BEEF CATTLE SCIENCE
HANDBOOK
VOLUME 19

International Stockmen's School Handbooks

Beef Cattle Science Handbook
Volume 19

edited by Frank H. Baker

The 1983 *International Stockmen's School Handbooks* include more than 200 technical papers presented at this year's Stockmen's School—sponsored by Winrock International—by outstanding animal scientists, agribusiness leaders, and livestock producers expert in animal technology, animal management, and general fields relevant to animal agriculture.

The *Handbooks* represent advanced technology in a problem-oriented form readily accessible to livestock producers, operators of family farms, managers of agribusinesses, scholars, and students of animal agriculture. The *Beef Cattle Science Handbook,* the *Dairy Science Handbook,* the *Sheep and Goat Handbook,* and the *Stud Managers' Handbook* each include papers on such general topics as genetics and selection; general anatomy and physiology; reproduction; behavior and animal welfare; feeds and nutrition; pastures, ranges, and forests; health, diseases, and parasites; buildings, equipment, and environment; animal management; marketing and economics (including product processing, when relevant); farm and ranch business management and economics; computer use in animal enterprises; and production systems. The four *Handbooks* also contain papers specifically related to the type of animal considered.

Frank H. Baker is director of the International Stockmen's School at Winrock International, where he is also program officer of the National Program. An animal production and nutrition specialist, Dr. Baker has served as dean of the School of Agriculture at Oklahoma State University, president of the American Society of Animal Science, president of the Council on Agricultural Science and Technology, and executive secretary of the National Beef Improvement Federation.

A Winrock International Project

Serving People Through Animal Agriculture

This handbook is composed of papers presented at the
International Stockmen's School
January 2–6, 1983, San Antonio, Texas
sponsored by Winrock International

A worldwide need exists to more productively exploit animal
agriculture in the efficient utilization of natural and human
resources. It is in filling this need and carrying out the public
service aspirations of the late Winthrop Rockefeller, Governor
of Arkansas, that Winrock International bases its mission to
advance agriculture for the benefit of people. Winrock's focus
is to help generate income, supply employment, and provide
food through the use of animals.

INTERNATIONAL STOCKMEN'S SCHOOL HANDBOOKS

BEEF CATTLE SCIENCE HANDBOOK VOLUME 19

edited by Frank H. Baker

Routledge
Taylor & Francis Group

LONDON AND NEW YORK

First published 1983 by Westview Press

Published 2018 by Routledge
52 Vanderbilt Avenue, New York, NY 10017
2 Park Square, Milton Park, Abingdon, Oxon OX14 4RN

Routledge is an imprint of the Taylor & Francis Group, an informa business

Library of Congress catalog Card Number   80-641058
ISBN 0-86531-509-4

ISBN 13: 978-0-367-01962-4 (hbk)

ISBN 13: 978-0-367-16949-7 (pbk)

CONTENTS

Part 18. ENVIRONMENT, FACILITIES, AND ANIMAL WELFARE

PREFACE

The <u>Beef Cattle Science Handbook</u> includes presentations made at the International Stockmen's School, January 2-6, 1983. The faculty members of the School who authored this nineteenth volume of the Handbook, along with books on Dairy Cattle, Horses, and Sheep and Goats, are scholars, stockmen, and agribusiness leaders with national and international reputations. The papers are a mixture of tried and true technology and practices with new concepts from the latest research results of experiments in all parts of the world. Relevant information and concepts from many related disciplines are included.

The School has been held annually since 1963 under Agriservices Foundation sponsorship; before that it was held for 20 years at Washington State University. Dr. M. E. Ensminger, the School's founder, is now Chairman Emeritus. Transfer of the School to sponsorship by Winrock International with Dr. Frank H. Baker as Director occurred late in 1981. The 1983 School is the first under Winrock International's sponsorship after a one-year hiatus to transfer sponsorship from one organization to the other.

The five basic aims of the School are to:
1. address needs identified by commercial livestock producers and industries of the United States and other countries,
2. serve as an educational bridge between the livestock industry and its technical base in the universities,
3. mobilize and interact with the livestock industry's best minds and most experienced workers,
4. incorporate new livestock industry audiences into the technology transfer process on a continuing basis, and
5. improve the teaching of animal science technology.

Wide dissemination of the technology to livestock producers throughout the world is an important purpose of the Handbooks and the School. Improvement of animal production and management is vital to the ultimate solution of hunger problems of many nations. The subject matter, the style of presentation, and opinions expressed in the papers are those of the authors and do not necessarily reflect the opinions of Winrock International.

ACKNOWLEDGMENTS

Winrock International expresses special appreciation to the individual authors, staff members, and all others who contributed to the preparation of the <u>Beef Cattle Science Handbook</u>. Each of the papers (lectures) was prepared by the individual authors. The following editorial, word processing, and secretarial staff of Winrock International assisted the School Director in reading and editing the papers for delivery to the publishers.

Editorial Assistance

Jim Bemis, Production Editor
Essie Raun
Betty Stonaker

Word Processing and Secretarial Assistance

Patty Allison, General Coordinator
Shirley Zimmerman, Coordinator of Word Processing
Darlene Galloway
Tammy Chism
Jamie Whittington
Venetta Vaughn
Kerri Alexander, Computing Specialist Assistant
Ramona Jolly, Assistant to the School Director
Natalie Young, Secretary for the School

Part 1

THE BEEF INDUSTRY AND ITS HERITAGE

1

THE MAJESTIC BULL:
OUR LIVESTOCK HERITAGE

Richard L. Willham

> "To admire our own successes, as if they had
> no past, would make a caricature of know-
> ledge."
>
> Bronowski (1973)

The BULL engenders, even today, a sense of ferocious grandeur, making him the fitting symbol for the proud past of all stockmen. Pure urbanites, divorced completely from all production agriculture, are still intrigued by the ritual act of charcoal broiling red meat, which reverts man back to the long forgotten bull cults of the Mediterranean basin and his common pastoral nomad heritage where the mobile food reserve, cloven-hooved mammals, were money. Our folk hero, the cowboy of the silver screen and now a part of high fashion, remains the CENTAUR of the most prestigious occupation in agriculture, cattle breeding. Today, the role of the stockman is even more exciting than ranging vast herds on recent buffalo land, by holding the water holes with armed cowboys. The future belongs to stockmen that use technology, spawned by science and the computer age, in the production of livestock products. Intelligent technology application is the new frontier of today for the livestock industry so blessed with rich tradition. It is fitting that the American Society of Animal Science incorporates the BULL in its logo, as did the International Livestock Exposition at its inception in 1900 in Chicago, then the hub of American agriculture.

The purpose of this paper is to catch a glimpse of the fantastic heritage of our livestock industry; stockmen who understand the lessons of the past can best weave appropriate technology into the current fabric of a romantic buy dynamic industry. The development is chronological—to better appreciate the subtitle interactions between man and his livestock and their profound impact on the cultural evolution of man and his humanities.

CHRONOLOGY

What follows is a chronological development of the symbiotic relationship between man and his livestock. Tannahill (1973), Burke (1978), and Thomas (1979) served as reference documents for this paper.

Prehistory

At least 4 million years ago, ape-man, for unknown reasons, moved from the forests to the rapidly developing savannas and changed his diet from a primarily vegetarian one to an omnivorous one. This move and diet change had a profound effect on the evolution of man culturally. Meat eating from hunting had far-reaching consequences. It cut down on harvest time and bulk consumption by two-thirds (Brownski, 1973). This allowed more time for man to foster social action and encouraged communication skills. Further, it created the need to spread out and migrate with the herds that were being hunted.

Then some one-half million years ago, the last large ice age of the Pleistocene descended. Man as a hunter and a gatherer was forced to adapt rapidly. The use of caves by clans of some 40 persons with maybe 10 hunters created the environment for more social action. The control and use of fire for warmth and cooking had a humanizing influence on man. The dynamic cave drawings signaled the increasing ability of man to be a shaper of the world rather than just another rather poorly equipped mammal in the existing ecosystem. Bronowski (1973) put it well, "This is my mark, this is man," when he spoke of the cave paintings.

With the retreat of the glaciers, starting some 30,000 years ago, some clans began to follow the herds and developed a transhumant way of life, north with the forage in the summer and back south in the fall for the winter. The reindeer and man, for example, developed a symbiotic relationship. The reindeer provided for the many needs of man for food, clothing, and shelter, and even some transportation, while man provided salt from his urine for the reindeer in the salt-deficient areas left by the receding glaciers. Probably wild herds of other cloven-hooved ruminants, such as sheep and goats, with their intense gregarious instincts were joined in a symbiosis with man, the developing herdsman.

Sometime during his life as a hunter and gatherer, man domesticated a carnivorous competitor, the Asiatic wolf. The dog provided man with his acute senses during the hunt and, in return, man provided a share of food and love. This transhumant development over time utilized the herds as a mobile food reserve, a use that was to persist throughout history. Root words for money reflect this use of animals (Leeds and Vayda, 1955).

Civilization

Sometime around 11,000 B.C. the ice retreated. Great fields of grass with seeds exploded into existence, especially in the Near East where spring is warm and moist from winds off the Mediterranean but summer is dry and hot from winds out of the steppes of Eurasia (Heiser, 1973). To exploit the grain required that man settle first to store, then to protect, and lastly to sow in what is called the Neolithic revolution. Exactly why man took up agriculture, which requires much effort, in exchange for a seemingly easier nomadic life of gathering, herding, and hunting is not known (Harlan, 1975). But he did and succeeded through exotic circumstances. By genetic accident, wild wheat crossed with goat grass to produce a fertile hybrid, Emmer, with 28 chromosomes. It spread naturally, but then it crossed again with another goat grass and produced a larger hybrid with 42 chromosomes, called bread wheat. It would not have been fertile without one specific mutation. Bread wheat will not spread with the wind because the ear is too tight and, if broken, the chaff flies off leaving the grain in place. Suddenly man and the plant, wheat, became symbiotic and have existed together since, even though the grain is indigestible before heating. Now the principal dependence of man for his food was a carbohydrate with relatively low-quality protein in the germ. This move from the nomad with his flocks to agriculturalist is beautifully recorded in the Torah where the story of Abraham resides. However, the transhumant life echoes through mythology and the many cattle cults of the Mediterranean basin (Cole and Ronning, 1974).

Farming and husbandry in a settled agriculture create a technology from which all science takes off (Bronowski, 1973). The most powerful invention of agriculture was the plow. "Give me a lever and I will feed the earth" was demonstrated by the farmers of the Near East. When man yolked the ox to the plow, he began to utilize power greater than his own. It increased the surplus he could win from nature. The surplus food production was what released men to create, invent, and build a civilization; something the nomad had no time to do. Cultures developed on great flood-plains of the Nile, the Tigris-Euphrates, the Yellow, and the Indus Rivers. Irrigation of crops required cooperation and developed administrative systems of empire. Law and government resulted. From cereal-based agriculture came cities and civilization as we know it. Trade developed, possibly to acquire salt and other limiting necessities. The animals domesticated for food and clothing and for a power source added a surplus to food production, especially when adapted to utilize the by-products of cereal agriculture. This is true only if livestock remained subservient to agriculture. One species did not.

The horse became a threat to the village surplus and to the existences of nations. Warfare was created by the horse, as a nomad activity (Bronowski, 1973). Centaurs of Greek mythology represent this fear by settled people. The entourage of states in the fertile crescent represent domination and then assimilation by waves of horsemen from the steppes of Eurasia, keeping alive the antagonism between the settled and the nomad, quaintly illustrated by Cain and Abel and Ishmael in the Torah. What Egyptian slaves ate when Rameses II stabled 70,000 chariot horses in Egypt remains a mystery.

Civilization resulted from a return to a plant-based diet made possible by agriculture and its plow. The domestication of animals guaranteed their use as a source of quality protein and fat. Their use often was ritualized so that the clan, or now the village and city, could utilize the products wisely; without storage possibilities, consumption had to be immediate. Animal sacrifice probably began to assure the use of the animal by the entire group, then it was for the gods, and later for the priests. The use of animals multiplied as the source of power for agriculture and transportation for trade. As it always does, a guaranteed food supply results in a population explosion. The human population was around three million in 10,000 B.C. and had risen to 300 million by 3,000 B.C., even before the beginning of most civilizations (Tannahill, 1973).

Classical

The seminal influence of Greek culture and the Hellenistic Age after Alexander the Great was in part due to the necessity of Greece to trade for grain with olive oil and wine and to the immigration of its citizens over the Mediterranean basin. Greek agriculture decimated its frail land resources in less than 400 years compared to 4,000 years for the Tigris-Euphrates flood plains. The Phoenicians, or alphabet givers, tied the Mediterranean basin together with their trade. The diet of the basin was grain, olive oil, wine, and mutton in small amounts. Cattle were responsible for agriculture power and bulk land transport. Ships carried grain to the city, Rome, from the frontiers of the empire. Spices to make the cuisine edible was the rule and was responsible for a flourishing trade with India and via the silk road with China. Rome commanded trade, and on couches around the table the rich enjoyed a variety of food from the Roman empire. To the plebes, Rome was the center of free grain and later bread and circuses. Even yet, nomadic culture helped shape the language. Latin for cattle is pecus from which comes our word pecuniary for money matters. Although the Phalanx of the Roman legion conquered the known world, horse warfare of the barbarians changed the Roman legions. The empire crumbled from within but outwardly from hordes of horsemen that pillaged. In the end, Byzantium remained to protect the heights of culture attained but lost. Europe and the Western basin were thrown on their own to survive.

Asia

Pastoral nomads continued to move in an intricate square dance over the steppes of Eurasia. Hunger on the steppes had far-reaching effects on the rest of the world. First the Huns under Attila swept into Europe in 400 A.D. followed by other horsemen, the Avars, in 500 A.D. The gigantic empires of the Khans were begun around 1100 A.D. The hordes of horsemen traveled and ate from the 18 head of horses used by each man. Blood and mare's milk, which is high in Vitamin C, along with the spoils of war, defined mobility; but conquest is not control. Empires come from settled agriculture, and in the end the Khans controlled but were assimilated into Chinese, Indian, and Russian cultures. In China, a competitor for grain with man, the pig, became the source of meat and a consumer of family waste. In India, the Aryan invasion signaled the beginning of the sacred cow (Harris, 1974). Cattle, as a power source, moved through oriental civilizations with rice culture. Intolerance to milk in adults in the Orient possibly resulted in an early interest in hygiene. India domesticated the jungle fowl.

The Arab world was united around 500 A.D. under Mohammed. Their holy wars by 711 A.D. had conquered much of the Iberian peninsula. Through this conquest, Europe was to benefit immensely because of Islam's development of the classics and science. Also, introduced were the agricultural system of ranching, the harvest of sparse vegetation by feral animals, and the weaving of wool (Grigg, 1974).

Just why so many inventions of the Orient had to wait exploitation from a maturing Europe remains a mystery: the stirrup that revolutionized European warfare from the battle of Hastings in 1066 on; the horse collar that let the horse become a power for agriculture; gunpowder that changed war dramatically.

Europe

When Rome fell, each area of the Western world was left to rely on its own resources. The Roman Catholic Church remained the only tie between the developing feudal systems of Europe. The feudal system was started initially to supply war horses for the lords. This system was a self-contained agricultural unit with guaranteed protection from Viking raids. Livestock production has always been central in European agriculture. The manors had flocks of sheep and poultry, cattle for oxen, and horses for war. Peasants had a cereal-based diet, with some meat consumption during fall slaughter; keeping livestock over the winter was impossible. Late winter was a time of suffering for both man and animal and the root of the stories of vampires, werewolves, etc. (Tannahill, 1973).

Crop rotation and the fallow became common. With this, beans were introduced that provided a protein source. The population increased and was more energetic. The unemployed nobility and many peasants participated in the crusades. The travel created an awareness of a larger world (with spices) again.

The utilization of the moldboard plow, a Scandinavian contribution, enhanced agricultural production on the heavy soils of Europe, but the ARD required eight oxen and resulted in a revolution in agriculture. Villages had great strips of tilled ground radiating from them. The length was what the oxen could plow without resting. Cooperation in ownership of the oxen was required. Within these villages, areas were designated as markets. They were recognized with a cross, the sign of market peace that was so necessary to trade.

Sheep were always an important part of the European economy. Peasants had several sheep and could clothe themselves, obtain milk and cheese and some meat in their diets --even when half the lamb crop went to the Lord. The Cistercian order of monks begun in 1100 A.D. was a response to the worldly ways of the older orders. They started monasteries on the poorest land and brought it to use through the production of sheep. Cistercian wool was the finest available in Europe. They also gathered farmers of the area and shared their developing knowledge on agricultural subjects. In Britain, the woolen industry and the production of sheep was a primary enterprise, so much so that the woolsack remains in the House of Lords today.

In contrast, Spanish sheep husbandry was transhumant with the movement of the flocks north and south. Large herds of the fine wool Merino breed were developed by the aristocracy and clergy. The owners formed the first livestock organization, the Mesta, to assure the smooth migration through the Iberian peninsula. The Mesta was second only to the Church in power in Spain.

As the cities of Europe grew in size, the town dwellers for a long time retained their cow, poultry, and pigs. In Medieval Europe, swine were the garbage collectors of towns. Milk was produced in cow sheds in the towns; foaming warm milk was the criteria of freshness for the buyer. Red meat was still recognized as the source of all strength and passion, an illusion of the pastoral nomad. The British guards, the beefeaters, were so named because, to keep their loyalty, they were fed great rations of beef by the Norman lords.

The little Ice Age of Europe created much hardship and an extremely poor diet. Although the chimney was invented to create warmth and, incidentally, changed the social structure of the population, the poor diets allowed the Plague to decimate nearly half of the population of Europe. Afterwards, workers were at a premium and created the need to develop labor-saving machinery such as mills run by water power, a thing the slave cultures of Greece and Rome had no need to exploit. Craft guilds flourished in the towns of Europe and were responsible for what few sanitary precautions were taken by the livestock slaughter sections of great towns such as Smithfield in London. Europe was saved from starvation through the importation of cereal grain from the developing Balkan countries. This cheap grain signaled the need for European agriculture to move more into livestock agriculture to compete in the markets. Wine making flourished; dairy products were produced in abundance.

In the late Middle Ages, the Knight in armour on their big horses, which represented Europe's response to the horsemen of the steppes of Asia, was shown to be beatable in battle (Agincourt in 1415). Oats, an excellent horse feed, was included in the crop rotation system and the Chinese horse collar gave the horse the ability to compete with the ox. The horse gradually replaced the ox and the development of cattle for meat or milk was initiated. Further, the horse made the European population more mobile than ever before.

Expanding

Spices again were to play an important role in world history. Personal fame, the glory of God, and a share in the spice trade was an unbeatable combination in fifteenth-century terms. Many Spanish conquistadors responded (Tannahill, 1973). Stockmen found rich rewards in the West Indies, founding a livestock-based aristocracy that was the envy of all Europe. Cattle, swine, and the conquering horse were taken to the New World, which lacked these species as natives, although the evolution of the horse was enacted on the North American continent eons before. The settlers brought their own stock and adapted them to the new habitat. The death losses on shipboard were close to 50%; the "horse latitudes" are named for the numbers thrown overboard in calms to save water. Corn, potatoes, and sugar were to revolutionize the old world; especially sugar with its vast plantations and slave labor acquired from the coast of Africa.

Englishmen brought to the Eastern seaboard their common law, Durham cattle, sheep from every point of embarkation, and innumerable swine that were let loose to run free in the forests. The colonizers were developers of the self-sufficient pioneer farm that cleared the land. Plantations in the South produced tobacco and then cotton. The Appalachian Mountains held the settlers until independence was won from England, and then the Ohio Valley with its fertile soil beckoned and the rush to the Mississippi River and beyond was on (Schlebecker, 1975). Meanwhile, the hacienda ranching life was developed in the Southwest and California with its vaqueros and their feral longhorns and the vast flocks of Mission sheep. Livestock agriculture in the new world became a rich mixture of Spanish exploitation and British tradition when they finally met.

Exploiting

Because of a renaissance in British agriculture brought on by the new enclosure laws, advances in the seed drill, the introduction of turnips as a root crop that could provide winter feed, Britain began the industrial revolution that changed the face of the earth yet again. Because the commons were enclosed, breeders of livestock could use male animals of their choice and could capitalize on this improvement through the demands of the Napoleonic wars. British livestock were adapted to the needs of the industrial revolution. For the first time since Rome, cities became markets of great dimension, and real commercial livestock production began.

Livestock breed formation started with a local type that was useful. A few men in the area inbred and selected the best of the type until a degree of uniformity was attained, and then they began to popularize the breed through the shows and fairs developed from the old Markets. Then to protect their development, herdbooks were started. These recorded the ancestry of the breed since almost complete reliance was put on heredity rather than environment in the Victorian age. To further foster the new breed, societies were formed to protect the purity of the breeds and encourage their multiplication. Developing nations such as the U.S., which was beginning to industrialize just after the Civil War, imported these breeds since they were already adapted to the needs of industrial society. Early maturity, and as a consequence heavy fat deposition at an early age, was the trademark. A source of quality protein and large amounts of concentrated energy were the needs. The British, because of the continuity brought about by the landlord-tenant laws and their relative isolation from war, developed large numbers of breeds. Only the Dutch on their isolated peninsula were able to compete with the Holstein-Friesian dairy breed (Willham, 1979).

Innovations of the industrial revolution coupled with its demand for food to feed the cities was synergistic with the development of livestock agriculture. New methods of preservation (canning, freezing, and chilling) revolutionized the food supply to the millions. The railroad, with its key symbol the "Iron Horse," allowed first the live animal to be moved from its area of production to the consumer, and then the chilled carcasses were moved to the consumer through the innovations of the great packing magnates of the hub of commercial agriculture, Chicago. The swine-packing industry moved from Cincinnati, where swine were driven in, to Chicago just prior to the Civil War. The dairy industry was revolutionized with railroad transport and microbiology.

Returning from the Civil War, Texas cowboys found the longhorn herds had swollen from 4 to 6 million head. These cattle driven north were to populate the great prairie where once roamed 40 million head of buffalo. Such a replacement of species was astounding. As with the Spaniard, the hide and horn were the products for lack of transportation. Then in 1867, a promoter named Joseph McCoy, from Chicago and a real entrepreneur, put things together. He rode to the railhead town of Abilene, Kansas, bought it for 5 dollars an acre and advertised in Texas with his remaining money ($5000) that he would double the price of a Texas steer at Abilene. The first ones arrived three months later. He bragged before leaving Chicago that he would bring 200 thousand head in ten years and actually brought 2 million head in four years (Cooke, 1977). This resulted in the saying "It's the Real McCoy." The large cattle spreads of the great plains and the golden age of the cowboy lasted about 20 years. The rough winters of 1886-87, the homesteader with his windmill, the introduction of barbed wire ended the romance. To keep this cowboy image in our folklore through the silver screen probably goes back to the centaur, the cattle-cults of the past in the Mediterranean basin, and our heritage from the pastoral nomad brought to us from the Moor invasion of Spain in 711 A.D. Things have a way of connecting (Burke, 1978).

A major factor of the industrial revolution was to increase dramatically the size of the middle class that provided services to both the rich and the factory worker. Food choices could be made by this group and they exercised this choice. They ate meat, drank tea and coffee, and increased their consumption of sugar. Cooking was improved through the development of the Rumpford closed-top range. Only after the industrial revolution do we see a livestock agriculture that bears any resemblance to what we now know it to be.

Contracting

Today we live in an advanced industrial society--a hallmark of which is that fewer than five percent of the population is engaged in production agriculture. However, the livestock industry still produces a biological product for the consumer. As biological products meat, milk, and eggs still vary in the amount of edible product and in eating quality. The old seasonal variation of the fall slaughter to preclude winter feeding still exists. However, the advent of the southwestern feedlots for choice beef, the dairy marketing cooperatives, and the vertical integration of poultry production both for eggs and broilers have all leveled the supplies dramatically. Some species still have an economic system of production that goes through cycles of "boom and bust"--to the consternation of producers. Lastly, the product produced still remains perishable, even with the innovations of preservation. That is, all the product produced is consumed at some price (Byerly, 1964).

Large scale, commercial livestock agriculture really began in the "Windy City" on the sand bar by Lake Michigan. William Ogden built Chicago, using the new 2 x 4 construction, and in 1841 the first bumper crop from the prairie paid off. In nine years, Chicago was transformed from a marsh to the largest grain and livestock market in the world (Cooke, 1977). Nowhere were more animals and crops so dramatically transformed and merchandized. First transport was by water, but Chicago soon became the biggest railroad center in the world. The giant meat packing industry (Swift, Armour, Morris, Cudahay, and Wilson companies--each with their prancing eight-horse hitches for advertisement), plus the abundance of immigrant labor, established the livestock market and drew livestock from the prairies of the Mid-west, for slaughter and distribution to the industrial East. It was boasted that all was used but the squeal. Use of inedible offal was possible because of the mass slaughter the provided tremendous quantities of the raw product necessary to produce products such as greases, oils, gelatin, glue, blood, hair, leather, feathers, bones, pharmaceuticals, fertilizers, and industrial products that paid slaughter costs. The carcass meat price was left to be divided among the producer, packer, and retailer (Byerly, 1964). The variety of processed meat products rivaled what Mom could do on the farm with fall slaughter swine.

Livestock markets developed in Chicago. Price to the producer depends on supply and demand, market news, quality of product, numbers available for efficient slaughter, and competition. The intricate network and interplay of factors affecting price make livestock agriculture industrial rather than peasant-based, as it was until the industrial revolution.

Today, Chicago is no longer the livestock hub. Slaughter has decentralized to where the livestock are fed and where labor problems are less troublesome. Carcasses or boned and boxed meat is more easily shipped than livestock once were.

Railroads really made livestock agriculture industrial. To move first the animals in stock cars and the milk in cans, and later the carcass or boxed meat, was the key to supplying industrial nations with livestock products. Further, grain movement to ports by rail and its sea shipment to Europe allowed the industrializing nations of Europe to concentrate their agriculture on livestock production and wine making. Livestock agriculture is today dependent on fossil fuel for its transportation.

The rise of the mammoth chain grocery organizations that have replaced the simple corner grocery store has dramatically affected the livestock industry. As the "Iron Horse" was the symbol of the early industrial revolution, the grocery cart, developed by a simple, little grocer as he calls himself, is the symbol of the rise of the supermarket (Wilson, 1941). The promotional tools of convenience, intense advertising, trusted brand names, and juicy, redbeefsteak wrapped in shiny cellophane have really separated the consumer from the product production chain. No more is the animal fed with family love, bathed in blood and guts at slaughter, and preserved for later use with salt, drying, and immersion in fat. Even with this opportunity to buy the produce of a nation and of the world, our consumption patterns are still governed by dietary customs developed over centuries.

Today the livestock industry has embraced scientific technology. Dietary requirements of livestock are better known than those of humans. Genetic improvement in production traits has been applied to most species of livestock. The product sold is produced after animal maintenance is satisfied. Thus, faster gains in general or more product per unit of time is essential. Since feed grains can dramatically increase the product-to-maintenance cost ratio, it will be fed until a certain price is reached. Beef and milk production is possible from sparse forage on the 20% of the land unsuitable for crop cultivation in the U.S. (Dale and Carter, 1955), but the amount produced will be much less than we consume currently.

Three of the principal livestock species are ruminants that are able to break down complex carbohydrates in their stomachs and utilize protein from the digestion of the rumen microorganisms. This symbiosis has resulted in beef and dairy cattle and sheep and goats being the most widely distributed species on the globe.

Since World War II, the dairy industry has had 46% as many cows, but 2.42 times more milk. Part of this increase in pounds of milk per cow is a result of concentrate feeding in larger, better managed herds, but the synergism of milk records, artificial insemination, and sire evaluation and the use of superior sires throughout the industry has accounted for much of the increase. Marketing cooperatives have helped stabilize prices. We have the safest milk supply in the world.

The beef industry is romantic, segmented, and quite cyclic in nature. Beef production may become simply an adjunct to owning land in the uncultivated, 20% where the cow can be the harvester of sparse forage. Because of the low reproduction rate, the maintenance of the breeding herd is the primary cost of production, only 30% goes for feed costs to the finishing animal. Improvement in growth and size of the beef animal in recent years has been a result of the demands of the commercial feedlots in the Southwest. British breeds have been selected for growth, and the importation of larger breeds from Continental Europe has increased the output of beef from concentrate feeding. Beef consumption is tied directly to disposable income, but people like beef (a vestige of the ancient Mediterranean cattle-cults or a part of our nomadic heritage and its myths).

Sheep and goats are primary in pastoral societies; they were a contributor to the rise of Europe and were important in feeding and clothing the Americas. Sheep are holding their own in the U.S.; polyester has had a role, but wool will soon be the preference to fossil fuel products.

Poultry and swine are monogastrics and are found in conjunction with feed-grain production or are used as scavengers in less-developed countries. Poultry production in the U.S. is the most industrialized, as compared to that of other species. Only a few companies scientifically breed the birds; crosslines are raised in hatcheries and made available for mass egg or broiler production. The profit per unit of production is minute, but volume has been steadily expanded (Byerly, 1964). Improvement in any livestock industry usually results in cheaper consumer prices.

Commercial swine production is tied to areas of feed-grain production, except in China and Polynesia where swine are scavengers. Swine face religious taboos as well since the pastoral nomads of the Near East despised their lack of gregariousness. Swine were the backbone of the American family farm. They were the "mortgage lifters" since one could get in and out of swine production quickly due to their reproductive potential. The ability of swine to put on fat has been exploited by man in the World Wars. Swine can be changed genetically at a rapid rate and have undergone type transformations at least four times since 1900. Today the meat type is in vogue and heavier market weights are being sought with the same fat content. The producer is moving toward being capital intensive, with complete confinement of his animals. Thus, it is no longer easy to get in and out of production quickly--one of swine's main assets.

We now produce our food using an oil-powered cereal agriculture supplemented with livestock products that are produced from the by-products of cereal agriculture and range land (quality protein, meat, and complete foods, milk, and eggs). Long gone is the age of animal power that for centuries contributed to the production of cereal. To do away with fertilizers that are estimated to produce over one-fourth of our current food is impossible. This is not the first time that the world has adapted to alternative energy sources. Deforestation of Europe in the production of glass produced reliance on coal and coke and aided in the development of power sources for the industrial revolution (Burke, 1978). Adapt we will.

Synthetic food production, especially of quality protein from cereal and bean production, is a reality. Costs are presently prohibitive; but more bothersome than this are the long established dietary customs of people in every country. Meat eating and milk drinking are a part of our pastoral nomad heritage and are not to be broken easily.

REFERENCES

Bronowski, J. 1973. The ascent of man. Little, Brown, and Co., Boston.

Burke, Janes. 1978. Connections. Little, Brown, and Co., Boston.

Byerly, T. C. 1964. Livestock and livestock products. Prentice-Hall, Inc., Englewood Cliffs, New Jersey.

Cole, H. H. and M. Ronning. 1974. Animal agriculture: The biology of domestic animals and their use by man. W. H. Freeman and Co., San Francisco.

Cooke, A. 1977. America. A. A., Knopf, New York.

Dale, T. and V. G. Carter. 1955. Topsoil and civilization. University of Oklahoma Press, Norman, Oklahoma.

Grigg, D. B. 1974. The agricultural systems of the world: An evolutionary approach. Cambridge University Press, Cambridge.

Harlan, J. R. 1974. Crops and man. Crop science society of America, Madison, Wisconsin.

Harris, M. 1974. Cows, pigs, wars and witches. Vintage Books, Division of Random House.

Heiser, C. B., Jr. 1973. Seed to civilization: The story of man's food. W. H. Freeman and Co., San Francisco.

Leeds, A. and A. P. Vayda. 1965. Man, culture, and animals. AAAS, Washington.

Schlebecker, J. T. 1975. Whereby we thrive: A history of American farming, 1607-1972. Iowa State University Press, Ames, Iowa.

Tannahill, Rey. 1973. Food in history. Stein and Day, New York.

Thomas, Hugh. 1979. A history of the world. Harper and Row, New York.

Willham, R. L. 1979. Our livestock heritage. A syllabus for An. Science 225 class taught at Iowa State University, Ames, Iowa.

Wilson, T. P. 1941. The cart that changed the world. University of Oklahoma Press, Norman, Oklahoma.

2

CATTLE PRODUCTION SYSTEMS IN THE WORLD

Henryk A. Jasiorowski

Cattle play a tremendous role in the contemporary world, providing valuable nourishment--milk and meat, valuable hides, and traction power. In many developing countries, cattle are a means of subsistence for millions of people. Especially in poor countries, cattle and often ruminants are valued because they can use feeds that are not consumed by people and monogastric animals. Hence the great role of ruminants that provide mankind with about 70% of the total animal protein consumed.

Worldwide, there are about 1.2 billion cattle and 130 million buffaloes. One should be conscious, however, that about 400 million are in the developed regions, i.e., Europe, North America, and Oceania. Thus, 60% of the world's cattle population is in the developing countries.

The conditions under which cattle live and produces vary widely which accounts for the great discrepancies in cattle productivity summarized in table 1.

TABLE 1. CATTLE PRODUCTIVITY IN DIFFERENT REGIONS OF THE WORLD IN 1980

Region	Kilograms of carcass per head of cattle/year	Milk production/ year/cow
Africa	17.4	491
North America	87.4	5007
South America	32.4	1001
Asia	10.4	678
Europe	80.2	3475
Oceania	58.7	3229
World	38.4	1927

A clear understanding of reasons of such great discrepancies in cattle yields requires an analysis of the world production systems, which in turn are conditioned mainly by the natural and economic circumstances.

Two classifications of the cattle production systems are discussed in this paper based upon 1) ecological systems and 2) production system inputs.

ECOLOGICAL SYSTEMS OF CATTLE PRODUCTION

Three principal systems of cattle production are linked with their ecology. These are: the nomadic, the transhumant (seasonal herd migration), and the landholder "settled" systems.

The nomadic system is probably the world's oldest cattle production system and is characterized by continuous migration of entire families and tribes that move with their animal herds to take advantage of seasonal changes in the growth of grass and the availability of water.

Nomads usually keep mixed herds (cattle, goats, sheep, donkeys, and camels), but in Africa they tend to keep mostly cattle. Pastoral tribes of nomads have no fixed residences, do not farm land, and on their migration routes often ignore political frontiers. They produce only what they need for their families; the size of herd is a source of pride, and its size means prestige to those in the nomadic community rather than wealth as it does in our culture.

The nomadic system of animal production continues in many parts of the world, and in Africa and Asia may be the dominant system. Examples of contemporary nomadic tribes are the Borans on the Kenyan-Somali border, who travel tens of kilometers daily--their camels, cattle, and other belongings with them--in search of water and grass. They often have to travel more than a week before finding water. Good examples of nomadism are the excellent cattle breeders in East Africa--the Masai tribes--whose diet consists almost exclusively of milk, cattle blood, and meat. The Baggara Arabs (from Western Sudan) travel "horizontally." In the dry season, they migrate as far south as Bahr el Arab, where they are forced to stop by the tsetse fly (which carries trypanosomiasis) and during the rainy season they go north to the fringes of the Sahara. On the other hand, the nomadic tribes of Ethiopia and East Africa travel "vertically," migrating to the mountains during the dry season and returning back to the semidesert regions during the rainy season.

The nomadic system is a very unreliable form of animal production; the lack of water or rain in the right place and at the right time may cause entire herds to perish. Managed correctly, however, this system can be ecologically useful and effective.

Although excellent animal breeders, nomads give little thought to the correct utilization of land, often allowing vegetation to deteriorate. This has caused a degradation of the entire regions of the Sahel and the Sudan and a further spread of the desert.

The seasonal herd migration/transhumant/pastoralist/ system seeks to make use of often distant pastures during the rainy season and in the dry season makes use of forage remaining after the harvest around the farmhouses. This system also allows the summertime use of mountain pastures inaccessible in other parts of the year. In the Sahel region of Africa, herds travel north away from their local dry season to the fringes of the Sahara when the rainy season moves in that direction. The men return with their herds to their homes in the south when the rains start falling there and it is time to farm the land.

This system of production is found mostly in Africa and Asia but is characteristic of some regions of southern Europe and in the Asian republics of the Soviet Union and in Mongolia. Modern technology, however, is being introduced even in this system of pastoral production, with the animals being carried to far-away pastures by trucks or trains in some countries.

In tropical regions, the seminomadic system usually fails to match the nomadic system in terms of production results. The nomads, who concentrate all their efforts on animal breeding alone, are usually able to produce better animals.

A mixture of the nomadic and settled forms of cattle production is the so-called shifting agriculture. Under this system, a plot of land is cultivated for a time, then it is abandoned and allowed to go fallow and the farmer moves on to farm another plot of virgin land. The practice is used in regions with ample rainfall, low population density, and with no shortage of unclaimed land or tribal land. The areas that have been cultivated and then abandoned offer good opportunity for cattle grazing. In some regions of West Africa, this system is used for the breeding of the N'dama cattle, which are resistant to trypanosomiasis, and the North African shorthorn cattle. In Brazil, the system serves as a means of gradual transformation of tropical forests into farming areas.

Settled systems of cattle production of the world differ widely, depending on soil and climatic conditions and the country's level of development. For example, they include milk production for the farmer's own family on southern Europe's one-cow farms, milk production in African villages with cattle grazing freely, as well as the operation of 2,000-cow dairy farms and large feeder operations with up to 100,000 animals annually.

Settled Systems of Alternating Farming

This system differs from the nomadic system of shifting agriculture chiefly in that the farmer leads a settled life--usually because of increasing population density, the shrinking of available land resources, and the emergence of the market. In West and Southwest Asia, the system is based on the cultivation-abandonment cycle and involves land around the village. The result is the appearance of natural vegetation (weeds, grass) in the abandoned areas that are used for cattle grazing. During the dry season, this vegetation is often burned to allow the growth of more valuable grasses and to prevent the growth of brush and trees, the aim being to keep the land suitable for pasturing.

In regions of the dry savanna in Africa, large cattle herds are kept under this system. The alternating system of using land for cultivation and grazing also is important in cattle production in Ethiopia, Tanzania, and Burundi. In Indonesia, the alternating system is employed for tree cultivation and grazing, with cattle grazing on the plantations in the night. In the mountain regions of Colombia, Venezuela, and Peru, however, where the alternating system is widespread also, cattle pasture does not play a major role.

From the ecological point of view, the alternating system of land use involving the destruction of tropical forests and their replacement with less productive vegetation is a negative phenomenon. This applies, for instance, to regions of Southeast Asia and Brazil. In West Africa, the process of tree felling, however, is helpful in controlling the spread of the tsetse fly.

Settled Subsistence Production System (In Regions Provided With Water by Irrigation Facilities and Rainfall)

Under this system, cattle are fed mostly waste fodder and forage remaining in the fields after harvest. Communal schemes are often adopted to allow all cattle in the village to graze in all fields after they have been cleared of crops. Working within this system, African farmers who keep no cattle themselves allow nomads to bring in their herds so that the animals may graze in their fields and deposit manure. In many treeless tropical regions, the manure left by herds is collected by children, mixed with straw, dried, and then used as fuel.

Cattle Production System Linked With Permanent Plantations

This system is rarely seen in temperate climates, but it can play a certain role in tropical regions. All waste products are important. Sugarcane plantations, for example, provide cattle feed in the form of sugarcane tops, molasses, and fibrous remains after the sugar extraction process. Pineapple plantations, cleared of fruit, also provide feed for cattle, as do the plantations of agave plants that are used for the production of sisal fiber.

Ranch System

This is a commercial form of nomadic pasturing adapted to the more populated conditions. Popular in regions with poor soil and little rain, this extensive system is used chiefly for meat production. Thus, it is found in the steppe regions of Central Asia, in the prairies of North America, in Central and South America, and also in Africa and Australia. The arrangements of the scheme depend on land quality and rainfall--the two factors that are crucial in determining the size of herd in relation to the land area. In wet equatorial regions, for example, there might be one head of cattle/ha, while in North Australia the proportion would be one head of cattle/250 ha. The ranch system is marked by wide fluctuations in production results, depending largely on rainfall, which may be very capricious in those regions.

Production System Based on Crop Rotation

Under this system, farmers lead a settled life and their cattle are fed on fodder cultivated in the fields or are allowed to graze in pastures in the summertime. This system is predominant in the temperate climate but also appears sporadically in the tropics (in Ethiopia, Nigeria, Uganda, Zambia, and Zimbabwe).

Production System Based on Large Industrial-type Farms

This system is apparently the latest stage in the evolution of cattle-production systems in the world. In their most developed forms, the farms are almost as independent of land cultivation as are modern chicken farms.

Large dairy farms in California provide a good example; the owner of a herd of 2,000 may have fewer than 20 ha of land and buy all fodder from outside, including maize and straw. Other examples are the large feedlot operation in Colorado that fattens some 100,000 cattle annually, buying all fodder they need and selling manure. Large industrial-type cattle farms also exist in the Soviet Union and Eastern European countries. They are characterized by high capital inputs.

PRODUCTION SYSTEMS CLASSIFIED ACCORDING TO TYPE OF INPUTS

These may be called the economic systems of cattle production. However, an increasingly popular classification of animal production systems is according to production intensity. We often describe the production systems as intensive, extensive, intermediate, or semiintensive. A system is called intensive when it yields high production of milk or beef per cow, land area, unit of labor, invested capital, and time expense.

Similarly, low productivity and low capital outlays for unit animal production indicate an extensive production system. High milk yields on commercial-type farms and fattening of young cattle with concentrates (as in the U.S.) are examples of intensive farming, whereas cattle fattening on the prairies of Australia, where animals reach the weight of 500 kg only at 4 or 5 years of age, is an example of extensive farming.

Cattle production systems can be easily classified in the conventional manner as intensive of extensive. For example, we could easily decide that the nomadic system is extensive. The matter, however, is not all that simple. If we assume that animal productivity is the principal criterion, we must also realize that there may be a wide range of reasons for high productivity. The best criterion therefore for assessing the intensity of animal production, including cattle production, is the relationship between the three principal production inputs: labor, land, and capital. If we accept that, we can classify the existing cattle production systems in the world in the following way:

Systems with land as the principal input. This is probably the oldest system of animal production. In its earliest form it included the hunting of bison in North America by Indians and early white settlers. Meat production was based on the use of large areas of land with low labor expense on hunting and even lower capital expense—on the purchase of weapons. The system also includes all forms of natural pasturing, nomadic pasturing including.

Systems with land and labor the principal inputs. Historically, this system has followed the above system as the population density increased and the first agricultural settlements appeared. The emergence of farming required the guarding of herds and their pasturing within certain areas only, which meant increased labor costs. From an economic point of view, this system evolves when land value (and costs) begin to climb but labor is still cheap.

Systems with land and capital the principal inputs. This system evolves as the next step after those noted above and appears with the growth of labor costs (because of higher pay in nonagricultural jobs, for example).

In cattle production, this system is applied wherever land is abundant and cheap while labor costs are high. The availability of capital is also vital.

Examples of this form of production include the cattle pastured in Australia, on the prairies of the United States, and on the steppes of South America. The shortage of water and the low quality of land require large land inputs per animal, while high labor costs encourage investment into permanent fences, watering places, roads, etc. In the final stages of the system's development, helicopters may be used to watch the animals. In the conventional classification, we called this system the ranch system. The system also may cover dairy cattle; for example, the schemes employed in Byelorussia where dairy cattle are grazed on large areas of low-quality pastures.

Systems with labor as the principal input. This system appears in areas with high population density where land is dear and capital hard to obtain. In the past, the system included milk production from cows that were kept by landless rural families. The grandmother and children pastured cows in roadside ditches and on boundary strips between fields. (A typical picture from Eastern Europe's prewar countryside is an example of this type of production.) The system also includes the present practice of cattle husbandry used by small and parttime farmers in many areas of the world.

In the developing countries of wet tropical regions, large dairy herds are kept close to large cities and fed with vegetation transported on the backs of workers. I watched this scheme in operation near Jakarta, Indonesia. Each worker was obliged to cut vegetation in ditches and other unused land strips and to carry significant quantities to feed five cows daily. The worker also milked the cows, and his children delivered the milk on bicycle to consumers in the city's center. This example provides a good illustration of cattle production based chiefly on labor outlays. It contrasts sharply with cattle breeding practices in Britain and the U.S., for example, where it takes only one man to care for 100 dairy cows.

Systems with labor and capital as the principal inputs. This form of production appears when land is expensive and the economic advance has made labor dear and capital readily available. It is employed, for instance, on specialized private farms in Poland where there are large investments in land development (fertilization, reclamation, etc.), but the lack of equipment requires that much work has to be done through manual labor. This system also includes cattle production on the Polish state farms and on some Soviet collective farms.

Systems with capital as the principal input. Such systems include large dairy farms where milking, feed administration, the removal of excrement, and other operations are mechanized or even automated. Other examples are the large feedlot operations where the cattle (kept indoors or outdoors, depending on the climate) are fed on large amounts of concentrated feedstuffs and all operations are fully mechanized.

In its most developed form, this system may operate with little land and is marked by low outlays of human labor. In recent years, the system has been gaining ground --especially in wealthier economics where labor and land are expensive and capital is easily available. This form of production is often found in vertically integrated opertions with units of machinery production, feed processing, and wholesaling. Companies that own large feedlot operations in the U.S., for example, may also own grain producing farms, slaughtering houses, retail shops, etc.

The capital-intensive system of animal production appears at times in integration with other systems. An example is the production of young cattle in the U.S. under the ranch system (land-intensive) up to the weight of 250 kg whereupon the animals are transported (sometimes over a distance of several thousand km) to feedlots, which work under the capital-intensive system.

Much attention in world literature is being devoted to the merits and demerits of the capital-intensive system of animal production. The system, as it was pointed out earlier, involves a large concentration of animals in one place, which poses some environmental problems. One such problem is dung--during the winter, each cow produces some 8 tons (10% to 12% dry matter content).

Economists claim that a larger scale of production spells higher profits, but their critics charge that they fail to include in their calculations the costs of environmental protection, the social costs of unemployment, etc. They also criticize the high grain needs of these systems of production. It is generally believed that, although capital-intensive systems of animal production may sometimes benefit wealthier countries, it is misguided to try to introduce them in the less-industrialized countries.

In further analysis of the division of cattle production systems into intensive and extensive types, the extensive systems can be said to include labor- or land-intensive systems, or simply those that produce low milk yields and low beef production per animal. Under European conditions, the intensive systems are often considered those that use large amounts of grain and feed concentrate, while the extensive systems are those that are based mostly on bulk fodder. This division is not precise, as was pointed out earlier. Thus, our discussion suggests that the expression "an intensive system" of milk or meat production does not mean much by itself, unless we specify production inputs. Under European conditions, the term "an intensive system" is usually a synonym for the capital-intensive system, but this is not necessarily so in the other parts of the world.

3

THE FUTURE FOR BEEF

W. T. Berry, Jr.

In 1981, a National Cattlemen's Association (NCA) Special Advisory Committee, along with officers and staff, analyzed the beef industry, and NCA published the committee's report in March of 1982. We entitled the report "The Future for Beef" and refer to it as the SAC report. The report has been acclaimed by both university and industry leaders as the best modern evaluation of the beef industry. In this presentation, I will discuss some of the highlights of the SAC report.

Several months into the study, the committee concluded that cattlemen, particularly the more innovative and able managers, would have real profit opportunities in the 1980s. However, not everyone in the cattle business, or willing to try, will succeed. The optimistic outlook was strengthened by the realization that, in spite of all the competition from other protein foods, beef remains by far the most preferred meat. The SAC report also concluded that the cattlemen who succeed are those who are more market-oriented, who use the appropriate technology, and who optimize production with costs and sales.

The business environment in which cattlemen operate is very complex. Profitability is dependent primarily on each cattleman's planning and management; however, it also is related to many outside factors—some of which the cattlemen can have some control over and others that he has little, if any, control over. Let's examine some of the more important outside factors.

Date and information. Cattlemen have been reluctant to call on outside help, but that is changing. They are recognizing the value of intelligent and responsible employees. They are subscribing to and using more market data and market-management information from Cattle-Fax and other private information sources. They are making greater use of outside consultants, subscribing to more newsletters and other sources of information that can contribute to better planning and management decisions.

Research and education. Research funds nearly always produce a return on investment, but research is slow, tedious, and oftentimes frustrating and disappointing. Even though research eventually leads to positive findings, all of agriculture is woefully behind in research funding. This severely limits aggressive, high-priority research projects. Setting priorities and adequately funding worthy research projects is a continual struggle.

State associations. State cattle-producer and feeder organizations are strong forces in bringing about progressive action in state government affairs and in research and education programs at the state level. They need the support of cattlemen.

Breed associations. Genetics will be a real key to continued improvement in beef's competitive position and in the profitability of cattle operations. Breeders of registered seedstock are a select and small group among all cattlemen, but they have a real impact on beef production and profits. The big challenge to the purebred industry is to anticipate trends and industry advances, and breed the types of animals that fit the merchandising system and contribute to improved efficiency. A big job for commercial cattlemen is to select the right seedstock.

National organizations. There are three national organizations that have great impact on the profitability of individuals in the beef industry. They are the U.S. Meat Export Federation, the Beef Industry Council of the National LiveStock and Meat Board, and the National Cattlemen's Association. Each of these organizations has its own funding, programs, and responsibilities. Recently there have been encouraging signs that, in the future, the activities of these organizations will become more unified in their services to the industry. The industry needs greater coordination of programs as well as accountability for the effectiveness of some of the programs. In the future, cattlemen will take an even stronger role in developing more cost-effective funding and structure of these beef organizations.

The U.S. economy. The U.S. economy--which includes unemployment, consumer buying power, inflation, interest rates, credit, and available financing--obviously impacts profitability. There is very little that individual cattlemen can do about the nation's economy, except at the ballot box; but I really believe that individual cattlemen's influence is being extended in a forceful manner through NCA. The NCA-PAC, the committee system, special task forces on finance and energy, plus the regular NCA staff, are effective and respected in our nation's capitol as they work together in both the legislative and administrative areas.

MATURING OF THE MEAT INDUSTRY

Today we are operating in a relatively mature meat industry. Meat is no longer a growth business such as it was in the 1950s, 1960s, and early 1970s. Average per capita consumption of all meat has leveled off at around 200 lb, retail weight. The mix of meat products has changed, but the total has shown relatively little change for more than a decade.

Per Capita Consumption of Meat (Retail Weight, Lb)

	Beef	Pork	Poultry	Total meat
1970	84	62	49	200
1976	95	54	52	205
1980	77	68	61	208
1982	76	55	62	196

When one considers all animal protein foods, including eggs, fish, milk and cheese, there has been relatively little change in average per capita consumption during the past 20 years. The total is about 430 lb.

Per Capita Consumption of Animal Protein Foods (Retail Weight, Lb)

	1960	1970	1980
Total meat	168	200	208
Fish, cheese, eggs	66	68	70
Fluid milk*	185	174	157
Total animal protein foods	419	442	436

*Poundage adjusted to make milk more comparable in moisture content.

Although the total has changed very little, the mixture of products has changed. Per capita meat consumption rose steadily until 1970 when it leveled off, but consumption of milk has continued to decline. The total is not likely to change significantly in the years ahead. But the mixture can change in direct proportion to consumers' preferences and buying decisions and the production and marketing skills behind each of the commodities.

In this mature meat industry, cattlemen are asking themselves:

Q: Will we lose more of the beef market?
A: That's possible.
Q: Can we hold our own?   A.: That's also possible.

Q.: Can we expand and do it profitably: A.: That, too, is possible--but much more difficult.

Future beef industry growth will depend on: 1) population growth, which is expected to average about 0.8% per year in the 1980s, 2) expansion of exports, which now account for only 1% of U.S. beef production, and 3) beef demand, the development of new beef uses and beef products, and improvement in the competitive relationship between beef and other meats. Keeping beef at the top of the protein mix will require the united efforts of the beef cattle industry as well as improved efficiency on the part of the individual cattleman.

Control Beef Supply?

Some persons suggest that we should make a stronger attempt to control beef supply. If it is in shorter supply it could become a specialty item, push up the prices, and increase profitability. But that is unrealistic. When this type of discussion starts, the question arises: Which cattleman is ready to go out of business? Which one is ready to cut down on production when volume production on a given set of assets usually means more efficiency?

As individuals, cattlemen can become better informed about market trends, beef supply information, placements, weekly kills, and much more background information. With this information they make short-term decisions that can contribute to profitability. It is usually a case of being just a few hours, days, or weeks ahead, in terms of key information, that helps a cattleman make better planning and marketing decisions and have a competitive edge.

But for the industry in total, supply reflects the actions of about 1.25 million cattle owners who respond to both economic and other incentives. Those other incentives keep many cattlemen producing cattle. They are almost immune to the pressures of cyclical market changes.

The Beef Production Machine Has Momentum

The Special Advisory Committee report concluded that for the rest of this decade the average per capita consumption (domestic production and imports) of beef will be about 80 lb, retail weight, which includes 73.5 lb produced in the U.S. and 6.5 lb imported. Keep in mind, however, that the 6.5 lb of imported beef per capita, valued at a total of about $3 billion, is offset by the successful exportation of $3 billion worth of beef, variety meats, hides and other beef by-products.

The 73.5 lb of beef produced annually in the U.S. are being produced on more than 1 billion acres of all types of forage by 1.25 million cattle owners.

Approximately 300,000 of these cattlemen produce 75% of our beef supply, and they are deriving sizable percentages of their incomes from beef production, but about 950,000 cattlemen produce 20% of our beef supply, and most of them are relatively immune to economic pressures in the cattle business. That is because they earn most of their incomes from other agricultural products, or they have high percentages of off-the-farm income.

A few of these small or part-time cattlemen are serious-minded about their business, but most are less interested and concerned and are slow to react to the economic pressures of the cattle business--so they keep on producing. Their breeding decisions, from the point of view of both the type of animal and the numbers of animals, have a significant impact on the U.S. beef industry.

Beef Is the Preferred Meat

More beef is consumed than either pork or poultry, and it sells at a higher price; beef commands a very substantial premium over other meats. In 1981, Choice beef averaged $2.39/lb, compared with $1.52 for pork and $.74 for chicken. Other meats present tough competition, but consumers are willing to pay substantially more for beef. The question is, how much more?

Market surveys clearly show that price is the single most important competitive problem for beef when consumers make their choices at the counters. Cattlemen sometimes wishfully think of higher prices but, realistically, beef prices must remain competitive to move approximately 80 lb of beef per capita per year on a regular basis. Profits, therefore, come from producing, marketing, and merchandising beef more efficiently.

Efficiency Is the Key

For both the individual cattleman and the industry, efficiency in production and marketing must improve--there is no real alternative. Many cattlemen feel that they are producing beef as efficiently as possible, but the Special Advisory Committee's analysis showed that most can improve their performance just by using technology and techniques now available. Records show that there are wide variations in efficiency and in actual cost of production.

Improvement in beef's competitive position will depend to a great extent upon the narrowing of the price spread between beef and its competitors, and it is possible to narrow that spread. There will be a slowing of efficiency growth in the pork and poultry industries while efficiency in beef production and processing will increase more rapidly.

To significantly and meaningfully reduce the price spread, cattlemen will not only have to make better use of current technology but also must take positive steps in financial planning, business management, and marketing strategy so that overall efficiency improves.

There will be further improvements in genetics, reproductive performance, disease control, and feed conversion, but some believe that the greatest potential for increased efficiency lies in better production and management of forage resources. The SAC report concluded that to further improve beef's competitive edge, cattlemen must take better advantage of ruminant animals' unique ability to convert otherwise wasted plant resources into meat.

Cattlemen are looking forward to new research to bring further advances in beef production, processing, and marketing. However, great strides in increasing efficiency can be achieved today by simply applying technology and systems already available and proven.

As mentioned earlier, profitability of the individual depends on the impact of outside forces to some extent, but the individual cattleman's success will depend mostly on his own abilities and efforts. He can, and should, do everything he can to help improve the climate in which he operates, but ultimately profitability will depend largely on his own relative efficiency and ability to anticipate and offset the business risks.

Risk management has been a popular subject on many recent programs, and there is a reason for that. Our business is cyclical in nature and much more volatile than it was in the past. The downside risk is much greater. Futures markets and other marketing tools will be more important.

Personal and small office computers will become part of the office procedure of medium-sized and even some small-sized cattle operations. This is an electronic age and even the small computers assist in communications and the basic accounting and financial planning.

Land Tenure

The cattle business has become so capital intensive that it is very difficult for new operations to develop enough cash flow from production to meet the debt service on land and operating requirements. This has resulted in more leasing to expend the size of the operation without involving greater mortgage obligations on the land. This trend of greater separation of land-ownership and cattle operations will probably continue. Land owners, as well as cattle operators, have different investment goals that make possible some of the more standard leasing and tenant contractual arrangements, as well as innovative joint-venture and leasing arrangements.

SUMMARY

This country is blessed with more than a billion acres of forage that is best converted by cattle into beef. As a superior, highly preferred product, beef is being produced by more than a million cattle owners, the majority of whom do not respond to economic pressures but produce about 20% of the beef supply. Supply control is impractical in such a segmented industry. The combination of over a million cattlemen operating on over a billion acres translates into a rather steady supply of about 80 lb of beef per capita for the foreseeable future. This large volume of a perishable product must be moved with aggressive merchandising at competitive prices.

In the 1980s the successful cattlemen who enjoy profits most of the time will be those who are the best managers and the most competitive. The profits will go to those who know their costs, who are the most efficient, who market most effectively, and who know how to manage their risks.

If beef producers take advantage of the potential for improvement in productivity, their own profits will grow, and beef's share of the mature meat market can hold and possible gain in the 1980s.

We can sum up this discussion with one last statement. To be profitable in the 1980s, "Cattlemen must be businessmen specializing in cattle rather than cattlemen functioning only occasionally as businessmen."

4

CREATIVE CATTLE CAPITALISM

Charles G. Scruggs

"Capitalism is the economic system in which goods and services are provided by the efforts of private individuals and groups who own and control the means of production, compete with one another, and aim to make a profit."

Basic beef cattle producers (beef cow owners) are essentially powerless to control their economic destiny.

Unorganized in an organized economy, beef cow owners flounder almost helplessly as "price takers" in one of America's most important and basic industries.

Beef is perhaps the "king of foods." Certainly beef is one of the most desired foods consumed by man. It is power packed with protein for good nutrition. Most people agree that no other food beats beef for enjoyable good taste and eating pleasure. And beef is a social status symbol.

Beef, with its health-giving protein, is the hope of a malnourished world for the future. The men and women who start and nurture the beef system (cow-calf producers) have been accorded a special status in history, folklore, economics, and politics. The words rancher, cowboy, cowman evoke images of all that is considered basic and good in the U.S.--and in much of the world.

With all these wonderful qualities and attributes, cowmen should be "kings of the hill," "masters of their own destiny." And indeed they are highly regarded most everywhere--except in the marketplace. Let me repeat my opening statement: Basic beef cattle producers are essentially powerless to control their economic destiny. They not only are at the bottom of the beef food chain; they very often are at the bottom in economic returns.

Why?

No one answer is possible. Books have been written seeking the answer. Meetings by the multithousands over the past 200 years have addressed this problem. Perhaps part of the answer is a paradox. Perhaps, because of one of their most highly acclaimed attributes--independence--basic beef cow producers have become victims of an organized economic system.

Cowmen have not focused and pyramided their economic and political power for their own benefit.

Can they?

Should they?

Will they?

Only time will answer the questions "should they?" or "will they?"

Can cowmen focus and pyramid their economic power and assets for their own good?

Yes.

Beef cowmen can begin to take control of their economic destiny by using basic American capitalism.

What do they have to work with? What tools do they have to begin to move from "price takers" to "price makers"?

Importantly, beef cowmen are "loaded"; they have money and assets. They have a lot of assets--really a staggering amount of assets--real wealth.

Beef cowmen control assets in the magnitude of $150 billion. (An estimated 50 million cows x an estimated $3,000 investment per cow unit in land, equipment, and animals.) And it's reasonable to assume that most of those staggering $150 billion are net assets. While it is true that most beef cowmen have mortgages on their land and cattle, the mortgages are for but a small part of the total assets. Debts cloud the titles, but real wealth is there. Nationally, the USDA estimates that farmers and ranchers have $8 in assets for each $2 in debts. Incidentally, this asset-to-debt ratio is far better than the ratio for most other American businesses, which have about $1.50 in assets for each $1 in debts.

Look at another comparison. On Dec. 31, 1981, _Forbes_ magazine estimated the market value of the largest 500 corporations in the U.S. at $918 billion.

It has been estimated that the 40 largest food retailing companies in the U.S. have a stock market value of only one-third of the assets of beef cow owners.

Thus, it seems clear that a lack of assets--economic power--is not a problem for cowmen. Cowmen have mammoth assets by any measure.

But their assets are diffused--spread out in the hands of from 1.5 million to 2 million individuals. Big ones, little ones, disinterested ones, middle-sized ones. The averge-sized beef herd in the U.S. is estimated at from 30 head to 45 head.

Over the last 100 ears, all sorts of cattle organizations have been established, grown, and thrived for a time, then faded from the scene or have become lethargic, "don't-rock-the-economic-boat," semisociopolitical organizations. Perhaps that will always be. For some strange reason, cowmen are big on talk in hotel lobbies at their conventions and at auction sales. But they have only moderate influence in Washington and state capitals--and little influence in business leverage.

This need not be. It's only a matter of decision.

If cowmen want to begin harvesting the just and fair returns on their staggering total assets, then they must focus and leverage their economic power to control their own destiny.

It's a generally accepted economic fact that the individual or corporation or group that controls a product from inception to the final consumer has the greatest opportunities to influence price and return.

Obviously, U.S. beefmen cannot reasonably expect to control all the steps, processes, and business organizations that move beef from conception to consumer. That being the case, cowmen then should seek control of the more attainable parts of the beef food chain (figure 1, The Beef Control Funnel).

Clearly, the most accessible and potentially controllable phase of the beef food chain is the processing industry. The assets and number of business organizations involved are the smallest in the beef chain. But, it should also be pointed out--quickly--that the beef processing business is a tough one. There are no automatic profits to be made. Operating beef packing firms requires business management skills of the highest order. The casualty list is long.

Yet, it is certainly acknowledged that a large beef processing firm has the opportunity to exercise more price-making leverage than does an individual rancher or feedlot. Such a firm also has the opportunity, through research, ingenuity, and good business judgment, to lever up--add value--to individual products. Plus that, there are opportunities with brand name products.

If the beef packing or processing industry seems to be a place where cowmen ought to try to make some economic impact, how can they do so?

There is an attainable, reasonable American way: Focused Creative Cattle Capitalism!

The idea is simple, perhaps deceptively simple. **Creative Cattle Capitalism means simply buying enough shares of stock in an established, profitable packing firm with the goal of acquiring control of that firm. Then, a second firm; then a third....**

Now, It's important to understand that control doesn't necessarily mean 100% ownership. In this day and time of giant business firms with millions of shares of stock owned by thousands and thousands of individuals and firms, effective control can often be gained through ownership of from 5% to 25% of the shares.

Let's look at an example to see what cowmen might have to do to take effective legal control--done the American way.

The example is hypothetical. But the numbers are based on those of a real company now operating successfuly as of

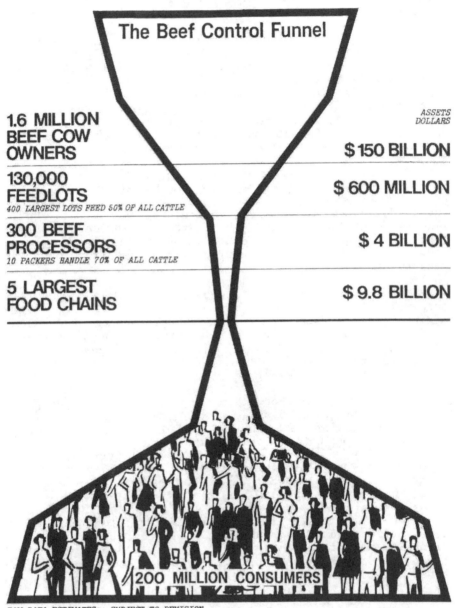

The Beef Control Funnel

ASSETS
DOLLARS

1.6 MILLION BEEF COW OWNERS — **$ 150 BILLION**

130,000 FEEDLOTS — *400 LARGEST LOTS FEED 50% OF ALL CATTLE* — **$ 600 MILLION**

300 BEEF PROCESSORS — *10 PACKERS HANDLE 70% OF ALL CATTLE* — **$ 4 BILLION**

5 LARGEST FOOD CHAINS — **$ 9.8 BILLION**

200 MILLION CONSUMERS

RAW DATA ESTIMATES: SUBJECT TO REVISION

Figure 1.

Dec. 31, 1981. The numbers are ratioed to protect the company's identity.

XYZ Co.

Assets	$ 100,000,000
Sales	1,150,000,000
Net profit	9,000,000
Market value of common stock	30,000,000
Shares outstanding	2,500,000
Price per share	$ 12

Goal: Buy 25% of the outstanding shares; (625,000 shares at $12 per share): Total investment is $7,500,000.

Now, how do cowmen find $7,500,000? How do they focus their economic power to assume control of the XYZ Co.?

First, how to pyramid the funds? Simply through ranchers buying shares of XYZ Co. in the open stock market. It seems reasonable that any cowman ought to be willing to invest a small portion of his assets if the investment makes good immediate economic sense (is a good buy) and offers to give some additional economic control of industry.

What is a reasonable amount? Five percent of the value of a cow would not be burdensome to most owners, and it could be handled out of cash flow. Thus, if cows are worth $500 per head, a cowman would invest $25 per cow in the stock of XYZ Co.

Too much? Make it 1% of the value of a cow for five years.

For purposes of illustration, let's assume that many cowmen would be willing to go the 5% route. That means that the owners of only about 300,000 cows theoretically could buy $7,500,000 worth of stock and end up with control of a $100 million corporation. Or to add another facet, if the $7,500,000 were invested in the stock of a Cattle Venture Capital Corp. it could be used as collateral with which to borrow another $15,000,000 for total buying power of $22 million. Such an amount invested in XYZ Co. would mean effective ownership of 75% of the stock outstanding--easily enough shares for control.

Good buy? Yes--you have acquired control of the assets of the $100 million XYZ Co. for 22 million. You easily could elect a majority of the members of the company's board of directors. You have acquired a good operating company; experienced personnel; a realiable, recognizable name; an established reputation and share of the market.

Obviously such a move as outlined above is more complicated and difficult than I have suggested. Yet, the pages of the **Wall Street Journal** report such actions day after day.

The secret--and it really is no secret--is the focusing and leveraging of assets. The secret is the focusing of assets by cowmen to reach a desired goal.

Those of you who understand business organization and finances can quickly suggest several other available alternatives by which cowmen can effectively purchase control of other phases or parts of the beef food chain, or restaurant chains.

Conversely, there are those of you who can and will just as quickly question the specific numbers recited and prove that such will not work.

But, our point seems abundantly clear: If cowmen want to begin to gain some control over the destiny of their food product--and the potential resulting profit--they must become as astute capitalists as they are cattlemen.

REFERENCES

The Concord Desk Encyclopedia, Concord Reference Books, Inc.

Forbes, "The Forbes 500," May 10, 1982.

A Global Food Animal Protein System--Pipedream or Possibility? C. G. Scrugs, International Stockmen's School, 1983.

Progressive Farmer, May 1973

5

NATIONAL CATTLE HERD STRUCTURES IN THE EUROPEAN ECONOMIC COMMUNITY AND IN THE UNITED KINGDOM

Harold Kenneth Baker

With the European Economic Community, beef production is an important sector of the agricultural industry. Indeed only the U.S. produces more beef than the EEC (table 1). Beef production had been rising up to 1980. There was, however, a slight fall in 1981, and a further fall is anticipated in 1982.

TABLE 1. BEEF AND VEAL PRODUCTION IN THE EEC AND OTHER SELECTED COUNTRIES[a]

	1979	1980	1981[b]	1982[c]
		(Thousand Tons)		
United States	9,925	9,999	10,343	10,262
Canada	946	971	1,000	990
Australia	1,770	1,537	1,435	1,375
New Zealand[d]	512	496	504	495
Argentine	3,092	2,876	2,975	2,970
Brazil	2,100	2,077	2,300	2,400
USSR	7,029	6,673	6,700	6,650
Eastern Europe	2,656	2,596	2,475	2,425
Japan	402	478	471	465
EEC - 10	6,886	7,168	6,953	6,867
Total	35,318	34,811	35,166	34,800

Source: MLC (1982).
[a]Carcass weight equivalent.
[b]Estimate.
[c]Forecast.
[d]Year ended September.

The major beef producing country in Europe is France with nearly 2 million tons of beef a year (table 2). Germany is the second largest producer, and the United Kingdom is third with a production of about 1 million tons of beef a year. Although overall the EEC is just about self-sufficient in beef, there are wide variations in self-sufficiency within the Community. The UK is just over 80% self-sufficient, while Italy imports even more and is only 60% self-sufficient.

TABLE 2. EEC BEEF AND VEAL PRODUCTION[a]

	1979	1980	1981	1982[b]
		(Thousand Tons)		
Belgium/Luxemburg	263	309	320	310
Denmark	256	246	240	232
West Germany	1,516	1,570	1,580	1,540
Greece	99	100	95	90
France	1,957	1,965	1,995	1,965
Irish Republic	426	539	410	410
Italy	880	923	865	905
Netherlands	411	430	450	435
United Kingdom	1,058	1,086	998	980
Total EEC Consumption	7,007	6,952	6,770	6,830
EEC Self-sufficiency (%)	98.3	103.2	102.7	100.6

Source: MLC (1982).
[a]Carcass weight equivalent.
[b]Estimate.

There are considerable structural differences between the national cattle herds in Europe and those in the New World. Generally speaking, in the Americas and Australasia, dairy production herds and beef herds are quite distinct and independent of each other. By contrast, in Europe there is a much closer relationship between dairy and beef production. Even within Europe, however, there are differences between the national herd structures.

As a broad generalization, cattle production in Continental Europe is dependent upon dual-purpose cattle that provide both milk and beef. These dairy cattle are typically either the Simmental or the Friesian. In the past, the Friesian has been intermediate in type between the traditional European dual-purpose animal and the extreme dairy-type Holstein from North America. Recently, however, the latter type has increasingly been gaining in importance and affecting the type of black and white dairy cattle used throughout Europe.

Compared with the U.S. beef cows are relatively unimportant; thus in West Germany they only account for 3% of the total cattle population (table 3). The greatest proportion of beef cows is in the United Kingdom where they account for nearly one-third of all cows. France has the next largest proportion, followed by Ireland and Italy.

TABLE 3.   DAIRY AND BEEF COW HERDS IN THE U.S. AND EEC, 1980
(THOUSANDS)

	Dairy cows	Beef cows	Total	Beef cows as % total herd
United States	10,779	37,086	47,865	77.0
Belgium	977	131	1,108	11.8
Denmark	1,066	65	1,131	5.8
Ireland	1,449	419	1,867	22.4
France	7,120	2,891	10,011	28.9
Italy	3,013	750	3,763	19.9
Luxemburg	69	13	82	15.9
Netherlands	2,356	–	2,356	–
United Kingdom	3,277	1,454	4,728	30.7
West Germany	5,469	170	5,639	3.0
Total EEC	24,796	5,890	30,685	19.2

Source: MLC (1982).

In Continental Europe, the beef cows are usually pure-bred and are a completely distinct population from the dairy cattle.   In the U.K., however, there is a greater inter-dependence between the beef and dairy cattle with many of the beef cows produced by crossing beef bulls onto dairy cattle.

This close relationship between beef and dairy herds appears to be confined to Britain and is biologically very efficient.   Dairy cows not required to breed replacements are crossbred to beef bulls, thereby raising the beef merit of their progeny.   Crossbred females from these matings can be reared as replacements for the beef suckler herd, sup-porting more crossbred beef cows than would be possible if the beef herd had to provide all of its own replacements.

The French cattle industry is similar to Britain's (in that many dairy females are crossbred to beef bulls) but stops short of using the resulting crossbred female calves as suckler herd replacements.   Suckler herds in France are composed almost exclusively of purebred females.

Within the EEC, there are considerable variations in the patterns of beef production.   These are influenced both by the structure of the industry and also by the way in which the different categories of animals are used.   Thus, the by-product calves from the dairy herd may be slaughtered as "bobby" calves at a few days of age, reared for veal, or carried through to slaughter as mature beef.   In most parts of the world, male calves are castrated, but within some European countries they may be kept intact males and used to produce "bull beef."   These bulls are normally slaughtered at between 9 and 24 months of age depending upon the inten-sity of feeding.   Within the EEC there is considerable variation in the production of bull beef.   In Germany more than 50% of the beef comes from bulls (table 4).   In con-trast only 1% or 2% of the Ireland and UK beef comes from bulls.

There are also different ways of using cows for beef production. For example, cow beef from dairy herds is, on the average, younger than that from suckler herds. This results from cow replacement rates in dairy herds (often about 25%) being higher than in beef herds. Quite heavy culling occurs in dairy herds during the first lactation because of poor production or reproductive failure, so some of the cow beef from dairy herds comes from relatively young cattle and is of good quality. By contrast, most culling in beef herds is among older cows and the beef is only of manufacturing quality.

TABLE 4. CATEGORIES OF BEEF PRODUCTION IN EEC COUNTRIES PERCENT BY WEIGHT (1977 EEC STATS)

	Cows	Bulls	Steers	Heifers	Veal and "bobby" calves
Belgium/ Luxemburg	29	31	9	20	11
Denmark	40	48	1	10	1
Ireland	21	1	53	25	1
France	35	12	18	14	21
Italy	26	48	5	7	14
Netherlands	42	15	1	11	31
United Kingdom	22	2	52	23	1
West Germany	30	51	2	13	4

Source: MLC (1982).

The various sources of beef can be seen in statistics from EEC countries categorized according to type of production (table 4). Cow beef makes the greatest contribution to production in countries where beef is exclusively a by-product of dairying, e.g., the Netherlands. The high level of cow beef in France is also a reflection of the export of young calves, many of them in Italy.

The various patterns of beef production clearly have an effect on the average carcass weight produced in the different countries. Table 5 shows the average cattle and overall carcass weights. The former are affected by the importance of cow and bull beef, but veal production clearly influences the overall average slaughter weight. Thus, the Netherlands has the lowest overall carcass weight and the highest proportion of veal production.

DETAILS OF UNITED KINGDOM CATTLE HERD

In the United Kingdom during the last two decades, there have been marked changes in the make-up of the national cow herd. Although the total dairy-cow numbers has remained fairly constant at around 3.2 million cows, the number of beef cows has shown a marked fluctuation. Thus from 1960 to 1974 the population of beef cows more than doubled from about 800,000 to nearly 2 million. However, since 1970, the number of beef cows has decreased to just over 1.4 million in 1981.

TABLE 5.   AVERAGE CARCASS WEIGHTS IN THE U.S. AND ECC

	Average carcass weight	
	Cattle (kg)	Overall (kg)
United States	288	272
Belgium	336	283
Denmark	231	225
Ireland	268	268
France	233	234
Italy	260	223
Luxemburg	336	283
Netherlands	286	199
United Kingdom	268	260
West Germany	300	277

The national cattle herd has consistently produced around 1 million tons of beef during the past decade. About 60% of this is produced as a by-product of the dairy industry, 34% comes from the national beef herd, and the balance of 6% is produced from feeding cattle imported from Ireland (figure 1). Trade from Ireland is much smaller and less important than it used to be. In the 1960s and early 1970s, 500,000 to 700,000 cattle were exported annually to the United Kingdom from Ireland; in 1980 the figure was about 180,000. This decline has been caused by the export of Irish cattle to other EEC countries (notably Italy) and non-EEC countries, particularly Libya,and by the retention of a higher proportion of cattle for feeding and finishing in Ireland.

Another significant development within the UK has been the build-up of an export trade in male calves from the dairy herd to Europe (table 6). Typically these calves are male British Friesians and many of these are used for veal production in France and Italy. Veal production is of little importance in the UK but, traditionally, the worst of the dairy progeny has been slaughtered as "bobby" calves, usually for pet food or other manufacturing purposes. The exported calves are generally considered to be of reasonably good quality for beef production, and the repeal of slaughterings on some of the poorer calves will have some effect on the finished commodity in terms of efficiency of production and carcass characteristics, although probably not as much as is often claimed by members of the meat trade!

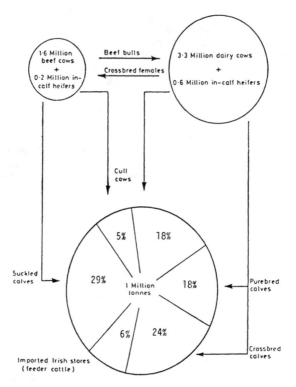

Source: MLC (1982).

Figure 1. Structure of the UK cattle industry and sources of beef, 1981

The UK Dairy Herd

The UK dairy herd provides a high proportion of the total cattle for beef production and hence its composition is important in the beef context. The breed structure of the UK dairy herd has changed considerably since 1955. In England and Wales the proportion of Dairy Shorthorn cows has fallen from 25% in that year to less than 1% in 1979. The proportion of the small-sized breeds (Ayrshire and Channel Island) has also fallen from 36% of the cow population to 8% while the Friesian has increased from 41% in 1955 to 89% in 1979. Similar changes have occurred in Scotland and Northern Ireland at the expense of the Ayrshire and Dairy Shorthorn, respectively. This trend is continuing so that the proportion of Friesian will now be probably over 90%. The main reason for this change was undoubtedly the milking ability of the Friesian, but there can be no doubt that the good beef quality of the British Friesian, particularly when crossed to beef breeds, augmented this change. The beef-cross calf from the Friesian has regularly commanded a high price premium and is an underlying reason for the current interdependence of the beef and milk industries.

TABLE 6.   THE NUMBERS OF EXPORTED CALVES AND "BOBBY" CALVES
           SLAUGHTERED IN THE UK (THOUSANDS)

	No. exported calves	No. calves slaughtered	No. calves slaughtered classified as:	
			"Bobby"[c]	Veal[c]
1970	38	256		
1971	27	259		
1972	29	154		
1973	29	141		
1974	13	416	378	40
1975	130	531	471	60
1976	250	295	262	33
1977	395	264	231	32
1978	421	157	128	29
1979	365	145	123	21
1980	299	146	123	22
1981	240[a]	118	93	25
1982	300[b]	100[b]	75[b]	25[b]

Source: MLC.
[a]Estimates.
[b]Forecast.
[c]No figures available on the split between "bobby" and veal
for years 1970 to 1973.

In the past few years, the move towards greater total
milk yield per cow has continued with an increased use of
North American Holsteins. In 1979, pure Holsteins accounted
for only 1% of the dairy cow population(Milk Marketing
Board, 1979).   However, it is estimated that about 20% of
the black and white calf crop now contains Holstein blood.
It seems likely that this trend will continue.

Incidence of crossbreeding on dairy herd. Crossbreeding, as a means of upgrading a new breed became an established practice in dairy breeding, and in cattle generally.
Thus the old-fashioned dual-purpose Shorthorn replaced the
Longhorn and many of the local British breeds. After the
last war, the Shorthorn itself was then almost entirely
replaced by the Hereford for beef and British Friesian for
milk. This latter breed is now being at least partially
replaced by the North American Holstein. Although the overall change in the breed structure of the dairy herd has been
dramatic, it has been achieved using only a proportion of
the breeding herd. Thus, parallel with normal dairy cow
replacement, and any current breed substitution program,
there has also developed the practice of using dairy cows--
those not required for dairy herd replacements--for crossbreeding with beef bulls.

44

Artificial insemination is very important in the
national dairy herd and accounts for about 80% of all matings. Survey data from the Milk Marketing Board on the use
of artificial insemination from 1963 to 1980 gives a good
indication of the type of matings in the national herd.
This data showed that the proportion of dairy cows being
mated to beef bulls has fluctuated around the 32% mark during the last two decades (figure 2). There has been a 5- to
6-year cycle in the levels of beef inseminations with peaks
of about 38 to 39%. These fluctuations have coincided with
the relative fortunes of the dairy and beef industries.

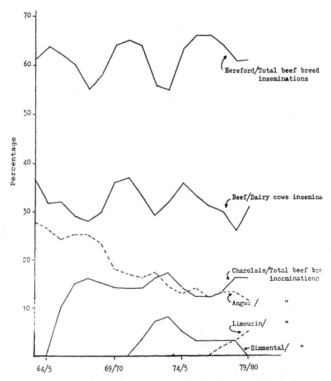

Figure 2. Breed demand for inseminations in dairy cows from
1963 to 1980

The use of the beef bulls on dairy cows has two major
objectives: first, to improve the beef potential of calves
not required as dairy herd replacements and, second, to
promote easier calvings. The latter is particularly important in dairy heifers. About one-third of the cows that
yield too low to breed as dairy cow replacements, and
three-quarters of the heifers, are mated to beef bulls. The
resulting calves are generally sold through auction markets
and for that reason, the ability of the beef bulls to color-
mark calves is important. Finally, beef production in the
UK is associated with the use of grass and grass products.

The combination of easy calving, color-marking, and the ability to finish on grass and (or) hay and silage has led to the popularity of the Hereford as the major crossing bull of the British dairy herds. This popularity is quantified in figure 2, which indicates that about 60% of the beef matings in the dairy herd are with Herefords. The Charolais has had an increasing impact on the beef crosses since its introduction in the early 1960s and has now replaced the Angus as the second important beef breed. The proportionate use of the Angus has declined from just under 30% to about 10%. Its now now is virtually limited to crossing with dairy heifers. After an initial period of increasing use, there has been an almost complete fall-off in the use of the Simmental. This is almost certainly because it has no distinct advantages over the Charolais and, at the same time, has failed to color-mark its calves. The Limousin has recently created considerable interest and the latest statistics show that it is competing with the Angus for third place. Like the Simmental, the Limousin fails to color-mark its progeny, but it has built up a reputation for producing a high-quality carcass, and many of its progeny are the subject of some form of cooperative, or contract, marketing.

Beef from the dairy herd. As indicated earlier, the 3.2 million dairy cows (about 90% of which are British Friesians or North American Holsteins) produce about 60% of the total beef produced in the UK (figure 1). Cull dairy cows provide 18% of total beef and another 18% comes from by-product, purebred dairy calves, typically Friesian steers. The remaining 24% of beef from the dairy herd comes from the crossbred calves resulting from the use of beef bulls on dairy cows. The most common of these is the Hereford cross Friesian steer with smaller numbers of Hereford cross Friesian heifers (many of these finding their way into beef herds as beef-cow replacements) and smaller numbers still of beef breed calves produced by crossing Continental beef bulls with Friesians.

The UK Beef Herd

Figure 1 showed that the national beef cow herd produces about 34% of the UK beef. Cull cows provide 5% and suckled calves 29%. Apart from the use of the dairy/beef crosses for feeding and finishing, a high proportion of the crossbred females from dairy cows enter the national beef herd as suckler cows to supplement the crossbred females such as the Blue-Grey, which are produced in hill and mountain herds. The Blue-Greys are traditionally produced by crossing Whitebred Shorthorn bulls with Galloway cows. It is important to recognize that the bulk of the commercial beef cows in the UK are, therefore, crossbred; in the lowlands the majority are Hereford cross Friesian plus a much smaller number of Angus cross Friesians. In the hills and tougher environments, there are more Blue-Greys supplemented by a number of other beef crosses, e.g., Hereford cross Blue-Greys or cross Angus.

Thus, over 90% of commercial beef cows are crossbred and well over one-third of these will be Hereford-cross Friesian. The only purebred commercial beef cattle are the Galloways and Welsh Blacks in the hills and mountains and a very small number of the native red breeds (Devon, Lincoln Reds, and Sussex) in the lowland. Pure breeding is normally restricted to the pedigreed beef herds, which in any case are relatively small and the average number of cows in a pedigreed herd is only about 20.

Continental breeds of cattle have been widely used, some of them in low numbers during the past decade. The Charolais has had a considerable impact since its original importation in 1961 and is numerically the most important of the Continental breeds. It is estimated that in 1980 about 40% of the calves in beef herds were sired by Charolais bulls and another 10% by other Continental breeds.

There has recently been an increase in the number and proportion of artificial inseminations undertaken in beef cows; the percentage of AI rose from 2% in 1964 to 4.7% in 1974. During this period, the number of beef cattle had more than doubled so that the number of artificial inseminations in beef cows increased five times. This was also associated with a big increase in Charolais inseminations.

Although there are now well over 30 beef and dual-purpose cattle breeds in the UK, the bulk of beef is produced as a result of breeding within and between three breeds: Friesian, Hereford, and Charolais. The most numerous types of feeding cattle are the pure Friesian steer, the Hereford-cross Friesian, and the Charolais cross out of Hereford-cross Friesian or Blue-Grey beef cows.

Part 2

A GLOBAL VIEW OF ANIMAL AGRICULTURE

6

A GLOBAL FOOD-ANIMAL PROTEIN SYSTEM: PIPEDREAM OR POSSIBILITY?

Charles G. Scruggs

Throughout the long ages of man, meat has been his primary food. Further, a surface study of archaeology seems to indicate that those civilizations that primarily depended on meat have been the most advanced. The telltale signs of man dating back as far as 10,000 years ago indicate that man was primarily a hunter--and thereby a meat consumer. Today, we marvel at the drawings of muscular bulls and heavy bison found on cave walls, drawings indicating the high regard early man had for meat animals.

Approximately 9,000 years ago, man began to domesticate animals. "At Zauri Chemi, dated 9000 B.C., many wild sheep had been killed when immature, as if the inhabitants had either fenced in the grazing grounds of wild sheep, penned herds, or even tamed sheep to the extent that they could control the age at which they were killed."

The earliest evidence of tame cattle is dated approximately 7000 B.C. By 2500 B.C. to 2300 B.C., the Egyptians were milking cows.

Again, a quick observation seems to indicate that civilizations that depended most heavily on cereals were often the least advanced. When animals were not present and man was forced to depend on cereals alone, the civilizations failed to flourish and often disappeared.

As man began to concentrate in cities, he still desired--indeed, seemed to prefer--meat. And meat animals were a lot easier to transport to the cities than were cereals. Meat animals had four legs. They could be driven to people. And so they have been for thousands of years.

As man moved into new territories, he often took his meat animals with him. Cattle and sheep have traveled with the armies of invaders and in the trains of the missionaries. These animals helped civilize the world. Some of the cargo on the earliest ships was live animals-- incredible, when you think about the tiny vessels and harsh long voyages made as long as 2,000 years ago. And even 200 years ago.

The earliest hunters and travelers discovered that by drying or salting they could preserve and transport meat more easily. In fact, it comes as something of a shock to us in this day and time to find that more pounds of salted or pickled meat were exported from the U.S. to Europe and the West Indies over a hundred years ago than are exported fresh today.

Then we made great progress--refrigeration was developed. This was an amazing advance--fresh meat was available year round! But refrigeration also became a shackle and a chain. In recent years, especially in the U.S., we have become chained to our refrigerators. We can go only so far as we can take our refrigerators!

On the other hand, refrigeration has also brought to the U.S. and Europe unparalleled adventures in good taste, variety, and quality of meat. Refrigeration has allowed us in the U.S. to enter the roast, steak, and hamburger era in a way never thought possible. But it has also allowed U.S. beef producers to go wandering off into the insignificant. We began looking inward and forgot the rest of the world, thinking, "Those dumb people overseas wouldn't know good beef if they saw it."

Instead of seeking wider markets, we have spent our time over the last 70 to 80 years arguing over such insignificant things as horns or lack of horns; or whether a few hairs of a different color are wrongly placed; or trying to make bulls and cows look like steers. Now we are rushing about and exchanging thousands of dollars on the difference of a centimeter or two in testicle size, trying to put legs on cattle that might look better on thoroughbred horses. And, oh, the hours we spend washing, puffing, combing tail hair so as to make a nice round-looking ball at the end of a show animal's tail! Insignificant fadism in the extreme!

We in the U.S.--the greatest food-animal producers in the world--have perhaps been so busy arguing over minute details of grade standards, which no one but a few experts understand, and over other inward-looking insignificant details, that we have missed the greatest world market for meat that has ever existed since the earliest Homo sapiens killed their first Bos primigenius (auroch or wild ox). In short, "we have been so busy fighting gnats that we let the herd get away."

There is a big world all around us--some four billion people--all hungry for meat! There are in this populous world of ours today approximately one billion people who exist at a malnourished level. Most of the people in the world today do not get the health benefit that we have proved comes from a diet in which food animal protein is the major ingredient. If, indeed, it is true that a good diet containing major elements of food animal protein makes for more intelligent, more ambitious, harder working people, doesn't it follow that U.S. livestock producers should seek to contribute that food animal protein to mankind--at a profit? Would not this contribution be greater than wars or multibillion dollar giveaway programs?

Think about it! What if the approximately 800 million Chinese could be induced to consume only one (1) pound of food animal protein per year above their present one pound? What would demand for 800 million pounds of meat do for beef, pork, sheep, and goat--food-animal--producers in the U.S.? Add to the Chinese the Russians, the Indians, and the Southeast Asians, and you have a need for food-animal protein that boggles the mind!

Thus, U.S. livestock producers must move from their present "cowboy" mentality to one in which they seek to become world marketeers of food-animal protein. We must move from being little more than herders to thinking of ourselves as food-animal producers whose goal is to sell meat protein to a hungry world--at a profit!

And this shift to world food-animal-protein suppliers can be made while we continue to supply the American consumer with desired beef, milk, pork, and mutton.

How?

Think big.

Think systems.

Let's be more specific. Here are what I believe to be the essential elements of developing a Global Food Animal Protein System.

Phase I. Search out a small or medium-sized country that has unfulfilled nutrition and food problems, heavy population density, and a reasonable amount of foreign exchange.

Study the food habits of the population in detail. Find out exactly how the people prefer to eat meat. Learn all there is to know about their overall food habits, preferences. Study their tariff laws, their customs, their religion--everything that could make an impact on their meat consumption. Learn to think as the natives think and react.

All the while your goal is a modest one: Increase per capita meat consumption, on the average, by one (1) pound.

The key to this phase is to drop the American habit of turning up our noses at the way other people like their meat. Just because we like roasts, there's no reason that all the rest of the world must eat roast the way we do. If the natives want barbecued tail bone, let's not call them stupid and say there is no market, and retreat to the nearest McDonald's. Let's sell them tail bones! The goal is a good old American custom: Study a market, decide what can be sold at a profit, then produce it well enough to earn a profit.

Phase II. Devise a production and processing system to deliver the product without waste of time or effort from conception to consumption. We in U.S. agriculture now do just exactly the opposite: We produce something and then try to peddle it in the form we want--or "to hell with 'em."

Phase III. Take the basic lessons learned in one country with modest goals, and apply them to other countries one by one--at a profit. Soon, we will have made a world contribution.

Let's explore some other dimensions of the idea of supplying food-animal protein to a malnourished world.

Let's look at Russia. Russia desperately needs meat. At present, they get their meager supplies from their own limited meat production system. They import some meat from Australia and some from their satellite European countries, but not nearly as much as the Russian population needs and wants. Instead of importing meat, the Russians are trying to do it the long, hard way: Import grains from all over the world, then process and feed them through their herds and flocks.

Has the U.S. ever seriously studied with the Russians (or the Hungarians, or the Saudi Arabians) a policy of importing ready-to-consume meat products (mutton and chicken) instead of raw grains and the resulting timelag to consumers? Wouldn't it be possible, technologically, for us to ship lean grass beef to Russia at a profit? Russians prefer beef that tastes much different from our own grain-fed beef. Mostly they get cull cow or bull beef--if they get any beef at all. Let's sell them what they are used to eating.

We can load forage-produced boneless beef on planes at Atlanta or Dallas, go up to 35,000 feet and quick-freeze it at no cost, and land it in Moscow 16 hours later. Too expensive? We don't know. So far as I know, we have never even tried it. Flying too much moisture? We can dehydrate beef fibers, reconstitute them in Istanbul or Pakistan. Remember our hunting, frontier-busting forebears? How did they transport beef supplies when traveling alone? Jerky beef. It needed no refrigeration. Wouldn't a malnourished native of Gambia be glad to have some jerky beef to mix with his root foods and maize?

American technological expertise is envied the world over. But we haven't used our skill to try to supply food-animal protein to a food-deficient world. Can we not preserve meat and milk through irradiation and/or treatment so that it can be held on a pantry shelf? Good U.S. agricultural policy, it seems to me, should be one that uses U.S. technical advantages to the benefit of the U.S.--and then for other citizens of the world. If so, we must begin to shift from a policy of shipping raw, unprocessed grains to a policy of exporting value-added, more nearly ready-to-eat food-animal products. The U.S. has the livestock. We have more feed and feedgrains than anyone. We have the nutrition knowledge. We have financing. Shouldn't we ship food-animal protein products and short-circuit the long food production chain that most countries now are trying to establish by buying our grain and soybeans?

What about the strategic military considerations of such a policy? Ship food protein ready to eat instead of supplying grain? That should be the U.S. goal. We must do this instead of supplying grains so the countries can build their own livestock infrastructure for future self-sufficiency.

The U.S. food-animal production industry has seen some quantum jumps in the last few years in development of all-tender, flavorful meat products. Meat protein cubes, dehydrated meat fibers, ready-cooked meals are just a few of the new forms and products pouring out of labs. More are about to emerge. However, we cannot hope to supply all the world with edible animal protein. Perhaps we should adopt this policy: Sell edible animal food protein to those countries whose needs are greatest and most immediate. To those countries with developed livestock industries, sell genetic and germ plasm materials to use in upgrading their production. This whole field of genetic engineering now in its infancy in the U.S. may shatter much of our previous livestock thinking. Livestock producer leaders must be equally bold in their thinking!

I have no doubt that if U.S. food animal producers--at one time called cowmen, sheepmen, dairymen--would set out to develop a Global Food Animal Protein System, they could do so rather easily. But to do so they must throw off the shackles of the past, think aggressively American, think profit, think systems, and above all think BIG.

* * *

"Populations of the lower- and middle-income countries still are increasing rapidly, and in many countries there is a growing number of affluent people whose diets are being upgraded to include more red meat, dairy products, and eggs. As more countries are unable to meet their national requirements for staple foods, their governments are looking for outside sources of supply, often on an urgent basis. It is dangerous politically for national leaders to let food shortages occur, driving prices up; they run the risk of unrest, violence, and even overthrow of governments."

Beyond the Bottom Line
The Rockefeller Foundation

REFERENCES

Fagan, B. M. Men of the Earth: An Introduction to World Prehistory. Little, Brown and Co.

7

FOUNDATION OF CIVILIZATION: FOOD

Allen D. Tillman

> "And he gave it for his opinion that whoever
> could make two ears of corn or two blades of
> grass grow on a spot of ground where only one
> grew before would deserve better of mankind and
> do more service to his country than the whole
> race of politicians put together."
> --Johnathan Swift
> The Voyage to Brobdingnag
> in Gulliver's Travels

I chose this quotation because it is apparent to me that man now has the knowledge and power to make two ears of corn or two blades of grass grow on a spot of ground that formerly would grow one or less.

Civilization is defined as "an advanced state of human society in which there is a high level of culture, science, industry, and government." The high level of civilization that we enjoy today has resulted from many technological developments in agriculture that increased the amount of food produced and the efficiency of human labor in producing it. Each innovation freed more people for the development of human society and of the culture, science, industry, and government found in it.

In discussing some of the developments, this paper is divided into major sections as follows: (1) a brief history of agricultural development worldwide; (2) the close relationship of agricultural and industrial developments in modern societies; (3) some characteristics of a successful agriculture at national levels, and (4) reasons for optimism about the world food problem.

HISTORY OF AGRICULTURAL DEVELOPMENTS IN THE WORLD

> "History celebrates the battlefield whereon we
> meet our death, but scorns to speak of the plow-
> ed fields whereby we thrive; it knows the names
> of the king's bastards, but cannot tell us the

origin of wheat. This is the way of human folly."
--Jean Henri Fabre

For convenience, I have divided this section into three parts, as follows: the gathering and hunting stage (2,000,000 - 7000 B.C.), the low-technology stage (7000 B.C. - 1750 A.D.); and the scientific stage (1750 A.D. - the present).

The Gathering/Hunting Stage

"Cultural man has been on earth for some 2 million years and for 99% of this period he has lived as a hunter/gatherer. Only in the last 10,000 years has man begun to domesticate plants and animals, to use metals, and to harness energy sources other than the human body. Of the estimated 90 billion people who have lived out a life span on earth, over 90% have lived as hunters/gatherers, about 6% by agriculture, and 4% have lived in industrial societies. To date, the hunting/gathering way of life has been the most successful and persistent adaptation man has ever achieved."
--Lee and Devore (1968)

The first ancestor of man, Australopithecus, appeared on earth about two million years ago. His main invention was the knife, which was made by putting an edge on a pebble, an invention that permitted him to kill animals, skin, and cut meat, thereby changing him from an herbivora to an omnivora. This change was dramatic, because the addition of dietary meat reduced the bulkiness of his diet by about two-thirds, permitting him to leave the trees and to become more mobile to better utilize the rapidly developing savannas. Also, meat required less time to gather, thus he had more time for social activities - improvement in communication skills and in his tools. And so man for the first time released the brake that environment imposes on his fellow creatures. It is significant that this basic tool was not changed very much for the next million years, attesting to the strength of the invention.

Homo erectus came onto the scene about one million years ago. His ability to walk upright freed his arms, which improved his ability to hunt, and his greater ability to adapt to many ecosystems permitted him to spread out from his place of origin, Africa. In fact, the classical discovery of Homo erectus was the Peking man, who lived in China about 400,000 years ago. He was the discoverer of fire, which was used for warmth and cooking. The Neanderthal man, who was discovered in Europe, appears to have led directly to us, Homo sapiens.

The test of the ability of Homo erectus to adapt came about 500,000 years ago when the Pleistocene Ice Age covered much of the earth. Clans of 40 or more moved to caves for protection and work, a move that required a new organization - the young and stronger men (usually 10 per clan - Willham, 1980) were the hunters, while the remainder were assigned duties in keeping with their abilities. Some of the dwellers even had time to paint on the cave walls. Bronowski (1973) felt that these are saying to all - "This is my mark, this is man." Man is now saying that he has the ability to shape the world and is not a mere creature to be shaped by the environment.

The great glaciers began to retreat about 30,000 years ago. Left in their wakes were great savannas that were soon filled with grasses of all kinds and with cloven-hoofed ruminants to consume these. In response, the clans came out of the caves and spread out over the plains following the animals, going north in the summer and returning south in the winter. This was the beginning of a transhumant way of life, which later became dominant in many areas of Eurasia. In fact, there are cases of this way of life even today-- East Africa, North Africa, Finland.

The dog was domesticated in about 20,000 B.C., and this greatly increased man's ability to hunt. Also , there developed a symbiotic relationship between man and animals: animals furnished meat, skins, and other products for man, while man furnished some protection and salt to the animals. Urine of meat-eating man contains salt, a valuable commodity to ruminants in many salt-deficient areas. Man developed oars in about 20,000 B.C. and the bow and arrow in about 15,000 B.C.; both inventions increased his efficiency as a hunter. Man domesticated the gregarious sheep and goats during the latter part of this stage, and benefited greatly by the increased food supply from these animals.

At the end of this stage, about 10,000 to 7,000 B.C., and for a long time after man had already established village agriculture, animals continued to be an important source of food for man. The live animal represents, until it is sacrificed, a reserve food supply. This fact is often forgotten or overlooked by planning economists, who plan for the aid programs that are given to the developing countries. Early man recognized the importance of animals: so much so that the root word used for money in many languages reflects the importance of animals to man (Leeds and Vayda, 1965).

Low Technology Agriculture (7000 B.C. - 1750 A.D.)

"The greatest single step in the ascent of man is the change from nomad to village agriculture."

--Bronowski (1973)

With the receding of the Pleistocene ice, there came great environmental changes in the Old World. The hot and dry winds off the Eurasian Steppes eliminated all but the hardier grasses in much of North Africa and some of Asia, thereby turning some of the lands into semideserts or deserts. As a result, wild animals migrated to the river valleys of the Euphrates, Tigris, Indus, Nile, and Yellow rivers. Agriculture began when the nomads decided to stay put to exploit plants. Whether the nomads planned to develop agriculture or were the benefactors of two genetic accidents is not clear. What is clear is the fact that the early ones came to hunt and to gather wild wheat. By a genetic accident (Harlan, 1975), wild wheat containing 14 chromosomes, crossed with wild oat grass, also containing 14 chromosomes, to produce a fertile hybrid, called Emmer. It contains 28 chromosomes, thus the grain was much larger than wild wheat. Therefore, man began to cultivate it. Emmer's grain is so tightly bound to the husk and chaff that it is easily dispersed by the wind. As a result, it spread over wide areas. It appears that by another genetic acident, bread wheat came to the settled agriculturalists: Emmer crossed with another wild oat grass to produce still another fertile hybrid, bread wheat, containing 42 chromosomes, which has a large grain. In contrast to Emmer, when bread wheat is broken, the chaff flies off, leaving the grain in place, thus it is not spread by the wind. With the advent of bread wheat, which man has to plant and cultivate, man and wheat developed a symbiotic relationship that remains up to this day (Heiser, 1978).

Farming and husbandry in a settled agriculture creates an atmosphere from which technology and science take off (Bronowski, 1973). The first tools used by village agriculturalists were the digging stick, which was invented about 7000 B.C. This later evolved to a crude plow, the footplow that used human labor, which appeared in about 6000 B.C. The cow was domesticated in about 6000 B.C., and when man yoked the ox to the plow he for the first time began to utilize a power source greater than the human muscle. This was undoubtedly the most powerful invention of this early period, making it possible for man to wrest a great surplus of agricultural products from nature. The surplus food released more men to create, invent, innovate, and to build great civilizations--something for which the nomads had never had time. Civilizations, with their specific cultures, developed on the flood plains on the Nile, the Tigris-Euphrates, the Indus, and the Yellow Rivers. Many of their activities, such as irrigation, required cooperation by many men; therefore there developed administrative systems that led to the building of empires. Law and government developed. The great cities of that period thrived on a cereal-based agriculture. Trade between cities developed in order for them to acquire necessities - salt, spice, metals, etc. (Thomas, 1979).

The animals found in the settled villages, up to about 3000 B.C., were goats, sheep, the oxen and the onager, a kind of wild ass. As long as the animals were the servants of agriculture, all went well. But some time after 4000

B.C., the horse was domesticated and the nomads learned to ride in about 2000 B.C. Thus, the nomad was transformed from a poor wanderer to a threat to the settled villages. Warfare of that period was intensified by the discovery of how to ride the horse, and warfare became a nomad activity. The nomads battered on doors of the settled villages from about 2000 B.C. until the early part of the 14th century A.D. Sometimes the nomads were successful and took over villages, but in all instances, the nomads were absorbed into the villages. Historians have made all of us aware of the great wars waged by the nomads, recording the names of the famous nomads--the Huns, the Phrygians, and the Mongols. The Mongols were defeated in about 1300 A.D., thereby ending the threat of their making nomad life supreme throughout sections of the Old World.

When the horse collar was discovered, the horse became an important draft animal, especially in northern Europe were great teams of horses turned the heavy sod. Without these teams of horses, which permitted the Vikings to produce great surpluses of grains, they could not have been such a military power and threat to much of northern Europe.

The low-technology stage continues right on through the European Renaissance (Thomas, 1979), during which time there were many contributions to agricultural innovations, each with its consequent improvements in food production and the efficiency by which man produced it. As there were many such improvements over time, for sake of time and space, let us summarize the advances:

- Man developed methods for the systematic exploitation of plants.
- Man developed methods for the cultivation of plants for the production of grain--wheat, barley, millet, and rice (Heiser, 1978).
- Man domesticated animals--dog, cow, sheep, goat, and the horse.
- Man developed systems of irrigation.
- Man developed some degrees of mechanization--the digging stick, the plow, the ox-drawn plow, the wheel, and others.

The developments in limited technology were not continuous but came in ebbs and flows throughout the period up to about 1750. There appeared to be a de facto technological ceiling upon agricultural production throughout the entire stage. At the core was the simple Malthusian element--population expansion ultimately pressed against the land, thereby producing malnutrition, famine, disease and, finally, a decrease in population. In China, Hung Liang Chi (Rostow, 1978) the predecessor of Malthus, wrote "during a long reign of peace--the government could not prevent the people from multiplying themselves." Rostow said it well--"During this period of limited technology, if war did not get you, peace did." And so ended the second era.

The Scientific State (1750 - present)

The scientific state began in about 1750 and continues until the present. In this period, the western nations for the first time broke the ceilings on agricultural and industrial technology so that invention and innovation came at a regular flow. The key to these, I feel, was the advent of the "scientific revolution which brought with it experimental science." Experimental science, for the first time, permitted and motivated the formulation of scientific laws to describe general and natural scientific phenomena. This led scientists to design experiments that would lead to the manipulation of nature to man's advantage. This exciting time saw the advent of scientific agricultural societies in which agriculturalists met together for the discussion of scientific discoveries and how these could be put to use by farmers. Innovations came at a faster pace, and as in the past, each invention or innovation increased the level of food production and its efficiency of production. The rates increased rapidly and we now have the development of high-technology agriculture. Some of the basic advancements that are characteristic of this stage are as follows:
   - Classification of soils, along with estimates of their fertilities.
   - Improved plants by gene manipulations (genetic engineering).
   - Improved animals by gene manipulations.
   - Scientific utilization of fertilizers.
   - Proper use of irrigation.
   - Proper use of fermentation and other means of food preservation (Tannahill, 1973).
   - Use of insecticides, fungicides, herbicides, vaccines, etc.
   - Use of modern techniques in farm mangement.

Many names stand out during this third era; however, it is significant to mention that King George III (better known to Americans for other reasons) gave much support to the newly developing agricultural research in England. Some feel that the innovations resulting from this agricultural research greatly increased agricultural production in Great Britain. In fact, some suggest that it was the English agricultural revolution that permitted that nation to defeat Napoleon at a time when agricultural imports were essentially cut off by France.

Agricultural research is so important that every modern nation has now developed a national agricultural production plan in which agricultural research is a powerful component. Those countries that are lagging in agricultural production are the ones that have been slow in developing good agricultural research, teaching, and extension programs.

How about the United States? Our country had its beginning in 1776, or sixteen years after the third stage began. When our Declaration of Independence was signed, the Revolutionary War was fought, and our people set out to build a nation, fully 90% of our people were directly engaged in agricultural production. If time and space permitted, it would be useful to point out the significant inventions, such as the first cotton gin, the first wheat thresher, the first steam or gasoline-propelled tractor and many others, all American inventions or innovations, and to note the effect of each upon food production and efficiency. However, I will end this section by saying that during the 200 years since Independence, our farm population has decreased from 90% of the total population to only 5%. However, these fewer people are producing food of improved variety and quality, providing nourishment for a vigorous population. Our national agricultural program has been one of the modern success stories.

THE CLOSE RELATIONSHIP OF AGRICULTURAL AND INDUSTRIAL DEVELOPMENT IN MODERN SOCIETIES

In reviewing agricultural developments worldwide, Rostow (1978) noted that the modernization of agriculture and industry have gone hand-in-hand in successful national development programs in the past. In fact, one finds that the modernization of agriculture must precede industrialization in most countries. There are many reasons for the close interrelationship; successful agriculture (Foster, 1978) provides:

- An ever-expanding supply of high quality foods to nourish its people, increasing their vigor.
- An adequate supply of high quality food, available at a reasonable price, to combat inflation which, if left uncontrolled, hampers national development.
- Capital for the expansion of the nonagricultural sector of the economy. This is very critical in the early stages of industrial development because at least 90% of the population in every country studied were farmers at this critical period.
- More food for the nonagricultural population. Therefore, all through the modernization process, more and more people are freed for work in the nonfarm sector.
- Educated and motivated people for work in the nonagricultural sector in the developed countries.
- Land for the nonfarm sector--for highways, railroad, airports, shopping areas, etc.
- A market for the nonfarm sector--tools, machines, medicines, insecticides, clothing, gasoline, etc. Again, this market is most critical in the early stages of industrialization.

SOME CHARACTERISTICS OF A MODERN AND PRODUCTIVE AGRICULTURE
AT THE NATIONAL LEVEL

In general, a successful national agriculture results
from a good agricultural plan or programs that provide farmers with relevant production information and assures an adequate infrastructure that is needed for both production and
marketing of agricultural products. In addition, farmers
must receive a fair return from their investment of land,
labor, and capital. Otherwise, production will be sporadic
rather than continuous. Some specific requirements are:
- Adequate government financial support for research, teaching, and extension (public service)
  for agriculture.
- Infrastructure: The national government also
  has to provide certain .components of the infrastructure needed by farmers--such as roads, harbors, railroads (in some cases). (Many of the
  infrastructures in the U.S. are now furnished by
  industry.)
Some inputs needed by modern farmers are as follows:
- Farm machinery and spare parts
- Farm tools and spare parts
- Power source--gasoline, electricity, diesel fuel
- Fertilizers
- Insecticides, fungicides, herbicides, vaccines,
  etc.
- Irrigation tools--pumps, pipe, valves, etc.
- Credit
- Others
Those who are familiar with the ready availability of
inputs on the American scene might well question why private
industry hasn't made these available in many developing
countries and in developing agricultural industries. In
many cases, private industry will not take the risk of production and distribution of many necessary inputs unless the
volume and price justifies the risk. In such situations,
the government has to subsidize these inputs until the industry is well along toward development. The writer has
spent over 10 years in four developing countries and found
that the unavailability of certain inputs, such as vaccines,
insecticides, and fertilizers, has served as a severe constraint on production. In the same vein, lack of marketing
services results in severe losses of the harvested produce.
Marketing services needed to assist production increases are:
- Farm produce collection and storage
- Processing and subsequent storage of the processed produce
- Wholesale distribution of produce
- Retail distribution of produce
- Marketing--farm markets, small stores, supermarkets, and others.

Even with all of the above components in place, farmers will not produce continuously unless the price for agricultural products pays for the costs of inputs and allows them a fair return on their investments.

REASONS FOR OPTIMISM ABOUT THE WORLD FOOD PROBLEM

The world food problem has two components--the demand side (population) and the supply side (production). Many reports today emphasize that the demand side is growing faster than the supply side. Let us analyze the population problem first.

Population

The demographic facts at first glance are frightening! If we plot population against time, we see that the population growth from about 8000 B.C. until about 1650 A.D. was almost a straight line function, a very slow growth rate. In 1650 A.D., there were about 0.5 billion people, and the rate of increase was only 0.3% per year. At that rate, the population would require 250 years to double. Instead we have had an exploding population since about 1700 A.D., and now some authorities estimate that we will have 6 to 7 billion people in the year 2000.

It is my belief that the population estimates have all been wrong, from Malthus down to the United Nations' recent estimates. For example, the U.N. estimated in 1976 that the world population by the year 2000 would be 7 billion. However, in 1979 that estimate was reduced to 6 billion. Therefore, in 3 years, we "lost" 1 billion people. There are many other examples of inaccuracies. Why is this so? Population projections require assumptions about the choices people have and make in regard to the size of their family. Therefore, it is easy to see why all past assumptions have been wrong.

Simon (1981) maintains that there is a built-in, self-reinforcing logic that forces the rate of population growth to respond to resource conditions. It does so by reducing population growth and size when food resources are limited, and expands it when resources are plentiful. Others have studied population changes in Europe for 1400 A.D. until 1800 A.D., and found that the population did not grow at a constant rate, and that it did not always grow. Instead, there were advances and reverses, provoked by many forces--famine and disease, however, were not the major forces. We know that increased incomes, associated with economic developments, reduce birthrate and population growth. For example, the populations in Singapore, Hong-kong, Japan, and other places have tended to stabilize within the last 20 years.

In summary, these facts lead me to suggest that population size tends to adjust to the production conditions.

Following a technological advance in agriculture, there is an increase in production followed by an increase in population size. However, the rate of population increase levels off as the new technology is "used up".

Food Production

If I am optimistic about the demand side of population problem, I have even more reasons to be optimistic about the supply side--food production. As pointed out by Schultz (1964), the world food problem exists because of low productivity in the poor countries. It is estimated that the amount of crops produced per ha of land in the poor countries was only about one-fourth of that in the industrial countries, and the production of animal products/animal unit was even lower. Therefore, the potential for greatly increasing production lies in the poor countries where it is needed.

Starting with the success of the Green Revolution in India during the 1965-70 period, there has been an awareness worldwide that a nation can increase its food supply if it has the will to do so.

Since the World Food Congress in Rome in the early 1970s, the FAO, the World Bank, the bilateral aid programs and other organizations have given priority in their aid programs to agricultural development in the poor countries. Up to 1975, the proportion of foreign aid dedicated to increasing food production was less than 10% of the total program of aid and the level has now more than doubled.

Some of the reasons for my optimism about future world food production are:
- The nature of the food/population problem as a whole is now becoming understood.
- The complexity of technology transfer from the industrial countries to the developing countries is also becoming understood; the international research and development centers have made excellent progress in these endeavors.
- The potential for increasing yields per land unit in the poor countries, most of which are located in the tropics, is enormous. We call this "payoff on research." I have made some estimates of the annual rates of return on monetary investments in agricultural research, both by commodities and by countries. If we consider rice research in the tropics, the figures run from 46% to 71%; however, if only Asia is considered the figures vary from 74% to 102%. the International Rice Research Institute, in the Philippines, has done an excellent job.
- Fertilizers and the knowledge of how to use them are now available to farmers in the poor countries. Many national governments are now subsidizing the price of this input.

- New and better adapted plant varieties are now available to most poor countries; many of these plants are resistant to certain diseases.
- Many national governments in the poor countries are now using aid and loans from the FAO, World Bank, bilateral sources and philanthropic organizations to support agricultural production by improving some or all of the following: (a) The infrastructure (such as roads, communication, and harbors), (b) The teaching, research, and extension structure facilities and activities. The research is now directed at solving their own production problems. (c) The credit structure. (d) Water resource utilization (e) Active intervention programs to increase the production of certain critical commodities. For example, in 1976-75, Indonesia initiated an active intervention program in rice production using subsidization of inputs, price stability, increased research, increased extension inputs, etc.

In 1969-71, the average production of padi was 2346 kg/ha in Indonesia. By 1978, this figure increased to 2921 kg/ha, an increase of 27.5% in 8 years, or about an increase of 3.5% per year. Preliminary figures for 1980 show a dramatic increase and give hope that Indonesia is about to gain self-sufficiency in rice production.

An interesting aspect of Indonesia's successful efforts on a single commodity, rice, is the fact that production improvements in other commodities have not improved, and some have decreased. To me, this is a god sign--meaning that with increased effort, Indonesia could increase production in any agricultural commodity chosen. Their leaders now know this and are putting forth successful efforts to increase production of other selected plants and animals.

REFERENCES

Bronowski, J. 1973. The Ascent of Man. Little, Brown and Co., Boston, Mass.

Burke, J. 1978. Connections. Little, Brown and Co., Boston, Mass.

Foster, P. 1978. Food as foundations of civilization. In Food and Social Policy. G.H. Koerseiman and K.E. Dole (Ed.) Iowa State University Press, Ames, Iowa.

Harlan, J.R. 1975. Crops and Man. Crop Science Society of America, Madison, Wisconsin.

Heiser, C.B. Jr. 1978. Seed to Civilization. W.H. Freeman and Co., San Francisco, CA.

Lee, R.B., and I. DeVore. 1968. Man, the Hunter, Aldine, Chicago, Ill.

Leeds, A., and A.P. Vayda. 1965. Man, Culture and Animals. AAAS, Washington, D.C.

Rostow, W.W. 1978. Food as foundation of civilization. In: Food and Social Policy. G.H. Koerseiman and K.E. Dole (Ed.). Iowa State University Press, Ames, Iowa.

Simon, J.L. 1981. World population growth. The Atlantic Monthly 248 (2):70-76.

Tannahill, Rey. 1973. Food in History. Harper and Row, New York, N.Y.

Thomas, Hugh. 1979. A History of the World. Harper and Row, New York, N.Y.

Willham, R.L. 1980. Historic development of use of animal products in human nutrition. Mimeo. Rpt., Iowa State University, Ames, Iowa.

8

WORLD LIVESTOCK FEED RELATIONSHIPS: THEIR MEANING TO U.S. AGRICULTURE

Richard O. Wheeler,
Kenneth B. Young

INTRODUCTION

Livestock producers both in the U.S. and worldwide have an important stake in the operation of the world grain economy. The continued availability of relatively low-cost grain in the world economy would tend to foster further livestock development in grain-deficit countries and provide competitive advantages in all countries for livestock and livestock production practices more dependent on intensive grain feeding. On the other hand, if world grain supplies become more restricted, grain-deficit countries would be more likely to import livestock products rather than grain, and livestock production practices less dependent on grain feeding would gain a competitive advantage.

Prior to the early 1970s, the world trend was toward increased grain supply and continued buildup of large stocks in the industrialized exporting countries, primarily the U.S. and Canada. These large stock levels helped to maintain relatively low prices and assured a stable supply for importing countries. For example, U.S. wheat export prices deviated very little from $170 per ton from the mid-1950s to the early 1970s. This long period of stable grain supply, sold at very attractive prices, encouraged the widespread use of grain feeding to increase livestock production. In addition, some grain exporters offered other inducements, including liberal credit arrangements and Public Law 480 assistance for countries unable to compete in the international grain market.

During this era, many developing countries adopted intensive grain-feeding practices for poultry and swine production, and cattle feedlots became a prominent feature of the U.S. agricultural system. On a worldwide basis, use of all cereals for animal feed increased from 37% in 1961-65 to 41% in 1975-77 (Harrison, 1981). Eastern Europe and the Soviet Union registered the largest increase in grain feeding of all countries--a change from 48% in 1961-65 to 69% in 1975-77. Use in Latin America increased from 32% to 41%.

At the present time, the world grain market has recovered from the 1972/73 shortfall and world stocks have been restored to the level of the 1960s. Nevertheless, the shortfall did mark a major shift in the supply-demand balance of the world grain market, a situation that had not occurred previously.

The structure of the world grain market has changed dramatically since the 1950s. World trade increased over 200% from 1960 to 1980. Currently, over 100 countries are dependent on grain imports from a few exporters. The U.S. dominates the world grain trade, accounting for about 60% of global coarse-grain exports and 44% of world wheat exports. About 40% of the total U.S. grain production is now exported and the annual rate of increase in exports reached 7% per year during the 1970s. Projections of future U.S. crop exports available from the Economic Research Service of USDA (1981) indicate that the growth in foreign demand will continue through the 1980s, although not as rapidly as in the 1970s (table 1). Annual average export demand for corn and rice is projected to increase about 4½% compared with 2% for wheat and soybeans.

TABLE 1. PROJECTED INDICES FOR U.S. CROP EXPORTS, 1981-1989

Com- modity	1981	1982	1983	1984	1985	1986	1987	1988	1989
					(1981=100)				
Corn	100	106	111	115	123	127	131	135	139
Wheat	100	96	99	101	103	105	107	110	115
Rice	100	109	113	117	120	124	127	131	135
Cotton	100	107	103	103	103	104	106	106	107
Soybean	100	100	101	104	107	111	113	116	119
Peanuts	100	123	140	147	150	153	157	160	163

Source: These projections have been calculated from Problems and Prospects for U.S. Agriculture, ERS-USDA (1981). They are not official USDA projections.

There is some question now about the U.S. ability to keep up the recent pace of expanding exports. Most of the available cultivated land is currently in production and we are losing about a million acres of cropland per year to nonagricultural uses. The rate of soil erosion has increased substantially with more intensive cultivation and use of marginal cropland formerly not used for crop production. The same problem is occurring in other countries and average crop yields are leveling off over much of the world.

PROJECTIONS ON WORLD GRAIN SUPPLY

A Winrock International study was completed in 1981 on world use of grain and other feedstuffs (Winrock International, 1981). The study was designed to evaluate the

interaction between the world livestock system and the feed-
and food-grain system.

Estimates of current world use indicate that poultry
consume 27% of all grain fed; swine--32%; draft animals--4%;
sheep and goats--2%; and cattle and buffalo, including dairy
animals--35% (table 2). Feed use in table 2 is expressed in
terms of megacalories of metabolizable energy

TABLE 2. ESTIMATED ANNUAL WORLD FEED USE FOR DIFFERENT TYPES OF LIVESTOCK, 1977

Livestock category	Livestock Output		Feed use				
	Meat	Other	Grain	Protein meal	By-products	Forage & other	Total feed
	(million metric tons)		(billion mcal ME)				
Poultry	22.8	23.3[1]	387.9	91.1	73.4	51.7	604.1
Sheep & goats	7.3	--	23.6	5.3	35.8	993.9	1,058.6
Cattle & buffalo	46.8	415.0[2]	507.3	42.8	204.2	4,101.0	4,855.5
Swine	41.0	--	460.9	56.0	213.1	157.2	887.2
Draft animals	13.4	--	57.9	5.9	23.5	1,214.9	1,302.2
All livestock	131.3	--	1,437.6	201.1	550.0	6,518.7	8,707.4

[1] Eggs.
[2] Milk.
Source: Winrock International (1981).

rather than metric tons due to variation in the quality and
variety of feed used in different countries. For example,
grain-feed use in the Soviet Union is reported on a "bunker
weight basis" generally containing excess moisture and
extraneous matter. The percentage of grain use in poultry
rations is estimated to be similar for developed, centrally
planned, and developing countries since the technology of
modern poultry production has been readily adopted all over
the world. Developing countries feed less grain and more
forage and by-products to swine. The ruminants in developing countries subsist almost entirely on forages. However,
nearly half of the world grain feeding occurs in developing
and centrally planned countries dependent on grain imports.

Grain feeding was projected to continue increasing
according to recent trends evaluated in the Winrock study.
There will be occasional setbacks for countries with severe
foreign exchange problems and domestic recession. Most of
the centrally planned countries have set target levels of
increased livestock production requiring additional grain
feeding. The Winrock study projected that total world feed
use of wheat and coarse grains would surpass direct human
and industrial consumption by 1985. Recent trends also
indicate that total world grain use is increasing at a
faster rate than world production. World grain demand has
been increasing steadily due to continued growth in world
population and rising per capita consumption of livestock
products dependent on grain feeding while the growth in
supply is slowing due to limitations on development of new

cropland and reduced productivity gains on existing cropland. This increased tightening in the world grain market implies that the grain export price should increase substantially by 1985.

GRAIN SUPPLY OUTLOOK FOR U.S.

Winrock International is currently initiating a study of both the potential for and implications of additional crop production in the U.S. The 1977 Natural Resource Inventory compiled by Soil Conservation Service of USDA shows a total of roughly 460 million acres of cropland available in 1977 containing about 70% prime land in the Class 1 and 2 categories. Heady and Short (1981) of Iowa State University have projected that the 1977 cropland base will dwindle to 353 million acres by the year 2000, but that there are 37.6 million acres of high-potential land and 90.1 million acres of moderate-potential land that could be converted to cropland. This land area for potential development is located primarily in the South Atlantic, South Central, Great Plains, and North Central regions of the U.S. However, other economists in the U.S. have serious doubts whether it would be feasible to convert this much additional land to crop production. Some limitations to development of additional cropland and current use of this land are shown in table 3; the data indicate that much of this potential cropland is currently used for pasture and timber production and that there are definite erosion hazards and probable high conversion costs to develop this land area for crop production. Such limitations imply that there will be a major increase in production cost to bring these new lands into crop production after we reach full capacity on existing cropland.

TABLE 3. ESTIMATED POTENTIAL CROPLAND AND LIMITATIONS TO DEVELOPMENT IN THE CONTINENTAL UNITED STATES

Type of limitation	Percent of potential cropland	Present use	Percent of potential new cropland
Erosion	59	Pasture-range	79
Drainage	23	Forest	17
Soil	7	Other rural	4
Climate	4		
No limitation	7		
Total	100	Total	100

Source: 1977 Natural Resource Inventory, Soil Conservation Service, USDA.

IMPLICATIONS FOR U.S. LIVESTOCK PRODUCTION

World population is projected to increase 50% between 1975 and 2000. There is increasing emphasis on livestock production to improve the quality of human diets, particularly in centrally planned and developing countries, and increasing pressure on cropland worldwide. In the short term, there may be temporary swings between shortages and surpluses in the world grain market. This is expected due to greater year-to-year variation in world crop production as a result of expansion of cultivation on marginal lands with increased drouth stress and other climatic variability. Stability of supply may also be reduced due to mounting pressure on exporting countries to reduce the carryover of grain stocks from year to year. Grain prices are projected to increase with gradual tightening of world grain supplies. Increased export volume will require eventual conversion of at least some pasture and timber land in most of the key exporting countries, with an associated rise in grain-production cost.

Some implications for U.S. livestock producers include increased grain-feeding costs eventually rising above the general inflation rate and the loss of some pasture and rangeland area converted to cropland as indicated in table 3. Higher grain prices will be translated into somewhat higher meat prices, particularly for poultry, swine, and fed cattle because these enterprises are highly dependent on grain feeding. However, it will become more profitable to utilize additional crop residues and by-products in livestock feeding, particularly for cow maintenance, to replace present grain use. The biggest deterrent to using these low-quality feeds is cost--primarily for labor, equipment, and interest. To date, the availability of a stable supply price for grain is analogous to the situation we had 10 years ago for oil and natural gas. It has not been cost effective to utilize many alternative sources of feed energy, although there is an abundant physical supply available in the U.S. The amount of corn crop residue physically available was estimated to be 231 million tons in 1977 (Ensminger and Olentine, 1978). This would support 117 million cows for a 4-month grazing period on a purely physical supply basis. The nutritive value of crop residues can be enhanced with special processing techniques, and some very promising research work has been done on ammonia treatment of straw.

Cattle, sheep, and goat producers could potentially utilize these alternative sources of feed to substitute, at least partially, for grain or to enlarge breeding herds even on a drylot basis if it became more economical to do so. If meat prices increase along with grain prices, some livestock producers may regain a competitive advantage over poultry and swine producers who are more vulnerable to rising grain prices. Under the current regime of depressed grain prices, poultry producers, in particular, have been gaining a significant competitive advantage over beef producers. Poultry meat prices have now declined to 30% of average beef prices compared with 80% a few years ago (National Cattlemen's Assocation, 1982).

A reduction in grain feeding of cattle in the U.S. would mean increased competition for use of existing pasture and range lands. With increased grain exports, there would be an associated reduction in the grazing land area, particularly in the Southeast and Great Plains regions of the U.S. Increased dependence on crop by-products and residues implies that more livestock will be produced in traditional cropland areas to utilize these waste products, as was the practice in the U.S. before the feedlot era began. This is the situation now in most developing countries where the bulk of livestock production is found in mixed crop/livestock systems (Winrock International, 1981).

Increased use of crop residues to reduce the amount of grain feeding would have a significant impact on the management system for livestock, especially cattle. Levels of annual offtake would decline as cattle would have to be nearly a year longer to reach market weight on a less intensive feeding program. The cattle operator would be forced to move cattle to crop-production areas and to lease crop residues from crop farm owners as is now done for wheat pastures. Additional use of feed supplements would be necessary as crop residues are generally lacking in total nutrient requirements. The cattle operator would also have to invest in additional fencing and equipment to utilize crop residues. Thus the overall implications are that major adjustments would be required in the U.S. cattle industry, including shifts in the location of production, the composition of herds, and feeding programs.

CONCLUSIONS

The general outlook for the international grain market points toward continued price variability for feedgrains, a gradually rising price level for grains as the world market continues to tighten, and eventual loss of some grazing land in the U.S. when additional cropland is needed to meet expanding export requirements. It is possible that the problem of price variability may be alleviated by additional government intervention such as paid acreage reduction or other methods of supply control on the market, but this does not appear likely in view of the current emphasis on curbing spending for most agricultural support programs.

Expected consequences of the grain-market outlook for cattle producers include continued fluctuations in feeder cattle prices and returns from cattle feeding during the next few years and a general trend toward higher feeding costs. Although price variability is nothing new to livestock producers, the sharpness and range of price movement will likely be increased as long as we continue to be the shock absorber for the world grain market. The U.S. is one of the few nations that exposes domestic producers to price fluctuations of the international market.

Short-term effects of the expected swings in prices will be of more immediate concern to most livestock producers than the longer term upward trend in feeding cost, particularly for those in a weak financial position. To some extent, producers may be able to reduce the financial risk through greater participation in the futures market or by direct contracting. However, their most urgent need to survive in the livestock business will probably be to secure alternative methods of financing to provide more flexibility on repayment of loans. Other possible options for reducing or spreading the risk of price movement include the development of programs for outside investors to assume partial ownership of livestock and other creative financing schemes to shift at least part of the risk from producers to other outside parties. There may also be an opportunity for further revision of the tax laws to encourage more outside investment in the livestock business.

Long-term implications of changes in the world grain market, as well as expected increases in transportation cost, are that the structure of the U.S. cattle production system will change. Projected world food-system trends suggest increasing prices for all livestock products due to rapidly rising consumption in most countries and upward pressure on grain prices. However, there may not be much increase in livestock-product consumption in the U.S. market because per capita consumption rates have stabilized. A continuing problem for beef will be competition from pork and poultry in the U.S. meat market. To recapture its former market share, beef will require more efficient production and marketing throughout the system.

Increasing grain prices may provide some opportunity for beef producers to improve their production-cost relationship relative to pork and poultry by changing to less intensive grain feeding. Additional research and development is needed on the utilization of crop residues and by-products to reduce cost in cattle production.

REFERENCES

Economic Research Service, USDA. (1981). Problems and Prospects for U.S. Agriculture, Washington, D. C.

Harrison, P. 1981. The inequities that curb potential. FAO Review on Agriculture and Development. Food and Agricultural Organization of the United Nations, Rome, Italy.

Heady, E. O. and C. Short. 1981. "Interrelationship among export markets, resource conservation, and agricultural productivity," Agr. J. Agr. Econ. 63:840.

National Cattlemen's Association. 1982. The future for beef. Special Advisory Committee Report, Englewood, CO.

Soil Conservation Service, USDA. 1977. 1977 Natural Resource Inventory. Washington, D. C.

Wheeler, R. O., G. L. Cramer, K. B. Young and E. Ospina. 1981. The World Livestock Product, Feedstuff, and Foodgrain System. Winrock International, Morrilton, Ark.

Winrock International. 1981. Report on Livestock Program Priorities and Strategy. Winrock International, Morrilton, Ark.

9

LIVESTOCK LEADERS
IN AN ERA OF CHANGE

L. S. Pope

You are fortunate to have participated in this landmark conference under new leadership. The planners of this conference have assembled an impressive array of livestock leaders and scientists from around the world. A massive amount of information has been presented--over 200 papers in all. New ideas have been presented, vital to survival in this turbulent era of change. As leaders in your own business or profession, you carry a dual responsibility--to improve your own operation and also to lead out with progressive and innovative new programs, setting the pace for others.

Seldom in our 205-year history has such leadership been necessary. The pace of change today is staggering; the problems confronting us are enormous. Economic survival of animal agriculture, as we have grown accustomed to it, is in jeopardy. Old landmarks have faded, and we sail uncharted waters. A single miscalculation or poor decision today can send a risk-prone operation into a tailspin. More and more, decisions from the outside are impacting on the profit-and-loss column. Today, each dollar of gross return to U.S. agriculture must service about $12 in debt load. In brief, our task is to operate and survive in a decade charged with risk and uncertainty.

Add to this the fact that we are confronted by new technology on a mammoth scale. Take, for example, the impact of computers on modern agriculture. It is almost impossible to grasp the magnitude of this one development in a world of rapid change. Financial decisions hinge on it, as do tax accounting and record keeping by sophisticated producers. Rations are formulated by computers to gain least-cost advantage, breeding plans are evaluated, sires selected and matings plotted to produce a specification product. Ingredients are purchased and formulated, sales arranged, market information dispersed, and transportation arranged through use of computers. These strange new machines, growing more accessible and economical and speaking a language of their own, now dictate decisions to an extent previously unknown.

No one possesses a crystal ball clear enough to predict the future with accuracy. At this conference, speaker after speakers has brushed back the horizons that might lead to increased livestock production. But each would agree that current projections are clouded by uncertainty. The future belongs to those willing to gamble and innovate, to seek and try new ideas, but at unprecedented risk. Let's point out a few of the emerging challenges that must be considered.

First, let's take a good look at the consumer, the final arbiter of our fate at the market place. Red meats are still king at the market place and on the table; they will remain so well into the next century. This year, the average American family of four will consume almost 3 t of food products, 2324 lb of which will be animal products. However, bear in mind that our yearly consumption of meat proteins has apparently leveled off at about 200 to 210 lb per capita annually. Total consumption is not likely to increase; rather, there is a good chance for a slow decline.

Part of this static condition lies in economics as credit-conscious housewives attempt to cut back on what appears to them to be an exorbitant cost of red meat in the diet. Also, middlemen costs continue to climb. There has been fundamental change toward "light" foods in America due to aversion to fat and the pressure of physicians who consider red meat to be a serious dietary mistake.

Further, we live in a highly mobile America with almost 20% of all families changing addresses each year. The fast-food craze and "eating away from home" have left their impacts--and still continue to do so. We are a population advancing in age, with the average life-expectancy now over 73 years. A younger generation, reared on hamburger diets, is now in the forefront of the working class of our society. According to a recent NCA survey, 64% of the households have significantly altered their eating habits in recent months by eating more poultry and pork (combined) than beef. Another startling statistic: in 70 percent of households today, teenagers do a significant amount of the spending!

In the case of beef, NCA spokesmen have wisely stated that we are now a "mature" industry. Dr. Max Brunk, economist at Cornell University and noted for his astute ability to observe trends, put it succinctly: "No longer can be depend on the growing affluency of the American consumer to provide a market. We must develop an effective market franchise with the consumer."

All this impacts on the future of the average stockman. The meat industry must be more creative and innovative. We must look for new products that can stand on their own. One possibility is "flaked and formed" or steak-like products, mostly made from the chuck, and resembling the true cuts of beef. They are especially valuable for quick preparation and serving by the busy housewife. Intermediate-value products may have enough appeal to move lower-priced beef into a better selling category. One might even let his mind wander--to the day of the "all steak" steer--where the entire carcass might appear as appealing steaks or roasts! The impacts on beef type or conformation, demand for bull carcasses, or even on imported beef are worthy of speculation.

Given the trends of the last few years and the plateau in beef consumption that now exists, it appears that production levels of about 80 to 90 lb per capita may be an optimum demand level. This year it is probable that consumption of poultry and turkey meats will exceed that of pork; the combination will exceed beef by half again. If so, this will be a remarkable shift form the 1960 to 1970 era where beef was king. We have witnessed perhaps one of the great changes in American dietary habits--the swing away from beef that may continue into the 1980s.

To better respond to consumer demands for less fat and more lean in beef, the industry has given much thought to changing beef carcass-grading standards. Surprising to some, it has met a storm of protest considering the relative modest changes proposed, i.e., reduced marbling and less outside fat cover. The reaction to the proposed changes is perhaps more a cause of concern than the change itself. This resistance to change appears to be strongly out of line with consumer demand. Failure to make even a modest move toward a leaner carcass reveals a deep conservative tendency in the industry which is disturbing. Some suggest that failure to change USDA carcass grades may lead to greater use of brand names in beef marketing. Already large supermarket chains such as Safeway have found a ready market among buyers with a leaner product than the present choice grade.

"Genetic engineering" has become one of the popular buzz words of our age. Actually, stockmen have long practiced genetic engineering through selective matings, even from the days of Bakewell and "breeding the best to the best." Today, the change comes through more sophisticated tools available. In plant genetics, for example, it is undoubtedly the hottest item in the scientists' laboratory.

No question, the possibility of creating new genetic combinations in the test tube, together with gene splicing and cloning, plus embryo transplants, could all but revolutionize the plants and animals of today's agriculture. Superior rates of gain and reproductive performance, increases in milk yield and disease resistance are among the tremendous opportunities. With dairy cattle, some experts feel that the only limiting factors to maximum milk production in the future may be structural unsoundness and lack of feed capacity of the cow to support exorbitant levels of milk production!

The opportunities are enormous. In operations coupled with the computer and other sophisticated techniques, stockmen of tomorrow may be dealing with entirely different kinds of animals, necessitating significant changes in nutrition and management. Consider the impact, for example, of the relocation of genes affecting reproduction into species with low reproductive performance, as in the case of the beef cow.

Equally startling results may be in the offing through improved nutrition, "targeting" of nutrients to change or alter body composition and increase muscle development. Increasing knowledge of the pathways and alternate routes of nutrients in meeting body needs brings with it exciting new dimensions. Although not as glamorous at present as new genetic frontiers, it may well have more useful impact in the future.

New understanding of muscle biology has paid off in improvements in meat technology. Widespread use of electrical stimulation to induce greater tenderness in the beef muscle is a prime example of the application to the production line of basic research from the 1970s. Today, tenderness can no longer be considered a prime factor in breeding and management; rather it can be introduced through technology. New products that can be prepared from the hot-boned carcass, cooked within hours after slaughter using energy-efficient techniques, may soon be adopted. The stockman must be aware that modern meat technology may soon alter, reshape, and indeed improve the product far beyond current feeding, breeding, or management techniques.

What of the future of large commercial feedlot confinement with swine, beef cattle, and dairy operations? This question must be carefully considered by the livestock industry, especially where enormous financial inputs are necessary. The trend has been toward bigness, but at what cost and risk? Will cattle feeding shift back toward older and more traditional areas, such as the Cornbelt? With shorter feedlot periods almost a certainty, together with the high costs and risks of cattle feeding as we know it today, plus exorbitant cost of money, the return to forage-type beef production may take on added new impetus.

While this conference has dealt mostly with improving production and increasing quality of product, stockmen must not lose sight of one of the biggest obstacles of all—profitable marketing. The era of "price taking" may be fast fading from the American agricultural scene—at least many bankers hope so. The possibility is strong that the 1980s will be recognized as the turning point in livestock marketing with much stronger group action than in the past.

Admittedly, this may run against the grain of independent-minded stockmen, long accustomed to making their own decisions. However, it seems inevitable that some sort of organized action will be required to match other supply sectors in our economy. Standing alone, the cattleman of today faces enormous risk; working in an organized fashion, he can gain strength from volume and numbers and can produce specification products to better meet market demands.

Coupled with this is the enormous risk in financing modern livestock operations. Hopefully, many young operators, heavily in debt and deeply leveraged, can weather the current storm. Cash-flow problems are a serious problem to stockmen who have much equity in land, cattle, and equipment but are short of cash to meet current demands. Financial agencies will insist on a tighter ship, with stronger fiscal accountability and more assurance of profit in the future. Use of the computer and system analysis will become a must in tomorrow's operations.

What about the environmentalists, the "do-gooders" who seem bent on continuing, even tightening, restrictions on modern methods of animals production? Today, no reasonable stockman shrugs them off. They will have their say, making an impact through the press, TV, and in the halls of Congress. The "animal rights" movement, as farfetched as it sounds to most stockmen, is a factor to be dealt with in this decade.

The economic realities of life may force some relaxation in the regulations on use of growth stimulants, chemicals, and confinement programs, but the prospects are not too encouraging. The significant decline in farm and ranch population has steadily eroded a once-powerful political base. Witness, too, the recent Congressional in-fighting this past spring between farm groups themselves and the fragile base upon which milk price supports currently rest. One would hope that agriculturalists of all colors and stripes would band together for the common good. More realistically, we can count on continued internal differences, to the advantage of groups that are consumer oriented. Also, we still operate under a national "cheap food" policy, increasing the difficulty of gaining true and effective legislation to serve our ends.

Basically, a question probably running through the minds of many at this conference is whether or not there will continue to be a place for a viable, profitable animal agriculture. The answer is not as obvious as one might think--or wish to believe. Obviously, there will be, but it will be a much different industry than we see today. The production practices we will use tomorrow, the tools available in breeding and management, the markets we must serve, the financial lifelines we must maintain--all will be vastly different. Given these circumstances, the time is right to fashion new approaches to meet the problems that will confront us tomorrow. Only resourceful stockmen, skilled in "risk management" and keenly aware of the fundamental and far-reaching changes underway can hope to survive. Fortunately, you represent a significant part of the vanguard that will lead the way toward a brighter future.

10

SOME CURRENT PROBLEMS
OF ANIMAL PRODUCTION IN POLAND

Henryk A. Jasiorowski

Our recent economic crisis and related food problems in Poland were given so much publicity in the West that it is impossible for me to lecture about animal production in Poland without starting with some general remarks.

Discussing the food problem on the world scale, we always recognize its two chief aspects: production and distribution. There is no exaggeration at all in the statement that worldwide food distribution poses far knottier problems than does food production.

The crux of the matter is that while world food output is generally adequate and even can be stepped up to a much higher level, there are many people in the world who lack the means to acquire even the merest food rations for themselves and their families.

Looking at the Polish scene from this worldwide perspective one can easily find that our recent food problems stem from diametrically different causes than is the case in other food-deficient countries.

In the past, we have managed to deal with the food distribution problem in a way that befits a socialist country. The main policy has been to maintain low food prices—which, however, has brought us many problems caused by low profitability of production and excessive meat consumption compared with the country's national income. In my opinion, these last two factors stemming from the country's sociocultural practices lay at the root of the food-related components of the latest crisis. Thus, although food distribution has caused us problems, those problems are the opposite of those found in food deficient countries.

In our climatic region the living standard of society can be measured among other things by the share of articles of animal origin in food consumption. The problem is illustrated in table 1 which shows clearly that food consumption in Poland was about average in Europe at the end of the 70s.

TABLE 1. AVERAGE DAILY CONSUMPTION IN THE YEARS 1977-79

Country	Calories		Protein		Fat	
	Total	% of animal origin	Total gr	% of animal origin	Total gr	% of animal origin
Poland	3,515	34.0	100.0	52.2	125.8	79.7
Hungary	3,553	35.7	95.7	51.5	136.0	81.7
France	3,412	36.2	97.9	61.8	149.7	69.0
Sweden	3,065	40.8	82.6	64.2	141.3	72.1
Italy	3,517	23.7	97.9	45.6	130.7	50.5
Greece	3,365	22.5	98.2	48.0	138.6	40.9
Yugoslavia	2,737	22.8	99.5	35.5	102.9	65.7

Source: FAO.

Regarding the share of products of animal origin in the national diet, we were decisively ahead of such countries as Italy, Yugoslavia, and Greece, but somewhat behind others such as France and Sweden.

These figures, although undoubtedly accurate, did not quite indicate public opinion on the matter. This was the fault of the distribution. While people in the starving countries lack the money for food, in our country (in the 1970s, at any rate) people were ready to spend far more on food than the national income and food supplies allowed.

Because the problem of food distribution (including the distribution of meat) has recently acquired the rank of an issue that is almost crucial to national existence and will be resolved through the price reform already announced, I think I need not deal further with this topic and can devote my attention to production issues.

FARMING

Poland has very diversified systems of farming. On one hand, there are large state and cooperative farms with the average from several hundred to several thousand hectares of land, but there are also small private farms that average about 5 hectares. In total, 75% of the agricultural land is owned and cultivated by private individual farmers and 25% of the agricultural land belongs to the state and coopera- tives. Thus, agricultural production in Poland is dominated by the small family-operated farms.

LIVESTOCK POPULATION

In 1980, Poland had about 12.6 million cattle, 21.3 million pigs, 4.2 million sheep, 83.4 million poultry, and 1.9 million horses. Is that a lot or not enough to produce an adequate amount of food of animal origin? It is important to record the official population of all livestock animals with the exception of horses. Any disruption in the growth trend always has been treated with great concern. Whether we have enough livestock animals or not we can only determine through comparison with other countries. Let us begin by comparing land acreage per capita in Poland with other countries.

Table 2 shows that we do not compare badly in terms of land resources with other countries. Of the countries around us, only Hungary can boast of a higher per capita acreage of arable land and pastures. The FRG's land resources per capita, for example, are 2.5 times lower than Poland's.

TABLE 2. LAND ACREAGE AND THE POPULATION

| Country | Population per square kilometer | Land acreage per capita | |
		Arable land and permanent plantations	Permanent pastures
Poland	114	0.42	0.11
Czechoslovakia	120	0.34	0.11
GDR	155	0.30	0.07
Sweden	18	0.36	0.09
FRG	248	0.12	0.08
Hungary	116	0.50	0.12
France	159	0.35	0.24

Table 3 shows the number of livestock animals per 100 hectares of agricultural and arable land.

TABLE 3. LIVESTOCK ANIMALS PER 100 HECTARES OF LAND (1980)

| Country | Animals per 100 ha of agricultural land | | Animals per 100 ha of arable land | |
	Cattle	Sheep	Pigs	Poultry
Poland	67.8	22.5	145.6	569
Czechoslovakia	72.4	12.9	149.4	941
GDR	93.3	32.9	254.5	1,080
Sweden	51.7	10.6	90.5	450
FRG	124.5	9.4	306.9	1,201
Hungary	30.8	46.2	165.9	1,388
France	79.2	38.9	65.7	1,148

In numbers of cattle, pigs, and poultry per 100 ha, we rank behind our immediate neighbors, Czechoslovakia and the GDR. Compared with Hungary, we have more cattle but fewer pigs and poultry. Because large state farms dominate in Hungary, and small private farms predominate in Poland, this

indicates that there is no clear relationship between livestock numbers per unit of acreage and the structure of agricultural production

The comparison of our land acreage with that of other
countries clearly shows that we still have untapped possibilities to step up livestock numbers. This statement is
given further support when the number of people employed in
agriculture is considered.

Another criterion for assessing the size of livestock
population is the number of animals per capita (table 4).

TABLE 4. THE NUMBER OF LIVESTOCK ANIMALS PER CAPITA (1980)

Country	Number of animals per capita			
	Cattle	Sheep	Pigs	Poultry
Poland	0.36	0.12	0.60	2.34
Czechoslovakia	0.32	0.66	0.49	3.12
GDR	0.33	0.12	0.72	3.07
Sweden	0.23	0.05	0.33	1.62
FRG	0.36	0.02	0.36	1.42
Hungary	0.18	0.27	0.78	6.53
France	0.45	0.22	0.21	3.72

In terms of livestock numbers per capita, our country
ranks well with only France having more cattle per capita,
and only the GDR and Hungary having more pigs than does
Poland. Our per capita sheep and poultry numbers also seem
adequate when compared with those of other countries.

Taking these findings into account, we must conclude
that it is not the size of our livestock population that
causes our meat shortage problems. This is not to say, of
course, that we can afford to ignore any declining trend in
animal population, but that topic I shall discuss a little
later.

MEASURES OF ANIMAL PRODUCTION

The most important measure of animal production is probably expressed as a per capita ratio. Again, comparisons
with other countries are necessary.

Table 5 shows that Poland produces quite a lot of milk
per capita, with only a few countries like France and GDR
ranking ahead of us. We produce more meat per capita than
do Sweden and the FRG--and as many eggs as they do. We do
not compare badly in wool production, either.

Judging by production, therefore, there should be no
shortage of products of animal origin in our country. If
the facts of our daily life seem to show something to the
contrary, the responsibility for the shortage must be placed
on the upset market equilibrium between supply and demand.

The only other possibility is that the statistics are wrong --even the international statistics--which is rather hard to accept.

TABLE 5. PRODUCTION PER CAPITA IN KILOGRAMS (1980)

Country	Milk	Meat	Eggs	Greese wool
Poland	478.5	83.2	13.8	0.39
Czechoslovakia	391.6	92.6	15.7	0.28
GDR	490.0	102.9	19.4	0.68
Sweden	417.9	64.1	13.4	0.07
FRG	402.4	76.4	13.6	0.07
Hungary	236.2	132.9	18.7	1.12
France	625.6	100.5	15.6	0.41

The next important measure of animal production is output per hectare of agricultural land.

Unfortunately, judged by this measure (output/hectare), our animal production no longer keeps pace with that of other countries (table 6). While things are not too bad with milk production, our meat and egg outputs/unit of agricultural land fall short of the results of other countries. We can conclude, therefore, that animal production in our country is more land-intensive than animal production in the remaining countries mentioned in table 6.

TABLE 6. PRODUCTION OF PRODUCTS OF ANIMAL ORIGIN IN KILO-GRAMS/HECTARE OF AGRICULTURAL LAND

Country	Milk	Meat	Eggs
Poland	896.4	155.8	25.9
Czechoslovakia	866.6	204.7	34.8
GDR	1,305.7	274.2	51.7
Sweden	934.1	143.3	29.9
FRG	2,050.0	389.4	69.5
Hungary	380.4	213.9	30.1
France	1,054.9	169.5	26.3

Source: FAO (1980).

As for the productivity of the animals themselves, table 7 shows that the gaps between the individual countries are substantial.

Table 7 shows the low productivity of livestock in Poland. Milk yields per cow are among the lowest in Europe and beef and pork production per animal is not much better. More important, during the past 10 years the productivity of our animals has failed to increase at a satisfactory rate. The average annual milk yield of our cows, for example, which was 2,441 kg in the years 1969 to 1971, increased to only 2,861 kg in 1980. Although this represents the average growth rate in Europe, we must remember that we started from a low level. In this period, average beef production per animal rose from 50.1 kg to 57.8 kg, while pork production declined.

TABLE 7. ANNUAL LIVESTOCK PRODUCTIVITY (1980)

Country	Average no. of kilograms of milk/dairy cow	Average no. of kilograms of beef/animal	Average no. of kilograms of pork/pig
Poland	2,861	57.8	79.7
Czechoslovakia	3,227	77.3	106.1
GDR	3,861	72.4	99.3
Sweden	5,281	79.0	117.0
FRG	4,538	101.0	118.3
Hungary	3,244	94.9	106.9
France	3,365	81.6	152.7

Although from 1969 to 1971 the average pork production per animal stood at 90.6 kg, in 1980 it was down to 79.7 kg. If these are accurate figures on the size of herd in those years, such reductions are truly alarming, especially when considered together with the sharp rise in intensive grain feeding in animal production in Poland during this period.

DEVELOPMENT OF ANIMAL PRODUCTION IN POLAND

Animal production in Poland grew rapidly after the war, reaching its highest growth rate in the 1970s (table 8).

TABLE 8. GROWTH OF MEAT AND MILK PRODUCTION IN POLAND (THOUSAND TONS)

	1938	1950	1960	1970	1975	1980
Beef	227.0	162.5	235.0	464.6	695.4	690.1
Veal	40.0	57.2	98.2	82.2	57.1	48.7
Pork	558.6	957.8	1,177.2	1,278.8	1,792.8	1,766.2
Mutton	12.1	4.6	27.5	23.8	21.6	26.0
Horse meat	2.0	5.0	26.6	35.2	43.3	53.8
Poultry meat	40.8	33.4	62.3	127.9	235.6	358.3
Total meat	896.6	1,245.3	1,646.2	2,031.5	2,867.5	2,943.1
Milk	10,000.0	7,760.0	12,123.7	14,498.6	15,882.7	16,606.9

Source: Statistical Yearbook (1980).

In the last decade, meat production per hectare of agricultural land increased from 11.2 t to 16.5 t, while milk production rose from 74.2 t to 87.1 t. As a result, meat consumption per capita increased from 53.0 kg to 70.6

kg annually, milk consumption from 262 liters to 264 liters, and egg consumption from 186 to 219 eggs a year.

These figures are truly impressive. But let us look at them against the background of the growth of grain, concentrates, and intensity of production.

According to the data of the Ministry of Agriculture, total consumption of feed concentrates for animal production in the years 1970 to 1980 increased from 13 million tons to 22 million tons (table 9). In this period, the supplies to agriculture of feed concentrates processed in central compounding facilities increased from 4.1 million tons to 13.7 million tons. This increase was achieved along with an increased importation of grain that reached 8 million tons annually in recent years. In comparison with annual grain production in Poland of around 20 million tons, the amount imported was very high.

TABLE 9. FEED CONCENTRATE CONSUMPTION (THOUSAND TONS)

	1960	1965	1970	1975	1976	1977	1978
Total consumption according to the Ministry of Agriculture	9.694	11.813	13.124	19.960	20.371	19.569	22.165
Including feed concentrates produced in central compounding facilities	–	–	4.125	11.078	10.853	12.405	13.729

Table 10 compares the growth of animal production and the growth of feeding of concentrates, providing a clear picture of the growing intensity of grain use in animal production in our country. If we calculate the growth of animal production in the 1970s as meat equivalent (total meat + 0.1 milk + 0.5 egg production), we find that 7.4 kg of feed concentrates were used to achieve a 1 kg increase in meat production during that period. This calculation is simplified, but nonetheless is highly alarming.

The growth of grain inputs per unit of animal production in Poland in the 1970s can be accounted for by: the decline of small producers (especially pig farmers); the growth of animal concentration; and the improper balance of feed concentrates, meaning their protein content.

However, the most important reason for the growth of grain inputs per units of animal production in Poland is probably the fact that, so far, the growth of production has been achieved to a greater extent by stepping up animal numbers than by increasing animal productivity.

As a result, a disparity developed between the intensification of the input, meaning the rapid growth in the consumption of feed concentrates and the extensive form of increasing animal production by stepping up the livestock population.

TABLE 10. GROWTH OF LIVESTOCK POPULATION AND ANIMAL PRODUCTION IN POLAND

	1960 $\frac{1978 \text{ production}}{= 100}$ %	1970 $\frac{1978 \text{ production}}{= 100}$ %
Beef and veal production	221	195
Pork production	150	138
Mutton production	95	113
Poultry meat production	575	280
Total meat production	180	146
Milk production	137	115
Concentrates and grain inputs	228	169

Source: Statistical Yearbook (1979).

During the last decade, there was a substantial increase in the country's livestock population: the number of cattle per 100 ha of agricultural land increased from 55.5 to 74.6 (including cows, from 28.8 to 31.9); the number of pigs from 61.8 to 113.9; and the number of sheep from 17.9 to 22.3.

If we now compare this increase with the growth of production, we find that only a very small share of the growth of milk, beef, and wool production resulted from increased animal productivity. The entire growth of pork and mutton production was due to the increase in the number of animals, with animal productivity actually slipping slightly (table 11).

TABLE 11. PRODUCTION PER ANIMAL OR FOWL

	1970	1978
Milk, kg/cow	2,384	2,766
Beef, kg/cow	90	121
Beef, kg/animal	50	56
Pork, kg/sow	849	804
Pork, kg/pig	95	81
Mutton/sheep	7.2	6.1
Wool/sheep	2.8	3.1
Poultry meat/hen	2.0	5.1
Eggs/laying hen	100	117

It is obvious that while we cannot continue further animal population increases or large grain importation, we should continue to boost animal production.

Because we are unable to apply the methods that have been used by highly industrialized countries (high levels of grain feeding), we must now search for our own ways to intensify animal production. Under our current conditions, this means the development of labor intensive systems.

Thus, for the near term, our plans are the following:

- We will emphasize production of ruminants, particularly dairy animals because of the grain shortage and the high cost of energy.
- We will use available supplies of feed grain first to feed poultry because they give the best returns in the production of eggs and poultry meat.
- Wherever possible, we will reinstitute the practice of feeding potatoes to pigs.
- Because it is probable that grain substitutes will be needed in the production of feed concentrates, the nutritive value of these substitutes must be monitored and the composition of rations must be constantly refined.
- Because highly intensive animal production in the sense of grain and capital inputs will not be affordable for some time, we must take care to keep the appropriate size of livestock population.
- We must avoid unjustified concentration of animals because it involves more grain-intensive production.
- Low-scale animal production, especially production based on low grain consumption and low capital expenditures, will deserve full support.

11

LIVESTOCK DEVELOPMENT PROSPECTS IN ASIA: IMPLICATIONS FOR THE U.S. LIVESTOCK INDUSTRIES

A. John De Boer

INTRODUCTION

International trade affects the U.S. livestock indus-tries in many ways--most directly through imports and exports of live animals, meat, and animal by-products. A less direct, but even more important factor, is the trade in feedstuffs. While trade in meat has increased substantially since World War II, the international trade in feedstuffs (primarily soybeans, soybean meal, and corn) has increased spectacularly. Houck (1979) reports changes in U.S. agri-cultural exports between 1953/54 and 1977/78 indicating that agricultural exports increased in value from $3 billion to $27 billion. In terms of percentage composition of these exports, oilseeds and oilseed products changed from 11% to 27%, feed grains from 7% to 21% and livestock products from 15% to 9%. This illustrates the fact that sustained income growth in many major economies has led to sharply increased levels of animal product consumption and increased levels of livestock production that depends upon imported feedstuffs. This trend has been most evident in Europe and Japan. The nine-nation European Community (EC-9) now accounts for 25% to 30% of U.S. exports followed by Japan at 15%. The Soviet Union is now a major factor in the market for both feed-stuffs and meat as it attempts to increase the quantity and quality of animal products available to its citizens. However, because of climatic instability, Soviet participa-tion in the market is much more erratic than the other major purchasers.

Against this background of increased agricultural trade, this paper examines the potential market for U.S. agricultural products in Asia and outlines some expectations of how such forces would react on the U.S. livestock indus-tries. Detailed discussion is limited to a few key countries.

LIVESTOCK DEVELOPMENT IN ASIA

The livestock industries of Asia represent a complex

series of interactions among environments, animals, cultures, and levels of economic development. In general, most developing countries of Asia have been unable to close the demand-supply imbalances for livestock products. The most recent figures on ruminant livestock populations are given in table 1 for selected countries of Asia. Tables 2 and 3 show percentage changes in animal numbers for cattle, sheep, goats, swine, and water buffalo on a regional basis over the 1970-79 period. Population growth of cattle and water buffalo, the two major species in terms of animal units, was virtually stagnant, while swine numbers showed moderate to strong growth. These numbers are in stark contrast to the projected increases in food demand over the 1970 to 1980 period (table 4). As a consequence, increases in imports of animal products have jumped sharply (table 5).

TABLE 1. RUMINANT LIVESTOCK POPULATIONS FOR SELECTED ASIAN COUNTRIES, 1979

Country	Cattle (000)	Sheep (000)	Goats (000)	Camels (000)	Buffalo (000)
Bangladesh	31,741	1,061	11,000	0	1,529
Burma	7,560	215	575	0	1,750
India	181,849	41,000	71,000	1,150	60,651
Indonesia	6,453	3,611	8,051	0	2,312
Korea Rep.	1,651	8	224	0	0
Nepal	6,850	2,360	2,480	0	4,150
Pakistan	14,992	24,185	27,804	830	11,306
Philippines	1,910	30	1,430	0	3,018
Sri Lanka	1,623	24	461	0	844
Thailand	4,850	58	31	0	5,500

Source: FAO Production Yearbook.

The outstanding feature of cattle and water buffalo production in Asia is the low offtake per unit. Table 6 shows production data for 1979, by species and region; the figures for large ruminants are heavily influenced by data from India. Overall, Asia has 37% of the world's total combined population of cattle and water buffalo--but produces only about 10% of the world's cattle and buffalo meat. Excluding the Indian cattle population from the total puts the combined Asian population of cattle and water buffalo at 20 million head, thus meat offtake per unit is still about 50% of world average. The same general pattern holds for sheep and goats. Although the offtake figures for the Far East (table 6) are quite high by developing country standards, the 46% of the world's small ruminant population held in Asia produces only 29% of the total world supply of sheep and goat meat. For nonruminants, an important distinction needs to be made between the modern and traditional sectors, since a dualistic production structure is evident (De Boer and Weisblat, 1978). The modern sector has productivity

TABLE 2. CATTLE, SHEEP, AND GOAT POPULATIONS IN 1979 AND ANNUAL PERCENTAGE CHANGES, 1970-1979

Region	Cattle		Sheep		Goats	
	No. head[1]	Annual % change[2]	No. head[1]	Annual % change[2]	No. head[1]	Annual % change[2]
Industrialized						
North America	123,192	- .1	12,654	-4.4	1,386	-5.3
Western Europe	95,064	.6	87,469	.4	9,842	.1
USSR, East Europe	153,558	1.9	186,733	.5	7,242	-1.1
Oceania[3]	36,203	1.6	197,264	-1.8	148	-2.1
Others[3]	17,600	2.0	31,711	-1.3	5,510	- .8
Total	425,617	.9	515,832	- .8	24,128	-1.0
Less industrialized						
Middle & South America	267,304	2.3	116,585	- .7	29,053	- .2
Central & Southern Africa	148,882	1.3	105,235	1.4	126,657	1.2
North Africa & Near East	35,854	1.9	140,772	1.3	62,250	- .1
South Asia	241,242	.6	91,671	1.7	115,766	2.0
Centrally Planned Far East	68,985	.2	109,891	2.3	77,333	1.7
Open Economies Far East	24,125	.2	3,969	1.5	10,734	2.0
Total	786,392	1.3	568,123	1.1	421,793	1.3
World Total	1,212,009	1.2	1,083,954	.1	445,919	1.1

Source: FAO Production Yearbook, 1979.

1 Total inventory, thousands.
2 Annual change between 1970-1979.
3 South Africa, Japan, Israel.

TABLE 3.   SWINE AND BUFFALO POPULATIONS IN 1979 AND ANNUAL
PERCENTAGE CHANGE, 1970-79

Region	Swine No. head[1]	Swine Annual change[2]	Buffalo No. head[1]	Buffalo Annual change[2]
Industrialized				
North America	68,126	0.0	0	0.0
Western Europe	109,253	2.6	86	3.0
USSR, East Europe	136,283	3.8	733	- .7
Oceania	2,770	-0.8	1	11.1
Others[3]	11,084	4.9	0	0.0
Total	327,516	2.4	820	- .4
Less industrialized				
Middle & South America	74,369	2.1	318	17.0
Central & Southern Africa	7,564	3.9	0	0.0
South Asia	312	5.8	3,790	- .3
North Africa & Near East	10,454	6.0	14,723	- .4
Centrally Planned Far East	316,941	2.6	32,421	.2
Open Economies Far East	24,520	1.1	78,485	1.3
Others[4]	1,783	3.6	0	0.0
Total	435,945	2.5	129,737	.8
World total	763,461	2.5	130,557	.8

Source: FAO Production Yearbook, 1979.

[1] Total inventory, thousands.
[2] Annual change between 1970-1979.
[3] South Africa, Japan, Israel.
[4] Oceania other than Australia and New Zealand.

levels almost comparable to developed country standards, while the traditional village subsistence production system achieves productivity levels far below that of the modern sector (APO, 1976). One reason for the low offtake figures for cattle is their widespread use for draft power, which results in large numbers of buffalo and oxen being held until an advanced age, in lower productivity from females used for work, and in offtakes approximating those of herds where no culling is practiced (about 10%). However, this sacrifice in output must be balanced against the major contribution made by draft animals. On a worldwide basis, draft animals total between 280 and 300 million head with a market value of $100 billion. They provide the equivalent

TABLE 4. PROJECTED POPULATION GROWTH, FOOD SUPPLY, AND FOOD DEMAND IN DESIGNATED LDCs IN ASIA

Country	Population Percent growth rate/year (1)	Food production[1] Percent growth rate/year (2)	Food demand[2,3] Percent growth rate/year[5] (3)	Food energy supply[3,4] Kilocal/caput/day (4)	Food energy supply[3,4] Percent of requirement[6] (5)	Protein[3,4] supply g/caput/day (6)
Bangladesh	3.5[7]	1.6[7]	--	1,840	80	40
Burma	2.2	2.4	3.3	2,210	102	50
India	2.1	2.4	3.0	2,070	94	52
Indonesia	2.5	2.0	2.6	1,790	83	38
Korea Rep.	2.7	4.8	4.7	2,520	107	68
Nepal	1.8	0.1	2.1	2,080	95	49
Pakistan	3.0	3.0	4.2	2,160	93	56
Philippines	3.2	3.2	4.2	1,940	86	47
Sri Lanka	2.5	3.6	3.1	2,170	98	48
Thailand	3.1	5.3	4.6	2,560	115	56

Source: FAO/UN. 1974. Assessment of the world food situation, present and future.
United Nations World Food Conference, Rome. (Taken from longer table.)

1 Food component of crop and livestock production.
2 Based on FAO commodity projections 1970-1980.
3 Total food.
4 1969-71 average.
5 Exponential trend 1952-72.
6 Physiological requirement plus 10% waste at household level.
7 1962-72.

of 150 million horsepower, the replacement of which by mechanical sources would cost nearly $250 billion. In the Asian region, the major role played by draft animals seems likely to persist.

TABLE 5. INCREASES IN IMPORTS OF MEAT AND MEAT ANIMALS PLUS ALL FORMS OF MILK PRODUCTS BY DEVELOPING COUNTRIES, 1975-1979

Continent	Dollar value of imports in 1979 (million)	Increases in imports from 1975-79 (percentage)
Meat and meat animals		
Africa (40 countries)	$ 843	+242.20
South America (13 countries)	553	+ 91.30
Asia (except China) (38 countries)	3,578	+ 46.90
Milk in all forms		
Africa (40 countries)	$ 677	+102.07
South America (13 countries)	204	+ 63.20
Asia (except China) (38 countries)	1,035	+ 97.10

Source: FAO Trade Yearbook, 1980.

TABLE 6. PERCENTAGE ANIMALS SLAUGHTERED, 1979

	Africa	Latin America & Caribbean	Near East	Far East	Other Developing Countries
Cattle	11	16	16	3	15
Buffalo	--	0	29	6	--
Sheep	29	17	35	37	52
Goat	32	28	33	45	29
Pigs	76	47	99	81	56

Source: FAO Production Yearbook, 1979.

Some reasons for the lack of growth in the ruminant sector relative to the nonruminant sector are advanced by De Boer (1982). Prior to the 1950s, most Asian livestock were produced under a scavenger system for subsistence consumption. This was reflected in low consumption levels and virtually no use of purchased inputs. Since so few inputs were purchased outside the farm production unit, the farmers' supply of animal products depended heavily upon individual and communal land resources. The local land resources provided the grazing, crop by-products, and other wastes used in animal feeding. Applied at a national level, this scenario implied that continued expansion of animal product supplies would come up against a land constraint in the absence of technological change in the animal feed sector.

Following the 1950s, the ruminant and nonruminant production systems began diverging. Nonruminant production was more influenced by modern production technologies and a trend began toward closer integration of the farm production unit with the purchased input sector. This trend was accelerated by two factors: 1) the dramatic increases in productivity achieved in pig and poultry production because of relatively straightforward technology transfers from temperate regions and 2) a gradual expansion of international trade in feedstuffs, including P.L. 480 commodities. These two factors encouraged the rapid growth of swine and poultry industries utilizing predominantly industrialized forms of production clustered around major urban markets, which were often points of entry for imported feedstuffs. These developments caused a gradual reduction in the importance of land resources as a limiting factor governing output of pork, poultry meat, and eggs. Supplies of these products became more dependent upon the availability of foreign exchange to finance feedstuff imports and the growth of domestic markets to absorb the output.

The ruminant animal industries in Asia experienced no comparable increases in productivity or in easily accessible feed supplies and thus continued to depend heavily upon domestic land resources as the major production input. This sector retained its location-specific character since large quantities of low-cost roughages were needed that could only be supplied economically from local land resources.

From the above discussion, several points seem clear. First, the resource limitations in many of the Asian countries meant that the increased demand for animal products that accompanied rapid economic growth had to be met by modern swine and/or poultry operations. This resulted in steady growth of feed imports. Second, where beef is a preferred commodity, the supply has not been able to keep up with income--generated growth in demand and beef prices has increased relative to food products, in general, and also relative to other animal products. Thus markets potentially exist for imported beef, usually of higher grade types. Third, the great variability of Asia makes it difficult to come to any general conclusions. Instead, it is more helpful to consider groupings of countries relative to trade in livestock products and feedstuffs. Some of these countries are discussed next.

MARKET PROSPECTS IN SELECTED COUNTRIES

Japan

The market of greatest interest is Japan, because the Japanese are our single largest trade partner, our single largest importer of agricultural products, and a potentially significant market for high-quality U.S. beef. To put the market situation in perspective, consider the following: Current Japanese beef consumption levels are about 2.7 kg per capita compared to U.S. figures of about 50 to 55 kg, West German consumption of 28 kg per capita, and United Kingdom levels of 19 kg per capita, despite the fact that income levels in the latter country are now substantially less than in Japan. In November 1977, retail prices for beef in Tokyo were 8.7 times those of Canberra, 7.2 times those of Washington, D. C., and 3.3 times those in Bonn, West Germany. Conversely, pork was only 30% to 90% more and broilers 30% more than in lower-priced capital cities. Part of the reason for such high retail prices relative to c.i.f. plus import duties is the excessively high profits made by wholesalers with import permits (Hayami, 1979; Longworth, 1978) and a high-cost domestic food distribution system (Longworth, 1978). Recent estimates by Saxon and Anderson (1982) gave the ratio of wholesale to border prices (the price at which a commodity can be landed in a country without duties) of beef as 2.71 compared to ratios of 1.00 for pork and 1.08 for broilers. The ratio for beef in 1977 reached 4.4, however. Despite these relatively low levels of protection accorded pork and chicken, Japan is 93% self-sufficient in both of these products compared to 78% for beef.

The structure of the Japanese beef industry also hinders adjustment toward a more rational use of resources. High levels of protection for rice create distortions that drive up costs of resources in all rural areas. In addition, about 60% of all domestically produced beef comes from dairy steers and culls, and the average herd is less than 5 head per farm. Rapid technological development has not occurred in the beef industry, but the poultry and pork sectors have moved swiftly. Therefore, beef prices have been supported at progressively higher levels to maintain some degree of income equality within the rural sector, as well as between the rural-urban sectors. This maintenance of relative income equality is a stated goal of Japanese economic policy.

Comparisons of the relative change in efficiency can be made by examining the situation in 1960 with that in 1976. In 1960, at the wholesale level, beef was cheaper than pork (287 yen/kg vs 346 yen/kg), but by 1976 beef was 1,462 yen/kg while pork was 660 yen/kg. Over the same period, beef consumption increased from 1.1 kg/person/year to 2.7 kg, while pork consumption climbed from 1 to 7.7 kg/year. This illustrates that Japan's traditional consumption habits based largely on marine and vegetable products have shifted rapidly in response to relative prices (Hayami, 1979).

In terms of overall impact of agricultural distortions, Bale and Greenshields (1978) estimated net social loss in consumption and production of 1975/76 distortions for 8 commodities at $276 and $111 million, respectively. The realization of the ambitious agricultural production targets set for 1985/86 will raise total net social losses to $7.6 billion, or about 2% of total Japanese Gross National Product. Hayami (1979) has illustrated how beef imports could be increased several times and Japanese beef producers could increase their incomes. If the import quota was abolished and a duty levied on imported meat at a rate permitting imports to increase from the current level of about 100,000 t to 300-400,000 t, then the resulting revenues would be sufficient to maintain the current price to producers through a deficiency payment mechanism, to leave a surplus for the treasury, and to lower prices to consumers by at least 35%.

Given the overwhelming evidence of welfare losses to Japan's economy as a whole, the resulting distortions in resource use, and the demonstrated fact that Japanese beef producers could be fully compensated for any losses resulting from less restrictive trade without any loss to the treasury, why does the system continue to resist efforts for reform from within and without Japan? First, the wholesalers licensed to import beef earn extraordinary profits without having to bid on the real value of the import licenses. These importers have a powerful lobby in the National Parliament. Second, the depressed rural areas where most cattle are produced have a disproportionate share of political power. These rural areas have rapidly lost population since World War II, but there has been no redistricting on the basis of population. Therefore, in many rural areas, one vote equals 14 to 18 votes in a major urban area. Third, major concessions in one sector of the highly protected agricultural sector could lead to demands for reform in other areas that are much more politically sensitive, such as rice, wheat, and sugar. Finally, producers resist the notion of a deficiency-payment check and would prefer to sell their cattle for a high price in the protected market. Given these factors, a major reform of the Japanese beef protection devices is unlikely. Most changes will be minor and superficial.

Recent changes in Japanese regulations regarding "non-price" barriers to meat imports should be taken within this context. The most current examples are concessions made by the Japanese Livestock Industries Promotion Corporation to ease the rules under which U.S. packers qualify to bid on import tenders. The most powerful industrial grouping in Japan, the Federation of Economic Organizations (Keidanren) recommended removal of import quotas in the agricultural sector. Such a move would help reduce Japan's domestic budget deficit, reduce international pressures against Japanese exports, and reduce wage pressures in the industrial sector.

Finally, how large a potential market are we talking about for the world's beef exporters? Even adjusting for differences in tastes and preferences, pricing beef in Japan at border prices would lead to increases in consumption of from three to five times. Assuming a 10 kg/capita increase

as the minimal change that would result from a trade liberalization of beef, total imports would rise by at least 1.1
million t. By comparison, Japanese imports had reached
about 110,000 t in 1973/74 when severe trade restrictions
were imposed to stop the flood of imported beef that had
reached 42% of total consumption. The additional imports of
1.1 million t can be compared with 1981 beef and veal production in the U.S. of 10.35 million t, Australian production of 1.42 million t, and Argentine production of 2.96
million t. Total world exports over the 1977-1979 period
averaged 4.3 million t (Simpson and Farris, 1982) so the
increased demand for the Japanese market would represent a
huge increase in the current international market. This
market is highly unstable because a relatively small number
of importers and exporters dominate the market and small
changes in demand or supply conditions can cause major price
instability. Such an increase in demand would undoubtedly
lead to higher short-run prices, which would reduce Japanese
(as well as other importer) demand.

The American meat industry could best expend its political and promotional muscle on market development and ensuring market access in some of the rapidly developing countries of Asia that have not yet erected the high barriers to
imported U.S. livestock products, particularly beef. The
most prominent examples are Korea, Taiwan, and Malaysia. A
number of other countries such as Thailand, the Philippines,
and Indonesia also have impressive records of economic
growth but also have a much more favorable resource base for
beef production than do the above countries. Current statistics on the overall economic performance of these countries are in table 7.

TABLE 7.   GROWTH RATES OF SELECTED EAST AND SOUTHEAST ASIAN ECONOMIES,
           1981 AND 1982 (EST.).   (% CHANGE IN REAL GNP)

	Hong Kong	Sing-apore	Mal-aysia	Thai-land	Indon-esia	Philip-pines	Japan	Tai-wan	South Korea
1981	10.4	9.7	6.9	7.8	9.6	3.8	4.1	5.5	7.1
1982 (est.)	8.0	10.0	7.2	6.9	6.5	4.0	5.2	7.5	7.0

Source: Far Eastern Economic Review, June 17-24, 1982.

Agricultural support is the critical issue facing
countries undergoing rapid structural transformation from an
agriculturally based economy to a manufacturing, service-
based economy. Despite the impressive evidence of the
economic costs incurred by Japan in its agricultural support
policies, both Korea and Taiwan are moving in the direction
of high levels of support for rice and beef.

Korea

A recent FAO mission to Korea was asked specifically to examine the potential of the Korean feed and animal sectors (FAO, 1981). Korea, like Japan, had relied heavily on draft animals in the traditional agricultural sector. Beef was largely a by-product of the draft-animal sector. Rapid industrial development and mechanization of agriculture resulted in sharp decreases in the rural labor force and a rapid decline of draft-animal numbers and a shift toward beef production as a by-product of the rapidly expanding dairy industry. The dairy sector is composed primarily of small-scale producers, due to land limitations. To help support these dairy farmers, high prices of feeder calves are needed. The FAO (1981) study team recognized the general government objectives of supporting rural incomes and devised a deficiency-payment scheme similar to that devised for Japan by Hayami (1979). This would result in domestic prices intermediate between those obtained by import quotas on beef imports and border prices for imported beef. However, the potential market for imported beef and veal, mutton, and lamb remains relatively modest compared with Japan. Projections from the FAO (1981) team were based on three different scenarios regarding Korean economic growth: a low assumption of 5% to 5½% per year, a medium assumption of 7% to 7½% per year, and a high growth assumption of 9% to 9½% per year. The actual and projected net imports for 1986 and 1991 are shown in table 8.

TABLE 8. ACTUAL AND PROJECTED TRADE IN LIVESTOCK PRODUCTS, KOREA (1,000 t)

		Total meat	Beef & veal	Mutton & lamb	Skimmed milk
Actual	1978	48.5	41.3	4.2	38.0
Low growth	1986	91.8	59.3	4.5	53.4
assumption	1991	28.7	57.3	5.9	49.4
Medium growth	1986	109.4	70.8	5.1	45.5
assumption	1991	52.5	68.4	7.2	43.4
High growth	1986	121.3	79.2	5.7	46.9
assumption	1991	69.3	69.8	8.4	49.4

Source: FAO (1981).

Even under the highest growth rates, beef imports would only be in the range of 70 to 80,000 t, an increase of only 30 to 40,000 t over current levels. This increase would be met largely from Australia and New Zealand and probably would represent a minor market for U.S. animal products, even with a more liberalized system of beef and dairy-product imports. A major U.S. export to Korea is live dairy cattle. Imports of 15,000 head were planned in 1980, with a value of about $30 million. One reason for the relatively slow growth of beef import demand, despite high rates of economic growth and a strong consumer demand, is the substantial expansion of domestic production that is forecast. Using "medium" projections, total cattle numbers are forecast to increase from 1.6 million head in 1978 to about 3.6 million head in 1991. This is heavily dependent on continued importation of beef and dairy cattle breeding stock.

The projected large increases in the domestic dairy, swine, and poultry industries imply rapid growth in demand for feed imports. This is borne out by the FAO report (1981). Concentrate feed requirements are projected to increase from 5.8 million t in 1978 to between 12.9 and 15.1 million t in 1991, with a medium projection of about 14 million t considered most likely. The "medium" projection feed import bill for 1986 is $1.2 billion, while for 1991 it is $2.0 billion. These figures are many times larger than the projected value of beef imports and indicate the feed grain trade will be far more of a factor for U.S. exporters than trade in meat.

In 1980, Korean imports consisted of 2.3 million t of corn, 1.8 million t of wheat, and 0.4 million t of soybeans with total value of $1.021 billion. Livestock and livestock products added a further $400 million to the import bill. In 1980 the Korean trade deficit was $4.8 billion and continued trade deficits will be a major factor constraining the growth of feedstuff and livestock imports.

Taiwan

The same general conclusions hold true for Taiwan, with several important exceptions. First, Taiwan's dairy industry is much less advanced than that in Korea or Japan; milk demand is projected to grow at a slower rate; and the climatic environment is less favorable for dairying. Therefore, a large supply of by-product cattle from the dairy industry will not be forthcoming. Second, pork is the preferred meat, with poultry a strong second. Pork is relatively cheap, consumption is high, and the long-term prospect for growth in beef demand is less than in Korea. In 1981, U.S. exports of animals and animal products to Taiwan were about $73 million. Feed imports in Taiwan have grown tremendously to support expanded pork and poultry production. The levels of coarse grain imports forcast in 1981/82 were 3.7 million t for Taiwan and 2.7 million t for Korea, despite Taiwan's having only about 35% of Korea's population.

Taiwan has a generally favorable trade balance and demand is constrained more by income levels and pork consumption levels that are already quite high.

Other Countries

Other Asian countries with medium to high rates of

economic growth include Thailand, the Philippines, Indonesia, Malaysia, Singapore, and Hong Kong. Singapore and Hong Kong are expanding markets for U.S. feedstuffs and livestock products, but populations are so small that they will not be major factors. The other four countries have several characteristics in common. They are, with the exception of Thailand, minor participants in the international livestock-feedstuffs market. They are all basically self-sufficient in meat, and animal product prices are at or below international levels. And, finally, they are all experiencing rapid growth in national incomes and population. Cattle numbers are virtually stagnant in these countries; water buffalo numbers are declining, and the commercial pig and poultry sectors are expanding rapidly. The increases in maize and soybean imports in the Philippines, Indonesia, and Malaysia reflect this trend. Beef prices have not risen dramatically because of the substitution of pork, poultry, eggs, and fish.

Cereal production in these countries (primarily rice) has also expanded rapidly and the availability of rice bran, wheat bran (from increasing wheat imports) and oilseed meals has helped ease the demand for imported concentrates. Continued rapid growth will increase the demands for animal feedstuffs and lead to increased imports, however.

The People's Republic of China

The livestock targets adopted in 1978 call for a 30% increase in the numbers of cattle, sheep, and pigs and a doubling of meat output by 1985. Consumption levels of meat in 1977 were about 7.5 kg per capita and consisted primarily of pork, although in the sparsely populated pastoral areas, meat-consumption levels of 50 kg per capita are common. To meet the livestock targets will require about 5 million more tons of grain (Tang and Stone, 1980), which is feasible from domestic production, but to attain the increased output would require an additional 6 to 19 million tons more of grains. This is likewise considered feasible in the coarse grains sector, but there seems little doubt that meeting the meat production target will require substantial imports of soybeans and/or soybean meal. Much of the increase in output must come from the swine and poultry sectors, and to get productivity increases, more oilseed meal must be made available. The probabilities of China entering the world meat market are considered very small.

SUMMARY AND CONCLUSION

The major impact of rapidly developing Asian economies on the U.S. livestock sector will be felt through the international grains and oilseeds market, rather than through the world trade in beef, pork, poultry and dairy products. The possible (but not likely) exception is Japan. The growth of the domestic livestock sectors will be concentrated in the nonruminant sector. Since the growth in animal product demand is closely tied to economic growth, the condition of the world economy will be an important factor governing livestock numbers.

The rapidly developing Asian economies are heavily dependent upon exports. Declines in export demand resulting from poor economic conditions in the industrial countries are quickly reflected back to these countries. This slows down the growth in demand for livestock products and, consequently, the demand for feedstuff imports. The growth prospects in these countries remain excellent but will remain heavily dependent on the economic policies of the major industrial powers.

REFERENCES

APO. 1976. Livestock Production in Asian Context of Agricultural Diversification. Asian Productivity Organization, Tokyo.

Bale, M. D. and B. L. Greenshields. 1978. Japanese agricultural distortions and their welfare value. Am. J. Ag. Econ. 60:59.

De Boer, A. J. 1982. Livestock development: the Asian experience. In: International Development Research Center, Ottawa, Canada (forthcoming).

De Boer, A. J. and A. M. Weisblat. 1978. Livestock component of small farm systems in South and Southeast Asia. Paper presented to International Conference on Integrated Crop and Livestock Production to Optimize Resource Utilization on Small Farms in Developing Countries. Lake Como, Italy.

FAO. 1981. Livestock and animal feed development in Korea. Food and Agriculture Organization of the United Nations Report No. ESC:CPCL/ROK. Rome, Italy.

Hayami, J. 1979. Trade benefits to all: a design of the beef import liberalization in Japan. Am. J. Ag. Ec. 61:342.

Houck, J. P. 1979. Agricultural trade: protectionism, policy, and the Tokyo/Geneva negotiating round. Am. J. Ag. Econ. 61:860.

Longworth, J. L. 1978. The Japanese beef market: recent developments and future policy options. Rev. Mktg. Ag. Econ. 46:167.

Saxon, E. and K. Anderson. 1982. Japanese agricultural protection in historical perspective. In: Livestock in Asia: Issues and Policies. International Development Research Center, Ottawa, Canada (forthcoming).

Simpson, J. R. and D. E. Farris. 1982. The World's Beef Business. Iowa State University Press, Ames, Iowa.

Tang, A. M. and B. Stone. 1980. Food production in the People's Republic of China. Research Report No. 15. International Food Policy Research Institute, Washington, D. C.

USDA. 1982. Outlook and situation. Economic Research Service WAS-28. United States Department of Agriculture, Washington, D. C.

Velasco, M. M. 1982. Livestock policy choices: the Philippine experience. In: Livestock in Asia: Issues and Policies. International Development Research Center, Ottawa, Canada (forthcoming).

Part 3

GENERAL CONCEPTS AFFECTING AGRICULTURE AND THE INDUSTRY

12

POLITICAL CHALLENGES FOR TODAY'S ANIMAL AGRICULTURE

George Stone

Not all of the challenges to today's animal agriculture are political. Some of the challenges are internal. They relate to the nature of our industry and to the nature of the people in the livestock industry.

We seem to have an allergy to change. And we have a disposition to go our own way as individuals, even if we could improve things for ourselves by working together.

ROLE OF FEDERAL GOVERNMENT

Part of the time, we want the government to leave us alone. Part of the time, we get very impatient when the government is too slow with helping us. This is not the first, nor the last, speech to be made in this nation on the role of the federal government in food and agriculture. We talk about that subject as if we were about to make an original choice. But, as a matter of fact, the choice was made long ago. As early as 1796 and as recently as 1977, and many times in between, the federal government has decided that the family farm should be fostered and encouraged.

Our society has decided that the federal government should take measures to help assure that land remains in the hands of family farmers.

Our society has decided that the federal government should be involved in the conservation and protection of land and water resources.

Our society insists on a federal involvement in environmental protection.

Our society has dictated a federal role in assuring that food supplies are safe and wholesome.

Our society requires federal supervision to see that pesticides and chemicals are safe for farmers and consumers.

Our society provides for federal supervision of the marketing system to try to keep it fair and competitive.

There are more kinds of government intervention in agriculture today than ten, twenty, or thirty years ago. There will probably be more federal involvement in farm and food policy in 1990 or the year 2000 than there is today.

The real question should be in regard to the nature, extent, and purpose of government involvement, and the degree to which farmers have a voice in decisions that will affect them.

Government intervention has often been justified when there was no other effective way to cope with problems. The government's role in agricultural research and education is widely accepted and advocated. Various federal farm credit programs had to be initiated because the private sector could not take the entire risk in financing agriculture. Commodity exchanges and boards of trade had to get federal supervision because they could not be left to police themselves. The federal rural electrification system had to be created because the private sector could not, or would not, do the job.

In many agricultural sectors, government involvement can be good or bad:

- Rules on farmers' handling of pesticides or chemicals can be reasonable or ridiculous, depending on how much farmer input there has been in the process.
- Feedlot pollution abatement rules can be workable and effective, or unrealistic and oppressive.
- Dredge and fill regulations can be a protection or a harassment for farmers.
- OSHA regulations can be a godsend or an aggravation.

Everything depends on how well farmers have involved themselves in the process and how much voice they have had in shaping these laws and regulations. That is not an easy task for us as farmers and livestock producers. The leading codification of laws affecting agriculture now runs to 14 volumes of about 500 pages each. That is 7,000 pages of laws.

MAJOR POLITICAL CHALLENGES

Having said this much in the way of background, let me now turn to what I perceive as some major political challenges for livestock agriculture. One thing that is quite obvious, but not generally appreciated, is that we do not function in a vacuum as livestock people. We depend upon consumer purchasing power and demand for our products, and we cannot expect to be stable and prosperous if there is high unemployment and weak buying power.

Recession and Unemployment

Much of the difficulty that has faced the livestock producer in the past three years has been attributable to recession and unemployment. High interest rates the past three years have diverted about 30 billion dollars a year of consumer purchasing power from food and other necessities. High interest rates have an impact on the cattle producer as well. Some calculations done at Oklahoma State University reveal that the interest cost per head of livestock sold would be $133 per head at 9 percent interest; $237 per head at 16 percent; and $297 per head at 20 percent interest. The difference between the two extremes of 9 percent and 20 percent is $164 a head, more than enough to wipe out any potential profit.

When national economic conditions are difficult and there is a pinch on consumer buying power, the tendency is for a reduction in the higher-priced meat purchases and, to some extent, for the consumer to buy other products. In a time of recession and unemployment, meat purchases are the first thing affected.

Specifically, because of the tendency for the American diet to be hurt by recession and unemployment, the Congress in its wisdom developed and implemented the food stamp program. Of course, one can expect that when times are tough, the food stamp program becomes costly. Each additional one percent of unemployment adds one million people to the food stamp rolls, not because it is a bad program, but because it is doing what it is supposed to do--maintain a healthful diet for the lowest-income people in our society.

About 27 percent of the food stamp benefits are spent by recipients to buy meat and meat products. This means that if the food stamp program makes available 12 billion dollars in food subsidies to low-income families, over 3 billion dollars will be used to buy meat. If the food stamp program is cut by 3.7 billion dollars, as it has been in fiscal budgets for 1982 and 1983, that means a 1-billion-dollar reduction in the demand for meat. That is a political decision and we live with it as livestock producers.

Competition Between Humans and Farm Animals

One of the political challenges that will become more serious as time goes on will be the competition for living space between humans and farm animals. Twenty years ago, a livestock economist at a midwestern land grant university was making the prediction that there would be little room in the world for livestock 100 years in the future. Human population was increasing so rapidly that the land would be needed for living space. There would only be room for enough livestock to provide meat flavoring for synthetic meat substitutes, he predicted. Although there are still 80 years to go to see if the good professor was right, we doubt that he was on the mark. Still, his kind of thinking has surfaced in some other forms.

One noted futurist has looked in his crystal ball and concluded that if we fed the grain to humans instead of to livestock, we could perhaps support a population more than three times greater than at present. Some world hunger activists were suggesting a few years back that if each of us would give up one hamburger a week, it would help feed the starving of the world. Still others have suggested that Americans should quit fertilizing their lawns and golf courses and send the fertilizer to the developing nations to help them grow more of their own food supply. These ideas are simplistic and, even if carried out, would have little measurable effect on hunger in the world. Basically, outside of famine caused by natural disasters, there is no actual shortage of food in the world, nor of land or fertilizer. There are more than adequate supplies of land, fertilizer, and food. What is lacking is the purchasing power to pay for the food. What is lacking is effective consumer demand. Wherever the cash is available to pay for the food, the food becomes available.

Politically Prescribed Diets

I rather expect that we will have increasing frustrations ahead with those who wish to try to influence the diets of the American people by political prescription.

For forty years, the Food and Nutrition Board of the National Academy of Sciences has been the widely recognized and respected source of dietary guidance. The Food and Nutrition Board has been the agency that has issued the "Recommended Dietary Allowances" or RDA that have been the basis for most nutritional education aimed at the consuming public. Just two years ago, the Food and Nutrition Board issued a report, entitled "Towards Healthful Diets," in which it advised that the average adult American whose body weight is under reasonable control should feel free to "select a nutritionally adequate diet from the foods available, by consuming each day appropriate servings of dairy products, meats or legumes, vegetables and fruits, and cereal and breads."

In the same report, the Food and Nutrition Board recommended that dietary change or therapy should be undertaken under a physician's guidance. Aware of some of the political headline hunting being done by self-appointed guardians of American diet, the Food and Nutrition Board plainly warned that it is "scientifically unsound to make single, all-inclusive recommendations to the public regarding intakes of energy, protein, fat, cholesterol carbohydrates, fiber, and sodium."

You are aware, of course, of the rash of studies and reports on diet and heart disease, diet and cancer, and other topics that have singled out animal fats and meat as the causes of human health difficulties. There was a surgeon general's report in 1979 and a study by the Senate Select Committee on Nutrition and Human Needs. Last summer, a special panel of the National Academy of Sciences on "Diet, Nutrition, and Cancer," issued a report which differed in important respects from the position of the Food and Nutrition Board and that had not, in fact, been submitted to the Food and Nutrition Board for review and evaluation. Like many of the other political diet studies, the cancer study issued sweeping recommendations that included a 25 percent reduction in the consumption of fats, fatty meats, and dairy products. The report recommended avoidance of smoked sausages and fish, ham, bacon, frankfurters, and bologna.

The cancer and diet panel admitted it did not know what percentage of cancer risks are attributable to diet or how much the risk could be reduced by modifying one's diet. Still, while admitting that there was considerable uncertainty about the scientific basis for its findings, the diet and cancer panel issued its recommendations anyway and, as you might expect, the press treated it in a sensational manner. It was entirely in order for the livestock industry to ask that a review be held to reconcile the contradictory advice that was originating at the same time from the National Academy of Sciences. In our belief, the practice of medicine should neither be carried out by politicians or advertising agencies. The practice of medicine should be left to the medical profession.

Livestock Marketing Revitalization

At a time when all other industry seems to be centralizing, livestock marketing is disintegrating--breaking up into bits and pieces--with no central system for determining price.

As a result, live cattle prices are being based on data reported in the Yellow Sheet or the USDA meat news--sources that may represent as little as 2 percent of the market volume. Farmers have been looking at other options, such as electronic auction markets, to restore some competition into the system. But, it will take time, considerable capital, and major organizational efforts to establish an effective producer-controlled system of that sort. We may need some federal help and encouragement to get the job done.

International Trade in Meat and Meat Products

Another political challenge we may face will be in regard to international trade in meat and meat products. We appreciate the desire of some in the livestock industry to expand foreign markets for U.S. meat and related products. Foreign market development ought to be pushed in any constructive way. There may well be some potential for gains, particularly as some of the developing countries increase their purchasing power and seek to upgrade their diets. However, care must be taken that nothing we do in seeking to expand world trade reacts to undermine our own meat-import control laws. We ought to recognize that if we attack the quotas, the nontariff barriers and protectionist devices of other countries, this will certainly expose our own Meat

Import Act of 1964, as amended by the Meat Import Act of 1979, to attack from abroad. You may recall that when the 1979 Act was adopted, it was criticized by some who look to the U.S. market to dump their oversupplies. While we now export almost one billion dollars' worth of meat and meat products, we import 2.2 billion dollars in meat and meat products, plus another 1.5 billion dollars in animals and animal products.

The expansion of U.S. meat exports will be a gradual, long-term proposition. It will take a long time before exports offset imports, and this will be particularly true if we go to a free market in meat trade. At any time that world meat supplies are excessive in relation to effective demand, the U.S. will tend to be the magnet for oversupplies. That situation will tend to prevail most of the time. So, in a free market situation, U.S. meat imports will tend to expand more rapidly than meat exports.

We have a good law in the 1979 Meat Import Act. It is a responsible measure that helps us retain a domestic livestock and meat industry. The countercyclical factor that determines the allowable level of imports is particularly important. When the U.S. cattle industry is in the liquidation phase and beef production is relatively high, the countercyclical factor will tend to reduce the allowable level of imports. When the cattle cycle is in the rebuilding phase and domestic production is low, the allowable imports will be increased. If we can hold foreign imports to the minimum figure of 1,250 million pounds or near to the figure, the situation will remain in control. But, in a free market situation, it is easy to imagine that without any controls, meat imports, now subject to the law, would quickly advance to new all-time record levels.

Animal Welfare Concerns

Another political challenge livestock producers may have to face would be from the animal welfare lobby. Up to a couple of years ago, few farmers were concerned about the animal welfare lobby--many had not heard of it. Thus far, there has been no serious effort in the Congress to delete the several words from the Animal Welfare Act that would end the exemption of farm animals and birds from that statute. However, there have been bills in the 96th and 97th Congresses that would regulate confinement feeding of animals. Some activist groups have emerged and have gotten some coverage from farm magazines and the media but, so far, it seems they have been open to dialogue with farm and livestock groups. We should be realistic enough to expect that concerns of nonfarmers about this area of farm production will increase but, if we can keep some lines of communication open with responsible citizen groups, perhaps the discussions can be kept on a reasonable basis.

National Farmers Union has had some concerns over the harmful effects of excessive concentration of poultry and animals. To be frank about it, our concerns have been more in terms of environmental, health, and economic effects of such concentration, rather than with humane treatment of livestock and poultry. Up to this point, here in the U.S., we have been able to avoid the confrontations between consumers and farmers that have been common in Western European countries over animal rights. Most of the difficulty arises with people who are almost totally lacking in knowledge about livestock production methods.

By attempting to carry on a dialogue, we may be able to avoid misunderstandings that tend to polarize viewpoints on their side or ours. If we can win some understanding, we may be able to keep the issue out of the political arena.

Sensible and Effective Regulations

In the time that we have had on the program today, it has not been possible to touch upon all the political decisions that will affect our business and livelihood. There is a whole array of potentially serious problems, if the hysteria for deregulation is carried to the extreme. Reducing needless paperwork burden is desirable, of course, but to eliminate needed regulatory measures is something else. The Packers and Stockyards Administration covers a whole range of competitive issues. What we need is sensible and effect regulation, not the elimination of regulation altogether.

Most of my adult life was spent in Oklahoma as a farmer and farm leader. We found at times that there were governmental decisions so bad that we had to fight them with all our might. But, we also found that we could reason with people and that if we got involved early enough and got a voice in shaping the decisions, we could avoid situations that otherwise might have become desperate. The key is to get involved.

Rest assured, the government is involved and will continue to be involved in our agriculture and our society. If we don't take part in the process, the results will be worse. If we curse the government, the government will not go away. We will just be destined to live under scarcely bearable regulations devised by someone who doesn't really understand livestock farming. Fortunately, we do have a choice and a voice, if we want it.

13

AN AGRI-WOMAN'S VIEW
OF THE POLITICS OF AGRICULTURE:
NATIONAL AND INTERNATIONAL PERSPECTIVES

Ruby Ringsdorf

As a city girl turned farm wife some 27 years ago, learning all about the various aspects of agriculture has been very interesting and educational as well as very rewarding for me. Because so many have left either the farm or rural area and relocated in the city, farm community or farm bloc no longer carries the clout it once did.

When we look at the statistics we see that in 1949 25% of the populations were farm folks--today we constitute about 2.7%. From a significant segment of society that was once courted, soothed, and cared for, we have become a minority of the voting public--our "clout" practically nil. Our collective ability to feed and clothe our nation and significantly affect this nation's economic status of favorable contributions to the balance of payments still exists, but now one producer feeds 78 people--not just a simple household. And therein lies the rub. Production capability and output efficiency do not vote. Contributions to the common good carry little political weight. This has created our current predicament. We, as producers, find ourselves in the unenviable position of providing a product with public utility overtones without even the restrictive protection that public utility status would bestow.

From a public relations standpoint we are viewed as either land-hungry, dollar-greedy, unpatriotic, rich farmers (witness the writing coming from some of our church groups--writings such as "Strangers and Guests in the Heartland"), or as not-so-smart farm folk in overalls or print dresses (note the celery farmer and his wife on the Baggies commercial), or as sheltered, dull, farm children, apprehensive of the outside world (M & Ms make friends and the hayseeds are readily won over with a hand full). We are none of these or all of these. Farm folk run the gamut--as do shop keepers, manufacturers, and other American entrepreneurs. We are business people. We love agriculture and perhaps are willing to sacrifice some "return on investment" in monetary measures for the "return on investment" in satisfaction and the personal pleasure that come from involvement in production agriculture. But agriculture is too expensive to keep as a hobby. We are commercial farmers. "Commercial" is not a dirty word--nor is "profit." The politics of agriculture has economic overtones and very real economic repercussions for the producer.

I would like to touch on a few of the political influences with which we have had to deal over the past few years. I apologize for the negative tone of this presentation, but other than our own attitude and the placement of a free marketer in the office of Secretary of Agriculture, there has been little to be positive about for commercial production agriculture.

Past grain embargoes and trade restrictions have worked to the detriment of the free-market system and have, as a by product, encouraged grain and oilseed production in other countries, in direct competition with the U.S. producer, thus diverting sales arrangements, agreements, and traditions to these countries and away from the U.S. We have become, in many cases, the residual supplier--the supplier of last resort. We have a reputation as an unreliable source of grain, a nation whose past political philosophy has allowed food to be used as a weapon and the farmer as a political pawn in the game of foreign relations. The 1973 embargo sent Japan to South America with the technology and economic necessity to develop a source of soybeans not dependent upon U.S. philosophical and political attitudes. Now, rather than competing with the protein supplied by the anchovies harvest of the coast of Peru, our U.S. soybean farmers must contend with a full-blown, mirror-image harvest in the southern hemisphere. We welcome healthy competition, but resent our own government policies encouraging such, while at the same time inadvertently and directly threatening the economic livelihood of our American producers.

The most recent embargo reinforced this scenario, left us a residual supplier to the U.S.S.R., and brought agripolitical America to its economic knees. The U.S. farmer cannot continue to produce in abundance without the reasonable expectation of an adequate return for the labor and capital involved; prices commensurate with this return are probable only if the world market is available to the producer. Ours is a global society. It makes no sense to subject the producer to the unpredictable risks of the world while confining his free marketplace to the narrow national confines of the U.S. and placing constraints on his participation in the world economy.

The political intrigue surrounding the long-term trade agreement with the U.S.S.R. and the threat to employ trade sanctions against our enemies have kept the farmer in the dark. Any long range capital plans, any cash flow projections are subject not only to the economic ebb and flow of supply and demand but to the implementation of foreign policy philosophies either to further our political ideologies in another country or to achieve some internal goal. our soybean farmers held their breath for possible repercussions when our government talked of imposing import taxes on Japanese automakers.

In 1979 we felt the political influences nipping at our heels as the Secretary of Agriculture Bob Bergland called for hearings on the "Structure of Agriculture." (Our official response as American Agri-Women is included as an Appendix.) We had women testifying at hearings held all over the country. Our basic premise was that agriculture does have some problems but we're basically sound. We were profoundly disturbed as we heard witness after witness talk of agriculture's problems and propose solutions bordering on political and economic systems inconsistent with our free-enterprise democratic society.

Other players in the political game are the food lobbyists. We, as producers, are part of the picture. With representatives of both general farm organizations and commodity groups in our state capitals and in Washington, we pay individuals to serve as our eyes and ears and also to pass along our concerns to the powers in political and economic decision-making.

Often working at cross-purposes with us, and just as often uninformed as to the probable repercussions of actions they propose, are various church groups, hunger organizations, social activists, and do-gooders. A recent example is the animal-rights lobby and their campaign against proven and protective commercial agricultural practices. In recent years we've dealt with land groups who favor land redistribution. There is also a strong school of thought that land belongs to everyone and should not be privately held nor used to make a profit.

We've had farm price-support programs since 1933. Although the concept may have initially been a necessary emergency measure, the fantastic productive capability of the agriculture sector today has been accomplished in spite of government interference via the farm programs, not because of them.

We would rather see the emphasis of government involvement redirected toward programs that will encourage a healthy, contributory agriculture. We see government commodity programs as an alternate source of credit rather than as a guaranteed price. The trend should be toward market development and a broader use of trade barriers and guarantees (supported by action) that the U.S. farmer is a reliable supplier.

Food is the primary need of all mankind. Efficiency in agricultural production has allowed portions of the labor force to enter industry, the arts, government, religion, and other human endeavor while depending on others to provide their daily requirements of food. The U.S. has always held individual and religious liberty to be a right of all its citizens and has chosen to protect this liberty by establishing a democratic republic and by fostering the private ownership of land and other resources. The ownership of this land is openly accessible to all who will work for it. American agriculture is the most efficient in the world, and its productivity has reduced the cost of food to the point

that only about 17% of our average disposable income is now
spent for this most basic human need. This monumental productivity has relieved 97% of our population of any need to
produce food, leaving less than 3% of our people directly
employed in farming. This separation of most people from
everyday agricultural production has raised a great deal of
public questioning of agricultural practices, farmland ownership, labor, and similar issues. These are political
issues and affect every individual in this country. They
deeply affect farmers and their methods of production and
could ultimately affect both the structure of American society and the total supply of food available to the people of
the world.

APPENDIX A

AMERICAN AGRI-WOMEN
POSITION ON STRUCTURE OF AGRICULTURE HEARINGS

Whereas, R. Bergland, Secretary of Agriculture has requested
input on the Structure of Agriculture in twelve areas, we,
American Agri-Women, submit positions on the following subjects:
1. Land ownership, tenancy and control.
    - Land ownership, control and tenancy belong
      in the private sector.
    - The majority of ownership will be limited
      to U.S. citizens.
    - Since unlimited leasing is not restricted
      in any other business, ownership and pro-
      ductive use of agricultural land need not
      be with the same party.
    - Agricultural landowners need the flexibil-
      ity to manage their land as their exper-
      tise guides them. A residency requirement
      for agriculture is archaic - just as it
      would be for any other business.
2. Barriers to entering and leaving farming.
    - Barriers to entering are basically deter-
      mined by the lending institution.
    - Leasing can be a vital avenue for entry
      into farming and therefore must be unre-
      stricted.
    - We object to government support of "life-
      style farmers" (non-commercial) because
      productive incentive will be lacking.
    - Capital gains taxes and environmental zon-
      ing restrictions are barriers to farming.
3. Production efficiency, size of farms, role of
    technology.
    - Each farmer should be able to determine
      his farm size. Technology should be uti-

lized to the fullest. Agriculture must not be singled out in use of size technology, and methods of production.

4. Government programs.
   - Much of the success of American agriculture is due to agricultural research. This represents only a small portion of the USDA budget. To feed a hungry world, we vitally need accelerated research programs.
   - Food assistance programs represent 56% of the total USDA budget. A cut in agricultural research, extension, soil and water conservation programs would not be in the national interest.
   - We oppose USDA funding of persons to testify at regulatory hearings.

5. Tax and credit policies.
   - Inheritance tax laws need changing, particularly the carry-over provision.
   - More tax credit incentives for conservation measures are necessary to justify the required capital outlay.
   - No taxes on transfers between spouses.

6. Farm input supply system.
   - Leave in the private sector. The current system works well. Leave it alone.

7. Farm product marketing system.
   - Leave in the private sector. The current American marketing system is the best in the world.

8. Present and future energy supplies.
   Develop multi-faceted energy supplies and aggressively work to decrease our foreign dependence.
   - Agriculture uses only 2.9% of U.S. energy supplies and we needn't apologize to anyone.

9. Environmental concerns, including conservation and the use of soil and water.
   - Farmers were the first environmentalists and conservationists as their income and livelihood depend on it. Let us continue with less regulatory interferences.

10. Returns to farmers.
    - Farmers need to develop collective bargaining. But farmers, like any other independent businessmen, accept the responsibilities and assume the risks.

11. Cost to consumers.
    - The American farmer produces the least expensive, most bountiful, and highest quality food and fiber for the consumer anywhere in the world.

- Farmers must be allowed to pass on their costs as in any other business.
12. Quality of rural life.
   - There is a difference between rural America and farming America. The economic quality of life of agripolitician America is dependent upon agriculture - just as the economic quality of life of metropolitan America is dependent upon the central business district.
   - Quality of life in rural America is as good as the people who live there and make up the rural community. As people from rural areas, we can attest to the fact that it is a good life if you work at it, and are allowed to work at it without government interference.

Whereas, we American Agri-Women, representing 24,000 people, present these proposals with a diversified history of farming success and failures in each of our 50 states; that the aforementioned proposals are based in fact and are the positions American Agri-women takes on the structure of agriculture.

Adopted November 3, 1979 by the governing body of A.A.W. at annual convention, San Diego, California.

14

CONFLICTS BETWEEN PRESENT AGRARIAN LEGISLATION AND RANGE MANAGEMENT

Martin H. Gonzalez

INTRODUCTION

On the 40% of the land on Earth classified as range-lands, management and conservation must follow similar principles--from deserts to subtropics, from sea level to the highest peaks, from the extremes of the Continents to the Equator, and from all latitudes and longitudes on Earth. These basic principles are applicable for proper use, improvement, and management of grazing lands to obtain a permanent, sustained production. Benefits obtained may be reflected in the vegetation, the soil, and the animal species using them--plus, of course, the benefits for people.

The extreme variations in ecological characteristics are not limiting factors for the application of basic range-management principles. The know-how and the tools exist, and are available in most parts of the world--what actually influences range management, particularly in most of the developing countries, is the great variation in legislative policies governing them. Unfortunately, such legislative policies seldom benefit the natural resources. In many cases because of social agrarian pressures, the extensive technological and scientific information related to ecology, range management and livestock production that is available world wide is not considered when new legislation is passed.

In many developing nations, legislation dealing with grazing and agricultural lands has been more for political reasons than for technological reasons. Mexico, as an example, has experienced the influence of such legislation for more than half a century with less than satisfactory results.

Social pressures are recognized in these countries, of course; but to improve the living conditions of one's own people, a balance must exist between the social and the ecological sectors when legislating the resources of a country and its inhabitants.

It has been evident, through the years, that many legislators are not aware of what our natural resources are nor how to protect them. Even when our legislators are aware and knowledgeable, political interests interfere with practical legislation.

We are living in an era when information produced by research and experience is extensive and available as ever before. That information should serve as the foundation for government's administrators to design, develop, and implement the programs and policies for proper utilization and management of the grazing resources of a country. However, that is not the case. In Mexico and other developing countries, science and technology are second to sociopolitical decisions when agrarian legislation is passed.

In many countries, the interest, and enthusiasm for needed range improvements are eclipsed by mainly two types of legislation concerning land tenure:

1. That which was adequate in the past but is now obsolete or archaic.
2. That which, from the beginning, was wrong and has not been modified because of:
   - Lack of basic technical knowledge in their formulation
   - Limited administrative ability
   - Political interests
   - Fear of change

One or several of these reasons tend to develop excessive beauracracy that in no way can benefit the proper development of a nation's resources. Although the damage caused to the grazing lands in many countries may be irreversible or rehabilitation may be very difficult and very expensive, the possibilities still exist for many countries to recover the productive potential of those lands.

This paper describes some of the most important impacts of the agrarian reform on Mexico's range and livestock resources since the agrarian revolution in 1910.

BASIS OF THE AGRARIAN REFORM IN MEXICO

The agrarian reform in Mexico was based on the Political Constitution proclaimed on February 5th, 1917. Article 27 of this Constitution establishes clearly the characteristics of the different land-tenure systems in the country. Several amendments have been made to the Agrarian Code in the Constitution to provide, theoretically, a better environment for the development of the agricultural and livestock industries in the country. However, these amendments as well as the whole agrarian code have lacked the technical basis to make them effective. Thus legislation has been made that was dominated by the theoretical points of view of lawyers and sociologists and by politicians with no experience in ranching and still less in management of resources.

To understand the problems facing rangelands in Mexico, one must know the different land-tenure systems in the country. According to the Constitution, all lands, including rangelands, fall within any of the following tenure systems: Ejido, Colony, and Private Property.

Ejido Landholdings

The national policy with respect to landholdings has had an important influence on agricultural and livestock production in Mexico. Before 1945, land distribution had been carried out chiefly through the ejido system. Each ejido is a legal entity--a community where the land is divided into family plots that are farmed on an individual basis. Holders of individual plots cannot sell or mortgage the land, but rights to the land may be passed along to their heirs. Each ejido member is assigned a piece of land to farm individually and has a "right" to use a certain percentage of the grazing land. Although each ejido landowner (ejidatario) owns and brands his own animals, grazing is communal and, supposedly, subject to certain ejido rules concerning the maximum number of animals that the "right of use" gives to each ejidatario.

The communal grazing of the ejido pastures presents many problems. In general, there is no control of the livestock operation: overgrazing reaches dangerous extremes and lack of adequate fences and water facilities hinders some management practices that could be applied. Under these conditions, only extensive technical and financial government assistance has made it possible for many of these ejidos to produce what they are capable of producing. At the present time, around 25% of the Mexican grazing lands belong to ejidos, but few of them are producing.

Private Landholdings

According to the Mexican Constitution, the size of a private ranching unit is determined by "the amount of land required to maintain 500 head of cattle, or the equivalent in other kind of livestock, according to the grazing capacity of the land." This statement, however, does not specify what type of "head," allowing room to many misinterpretations. This term is the biggest error in the legislation, but its use has remained unchanged. The "head of cattle" concept is ambiguous. The "animal unit" concept, as we know it today, is never mentioned in the Constitution or in the Agrarian Code. On the other hand, the failure has been in determining the "capacity of the land" to set the legal boundaries of a ranch (a point that is discussed further below). Any ranch exceeding those limits was, and is, subject to seizure, and the area exceeding the limit is subject to being confiscated by the government or parceled out to ejido solicitors.

Colonies

The "colonies" are basically several privately-owned properties particularly in the grazing areas. They suffer the same overgrazing and management problems of the ejidos and the private ranches. However, they are more flexible and usually the members work in harmony.

Criteria Used for Range Legislation

Some of the criteria have been changed for determining the legal size of the private ranch--that is, to delineate the area required to support 500 head of cattle--but these changes have not brought sufficient improvement. The "aridity index," "grazing index," and "grazing coefficient" --terms adopted by the agrarian legislators in different times as an effort to define the productivity of the Mexican rangelands--have not contributed significantly to the solving of these problems. This is because none of these terms had a quantitative basis for reflecting forage production, and the criteria were based only on climates.

The aridity index is a relation between rainfall and temperature and the grazing index and the grazing coefficient are to determine the amount of land needed to feed 500 head of cattle in the natural state of the range. However, the classification of the grazing zones in Mexico at that time (1942) included only eight extensive zones as shown in figure 1. This classification was not in agreement with the multiple ecological conditions of Mexico or even within each of the eight zones.

CARRYING CAPACITY REGIONS — 1948

1 South and Central Gulf of Mexico.

2 South and Central Pacific Littoral

3 Indirect watersheds with drainage to the Pacific Ocean.

4 Central watersheds with drainage to the Pacific Ocean.

5 North portion of the Gulf of Mexico watersheds

6 Rio Conchos Basin and closed — basins of Durango and Zacatecas.

7 Northwest watersheds.

8 North - Central region.

Figure 1. Subdivisions of Mexico into 8 carrying-capacity regions. Agrarian Dept. 1948.

Use of the indexes, the grazing coefficient, and the grazing zone classifications led to error because:
- Soil and topography factors were not considered-- these are fundamental elements in any determination of forage production.
- The dynamics of the vegetation was ignored.
- For each one of the eight divisions, three grazing-capacity figures were given according to the quality of the grasses: good, medium, or bad without giving adequate taxonomic information.
- The "grazing coefficients" were given in terms of hectares per head of cattle, regardless of age or size of the animals, and only one coefficient was given for large heterogeneous regions (only climate was used as a decision guide).

The New Range Management Approach

It was not until 1965 that the government of Mexico integrated a federal commission (COTECOCA) within the Department of Agriculture to study the actual capacity of the country's rangelands. This new dimension to the Agrarian Department was based on the scientific studies and experiences in the Southwest of the U.S., which were adaptable to the conditions in Mexico and similar countries.

The most important contribution was the adoption of the "animal unit" concept and the application of the ecological principles of range management. It was established that all legislative action should have technical and ecological (besides socioeconomic) bases for the improvement and preservation of the 62% of the nation's area classified as rangelands.

However, in spite of the demonstrated facts and advantages of the new approach, a constitutional amendment has not yet been made in the legislation. The amendments are expected in the very near future.

With the advent of basic principles in range ecology and management, a significant change took place in the approach to determine the productivity of Mexico's rangelands: more detailed inventory and mapping of vegetation was developed; including determination of range sites and their productivity. This last phase is still in progress, and COTECOCA-SARH is now officially recognized as the federal agency responsible for providing technical information to the Agrarian Reform Department, especially as related to the factors that determine the legal size of a ranch. Unfortunately, much valuable information provided by COTECOCA is not fully accepted by the Agrarian Reform Department.

PROBLEMS IN RANGE MANAGEMENT

Both of the major land-tenure systems in Mexico (ejido and private property) reflect similar cause of vegetation and soil deterioration: overgrazing; lack of an adequate management program; limited infrastructure; and inadequate technical assistance. But each system has more specific problems.

The Ejido

Within the spectrum of ejido land management, the following aspects of their range management are causing the most serious problems:
- Too many solicitors assigned to an ejido.
- When too many ejido solicitors are assigned to good, productive range, they often plow up the grass cover to do some rainfed cropping, even when there is insufficient rainfall. This accelerates the desertification process.
- No technical base is used in determining the number of animals grazing the range. This results in overgrazing, mostly by unproductive animals such as burros, horses, and mules. At least two-thirds of the livestock on ejido lands in central Mexico are estimated to live at a maintenance level--not a productive one.
- The range is overgrazed because the land is rented to ranchers or because the most influential men in the ejido abuse it.
- Due to rural social pressures, ejidos sometimes are located in arid lands with the potential for improvement so low that the investment in infrastructure is not justified.
- The "caciques" (economically powerful men in the ejidos) oppose any collective action for improvement or innovations in management because their private interest would not be served. Usually these people dominate the council or assembly in each ejido that makes the final decision for every communal program.
- By renting the land to other people, the ejido loses its social control as well as control of animal numbers, which seriously damages the land.

All of these existing problems make range improvement programs difficult to implement. The worst obstacle is the resistance to reducing the number of animals (stocking rate) in the ejido. Unless this obstacle is removed, no new program will be feasible.

A recent innovation in the ejido system is the formation of production groups made up of members who want to do things better and improve their operation. The National Rural Bank (BANRURAL) organizes them and provides long-term, low-interest credit for infrastructure, range improvement practices, and purchase of livestock. This credit is paying good dividends because, in addition to the loan, the participants receive technical assistance, mainly from the Rural Bank.

This paper is not the appropriate place to suggest changes in the procedures for the allocation of ejido centers or for the number of people to be assigned to each of them since the process of repartition of land in the agrarian reform has almost been concluded. However, from the standpoint of the management of those poorly distributed lands, a series of problems exist that must be solved by putting aside political interest so that the area can regain its productive status.

There will never be sufficient land in the country to meet the demands of an ever-growing number of solicitors. Other than creating agro-industries as a source of jobs for the rural people, the solutions could be:

1. Protect rangelands against plowing and attempts to convert them into unproductive, dry-land farming areas where ecological conditions are not adequate.
2. Eliminate unproductive animals from the range and stock it according to the technically determined condition of the range.
3. Organize production units in the ejido and select people who are willing to accept responsibilities under a well-planned and supervised investment and management program.
4. Strengthen technical assistance from BANRURAL and the extension service.
5. Enforce the prohibition of renting the land to outsiders.

Private Property

One of the factors that has limited the development of the livestock industry in Mexico is the lack of security in land tenure for the private rancher. Even when the government has completed (theoretically) the repartitioning phase, the rural demographic pressure is so tremendous that a great many solicitors are demanding that new ejidos be formed. Unfortunately, these demands get support from many official agencies and politicians. A constant, restless atmosphere among ranchers is created. A least 40% of the rancher's time is spent dealing with agrarian authorities: meetings, inspections, lobbying, filling forms, discussions with lawyers, etc. These inconveniencies and the almost complete absence of incentives have drastically reduced investments for infrastructure, for range improvements, and for livestock on ranches of the private sector.

Some factors that negatively influence the private ranching industry in Mexico from the legislative view are:

1. The legal size of a private property has not been clearly defined.

- The "500 head" concept lacks uniformity and is subject to many misinterpretations.
- The carrying capacity figures determined by COTECOCA-SARH are not fully accepted nor respected by the Agrarian Department.
- Frequent inspections by personnel of the Agrarian Department to confirm the "correct" number of animals on the ranch are often for other reasons.

2. Conversion of some areas of the ranch to intensive forage production when conditions are adequate is not practical because that puts that area of the ranch into another land category so that it then is more vulnerable to ejido solicitors. (This is true also for other types of investments, particularly in infrastructure.)

3. Ranchers fear being invaded by ejido solicitors, even when the ranch has all the certificates of ownership and of legal tenure in order. Many of these invasions are sponsored or even directed by corrupt agrarian leaders. When the invasion succeeds, a former productive unit is broken up and the over-all production is severely reduced by the ejido problems mentioned above.

Obviously, as long as the government does not determine the technical size of the private property, the rancher will not have incentives to invest and to improve livestock production. The present status of ranching is as follows:

1. Legal ownership certificates are available to ranchers according to the Constitution. It is just a matter of applying for them and respecting them.

2. Cultivated pastures as a complement to the range livestock operation should be encouraged. Many ranches are in a position to do this to improve the overall ranch production. This additional investment would create more jobs for rural people and help the overall production of the country.

3. Legislation is needed for the marketing of livestock products—both for national consumption and for exportation. A clear, long-term price policy is needed in Mexico in order to provide incentives and to offer security to plan any range livestock operation.

Mexico has suffered from inadequate legislation for the range livestock industry. The principal problem has been related to the tenure of the land dedicated to grazing and (or) cropping. If this problem is not solved soon, numerous ranchers are going to lose interest in this type of operation. The country has the capability to correct the system and let a prosperous livestock industry flourish.

The problem is not just a matter of increasing production, it is a matter of survival for the range resources of the country and, consequently, for its own people.

15

REGULATION OF AGRICULTURAL CHEMICALS, GROWTH PROMOTANTS, AND FEED ADDITIVES

O. D. Butler

Agricultural chemicals, from fertilizers to pheromones, help make U.S. agriculture the most productive in the world. Discovery, testing for efficacy and safety, manufacturing, marketing, and proper use all represent the ultimate in biological sciences, in ingenuity, and in exercise of the free enterprise system.

Some say that in this case the enterprise system is not very "free." Thalidomide, DDT, aldrin, dieldrin, arsenic, and many others did not pass safety tests. The thalidomide tragedy may have aroused the most fear in public minds, but the diethylstilbestrol use in the 1950s for sustaining pregnancy in women, which apparently resulted in increased incidence of cancer in their daughters 20 or more years later, would have to be rated a close second in the world, and first in the U.S.

Public demand expressed through members of Congress the last couple of decades caused ever-more-strict federal regulations on development and use of agricultural chemicals. During the past year, however, the Food and Drug Administration, the USDA, and the Environmental Protection Agency (the major responsible agencies) have shown good evidence of more reasonable postures concerning laws, regulations, and interactions with manufacturers and users of agricultural chemicals.

President Reagan's appointment of a cabinet-level committee chaired by Vice-President George Bush with a mission for reducing burdensome regulations, gave an unmistakable signal to the agencies. Now we see Congress considering revision of the Federal Insecticide, Fungicide, Rodenticide Act (FIFRA), and the Food Safety Laws, especially the extremely strict 20-year-old Delaney anti-cancer clause. This clause was made obsolete by almost unbelievable advances in assay procedures that now detect parts per trillion of materials in foods that were considered to have zero residue with the parts-per-million capability of assays in the 1960s. Assays are now as much as a million times more sensitive.

Strict laws that were formerly written to ban toxic substances on the basis of risk alone are being reconsidered. A couple of reasons derive from the issue of essential elements--such as selenium required by the body at a low level, but toxic at higher levels, and nitrite used for centuries in meat curing to give the characteristic color. Derivatives--for example, nitrosamines that may be developed during cooking of bacon--have been shown to cause an increased incidence of cancer in susceptible laboratory animals. More recently, the finger of suspicion has been pointed at nitrite itself, in a highly disputed experiment with laboratory animals. Nitrite produces color, but more importantly, it protects against the deadly botulism bacteria, so use of nitrites has not been banned, but has been strictly limited. Critics of the regulations point out that many natural foods contain nitrites and that human saliva does also. Avoiding cured meats would reduce nitrite consumption by a very small and negligible amount, critics say. But the "scare" stories certainly reduce demand for ham, bacon, and hot dogs.

What are producers' primary concerns about agricultural chemicals? I believe that you should have a general idea of how they are discovered, tested for efficacy and safety, and used in a safe and effective way. You should also know the direct cost of materials, as well as the indirect cost, if consumer concerns affect demand for products marketed.

Good basic biological research done primarily by public institutions, such as the Land Grant Universities, usually provides the foundation for development of an effective product. The need for products to control pests or diseases usually is expressed by producers reinforced by producer organizations, by extension specialists and research workers who interact with producers, and by supplier representatives.

Because of the similarity of all living cells, there must be a good understanding of the biology of both species affected to be able to kill a parasitic living organism without consequent toxic effect on the host. Then, for food producing plants and animals, there must be great concern about residues that might have an effect on consumers.

Animal producers are served well by a group of competing companies seeking profit by manufacturing and marketing drugs, biologicals, pesticides, and related materials. Most of the companies belong to an industry trade association, the Animal Health Institute (AHI), headquartered in the Washington area. It serves the industry the same as the many other trade associations there, trying in every way to protect the opportunity for the industry to produce products that customers will buy and use because of benefits and thereby earn a profit for investors.

Almost inevitably it seems, any position taken or change advocated by the AHI is opposed by one or more organizations that classify themselves as consumer protectionists. Lawmakers and regulators usually have to make decisions between opposing viewpoints without the benefit of absolutely conclusive evidence. In the last decade such controversy has been a major stimulant to the formation of the American Council on Science and Health (ACSH) and Council for Agricultural Science and Technology (CAST), both of which I support.

"The American Council on Science and Health (ACSH) is a national consumer education association directed and advised by a panel of scientists from a variety of disciplines. ACSH is committed to providing consumers with scientifically balanced evaluations of food, chemicals, the environment, and human health." This is quoted from their March 1982 publication, "The U.S. Food Safety Laws: Time for a Change?"

The Council for Agricultural Science and Technology (CAST) is an organization sponsored and managed by twenty-five scientific agricultural societies. Its major purpose is to assemble and report the scientific information on important issues of national scope for the benefit of lawmakers, regulators, and the general public. It is not an advocacy organization. Most of its task force reports, now numbering about a hundred, were prepared at the request of members of Congress, some by government agencies, and some because the 47 officers and directors, all representing the scientific societies, decided that there was a need to assemble and print the scientific evidence on an important issue. CAST celebrated its tenth birthday anniversary in July 1982 at a directors' meeting at its headquarters. I have the privilege of serving as president of CAST in 1981, as did Frank Baker, the Director of this International Stockmen's School, in 1979. (I want to especially recommend CAST task force reports mentioned in the references.)

Some of the scientific societies work directly with regulatory agencies. I served as chairman of the Regulatory Agencies Committee of the American Society of Animal Science for about 10 years until 1981. The Institute of Food Technologists, like the American Society of Animal Science, has been very active in identifying and nominating qualified scientists to serve on CAST task forces and has also produced independent papers on various aspects of food safety.

Drug manufacturers have been very critical of the Food and Drug Administration (FDA) for taking so long to consider new animal drug applications (NADAs) before approval. A recent report entitled "The Livestock Animal Drug Lag" by the AHI describes the problem and suggests solutions. U.S. manufacturers have been able to obtain approval to market their products in the United Kingdom and European countries in a fraction of the time required for U.S. approval. An example is albendazole, a broad spectrum anthelmintic effective against gastrointestinal roundworms, lungworms, tapeworms, and liver flukes in cattle. Approval was obtained in 5 months in England in 1978. The same application filed in the U.S. in 1977 is still pending, though strong producer pressure resulted in limited approval in 1979 under a special investigative New Animal Drug authorization in a limited number of states. After the Food and Drug Administration banned hexachloroethane for liver fluke control, cattlemen had no approved drug. Texas and Florida producers, with pastures along streams and low-lying areas that have snails (the intermediate fluke host), just had to have an effective drug. Cattle producer organizations rallied to the cause and helped obtain the limited approval.

The AHI sponsored a Forum on Regulatory Relief in Alexandria, Virginia, in June 1982. Dr. Arthur Hull Hayes, Commissioner of the FDA, announced there that "I've decided that all activities in the Review of Animal Drug Applications, including issues of Human Food Safety, will be consolidated within the Bureau of Veterinary Medicine." That is certain to allow faster decisions. The Bureau of Food review has been blamed for much of the delay in the recent past.

Dr. Hays gave a definition of safe as "a reasonable certainty of no significant risks based on adequate scientific data, under the intended conditions of use of a substance." More and more we are realizing that there is no such thing as absolute safety, or zero risk. His speech gave some reassurance concerning "sensitivity of method" regulations that have been under consideration for several years by FDA. The bureau now seems willing to accept foreign data in support of New Animal Drug Applications under certain restrictions and also to consider cross-species approvals. It is not a good investment for drug companies to spend several million dollars to obtain approval of an anthelmintic for goats, for instance, that is very important in Texas (which has about 95% of U.S. goats) because the market is so limited. Other minor species, even sheep, fall in that same category. I believe it will be necessary for publicly supported institutions like the Texas Agricultural Experiment Station to assist in developing drugs and obtaining approval for use in such minor species.

The FDA is also considering some liberalization of restrictions on feed manufacturers. Dr. Lester Crawford, recently reappointed to the position of Director of the Bureau of Veterinary medicine (BVM) of FDA spoke to the American Feed manufacturers' 74th Annual Convention at Dallas in May 1982. He reported that "The Subcabinet Working Group, chaired by USDA Assistant Secretary Bill McMillan, has proposed the total elimination of FD 1800s, the notorious application required to authorize manufacturing and sale of medicated feeds. Instead, the BVM would have authority to deny registration of feed manufacturers that lacked adequate facilities and controls to assure safety."

Even the Environmental Protection Agency (EPA) is trying to "simplify the regulatory burden on industry and reduce unnecessary costs." So said John A. Todhunter, Assistant Administrator for Pesticides and Toxic Substances, at the 1982 Beltwide Cotton Production Mechanization Conference, January 1982, at Las Vegas. He described a reassuring response to Vice-President Bush's task force, especially that FDA has instituted a plan to improve the quality of scientific assessment, including a peer review system for major scientific studies and reports. There is, therefore, hope for maintaining availability of the herbicide 2, 4, 5T and even reapproval of compound 1080 for predator control. The states also have regulatory authority and enforcement responsibility. We are all aware of Governor Brown's reluctance in California to institute effective control measures for the Mediterranean fruit fly because of the political pressure of environmentalists.

Food Chemical News, a weekly publication, keeps you up-to-date on what is happening in Washington.

For those of you mixing your own feed, and for feed distributors, I recommend the annual Feed Additive Compendium, a guide to use of drugs in medicated animal feeds with monthly, up-to-date supplements.

In conclusion, I want to make a plea to agricultural producers for closer adherence to label requirements and restrictions on use of agricultural chemicals. The Agricultural Extension Service in every state has a responsibility for assisting producers in the proper use of chemicals. More attention is being devoted to that. Very few people deliberately break the laws, but many are not aware of the precautions necessary to prevent cross contamination of products and elimination of residues in feeds and foodstuffs. The USDA state producers' effort to eliminate sulfa drug residues in pork is an example of the kind of cooperation required to maintain availability of chemicals so important to modern food production. Let us resolve to intensify the effort for safe use of agricultural chemicals in order to gain greater public confidence in the safety of our abundant food supply.

REFERENCES

American Council on Science and Health (ACSH). 1982. U.S. food safety laws: Time for a change? 1995 Broadway, New York, N.Y.

Council for Agricultural Science and Technology (CAST). 250 Memorial Union, Ames, Iowa 50011.

CAST. 1977. Hormonally active substances in foods: a safety evaluation. Report No. 66.

CAST. 1981. Antibiotics in animal feeds. Report No. 88.

CAST. 1981. Regulation of potential carcinogens in the food supply: the Delaney clause. Report No. 89.

CAST. 1982. CAST-related excerpts from U.S. House of Representatives hearing on the Federal Insecticide, Fungicide, and Rodenticide Act (FIFRA). Special Pub. No. 9.

CAST. 1982. CAST-related testimony on the food safety amendments of 1981. Special Pub. No. 11.

Feed Additive Compendium. Miller Publishing Co., 2501 Wayzata Boulevard, P.O. Box 67, Minneapolis, Minnesota 55440.

Food Chemical News. 1101 Pennsylvania Ave., S.E., Washington, D.C. 20003.

16

THE CURRENT STATUS
OF THE FAMILY FARM
IN AMERICAN AGRICULTURE

George Stone

Almost everyone knows what a family farm is, but hardly anyone is able to define it on paper to the satisfaction of other people. A particular farm may meet most of the criteria that might be suggested, but there will always be some differences of opinion on such things as size, ownership, and control. There is not time for a debate on the fine points of a family-farm definition. But if I am to talk to you today on family-farm agriculture, you are entitled at least to know what I think I am talking about.

FAMILY FARM DEFINED

I like the National Farmers Union's definition of a family farm. It says:
"A 'family farm' is, ideally, one which is owned and operated by a farm family, with the family providing most of the labor needed for the farming operation, assuming the economic risk, making most of the management decisions, and depending primarily on farming for a living."
That is probably as well as it can be explained in less than fifty words.

PUBLIC POLICY RELATED TO FAMILY FARMS

Our national public policies have endorsed and advocated a family-farm structure of agriculture for almost 200 years, dating back to the Ordinance of 1785, the Land Act of 1796, the Pre-emption Act of 1841, the Homestead Act of 1862, the Reclamation Act of 1902, and a half dozen major statutes in this century as recently as the Food and Agriculture Acts of 1977 and 1981.
The 1977 Act includes a declaration by the Congress that it "firmly believes that the maintenance of the family-farm system of agriculture is essential to the social well-being of the nation"...and that "any significant expansion of nonfamily owned large-scale corporate farm enter-

prises will be detrimental to the national welfare." The 1977 Act also mandated that the Department of Agriculture should issue an annual report on the "Status of the Family Farm." This has been done, supplying some continuing data on the structural trends and changes in agriculture.

In March 1979, speaking at the national convention of the Farmers Union, Secretary of Agriculture Bergland called for a national dialogue on the structure of agriculture, declaring:

"We are at a point in our history where a broad-based public discussion of the issues that shape national policies is needed to promote the kind of agriculture and rural living this nation wants for the future."

In that Kansas City speech, Secretary Bergland observed that "we really don't now have a workable policy on the structure of agriculture," and warned:

"We can act now to insure the kind of American agriculture we want in the years ahead. Or we can let matters take their course, with the probable result that we will wake up some morning to find that we have forfeited our last chance to save those characteristics of the farm sector we believe are worth preserving. I, for one, do not want to see an America where a handful of giant operators own, manage, and control the entire food production system. Yet that is where we are headed, if we don't act now."

In late 1979 and early 1980, Secretary Bergland conducted this national dialogue at a series of regional hearings. Numerous economic papers and a comprehensive report were eventually published. It was a worthwhile exercise in stimulating Americans to think about what they want in an agricultural system and what they want their federal government to do to assure such a system. But, while there may be better public understanding of our agricultural system, there is little in the way of agricultural legislation or administrative decision-making that can be attributed to the Bergland study.

ASSESSING CURRENT FARMING CRISIS

We find ourselves here, early in 1983, still trying to assess the current situation of family farms. Through our 80-year history, National Farmers Union has been totally dedicated to the family-farm system. We believe that the family farm represents the best choice for the American people on every score:
- Assured abundance
- Efficient production
- Best care and use of land and water resources
- Rural employment
- Quality of life in rural communities

- Highest export earnings
- Most favorable balance of trade

During the 80 years of Farmers Union, the family farm has proven its durability and staying power. Family farms have survived wars, natural disasters, and a total of 14 recessions, panics, and depressions, including the most recent. During most of this century, it was usually assumed that the family farm would survive as the dominant form of agricultural structure. Now, although it is readily acknowledged that the family farm is the most efficient agricultural production unit, it is no longer that certain that it will survive much longer.

In a spirit of candor, one must admit that there have been other times when the survival of the family farm appeared to be in doubt. Calamity seemed at hand. Yet, while some farm operators were lost in the crises of the past, and their loss was regrettable, the system as a whole survived and continued to produce for the nation.

Having expressed the caution that at times in the past things have appeared worse than they turned out to be, there are signs that the current challenge to the survival of family farms is the most dangerous, at least since the years of the Great Depression.

Not since the early 1930s has the nation had three such bad years in agriculture in succession. Many signs indicate the magnitude of the farm crisis. Net farm income dropped from $32 billion in 1979 to $19.8 billion in 1980, $18.9 billion in 1981, and while we do not yet have final figures, appears destined to be still lower in 1982. In terms of purchasing power, the farm parity ratio in much of 1981 and into 1982 has been the most unfavorable suffered by farmers since 1933. In 1933, U.S. farmers had $3 billion in net income, but only $9.1 billion debt. That was a ratio of $3.10 in debt for each dollar of net income. Today, we have something over $11.00 in debt for each dollar of net income.

In 1981, for the first time in recorded history, U.S. farmers paid out a total of $19 billion in interest outlays on their debt, a sum that exceeded their net income for the year. At the worst of the Depression of the 1930s, the interest rates paid by farmers averaged 6.4%, while recently the rates paid by farmers on loans to commercial banks averaged over 18%, as reported in the Federal Reserve Bulletin.

In these last three years, farmers have been substituting credit for income at an alarming rate. Years ago, farmers were able to generate much of their capital needs internally. In 1970, for example, farmers depended on borrowed capital for only 5% of their cash operating funds. By 1980, the proportion was up to 21% and, in 1981, it was almost 23%.

Another important measurement is the liquidity ratio of farmers. In 1950, as an example, U.S. farmers had $13.8 billion in cash assets such as deposits, currency, and savings bonds. Against this, they had $12.4 billion in debts. That was a liquidity ratio of 111%. In 1960, there were cash assets of $13.9 billion and $24.8 billion in debts, a liquidity ratio of 56%. By 1970, there were $15.6 billion in cash assets and $53 billion in debts, a liquidity ratio of 29%. In 1981, the cash assets totalled $19.9 billion against a total debt of $194.5 billion, a liquidity ratio of about 10%. Behind all those statistics are human families trying to earn a living in a productive and useful endeavor.

FACT-FINDING HEARINGS

To document the human side of the farm crisis, the National Farmers Union held a series of nine regional fact-finding hearings in March and April of last year. We heard testimony from 230 witnesses, including farmers, farm wives, main street businessmen, cooperative officials, teachers, bankers, and community leaders. The summary report, which we published on these hearings, is entitled "Depression in Rural America." It did not deal just in generalities or endless statistics, but told the personal story of families beset with hardship and despair because of conditions over which they had no control. The report showed how the desperate economic conditions were affecting the lives and survival of working farmers, their business communities, and the fabric of life in rural America.

The purpose of the hearings and the report was, of course, to mobilize opinion and develop a sense of urgency about farm legislation that would help family-scale farmers survive.

Of course, there are some who say that the federal government should not intervene on behalf of family farmers —that we should just let nature take its course. The theory is, if we just let the decline in family farms continue, then, after a while, just the efficient farmers will remain and they will be able to prosper.

But those who have been involved in agriculture for a lifetime have yet to see such a scenario work. In 1960, for example, there were about four million farming units and they were earning farm income and purchasing power equal to 80% of parity. In the decade of the 1960s, the nation lost one million farmers, but farm income did not go up. We have been as low as 57% and 58% of parity. At this rate, it might be asked, how long will it take to get to 100% of parity?

The truth is you won't reach some sort of ideal economic situation for farmers by that route. The truth is that the economic hardship is not weeding out small, marginal, or innefficient farmers. The farmers who have been hurt most in these past three years of low farm prices and high interest rates have been the good, efficient operators in the middle of the scale in farm size.

Further, the projections are that this kind of attrition of our best farmers will continue. In 1980, USDA econ-

omists did a projection of what will happen to farm size and
structure by the year 2000. The report projected that the
number of farming units would drop by 30% to 1.8 million by
the year 2000, with most of the decline in middle-sized
farms. Small farms, with less than $20,000 annual gross,
will still make up 50% of the total farming units, with
large farms, with $100,000 or more in gross sales, edging
out the middle-sized operators.

Along with these structural changes, the USDA officials
foresee an increase in concentration in both farmland ownership and production. The USDA specialists project that it
will take $2 million in capital assets to run a farm capable
of grossing $100,000, and that these large capital require-
ments will tend to concentrate farm wealth in the hands of a
relatively few.

Young beginning farmers will have increasing difficulty
entering the industry. USDA projects that there will be
fewer than 300,000 farmers under the age of 35 years in the
year 2000, a drop of 200,000 from the current level. The
number of individual ownerships and partnerships in farming
will decline by the year 2000, while the number of corporate
farms and multiownership units will increase, the report
indicates.

In another report associated with the farm structure
dialogue, entitled "Another Revolution in U.S. Farming,"
USDA economists predict that there will be further declines
in the number of farms, but not at as sharp a rate as in the
1950s and 1960s. However, there will be increasing concentration of production among the largest producers, along
with strong pressures for the separation of ownership and
use of farming resources. Because of taxes and other factors, off-farm investors can get higher overall returns by
investing in farmland than they can by investing in common
stocks of business and industry.

In the Farmers Union, we view the separation of land
ownership and farming operations with a great deal of concern. It seems to us that such a trend will have the tendency to create a new generation of sharecroppers--people
who have little control over their own destiny. That is why
Farmers Union in the past several years has taken the
leadership in seeking to limit absentee ownership of agricultural land. The threat has come from three different
sources:
- Investments in U.S. farmland by American business corporations, conglomerates, and off-farm
  investors. This has included efforts by individual business firms and such spectacular
  schemes as the Ag-Land Trust Company.
- Investments by foreign corporations and investors.
- Proposals to invest pension fund assets in U.S.
  farm cropland.

In regard to the threat of domestic corporations and
investors to take over farmland, several midwest states in

the past several years have enacted limitations on corporate ownership of farmland and corporate farming. Most of the states in the Mississippi Valley now have restrictions of some sort on corporate farm ownership. Largely because of the vigorous campaign by the Farmers Union and the opposition raised at a Congressional hearing in Washington, D.C., the Ag-Land Trust proposal was dropped.

Because American farmland had become a magnet for foreign corporations and investors, the Farmers Union successfully won adoption of the Agricultural Foreign Investment Disclosure Act of 1978, under which foreign persons or foreign-controlled firms acquiring U.S. farmland must report such holdings to the USDA. Although there may be some evasion of the disclosure law, we now have some hard data on the extent of foreign holdings. It is now clear that there has been more foreign investment in American farmland in the past five years than in the previous fifty years. The latest annual disclosure report by USDA shows that almost 5 million acres of land were acquired by foreign persons during 1981. The foreign acquisitions were equal to about 25% of the total of 18.1 million acres of farmland sold during the year.

The third proposal, that of the American Agricultural Investment Management Corporation of Chicago, proposed to facilitate the investment of nonprofit pension funds in U.S. farmland. We became concerned because pension funds represent a huge pool of capital earning very modest returns on the order of 3% to 4% a year. We thought the opportunity to invest and take advantage of the rapid appreciation of farmland values would be irresistible, even if the profit from farming were modest. As a matter of fact, pension fund assets now total about $700 billion and are expected to rise to $1.5 trillion by the year 1990. Obviously, there would be enough capital to buy all of the farmland in the nation.

We don't expect that to happen. But even if only 3% or 4% of the pension funds were invested in farmland, that would total $18 to 24 billion a year--about the total of farmland sales values in recent years.

The ability of beginning farmers--or existing farmers seeking to expand their operations--to bid for land would certainly suffer by the presence of institutional investors who would not have to pay for the land from their agricultural earnings. Young farm couples who hope to acquire a viable farming unit would be virtually fenced out of the competition for farmland by vast amounts of absentee capital. Of course, the promoters of the pension fund scheme claim they would be doing farmers a huge service by relieving them of the necessity to own farmland. Such a separation of land ownership and farming operations would enable farmers to use all of their limited capital in production.

This is a phony argument. The operating farmer pays a land cost whether he owns or rents. He pays land costs whether he is a cash renter or a share renter. We take this attitude in the Farmers Union because we believe that public policy, whether federal or state, ought to be helping families become owners of the land they farm, not separating them from that possibility.

In conclusion, we regard our efforts to keep farmland in the hands of operating farm families as very important. But, it should be pointed out, the challenge of these outside forces is most damaging because of the weak economic position of our farmers. If farm prices and income were maintained at a more satisfactory and stable level, farmers would be able to withstand more easily the competition of outside investors. Low farm prices and income, accompanied by high interest rates, compound the problems of farmers in sustaining themselves in land ownership and farming. Because this is true, we cannot simply go on as we are and let nature take its course. We must act positively on farm income and other measures to assure that we continue to have a predominantly family-owned, family-operated farming system in our nation.

17

OREGON WOMEN FOR AGRICULTURE TALK ABOUT THE STRUCTURE OF AGRICULTURE

Ruby Ringsdorf

Let it be known that Oregon Women for Agriculture do believe that it is the inherent right of every child born today to have adequate nourishment; and that the American farmer will continue to feed the hungry if not strangled with bureaucratic rules and regulations.

We furthermore feel that it is neither our duty, nor even our right, to enter into the internal policies of a foreign country whose political system, or local corruption, are preventing food from reaching their hungry masses. Neither are we prepared to let those countries' Marxist-oriented political ideologies creep into and destroy our free enterprise system. The free enterprise system is the propelling factor that has made American agriculture the envy of the world!

LAND OWNERSHIP, CONTROL, AND TENANCY

Fifty-five percent of the land in the state of Oregon is already publicly owned (L.C.D.C.). The number of all commercial farms (farms with sales of $2,500.00 or more) in Oregon increased rather than decreased from 1969 to 1974.

The number of commercial farms in Oregon with sales greater than $40,000.00 increased 67% from 1969 to 1974 and comprised 30% of all of Oregon's commercial farms. At the rate of inflation over the past 10 years it is surprising that this percentage is not greater. It doesn't take much of a farm to produce $40,000.00 in sales today, but the net probably isn't enough to keep the family dog in dog food for the year.

According to an Oregon State study (EM N:23), family farms, nonincorporated, comprise 96.4% of all commercial farms with 3.6% being corporate farms. Of the corporate farms, 87.3% are family farms (94% have 10 or fewer shareholders, 44% are controlled by one stockholder).

Many family farmers own some land and lease more from retired farm relatives or neighbors in order to make their units more economical.

BARRIERS TO ENTERING AND LEAVING FARMING

Inflation, high interest rates, inheritance tax laws, FHA regulations, EPA regulations, and our national cheap food policy are all barriers to entering and leaving farming. Inflation is not only driving up the cost of land to an impossible price for a beginning farmer (the interest alone for each acre of ground is much more than what he would have to pay to lease the ground), but also the cost of the equipment needed to start farming. Inheritance tax laws, especially the carry-over provision, make it almost impossible for children to inherit a farm without selling a portion of it or splitting it up to meet the tax obligation so that they no longer have an economical unit.

EPA regulations are becoming increasingly more difficult to cope with and discourage many young people from even thinking of farming. FHA regulations restrict the amount of money available to a young farmer for a small acreage because it is not an efficient, economical unit that could produce enough net income to service debt and provide a decent living for his family without off-the-farm income. And yet, if the unit is large enough to do both, the cost is far more than FHA is allowed to cover for one farm.

It is difficult for a farmer to retire and leave farming. The land cannot be sold because of strict zoning laws, because of capital gains tax on the appreciated land value (often the only net savings realized from the farmer's investment in time and labor), and because his acreage is no longer large enough to be an economic unit for a family farmer. He stays on the land, rents to others, and does the best he can, too often becoming another rural-poor statistic.

PRODUCTION EFFICIENCY AND SIZE OF FARMS

Production efficiency and size of farm are tied together. Our Oregon State study shows that the average size of all commercial farms in Oregon increased very little from 1969 to 1974. It would seem the trend to larger farms has already peaked because of production efficiency.

Size and number of farms in Oregon vary greatly from one geographical area to another. In the Lake Labish area near Salem, where land sells for $10,000.00 per acre and up, a 20 A farm is considered large. In the southern end of the Willamette Valley, in the grass seed capitol of the world, a thousand acre farm is not considered large, and in the cattle grazing lands of eastern Oregon a 5,000 A ranch is not large. A dairy farmer can have only 100 A but milk 500 cows and be considered a large farmer, but another dairy farmer can have 50 cows on 500 A and be considered a small farmer.

Because Oregon produces over 170 different marketable commodities, it is unfair to use a gross dollar amount figure for sales to define large and small. Different crops show different net results. It is impossible for a 20 A berry farm to net more than a 500 A wheat ranch. Most often a farmer cannot convert his acreage to higher value crops because of soil types, marketing limitations, increased risks, and increased operating and capital requirements.

The efficiency of the American farmer is the envy of the world. After American farmers feed the U.S. they export 60% of their wheat and rice, 50% of the soybeans, one-fourth of their grain sorghum and one-fifth of their corn. The U.S. provides half of the world's wheat (Oregon Grange Bulletin 9-4-78).

Agricultural products are the second largest category of U.S. exports. Agricultural exports returned $23 billion to our country in 1976. In 1975 agricultural exports provided the foreign exchange to cover 83% of our petroleum imports.

American farmers provide all this despite the fact that the number of U.S. farms and farm workers has decreased by two-thirds since 1940.

One American farmer can now produce enough to feed 60 people. From 1950 to 1978, farm productivity increased at an annual rate of 5.3%—more than twice as much productivity as compared to any other nonagricultural business (Oregon Grange Bulletin 9-4-78).

GOVERNMENT PROGRAMS

Much of the success of American agriculture is due to agricultural research. This represents only a small portion of the USDA budget. Food assistance programs, including Food Stamps and Child Nutrition (programs benefiting from past research programs), represent 56% of the total USDA budget. A cut in agricultural research, extension, and soil and water conservation programs would not be in the national interest.

As to having a national or world grain reserve, why not establish a worldwide monetary food fund (a required UN fund, a contribution fund with participation by churches and other interested groups, internationally funded and administered) to be used for international food crises. Reserves have a history of depressing prices to producers and stabilizing prices at the lower levels. Reserves have also acted as a disincentive to production so that farmers change (if possible) to producing crops that will hopefully yield more net return. Any grains in reserve should be isolated from world markets and used as aid rather than trade. The cost of this grain reserve should be shared by all people internationally.

Another program that is under consideration for USDA funding is the plan to pay the expenses of low-income and nonprofit groups that testify at regulatory hearings. We are very much opposed to such a plan; even though Women for Agriculture would qualify for funding.

TAX AND CREDIT POLICIES

Inheritance tax laws need changes, particularly the carry-over provision. Prior to enactment of the carry-over provision, beneficiaries inheriting appreciated property received a stepped-up tax basis on property at the time of inheritance, and each generation of a farm family was subject to capital gains tax only on the appreciation that occurred while they owned the property. This procedure was radically changed by the carry-over provision that bases capital gains on inherited property on the descedent's acquisition price and not the market value at the time of transfer.

There should be more tax credit incentives given for conservation practices, since these can be very costly for one individual.

FARM INPUT SUPPLY SYSTEM

Government regulations and inflation have had a strong influence on the farm-input-supply system. Labor costs have spiraled, inventory taxes have prevented smaller manufacturers and suppliers from keeping a full inventory; land costs have spiraled and it is becoming more difficult to obtain necessary capital.

FARM PRODUCT MARKETING SYSTEM

There is a lot of talk about direct marketing from farmer to consumer. This works only in agricultural areas. Most crops sold in this manner are perishable, thus limiting choice, variety, and quality. It takes much more time and energy to drive all over to pick up vegetables here, eggs there, milk at that place, and fruit at still another stop. We already have the most energy-efficient distribution system.

Farmers are being told, "You can get a better price for your produce than the processor gives and the consumer can get it for less. Let's cut out the middle-man." Just who is the middle man? If we cut out all the middle men in the food-processing chain, our unemployment rate would probably be closer to 30% or 40% than the 8% or 9% it is now. Also, at the same time we are hearing rumblings about vertical integration (selling your own produce as a finished product and cutting out all middle men). This is what multinational corporations are accused of doing. Direct marketing is the same.

Probably close to 100% of Oregon's fruits (not including tree fruits) and vegetables are sold by forward
contracting or contractual arrangements with a processor.
This method is preferred over freedom of decision-making at
harvest time. There are not many farmers who would care to
wait until harvest time to search for a market for their
perishable products. They would be at the mercy of the processor who would then know the farmers have no choice but to
take whatever price is offered. If an equitable price is
not offered at planting time, the farmer has the freedom of
decision to plant or not to plant.

Contractual arrangements and forward contracting on
seed and grain crops isn't bad either. It can certainly
provide some freedom from fear. The farmer who stays in
business all his life usually is the one who contracts ahead
whenever he feels the price is such that he can make a fair
return. It takes some of the gamble and risk out of
farming.

PRESENT AND FUTURE ENERGY SUPPLIES

In the early 1920s we had 25 million horses to pull the
plow, the wagon, and the carriage. We fed about one-fifth
of our grain and roughage to those horses. (Today it would
take one-third of our crop land plus 20 years to breed
enough horses and mules for today's needs.)

It is time we look into using biomass or agricultural
products as a future energy source. The liquid energy that
we import is priced at $1.50 to $2.50 per gallon in most
major industrial nations. We are nearing these world prices
now, which will make the production of biomass fuel profitable. There are tons and tons of grass straw in Oregon
alone that can be used for fuel pellets or biomass conversion.

We feel it is unfair for American farmers to be told to
conserve fuel and energy when farming uses only 3% of the
total energy consumed in this country. It takes more energy
in the home for food preparation than it does for agricultural production (including fertilizer and other energy-
intensive inputs).

Productivity per man-hour in agriculture has been
increasing about twice as fast as the rate of productivity
per man-hour in manufacturing.

ENVIRONMENTAL CONCERNS, INCLUDING CONSERVATION AND THE USE OF SOIL AND WATER

Oregon farmers are also environmentally concerned. We
are also concerned because we who are engaged in agriculture
are such a minority. Even though Department of Environmental Quality tests gave proof to the fact that smoke from
field burning, a practice used in the Willamette Valley to

sanitize our seed crop grass fields, really had little or no effect on the quality of the air in the Valley, the EPA would not allow easing of regulations that are now strangling the Grass Seed Industry in Oregon. Why? Because all other sources of pollution are increasing yearly!

Bureaucratic rules and regulations often conflict with each other. In the Silverton Hills area, for example, 25 to 30 years ago farmers were losing tons of top soil every winter as a result of water run-off on cultivated fields. They discovered that their soil and climate was suitable for perennial grass seed production, which held the soil on the hillsides. Now EPA has restricted field burning because of air quality. Without burning, the grass fields become diseased and seed production is no longer economically feasible. We are again faced with soil erosion and resulting probability of water pollution.

Oregon has been working on Water Quality Management Programs, or non-point-source pollution. There are six major agricultural pollutants: sediments, nutrients, salts, organics, pesticides (including herbicides) and disease-producing organisms. So far only sediments have been found in our streams and these have been coming primarily from nonagricultural sources.

America's agricultural engineers are coming up with new and better seed drills that utilize low-till and no-till methods. Oregon farmers have been using grassland drills for many years, but we cannot use a grassland drill in a grass field that has not been burned. Here again, air quality versus water quality.

RETURNS TO FARMERS

Because our government has long endorsed the cheap-food policy, returns to farmers for the last 35 years have not kept up with the rest of the economy.

The American farmer was forced to become more and more efficient and only those who were efficient, excellent managers stayed in business. (Often even the most efficient were wiped out because of extraordinary conditions, such as weather. The margin in good years was so slender that a poor year wiped them out.)

Naturally, when the returns per acre became less and less, we had to expand our acres if we were going to live off the land. Now, suddenly, we are all called "commercial farmers" and that seems to be the wrong kind of farmer to be. "Commercial farmer" seems to be a dirty word in many circles. At the Rural America Conference in Washington, D.C., in June of this year, commercial farmers were being blamed even for dope addiction because farm mechanization caused these people to be out of work.

The people in Rural America meetings defined the family farm as a unit that produces food only for the people on the farm unit and who do not sell any farm product for profit.

In 1945 our fathers sold rye grass seed for 12 cents a lb. (They also sold wheat for $100 a ton or $3.33 per bu.) A tractor at that time cost around $3,000 and the first self-propelled combines came out for around $5,000. Today a tractor costs between $30,000 and $60,000; a combine from $65,000 to $70,000. Guess what the price of rye grass seed is today? Twelve and one-half cents per lb. Last year it was only 10 cents per lb so there are big headlines in the newspapers that agricultural wholesale prices increased almost 25% over a year ago! It is usually buried in the back pages when our farm product prices decrease or simply stay at the level of 30 years ago.

COSTS TO CONSUMERS

We as farmers are always hearing from our city friends about farm subsidies and how we are being paid to keep land out of production, etc. The truth is, the American farmer has been subsidizing the consumer for the last 40 years.

The trend toward greater efficiency in farming has benefited the consumer most. People have never had a greater variety of safe and nutritious food at so low a cost. The consumer can purchase more food for his hour of labor than at any time in history or in any other country. When the housewife complains about high food prices, she is often paying for a maid-in-a-box by buying prepared and semiprepared foods that cut down on food preparation time at home. It all depends on your priorities--time versus cost.

We also feel that if it is a policy to serve the public interest, by drastic and disruptive actions (such as export controls and import floods), then the general public, not just the producer, should pay or help pay for this policy.

QUALITY OF LIFE IN RURAL AREAS

This is interesting! Rural areas, not agricultural areas! Rural America: Educational Needs of Rural Women-- these programs are not referring to farmers or farm women. Rural America means the urban population that has moved to the less-populated areas of the U.S. from all socioeconomic levels and who most likely have never farmed and never will. The problems of these people should come under the jurisdiction of H.E.W., not USDA.

We farmers have noticed that this migration to the rural areas is causing problems. They tell us, "We are moving out to the country because we want peace and quiet and we want to be one with nature." And then they complain about the noise and dust from farm machinery and activities; so now we have noise pollution and dust-control laws. Now rural residents are suing because of smells coming from swine production and cattle feeders, forcing farmers to put in costly equipment to take care of the smell.

The environmentalists insist we use too much commercial fertilizer, but they complain about the smell when the local dairy farmer puts the liquid manure on his fields with an irrigation gun. There are over 100 known toxic substances occurring "naturally" in the environment, yet they complain any time they see our spray rigs come out of the yard.

Rural America says there are 131 rural towns in the U.S. without a doctor, that many people have to drive from one to two hours to a medical center. This is not always bad. We would rather drive for two hours to a good medical center with full facilities than fifteen minutes to a small facility where sometimes a local doctor tries to do only what a very specialized doctor should do.

Rural America also says one-half of the maternal deaths occur in rural areas. We believe that. Much is due to the back-to-nature trend, which is currently popular along with do-it-yourself childbirth. This is fine if everything proceeds normally. In the home we do not have the back-up facilities to aid a difficult delivery; it is too late then to rush to the hospital. Many of these same people have no prenatal care of any kind and their diet is often very inadequate because of their chosen life style and eating habits.

If we talk about the quality of life of real farm and rural people, we are talking about something entirely different. Genuine farmers seem to have fewer divorces per thousand population and they generally have a very strong family unit because everyone learns to work together.

A farmer has a lot of respect for his Creator; he is too closely involved with growing and living things to think that we are here purely by chance.

Generally speaking, the quality of life depends on the individual involved. The socialite who grew up in the large city might be quite disenchanted and bored living in a small rural community. She might complain bitterly about the lack of culture! On the other hand, the people who grew up in that same small area are quite content and feel that they have the good life. If they were to move to the large city they might then cry bitterly about the hustle and bustle, the unfriendliness, the foul air, the crime rate, etc., etc.

The kind of life we want in America can be found by any one who intends to earn it. If we expect it to be given to us, we will never find it.

18

LIVESTOCK PRODUCTION
ON NEW ENGLAND FAMILY FARMS

Donald M. Kinsman

INTRODUCTION

The Northeast region of the United States encompasses the 12 states of Delaware, Maryland, New Jersey, New York, Pennsylvania, West Virginia, Connecticut, Maine, Massachusetts, New Hampshire, Rhode Island, and Vermont; with the last 6 named constituting the New England states. Table 1 cites the basic facts and figures for the total Northeast region, which contains 128 million acres of land or 5.6% of the land area of the U.S. The 6 New England states represent 40 million acres (about one third of the Northeast) or 1.7% of the U.S. land area, yet the Northeast contains 21.6% of the nation's population or about 49 million people. New England has 12.3 million people or 5.4% of the U.S. total. The average annual precipitation is 40 to 46 in., and the mean temperature variation is from 20°F to 40°F in January to 70°F to 80°F in July. Temperature and snowfall vary considerably with elevation, which is dominated by the Appalachian Highlands. The Atlantic Ocean serves as a moderating influence along the coast. The frost-free period ranges from 90 to 150 days.

Sixty-four percent (64%) of the area is forested, compared with a U.S. average of 32%. The Northeast contains 11.4% of the nation's forest lands; New England represents 4.5%. The land suitable for agricultural production is primarily gray and brown podzolic soils, and agriculture on these lands is intensive.

Being the most highly urbanized region in the U.S., over 12% of the land is city, urban, and industrial compared to 9% U.S. average. Grassland pasture represents 3.2% of this area versus 26% for the U.S., and 15% is crop land as compared to 21% for the nation.

LIVESTOCK PRODUCTION

Against such a background, one might wonder about the livestock potential for this Northeast region of which New England is a microcosm. Approximately 6.1% of the U.S.

TABLE 1.   COMPARATIVE AGRICULTURAL DATA* FOR NEW ENGLAND,
THE NORTHEAST, AND THE U.S.A.

	New England		Northeast		
		% of U.S.		% of U.S.	U.S.A.
States	6	–	12	–	50
Land area (million acres)	40	1.7	128	5.6	2,264
Population (million)	12.3	5.4	49	21.6	227
Number of farms (thousand)	26.4	1.1	179	7.7	2,333
Average farm size (acres)	171	–	183	–	450
Cropland (million acres)	2.2	0.5	19	4.2	456
Forestland (million acres)	32	4.5	82	11.4	718
Livestock numbers (thousand):					
Cattle (beef & dairy)	753	0.6	5,750	5.0	115,013
Sheep	42	0.3	450	3.6	12,492
Swine	106	0.2	1,687	2.6	64,520
Agricultural cash receipts ($ billion):					
All commodities	$1.4	1.1	$8.0	6.1	$131
Livestock products	$0.98	1.4	$5.4	7.8	$69
Dairy products	$0.55	3.7	$3.1	20.7	$15
Cattle & calves	$0.08	0.2	$0.7	2.0	$35

*Data selected from USDA reports.

agricultural commodity value is produced in the Northeast.
The major portion of the livestock receipts is from dairy
products--3.1 billion dollars.   Beef cattle and calf receipts account for $700 million annually or approximately
2.0% of the U.S. total.   All cattle in the Northeast represent 5.0% of the U.S. total, sheep 3.6%, and hogs 2.6%; for
New England alone, those percentages are an infinitesimal
0.6%, 0.3%, and 0.2%, respectively.   Therefore, it goes
without saying, the Northeast is a meat deficit area, and
New England in particular produces but 4% of its meat consumption, thus it imports 96% of its meat supply, chiefly
from other sections of the nation.   Livestock production,
exclusive of dairy, in New England must be considered as
consisting primarily of small-livestock farm operations.
Using the USDA definition of a small farm ($20,000 or less
gross sales), 42% of all farms in New England are considered
small.   Most producers are part-time farmers.   These part-
timers are on family farms and do contribute to their

owners' well-being in addition to producing meat for a ready market.

CAN NEW ENGLAND LIVESTOCK FARMS CONTINUE?

Although dairy farms in New England do continue to decrease in numbers and dairy cow numbers diminish accordingly, the milk production per cow and per herd or farm continues to climb. This, in turn, maintains the milk supply, perhaps at too high a level, but also makes available land, facilities, and expertise for other agricultural pursuits. To protect the better agricultural land, several northeastern states, namely Connecticut, Massachusetts, and New Jersey, have instituted Farm Land Preservation Acts that set aside for perpetuity the best agricultural lands to remain forever available for food production. This needs to be done nationally before any more of our precious, highly productive agricultural land comes under control of the developers and their paved jungles. We in New England have felt the pressure first and gladly share our experience with all to maintain a viable agriculture. Our aim is to achieve greater self-sufficiency in producing more of our food requirements as we recognize the danger of possible isolation in the paths of energy crises, weather, transportation strikes or failures, and our climate restrictions. With the courage, fortitude, industry, and imagination of our early forefathers, we are moving toward narrowing the gap between dependency and self-sufficiency.

Multiple land use is important to livestock production in New England. In the one instance, where our forestlands account for twice the proportion of total acres compared with the U.S. average, we have a large potential to utilize forages within these forestlands for grazing. Secondly, with the decrease in dairy numbers, there is a tendency to replace them with beef cattle, or sheep, or even hog operations--thus utilizing existing land, facilities, and labor for a combined or replacement operation. Additionally, the availability of inexpensive by-product feeds and the use of unconventional feedstuffs encourages these livestock operations. Because the ruminant especially can be maintained on lower quality or by-product feeds, New England as a natural, cool-weather grass country can produce forage-fed beef and lamb with a minimum of purchased feed through wise grazing and forage harvesting management.

Furthermore, the markets are prevalent in New England, with its population density (albeit concentrated in large cities along the sea-coast), thus allowing rural production within easy access to the consumer. This proximity also permits direct marketing from producer to packer or to consumer. This proximity also permits direct marketing from producer to packer or to consumer, development of a freezer trade, and utilization of farmers' markets as well as the existing auction markets. New England does lack a major

terminal market within its confines. Livestock marketing
pools are becoming more prevalent. Specialty marketing,
catering to the natural or organic food interests, also is
practiced to some degree. Some cater to specific ethnic
demands, often of a seasonal nature, such as Easter lambs.

THE TIME IS NOW

In general, New England does have much in its favor for
the production and marketing of livestock for meat purposes. New England farmers historically have been excellent
livestock men. The first U.S. meat packer was Captain John
Pynchon (established in 1645) in Springfield, Massachusetts. The Brighton (Massachusetts) Stock Yards were
developed to feed George Washington's Continental Army in
1775--the nation's first and oldest terminal market. "Uncle
Sam" Wilson followed suit supplying the U.S. troops with
meat during the War of 1812. The first agricultural and
livestock show or fair in the U.S. was held in Pittsfield,
Massachusetts, in 1810. Many of the early imports of livestock from Europe funneled into the New England states, and
in time this breeding stock of all species was disseminated
west. The first major U.S. importation of Merino sheep was
to Weathersfield Bow, Vermont, in the early 1800s by Hon.
William Jarvis, then U.S. Consul to Lisbon, Portugal. By
1865, there were 1.5 million sheep in Vermont alone. These
sheep were the foundation of the great Merino flocks of Ohio
and now Texas. In 1875, Herefords from the Bodwell and Bur-
leigh herd of Vassalboro, Maine, sold to the Hon. William
F. Cody of Scout's Rest Ranch, North Platte, Nebraska, and
to other prominent breeders of that day.

Although there are some large-acreage livestock farms
in New England, most are small, family farms. The average
farm size in New England is 171 acres compared to a U.S.
average of 450 acres. Some of these are registered, purebred breeders supplying breeding stock to the area and
throughout the U.S. A limited few even sell breeding stock
or semen internationally. New England has long been a seedstock producing area, and its livestock compete very successfully in the show and sales rings of the nation's major
expositions and sales.

The New England farmer has often been faced with the
quandary of how to make a living under ofttimes less-than-
desirable conditions. Frequently he has survived by living
on "not what he earned, but what he did not spend." Through
the vagaries of climate, weather, topography, land capabilities, and pressures, the New England livestock producer has
developed a unique capability in growing, managing, harvesting, and preserving forage in the form of grass or
legume hay, haylage or silage, and corn silage where
possible. The producer has realized maximum TDN per acre
through wide pasture management and has obtained maximum
livestock production on his precious land. New England has

been forced to take the lead in forage production of meat
and dairy animals as its ability to raise grain and protein
supplement has been very limited to practically nonexistent. This has perhaps been the salvation of the New
England livestock producer and especially the family farm
where homegrown labor and homegrown feed have been the major
resources for survival.

Some New England farmers combine livestock operations
with other agricultural pursuits such as:

Major Enterprise	Supplementary Enterprise
1. Cucumbers (pickles)	Hogs
2. Dairy	Sheep, feed cattle
3. Forestry, firewood	Beef cattle, sheep
4. Hogs	Sheep
5. Landscape and bedding plants	Beef cattle
6. Maple products (syrup, sugar)	Sheep, beef cattle
7. Orchard	Sheep, beef cattle
8. Poultry	Beef cattle, sheep, hogs, veal
9. Tobacco	Beef cattle
10. Vegetable gardening	Hogs, sheep, cattle
11. Vineyard	Beef cattle

Generally, these are family farm operations that have
diversified to utilize surplus feed, labor, facilities, or
alternatives that best fit the existing situation and provide additional homegrown products for family use as well.

NEW ENGLAND, WHAT'S AHEAD?

With the advent or resurgency to greater self-sufficiency, we have already noted a greater number of "backyard"
meat animals being produced for the home meat supply. More
family farms and part-time farmers are turning to this pro-
gram--not only for their own meat requirements but also for
producing "a few extra to sell." Dairy farms are replacing
some of their cull cows with dairy steers or other livestock
to utilize homegrown roughage that can be marketed through
these animals. Some operators are expanding their programs
to satisfy the continuing and expanding demands for fresh
and processed meat of all species, as well as from the
fast-food chains. Most of these increasing needs require
leaner meat that favors a forage-fed program. The challenge
is to develop animal-forage management systems that will
maximize the utilization of forages through grazing. Additionally, with the recognized growth efficiencies and
greater muscle production of intact males, and with the
great availability of dairy bull calves in this region, New
England has the opportunity to utilize these surplus (to the
dairy herd) bulls and feed them out for a specialized market. These bull calves provide an alternative veal produc-

tion system that presently serves as a viable program for the small family farm that may feed out 100 to 500 vealers in confinement systems. Over one million dairy bull calves are produced annually in the Northeast with approximately 161,000 of these being New England-reared.

The future belongs to those who prepare for it. New England livestock producers, though small and often diversified, are facing the future with courage, adaptability, innovation, and confidence that they will continue to do a respectable job in maintaining their families and farms and contributing to the nation's meat supply.

(Statistical data presented herein has been derived from "Beef Research Program for the Northeast," [in progress, 1982], of which the author is a member of the Steering Committee.)

19

LIVESTOCK PRODUCTION
ON FAMILY FARMS IN INDONESIA

Allen D. Tillman

Indonesia is of great strategic importance to the United States and the Free World. Located at a major cross-roads in the world, it forms a barrier between the Indian Ocean and the China Seas. The Straits of Malacca, shared with Malaysia, is a busy seaway because it is the shortest sea route from the Suez Canal to China, Hongkong, and Japan. A major portion of the Mideast oil bound for Japan and other countries in the Orient is transported through these straits.

The purpose of this paper is to discuss some of the general features of Indonesia, to introduce its people, and to explain the agricultural production systems and the roles of farm animals in these systems.

GENERAL FEATURES

Geographical Setting

The geographical setting of Indonesia is shown in figure 1. The country forms a barrier between Australia and Asia and blocks all southern sea routes to the mainland of Asia. A striking feature of Indonesia is its size--it stretches southeastward from the mainland of China from about 15° to 140°, a distance of about 5100 kilometers (3100 miles). If a map of the United States were superimposed on that of Indonesia, both New York City and San Francisco would lie within Indonesia's borders. Also, it stretches from about 5° north to 11° south, a distance of about 1600 kilometers (1000 miles). Indonesia is the third largest country in Asia, ranking in size and population behind China and India. Her population in 1980 was about 148 million, making it the fifth largest nation in the world in numbers of people.

Physical Environment

The physical environment of Indonesia is greatly influenced by the seas (figure 1). In fact, all Indonesians

154

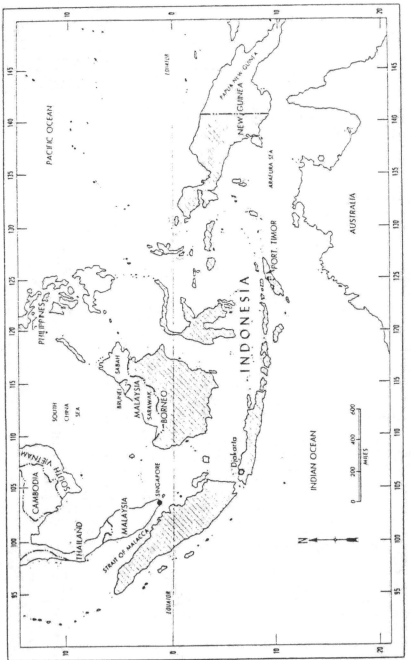

Figure 1. Indonesia, Geographical Setting

think of their country as <u>tanah air kita</u> (our land and water). The seas serve as a communication medium and as a means of isolation: The calm, shallow seas to the north foster seaborne transportation between the islands and easy communications between the islands and their coastal cities. The coastal cities of the larger islands of Sumatra, Kalimantan, and Sulawesi developed only when river transportation into the interior was available. Otherwise the interiors remained undeveloped and the people lived under primitive conditions. Even today the interiors of all islands except Java, Madura, and Bali remain undeveloped and primitive. The lack of good roads is a major constraint upon further development.

The shallow seas of Indonesia are teeming with fish and other forms of sea life that provide a rich food resource for domestic consumption and for export. Oil discovered under these shallow waters and inland has made Indonesia a major oil-exporting country.

There are about 13,667 inhabitable islands in Indonesia that vary in size from coral atolls to one the size of France--Kalimantan. The five largest islands (Kalimantan, Sumatra, Irian Jaya, Sulawesi, and Java) constitute 90% of the land mass, which is 1,934,944 km^2.

The climate of Indonesia is tropical and influenced locally by location and the monsoons. There are two seasons in most of the country--wet and dry. Two monsoons bring rains to the country: the northwest monsoon brings rains to Java, some of Kalimantan, South Sulawesi and to the islands in the chain that extend from Java to the southeast. However, those islands nearest Australia receive little, if any, rain from the southeast monsoon. Some areas receive rains from both monsoons--west Java, parts of Kalimantan, southern Sumatra, and northern Sulawesi. The northern part of Sumatra has a rainfall pattern similar to that of Malaysia and Thailand. Table 1 shows average rainfalls in the capital cities of each of the provinces.

There are three major soil types in Indonesia that have influenced agricultural development. A major soil type on the outer islands of Sumatra and Kalimantan is the red-yellow podzolic soil, which is infertile and hard to cultivate. Because of these poor characteristics, the natives have developed a "slash and burn" (swidden) agricultural system, a system which requires much land in order to produce their food needs. In general, the Javanese people have avoided these lands and remained on overcrowded Java.

The androsols and regosols are volcanic soils that are found in many areas of Java, Bali, and a small area of northern Sumatra. These soils are fertile and easier to cultivate. Some feel that this is a primary reason for the overpopulation of Java, Madura, and Bali. The hydromorphic alluvial soils are found on all coastal areas--the sites of the mangrove forests. These forests constitute a valuable resource and should be maintained rather than used for the production of rice.

Table 1. Some Agroclimatic Data for Indonesia

Province	Av. Annual Rainfall (mm)	Average Temperatures		
		Min. C°	Max. C°	Av. C°
Sumatra:				
Aceh	1505	23.1	32.4	26.3
North Sumatra	1984	23.1	31.4	26.1
West Sumatra	3080	22.3	-	-
Riau	2454	23.0	30.7	26.1
Jambi	2535	23.0	31.0	26.3
Bengkulu	2501	21.6	-	-
South Sumatra	2558	23.5	31.3	26.6
Lampung	2784	-	-	-
Java:				
West Java	1907	18.7	27.9	22.7
Central Java	1851	24.0	31.9	26.9
East Java	1429	23.8	31.3	27.5
Kalimantan:				
West Kalimantan	3382	-	-	-
Central Kalimantan	3377	-	-	-
South Kalimantan	2254	23.3	32.0	26.8
East Kalimantan	2570	20.9	-	-
Sulawesi:				
North Sulawesi	2765	22.6	30.0	25.8
South Sulawesi	3001	22.7	31.4	26.4
Southeast Sulawesi	1600	-	-	-
Bali	2031	25.2	30.3	27.8
West Nusatenggara	1274	21.5	-	-
East Nusatenggara	1146	22.8	-	-
Maluku	3101	-	-	-
Irian Jaya	2391	-	-	-
East Timor	1475	22.6	-	-

Source: Bureau Pusat Statistics, Indonesia (1971).

THE PEOPLE

Indonesia became independent in 1949 after more than 300 years under Dutch Colonial rule. The new nation appeared to have the potential to become prosperous. However, progress over the past 33 years has been slow and poverty persists.

President Sukarno, who was a popular revolutionary leader, was the first president and governed from 1949 to 1966. By incomparable rhetoric and a compelling personality, Sukarno was able to enlarge the country by wresting West Irian from the Dutch and to unify the many islands under one government. However, during his administration there was (a) a rapid increase in population, (b) the world's most inefficient bureaucracy, (c) many costly foreign adventures, (d) a general neglect of social and economic needs, and (e) mismanagement of resources. These led to economic disaster. In other words, President Sukarno brought charisma to Indonesia along with many serious economic problems.

President Sukarno was replaced by General Suharto in 1967 who has governed the country ever since. The new president immediately initiated economic reforms and by 1970 the general economic conditions were improved. Under his leadership, Indonesia established an economic planning body (Bappinas) in which the best economists (trained under a Ford Foundation program in Indonesia) provided effective economic planning for their five-year plans. At the present time, about 90% of the membership of Bappinas are graduates of the Ford Foundation program, and most of these have advanced degrees from American Universities.

There has been tremendous progress in economic development and in sociological programs under President Suharto's leadership. However, many complex sociological problems from the Colonial period remain and will require more than 33 years, the age of this young Republic, to solve. For example, the problem of overpopulation and poverty on Java, Madura, and Bali, has only become worse. It is doubtful any Government of Indonesia (GOI) program or series of programs can bring fast relief. Tables 2, 3, and 4 reveal the magnitude of the problem: Table 2 shows the population of the provinces (states), while table 3 shows the population of the major island groups. The Java-Madura-Bali area contains 63.7% of the people (94 million) on only 7.2% of the land while the outer islands contain only 36.3% of the population (54 million) on 92.9% of the land mass. The relative density of people per square kilometer is 685 vs 30 for Java-Madura-Bali and the outer islands, respectively. The question is, "Why don't the Javanese people in these thickly populated areas move to the outer islands?" There is no easy answer. The GOI is attempting to alleviate this problem by the transfer of people from Java, Madura, and Bali to transmigration areas in Sulawesi, Sumatra, and Kalimantan in a very costly program supported by World Bank. In spite of a major effort, there is reasonable doubt as to whether the transmigration scheme is likely to be effective.

Sociological problems in addition to the overpopulation on the islands of Java, Madura, and Bali remain: Indonesia is the classic example of a pluralistic society. There are more than 300 tribal or ethnic groups who identify as ethnically different. Also it is estimated that about 250 different languages are spoken. However, in this regard, Indonesia is fortunate in that most Indonesians spring from Deutro-Malay people and that most of the dialects belong to the Malay-Polynesian family. However, there are many conflicts in the Indonesian society--conflicts between Christians and Muslims, conflicts between Muslim groups, conflicts between the Javanese rulers and the powerful Chinese minority that controls the commerce and has the skills and drive necessary for economic development of the country. These conflicts are expressed locally, regionally, and nationally, and all present continual problems.

The economy is predominately rural with some 60% of the labor source being engaged in agriculture. Most live on small family farms of less than one hectare. Agriculture accounts for about 30% of the Gross National Product (GNP), and the proportion appears to be decreasing each year; the service and mining sectors are the fastest growing segments. The overall GNP is growing at 8% per year, while agriculture is growing at a 2.9 to 3.5% rate. The average income is about $370 per year, but the small farmers are below the average of the lowest income group.

The GOI owns the National Petroleum Company that is the economic giant in Indonesia. It is generating enormous incomes which, if properly used, could be used to build the infrastructure necessary to develop a viable agricultural production system, especially if the price of petroleum products remains high and the supply lasts.

AGRICULTURAL AND LIVESTOCK PRODUCTION SYSTEMS IN INDONESIA

Agriculture

Indonesia has about 14.4 million farms: 6.6 million (45.8%) contain less than 0.5 hectare and 7.8 million (54.2%) contain more than 0.5 hectare. In addition, there are about 2.5 million hectares in estate crops (rubber, tea, coffee, palm, etc.) that produce crops for export. Most of the farms are found on the islands of Java, Madura, and Bali, where the average farm size is about 0.4 hectare--or one acre. In addition, small farms are found in northern Sumatra and in some parts of Sulawesi. As the small farms must produce food for the family of four to six or more people and produce products that can be sold to obtain monies for school fees, clothes, and various celebrations (selamatans), the system is classified as subsistence farming in which the utilization of land and labor are quite intensive.

Rice is the primary crop and is grown almost everywhere during the rainy seasons. If irrigation is available, three crops of rice can be grown. However, most farmers in these areas produce two crops of rice, timing the planting of both crops to coincide with the rains. The second rice crop is harvested at the end of the rainy season and immediately afterward one or more of the secondary crops is planted-- corn, peanuts, soybeans, etc., and these grow during the dry season. Farmers in the upland areas who do not have irriga- tion plant their rice to fully utilize the heavy rains. By using dikes and drainage ditches, they are able to trap the rain waters for flooding the rice. However, some upland rice is grown without flooding.

To reduce the time that rice has to grow on the rain- scarce lands, it is first grown in a nursery until it attains a height of 8 to 10 inches. The plants are then transplanted to specially prepared fields that have been "puddled": the land is plowed, flooded, and while still flooded is repeatedly harrowed until all clods of dirt are broken and become pulverized. When the muddy waters have set for a while, the plants are transplanted from the nursery to the fields. Transplanting is done by hand, as is the subsequent cultivating and harvesting.

Large ruminants (cattle and swamp buffalo) are used for land preparation. In the thickly populated areas, cows usually pull the moldboard plows and the harrows that are used in the puddling operation. Cows may be used for farm transport, but road transport is mostly by mature bulls. The cows serve as reproduction units and produce a calf at about 12- to 18-month intervals. The large and small rumi- nants (sheep and goats) are also important to the farmers for utilizing the enormous quantities of crop residues that are produced in the intensive farming system. The crop residues on the typical farm unit consist of rice straw, peanut vines, sweet potato vines, soybean stalks, corn stalks, and cassava stems. In many areas, one hectare of land will produce enough crop residues to provide for the forage needs of one pair of large ruminants.

Very few tractors are being used for land preparation in the Java-Madura-Bali area, and there is no indication that the numbers will increase soon. Animals still play a major role in the land preparation, which accounts for the great concentration of animals in the overcrowded areas (tables 2, 3, 4). In the Java-Madura-Bali area, each km^2 contains 685.2 people, 31.2 cattle, 7.8 buffaloes, 4.9 pigs, 0.9 horses, 428 chickens, and 56.5 ducks in comparison to 29.9 people, 1.2 cattle, 0.7 buffalo, 1 sheep-goat, 1.1 pigs, virtually no horses, 17.3 chickens, and 3.6 ducks in the outer islands. Expressed another way, the thickly popu- lated areas contain 685 people and 50 animal units versus 30 people and 2 animal units on each km^2. These figures show that farm animals are so vital to the small farmers that they are found in very close association with the people.

Table 2. Areas and Populations of People, Buffalo, Cattle, Goats, Sheep, Chickens and Ducks by Provinces in Indonesia in 1980 [1]

Name of Province	Area Km²	Human Number 000	Human Density No/km²	Cattle [4] Number 000	Cattle [4] Density No/km²	Buffalo [4] Number 000	Buffalo [4] Density No/km²	Goats/Sheep [3][6] Number 000 [5]	Goats/Sheep [3][6] Density No/km²	Swine [4] Number 000	Swine [4] Density No/km²	Horses [4] Number 000	Horses [4] Density No/km²	Chickens [4] Number 000	Chickens [4] Density No/km²	Ducks [4] Number 000	Ducks [4] Density No/km²
D.I. ACEH	55,392	2,595	46.8	164	3.0	170	3.1	153.7	2.8	11.7	0.2	5.1	0.1	2,587.1	46.7	995.7	18.0
North Sumatra	70,787	8,554	120.8	145	2.0	105	1.5	215.0	3.0	532.0	7.5	9.0	0.1	5,762.9	81.4	886.1	12.5
West Sumatra	49,778	3,607	72.5	156	3.1	99	2.0	70.0	1.4	1.6	0.0	7.0	0.1	1,747.7	35.1	629.4	12.6
Riau	94,562	2,091	22.1	10	0.1	15	0.2	90.0	1.0	25.0	0.3	2.0	0.0	1,164.2	12.3	129.2	1.4
Jambi	44,924	1,299	28.9	13	0.3	31	0.7	57.0	1.3	5.0	0.1	1.0	0.0	1,019.0	22.7	227.3	5.1
South Sumatra	103,688	4,448	42.9	88	0.8	35	0.3	100.0	1.0	28.0	0.3	0.5	0.0	2,091.7	20.2	562.9	5.4
Bengkulu	21,168	670	31.7	21	1.0	29	1.4	28.0	1.3	-	-	1.0	0.0	618.4	29.2	115.9	5.5
Lampung	33,307	3,587	107.7	73	2.2	22	0.7	196.0	5.9	21.0	0.6	1.0	0.0	2,217.4	66.6	280.0	8.4
DKI Jakarta	590	7,345	12449.1	-	-	-	-	-	-	-	-	-	-	2,168.7	3,675.8	-	-
West Java	46,300	25,526	551.3	118	2.5	465	10.0	3281.0	70.9	27.0	0.6	20.0	0.4	15,130.7	326.8	2,756.8	59.5
Central Java	34,206	25,814	754.7	1,022	29.9	368	10.8	3292.0	96.2	64.0	1.9	53.0	1.5	19,610.8	573.3	2,675.0	78.2
D.I. Yogyakarta	3,169	2,938	927.1	174	54.9	20	6.3	447.0	141.1	12.0	3.8	4.0	1.3	3,159.4	997.0	252.5	79.6
East Java	47,922	30,121	628.5	2,570	53.6	211	4.4	2876.0	60.0	22.0	0.5	49.0	1.0	18,309.8	382.1	1,621.1	33.8
West Kalimantan	146,760	2,545	17.3	55	0.4	-	-	25.0	0.2	256.0	1.7	-	-	581.4	4.0	180.5	1.2
Central Kalimantan	152,600	882	5.8	11	0.1	1	0.0	3.0	0.0	45.0	0.3	-	-	464.4	3.0	45.2	0.3
South Kalimantan	37,660	2,141	56.9	23	0.6	8	0.2	17.0	0.5	3.0	0.1	1.0	0.0	1,550.4	41.2	638.9	17.0
East Kalimantan	202,440	924	4.6	-	0.0	-	-	8.0	0.0	26.0	0.1	-	-	612.8	3.0	15.2	0.1
North Sulawesi	19,023	2,163	113.7	158	8.3	9	0.5	36.0	1.9	158.0	8.3	8.0	0.4	-	-	51.4	2.7
Central Sulawesi	69,726	1,151	16.5	108	1.5	357	5.1	74.0	1.1	38.0	0.5	6.0	0.1	670.1	9.6	86.8	1.2
South Sulawesi	72,781	6,533	89.8	567	7.8	11	0.2	174	2.4	139.0	1.9	163.0	2.2	2,857.7	39.3	1,191.9	16.4
Southeast Sulawesi	27,686	899	32.5	15	0.5	13	0.5	22	1.9	1.0	0.0	6.0	0.2	-	-	48.1	1.7
Bali	5,561	2,645	475.6	366	65.8	11	2.0	52	4.0	555.0	99.8	4.0	0.7	2,748.0	494.2	480.9	86.5
West Nusatenggara	20,177	2,747	136.1	147	7.3	204	10.1	128	6.3	12.0	0.6	85.0	4.2	1,667.7	82.7	270.8	13.4
East Nusatenggara	47,876	2,864	59.8	403	8.4	140	2.9	266	5.6	638.0	13.3	194.0	4.1	2,019.9	42.2	41.6	0.9
Maluku	74,505	1,359	18.2	18	0.2	-	-	62	0.8	24.0	0.3	-	-	-	-	-	-
Irian Jaya [7]	421,981	1,153	2.7	10	0.0	-	-	2	0.0	3.0	0.0	1.3	0.0	603.6	1.4	18.8	-
East Timor [7]	30,775	1,515	49.2	-	-	-	-	13.3	-	225.7	0.5	-	-	-	-	-	-
Total	**1,934,944**	**148,116 [2]**	**76.2 [8]**	**6,436**	**3.3**	**2,313**	**1.2**	**11688.0**	**6.0**	**2873.0**	**1.5**	**620.9**	**0.3**	**90,065.7**	**46.5**	**14,202.0**	**7.4**

Footnotes: [1] From Biro Pusat Statistics. [2] This is a projection; a preliminary report of 1980 census indicated 147.4 million people.

[3] Estimated – 1977 figure, [4] 1978 figure, [5] 1976 figure, [6] Goats and sheep figures are not separated.

[7] Data are not available for most livestock. [8] 1980 census figure used for calculations.

Table 3. Number and Density of People, Livestock and Poultry on the Densely Populated 1/
Areas (Java, Madura, Bali) Versus the Less Densely Populated Outer Islands 1/

Island Group	Area km²	Human		Cattle		Buffalo		Sheep/Goats		Swine		Horses		Chickens		Ducks	
		Number x 000	Density per km²	Number x 000	Density per km²	Number x 000	Density per km²	Number x 000	Density per km²	Number x 000	Density per km²	Number x 000	Density per km²	Number x 000	Density per km²	Number x 000	Density per km²
Sumatra	473,606	26,851	56.7	670	1.4	506	1.1	909.7	1.9	624.3	1.3	25.6	0.1	17208.4	36.3	3,826.5	8.1
Kalimantan	539,460	6,492	12.0	90	0.2	9	.02	53.0	.1	330.0	0.6	1.0	–	4764.5	8.8	879.8	1.6
Sulawesi	189,216	10,746	56.8	848	4.5	378	2.0	336.0	1.8	336.0	1.8	183.0	1.0	5342.9	26.8	1378.2	7.3
E. Nusa Tenggara	47,876	2,864	59.8	403	8.4	140	2.9	266.0	5.6	638.0	13.3	192.0	4.1	2019.9	42.2	41.6	0.9
W. Nusa Tenggara	20,177	2,747	136.1	147	7.3	204	10.1	128.0	6.3	12.0	0.6	85.0	4.2	1167.7	82.7	270.8	13.4
Maluku	74,505	1,359	18.2	18	–	–	–	62.0	0.8	24.0	0.3	–	–	–	–	–	–
Irian Jaya	421,981	1,153	2.7	10	–	–	–	2.0	–	225.7	0.5	1.3	–	603.6	–	18.8	–
E. Timor	30,775	1,515	49.2	–	–	–	–	13.3	–	–	–	–	–	–	–	–	–
Total	1,797,596	53,727	29.9	2,186	1.2	1,236	0.7	1,770.0	1.0	2,190.0	1.2	490.9	0.3	31107.0	17.3	6,415.7	3.6
Java, Madura	132,187	91,744	694.0	3,884	29.8	1,064	8.0	9,896	74.9	125	0.9	126.0	1.0	56210.7	425.2	7,305.4	55.3
Bali	5,561	2,645	475.6	366	65.8	13	2.3	22	4.0	555	99.8	4.0	0.7	2748.0	494.2	480.9	86.5
Total	137,748	94,389	685.2	4,255	31.3	1,077	7.8	9,918	72.0	680	4.9	130.0	0.9	58958.7	428.0	7,786.3	56.5
Grand Total	1,934,944	148,116	76.5	6,436	3.32	2,313	1.2	11,688.0	6.0	2,870.0	1.5	620.9	0.3	90065.7	46.5	14,202.0	7.4
% on Java, Madura, Bali	7.1	63.7		66.1		46.6		84.9		23.7		20.9		65.5		54.8	
% on outer islands	92.9	36.3		33.9		53.4		15.1		76.3		79.1		34.5		45.2	

1/ From table 2

Table 4. A Comparison of the Populations of People, Their Livestock, and
Densities in Indonesia, Java, Madura and Bali, and the Outer Islands [a]

Items	Indonesia		Java, Madura, Bali			Outer Islands		
	Numbers 000	Density km²	Numbers 000	Density km²	% of Total	Numbers 000	Density km²	% of Total
Land Area, Km²	1934.944		137.748		7.1	1,797.196		92.9
Populations								
Human	148,116	76.5	94,389	685.2	63.7	53,727	29.9	36.3
Cattle	6,436	3.3	4,250	31.3	66.1	2,186	1.2	33.9
Buffalo	2,313	1.2	1,077	7.8	46.6	1,236	0.7	53.4
Sheep/Goats	11,688.0	6.0	9,918	72.0	84.9	1,770.0	1.0	15.1
Swine	2,873.0	1.5	680	4.9	25.7	1,967.3	1.1	74.3
Horse	620.9	0.3	130	0.9	21.1	490.9	–	79.1
Chickens	90,065.7	46.2	58,958.7	428.0	65.5	31,102.0	17.3	34.5
Ducks	14,202.0	7.3	7,786.3	56.5	54.8	6,415.7	3.6	45.2

a Data from table 2

Farm managers in Indonesia, as everywhere, attempt to utilize their resources in the most efficient manner. Even though the large ruminants have traditionally been the source of draft power for land preparation and transport, in some areas animals used in road transport are being replaced by motorized trucks. Since gasoline prices are being subsidized by the GOI, this replacement is rapid. In many thickly populated areas on Java, draft animals are being replaced by human labor for land preparation. When 5 to 6 people, equipped with the Javanese hoe, can be hired for the same price as a worker with a pair of cows to pull a plow, farm managers are choosing to utilize the former. In addition, small tractors are beginning to be used by some of the largest of the small farmers on the outer islands, particularly in south Sulawesi.

Animal Production

The populations of species of animals are shown in table 5. When the FAO figures are used to convert animal numbers to animal units, it is found that Indonesia contains about 10 million animal units for 148 million people, only 0.07 animal units per person. This is one of the lowest figures in the world. Because of the combination of low animal numbers and the inefficiencies of reproduction and production, the daily intake of animal protein is less than 5 grams.

TABLE 5. LIVESTOCK POPULATION IN INDONESIA EXPRESSED IN ANIMAL UNITS

Kinds of animals	Number (000)	Factor to obtain units	Animal units
Cattle	6,436.0	0.8	5,148.8
Buffalo	2,313.0	1.0	2,313.0
Horses	620.9	0.8	496.7
Swine	2,870.0	0.3	861.0
Goats/sheep	11,688.0	0.1	1,168.8
Poultry	104,267.7	0.001	104.3
Total	--	--	10,092.6

Source: Table 2.

Most of the large ruminants are owned by the small farmers and used for draft power. Secondary uses are for the production of meat, hides, milk, and manure (table 6). It is estimated that between 1.9 and 2.0 million pairs of cows (both cattle and buffalo) are used for land preparation while the remainder represent young stock and cows used for other purposes. Most owners keep the animals in covered stalls near their homes and bring crop residues and supplemental feeds to the pens in a zero-grazing scheme. As the crop residues are available at different times of the year and in different amounts, the nutritive states of the animals varies depending upon the available quantities and qualities of the crop residues. Many farmers grow tree legume shrubs (as "living" fences or otherwise) from which they cut leaves for the animals throughout the year. The leaves of <u>Leucaena</u> <u>leucocephala</u>, <u>Sesbania</u> <u>grandiflora</u>, <u>Gliricidia</u> <u>sepium</u>, and others are quite high in protein. Therefore, protein is not a limiting nutrient in many management situations. However, there are indications that energy and, perhaps, some of the minerals are limiting nutrients in a number of systems.

TABLE 6. MEAT, MILK, AND EGGS PRODUCED IN INDONESIA

Animal or product	Number of animals	Yield per animal kg/animal	Total production metric tons
Beef and veal	897,000	156	161,000
Buffalo meat	211,000	160	34,000
Mutton & lamb	2,084,000	10	21,000
Goat meat	3,746,000	10	38,000
Pork	2,109,000	55	116,000
Horse meat			1,000
Poultry meat			102,000
Total meat			473,000
Cow's milk	40,000	1,725	69,000
Eggs			80,000
Cattle/buffalo hides			28,122
Sheep skins			4,168
Goat skins			7,491

Source: FAO Production Yearbook.

There are about 50 large cattle ranches in Indonesia, and these are on the outer islands. All are devoted entirely to animal production, but these constitute a total land area of about 85,000 hectares and contain less than 30,000 cattle. Most ranches are owned by GOI or by parastatal units that are responsible to the GOI. The primary purpose of the largest of these ranches is to produce animals for draft power in the transmigration areas. The large ranches are expected to serve as models, as well as training sites, for the training of future ranchers for the remote areas of all of the larger outer islands. Such training is needed because most Indonesians lack both the capital and experiences needed for successful ranching operations.

In general, there are three major breeds of beef cattle in Indonesia--Ongole, Banteng, and Madura. The Ongoles (Bos indicus) and their crosses are the most numerous of the breeds, and about 4.5 million of these are estimated to be scattered throughout the country. The breed came to Indonesia from India many years ago and was kept relatively "pure" on the island of Sumba. The GOI has, by decree, forbidden crossbreeding of the purebred Ongoles on Sumba, thus the breed will be maintained there. The second most numerous breed is the Banteng (Bos bibos or Bos sandaicus), which is found on the islands of Bali, the East and West Tenggaras, and Sulawesi. There are about 1.5 million Bantengs in Indonesia and 360,000 on the island of Bali, which by GOI decree must be kept in the "pure" state. The Banteng, probably belonging to the buffalo family, was found to exist in the wild state on several places in Southeast Asia, including Indonesia. Its period of domestication probably extends back to prehistoric times. The F_1 hybrid male, resulting from a union of the Ongoles and the Banteng, appears to be sterile. However, the F_1 female hybrid is fertile and will breed either to Ongole or Banteng bulls. In Sulawesi, where crossing is common, one sees only F_2 crosses with either three-fourths Banteng and one-fourth Ongole breeding, or the reciprocal. Madura cattle are found on the island of Madura and probably resulted from an early crossing of the Banteng with native cattle, some of which were imported by the Arab traders; thus it appears likely that some Bos taurus crossing is found in the breed.

There are very few purebred dairy cattle in Indonesia, and these are primarily Holstein-Friesians. In fact, the number of dairy cattle appear to be about 100,000, and almost all are Friesian crossed with indigenous cattle, called Grati. The Grati animals are used for milk production; only a few, if any, are used for draft purposes.

The swamp buffalo, used for draft purposes throughout Indonesia, is usually found in the alluvial swamp lands where the soils are heavy and difficult to plow. As these animals are heavier than the cattle, have larger feet, and have an ideal temperament for heavy work, they are in demand for such areas.

Relative weights of the cattle and the swamp buffalo at different ages are shown in table 7. The Banteng, Madura, and Ongoles are smaller at birth and grow slowly, while the Grati and the swamp buffalo are larger at birth and exhibit faster growth rates. For example, at three years of age the relative weights for the Banteng, Madura, Ongoles, Grati, and the swamp buffalo were 320, 250, 330, 400, and 430 kg, respectively.

The heat tolerances of the Banteng, Madura, and Ongole cattle are the highest while those of the Grati and the swamp buffalo are lower. All animals are fairly resistant to diseases and appear to be adapted to Indonesian conditions. Also, the three beef cattle breeds appear to be able to utilize poor quality forages quite well, while the Grati

TABLE 7. WEIGHTS (kg) OF CATTLE AND SWAMP BUFFALO AT
DIFFERENT AGES

Item	Cattle				Swamp buffalo
	Bali	Grati	Madura	Ongole	
Birth weight	16	22	14	20	30
One year old					
Males	110	150	90	120	170
Females	90	130	70	90	150
Two years old					
Males	220	280	180	230	300
Females	170	240	130	180	270
Three years old					
Males	320	400	250	330	430
Females	240	330	180	250	380
Four years old					
Males	400	500	300	430	550
Females	300	400	220	320	480
Adult					
Males	400	500	300	430	600
Females	300	400	220	320	530

Source: FAO/World Bank Team (1978).

animal must have higher quality forages if high milk production is obtained. The swamp buffalo appears to be the animal of choice in the swampy areas because it can survive on
poor quality forages and in wet places.

The relative growth performances of male cattle and the
swamp buffalo when fed a typical American feedlot ration
containing high levels of concentrates are shown in table
8. The animals were fed the diet for 154 days. An attempt
was made to obtain animals of the same physiological ages,
but this was difficult. The Banteng and Madura, which are
smaller than the other animals, apparently matured faster
and therefore were fatter at the end of the feeding period.
The Grati cattle, which are much larger than the Banteng or
Madura animals, were still growing at the end of the experiment. The Ongole growth performance was between these two
groups. The swamp buffalo, even though larger, was quite
fat at the end of the trial indicating that it, too, matures
early and readily fattens.

Some of the production characteristics of Banteng,
Madura, and Ongole on large Indonesian ranches have been
estimated and are shown in table 9. Since these are
estimates based upon limited data, they are in need of
verification by good research and observations.

Currently there is much world interest in the genetic
capabilities of both the Banteng and the swamp buffalo. It
is possible that they possess genetic capabilities to help
feed the world's populations. Studies to this end should be
initiated as soon as possible.

The small ruminants (sheep/goats) number about 11.5 million and most are found on Java-Madura-Bali. Goats are more numerous than sheep and have few distinct breeds. However, those that have been identified are shown in table 10. All are small and productive.

TABLE 8. SOME CHARACTERISTICS OF LOCAL INDONESIAN CATTLE

Items	Bali	Madura	Ongole
Fertility	+++	++	+
Calving difficulties	-	-	-
Calf mortalities	+	++	+
Birthweight, kg	13-15	12-18	20-25
Weaning weight, kg	70	60	85
Daily gain, kg	0.35	0.25	0.30
Feed conversion	++	+	++
Age at puberty, month	18-24	20-24	24-30
Disease resistance	++	+++	+++
Heat tolerance	+++	+++	+++
Male libido	++	+	+
Grazing ability	+++	+++	+++
Mothering ability	+++	+++	+++
Milk production	+	+	+
Mature weight, kg			
Male	375	275	400
Female	275	250	300
Dressing percentage	56	48	45

+ = poor, ++ = good, +++ = excellent
Source: Directorate General of Animal Husbandry (1977).

TABLE 9.   PRODUCTION PERFORMANCES OF BREEDS OF CATTLE
            INDIGENOUS TO INDONESIA FED A HIGH CONCENTRATE
            DIET

Item	Breed of Cattle				
	Madura	Ongole	Bali	Grati	Buffalo
Average daily					
gain, kg	0.60	0.75	0.66	0.90	0.73
Range	0.18-	0.34-	0.32-	0.56-	0.47-
	-0.89	-1.03	-0.87	-1.62	-1.06
Dry matter intake,					
kg/day	5.33	6.42	6.02	7.97	5.80
g/kg 0.75/day	72.6	72.3	76.8	84.9	76.6
Feed conversion					
kg feed/kg gain	9.22	8.56	9.12	8.85	7.95
Organic matter					
digestibility, %	70.6	72.6	68.6	73.7	68.3
Dressing percentage	63.2	57.9	56.6	57.3	51.8
Rib eye area, cm2	63.4	54.5	63.7	61.0	46.2
Muscle: Bone ratio	5.34	4.23	4.44	3.89	3.34
Relative rates of					
fat deposition	0.49	0.34	0.52	0.33	0.57

Source: Moran (1979).

TABLE 10.   WEIGHTS (kg) OF MATURE SHEEP AND GOATS IN
            INDONESIA

Breeds	Male, kg	Female, kg
Sheep		
Thin-tailed	40	30
Priangan	60	35
Fat-tailed	40	35
Goats		
Kacang	35	30
Grade Ettawa	60	50
Gembrong	45	38

Source: Hardjosubroto and Astuti (1979).

    Indonesian sheep have attracted much attention because
they show no photo-periodism, which makes it possible to
breed throughout the year.   Under good feeding and management conditions, two lamb crops per year are possible.
Indonesian sheep, like buffalo, are a neglected genetic
resource that deserves further study.
    Approximately 85% of the sheep and goats are found in
the thickly populated areas and are owned by small subsis-

tence farmers or by landless peasants who are provided a home in the villages. Some of the farmers keep their sheep and goats in covered shelters, usually with cattle, and bring feed to them in the usual "cut and carry" system. Others, particularly the landless peasants, keep the animals penned at night, but during the day they are permitted to graze on public lands such as roadsides and in parks. Older people and young children are usually the shepherds.

A delicious sheep and goat meat, called kambing, is in demand for making sate in both indoor and outdoor cafes. It is prepared by wrapping pieces of meat around skewers and cooked over burning charcoals.

The swine industry is small in Indonesia because 90% of the population is Muslim. There are three different types or breeds of pigs in Indonesia--the Java, Bali, and Sumatra pigs. The Java pig in reality is a crossbred animal that has varying levels of European breeding (primarily York-shire). In the large and improved operations, the crosses are essentially purebred animals under modern management.

The Bali pig resembles the Chinese pig in that it is small, swaybacked, and has excellent reproductive capa-bilities. Because it is a scavenger, it grows slowly, yielding only a small quantity of overfat pork. The indige-nous Sumatra pig resembles the East Indian pig. It is black, small, and grows slowly. Like its Javanese counter-part, it can be used for crossbreeding with the larger European breeds.

The chicken industry in Indonesia exists in two forms--the kamgpung (village) industry and the large commer-cial units. The village industry is based upon the native chickens, which are descendants of the red jungle fowl that are still to be seen in many parts of Indonesia. The birds are scavengers around the home, consuming scraps from the table, loose rice grains, insects, and other feeds. As feed is limited, they grow slowly and produce only a few eggs. The mortality rate among the young is excessively high; how-ever, some do survive and provide meat and eggs for home use or for sale, with only a few inputs.

The commercial industry is a rapidly growing industry and is based upon using hybrids for both broiler and egg production. The units are usually large with modern feeding and management. The commercial chicken industry is the fastest growing segment of the livestock industry in Indo-nesia.

Ducks are very important to the Indonesian people. In the country are four distinct breeds--Alabio, Tegal, Bali, and Manila. Tegal ducks are most numerous in Java, but Bali ducks are most numerous in Bali. Both are kept by small-holders in flocks of 40 to 200 female, penned at night, and shepherded to rice paddies or lagoons during the day. They consume waterplants, small fish, and other forms of water life in the flooded fields. They are used as scavengers on all newly harvested fields to feed on the shattered grains.

In Kalimantan, the flocks are found in the floodplains and are larger. Usually the ducks are kept in one-story houses that float. The birds are penned at night but are permitted to feed in the shallow waters during the day. Alabio ducks are usually used, and the rate of egg production is quite high with some flocks reporting averages of 240 eggs per year. As these are unselected populations, this is remarkable and is another neglected genetic resource.

Additional General Remarks

Most farm livestock is owned by the smallholders who find that animals are vital for the production of meat, milk, eggs, hides, and other animal products for home use and sale. In addition, the large ruminants provide draft power for land preparation and provide a use for the large quantities of crop residues that are by-products of the system. As the large ruminants and some small ruminants are kept in covered shelters, all of the manure produced by the animals is preserved for home use, particularly for growing home vegetables and fruits.

Indonesian farmers regard the animals not used for draft purposes as "living" savings accounts. Therefore, the marketing of young stock for meat depends more on the farmer's needs for ready cash than upon the size and condition of the animal. Under such conditions, price changes have very little influence upon production. For example, since 1968, the demand and consequent price for meat have increased several times as fast as has the price of rice, the basic staple. However, this price increase has had very little influence upon numbers of cattle, buffalo, and sheep/goats. In fact, the numbers of these animals have remained the same or have actually decreased. The numbers of producing females appear to actually have decreased while consumption has risen. If this is so, the shortfall is being met by the slaughter of old road-transport bulls (being replaced by motorized trucks) and young, often pregnant, females. In the latter case, the price offered for these animals to be transported live to the transmigration areas is lower than the price obtained if slaughtered for meat. Therefore, the farmers are slaughtering these animals and this is happening at a time when the GOI and some private ranchers are importing young cows from Australia to build up cattle numbers on the large ranches on the outer islands.

REFERENCES

Cockrill, W. R. (Ed.) 1974. The Husbandry of the Domestic Buffalo. FAO, Rome.

de Guzman, M. R. 1975. The Asiatic Water Buffalo. Food and Fertilizer Technology Center (ASPAC), Taipei.

Devendra, C. 1980. Goat and sheep production in the Asian region. World Animal Review 32:33-42.

FAO. 1979. Land Use. FAO Production Yearbook, FAO, Rome.

Hartadi, H., S. Reksohadiprodjo, S. Lebdosukyo, A. D. Tillman, L. C. Kearl and L. F. Harris. 1980. Tables of Feed Composition for Indonesia. IFI, Utah State University, Logan, Utah.

Hardjosubroto, W. and M. Astuti. 1979. Animal Genetic Resources in Indonesia. Proc. Workshop on Animal Genetic Resources. TSU Kuba City, Japan.

Hutasoit, J. H. 1974. Animal husbandry development perspectives in Indonesia. Min. of Agriculture, Jakharta.

Mason, I. L. 1978. Sheep in Java. World Animal Review 27:17-22.

Moran, J. B. 1979. Performance of Indonesian beef breeds when fed high concentrates. Mimeo. Report. Center for Animal Research and Development. Ciami (Bogor) Indonesia.

Payne, W. J. A. and D. H. L. Rollinson. 1973. Bali cattle. World Animal Review 7:13-21.

Tillman, A. D. 1981. Animal Agriculture in Indonesia. Winrock International, Morrilton, Arkansas.

Part 4

WATER, WEATHER, AND CLIMATE

20

WORLD AGRICULTURE IN HOSTILE AND BENIGN CLIMATIC SETTINGS

Wayne L. Decker

CHARACTERIZATIONS OF CLIMATE

Health, nutrition, and suffering of the human population are determined, in part, by weather and climate. Regional wealth and the levels of economic development are impacted by the natural resources, including the climate resource. But agriculture and the associated food production industries are more directly affected by weather and climate than any other sector of the economy. Climate determines production potential of both grain and livestock producers, identifies strategies available to the producer for resource allocations and marketing, and determines the feasibility of plans for exports and imports of commodities. An improved understanding by agriculturalists of the nature of the climatic resource is essential if the impacts of climatic risks are to be minimized.

Climate is defined by the space and time distribution of weather events: temperature, precipitation, wind, humidity, and sunshine. In spite of the unpredictability of weather events, climate occurs systematically in both the space and time scale. As a result of these consistencies, climatic zones are easily recognized. For example, in the tropics and subtropics some regions are consistently rain-free in summer, others are smaller areas with an even seasonal distribution and abundant rainfall. In the temperate latitudes, the continental regions also demonstrate regional consistencies in climate. The west coasts of continents are mild with abundant winter rainfall, while the continental interiors tend to exhibit summer maximum of precipitation and marked seasonal temperature variations (Mather, 1974).

In most climatic regions, there are periods during the year with hostile climates for agriculture. These climatic hostilities are associated with temperature stresses (both high and low) and with deficiencies of rainfall. In many of these regions, there are periods of the year during which the weather is consistently dry, thus producing a hostility. This climatic hostility can be removed by irrigation, or avoided by adopting an enterprise with a low water need. For the hostile climates produced by temperature stress, shelters may be constructed to protect animals from the critical temperatures, and crop production can be scheduled to avoid the consistent occurrence of high or low temperatures.

Many regions have climates that are consistently favorable for agricultural production. These climates are usually characterized by dependable water supply (rain or irrigation supply). Benign climates are also characterized by moderate temperatures without a high probability of extreme temperatures during critical times for sensitive plants.

CLIMATIC CHANGE

Climatic change and the impact of climatic change on man have become controversial issues in recent years. Articles on the subjects appear regularly in technical and popular magazines, and both paperback and hardback books have been published dealing with climate change. The written opinion concerning climatic change and its impact are as different as day and night. Even scholars of climatology are confused by the diversity of opinion.

The evidence to support the existence of major climatic changes through geologic time is well documented. Long periods of geologic history are characterized by mild climates, i.e., benign climates. These periods were interrupted by relatively short intervals when glaciers extended into the middle latitudes (the ultimate in hostile climates). It is generally accepted that the current climate of the world is more like that of the glacial period than the warmer "climatic optimum." Climatic researchers do not agree about the mechanisms causing these major climatic changes. Current thinking focuses on long-term oscillations in the slope of the terrestrial axis, but continental uplift and the associate volcanic activity appear to be necessary conditions for the glacial climates.

Variations in the climate of the earth also have been documented from historical records. The rise and decline of civilizations during the past 4,000 years appear to be related to changes in climate. Plagues, famine, and migrations have been linked to shifts in climate. In modern history, the period corresponding to the North American settlement and the establishment of the United States was a period of climatic stress, frequently called "the little ice age." Again, meteorologists do not agree on the physical processes that caused the climatic variations in historical time. Volcanic activity, variability in the solar output, and combinations of both these factors are mechanisms receiving prominent attention.

It was not until the late 1800s that a worldwide network of weather observing stations was established. Although records of weather observations can be traced into the 18th century at selected points, networks of observational stations did not generally exist until the early and mid-nineteenth century. In the United States, for example, it is difficult to find documented weather records prior to the establishment of the Weather Bureau in the Department of Agriculture in 1890. For this reason studies of climatic change based on meteorological observations are confined to the most recent 90 years.

Attempts have been made to establish the trends in climate from the meteorological observations. The best documented estimate of the trend in climate is shown in figure 1 from Waite (1968). During the first 40 years of this century, the average air temperature near the earth's surface increased, but about 1940 this trend was reversed. Figure 1 verifies that these trends in temperature apply to regions of different size and are the most pronounced in the polar and subpolar regions of the northern hemisphere.

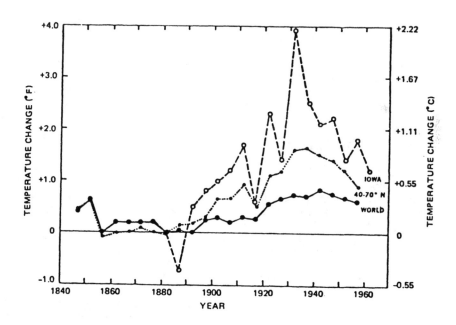

Figure 1. **Worldwide trend in mean annual temperature** (Waite, 1968)

For agriculture it is more instructive to look at the climatic variabilities associated with precipitation. In figure 2 and figure 3 the historical records of summer rainfall are summarized for the prairie provinces of Canada and the winter wheat region of Russia. It is difficult to identify trends from these data, although many interesting pulsations in the annual rainfall statistics, lasting for a decade or so, are apparent.

The atmospheric mechanisms producing the climatic trends and fluctuations that extend for a decade or a few decades have probably all been identified. These mechanisms include ocean-atmosphere interactions, volcanic activity, man's interference (CO_2 and particulate matter), and solar activity. The analytical contribution of each individual mechanism and the interaction between the mechanisms have not been defined; a major objective of the meteorological community to mathematically describe these processes based on known physical relationship efforts.

Efforts to research the physical causes for climate change led Dr. B. J. Mason, Director General of the British Meteorological Office, to observe in a recent article in the New Republic (1977):

> The atmosphere is a robust system with a built-in capacity to counteract any perturbation. This is why the global climate, despite frequent fluctuations, is fairly stable over periods of 10,000 years or so. Sensational warnings of imminent catastrophe, unsupported by firm facts or figures, not only are irresponsible but are likely to prove counterproductive. The atmosphere is want to make fools of those who do not show proper respect for its complexity and resilience.

FLUCTUATIONS IN CLIMATE

Climate variability adds an additional stress to the agriculture system and adds a component to climatic hostility. When the weather of one or more years departs markedly from the expected, a farm management strategy that has been successfully used becomes inappropriate for the agricultural enterprise for a particular year or growing season. Several years of drought (such as the 1930s on the U.S. Great Plains) is hostile to the farm enterprise adopted to nondrought enterprises. On the other hand, periods of years with benign climates often lure farmers into strategies not adapted to the hostile and stressed condition that follows. The type of fluctuation most often used by climatologists deals with the variation of climatic events between years. This variability refers to the variation in annual or seasonal temperatures and precipitation totals. McQuigg (1973), for example, demonstrated the low variability in climate between 1955 and 1970 for the major agricultural production regions of the U.S. McQuigg simulated

CANADIAN PRAIRIE PROVINCES

MAY TO JUN BINOMIAL RUNNING MEAN TOTAL PRECIP

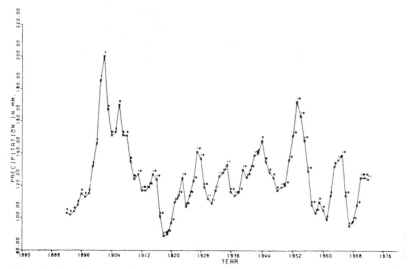

Figure 2. The year to year variability in the total precipitation for May and June (smoothed by a binomial technique) in the spring wheat producing area of Central Canada.

SOVIET REGION WEST OF VOLGA

MAY TO JUN BINOMIAL RUNNING MEAN TOTAL PRECIP

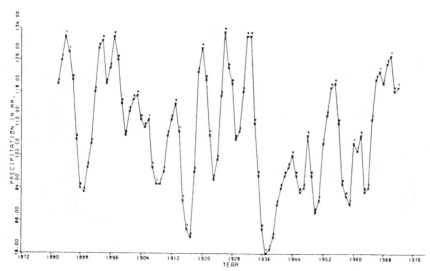

Figure 3. The year to year variability in the total precipitation for May and June (smoothed by a binomial technique) in the winter wheat producing area of the Soviet Union (west of the Volga River).

yields of grain throughout this century from climatic data at a constant technology. He showed (figure 4) that the yields simulated from climate during the period extending from the late 50s through the 60s were remarkably constant and relatively high, i.e., the climate of the U.S. Corn Belt was benign. The year-to-year variability in climate in the U.S. has been greater in the 1970s and the early 1980s than the preceding decade and a half.

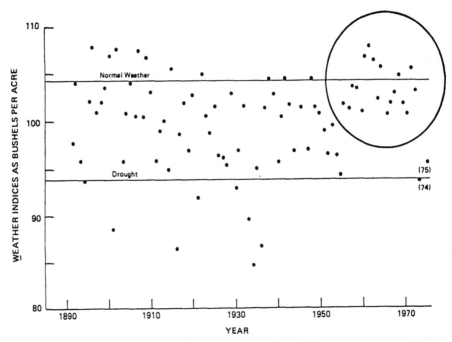

Figure 4. **Year-to-year variability in climate expressed in simulated corn yields for the U.S. (NAS, 1976)**

To examine how climatic fluctuation varies through time, the variances of climatic elements by decades have been computed for major agricultural production areas in the world. Figure 5 shows the 10 year variances in the May plus June precipitation in the Canadian prairie region, while figure 6 presents the same values for the Soviet winter wheat region. The May and June rainfall totals are vital to wheat production in these two regions. In both cases, the variability in May and June precipitation during the most recent decades were below average; however, the tendency for a lower variance does not appear to depart from that expected from the normal variability. The high year-to-year variability in the May-June rainfall in Canada just after the turn of the century is quite striking. There was also a period of high variability for the May and June precipitation in the Soviet Union between 1925 and 1950.

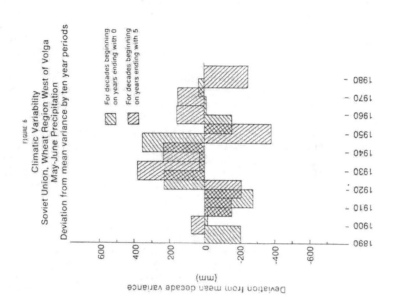

FIGURE 6
Climatic Variability
Soviet Union, Wheat Region West of Volga
May-June Precipitation
Deviation from mean variance by ten year periods

For decades beginning
on years ending with 0

For decades beginning
on years ending with 5

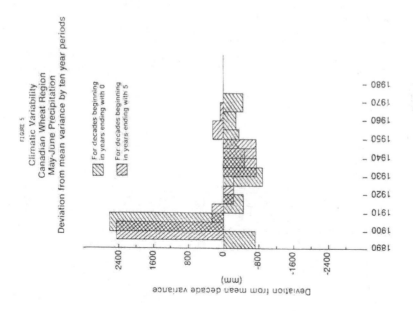

FIGURE 5
Climatic Variability
Canadian Wheat Region
May-June Precipitation
Deviation from mean variance by ten year periods

For decades beginning
in years ending with 0

For decades beginning
in years ending with 5

Figures 5 and 6 show that climate did not experience unusually low variability everywhere in the world during the period 1955 to 1970. A second, and more important, lesson concerns the recognition that periods of 1 or more decades in length with high (and low) year-to-year variability in climate do occur.

A report by the National Research Council (1976) identifies sudden and unexpected climatic fluctuations as the greatest climatic hazard to agricultural production. The managers of agricultural systems are forced to use practices that are not adapted to the regional climate. The farm manager has difficulty in finding the "best" strategy during periods with fluctuating climates. A large year-to-year variability forces the manager to choose management options that are "out of step" with the season's weather.

RESPONSES TO CLIMATE CHANGES AND FLUCTUATIONS

The policy makers and agricultural managers could easily respond to weather and climate fluctuations, if events could be anticipated before they occurred. Although several climatologists, with impressive credentials, regularly make seasonal forecasts, most meteorologists agree that the science of meteorology has not advanced sufficiently for these forecasts to be considered valid. The thirty-day outlook, which projects the general trends in mean temperature and precipitation, is correct only about 55% to 60% of the time. Seasonal forecasts (cold winter, dry summer, etc.) have an even lower accuracy. Complex computer models, which simulate the atmospheric circulation, offer the best promise for the development of rational forecasting schemes. As these atmospheric models are improved, weather and climate forecasts will have improved accuracy and be extended for longer periods. The improvement in forecasting skills will be slow. In the next two decades there appears little hope for major and sudden breakthroughs in our understanding and interpretation of atmospheric circulation.

Weather modification provides an additional strategy for removing weather risks. Modification of the surface energy budget through changes in the surface color, drainage of the land, or shaping of the soil surface through tillage, offers many important options for improved technologies for agriculture. Increased rain through cloud seeding is more often considered as a weather modification option. There are several difficulties that reduce the potential for cloud seeding to respond to fluctuations in climate:

- Proof of small increases in amounts of rain are almost impossible to obtain because variations in area and time affect the amount of rain falling over a region.
- Rain-making only augments the natural rainfall, so it is not a "drought stopper."

- The possibility exists that the manipulation in the clouds will reduce the rainfall from some clouds within a given weather system.

Cloud seeding appears to be most favorable for use in the mountainous regions—to increase the winter snow pack. Increased snow in the mountains improves the water supply for the agriculture of the adjacent semiarid and arid regions.

Disaster insurance spreads the risks of "bad" weather. An international program for grain and food storage should stabilize supply between the "lean" and "bountiful" years. A marketing cooperative, or even the individual farmer, may establish an "ever-normal granary" by withholding grain from the market. For the individual farmer, the best opportunity for spreading the risk is through disaster insurance. A national food policy must include an insurance program using both governmental and private agencies as underwriters. Insurance programs may stabilize farm incomes, but, in the long run, will not provide increased productivity of food for the expected increase in world population.

Technologies developed through private and public research provide an additional response to provide for a stable food supply and farm income under a fluctuating climate. Meteorological science cannot be expected to deliver completely reliable warnings of pending shifts and fluctuations in climate. The meteorologists will not save us from the adversity of a variable and often hostile climate. Agricultural strategies and technologies must be developed to respond to the expected climatic variabilities. These developments will emerge from integrated, interdisciplinary agricultural research.

REFERENCES

Mason, B. J. 1973. Bumper crops or droughts. Mimeo. NOAA, U.S. Dept. Commerce, Washington.

Mather, J. R. 1974. Climatology, Fundamentals and Applications. McGraw Hill, pp 112-131.

National Research Council. 1976. Climate and Food. Report on Climate and Weather Fluctuations and Agricultural Production. National Academy of Science, Washington.

Waite, P. J. 1968. Our weather is cooling off. Iowa Farm Science 23:13.

21

THE IMPACTS OF CLIMATIC VARIABILITIES ON LIVESTOCK PRODUCTION

Wayne L. Decker

CLIMATE AND LIVESTOCK

Regional climates contain factors that need consideration in determining the kind of profitable livestock production for a region. The variability of weather and climate provides a component in determining the profitability of the livestock enterprise. Climate imposes both direct and indirect effects on commercial animal agriculture. The direct effects include weather events producing physical injury (lightning, wind, flood, temperature extremes), occurrences associated with physiological stress (such as heat and humidity), and weather events promoting insect or disease episodes. Indirect climate events are those that impact on availability of forages and the supply of feed grains. These indirect impacts of the weather and climate are generally imposed by chronic deficit in water and occasional droughts.

LOSSES IN ANIMAL PRODUCTION DUE TO CLIMATE EVENTS

Animals, grown commercially on farms and ranches, are normally subjected to ambient environmental conditions. On many occasions, the atmospheric conditions are less than ideal and the animal is subjected to stress. This stress, which is usually related to the heat and energy balance of the animal, reduces the production. The stress causes declines in egg or milk production and reduced weight gains for swine, beef, or broilers.

Over the years there have been repeated attempts to mathematically define the impact of stress imposed by atmospheric conditions (temperature, humidity, wind, etc.) on animal production. The experimental basis of these efforts comes from two sources: (1) barns or chambers with controlled environmental conditions (Brody, 1948) and (2) field experiments measuring animal performance as related to observed ambient environmental conditions. From the observations obtained through the experimentation, functional relationships between the weather event (or events) are derived. The resulting mathematical formulas are usually obtained through standard statistical procedures. Strickly speaking, the expressions are only applicable to the experimental conditions from which the relationship is derived, so each relationship must be tested against independently collected data.

Literature has many examples of relationships between the performance of domestic animals and weather and climate events. Two mathematical expressions for relating cattle performance to environmental conditions are discussed below.

Milk Production

Using data obtained from controlled experiments, Berry et al. (1964) related the decline in milk production to an index involving both atmospheric temperature and humidity. This index, which is called the temperature-humidity-index (THI), is shown in equation (1).

$$THI = T + .36 \; T_d + 41.2 \tag{1}$$

where

T is the temperature in °C,
T_d is the dew point temperature in °C.
The relationship between THI and the decline in milk production (MD) is shown in equation (2).

$$MD = -2.37 - 1.74 \; NL + .0247 \; (NL)(THI) \tag{2}$$

where

NL is the normal production of a cow under thermoneutral conditions.
THI is the temperature-humidity-index.
Since negative values for MD do not make sense under the definition in this equation, decline in milk production is assigned the value of zero for all negative values in equation (2). This means that a zero production decline is expected until a critical value of the THI is reached. For higher values of THI, the decline in production decreases linearly. This critical value of THI is between 70 and 74 for normal production levels of between 25 and 30 pounds of milk per day.

Meat Production in Cattle

Bolling and Hahn (1981) and Bolling (1982) report the results of a regression analysis relating climatic variables to rates of gain for beef animals. The functional relationship, which best explains the reduction in weight gains, contained terms related to cold stress, heat stress (THI), precipitation, and wind. The results of the regression analysis are shown in figure 1, as taken from the work of Bolling (1982). The author concluded:
"Analyses consistently suggested that stress resulting from the direct effects of cold, combined with the effects of precipitation, have a greater impact on feedlot cattle in Nebraska than does any

other type of atmospheric stress studied. Interestingly, heat stress rarely appeared to be a significant factor affecting cattle performance."

Of course, this conclusion concerning heat stress is counter to the one for milk production. This difference may be due to the difference in climate of the regions where the milk production and rate of gain experiments were conducted.

Hahn et al. (1974) noted that beef cattle were able to overcome heat stress. In the Missouri Climatic Laboratory, beef cattle were stressed by being subjected to 5 weeks of temperatures of 30°C. A marked decrease in rate of gain as compared to a control group resulted; when the animals were returned to optimal conditions, the stressed animals outgained the control group. This result, which Hahn calls "compensatory growth," is demonstrated in figure 2. No such compensation occurred after animals were subjected to a greater stress (35°C). Hahn (1976) indicates that similar compensatory growth occurs with swine and broiler chickens. It is significant that (1) the research indicates that compensatory growth does not occur after exposure to a high degree of heat stress, (2) the laboratory experiments on compensatory growth were done at constant temperature and may not apply to temperatures experiencing a diurnal range, and (3) no experiments have been made to discover whether "compensatory growth" occurs after cold stress.

MORTALITY OF ANIMALS DUE TO CLIMATIC STRESS

Hostile climates do cause mortality to domestic livestock. This hostility occurs as a result of both heat and cold stress. In summer, high temperature and humidities (THI values of 80 or higher) produce stress that can lead to death. This condition is, of course, aggravated by other stress factors associated with handling and/or shipping. In winter, the cold stress can cause tissue to freeze and the animal to die. For U.S. cattlemen of the open ranges in the High Plains and eastern slopes of the Rocky Mountains, the cold stress may be further aggravated by high winds with snow. These blizzards cover the winter food supply, make access to the herds difficult (if not impossible), and bury herds under the drifting snow.

Bolling (1982) presents analyses that document the weather impacts on cattle mortality under feedlot confinement. Strong winds and cold stress were the "best" predictors of mortality under Nebraska conditions. The author was apparently unable to document the mortality due to stress imposed by hot and humid weather. Of course, one would not want to use these results to estimate mortality under range or pasture exposures.

The National Weather Service has established policies for issuing weather advisories to stockmen. These advisories are issued by Weather Service Forecast Offices located

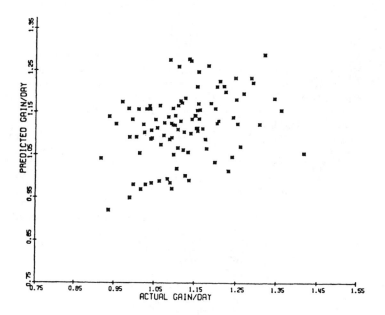

Figure 1. Relationship between measured gain of beef cattle and that predicted from weather data using a statistical relationship (Bolling, 1982)

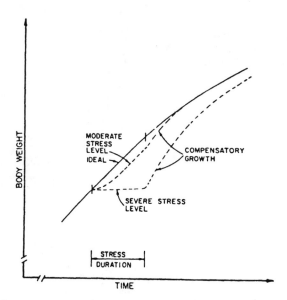

Fig. 2. Schematic to Illustrate the Principle of Equifinality in Animal Growth (Hahn, 1982)

in each state. These advisories are:
- Heat stress advisories are issued whenever high temperatures combine with high humidities to present danger to livestock. The THI is used as the index for danger to livestock. Two categories are recognized: (1) livestock danger (THI 79 to 83) and (2) livestock emergency (THI higher than 83). When air temperatures are below the body temperature, wind provides cooling. Because of this, wind is mentioned in the stockman's advisories when velocities higher than 20 mph are expected.
- Winter storm watches are issued when there are strong indications of a blizzard, heavy snow, freezing rain, or sleet. A blizzard is defined as a condition with high winds (in excess of 35 mph), with falling or blowing snow (visibilities less than 3 miles). A winter storm warning or blizzard warning is issued when the storm's development is a virtual certainty.
- Chill indices are released routinely. This index combines wind and termperature in an effort to approximate the equivalent temperature under light wind (4 mph) conditions. At Columbia, Missouri, in the winter of 1981-82, there were 398 hours (11% of the hours) with chill factors below 0°F, 181 hours (5%) with chill factors below -10°F and 72 hours (2%) with chill factors below -20°F. The number of days with chill factors below 0°, -10° and -20° sometime during the day was 30, 18, and 7 days, respectively.

REFERENCES

Berry, I. L., M. D. Shanklin and H. D. Johnson. 1964. Dairy shelter design based on milk production decline as affected by temperature and humidity. Trans. Amer. Soc. Agr. Engr. 7:329.

Bolling, R. C. 1982. Weight gain and mortality in feedlot cattle as influenced by weather conditions: Refinement and verification of statistical models. Progress Report 82-1. Center for Agricultural Meteorology and Climatology, Univ. of Nebraska, Lincoln.

Bolling, R. C. and G. L. Hahn. 1981. Climate effects on feedlot cattle: Growth and death losses. Proc. 15th Conference on Agr. and Forest Meteorology: 86-89. Amer. Meteorological Soc., Boston.

Brody, S. 1948. Environmental physiology with special reference to domestic animals, I: Physiological background. Res. Bull. 423. Mo. Agr. Exp. Sta., Columbia.

Hahn, G. L. 1982. Compensatory performance in livestock: Influences on environmental criteria. Livestock Environment, Proc. 2nd Internatl. Livestock Environment Symposium. Amer. Soc. Agr. Engr., St. Joseph, Michigan. (In Press.)

Hahn, G. L. 1976. Rational environmental planning for efficient livestock production. Proc. 7th Intern. Biometeorological Cong., College Park, MD. pp 106-114.

Hahn, G. L., N. F. Meador, G. B. Thompson and M. D. Shanklin. 1974. Compensatory growth of beef cattle in hot weather and its role in management decisions. Livestock Environment, Proc. Internatl. Livestock Environment Symposium, pp 288-295. Amer. Soc. Agr. Engr., St. Joseph, Michigan.

22

THE CLIMATIC RISK TO LIVESTOCK PRODUCTION IN DEVELOPING COUNTRIES

Wayne L. Decker

DEVELOPING COUNTRIES

Developing countries comprise a unique community of nations. These nations are often identified as the "under-developed" or "less-developed" countries of the world. Generally, these nations have large and growing populations with the economically poor and undernourished segments forming a high percentage of the total population. Health problems associated with these disadvantages produce shorter life spans and higher infant mortality rates, as compared to the world averages.

The developing countries make very small contributions to world trade. The term "developing" applies to the aspirations of these nations to become more involved with export trade and to improve the average income of their citizens. These countries represent a wide spectrum of levels of economic activity. Some are so poor that they are classified as "the least developed of the developing countries." Other developing countries occupy marginal positions between the developing and developed countries (Argentina, Mexico, etc.). There are, of course, also borderline developed countries (Israel, Spain).

The human and national problems associated with the developing countries are many and varied. Some of these countries have poor distributions of land among the population, which suggests that land reform could solve the problems of development. Other developing countries have rich mineral or petroleum deposits and well-planned use of this wealth appears to offer promise for development. But many of these countries have no latent sources of wealth and appear to be destined to a meager existence for many decades to come.

Invariably human or social factors appear to be hampering the development of the poor countries. These social features are often associated with the culture, family or caste structure, and religious belief. In most cases, the developing countries have populations that are currently too large for the national resources and (or) have population growth outreaching the national potential for food production.

The nutrition of the developing countries is tied to the human consumption of grain or other plant-derived protein and energy sources. The dietary use of animal products that are derived form hand feeding or grazing (red meat, poultry, milk, eggs) is often associated with development. Since many of these animals are ruminants, the maintenance of adequate grazing land is important.

NATURAL RESOURCES AND THE DEVELOPING COUNTRIES

The developing countries are either poor in natural resources or lack the human and economic resources to develop the existing resources. The soils of many of these countries are unproductive with low fertility and/or poor physical conditions. In addition, some developing nations have inadequate water to sustain the agricultural system. The water supply and soil properties are determined by the past and present climatic resource.

For most developing countries the climate resource is one of the primary limitations to economic development. These climatic deficiencies lead to limitations in agricultural production, and agricultural production is the basic requirement for development. There are, of course, management alternatives that may be used to remove part of these climatic limitations. These alternatives offered by modern technology include irrigation, genetic improvement of existing crops, and adoption of new crops.

In the simplest terms, climate restricts agricultural production through limitations in the length of the growing season. In the temperate and polar regions, this limitation occurs because of cool temperatures in the fall, winter, and spring. In these areas, the length of the production season is associated with the period when temperatures are warm enough to allow plant development. Although freezing temperatures (0°C) cause a termination of growth, most agricultural crops do not grow and develop until temperatures are above 5°C. The length of the growing season is also limited by the availability of water. Precipitation falls on the land with a portion of the water penetrating the surface to be stored in the soil. Evapotranspiration (water vaporized through evaporation at the soil surface and through transpiration by vegetation) removes water from the soil reservoir. So long as the precipitation is sufficient to prevent the reservoir of soil water from being depleted, the growing season continues. Obviously, the growing seasons associated with soils with high water retention will be longer than for other soils.

To further understand the growing season as indicated by water supply, it is necessary to look at the factors determining its length. The evaportranspiration component, although determined by the climate, is not a simple relationship. The maximum evapotranspiration, called the potential evapotranspiration, is the rate of vaporization of water with nonlimited soil water. The magnitude of the potential evapotranspiration is determined by the weather conditions. The actual evapotranspiration is never greater than the potential evapotranspiration. The amount that the actual is below the potential rate of evapotranspiration is a function of the level of soil moisture and the stage of crop development.

Table 1 shows examples of the water supply and demand for three locations (FAO, 1980) in southern Asia. For Bogor, Indonesia, located near the equator, there is always a surplus of water because the rainfall exceeds the potential evapotranspiration throughout the year. The growing season at Bogor continues through the year. The two locations from India exhibit growing seasons that are shortened by deficiencies in rainfall. For Satna, the season is about 5 months long and at Rajkot only 3 months long. The length of the growing season cannot be directly reproduced from the values in table 1 because the actual evapotranspiration during periods with soil moisture deficiencies is less than the potential evapotranspiration.

TABLE 1. AVERAGE PRECIPITATION AND POTENTIAL EVAPOTRANS-PIRATION WITH THE RESULTING LENGTH OF GROWING SEASON FOR THREE LOCATIONS IN SOUTHERN AND SOUTH-EASTERN ASIA

Month	Bogor, Indonesia (7°S, 106°E) Precip (mm)	Evapotr (mm)	Satna, India (24°N, 81°E) Precip (mm)	Evapotr (mm)	Rajkot, India (22°N, 70°E) Precip (mm)	Evapotr (mm)
Jan	392	112	36	61	1	146
Feb	375	104	23	86	0	163
Mar	391	119	14	153	1	264
Apr	401	115	10	193	3	309
May	346	108	1	232	7	361
June	236	100	126	197	99	235
July	217	111	356	136	293	171
Aug	280	123	320	122	143	158
Sept	284	132	176	123	93	156
Oct	435	137	47	112	25	176
Nov	410	124	13	74	5	156
Dec	308	119	6	59	4	134
Annual	4076	1403	1137	1549	674	2430
Growing season	365 days		157 days		96 days	

A common characteristic of the locations shown in table 1 is that the season of maximum rainfall occurs during summer (Bogor, Indonesia, is in the southern hemisphere). The occurrence of the maximum seasonal rainfall during the time of the year when the noon sun is high in the sky is a common characteristic of the climates of most regions located in the tropics and subtropics. In most cases, these tropical and subtropical regions have a distinct wet season in summer and a very dry (in many cases virtually rain-free) winter period. The amount of rainfall is also associated with geographic position in the tropics and subtropics. As one moves toward the pole from the tropics into the subtropics, the total annual rainfall decreases. Similarly, there is a decrease in annual rainfall as one moves from the continental east coasts toward the interiors of the continents. Because of the seasonal distribution in rainfall, as the annual total of rainfall decreases the length of the summer wet period (the length of the growing season) becomes shorter. The values shown in table 1 for south and southeast Asia are illustrative of this principle. This tendency is duplicated in Africa, South America, and Central America. (The generalizations concerning the rainfall amount and length of the rainy period do not hold for areas where mountain barriers provide an added mechanism for rain production. Northern India is a good example of a region where mountains enhance the rainfall.) The summer maximum in precipitation is important to the type of agricultural development in the tropics and subtropics. The cloudiness associated with the rainy period reduces the solar energy reaching the earth's surface during the time of the year when the sun is highest in the sky. Less energy is available for evaporation and transpiration and cooler temperatures occur than would occur if the skies were cloudless. The cloudiness and the associated low-light intensities also limit growth by reducing photosynthesis.

On the whole, the summer maximum of precipitation enhances agricultural production. An even distribution of rain throughout the year with the same total annual rainfall would produce a cloud pattern leading to higher evaporative demands in summer causing an intensification in the aridity of the region and leading to shorter growing seasons.

There are also subtropical regions in the world where the season of maximum rainfall is in winter. These regions are found along the subtropical west costs of the continents (California in North America, European and African Mediterranean, and Chile in South America). The climates of the summer dry regions are characterized by very hot summers with cool and relatively damp winters. The agriculture of these areas must adapt to these climatic characteristics. Crop and forage production during summer can only be sustained with irrigation.

In the tropics, there are limited regions where the maximum seasonal precipitation occurs in winter. Such regions are favorably located with respect to the prevailing winter circulation. For example, the northeast trade winds bring winter precipitation to eastern Malaysia and the eastern shores of Thailand and Vietnam.

THE IMPACT OF CLIMATIC VARIABILITY ON FOOD PRODUCTION IN THE DEVELOPING COUNTRIES

In addition to the restraints introduced by the normal seasonal climatic patterns, there are year-to-year variations in the climate. When the seasonal precipitation is late in coming, or deficient in the amount, a failure in food production for that year is likely. Figure 1 shows (Steyaert et al., 1981) an analysis of rice yields in the northeast region of Thailand from 1954 through 1974. The dashed line with a constant slope in this figure represents an estimate of the trend in yields due to improved technology over this 20 year period. The fluctuations in yields

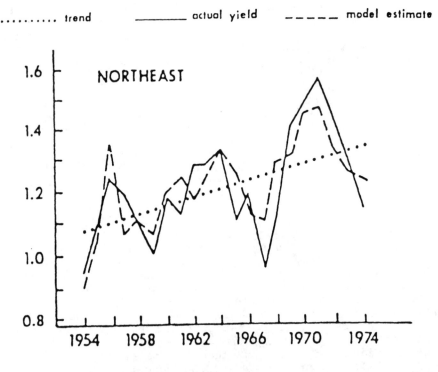

.... trend ——————— actual yield — — — — model estimate

Figure 1. Relationship between actual yield and model-estimated rice yields for the northeast region of Thailand (L. T. Steyaert et al., 1981).

around the trend represents the impact of weather variations on yield. It is apparent that the variation in rice yields due to weather ranges from 20% to 25% above the expected on some years and 20% to 25% short of expectations on other years. In many countries, this amount of variability induces critical shortages of food.

From a report of the National Research Council (1976) the following generalization was made (the words in parenthesis are the author's):

Climate is not fixed; it has varied throughout history. Trends in climate over decades or a century are not as significant to food production as fluctuations during a season or a few years. A field of corn, wheat or forage cannot escape hail, drought, sun or drying wind. Fluctuations in climate--from wet to dry, cold to warm, windy to calm--add uncertainty or risk to the production of food and (these fluctuations) have (the) greatest impact when they occur either in the "bread basket" regions or regions with large populations (i.e., the developing countries).

Animal production also is impacted by climatic fluctuations. Heat and cold reduce weight gain in animals, reduce milk and egg production, and may lead to premature death of animals. But an even larger impact by climatic fluctuations on livestock production occurs because of the climate-induced variation in forage production. During years favorable for forage production, animal herds increase in size and meat production increases. The dietary demands and economies become adapted to the availability of meat supplies. Droughts, on the other hand, reduce forage growth and cause the herds to consume all the available forages. With the forages gone, the animals either die of starvation or are marketed early. The barren land is exposed to wind erosion (and water erosion, too, when the rains return).

Such was the scenario of events that occurred in Subsaharan Africa during the 1970s. Figure 2 shows a plot of the annual rainfall for this area during the period of historical climatic record from the work of Puterbaugh (1981). (Like most subtropical regions this area receives virtually all of its rainfall during the summer season of the northern hemisphere.) The variations in rainfall over the period of record for this region of hostile climates is quite apparent, and the shortage of rainfall during the recent 1970s is obvious. Animal production in these regions consists of ruminants (cattle, sheep, goats) raised under nomadic conditions. It is well-documented (Glantz, 1977) that the herds of the nomadic tribes were sharply reduced by the lack of forages during this period.

There are management strategies available to the individual farmers and governments to limit the climatic risk to agriculture. The reduction of the farmer's vulnerability

to climate can be accomplished by: 1) increased water availability or a more efficient use of the existing water supply, 2) development of drought resistant varieties (including drought-tolerant or resistant strains of forages), 3)
modification of the plant environment (tillage methods for
row crops), and 4) improved farming methods through extension or adult-education programs. The developing country
occupying a region of the world with hostile climates frequently lacks the natural and human resources to adopt
strategies to overcome the climate hostility.

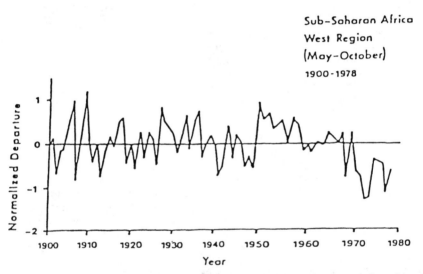

Sub-Saharan Africa
West Region
(May-October)
1900-1978

The normalized departure for a given station is the seasonal value minus
the long term mean for that station divided by the standard deviation of
the seasonal rainfall at that station. The regional values for each year
are obtained by averaging the normalized departure for all stations in
the region.

**Figure 2. Year to year variabiities in the normalized rain-
fall deprtures for the western portion of sub-
Saharan Africa (T. L. Puterbaugh, 1981).**

Since new technologies to combat the hostile environment
will be adopted slowly (or not at all) in the developing
countries, it is important that policy planners and decisionmakers in these countries be aware of pending crises in
food production induced by climatic variabilities. Although
seasonal weather outlooks are not reliable, it is still
possible to develop "early warning" systems that identify
the potential for food disasters 3 to 6 months prior to the

actual existence of the food shortage. Through these methods the weather events during the critical periods for crop and forage development are monitored, and warnings are issued when the weather begins turning bad for agriculture. This evaluation or assessment process is shown schematically in figure 3 as adapted from Steyaert et al. (1981). Using historical crop and climatic data, statistical analyses are developed that define the mathematical relationships between weather events (precipitation, temperature, etc.) and the crop or forage conditions or yields. These equations use as variables weather events occurring at critical times in the growth of the crop (say at flowering) and at times several weeks prior to the harvest. Thus calculations can be made of expected yields and the estimated yields transposed into assessment statements. Using these statistical models, "real time" weather data provide an "early warning" for shortfalls in production for policymakers.

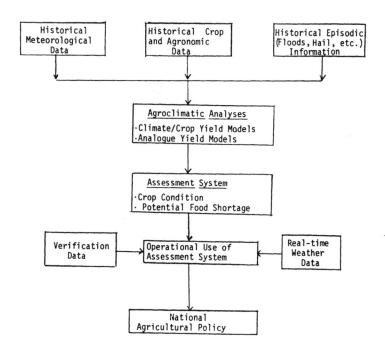

Figure 3. A schematic diagram of a system for evaluting the impact of weather and climate variations on food production (L. T. Steyaert et al., 1981).

In regard to developing countries, the National Research Council 91976) reported:

The widening gap in most developing countries between demand and production, coupled with rising prices of imported food and petroleum makes accelerated development of food production a policy priority. To this end, the first step for every nation (but most especially by those chronically short of food stuffs) would be to develop plans to meet short-term climatic fluctuations. The developing countries of south and southeast Asia can be used as an example for an early-warning system of climatic-induced food shortages. Rice is critical to the diets of this entire region. Failure of a rice crop has serious consequences, and an early warning of a pending shortfall will assist in planning a response. Steyaert et al. (1981) has analyzed the impacts of wet season climatic variabilities on the rice production for each state in India. For the State of Bihar, which is located in northeast India and produces over 10% of all the rice grown in India, an equation for estimating rice production was calculated using as a predictor an index of water supplied by rainfall during the July through October period. (Monthly indices are computed from the monthly values the rainfall minus potential evapotranspiration with the estimate of climatic index used in the equation being the 4-month total [July through October].) Similar relationships were developed for the other states in India. Using these relationships, the climatologist can begin at the end of July to provide assessments from estimates of production for each state. A final estimate is made at the end of October prior to the November harvest of rice.

REFERENCES

Food and Agricultural Organization. 1980. Report on the agroecological zone project. Results for Southeast Asia, World Soil Resources Report 48:4.

Glantz, M. H. 1977. Politics of Natural Disaster: The Case of the Sahel Drought. Praeger Pub., New York.

National Research Council. 1976. Climate and Food. Climatic Fluctuation and U.S. Agricultural Production. National Academy of Science, Washington.

Puterbaugh, T. L. 1981. Regional comparisons of seasonal African rainfall. M.S. Thesis. Univ. of Missouri.

Steyaert, L. T., V. R. Achutuni and A. V. Todorov. 1981. Proposed agroclimatic assessment methods for early warning of drought/food shortages in south and southeast Asia. Final Report to the U.S. AID, NOAA and Univ. of Missouri.

23

HIGH TECHNOLOGY AGRICULTURE AND CLIMATIC RISKS

Wayne L. Decker

CLIMATIC RISKS AND UNCERTAINTIES

Risk and uncertainty may seem to be synonymous terms, but there is a real and important difference in their meanings. Risk is interpreted in a probabilistic sense, where the probability of the unfavorable event can be calculated given all the facts about its occurrence. The probability of many unfavorable weather and climate events is calculable from historical climatic records. For example, the probability or risk of freezing temperatures after a given spring date can be calculated. Similarly, the chance for a temperature-humidity-index (THI) greater than 72 can be estimated for a given place and time period. Uncertainty, on the other hand, concerns adverse events for which the probability of occurrence is unknown or impossible to assess. The price of petroleum 10 years from now is uncertain, and one cannot assign a probability of the price being twice its current value. Similarly, the probability of the average air temperatures being warmer by a fixed amount by the year 2000 cannot be calculated because climatic trends over a decade or two are uncertainties. But most climatic events that adversely impact on plant and animal production do have estimable probabilities of occurrence, providing all the facts about the occurrence are known. So it is proper to speak of climatic risks.

If the risk from the climatic event is greater than desired, it is often possible to either avoid the risk or to ameliorate its consequence. Although there are probably examples of risks that can be avoided or ameliorated at no expense, either labor, capital, or both generally are required to change the climatic risk. Thus, the producer must examine cost and expected returns in reaching a decision on the most profitable management options to adopt in response to climatic risks.

DOCUMENTATION OF THE IMPACTS OF CLIMATE AND WEATHER EVENTS

Episodic weather events represent one type of climatic risk. These events are sudden (often violent), producing a finite damage to farm structures, crops, and animals. Episodic events are seldom, if ever, favorable to agricultural production. Examples of episodic events are killing freezes, hail, wind, floods, lightning, etc. The risks from these events are usually estimated from historical climatic data or other forms of records. For example, the probability of a flood resulting from a given height or level of a river can be estimated from climatic or river-flow data; or the probability of hail during the flowering time of a particular grain crop is available from hail actuarial statistics. The consequence from many episodic weather events is softened through insurance, with the insurance rates established from calculated risks.

Episodic weather events also impact the farm enterprise. These impacts may be either favorable (i.e., good weather) or unfavorable. Careful analyses are required to mathematically define the effect of weather on crop yields, animal gains, or milk and egg production. In the simplest form, this calculation is accomplished by evaluating the equation:

$$Y = f(W_1, W_2 \ldots W_n)$$

where

Y   is the agriculture production that relates to the "n" different weather variables (W), and

$f(W_i)$ is the functional relationship, which may be linear or nonlinear for each weather variable, and for each functional relationship, coefficients must be estimated from experimentation.

An example of this equation is found in a recent paper by Balling (1982) where an equation was developed relating the rate of gain in cattle on a Nebraska feedlot to an index of cold, an index of heat, and number of days with precipitation. He used feedlot experiments to estimate the coefficients appearing in the equation. Using Balling's adaptation of this equation, the rate of gain in other feedlots can be simulated if the climatic indices are known for the different feedlots.

A word of warning on the use of relationships derived from experimental data is appropriate--the coefficients estimated from the experimental data are often specific to the experiment from which they were derived. It is very important that each mathematical expression be evaluated or tested using independent data. For example, it should be assumed that the equation derived by Balling applies only to cattle in the feedlots for which it was derived and for the seasons used in the experiment. Independent data from other

feedlots and seasons must be used to validate the relationship.

MODELING RESPONSE TO CLIMATIC RISK

For every weather-sensitive farm and livestock operation, the manager (farmer or rancher) must make decisions
relative to the management options that will lead to
avoidance or amelioration of the climate risk. The decision
criteria may be complex or quite simple, and in some cases
the manager may not be consciously aware that he is making a
management decision. The options are schematically por-
trayed in figure 1 (adapted from a report by the National

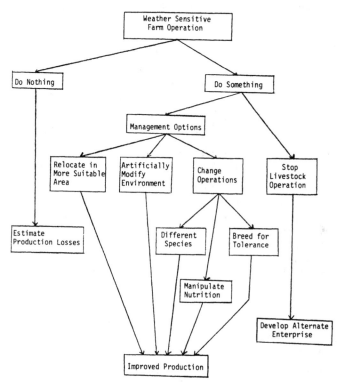

**Figure 1. Branching pattern for alternate options of the
manager in response to the impacts of the
environment**

Academy of Science, 1976; and from Hahn, 1977), which shows
that the manager may not be in a position to respond to the
climatic risk, but he must accept the consequence of his
failure to act. His inaction may be due to ignorance about
the techniques of overcoming the climatic risk, to the lack
of capital, or to the profitability of the modification. On

the other hand, the consequence of the climatic risk may be
so severe that the best choice for the manager is to divert
his land, capital, and labor into a different operation.
For the livestock producer this option would be to divert
his production system into some other farming activity.

Management options that the manager should consider are
to either avoid the climatic risk or alter the environment
into something less hostile. For crop production, the
environmental modification often involves new tillage
methods, frost protection, or irrigation. In animal production, the modification concerns the use of barns, windbreaks, shades, or heating and air conditioning. Options
also are available to he manager concerning changes in his
current operation without environmental modification, such
as selection of improved genetic stock, altering the nutritional inputs, and raising a different species of crop or
animal.

The simple statements contained in figure 1 represent
responses to a series of complex evaluations. How does the
livestock operator evaluate the consequences of the
available management options? One method of making an
evaluation is to adapt the technology and discover its
impact on production. Another method is to pool resources
or encourage governmental agencies to conduct extensive
experiments concerning the improvements in production for a
wide spectrum of options replicated through several years
and under varying climate conditions. These methods of
evaluation are either too expensive, require too much time,
or both. A more realistic approach is to use either
existing or new experimental data to evaluate our equation
and validate the relationship with independent data. The
increased production can then be simulated for other
locations and climates using the solution obtained from
relationships obtained for the equation noted previously.

An example of such a simulation analysis concerns the
milk production gains from the use of barn cooling. Berry,
et al. (1964) using climate-control chambers, developed a
general relationship for losses in milk production because
of hot and humid weather. This relationship for our equation is a simple linear equation with the THI[1] as the
weather variable (THI [Temperature Humidity Index] combines
temperature and humidity to produce an index related to comfort). Hahn (1981) used this relationship to simulate the
expected production increases for various schemes of environmental control. Figure 2 shows the general relationship
suggested by Hahn for estimating the improved production
from various types of control. Specifically, Hahn reports
that for a dairy herd with the production potential of 50 lb
per day, evaporative cooling will give no additional milk
production along the U.S. - Canadian border but it will
increase production more than 400 kg per season in southern
Texas.

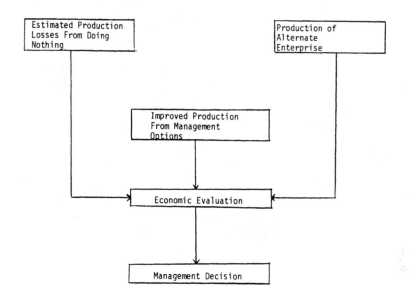

Figure 2. Branching pattern for the economic evaluation of management options in response to environmental events

The producer must integrate into his management decision considerations of the costs and benefits from the management options dealing with environmental control. This requires further branching to the schematic diagram of figure 1, as shown in figure 2. The economic evaluation specified must include the market price of the commodity, level of production, cost of feed, energy cost, and initial costs. Osburn and Hahn (1970) have completed such an analysis and defined the areas of profitability in the U.S. for evaporative cooling and air conditioning of dairy facilities. Using cost figures and milk prices appropriate for 1970, the break-even line for evaporative cooling passes through the central U.S.

TECHNOLOGY AND CLIMATE RISKS

Risks from weather and climate can be reduced through adoption of new technologies. The method just discussed can be used to calculate the profitability of these technologies. There are technologies other than amelioration of hot-humid climates for which economic feasibilities can be estimated, such as the feasibility of grain and forage drying facilities, irrigation systems, and size of farm machinery (implements for tillage and harvest).

The producer should be aware that technologies reduce risks but none completely "weather proof" the system. Large combines and pickers hasten harvest so that grain can be quickly removed from the fields reducing weather-related harvest losses. Yet, these same machines may be unusable during infrequent periods with exceptionally heavy rain. Fans and evaporative coolers are essential for climate control for confinements of cattle, swine, and poultry. But when a severe storm episode knocks out electrical power for several hours or a day or two, heavy losses in animal production or animal deaths occur. The risks from the episodic events impacting on production need to be incorporated into the analysis.

Technology will likely continue to be used to reduce weather risk and (or) labor cost in agricultural production. This trend leads to more intensive agriculture. The costs of energy and capital equipment require that a careful analysis be made concerning the profitability of these technologies for the particular scale of production.

REFERENCES

Balling, R. C. 1982. Weight gain and mortality in feedlot cattle as influenced by weather conditions: Refinement and verification of statistical models. Report 82-1. Center for Agr. Meterology and Climatology. Univ. of Nebraska, Lincoln. p 47.

Berry, I. L., M. D. Shaklin and H. D. Johnson. 1964. Dairy Shelter design based on milk-production decline as affected by temperature and humidity. Trans. Amer. Soc. Agr. Engr. 7:329.

National Research Council. 1976. Climate and Food, Climatic Fluctuations and U.S. Agricultural Production. National Academy of Science, Washington.

Hahn, G. L. 1977. Livestock environment selection for hot climates. Working Paper FAO/SIDA. Storage and structures in developing Countries. Egerton College, Kenya.

Hahn, G. L. 1981. Housing and management to reduce climatic impacts on livestock. J. Anim. Sci. 52:175.

Hahn, G. L. and D. D. Osburn. 1970. Feasibility of evaporative cooling for dairy cattle based on expected production losses. Trans. Amer. Soc. Agr. Engr. 12:289.

24

WATER MANAGEMENT—A CRITICAL NEED: AN ARIZONA CASE STUDY

B. P. Cardon

One might say that a desert environment is too fragile and should not be a major habitat for man. Experience, however, indicates otherwise. Since the beginning of man's history, he has been drawn to the desert and arid lands. I have just recently returned from a short visit to Israel. While there, we examined agricultural development from the Golan Heights in the north, through the Negev Desert in the south. Historical records indicate that during many centuries large numbers of people have lived and prospered in this area. Insofar as we can tell, the climate has not changed appreciably during that 5,000 or more year period in man's history there.

Our Southwest experience is another example. Not only do we have records of large numbers of people living here during the past 2500 years, but during recent history, as the western United States was settled by migrations of people from the East, many of them settled in this southwestern area. So it is not a matter of "should" people live in this environment, but that they do and they must be accommodated.

Food is perhaps the most critical of all the human needs in an arid land environment. Because of the shallow, natural vegetative base caused by limited rainfall, or the seasonality of that rainfall, productive agriculture is placed in a most precarious position. This limitation on water supply, as well as the arid environment and high temperatures prevalent in the area, make water management one of our most important activities.

My hope today is to reiterate problems relating to agricultural water management in general and in Arizona in particular. Obviously most of what we discuss here applies equally well to any arid lands in the world. While I'll be speaking of other uses of water, they are really peripheral to agricultural use.

Let's look at our water situation here in Arizona. What is the surface water supply (chart 1)? We are in the Colorado River Basin. It is claimed that this is the most developed river basin in the world. Currently it supplies to Arizona 2.3 million acre feet (MAF) of water (net basis). Of this about 900 thousand acre feet are directly from the Colorado and 1.4 MAF are from its tributaries. Until recently, our use of Colorado River water has been restricted to lands along and adjacent to the river.

Colorado River Basin

Chart 1

I'm sure most of you have heard of the Central Arizona Project. Briefly this is a U.S. Reclamation Project that will transport in excess of 1 MAF of Colorado River water to central Arizona for use by cities, industries, and agriculture. The first water delivery into the Phoenix area is scheduled for 1985. It is hoped that water will reach Tucson by the late 1980s or very early 1990s. The canal to carry the water is almost completed to the vicinity of Phoenix (chart 2). Currently, possible routes to Tucson are being studied.

At times, some visionaries have dreamed of waters from other areas to augment the limited supplies of the Southwest. One such dream involves the waters from the Pacific Northwest (chart 3). But nothing definite has been accomplished or planned for the future.

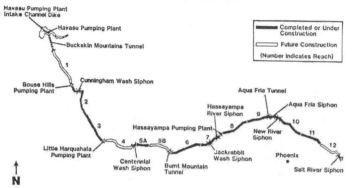

Chart 2

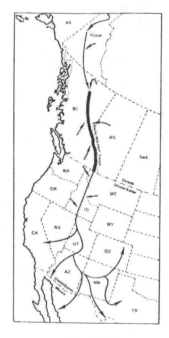

Chart 3

For a moment, let's go back in history--not to the ancient people but to the early settlers that moved into this area in our time. These first settlers coming in during the early part of the 19th century were cattlemen. As they moved herds into this area, water was a problem, and initially the cattle were concentrated along the few streams and permanent or seasonal bodies of water that were viable. Eventually, however, catchment basins were constructed that would collect runoff from rain and store it to supply drinking water to the cattle during the dry periods. Through the years this process was improved upon and expanded to include windmills and a few pumps driven by electric or gasoline motors to supply ground water to augment the limited surface supply. But other than developing a more stable water supply, the industry is run today pretty much as it was during the 19th century when cattle first moved into the area. A ride across any ranch in southern Arizona will confirm this.

The next group to arrive were the farmers who moved into the Southwest in substantial numbers about the middle of the 19th century. Originally, they settled along the riverbanks and diverted water out of the river streams for irrigation. Eventually dams were constructed to create reservoirs and agriculture became more stabilized and expanded further from the rivers. Underground water aquifers were discovered and pumps were developed to bring this water to the surface to be used for irrigation. Fortunately for Arizona, there were a substantial number of these aquifers and, in general, the water in them was close enough to the surface to permit the economic support of crop production.

The amount of underground water pumped to the surface for irrigation purposes, as well as the increase in total acres farmed, increased steadily up to the late 1950s or early 1960s (chart 4). Since then, the amount pumped has stabilized. The reason for this change in rate of increase was apparent. The level of water in some aquifers was being lowered to where the cost of pumping could not be offset by the crops produced. This was particularly true in central Arizona. Due to the high price of cotton during the early 1950s, and the economic availability of underground water, much of the desert land there was brought under cultivation into cotton. When the price fell in the late 1950s, economics forced the abandonment of these farms. That land is still uncultivated.

Of course, during all of this period, the miners were active in the area looking for precious metals, particularly copper. Initially, because of the cost of transport, only copper ore of high mineral content could be economically mined. Prior to World War II, however, a flotation system was developed in which low levels of mineral in ore could be concentrated. This eventually permitted the economic use of ore with low copper content. This process uses a substantial amount of water and now the miners were competing with others users for the small amount of water available in Arizona.

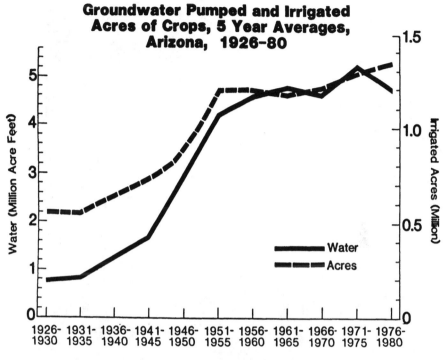

Groundwater Pumped and Irrigated Acres of Crops, 5 Year Averages, Arizona, 1926-80

Chart 4

As agriculture and mining developed, the population of Arizona grew (chart 5). In fact, the population grew faster than these two industries so that additional industries were attracted to the area. The people needed water and they also wanted a pleasant environment that required still more water. All of this meant more competition for the limited water available.

As the use of underground water increased, it was apparent to all that we were using more water than we could reasonably expect to be available on a continuing basis—we were mining water out of the ground. The question then became how much was available on a dependable basis and who would get it. It was this competition for a limited supply of water that forced a reevaluation of the right to use groundwater and the eventual development and passing of the Groundwater Law of 1980. This legislation is really a landmark law. It is the first time any state has faced squarely its groundwater problem.

The Groundwater Law of 1980 passed by the Arizona Legislature is a very complex document. This is understandable when one considers the subject addressed, competition that has developed among users, and the complexity of

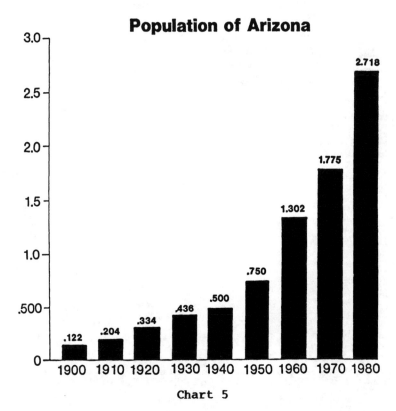

Population of Arizona

Chart 5

previous statutes and legal actions taken during the past 50 or even 100 years. I will not mention all provisions of the law--only those of major importance.

The law created a Department of Water Resources with a water engineer as head (chart 6). This Director of the Department of Water Resources has a great deal of responsibility and authority under the law. The usefulness of the law to the people of Arizona will depend, to a large measure, on the abilities, leadership, and industry of this individual.

The law established certain areas of the state, called the Active Management Areas (AMA) where the provision of the law applied (charts 7, 8, 9, and 10). These areas were selected because of competition for water use between different types of users, such as agriculture, municipalities, and mines. Where a single type of use prevailed, as for agriculture, in general the area has not controlled.

The second item of importance was that the right to use groundwater was shifted from the right to use a given supply, or well, to the right to use water for a particular purpose (chart 11). Agricultural land to be irrigated, a "grandfather right," depended upon the history of farming on that land. If the land was in an Active Management Area and

Administrative Structure

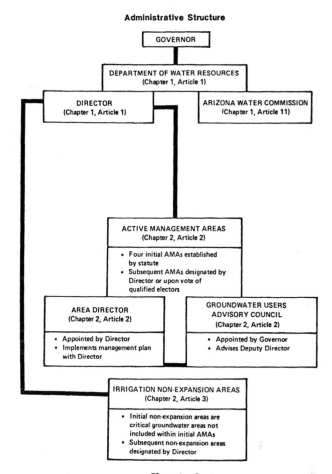

Chart 6

had not been farmed during at least one year of a particular time period (the period chosen was 1975-1980) it could never be irrigated legally again.

Because the right to use groundwater shifted from the source to the use, conservation measures could be established and enforced on all users.

What does this law really mean to agriculture in Arizona (chart 12)? First, as the population grows and we work to balance the underground water supply, it is inevitable that less water will be available for agricultural purposes. What is the solution for agriculture? There are some that suggest that crop production be phased out, or sent back to Iowa where some say it belongs! Obviously I don't agree, but the parameters of water use and the future of irrigated agriculture in Arizona have become uncertain.

212

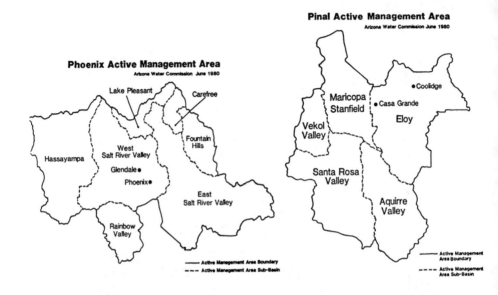

Phoenix Active Management Area
Arizona Water Commission June 1980

Lake Pleasant
Carefree
West Salt River Valley
Fountain Hills
Hassayampa
Glendale●
Phoenix●
East Salt River Valley
Rainbow Valley

—— Active Management Area Boundary
--- Active Management Area Sub-Basin

Chart 7

Pinal Active Management Area
Arizona Water Commission June 1980

●Coolidge
Maricopa Stanfield
●Casa Grande
Eloy
Vekol Valley
Santa Rosa Valley
Aquirre Valley

—— Active Management Area Boundary
--- Active Management Area Sub-Basin

Chart 8

Prescott Active Management Area
Arizona Water Commission June 1980

Little Chino

Upper Aqua Fria

Prescott●

—— Active Management Area Boundary
--- Active Management Area Sub-Basin

Chart 9

Tucson Active Management Area
Arizona Water Commission June 1980

Tucson●

Avra Valley
Upper Santa Cruz

—— Active Management Area Boundary
--- Active Management Area Sub-Basin

Nogales●

Chart 10

Pumping Rights of Agriculture in an AMA

- Restricted water use.

- Must practice conservation, initially with:
  - Reasonable leveling
  - Cement lined ditches
  - Tailwater pump back system

- No new farmland in an active management area.

- In the city service area can only sell water to another agriculture user. Agriculture can sell land to another user, but city now sells his water.

Chart 11

Right to Pump Groundwater Tied to Use Not to Well

- Agriculture. Water duty of land irrigated.

- Industry and Mines. Permit granted for use.

- Municipalities. Can pump the amount permitted (gal/cap/day) by water engineer.

Chart 12

I would like to outline briefly some of the areas of research underway at our College of Agriculture to resolve some of these uncertainties. I have not necessarily listed them in a priority sequence, for all are almost equally important.

The first area is a focus on conventional crops. These crops were, in many cases, domesticated and selected in a water-rich environment. Our research efforts are aimed at developing cultivars that will produce, yet use less water. We are also focusing on adapting conventional crops to using water of lower quality, i.e., high salt content.

I will not attempt to review all research done in this area, but will mention some examples. One of our research scientists has developed a barley strain that will grow in seawater. The yield is low, but if irrigation water of two-thirds seawater mineral content is used, the yield is almost normal. Another strain of winter barley will produce almost a normal yield using only 50% of the irrigation water required by current commercial strains. Experiment Station personnel have also developed a short season cotton variety that can be planted later, harvested earlier, and needs only

80% of the irrigation water required by commercial varieties. Work is also being done with several alfalfa varieties.

The second area of research is to identify new species that use less water and can be adapted to human use and utility. Our Experiment Station, along with other research organizations in the Southwest, has identified several promising species. The best known is the Jojoba (Simmondsia chiensis) "coffee bean" shrub. I'm sure most of you are familiar with this shrub and the plantings now under trial throughout the arid lands of the world. The oil or wax from the Jojoba bean was first sought as a lubricant; more recently, it has been used as a very desirable base for cosmetics.

Another species, buffalo gourd, (Cucurbita foetidissimo) also is receiving attention. Preliminary work at our station indicates that the seed from the gourd will compete favorably with soybeans in the yield of high-quality oil and protein meal per acre. In addition, we have produced over 1 t/A of pure starch from the roots of the plant. Water needs of the plant are substantially less than those for conventional crops.

The common gopher weed (Euphorbia lathris) has been proposed as a crude oil source that would substitute for crude petroleum oil. Its water needs when growing "wild" are very low and our Office of Arid Land Studies has an intensive development project evaluating this prospect.

In the past, irrigation water was been relatively plentiful and cheap; consequently, the farmers have used it in abundance. Our main area of focus in research today is a critical evaluation of methods of water application.

Flood irrigation developed under conditions of low water costs and high relative availability. Current research shows that flood irrigation generally uses several times the consumptive needs of the crop grown. Lining the delivery ditches with cement and installing a pump-back system to prevent runoff from the irrigated area can save substantial amounts of water.

More specifically, our research relating to "sprinkle" and "drip" irrigation shows that additional amounts of water, in some cases more than 50%, can be saved by these newer methods of water application. The water saved, however, must be balanced against substantial capital investment, as well as increased soil nutrient and salt-management requirements. Needless to say, we anticipate that the cost of water will continue to increase, so the capital and increased management requirements will become increasingly more important in the future.

Philosophers, and perhaps researchers, dream about the future and what it should entail. Farmers, however, are trapped in what I call the "economics of the moment." Farmers must always be competitive and must look to the economics of production in all of their management practices. With respect to irrigation water, economics are working against the farmer. There will be less water in the future and it will cost more. So our ultimate testing depends on how well we manage the water that is available to us.

Part 5

GENETICS AND SELECTION:
INTERNATIONAL PERSPECTIVES

25

BEEF AND DAIRY CROSSES
FOR BEEF PRODUCTION

Harold Kenneth Baker

INTRODUCTION

The British beef cattle scene at first sight seems very diverse and fragmented. Many different breeds and crosses are used in a wide range of environments and within many different production systems. There is also a well-established crossbreeding program in the national dairy herd that produces both feeding and breeding cattle for beef production.

While the national beef herd contains a wide variety of breeds and crosses, there have been no discreet experimental programs to compare the relative reproductive performance of different types of cattle. However, a considerable amount of comparative data has been collected through Meat and Livestock Commission's (MLC) large-scale, commercial, on-farm, beef recording schemes on both breeding and feeding farms. The livestock performance records from these commercial firms have been accumulated into a data bank that provides a large amount of robust comparative information for both reproductive characteristics in breeding cattle and growth rate and stocking performance for rearing and finishing stock. However, the records do not give any viable data on either feed efficiency or carcass characteristics. To provide this essential information, the MLC has just concluded a large-scale experiment with feeding cattle from both the beef cow and dairy herds. The data from the commercial recording schemes and these experimental programs are used in the subsequent comparative data between different breeds and crosses.

The sources of total beef production in the United Kingdom are given in table 1. This shows that about one-third of the total beef production is derived from the beef herd and 60% from the dairy herd.

BEEF COW PRODUCTION

Within the U.K., the beef herd is predominantly crossbred. While no reliable census data are available on breed structure, probably over 90% of the beef cows are crossbred and almost 100% of the bulls are purebred.

TABLE 1. SOURCES OF BEEF PRODUCTION IN U.K. (1981)

		%	%
Beef herd			
	Cull cows	5	
	Suckled calves	<u>29</u>	34
Dairy herd			
	Cull cows	18	
	Purebred calves	18	
	Crossbred calves	<u>24</u>	60
Imported Irish stores			<u>6</u>
			100

Source: MLC (1981).

In the past most beef cows were produced by crossing beef breeds. The most popular of these crosses was the Blue-Grey, which resulted from the breeding between the Whitebred Shorthorn bull and Galloway cows. More recently a high proportion of beef cows have been produced by crossing beef bulls with Friesian dairy cows. The most popular of these crosses is the Hereford cross Friesian, and it is estimated that about one-third of all beef cows will be of this type.

Effect of Cow Type in Beef Herd

Comparisons are restricted to different crossbred types as purebred cows are too few in commercial beef herds to provide a reasonable amount of data. Records from commercial herds are given in table 2 to show the relationship between cow size and calf performance.

The larger beef-type cows produce heaviest calves at 200 days of age. However, the cross between the small British breeds of bull and the dairy cow is popular as a suckler cow. Their relatively small size and good milking potential result in high calf weights per unit of cow size when compared with the large Charolais and red-breed crosses.

Calving interval also is related to cow size (table 3) with the large Charolais x Friesian having an interval 8 days longer than the small Angus x beef cows.

Dystocia and perinatal calf mortality have only been reliably recorded in herds containing Hereford x Friesian and Blue-Grey cows (table 3). Both assisted calvings and calf deaths were greater in the larger Hereford x Friesian cow by 1.2% and 0.8%, respectively.

TABLE 2.   WEIGHTS OF SUCKLER COWS AND RELATIONSHIPS WITH
CALF PERFORMANCE

Cow breed type	Cow wt* (kg)	Calf 200-day wt (kg)	Calf at 200 days per 50 kg cow wt (kg)	Mean calving interval (day)
Shorthorn crosses	443	192	22	N.A.
Angus x Friesian	449	193	21	374
Blue-Grey	450	194	22	367
Angus x beef breed	453	190	21	372
Hereford x Friesian	472	203	22	376
Hereford x beef breed	485	194	20	372
Red-breed crossess	552	204	18	380
Chrolaix x beef breed	628	239	19	378

Source: MLC (1979).
*Average of spring and autumn weighings

TABLE 3.   ASSISTED CALVINGS AND CALF DEATHS IN SUCKLER
HERDS (COW MATINGS)

	Breed of cow			
	Hereford x Friesian		Blue-Grey	
Breed of sire	% assisted calvings	% calf deaths	% assisted calvings	% calf deaths
Charolais	10.1	5.1	9.0	4.4
Simmental	9.7	4.7	8.9	3.7
South Devon	8.4	4.4	6.3	3.7
Lincoln Red	6.0	3.5	4.8	2.4
Devon	6.1	3.2	5.0	2.0
Sussex	4.0	1.7	2.7	1.3
Limousin	7.9	4.9	6.4	3.9
Hereford	4.2	1.8	3.3	1.3
Aberdeen-Angus	2.0	1.5	1.1	1.1

Source: MLC (1979).

Effect of Sire Breed

Hereford and Angus bulls predominated as terminal sires
in beef herds until Continental breeds became available.

For many years, the use of Continental breeds was restricted by high bull prices. However, there has been a steady increase in the use of Continental breeds, and it was estimated that in 1980 possibly 40% of calves from the beef herd were sired by Charolais bulls and a further 10% by other Continental breeds.

The largest breeds obviously produce the heaviest calves at weaning (table 4). It was traditionally believed that the relative order of performance would change under different environments. However, as indicated, the order of performance remains consistent under a wide range of environmental conditions, although the magnitude of difference decreases on moving towards the harsher conditions.

TABLE 4. COMPARATIVE EFFECTS OF SIRE BREED ON CALF 200-DAY WEIGHTS

400-day wt of purebred bull (kg)		Wt at 200 days			Within herd comparison to Hereford x calves (kg)
		Lowland (kg)	Upland (kg)	Hill (kg)	
Charolais	551	240	227	205	+75
Simmental	532	232	222	198	+13
South Devon	520	231	221	200	+12
Devon	460	225	215	191	+ 9
Lincoln Red	510	222	214	189	+ 8
Sussex	454	215	207	186	+ 5
Limousin	475	215	204	186	+ 8
Hereford	424	208	194	184	0
Angus	387	194	182	176	- 4

Source: MLC (1979).

These differences in preweaning growth are carried through into the finishing period of mixed diets in winter or grazed grass in summer (see below).

As with cow-breed type, bull-breed size also is associated with higher levels of dystocia, assisted calvings, calf mortality (table 3) and longer calving intervals (table 5).

While the larger terminal-sire breeds have higher levels of calf mortality and also result in a longer calving interval, when they are compared in herds of a given size, they still produce greater productivity because of the much heavier 200-day weights of their progeny. This is shown in table 6 where the effects of Charolais, Hereford, and Angus bulls are compared on Hereford x Friesian and Blue-Grey cows, which are themselves the most efficient of the various cow types (table 2).

TABLE 5. EFFECT OF BREED OF SIRE ON CALVING INTERVAL

Sire breed	Days to next calving	Days from previous calving
South Devon	376	374
Charolais	375	373
Lincoln Red	374	372
Devon	374	372
Sussex	372	371
Hereford	371	372
Angus	370	371

TABLE 6. EFFECT OF SIRE BREED ON THE PRODUCTIVITY OF HEREFORD x FRIESIAN (H x F) AND BLUE-GREY (BG) SUCKLER COWS

Sire breed	Charolais		Hereford		Angus	
Cow breed	HxF	BG	HxF	BG	HxF	BG
Live calves per 100 cows calved	94.9	95.6	98.2	98.7	98.5	98.9
Calf weaning weight* (kg)	294	278	249	234	236	220
Calving interval for cows that calved (days)	378	369	372	367	370	366
Calf weight per 50 kg cow weight (kg-year)	26.9	29.3	23.9	25.0	23.1	24.0

Source: MLC (1979).
*Calf weight adjusted to 250 days.

Feeding Cattle from Beef Cows

The relative performance, and carcass characteristics, of weaned calves from the beef herd fed to slaughter at constant levels of external fatness have been studied in the MLC's beef-breed-evaluation program. The cattle were compared under two systems, which equated to either winter finishing indoors, or to grass finishing during the summer. The results from the two feeding systems were broadly similar and only the winter finishing results are given in this paper. The cattle were produced by crossing nine different beef-bull breeds onto either Friesian x Hereford or Blue-Grey cows. The calves out of Hereford x Friesian cows were

12 kg heavier than those out of Blue-Grey cows at the start of the feeding experiment. The daily gains and age at slaughter were similar and the results of using these two types of cows are combined in the data given below.

The results confirm large differences between the crosses out of different breeds in the duration of the feeding period, weight at slaughter, and total feed consumption (table 7). For example, the Charolais crosses needed about 40% more feed in total than the Hereford crosses.

Charolais crosses had the highest slaughter weights and were older at slaughter. Simmental and South Devon crosses were broadly similar to the Charolais in performance. The very early maturing Aberdeen-Angus crosses had the lowest slaughter weights; Hereford crosses were only slightly older at slaughter than Aberdeen-Angus crosses, but were heavier. Devon, Lincoln Red, and Sussex crosses were included in the early maturing category, while Limousin crosses were late maturing.

Size and spread of maturity had significant effects on feed consumption. Big, late-maturing crosses ate much more feed in total than did the smaller, early maturing crosses. Being bigger, they ate more per day, and being later maturing they had to be fed for longer. However, the extra feed was about balanced by extra production, so differences in feed conversion efficiency were generally small.

Carcass Differences

Important carcass characteristics are shown in table 8. Sire breed averages are given for killing-out percentage, yield, and distribution of salable meat in the carcass and retail value. Retail value is calculated by accumulating the value of trimmed deboned primal joints, expressed as pence per kg carcass weight, using 1980 prices from MLC's retail price surveys. It provides a standardized comparison of the estimated retail values of the carcasses from different breed types.

Killing-out varied between breeds by up to 2.5%: Charolais and Limousin crosses had the highest values. Differences in yield of salable meat occurred, even though breeds were compared at the same level of external (subcutaneous) fat. The differences were due to variations in bone content or in the quantity of seam (intermuscular) fat that was removed during joint preparation. The high salable meat yield of the Limousin crosses (and, to a lesser degree, the Aberdeen-Angus crosses) was due to a low bone content, while the Charolais and South Devon crosses had low levels of fat trim.

Breed differences in distribution of salable meat in the higher-priced cuts were small, ranging from 43.9% for Sussex crosses to 46.4% for Limousin crosses. Breeds with better conformation tended to have more salable meat in the higher-priced cuts.

TABLE 7. EXPERIMENTAL COMPARISON OF WINTER-FINISHED CROSSBRED SUCKLED CALVES

Sire breed	No. of cattle	Start weight	Finishing period	Daily gain	Slaughter weight	Total feed (t DM)	Kg feed DM per kg gain
		(kg)	(kg)	(kg)	(kg)		
Aberdeen-Angus	60	319	97	0.77	393	0.83	10.5
Charolais	61	363	157	0.84	494	1.37	11.0
Devon	61	328	116	0.78	419	0.98	10.2
Hereford	59	322	113	0.78	410	0.96	10.2
Limousin	48	332	156	0.78	454	1.21	10.6
Lincoln Red	60	336	108	0.85	428	1.01	10.6
Simmental	62	359	153	0.86	490	1.34	10.7
South Devon	59	335	150	0.77	451	1.21	10.7
Sussex	59	324	136	0.76	428	1.13	10.6

TABLE 8. WINTER-FINISHED CROSSBREDS: SIRE BREED AVERAGES
FOR CARCASS CHARACTERISTICS

Sire breed	Killing-out (%)	Salable meat in carcass (%)	Salable meat in higher-priced cuts (%)	Retail value index* (p/kg)
Aberdeen-Angus	52.5	72.5	44.1	222
Charolais	54.8	72.7	44.8	223
Devon	52.7	71.6	44.0	219
Hereford	52.3	71.9	44.1	220
Limousin	54.7	73.8	45.4	226
Lincoln Red	52.3	70.8	44.3	216
Simmental	53.0	72.0	44.8	221
South Devon	53.2	72.0	44.3	220
Sussex	53.1	72.6	43.9	222

*At 1980 retail prices; pence/kg.

When differences in yield and distribution of salable
meat were combined into an estimate of retail value, a 5%
difference (10p per kg at 1980 values) was apparent between
extreme breeds (Lincoln Red 216p U.K. per kg, Limousin 226p
per kg). This is equivalent to 23 (U.K.pound) on a 230 kg
carcass.

BEEF FROM THE DAIRY HERD

Effects of Cow Type

The predominance of the Friesian breed during the 1960s
and 1970s reduced the commercial relevance of cow-breed effects. Purebred Ayrshire and Channel Island cattle perform
so badly in beef systems that most of those not required for
breeding are slaughtered as young calves. To some extent
crossbreeding these breeds to Charolais has increased their
acceptability, but they are still inferior to beef crosses
out of the Friesian.

To a large extent, the present structure of the U.K.
beef industry, involving the interdependence of dairy and
beef production is due to the beefing quality of the British
Friesian. The recently introduced North American Holstein
is of inferior quality for beef production and the increasing level of inclusion of Holsteins in the U.K. dairy herd
may adversely affect this structure. The present subjective
standards applied at all levels of beef industry discrimi-
nate heavily against the Holstein or Holstein type and, if

continued, may lead to a larger proportion of the calves out of Black and White dairy cow being rejected for beef production and slaughtered.

The Holstein has a lower yield of salable meat, although this difference is less than would be supposed on the basis of differences in shape between the two breeds. Table 9 summarizes current findings from U.K. trials: compared with Friesians, Holsteins have 1% to 2% less salable meat in the carcass, and a slightly lower proportion of salable meat in the higher-priced cuts. In addition, killing-out percentage is about 1% lower than with Friesians.

TABLE 9. CARCASS COMPOSITION OF FRIESIANS AND HOLSTEINS

Trial	Beef system	No. of cattle	Salable meat* % of carcass	Salable meat in higher-priced cuts %
Friesians				
Various	18 and 24 mo	70	70	45
Holsteins				
MMB/MLC	18 mo	30	69	44
ADAS	18 mo	10	68.5	44.5
MLC	18 mo	18	68.5	43.5
MLC	24 mo	12	67.5	44.5

*Yield of trimmed, deboned cuts.

All these differences have important financial implications that are given added emphasis by differences in the numbers of Friesians and Holsteins that do not meet the certification standards necessary for variable premium payment under the Beef Premium Scheme. Undoubtedly, the market already discriminates against the poor shape of the Holstein. Under British conditions, it is estimated that to give the same level of profitability, the Holstein calf is worth about 25 (U.K. pound) less than the British Friesian.

Effect of Sire Breed

Dairy beef systems also have been included in the MLC Breed Evaluation Program. Growth and feed intakes have been recorded for the pure Friesian and crosses by the important beef breeds out of the Friesian in production systems lasting 16 and 24 months. The 16-month system, in which growth rate is essentially linear, approximates to the 18-month system of grass feeding with yard finishing that is commercially popular in the U.K. In the 24-month system, yearlings are subjected to a store period lasting 9 months during which growth rate is restricted by feeding a low-quality feed. This system is similar to the commercial 2-year grass-finishing system in the U.K. Results from both these systems were similar. In terms of growth rate and slaughter weight, the breed differences recorded in these trials are very similar to those recorded on commercial farms by MLC and confirm the relevance of the results to the British meat industry. The average performance and carcass results are from these two systems and are given in tables 10 and 11.

TABLE 10. PERFORMANCE CHARACTERISTICS OF 2-YEAR BEEF FROM DARY HERD

				3-month slaughter	
	Daily gain kg	Feed con-sumption kg	Feed effi-ciency g LWG/KG DOM	weight at slaughter kg	Live Age at slaugh-ter days
Friesian (F)	0.70	6212	148	542	738
Angus x F	0.63	5153	151	454	683
Charolais x F	0.79	6418	160	616	743
Devon x F	0.67	5358	158	496	705
Hereford x F	0.68	5182	165	495	698
Lincoln x F	0.69	5700	151	507	699
Simmental x F	0.72	6363	149	569	744
South Devon x F	0.72	6043	155	554	733

TABLE 11. CARCASS CHARACTERISTICS OF 2-YEAR BEEF FROM DARIY HERD

	Carcass wt (kg)	Salable meat in carcass %	Salable meat in high-priced joints %	Efficiency of salable meat gain (g meat gain per kg DOM intake)
Friesian (F)	266	69.7	43.8	51.1
Angus x F	221	71.3	44.0	42.9
Charolais x F	317	70.8	44.0	59.7
Devon x F	242	70.2	43.8	54.8
Hereford x F	245	70.1	43.9	57.8
Limousin x F	239	69.3	44.0	49.8
Simmental x F	286	70.3	44.3	53.4
South Devon x F	279	70.5	43.3	56.0

Both in terms of feed efficiency expressed as live weight gain per unit of feed, and in terms of salable meat per unit of feed, the Charolais and Hereford x Friesian were superior to the other breeds and crosses. While the Charolais was more efficient than the Hereford in converting feed to meat, it is unlikely that the Charolais will replace the Hereford as the dominant cross on Friesians. The Hereford gives much easier calving than the Charolais, particularly on heifers. Difficult calvings may not only result in the loss of the calf but also influence the milk yield of the dam and increase the calving interval, factors of supreme importance to the dairy producer.

PRODUCTION SYSTEM DEVELOPMENT

The above data is now being used to assist in development of systems for beef production. Within Britain there will continue to be scope for different types of beef production and, indeed, the available resources (feed and housing), cash flow requirements, and preferences encourage a degree of flexibility. Thus table 12 indicates how different breed types from the dairy herd may be used within a two-year feeding system. In this type of production the earlier maturing types tend to fit better as they are sold in midsummer before grass quality begins to fall, and they release grazing for the next generation of young cattle. The late maturing cattle are a better investment when used in production systems with higher lifetime growth rates, lower slaughter ages, and less dependence on forage.

TABLE 12. PRODUCTION MODEL FOR DAIRY-BRED STEERS IN GRASS FINISHING

	Angus x Friesian	Hereford x Friesian	Friesian	Charolais x Friesian
Slaughter age (mo)	20	21	22	23
Slaughter weight (kg)	430	475	520	570
Carcass fat class*	4	3H	2 or 3L	2
Final grazing period (days)	110	120	170	190
Total gain (kg/ha)	1100	1100	1100	1100

*Fat class on a scale 1 (leanest) to 5 (fattest).

A contrasting example is given in table 13 where the information from the data bank is used to calculate the grassland acreage for different breeding systems to produce comparable quantities of meat.

TABLE 13. TOTAL FEED REQUIREMENT OF SUCKLER COWS AND CALVES TO SLAUGHTER TO PROVIDE ONE TON OF SALABLE MEAT

Sire breed	Angus		Charolais	
Cow breed	Blue-grey	Hereford x Friesian	Blue-grey	Hereford x Friesian
Cow/calf requirement to* transfer per cow (ha)	0.67	0.69	0.69	0.72
Cows required	6.9	6.8	5.4	5.4
Total cow/calf requirement* to transfer (ha)	4.6	4.7	3.7	3.9
Calf requirement per head – finishing (ha)	0.14	0.14	0.23	0.23
Calves required	6.8	6.5	5.1	5.0
Total finishing requirement (ha)	0.9	0.9	1.2	1.2
Total system requirement (ha)	5.5	5.6	4.9	5.1

*Transfer from calf production unit.

From the example it can be estimated that a total production system based on Angus terminal sires used on Hereford x Friesian cows would require 14% more land than a system based on Charolais sires on Blue-Grey cows. The data can also be used to provide financial comparisons both within and between systems and this is now being routinely done in the MLC 'Beefplan' service for commercial producers (table 14).

Thus table 14 shows the returns per head that can be anticipated for the winter-finishing of suckled calves. Of course, the producers clearly have a shrewd idea of the relative performance of the different breeds and crosses. However, reference to this type of detailed information does enable farmers to assess what financial premiums, or discounts, they can apply to stock when purchasing them.

TABLE 14. CALCULATED GROSS MARGINS FOR WINTER-FINISHED
SUCKLED CALVES (UK POUND)

	Sire breed				
	A.Angus	Hereford	Simmental	Charolais	Limousin
Net output	97	109	152	153	141
Variable costs	43	49	69	71	63
Interest on working capital	12	14	22	23	14
Gross margin less interest on working capital	42	46	61	59	57

26

PEDIGREE TESTING AND SELECTION
FOR BEEF PRODUCTION IN GREAT BRITAIN

Harold Kenneth Baker

INTRODUCTION

About 3 million steers and heifers, plus just over 900,000 cull cows, are slaughtered for beef each year within Great Britain. About 1.2 million of the steers and heifers come from commercial beef herds and a further 0.6 million are beef-cross dairy calves. Therefore, about 1.8 million calves that are slaughtered for beef are sired by bulls from beef breeds. In addition, about 0.4 million beef-cross dairy-calf heifers are transferred into the national beef cow herd as breeding replacements.

The one-million beef matings in the dairy herd are dominated by artificial insemination, which accounts for about 80% of the total. These 800,000 matings are accomplished by an AI stud that has an annual intake of some 40 to 50 bulls. Within the beef cow herd, AI is much less important, and well over 90% of all matings will be through natural service. The net result of this system of breeding is that nearly 7,000 purebred beef bulls enter the national stud each year. When bull licensing was abolished, there were fears that there would be a "mongrelization" of our breeds, and that crossbred bulls would be used in abundance. This has not, in fact, happened, and hardly any crossbred beef bulls are used, except for grading-up purposes in pedigreed herds.

With the introduction of Continental beef breeds, some 30 breeds of cattle are available for beef production, but only about 10 of these are used in significant numbers and are capable of supporting a structured, national breeding program.

In addition, the size of purebred herds is small. There are now no reliable national statistics, but an examination of the 1,400 pedigreed herds in the MLC's pedigree-recording schemes gives the following numbers of cows per herd (table 1). It can be assumed that the relatively small number of nonrecorded herds will be smaller rather than larger.

Table 2 shows the distribution of recorded herd size by breed. The average herd size of the more recently imported

TABLE 1. AVERGE HERD SIZE IN THE MLC PEDIGREED
BEEF-RECORDING SCHEME

	Avg number of cows/herd	Avg number of bulls used/year
Aberdeen Angus	12	1.1
Devon	25	1.2
Hereford	24	1.1
Lincoln Red	47	1.9
South Devon	25	1.3
Sussex	37	1.4
Welsh Black	25	1.2
Charolais	9	2.1*
Limousin	7	1.7
Simmental	7	2.0*

*Includes AI usage.

Continental breeds is small, although the situation is changing quickly, particularly as grading-up schemes mature. If it is assumed that a 50-cow herd is the minimum size for making genetic improvement for performance, the proportion of herds able to sustain a breeding program in isolation is relatively small in the major breeds.

TABLE 2. DISTRIBUTION OF HERD SIZE BY BREED

	Herd size (no. of cows)				
	Below 10	10 - 30	30 - 50	50 - 100	100 plus
	% of herds				
Aberdeen Angus	49	27	17	7	1
Charolais	73	23	3	1	0
Devon	36	36	13	12	2
Hereford	34	35	17	10	2
Limousin	80	19	1	0	0
Lincoln Red	31	12	16	24	16
Simmental	80	15	5	0	0
South Devon	41	29	13	14	3
Sussex	27	22	20	28	2
Welsh Black	40	37	13	8	3

Table 1 also shows the average number of bulls used per year. The overall herd size is 21 cows, and 1.3 bulls are used per year including AI. Average herd life of cows is

6.7 calvings, and bulls are used an average of 3.3 years. There is a very wide range in the length of time bulls are used in herds with an approximate 30% used only for 1 year and a significant proportion (17%) used for longer than 5 years.

Within all commercial beef production systems in Britain, there are strong correlations between growth performance and profitability. Thus table 3 shows the influence under British conditions of performance and profitability.

TABLE 3.   RELATIONSHIP BETWEEN GRASS MARGIN AND GROWTH RATE

System	Grass margin increase (British lb/head) for each additional .1 kg live weight gain per day
	British pound
Cereal beef (Friesian steers)	13
18-month grass/cereal beef (beef x Friesians)	14
20 to 24-month grass beef (beef x Friesians)	15
Suckled calves--autumn born (beef x's)	23
--winter/spring born (beef x's)	17

To date, therefore, the improvement programs, particularly on farm recording, have concentrated on selection for live weight gain. Increasingly, however, levels of fatness, feed efficiency, and ease of calving will become important.

PEDIGREE ON-FARM RECORDING

The base line of British pedigree testing lies with the on-farm recording scheme. About 1,400 pedigree breeders from 23 breeds record about 24,000 growing cattle, and it is estimated that from these are selected 60% to 70% of all bulls that enter the national stud. The proportion of bulls registered that are performance-recorded varies between breeds: among the Continental breeds (Charolais, Simmental, etc.) it is more or less 100%; among the native red breeds (Devon Lincoln Red, and Sussex) it is over 80%; and in the Hereford and Angus breeds it is only 40% to 50%.

Within this scheme, growing cattle are weighed four times a year until they are 20-months old, thereby establishing a cumulative growth record. Weights are adjusted to the nearest 100-day point. Special attention is paid to the 100- and 200-day weights as indicators of mothering ability in the dam, and to the 400- and 500-day weights as indicators of the genetic potential for live weight gain in the recorded animal itself. Weight for age averages and ranges for the main breeds at 400 days in 1980 are given in table 4.

TABLE 4. BREED AVERAGES AND RANGES FOR 400-DAY WEIGHTS

Breed	Average weight	Range
	(kg)	(kg)
Charolais	556	470 - 780
Simmental	544	439 - 730
South Devon	528	419 - 715
Blonde d'Aquitaine	504	377 - 699
Lincoln Red	496	402 - 625
Sussex	458	367 - 601
Devon	450	346 - 605
Limousin	475	382 - 658
Welsh Black	430	322 - 623
Hereford	422	344 - 603
Beef Shorthorn	413	399 - 534
Aberdeen Angus	388	308 - 536

Some of the variation in weight between cattle born at different seasons or reared in different herds is due to management but part is due to real genetic differences between individual cattle. Those that are heavier than their herd contemporaries, as well as being above the breed average, are likely to produce the fastest-gaining offspring.

There is a wide range of weights at each age and this gives ample scope for selecting heavier cattle to maintain or improve performance. The most useful function of farm records is to analyze individual herd performance. A computerized performance analysis is therefore produced for each herd to show:

- Calf ranking list: Identifies bulls and heifers showing superior or inferior weight for age at 400 days or other 100-day intervals on request.
- Cow summary: Lifetime records of all cows corrected for the number of records so that all can be compared on an equivalent basis. This summary includes the calving intervals of the cow.
- Cow ranking list: Identifies dams with a progeny record of superior or inferior 200-day weights or other chosen preweaning, weaning, or post-weaning weights.
- Sire summary: Provides a progeny test of any or all bulls in the herd.

Each year about 120,000 weights are processed and over 1,800 herd analyses are prepared for individual breeders.

This comprehensive system provides breeders with very full information, but frequently comparisons are difficult to make because of small herd sizes, widespread calving patterns, and differing herd management within and between herds. However, analysis of the sire summaries has shown that the heritability of growth rate within these herds is reasonably good and has varied within the different breeds from 27% to over 50%.

It might be expected that during the past decade there would be a steady increase in the average breed weights, e.g., at 400 days of age. However, table 5 shows that during the early 1970s there was a tendency for the average 400-day weights of bulls of most breeds to decline. This was partly the result of better record keeping, but it also reflected the adoption by breeders of more economical rearing methods. The 1978 to 1980 averages suggested that this trend had been reversed, and this is confirmed by the 1979 to 1981 values; for most breeds the latest averages are higher than the previous average. The exceptions are the Devon, South Devon, and Sussex where a further decline in the average weight of bulls probably reflects further adjustments in bull-rearing methods because the heifer averages remained the same for the Devon breed and increased in the South Devon and Sussex breeds. The rather large fall in the average weight of Sussex bulls may have been influenced by a substantial increase in the number of bulls recorded at 400 days of age.

CENTRAL PERFORMANCE TESTING

While the on-farm recording scheme provides the individual breeder with information for within-herd selection, this is partly negated by the small size of herds. Central Performance Testing, therefore, has an important part to play in Britain where bulls from small herds, in different environmental conditions and with varying managements, can be more accurately assessed.

The MLC operates five Bull Performance Testing Centers, and some 500 bulls are tested each year. Only bulls with above-breed-average 100- or 200-day weights are accepted for testing. They enter the centers in breed groups at around 6 months of age and are tested for 7 months. The bulls remain the property of the breeders who pay a test fee equivalent to the cost of feed, bedding, and veterinary charges. The bulls start their test in groups and are then individually housed and their feed recorded for the latter part of the test. They are fed ad libitum on pellets containing dried grass and rolled barley, supplemented with minerals and vitamins.

Considerable attention has been paid to the techniques of performance testing during the last 10 years and a series

TABLE 5. TRENDS IN AVERAGE BULLS 400-DAY WEIGHTS SINCE 1968/70 (kg)

	1968/70	1970/72	1972/74	1974/76	1977/78	1978/80	1979/81
Aberdeen Angus	420	411	397	384	382	388	398
Charolais	566	595	579	554	553	556	559
Devon	495	488	477	471	463	450	435
Hereford	444	454	441	425	425	422	427
Limousin	-	-	455	450	462	475	488
Lincoln Red	475	494	498	511	496	496	502
Simmental	-	-	532	528	531	544	555
South Devon	510	526	530	523	526	528	512
Sussex	460	466	442	451	447	458	421
Welsh Black	438	460	475	439	428	430	444

of refinements and extra measurements have been progressively introduced. Test measurements now include:
- Growth rates and weight at 400 days of age.
- Feed conversion efficiency.
- Backfat thickness.
- Withers height.
- Type classification.

Weight for age. Within Britain the 400-day termination point used by MLC has been a point of major discussion. It was chosen for several reasons:
- Four hundred-day weights are very highly correlated with subsequent weights.
- The bulls' weights at this age are similar to the slaughter weights of their commercial crossbred progeny.
- Unnecessarily long test periods add excessively to the direct costs of testing program and reduce through-put in expensive facilities.

To check the reliability of 400-day weights, tests have been extended experimentally to 500 days. These produced very high correlations between both the weights and the rankings of the bulls concerned at 400 and 500 days. Further supporting information on weight-for-age relationships is available from an ongoing mass of farm data on weights to 500 or 600 days. The use of 400-day weights as a measure in performance tests is also supported by subsequent farm 500-day weights.

Feed conversion efficiency. In 1971, complete diets based on dried grass and rolled barley were introduced so that accurate measurements of feed intake could be obtained. The diet was designed to incorporate as much fiber as possible and to be fed ad libitum in order that the bulls' performance should not be subjected to external influences. Diet quality has been designed to allow high rates of growth so that the bulls can express their growth potential.

Feed conversion is measured over stated live weight ranges during the latter part of the test. This avoids the period when most of the compensatory growth-effects occurs and confines the assessment to a period when the animals are likely to be depositing fat so that the measurement becomes a more sensitive measure of efficiency.

Subcutaneous fatness. In the traditional British breeds, the percentage of fat in the carcass is the most important variable constituent at a given weight or age and it largely determines the yield of salable meat. The development of ultrasonics has allowed measurements in live bulls of subcutaneous fat thickness. This technique was routinely incorporated into MLC performance tests in 1973.

Withers height. To provide supporting information on bulls' skeletal development at the yearling stage, withers height measurements have been taken since 1974.

Type classification. Conformation assessments have been a regular feature of performance tests. A system of type classification was introduced in 1975. Emphasis is placed on describing structural soundness and functional characteristics to supplement the physical characteristics that are routinely measured in the performance test. Strong or weak points can be identified so that these can be used to aid selection, and a permanent record is obtained that describes bulls that have been tested. Differences, if any, between bulls of overall similar merit will also be identified.

Interpreting Test Results

Typical test results are given for Simmentals (table 6).

TABLE 6. PERFORMANCE TEST RESULTS FOR SIMMENTAL

		1978/79	1979/80
Number of bulls		59	85
400-day weight (kg)	mean	623	619
	range	546 - 729	532 - 718
Feed conversion efficiency	mean	6.7	6.4
(kg feed/kg gain)	range	4.6 - 9.5	4.8 - 9.6
Withers height* (cm)	mean	129	128
	range	121 - 135	119 - 138
Backfat thickness* (mm)	mean	2.4	2.4
	range	1.1 - 4.3	1.2 - 4.9

*At 365 days of age.

Breeders are encouraged to look for bulls with a high 400-day weight matched by a good withers height measurement, with test gain as good as or better than pretest gain. Lower gains on test than pretest should be questioned. Better than average feed conversion efficiency is also required. The best values will often be associated with high test gains and low backfat readings, but good values may also be associated with high test gains and above-average fatness. This may indicate superior appetite with surplus energy stored as fat. However, high backfat reading associated with average or below-average test gain and below-average withers height is considered undesirable. High backfat readings will sometimes be associated with several plus factors and may not necessarily be a bad sign.

Examination of results from tested bulls through the study of crossbred progeny from the dairy herd and of purebred progeny in pedigreed herds have confirmed high heritability for growth rate. The best bulls are now invariably used as pedigreed herd sires, or are purchased for use in AI, where they obviously have a major effect on commercial beef production through their large numbers of progeny. The average and below-average bulls tend to be used in commercial suckler herds as terminal sires. This practice is quite acceptable as the average performance of tested bulls is well above the breed averages. Finally, the prominent featuring of performance-tested bulls in breed-society shows and sales for all the participating breeds is an indication of their acceptance by breed societies.

Cooperative Breeding Scheme

Each breed which participates in central performance testing also has to participate in a cooperative breeding project known as Young Bull Proving Scheme. This scheme is operated jointly by the breed societies and MLC.

Each year a group of bulls, each at least 1 1/2 deviations better than the test mean, are nominated to the relevant breed society. The society then selects a number of bulls from this group (depending upon the breed's numerical strength) to form the team for the scheme. From these bulls semen is collected and offered to breeders at no cost. The aim is to obtain at least 40 progeny per bull across a wide range of herds. The calves are then compared to calves sired by other stock bulls used in the breeders' herds on a contemporary comparison basis. The scheme is diagrammatically shown in figure 1.

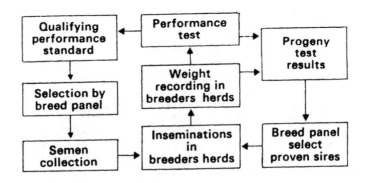

Figure 1

The twin objectives of this scheme are to spread the influence of the best bulls and also to obtain a progeny test within breeders' herds. The semen from bulls, which have their early promise confirmed by above-average progeny weights, is offered to breed societies for further use.

Progeny testing. In addition to the Young Bull Proving Scheme, a continuing progeny-testing program of AI beef bulls on commercial cattle is carried out by Milk Marketing Board. This tests crossbred progeny from dairy cows at a central progeny station (Warren Farm) and calves from commercial suckler herds. The calves are tested through on-farm recording in association with the MLC. Differences in daily gain among these highly selected AI bulls tend to be small, but there are important variations in carcass weight and carcass conformation and, bearing in mind the large numbers of commercial progeny, the test program is highly cost-effective.

Ease of calving. An accurate progeny test of growth performance can be accomplished using 20 or so calves. However, even a moderately accurate rating of difficult calvings requires hundreds of calvings to be surveyed. Effectively this rules out herd testing, because the results are more misleading than they are helpful, and information on ease of calving from individual bulls can only be contemplated from surveys of results from the use of AI bulls, particularly on dairy cows. This is done, and even for the larger breeds some of the AI organizations have semen available from easy-calving bulls which is particularly important for the dairy producer.

Dairy bull assessments. Because purebred and beef-cross dairy calves provide together over 40% of the beef produced in the U.K., it is considered important to monitor the beef characteristics of black and white cattle. Within the national dairy-progeny-testing scheme all Friesian/Holstein female progeny are assessed for beef shape as well as for overall breed conformation. The value of these assessments are being tested in a trial where steer progeny from a group of bulls representing the complete range of beef shape are being reared and slaughtered for carcass evaluation.

On-farm performance testing. The most recent development has been the introduction of an on-farm performance-testing scheme for breeders with herds of 40 cows or more.

The objective is to develop an improved method of within-herd selection. The testing protocol requires that bulls are reared in groups of not less than 10 and all bulls within the group shall be born within a 2-month period, preferably less. Breeders with promising bulls born outside the main calving period are encouraged to submit them for testing at MLC performance test centers.

Bulls within a test group are housed together or in a similar accommodation and treated alike in terms of exercise, etc. In particular the feeding regime should allow equal opportunity to all bulls though feeds can be selected by the breeder according to normal rearing practice within the herd. The test procedure can start at any time before bulls are 7 months of age. Bulls are treated alike in the pre-test period, i.e., receive the same suckling management, are weaned at the same age or date, and are given similar postweaning feeding.

The tests are normally restricted to indoor feeding, and the test period lasts at least 6 months so that bulls are 400 days old by the end of test. Tests can be extended further to produce 500-day weights.

The bulls are weighed monthly, and measurements for withers height and ultrasonic scanning of backfat thickness are done at about a year of age. The final test reports are modeled on those published at central performance-test centers.

In 1981/82 for the first time the number of bulls undergoing on-farm performance testing within this scheme exceeded the number of central test; 435 bulls representing six breeds were tested in 33 farm groups. It is envisaged that this scheme will become relatively more important within the national British testing program, and that the breeders and breed societies may see these on-farm testing herds as the central point for future bull selection and breed improvement.

27

FRIESIAN CATTLE STRAIN COMPARISON: BEEF PERFORMANCE

Henryk A. Jasiorowski

In Europe, dairy herds are the principal source of calves for fattening, thus traits affecting beef performance have high economic value--in a comprehensive selection index for all-purpose dairy-beef cattle this value was estimated at about 40% (Fewson, 1978).

Within the world population of Friesian cattle, there are a series of local strains, differing in production and exterior traits that result from different economic and environment conditions and different methods of selection applied in various countries. Probably the greatest differences are between the Holstein-Friesian strain from North America that is selected for progress in milk traits and the Dutch Black-and-White strain that is utilized as a dual-purpose milk and beef cattle. Compared to the Dutch cattle the American Holstein-Friesian strain is about 100 kg heavier and is taller by an average of 10 centimeters (Oldenbroek, 1977). The Dutch strain is characterized by a more compact build and shorter legs, which gives the appearance of better musculature. These traits are considered an important indicator of carcass quality and value.

The beef utility of European strains of Black-and-White cattle has been compared with that of the Holstein-Friesian strain in studies in several countries. Generally, these studies show that the weight gains of Holstein-Friesian strain cattle during fattening are higher than those of the European types (Oldenbroek, 1977). However, the post-slaughter evaluation is usually to the disadvantage of Holstein-Friesian cattle; their carcasses are long and flat in comparison with those of the Black-and-White cattle, giving an impression of a poor meat fill (Cook et al., 1979; Oldenbroek, 1977).

The lower dressing percentage and poorer muscle marbling are probably the principal reasons why a preslaughter evaluation places the Holstein-Friesian cattle in a lower class than that of the European Black-and-White cattle (Weniger et al., 1967; Oldenbroek, 1977; Cook et al., 1979).

The important role of Black-and-White cattle in European beef production is the main reason that so much interest is shown in the beef features of different strains and breeds that are considered to be dairy breeds in North America. Thus, the testing of beef performance of different Friesian strains has been included in special international studies.

EXPERIMENTAL PROCEDURE
(The general principles of comparison of 10 Friesian cattle strains are described in another paper that I have prepared for this school.)

The experimental procedure for meat performance evaluation was similar to that used in the evaluation of the dairy performance, i.e., the experiment was done in the field under intensive feeding conditions. The majority of young bulls were to be fattened to 450 kg live weight. The system of nutrition was based mainly on corn silage, with green forage in the summer. Only the minimum amount of concentrates was given. Because of the field conditions under which this part of the experiment was done, the data could be collected only on body weight at the age of 12 months, final body weight before slaughter, and slaughter value score on live animals.

TABLE 1. LEAST SQUARE MEANS FOR LIVE WEIGHT, AGE AT SALE AND DAILY GAINS OF F_1 BULLS/FIELD CONDITIONS

Country (strain)	No.	Live weight in kg at birth	12 mo	at sale	Age at sale (days)	Daily gains in g 0-12 mo
e	5972	37.2	302	471	599	751
U.S.	560	38.2	304	471	591	756
Poland	1038	36.1	301	465	598	752
Canada	580	37.9	304	473	588	757
Denmark	562	37.6	301	471	604	750
Great Britain	498	36.8	298	469	605	745
Sweden	561	37.4	307	474	595	765
Gorman Fed. Rep.	583	37.4	299	474	604	745
Holland	519	36.5	295	468	614	734
Israel	536	37.3	308	474	592	771
New Zealand	535	36.8	304	473	596	764

Source: M. Stolzman, H. Jasiorowski, Z. Reklewski, A. Zarnecki (1981).

The intensive feeding test was performed on 180 heads of F_1 males representing all paternal strains (28 per strain); 50% of them were fattened to 450 kg and 50% to 550 kg of live weight.

The male calves for this test were purchased at the age of 8 weeks. Until 109 days old the calves were reared on milk replacer, concentrate, and hay. The experimental fattening started at the age of 110 days and terminated when the animal reached 450 or 550 kg of body weight.

During fattening the bulls were individually fed concentrates ad libitum with an addition of 1 kg of hay daily. Animals were starved for 24 hours before slaughter. After slaughter, the carcass quality and composition were determined.

GROWTH AND BEEF PERFORMANCE OF MALES UNDER FIELD CONDITIONS

Data presented here show preliminary results based on 5,972 F_1 bulls fattened in 73 herds. Table 1 shows the least square means for live weights at birth, at 12 months, and at sale, and daily gains from 0 to 12 months. The bulls were sons of 218 sires; each sire was represented by at least 10 sons. In the birth weight, the greatest difference was between the F_1 progeny of U.S. sires and Polish bulls; F_1 Holsteins were 6% heavier. The birth weights of British, Dutch, and New Zealand F_1 bulls also were below the average.

The level of daily gains achieved (754 g/day) is characteristic of those obtained under semiintensive feeding conditions on the Polish state-owned commercial farms where 1.5 to 2 kg of concentrates are fed daily.

At such feeding levels, the differences between paternal strains were small and not significant; not exceeding 4% in live weight at 12 months, at sale, and in daily gains. The average age of animals at sale was 599 days and the live weight was 471 kg.

GROWTH AND BEEF PERFORMANCE OF MALES UNDER INTENSIVE FEEDING CONDITIONS.

As mentioned earlier, 28 F_1 bulls of each paternal strain were fattened under intensive feeding conditions. The slaughter weight was 450 and 550 kg, with other treatments (feeding, etc.) being equal. For the purposes of this presentation, it was considered that the differences between the groups of different slaughter weight are not so important as differences between paternal strains, thus data were combined for both slaughter weights.

Table 2 shows the average daily gains. The highest daily weight gains (124 g) were obtained by bulls with sires from the U.S. Significantly lower weight gains were obtained from New Zealander, Danish, British, Swedish, Dutch, Polish and West German strains. The Israeli and Canadian groups produced daily weight gains significantly greater than those obtained by the Danish and New Zealander strains. The lowest gain was from the New Zealander group--124 g lower than that of the American group. The

TABLE 2. LEAST SQUARE MEANS FOR DAILY GAIN/INTENSIVE TRAIL, BULLS SLAUGHTERED AT 430 AND 550 KG OF LIVE WEIGHT.

Strain	n	Mean daily gain g		Mean net gain g	
		$\bar{x}$	SD	$\bar{x}$	SD
U.S.	27	1204.4*	18.9	648.0	9.6
Canada	29	1166.8*	18.6	640.1	9.4
Israel	28	1164.2*	18.9	646.2	9.6
Poland	29	1141.7	18.3	636.0	9.3
FRG	28	1140.0	18.6	616.4	9.4
Netherlands	28	1126.2	19.3	630.5	9.8
Sweden	28	1125.1	18.6	635.2	9.4
Great Britain	26	1112.0	18.8	626.2	9.5
Denmark	27	1086.2	18.3	603.1	9.3
New Zealand	30	1080.3	17.9	597.0	9.1
Average	280	1134.8	5.9	627.9	30

Source: Reklewski Z., H. Jasiorowski, M. Stolzman, A. de Laurance (1981).
*Significantly different from some other groups

mean daily net gains show a similar order. The best net gain was obtained by the American strain (648 g); the Israeli and Canadian strains were only slightly less.

The New Zealander and Danish strains showed net gains significantly lower than those obtained by the American, Israeli, Polish, Swedish, and Dutch Strains.

Carcass dissection data are shown in table 3. The highest dressing percentage was obtained by the British strain (58.4%); this group was significantly better than the Israeli, German, American, and Canadian groups. The lowest Black-and-White dressing percentage was shown by the American group, which was significantly less than the Swedish and Dutch groups.

There were no significant differences between groups in valuable carcass cuts, except for the New Zealander group which was significantly poorer than most of the others.

No differences were observed between the strains in terms of the percentage of lean meat in carcasses

The highest fat deposition in the carcasses was shown by the New Zealander and Danish groups, which differed significantly from the American, Swedish, German, and Israeli strains.

The lowest content of bones in a carcass-side was shown by the Polish strain; significantly higher proportion of bones was demonstrated by the American, Canadian, German, and Danish strains.

TABLE 3. LEAST SQUARE MEANS FOR CARCASS DISSECTION RESULTS: INTENSIVE FEEDING TRIAL, BULLS SLAUGHTERED AT 450 AND 550 KG OF LIVE WEIGHT

Country	n	Dressing % x̄	Dressing % SD	Valuable cuts % x̄	Valuable cuts % SD	Lean in carcass-side % x̄	Lean in carcass-side % SD	Fat in carcass-side % x̄	Fat in carcass-side % SD	Bones in carcass-side % x̄	Bones in carcass-side % SD
Great Britain	28	58.4*	0.28	62.6	0.19	65.0	0.48	18.4	0.48	16.7*	0.25
Netherlands	26	58.3*	0.29	62.6	0.20	64.7	0.49	18.3	0.47	17.0	0.26
Sweden	28	58.2*	0.28	62.6	0.19	65.3	0.48	17.5	0.48	17.2	0.25
Poland	29	57.8	0.28	62.2	0.18	65.1	0.47	18.4	0.49	16.5*	0.25
New Zealand	30	57.6	0.27	61.6*	0.18	65.3	0.48	19.7*	0.45	17.0	0.25
Israel	27	57.4	0.29	62.5	0.19	64.7	0.48	18.0	0.47	17.3	0.25
FRG	28	57.4	0.28	62.8	0.19	64.6	0.48	17.6	0.48	17.8*	0.25
Canada	26	57.4	0.28	62.5	0.19	63.9	0.48	18.3	0.47	17.9	0.25
Denmark	29	57.3	0.28	62.4	0.18	64.0	0.47	18.6*	0.46	17.4	0.25
USA	27	56.9	0.29	62.4	0.19	65.1	0.48	17.1	0.47	17.8	0.26
Average		57.7	0.09	62.4	0.06	64.6	0.45	18.2	0.44	17.3	0.08

Source: Reklewski A., H. Jasiorowski, M. Stolzman, A. De Laurance (1981).
*Significantly different from some other groups.

Ranking for beef performance can be done only for the intensive feeding trial. The field test was done at a low feeding level so that the bulls had no opportunity to show their genetic potential.

Beef performance can be characterized by many features; for this study the average daily gain of lean meat was the characteristic selected and the relevant data are presented in table 4. In this respect, the New Zealander strain received a much lower ranking than did most of the other groups, except for the Danish group, which had a low ranking also.

TABLE 4. RANKING OF STRAINS ACCORDING TO BEEF PERFORMANCE

| Strain | Mean daily gain of "lean meat" (g) | | |
	n	$\bar{x}$	SD
U.S.	27	417.0	7.6
Israel	27	414.0	7.6
Sweden	28	409.5	7.5
Poland	29	409.2	7.3
Canada	26	405.	7.0
Netherlands	26	405.3	7.7
Great Britain	28	403.2	7.5
FRG	28	397.6	7.5
Denmark	29	383.3*	7.3
New Zealand	27	375.5*	7.2
Average	230	402.1	2.4

Source: Reklewski Z., H. Jasiorowski, M. Stolzman, A. de Laurance (1981).
*Significantly different from some other groups.

The first two places fell to the U.S. and Israeli strains. The ranking of the strains for slaughter value did not change greatly when the classification was based on the intake of nutrients per 1 kg of "lean meat" gain.

The highest daily gains were obtained by the American, Canadian, and Israeli strains, which for many years have been selected for milk yield exclusively. This is probably related to the correlation that exists between milk production and the body weight. In general, the Holstein-Friesian strains (U.S., Canada, and Israel) showed very good beef performance.

REFERENCES

Cook, K. N., N. Newton, M. Jennifer. 1979. A comparison of Canadian Holstein and British Friesian steers for the production of beef from 18-month grass/cereal system. Animal Production 28:1.

Fewson D. 1978. Weiteretwicklung der Zucht planung bein Rind Zuchtungskunde.

Oldenbroek, I. J. 1977. Vergleich nordamerikanischer Holstein Friesians mit niderlandischer Schwarz und Rotbunten - Tierzuchter.

Reklewski, Z., H. Jasiorowski,, M. Stolzman, and A. de Laurance. 1981. Estimation of the slaughter value of various strains of Friesian cattle. EAAP Annual Meeting, Zagreb.

Stolzman, M., H. Jasiorowski, Z. Reklewski, A. Zarnecki, and G. Kalinowska. 1980. Friesian cattle in Poland. Preliminary results of testing different strains. World Animal Review, 38:8.

Weniger, H., R. Riechers, W. Dieteret, J. Zeddies and Engelke. 1969. Deobachtungen und Betrachtungen uber den Einsatz von Holstein Friesian Bullen in der Zucht des Deutschen Schwarz-buntes Rindes VI. Tierzuchter.

28

RESULTS FROM LARGE-SCALE BEEF BREED EVALUATION TRIALS IN NEW ZEALAND AND EVIDENCE FOR GENOTYPE BY ENVIRONMENT INTERACTIONS

R. L. Baker

INTRODUCTION

Over the last decade the world has witnessed a revolution in animal breeding. Many of the advanced agricultural countries, and to an increasing extent many of the third world countries, actively promoted importation of new breeds of livestock designed to improve animal productivity. North and South America, Britain, South Africa, Australia, and New Zealand all joined the exotic breed bandwagon.

Why this sudden awakening of interest in new breeds? A number of reasons can be cited. Greater travel and communication have led to growing knowledge of other countries-- their people and their livestock. Widespread adoption and publication of performance records has focused attention on breed differences. Advances in the diagnosis and prevention of disease have overcome many of the serious obstacles to movement of stock.

At least with cattle, development of semen storage techniques has paved the way for convenient and inexpensive transport of genetic material. But perhaps the most important single factor has been the universal quest, dictated by economic pressures, for greater productive efficiency. The suitability of traditional breeds--under changing patterns of production and, particularly, of consumer demand--has rightly been called into question.

New breeds can confer valuable flexibility in adapting to changing needs. Undoubtedly the worldwide shift in consumer demand from fat to lean beef has been one reason for the interest in the larger European cattle compared to the traditional British beef herds. One must have some sympathy for the past breeders of Hereford, Angus, and Shorthorn cattle. Their very success in changing their animals to meet consumer requirements of the time, i.e., early maturity at relatively light carcass weights, left them at a distinct disadvantage in view of modern demands relative to European breeds that have evolved less rapidly from the muscular draft cattle of old.

Opinion has been sharply divided in New Zealand and many other countries as to the merits of the new versus the old. Surely, however, none can dispute 1) the possibility that new breeds may possess superior productive attributes compared with present stock, and 2) the need for factual performance information on which to base rational choice among breeds.

This philosophy underlies the New Zealand Government policy that resulted in the introduction of certain promising new breeds of both cattle and sheep (with stringent quarantine safeguards), so that their potential contribution to improving livestock productivity could be assessed. A logical approach to exotic breed evaluation involves:

- Specification of local improvement needs in relation to local farming systems.
- Selection of the most suitable overseas breeds.
- Importation with minimum disease risk.
- Evaluation, in comparison with local breeds.
- Development of effective methods of utilization.
- Commercial exploitation, if clear-cut advantages are established.

The results of ten years of research using such an approach with cattle are reported here. While several new breeds of sheep also were imported into New Zealand, a suspected case of scrapie in one of the animals resulted in complete slaughter of all of these imported animals and their derived progeny.

THE NEW ZEALAND LIVESTOCK INDUSTRY

It is important at the outset to document some of the important differences between livestock production systems in New Zealand and North America. The New Zealand livestock industry is based on an intensive year-round grazing regime, with pasture being supplemented by hay and silage in periods of winter and (sometimes) summer feed shortages. Forage or root crops are sometimes fed, but supplementary grain feeding is uncommon. This pastoral system is dictated by a temperate climate favoring some pasture growth through most of the year; terrain that is frequently hilly but can support improved pastures through aerial top-dressing and most stock management; large farm size; high costs of buildings, of concentrate feeds, and of farm labor; and above all, the need to keep production costs as low as possible to be able to compete on distant markets. The national economy is almost totally dependent on the export of animal and (to a lesser extent) forestry products. About 60% of New Zealand's total beef output is exported, with considerable diversity of beef markets and their requirements.

Reliance on pasture feeding has three important consequences: 1) animal performance is naturally lower than

under more intensive production systems; 2) maximum utilization of seasonally varying pasture supply is achieved by
concentrating calving and lambing in the spring and slaughtering surplus beef animals (usually steers) at 18 to 21
months, prior to their second winter--actual slaughter age
is governed more by feed availability on the farm than by
degree of "finish"' and 3) individual feed intakes, hence
feed conversion efficiency, cannot be measured directly.
Since maintenance needs are relatively higher than those
under intensive systems, size of the breeding animal assumes
greater significance. This poses real problems in performance comparisons among breeds differing widely in body
size.

Some other salient features of New Zealand's suckler
beef industry are: large self-replacing herds that are
closely integrated with sheep production; breeding herds
usually maintained on the less productive hill country and
supplying breeding or fattening stock (mainly steers) for
more favored lowland farms; and little use of artificial
insemination.

The dominant breeds in a population of about 2 million
beef cows are the Angus (65%) and the Hereford (20%), other
established breeds being the Shorthorn, Galloway, Red Devon,
and Red Poll. Introductions since the mid-1960s, largely
through semen imports, include the Simmental, Charolais,
Limousin, Maine Anjou, South Devon, Blond d'Aquitaine,
Murray Grey, Brown Swiss, and Italian White breeds, all of
which are now represented by local breed societies. A few
Bos indicus breeds and derived breeds have been introduced,
including, for example, Brahman, Sahiwal, Santa Gertrudis,
and Brangus.

Traditionally, the dairy industry has been sharply
divorced from beef production (apart from cull cows for
slaughter), with most male and surplus female dairy calves
being slaughtered for export as 3 to 4 day old veal (bobby
calves). The Jersey and the Friesian (Holstein) contribute
almost equally to New Zealand's 2.1 million dairy cows,
along with a few of Milking Shorthorn and Ayrshire. Concomitant with a swing from Jersey to Friesian cows and high
beef prices, increasing numbers of dairy-bred stock sired by
Friesian or Hereford bulls are reared for beef on dairy or
fattening farms. The potential of beef x dairy crossbred
females as breeding cows in suckler herds has so far been
little exploited. In contrast to beef animals, about 50% of
all dairy cows are artificially inseminated.

EVALUATION TRIALS

Three major New Zealand experiments are summarized
here; the first involves the two main local beef breeds, the
others deal with a range of local and exotic sire breeds
crossed with beef cows or dairy cows, respectively. Results

of these and of other New Zealand breed evaluation trails have previously been reviewed by Carter (1975) and Baker and Carter (1982).

Reciprocal Crossbreeding in Angus and Hereford Cattle

The first formal evaluation of the relative merits of the two main beef breeds in New Zealand was begun in 1968, some 100 years or so after the original importation of these breeds from Britain.

In a crossbreeding experiment, about 50 Angus and 50 Hereford bulls sired some 1200 purebred and reciprocal crossbred calves born from 1969 to 1972 (Carter, 1975; Baker and Carter, 1976). Base Hereford cows were 5 to 10% heavier than were Angus, with little breed difference in calf-weaning rate. Relative to Angus calves, straightbred Herefords were born 8 days later and 4 kg heavier, reflecting longer gestation periods (the crosses being intermediate). Superior maternal performance of the Angus dam was shown by 3% heavier weaning (5-month) weights of her calf relative to the Hereford. Postweaning weights and carcass weights were higher (about 8%) for the Hereford. Comparison of the average of the purebreds with the reciprocal crosses indicated heterosis of about 3% of 6% in weights at birth, weaning, and 20 months.

Table 1 summarizes reproductive performance (to 1979) of the heifers born 1970 to 1972, reflecting 1075 matings of 274 females first joined (bred) at 15 months of age. Compared to contemporary straightbred females, crossbred heifers and cows were 8% heavier and, when mated to the same bulls, weaned calves were 8% heavier. The superiority of crossbred females for calf weaning rate was consistent with overseas evidence averaging 13%, but greater at a younger age. The overall superiority of crossbred over straightbred dams amounted to about 22% more weight of weaned calves, and about 18% more 20-month steer carcass weight, per cow mated. As reflected in total lifetime calves weaned per heifer joined (table 1), crossbred cows have survived longer than purebred cows. Angus cows have shown greater longevity than Herefords and, in contrast with the results for the foundation cows, have shown superior weaning percentage (10% advantage in cows 3 years and older).

TOPCROSS COMPARISON OF LOCAL AND EXOTIC SIRE BREEDS

Beef Breed Evaluation Trial (Beef herds)

The Ruakura Beef Breed Evaluation (BBE) trial aims to compare the performance of the main local beef (Angus, Hereford) and dairy (Friesian, Jersey) with seven 'exotic' (South Devon, Charolais, Limousin, Blond d'Aquitaine, Simmental (four strains), Maine Anjou, Chianina) breeds of cattle in crosses to Angus and Hereford cows. Assessment is

TABLE 1. JOINING WEIGHTS AND CALVING PERFORMANCE OF ANGUS, HEREFORD, AND CROSSBRED COWS, 1972 to 1979

Cow Age	Trait	Angus (A)	Hereford (H)	Crossbred (AH+HA)	Heterosis[a] (%)
2 yr	Joining wt (kg)	215	210	230	8
	Calving %	72	74	85	16
	Weaning %	63	61	72	16
3 yr+	Joining wt (kg)	342	351	374	8
	Calving %	85	76	89	11
	Weaning %	80	70	85	13
Total calves weaned per heifer joined		4.0	3.2	4.2	17

[a]Advantage of crossbred over average of straight-bred cows.

based on both meat production of the steers and overall reproductive and maternal performance of the female progeny.

Semen from the same bulls was used at each of three locations that represented a wide range of environments and of animal performance. Tokanui (Waikato, near Hamilton – 500 Angus cows) has the most favorable conditions; Goudies (near Rotorua – 850 Angus and 500 Hereford cows) the most severe; Templeton (near Christchurch – 300 Angus cows) is intermediate. From 1971 to 1977, 191 different sires (averaging 20 [range 12 to 32] of each breed), were selected to generate test progeny by AI.

Steers were slaughtered at about 20 months, except that a balanced half from Goudies were retained to 31 months. Most animals were slaughtered through commercial plants. The carcass information varied, but included weight, grade, and an estimate of subcutaneous fat depth for all steers. All yearling heifers were bred to Angus or Shorthorn bulls, while subsequent mating seasons (mainly by AI) included Blond d'Aquitaine, Limousin, Simmental, Maine Anjou, and Charolais breeds as terminal sires. Females were culled if not impregnated after two mating seasons.

This study has been reported by Carter, et al. (1975); Baker, et al. (1981); and Baker, et al. (1981).

Dairy Beef Trial

Friesian cows in cooperating dairy herds were artificially bred to ten bulls of each of three different breeds in four successive years (1973-76), the Friesian breed being repeated each year (Everitt et al., 1978a). The nine sire breeds were the locally established Friesian, Hereford, Angus, and Red Devon, and the 'exotic' Limousin, Simmental,

South Devon, Blond d'Aquitaine and Charolais. Steer progeny were allocated (after weaning) among a number of grazing farms for rearing to slaughter (commercially) at about 20 months of age. Weaned heifer calves were transferred at about three months of age to a large hill-country farm (Tahae) and reared as normal herd replacement. They were joined with Angus bulls at 15 months of age; those subsequently having a live calf were joined again to calve when 3-years old.

Results of this trial have been reported by Everitt et al. (1978a, b); Dalton et al. (1980); Everitt et al. (1980); Jury et al. (1980).

SYNOPSIS OF RESULTS IN TOPCROSS COMPARISONS

Calving Performance (base cows)

Table 2 summarizes information on difficult births (observed or inferred), calf survival, birth weight, and gestation length, as related to Angus, Hereford, or Friesian cows. Relative to the local sire breeds, all of the exotic crosses were more prone to difficult births. Calving difficulties tended to be more frequent among Hereford and Friesian than among Angus cows, especially when mated to Charolais, Blond d'Aquitaine, Simmental, or Maine Anjou sires. Friesian sires caused serious dystocia only with Hereford dams, while South Devon sires induced more dystocia with Friesian than with Angus or Hereford dams. Breed rankings for calf losses to weaning (including calves that died at birth) were similar to those for calving difficulty, except for high survival rates of Friesian x Hereford, Maine Anjou x Hereford and South Devon x Friesian calves relative to associated dystocia levels. In general, exotic sire breeds were more similar to local breeds in terms of calf losses than for difficult births.

Calves sired by exotic breeds were carried longer than those sired by local breeds. The Chianina and Blond d'Aquitaine in the beef herds, and the Limousin when crossed with Friesian dams represented the extremes; their calves had gestation lengths about 9 to 10 days longer than those of Angus crosses. Birth weights likewise were higher for exotic crosses than local crosses. the heaviest calves were sired by the Chianina, Maine Anjou, and Charolais breeds, while the lightest were Jersey crosses.

Growth and Carcass Performance

Live weights at 20 months of age (all progeny) and carcass weights of steers are summarized in table 3. The carcass weight rankings from the BBE trial are pooled across the two dam breeds and two slaughter ages (20 and 31 months) since no marked interactions with sire breed were apparent. Breed rankings were broadly similar out of beef and Friesian

TABLE 2. CALVING PERFORMANCE (3 YEARS AND OLDER) OF BASE COWS[a]

Sire breed	Calving difficulty (%)			Calf losses to Wng (%)			Birth wt (kg)		Gestation length (days)	
	A dam	H dam	F dam	A dam	H dam	F dam	A&H dams[b]	F dam	A&H dams[b]	F dams
Angus	4	4	8	6	6	7	28	32	279	280
Hereford (H)	3	2	5	5	6	5	30	34	283	286
Friesian (F)	4	14	8	4	4	8	31	35	279	281
Jersey	2	5	–	4	5	–	26	–	281	–
South Devon	6	5	15	6	5	8	33	38	285	287
Charolais	15	24	26	12	18	18	34	40	284	285
Limousin	5	11	9	6	11	8	31	37	286	289
Blond d'Aquitaine	8	17	16	6	9	14	33	37	288	290
Simmental	8	16	14	7	12	10	33	38	285	287
Maine Anjou	13	24	–	9	8	–	35	–	286	–
Chianina	11	10	–	8	7	–	36	–	289	–
Red Devon	–	–	8	–	–	10	–	35	–	285

aApproximately 5000 calvings of Angus and Hereford dams and 2200 of Friesian dams (excluding abortions and twin calves).

bIncludes data from 2-year-old dams.

dams. Progeny of exotic breeds grew faster than Hereford (or Angus) cross calves, although difference was small for Limousin crosses. The fastest growing calves were sired by Charolais, Simmental, and Maine Anjou bulls, followed very closely by Friesian bulls. High dressing percentages for Blond d'Aquitaine, Limousin and Charolais cross calves improved their carcass weight rankings relative to 20-month live weights. Conversely, Jersey crosses (from beef cows) and purebred Friesian steers, with lower dressing percentages, showed relatively low carcass-weight rankings.

In general, the exotic crosses were leaner and a smaller proportion graded "prime" than was the case for Hereford and Angus crosses. Jersey crosses, straight bred Friesian steers, and to a lesser extent Friesian crosses, were downgraded primarily for undesirable conformation or fat cover. In the BBE, grading of all exotic crosses and the Friesian crosses was improved considerably when slaughtered at higher carcass weights. Through either better feeding to 20 months of age (e.g., at Tokanui) or deferred slaughter to 31 months of age (Goudies), the overall proportion grading prime increased from 21% to 62%, as carcass weights increased. In contrast to the BBE results, few of the steers out of Friesian dams graded "prime", most being classed as "manufacturing grade." The Hereford x Friesian steers graded best, and the straightbred Friesian worst, leading Everitt et al. (1980) to suggest that the Friesian male is eminently suited to bull-beef production, in New Zealand where producer payment is related solely to carcass weight, disregarding the subjective vagaries of carcass grade. Similar conclusions may apply to crosses of the larger, leaner 'exotic' breeds.

Carcass weight produced per cow calving (table 3) is calculated by multiplying carcass weight by percent of calf survival from birth to weaning. Weaning to slaughter losses were relatively small and constant across breeds in the BBE trial, but there was some evidence that Charolais-Friesian crosses had lower survival from weaning to slaughter (Everitt et al., 1980).

In matings to Angus and Hereford cows, the Blond d'Aquitaine and the Friesian, closely followed by the Maine Anjou and South Devon, produced the greatest carcass output. The Charolais was penalized by its high calf losses. The Limousin excelled as a crossing sire with Friesian cows, but not with Angus or Hereford cows. The South Devon and Simmental breeds were also productive as terminal sires with Friesian cows, although Everitt et al. (1980) favor the Hereford x Friesian cross because of its superior carcass grading.

Performance of crossbred Females

Reproductive and maternal performance of the female progeny is summarized in table 4. The BBE data relates to 2163 females born 1973 to 1977, with a total of 6187 matings and 4537 calvings (Baker, Carter, and Muller, 1981). The

TABLE 3. GROWTH PERFORMANCE AND CARCASS WEIGHT RANKINGS (%)
Note: (S) DENOTES STRAIGHT-BRED CALVES, ALL OTHER BREED GROUPS ARE CROSSBREDS

Sire Breed	20 mo. wt (both sexes)		Per head		Per cow calving	
	A&H dams	F dams	A&H dams[a]	F dams	A&H dams[a]	F dams
Angus (A)	94 (S)	99	89 (S)	99	88 (S)	97
Hereford (H)[b]	100b	100	100b	100	100b	100
Friesian (F)	107	107 (S)	106	104 (S)	107	101 (S)
Jersey	96	–	91	–	92	–
South Devon	106	108	106	105	105	103
Charolais	108	111	110	111	101	95
Limousin	101	104	103	107	101	104
Blond d'Aquitaine	106	107	109	106	108	96
Simmental	108	108	107	107	104	102
Maine Anjou	109	–	110	–	105	–
Chianina	105	–	107	–	104	–
Red Devon	–	104	–	101	–	95

a20- and 30-month carcass weights combined.

bHereford x Angus plus Angus x Hereford crosses in BBF.

TABLE 4. PERFORMANCE OF CROSSBRED FEMALES

NOTE: (S) DENOTES STRAIGHTBRED FEMALES; 'PRODUCTIVITY' ESTIMATES OF CALF WEANED PER COW MATED

Sire breed	Age at puberty (days)		Weaning % ranking[a]			Productivity ranking %[a]			Cow weight ranking(%)[a]		
	A&H dams	F dams	A&H dams (Goudies)	A dams (Tok)	F dams	A&H dams (Goudies)	A dams (Tok)	F dams	A&H dams (Goudies)	A dams (Tok)	F dams
Angus	413(S)	396	83(S)	96(S)	102	80(S)	96(S)	104	93(S)	94(S)	99
Hereford(H)[b]	392[b]	373	100[b]	100	100	100[b]	100	100	100[b]	100	100
Friesian(F)	353	323(S)	107	112	88(S)	125	129	99(S)	104	102	103(S)
Jersey	340	–	103	105	–	110	113	–	93	94	–
South Devon	402	411	94	100	86	101	108	97	107	105	108
Charolais	430	408	84	93	91	88	98	94	106	104	113
Limousin	436	387	78	104	75	78	105	71	98	96	102
Blond d'Aquitaine	443	418	80	101	74	81	108	77	102	106	110
Simmental	414	375	81	105	92	87	121	96	103	102	105
Maine Anjou	404	–	94	103	–	99	109	–	105	108	–
Chianina	–	–	70	93	–	88	98	–	102	106	–
Red Devon	–	367	–	–	103	–	–	110	–	–	104

[a]Two- to five-year-old cows at Goudies and Tokanui; two- and three-year-old cows only out of Friesian dams.

[b]Hereford x Angus plus Angus x Hereford crosses at Goudies.

dairy-beef trial covers 881 yearling heifers, 1532 matings, and 1194 first or second calvings at Tahae (Dalton and Jury unpublished). Only 51% of the 881 yearling heifers originally mated survived to the second calving. Sire breed, differences in age at puberty, and weaning rate were broadly similar in the two experiments, particularly in comparisons of the Goudies and Tahae locations. Late sexual maturity of the Limousin, Blond d'Aquitaine and Charolais crosses adversely affected calving success at two years, especially under harsh conditions (Goudies, Tahae), and this contributed largely to their relatively low average lifetime performance. In the BBE trial, breed differences in productivity (weight of calf weaned per cow mated) were greater at younger ages than at older ages, although breed rankings were broadly similar (Baker, Carter, and Muller, 1981). Performance of all 'exotic' crosses was relatively much better under the favorable conditions at Tokanui than in the harsh Goudies environment (table 4), indicating that these crosses need high levels of feeding to express their reproductive potential. The relative productivity superiority of the Tokanui environment over the Goudies environment was greatest for Simmental, Limousin and Blond d'Aquitaine crosses and straightbred Angus; least for Hereford-Angus, Friesian, and Jersey crosses; and intermediate for the remaining crosses.

Except for the Limousin and Blond d'Aquitaine crosses, performance of all the dairy-bred females at Tahae was satisfactory; highest productivity was achieved by Angus and Red Devon crosses. These findings are conditioned, however, by the culling policy and lack of data for more than two calvings.

In comparing the relative productive efficiency or profitability of different breeds and crosses, allowance should be made for the higher weight-related maintenance feed costs of larger animals. Under grazing conditions, annual cow requirements comprise about 70% of total feed used when progeny are slaughtered at about 20 months of age (Carter, 1982). Taking productivity per kg cow live weight as crude measure of calf production efficiency, the results in table 5 show a substantial and quite similar superiority of Friesian and Jersey crosses from beef dams, and of Red Devon and Angus crosses from Friesian dams, relative to other breed types.

CONCLUSIONS AND IMPLICATIONS

One of the most significant findings is the high level of heterosis manifest for cow productivity, combining the effects on reproductive rate, calf survival and preweaning growth. The value of 22% for the Angus-Hereford cross reported here is higher than than (14.8k%) found in crosses among the Angus, Hereford, and Shorthorn breeds in the

TABLE 5. PRODUCTION EFFICIENCY IN TERMS OF WEIGHT OF CALF WEANED PER COW JOINED PER KG OF COW LIVE WEIGHT FOR CROSSBRED COWS

Sire breed	A&H dams (Goudies)	A dams (Tokanui)	F dams
Angus (A)	86(S)	102 (S)	105
Hereford (H)	100	100	100
Friesian (F)	120	126	96 (S)
Jersey	118	120	-
South Devon	94	103	90
Charolais	83	94	83
Limousin	80	109	70
Blond d'Aquitaine	79	102	70
Simmental	85	119	91
Maine Anjou	94	101	-
Chianina	86	93	-
Red Devon	-	-	106

U.S.A. (Gregory and Cundiff, 1980). This supports other evidence (Barlow, 1981) of greater expression of heterosis, for reproductive traits at least, under less favorable environments. Performance, and presumably nutritional levels, were higher in the American as compared to the New Zealand trials.

In summarizing the results of the breed evaluation trials, it is clear that no one breed is "best" for all important beef production attributes and for all environments. For example, rapid progeny growth is important to the finisher, but may entail considerable real costs to the calf producer in terms of increased calf mortality and greater calving difficulty. All crossbred groups in the BBE trial nevertheless produced more carcass weight per cow calving than did the straight Angus; and all except the Jersey cross out-produced the Hereford-Angus cross. Thus, despite the fact that large exotic breeds were associated with higher calving losses than were local breeds, such losses can be more than offset by superior growth potential and high carcass weights. In addition, calving problems can be appreciably reduced by using 'exotic' sires on only mature cows, by judicious precalving cow management, and by choosing progeny tested sires showing low incidence of dystocia.

Breed x environment interactions (i.e., change of breed rankings in the different environments) are particularly evident for reproductive and maternal performance. Successful mating of yearling heifers is clearly contingent on their attaining puberty before the end of the limited spring mating period, and this in turn is influenced both by breed and by nutrition. Thus, the later sexually maturing exotic breed crosses are at a greater disadvantage relative to local breeds, (especially dairy crosses) under conditions of feed restriction (Goudies, Tahae) as compared to a more favorable environment (Tokanui).

The importance of the genotype-by-environment inter-action for overall reproductive and maternal performance in terms of weight of calf weaned per cow joined was demonstrated by comparisons of breed rankings in the harsh Goudies environment with those in the more favorable Tokanui environment (table 4). The Friesian and Jersey crosses excel at both locations. Use of crossbred cows of Chianina, Limousin, Blond d'Aquitaine or Simmental parentage clearly cannot be recommended under tough farming conditions in New Zealand. But when given favorable conditions, all of the exotic-cross cows either matched or out-produced the Hereford x Angus. In particular, the Simmental cross ranked second at Tokanui, but only seventh at Goudies; conversely the Hereford-Angus cross raked fourth at Goudies, but eighth at Tokanui.

The importance message here, supported by evidence from overseas is that breeds and crosses must be chosen to suit the environment and management system. In general, breed rankings for most productive traits recorded in these New Zealand trials concur with those from similar large scale comparisons in the U.S. (Cundiff, 1982) and Europe (Vissac, Foulley, and Menissier, 1982). Table 6 shows two composite traits that permit evaluation of breeds in terms of their merits as terminal sires or as sires of crossbred cows, using data from the New Zealand BBE trial and for those breeds in common in the "germ plasm evaluation" at Clay Center, Nebraska (U.S. Meat Animal Research Station). The Friesian and the Blond d'Aquitaine were not tested in the Clay Center program, but they did, however, test seven breeds that had not been evaluated in New Zealand (i.e., Red Poll, Brown Swiss, Gelbvieh, Tarentaise, Pinzgauer, Sahiwal, Brahma). In both cases these evaluation trials involved mating sire breeds to Angus and Hereford cows.

The steer carcass weight per cow calving from Clay Center results were calculated using the carcass weight measured at a standard age (465 days) and the total calf losses from birth to weaning. In the Clay Center study, the calf losses for Hereford-cross calves (2%) were lower than in our N.Z. trials (5%). The Limousin illustrates well the impor-tance of the criteria of evaluation on breed ranking. In the N.Z. trials, the Limousin ranked relatively poorly for carcass weight--which is the main basis of payment to New Zealand beef producers. By contrast, under high-energy feeding at Clay Center, Limousin crosses excelled in effi-ciency of food conversion to lean meat, which just could not be evaluated under our pasture feeding conditions.

The first-cross cow performance rankings under good New Zealand conditions are reasonably consistent with the Clay Center results. Two exceptions are the Chianina crosses, which have excelled at Clay Center but have performed very poorly in New Zealand. Conversely, Jersey crosses have per-formed relatively better in New Zealand than at Clay

TABLE 6.   COMPARISON OF NEW ZEALAND AND CLAY CENTER (U.S.)
BREED EVALUATION RESULTS

Breed or cross	As terminal sires: steer carcass wt per cow calving		First-cross cows(2yr+): weight of calf weaned per cow joined		
	N.Z.	U.S.A.	N.Z. (harsh)	N.Z. (good)	U.S.A.
Angus purebred	88	93	80	96	-
Hereford-X	100	100	100	100	100
Friesian-X	107	-	125	129	-
Jersey -X	92	88	110	113	104
South Devon-X	105	98	101	108	105
Charolais-X	101	98	88	98	102
Limousin-X	101	98	78	105	100
Blond d'Aquitaine-X	108	-	81	108	-
Simmental-X	104	97	87	121	109
Maine Anjou-X	105	103	99	109	112
Chianina-X	104	101	88	98	114

Center. The Clay Center results, however, include more calvings (2 through 8-year olds) than in the New Zealand results to date (2 through 5-year olds), suggesting that the superiority of the Jersey crosses may decline as they get older. In fact, the early Clay Center results (after only 3 calvings) ranked the Jersey crosses higher than their present rankings in a more complete assessment of lifetime production.

An important deficiency of the New Zealand results to date is the lack of information on feed consumption for steer finishing or, more importantly, for cows. This information is required to make final definitive decisions on the total herd productivity or profitability when the herd is composed of different breeds or crosses. Research is being initiated to attempt to answer this question in New Zealand, but measuring feed intake is difficult and expensive under extensive grazing conditions. For the present, dividing weight of calf-weaned-per-cow-joined by kg-cow-live weight is a crude way to adjust for weight-related maintenance costs; this statistic still suggests worthwhile superiority of a number of the crossbred cow types relative to the more traditional purebred Angus or Hereford-Angus cross (table 5).

It would take another paper to document the steps now being taken in New Zealand to utilize those breeds that we have identified as showing promise to increase beef cattle productivity. In brief, however, we are evaluating a number of 2- and 3-breed synthetics (composites), trying to encourage greater use of dairy-beef cross cows mated to terminal sires, and evaluating a simple two-breed rotation.

In terms of commercial application of these breeds, crosses, and different, more efficient breeding systems, we still have a long way to go. For many years, the commercial development of these new exotic cattle breeds was based on selling stock at inflated prices to our Australian neighbors. This bubble has now burst, and those breeders of exotic cattle who have stayed in the game are starting to ask relevant questions about how to set up well-based genetic improvement programs and where their breed may fit into the overall industry scene. There seems little doubt that profitable beef production in New Zealand in the future must rely increasingly upon crossbreeding systems designed to ensure efficient utilization of breed resources.

REFERENCES

Baker, R. L. and A. H. Carter. 1976. Influence of breed and crossbreeding on beef cow performance. Proc. of Ruakura Farmers' Conference. pp 39-44.

Baker, R. L. and A. H. Carter. 1982. Exploitation of breed differences: Implications of experimental results with beef cattle in New Zealand. Proc. 1st World Congress on Sheep and Beef Cattle Breeding, Massey University, N.Z., Nov. 1980, Vol. 1. (in press).

Baker, R. L., A. H. Carter and C. A. Morris. 1981. Evaluation of exotic breeds of beef cattle in beef herds. Aglink, FPP81 (2nd revise). Media Services, MAF, Private Bag, Wellington, New Zealand.

Baker, R. L., A. H. Carter and J. P. Muller. 1981. Performance of Crossbred cows in the Ruakura beef breed evaluation trial. Proc. of N.Z. Society of Animal Production 41:254-266.

Barlow, R. 1981. Experimental evidence for interaction between heterosis and environment in animals. Animal Breeding Abstracts 49:715-737.

Carter, A. H. 1975. Evaluation of cattle breeds for beef production in New Zealand - A review. Livestock Production Science 2:327-340.

Carter, A. H. 1982. Efficiency of production in the pasture-animal grazing complex. Proc. 1st World Congress on Sheep and Beef Cattle Breeding, Massey University, N.Z., Nov. 1980, Vol 1. (in press).

Carter, A. H., J. P. Muller and R. L. Baker. 1975. Exotic breeds for beef herds: Preliminary results. Proc. of Ruakura Farmers' Conference. pp 19-24.

Cundiff, L. V. 1982. Exploitation and experimental evaluation of breed differences. Proc. of 1st World Congress on Sheep and Beef Cattle Breeding, Massey University, N.Z., Nov. 1980, Vol. 1. (in press).

Dalton, D. C., K. E. Jury, G. C. Everitt and D. R. H. Hall. 1980. Beef production from the dairy herd. III. Growth and reproduction of straight-bred and beef-cross Friesian heifers. N.Z. Journal of Agricultural Research 23:1-10.

Everitt, G. C., K. E. Jury, D. C. Dalton and J. D. B. Ward. 1978a. Beef production from the dairy herd. 1. Calving records from Friesian cows mated to Friesian and beef bred bulls. N.Z. Journal of Agricultural Research 21:197-208.

Everitt, G. C., K. E. Jury, D. C. Dalton and J. D. B. Ward. 1978. Beef production from the dairy herd. II. Growth rates of straight-bred and beef-cross Friesian steers and heifers up to 4 months of age in several environments. N.Z. Journal of Agricultural Research 21:209-214.

Everitt, G. C., K. E. Jury, D. C. Dalton and M. Langridge. 1980. Beef production from the dairy herd. IV. Growth and carcass composition of straight-bred and beef-cross Friesian steers in several environments. N.Z. Journal of Agricultural Research 21:11-20.

Gregory, K. E. and L. V. Cundiff. 1980. Crossbreeding in beef cattle: Evaluation of systems. Journal of Animal Science 51:1224-1242.

Jury, K. E., G. C. Everitt and D. C. Dalton. 1980. Beef production for the dairy herd. V. Growth of steers and heifers in different environments to 600 days of age. N.Z. Journal of Agricultural Research 23:21-25.

Vissac, B., J. L. Foulley and F. Menissier. 1982. Exploitation of breed differences: Implications of experimental results with beef cattle in Europe. Proc. 1st World Congress on Sheep and Beef Cattle Breeding. Massey University, N.Z., Nov. 1980, Vol. 1 (in press).

29

IMPLICATIONS OF NEW ZEALAND BEEF-CATTLE-BREEDING RESEARCH WITH SELECTION AND ASSOCIATED FACTORS

R. L. Baker

INTRODUCTION

In my previous paper I discussed New Zealand results on breed evaluation and crossbreeding with beef cattle. The other important way in which improvement through breeding can be achieved is selection within a population, whether it is a straightbred or a synthetic herd. In this paper I would like to share with you some of our research results pertaining to selection in beef cattle. Some of our results are compatible with similar North American studies; others that are not may challenge you to think about some currently accepted practices.

MEASUREMENT OF GROWTH

Because of its predominant influence on profitability of beef production, growth rate has received major emphasis in experimental studies in New Zealand as elsewhere. A basic problem here is choice among alternative measures of growth performance, such as daily gains over specified periods, live weights at specified ages, age-corrected weights at specified times, or some combination of these.

In a selection experiment and associated progeny test herd with grazing Angus cattle, Carter (1971) studied the relationship of progeny growth performance with live weight records of their sires; results are summarized in table 1.

The conclusion that yearling (14-month) weight is a more effective within-herd selection criterion than weaning weight or postweaning gain for improving progeny live weight and carcass production has been borne out by subsequent studies (Baker and Carter, 1975). These findings strongly suggest that breeds and crosses should be compared in terms of "weight-for-age", rather than for weight gains over limited periods, at least for grazing cattle.

TABLE 1.   REGRESSION OF PROGENY PERFORMANCE ON SIRE'S OWN
PERFORMANCE[a]

Progeny performance	Sire trait		
	W	G	Y
Weaning (5-month) weight = W	0.12*	-0.02	0.09
Weaning to 14-month gain = G	0.11*	0.01	0.12*
Yearling (14-month) weight = Y	0.22*	-0.02	0.21*
Preslaughter (20-month) weight; steers only	0.33*	-0.14	0.30*
Repeatability of progeny tests	0.36*	0.36*	0.43*

*P<0.05
[a]Based on 55 Angus sires with an average of 18 recorded
progeny in selection herds and 29 of the same sires
averaging 24 progeny each in a test herd (adapted from
Carter, 1971).

SELECTION FOR GROWTH

Practical New Zealand evidence on the responses achieved to growth selection come from two long-term experimental
selection programs (both closed to outside breeding with
calves weaned at 4 to 5 months of age) and a large open-
nucleus group breeding program.

Commencing with calves born in 1962, a selection at
Waikeria with Angus cattle has been based on "corrected" 14
-month live weight in one line and on actual weight gain
from weaning to 14 months in a second line. Five yearling
bulls are used in each 125-cow line each year, some of these
bulls being subsequently mated in a 180-cow progeny test
herd (Carter, 1971).

At Waikite, selection was initiated in 1971 (Baker,
Carter and Hunter, 1980) for (1) 13-month weight with bulls
and heifers mated as yearlings in both an Angus (AS1) and a
Hereford (HS1) line; and (2) for 18-month weight, first mating at 2 years of age, in an Angus line (AS2) genetically
similar to AS1. An Angus control herd (AC0) also was established with sires chosen at random and first mating at 2
years of age. Each line comprised about 125 cows that were
2 years and older; 6 bulls are used annually in AS1 and HS1,
5 bulls in AS2 and 10 to 12 bulls in AC0. Semen from the
first set of selected sires (born in 1970) has been stored
for future use to allow another estimate of genetic progress.

The largest open-nucleus group breeding scheme with Angus cattle commenced in 1969-70 with the establishment of a
450 cow nucleus herd chosen on the basis of having weaned

large calves in the contributing herds. This program is designed to produce large numbers of high-breeding value bulls for use on Angus cows in the Lands and Survey Department's Rotorua Land Development District. The district has 53 farm settlement blocks, with some 335,000 breeding ewes and 26,000 breeding cows, on 62,600 ha of pasture. The nucleus (on the Waihora block) now consists of about 850 cows, 250 rising 2-year heifers, 450 heifer calves, 360 rising 2-year bulls, and 400 bull calves (Nicoll, 1978). While not an experimental study as such, this breeding project has had considerable input from geneticists in New Zealand and is an excellent demonstration of what can be accomplished in practice. It will be impossible to ascertain to what extent the high performance levels in the nucleus herd to date have resulted from the original screening or from subsequent performance selection. However, trials have been established to compare the performance of progeny sired by industry bulls and by bulls bred in the program. In addition, semen from the foundation sires has been stored for future use.

In the absence of a random-bred control herd at either Waikeria or Waihora, a convincing proof that selection has been effective comes from progeny test comparisons in other herds.

Bulls from Waikeria and Waikite, along with some bred in the Waihora bull breeding program, have been compared with a wide range of industry bulls in a number of progeny tests (Morris and Baker, 1978). Results in table 2 confirm a clear cut genetic superiority for growth of selection-bred bulls from the above three herds as compared to a broad sample of industry bulls believed representative of those available to commercial producers. Because genetic responses to selection are cumulative over time, the superiority of Waikeria bulls over Waikite and Waihora bulls is not surprising.

TABLE 2. PROGENY PERFORMANCE RELATIVE TO THE AVERAGE OF 122 INDUSTRY ANGUS BULLS FROM 60 HERDS

Source	No. bulls	Weaning weight (kg)	Yearling weight (kg)
Waikeria	25	+5	+11
Waikite	30	+2	+10
Waihora	34	+1	+5
Waihora	34	+1	+5

In the selection experiment at Waikite, direct and correlated responses to growth selection have been measured as deviations of adjusted selection line means from contemporary adjusted control mean (Baker, Carter and Hunter, 1980) and are shown in table 3.

TABLE 3. DIRECT AND CORRELATED RESPONSES TO SELECTION AT WAIKITE[a]

Line	Birth Wt	Weaning Wt	13-month Wt	18-month Wt
HS1	0.08 ± 0.04	0.88 ± 0.45	<u>1.65 ± 0.56*</u>	3.10 ± 1.39*
AS1	0.25 ± 0.05**	0.98 ± 0.38*	<u>2.56 ± 0.55**</u>	2.71 ± 1.01*
AS2	0.04 ± 0.06	0.34 ± 0.32	<u>1.22 ± 0.65</u>	<u>1.48 ± 0.96</u>

[a]Responses expressed in terms of kg/yr ± standard error.
Direct responses are underlined.

The results demonstrate that (1) selection has indeed achieved increases in 13-month and 18-month weights and these responses are continuing; (2) selection on yearling weight has been equally effective in Angus and Hereford lines; and (3) selection on yearling weight has improved both yearling and 18-month weight more than selection on 18-month weight.

A similar pattern emerges for weaning-weight responses. Birth weights also have tended to increase in response to the selection applied on later weights, a trend also observed in the Waikeria herd.

SELECTION FOR GROWTH--EFFECTS ON COW PERFORMANCE

The effectiveness of weight-for-age selection in improving progeny growth rate and slaughter weights is surely beyond dispute, as is supported by much North American research and application. But what have been the effects on that other vital component of herd productivity, calving performance?

Our selection trials have been research oriented, based solely on growth without reference to other traits except obvious physical abnormalities. The aim was to assess both direct responses to the selection and indirect effects on other important traits; production of bulls capable of increasing production in commercial herds was incidental.

Trends in cow performance over the 18 years of applied selection at Waikeria are condensed into table 4 showing performance for heifers (calving at 2 years) and cows (3 years and older) separately, mean breeding weights, calf losses, and marking percentage for three consecutive 6-year periods. Average birth weights are also indicated (Carter, Baker and Hunter, 1980).

As expected from the growth selection responses already established, weights of heifers have increased as have those of cows to a proportionately lesser extent. Calf losses here include animals born dead and those that died the first 6 weeks after birth. The main difference between heifers and cows is in the very much higher proportion of calves born dead (20.5% and 2.4%., respectively) due primarily to difficult births. In some earlier years, dystocia levels among the heifers were disturbingly high as birth weights increased in response to selection. The situation has subsequently improved, as shown by the decline in total calf losses, despite the continuing increase in birth weights.

The important conclusion from table 4 is that, over the course of the experiment, the percentage of calves marked has steadily increased. The improvement is comparable with that expected--had all selection emphasis been devoted to improving reproductive rate. It cannot be discounted that improved herd management has contributed to better calving rates, but there is certainly no evidence of genetic decline in cow performance.

TABLE 4. WAIKERIA SELECTION HERD: FEMALE PERFORMANCE OVER THREE 6-YEAR PERIODS (1962-79)

Period	1962-67	1968-73	1974-79
Heifers (total 2,037)			
Breeding weight (kg)	261	275	282
Calf losses (%)	16	15	12
Calves marked (%)	67	71	74
Cows (total 5,392)			
Breeding weight (kg)	430	445	447
Calf losses (%)	5	5	4
Calves marked (%)	82	88	89
Calves			
Birth weight (kg)	30.5	31.5	32.1

Supporting evidence from the Waikite experiment is presented in table 5. Although this experiment is at an earlier stage, the results covering 5 calving years again suggest favorable effects of growth selection on cow performance. The obvious superiority of Angus over contemporary Hereford cows is in accord with other New Zealand and North American findings.

TABLE 5. WAIKITE: RESPONSE IN BREEDING WEIGHT (J) AND
          WEANING PERCENTAGE (W%) TO GROWTH SELECTION

Herd	2 year		3+ year	
	J (kg)	W%	J (kg)	W%
HS1	248	65	363	78
AS1	259	70	355	84
AS2	-	--	345	84
AC0	-	--	363	79

MATING HEIFERS AS YEARLINGS

Mating heifers as yearlings tends to be the norm rather than the exception in the U.S. (Cundiff & Willham, personal communication), but it is not common in New Zealand where heifers are usually first mated as 2 year olds to calve as 3 year olds. Another obvious way to increase the calf crop is to mate heifers as yearlings, as is routine in dairy herds in New Zealand. Carter and Cox (1973) showed that, over a range of herd environments, heifers bred as yearlings marked 65% to 70% of calves at 2 years, subsequently performed as well as females first bred at 2 years, and increased herd productivity by weaning 20 more calves per 100 cows 2 years and older.

The above results did not, however, take account of the obviously higher feed costs with yearling mating. When first mating is deferred to 2 years of age, heifers can be reared at lower nutritional levels. Assuming they are bred at 340 kg live weight, their total feed requirements up to this stage are estimated to be about 60% of those for yearling-mated heifers, allowing for the latters' higher rearing costs to the yearling stage and subsequent requirements for their pregnancy and lactation. More replacements will, however, now be needed at a replacement rate of 25%. Given the same total amount of feed for the herd, it then follows that the number of cows 2 years and older must be reduced by 15% if heifers are mated as yearlings. The extra calves from the heifers lead to an increase of 5% in herd calf crop and an increase of 9% in total weight of slaughter progeny. Mating heifers as yearlings thus represents a net gain of nearly 9% in both feed efficiency and profitability (Carter, Baker and Hunter, 1980).

Yearling Versus Two-Year-Old First Breeding

The data routinely collected in the Waikeria and Waikite selection herds allow comparison of lifetime calf-weaning performance of heifers first bred at 1 versus 2 years of age. This has recently been summarized by Carter, Baker and Hunter (1980), basically updating the material presented by Carter and Cox (1973), and the results are presented in table 6.

At Waikite, the two contemporary Angus herds of similar genetic background differ solely in the age of selection and first mating. At Waikeria the data are for two consecutive 6-year periods; the figures in parentheses, relating to performance of two subgroups recorded in the same years, suggest, however, little effect of year of calving on the comparisons.

Despite their very different environments, reflected in cow live weights, neither location provides any evidence of adverse effect of yearling mating on subsequent calf production of cows 3 years and older. The 2-year-old calf crop is indeed a bonus.

TABLE 6. LIFETIME BREEDING WEIGHTS (J) AND WEANING PERCENTAGE (W%) OF FEMALES FIRST BRED AT 1 YEAR VERSUS 2 YEARS

Age first bred	No. heifers	Calving age					
		2 years		3 years		4 + years	
		J(kg)	W%	J(kg)	W%	J(kg)	W%
Waikite							
1 year	315[a]	251	69	315	76	361	86
2 year	229[a]	-	-	295	82	354	84
Waikeria							
1 year	619[b]	260	66 (71)[d]	364	79 (76)[d]	463	86 (82)[d]
2 year	419[c]	-	-	341	- (75)[d]	446	85 (83)[d]

[a]Born 1970-77
[b]Born 1960-65
[c]Born 1954-59
[d]Born 1959 or 1960 with first (1962) and later calvings in the same years.

Yearling Mating Weights and Effects on Lifetime Production

The relationship between yearling mating weights and lifetime calving performance has been investigated, measured in terms of average weaning percentage or of total calves weaned per heifer bred. The data have been summarized for some 1600 females born in 1960 to 1972 at Waikeria, which

were joined as yearlings and having at least five calving
opportunities over the period from 1962 to 1979 (Carter,
Baker and Hunter, 1980).

Apart from the lightest heifers (below 215 kg), there
is no obvious relationship between breeding weight and lifetime performance. This contrasts with earlier, less complete, results presented by Carter and Cox (1973) that appeared to indicate some decline in lifetime calving performance of the heaviest heifers (over 300 kg). This was presumed due in part to possible adverse effects on later reproductive performance of high milk intake during suckling,
leading to above-average weaning or yearling live weights.

Similar analyses for females in the harsher Waikite environment demonstrate the same general picture. However,
the "threshold" breeding weight, below which lifetime calving performance is reduced, varies with breed and breeding
age as follows: Hereford yearling-mated, 215 kg; Angus
yearling-mated, 195 kg; Angus first bred at 2 years, 250
kg.

Feed Allowances for Heifers

The results quoted have been collected under pasture-
grazing conditions, with no high-energy supplements. Some
recent research at the Whatawhata Hill Country Research Station has been studying the pasture feeding levels required
to ensure adequate reproductive performance in yearling
heifers (Smeaton and Winn, 1981).

TABLE 7. AVERAGE WEIGHT AT START OF MATING AND REPRODUCTIVE
PERFORMANCE IN 15-MONTH ANGUS HEIFERS.

Year	Yearling Nutrition	Autumn-winter pasture allowance ((kg DM/100 kg LW)	Weight at start of yearling mating (kg)	Proportion in-calf 6-wk mating(%)	Day of year con-ceived
1978	High	8	289	74	326 (Nov22)
	Low	3	253	85	326
1979	High	6	281	81	323
	Low	2	245	74	333
1980	High	3.5	267	78	323
	Low	2	234	45	334

The results (table 7) show that yearlings whose group-
average weight was about 250 kg or more at the start of

mating at Whatawhata attained acceptable in-calf rates (74 to 85%) without experiencing any increase in calving diffi- culty as compared with older cows, and over 83% conceived as 2 year olds. In terms of feed allowances, heifers weaned at Whatawhata at 150 kg at 6 months of age could be grown out to an average of 250 kg at 14 months by rotationally grazing them to a residual pasture level of 500 kg DM/ha in the autumn and winter, and to 1500 kg DM/ha in the spring through to the end of mating.

Conception Dates and Calving Intervals

Yearling mating may delay the date of conception at two years of age, but this delay can be made up by 3-year-old mating if subsequent nutrition is adequate. Data from many New Zealand analyses of research herds (Invermay, Massey, Ruakura, Whatawhata) now show that females calving later than average have shorter subsequent postpartum anoestrus intervals (PPAI) and intercalving intervals (CI) by about seven days, for each ten days later to calving. Unpublished Beefplan data analyzed by Morris (1980) with Angus and Hereford herds (mainly bull breeding) are consistent with this (table 8).

TABLE 8. REGRESSIONS OF CALVING INTERVAL ON FIRST CALVING DATE IN BEEFPLAN ANGUS (A) AND HEREFORD (H) SPRING-CALVING HERDS (95 A AND H HERDS; ONE PAIR OF YEARS).

	Calving date Year 1		Calving inter- val (Days)		Regression*	
	A	H	A	H	A	H
Overall (168 herds)	Aug 21	Sept 1	369	366	−0.70	−0.5
South Auckland region	Aug 21	Aug 25	367	366	−0.61	−0.4
South Island	Aug 24	Oct 4	366	371	−0.78	−0.6
Cow age in Year 1						
2 years	Aug 23	Sept 2	375	370	−0.75	−0.5
3 years	Aug 22	Sept 1	372	368	−0.76	−0.6

*Days change in calving interval per day later calved.

Two conclusions from table 8 are:
- Two and 3 year olds that calve slightly later are not necessarily a problem.

- Commonly given advice to breed heifers 3 weeks earlier than the cows will not necessarily be useful. This practice also extends the calving period by 3 weeks at a time when the availability of labor for calving and lambing can be critical.

In Angus herds, 15 days of the supposed 21 day "advantage" is lost by the next calving. The spring months for yearlings can be critical in terms of age at and date of first estrus, especially with low postweaning nutrition. A 3 week earlier mating can make a difference of 20 to 40 % in the number of heifers achieving first estrus by the time the bulls join them (Smeaton and Winn, 1981).

CENTRAL BULL PERFORMANCE TESTS

Central performance tests of beef bulls are used in many countries as the basis for selecting sires originating in different herds. Commonly, bulls from several herds are moved to a central location at 6 to 10 months of age (soon after weaning) and grazed as a single group or fed-out in feedlots for 150 to 300 days. The bulls are then ranked for final weight gain, and these rankings are used as a basis for making selection decisions.

The New Zealand Dairy Board has conducted four central performance tests (1972 to 1975) using Hereford bulls. A total of 100 bulls were performance tested from 16 different herds of origin. The bulls started the performance tests at an average age of 297 days and at an average weight of 275 kg. They remained on test for 273 days, gained in weight at 0.95 kg/day with a final weight of 535 kg. Sixty-six bulls were representatively sampled (including bulls with high and low performance test rankings) from the 100 bulls that were performance and progeny tested; these 66 were used to inseminate Jersey or Jersey-Friesian cross cows in 60-70 different dairy herds. Progeny (about 20 per sire) were reared on foster dams or artificially, then assembled in one location at about four months of age, and kept together for 14 to 15 months prior to a slaughter.

The correlation between central performance test and progeny test results was 0.15 for final live weight (at about 550 days of age) which was not significantly different from zero. The effective heritability, from offspring-sire regression, was 0.06 for final live weight. This is significantly different from the value of about 0.35 to 0.45 found in New Zealand and elsewhere for "within-herd" estimates of the heritability of 18-month live weight. It is concluded that central performance tests, as presently organized in New Zealand, are of limited value for ranking breeding values of bulls for growth traits. It is suggested that validation is needed for the effectiveness of similarcentral performance testing programs in many other countries around the world.

Morris, Jones and Hopkins (1980) have demonstrated the usefulness of sire-reference, progeny-test schemes to provide accurate rankings of sires across herds. They also showed that this sort of progeny test selection, in combination with within-herd selection programs, can be more effective than individual within-herd selection programs in producing annual genetic change for growth. The development of sire reference progeny test schemes in New Zealand is discussed next.

SIRE-REFERENCE PROGENY-TEST SCHEMES IN BEEF CATTLE

Industry sire-reference schemes, in which common "reference" sires are used in different recorded (Beefplan) herds were established in 1977 or 1978 for the Angus, Charolais, Simmental and Hereford breeds. The Angus, Charolais, and Simmental programs are operated under direction of Ruakura, while the Hereford scheme is under direction of the New Zealand Dairy Board.

These programs were initiated for four main reasons:
- Questionable technical merits of the central beef bull performance tests (Dalton and Morris, 1978).
- Eighty percent of pedigree Angus bulls used for breeding in New Zealand are not homebred (Cheong, 1977).
- Bull breeders seem more concerned about progeny-test results than about within-herd performance data.
- Because annual genetic progress from a multiple--herd reference sire testing and selection scheme can be at least as great as within-herd mass selection (Morris et al., 1980), the establishment of valid across-herd testing seemed worth while.

The Angus sire-reference scheme, begun with first matings in 1977, is the largest and most active of the four programs. By supervising the establishment of random mating groups of cows, home sires are progeny tested against the artificial insemination (AI) reference sires. A total of 27 reference sires has been used; repeat sires were arranged across years with 90 herd-years of testing so far set up in 31 separate herds. In five seasons, 4754 straws of semen have been ordered from reference sires and used at the rate of 1.5 straws per cow. Three research herds and a government nucleus breeding scheme (Waihora) also have been connected via reference sires. Progeny records from 503 sires were analyzed for weaning weight and 450 sires were analyzed for yearling weight using 3 years of data, plus current and back data from the research heads and Waihora. These include eight AI sires from other countries. Thebreeding values in New Zealand were 43 and 74 kg, respectively, for weaning and yearling weight as compared with weaning weights of 39 kg in Canada, 36 kg in the U. S. and 39 kg in Australia. The averages of all AI sires tested in New Zealand were +3.2 and +4.5 kg, respectively, above all non-AI sires. The reference sires were +1.8 and +5.7 kg above the mean of all AI sires.

Although still in their infancy, it does appear that sire-reference schemes in New Zealand are identifying the breeding values (BV) of large number of bulls, with the objective of providing the individual bull breeder or commercial producer with comparative information across herds, at least for weaning and yearling weights.

Two main factors are of concern at present. The first is the long delay experienced by the organizers in obtaining completed performance records from participants. The second is the high price and/or registration fee charged by the owners of some reference sires that provides a deterrent to greater membership in the scheme or more semen used per member.

The registration status of bulls is still a major barrier. In practice, the present situation enables breeders of unregistered bulls to have access to a wider sample of high BV bulls and semen than does the registered breeder, and generally at a lower average price.

Another factor delaying wider participation in these schemes is a reluctancy by farmers to become involved with an AI program. Progress is being made in New Zealand, however, with developing estrus synchronization techniques (Smith, 1977, 1978). An obvious advantage of synchronization is in the facilitation of the use of AI in beef cows and in the reduction of the costs of the AI program through savings in time, labor, special feed requirements, technician, and semen costs. Research in New Zealand with the P.R.I.D. (a stainless steel spiral coated with elastic rubber that has been impregnated with progestrone) has produced the most satisfactory synchronization results to date (Smith, 1978).

In the U. S., where an Angus SRS has been in operation for a longer time, AI sires have been used for both reference sires and weight-selected sires. Genetic progress estimates for 14 years and 564 sires were obtained for WW (0.39 kg/year) and YW (1.07 kg/year) by Willham (1982). In addition, a negative selection response had been achieved for birth weight (-0.1 kg/year). Closed research weight-selection herds in New Zealand have achieved twice the genetic progress in YW as the above (Baker et al., 1980), but it is not yet known what rate the industry is achieving as a whole.

DISCUSSION AND CONCLUSIONS

There is now ample evidence that growth traits in beef cattle are moderately heritable and will respond to selection (reviewed by Koch, Gregory and Cundiff, 1982; Dalton and Baker, 1980). New Zealand and Australian studies would suggest that adjusted weight per day of age at the yearling stage (Carter, 1971), or at 18 months (Seifert, 1975), are better measures of growth potential in beef cattle than is postweaning gain to these later ages.

Much less is known about genetic associations of growth traits with other traits of importance in beef cattle. In their review (made entirely of North American selection studies with beef cattle) Koch, Gregory and Cundiff (1982) concluded that selection for postnatal growth rate increased efficiency of gains but also increased birth weight that resulted in more difficult births (4%/kg) and higher calf mortality (1.2%/kg) among 2-year-old heifers. Barlow (1978) concluded that if cow condition is to be maintained in a herd following selection for growth, then food intake is expected to increase by 0.2 to 0.4% for each kg increase in weight at the selection age. If the number of cows in the herd is kept constant, with no increase in land area, then cow condition will decline. Barlow suggested this could result in herd weaning percentages falling to about 0.2 fewer calves weaned per 100 cows bred for each 1 kg of weaning weight improvement, or for each 2 kg of yearling weight improvement.

Neither the New Zealand research results reported here, or some similar preliminary analyses of changes of reproductive performance in the Canadian Shorthorn lines selected for yearling weight (Newman, personal communication) would support these findings, since in all cases there was an increase in cow reproductive performance in these herds selecting for growth. The high level of feeding heifers in many North American management systems is questioned relative to much more moderate levels in New Zealand. In fact, in herds selecting for growth, it may predispose heifers to calving difficulties (especially at the 2-year-old calving) and possibly to reduced lifetime reproductive performance.

Morris (1980a) reviewed the overseas and New Zealand data on first calving performance and subsequent production from females first bred at 1 versus 2 years of age. The overseas and New Zealand data are reasonably consistent, showing a rather small difference in first-calf weaning percentage (per heifer bred) and a lifetime advantage to the yearling mating system. The New Zealand data summarized here show that as long as heifers are over 210 to 215 kg individually, i.e., with a group-average of about 250 kg, then reasonable two-year-old calving results can be obtained, with no subsequent detrimental effects. Unfortunately, the advice of many people in North America seems to be to recommend very high minimum mating weights of 272 or 295 kg (e.g., 600 or 650 lbs) so that all heifers below that weight automatically achieve 0% calves weaned (i.e., heifers not mated) rather than letting the bull find out which heifers are cycling.

Likewise our New Zealand research results do not support the common advice in North America that heifers should

be bred 3 weeks or so ahead of the mature cows or that late-
-calving heifers will always be late-calving cows. This
whole subject was comprehensively reviewed and discussed by
Morris (1980b). The association of postcalving anestrous
interval with first calving date for cows and heifers was
mainly negative in studies in different countries around the
world. Some positive associations did exist in the literature (one from Britain, another in the U. S.), which Morris
suggested might be due to different feeding levels in these
studies. A restricted precalving diet, with no increase in
feed offered immediately after calving, as often applies to
early calvers under range conditions, would penalize the
heifers more than late calvers (where more abundant feed is
usually available) in terms of their ability to come into
estrus quickly. This would result in the negative association. When the feeding level is increased immediately after
calving, no such penalty applies and a positive association
may be found, i.e., early calves would have shorter anestrous intervals. Thus a recommendation to breed heifers a
few weeks before the cows is not necessarily valid in unsup-
plemented pasture conditions.

One of the main purposes of a central performance test
is to allow genetically superior bulls to be identified. if
the central performance test is successful in doing this,
then there should be a positive correlation between performance-test and progeny-test results. The four central performance tests analyzed here were shown, by subsequent progeny tests, to give a poor indication of the genetic merit
of individual bulls. In contrast, on-farm performance tests
have been checked in a similar manner under New Zealand
farming conditions (Carter, 1971; Baker, et al., 1976), and
were found to be as accurate as expected from the relevant
heritabilities.

Reasons for the worse-than-expected relationship between central test performance results and breeding values
estimated from progeny testing have not been fully examined,
but almost certainly relate to pretest environment effects
and age at the start of the performance tests. Dalton and
Morris (1978) reviewed New Zealand experiments that compared
the subsequent effects of differences induced by early feeding treatments and on calves. From trials with a variety of
different treatments and objectives, over 70% of the original live weight differences induced by early feeding persisted to the end of 12 of the 21 trials. From this review,
Dalton and Morris questioned the merit of central tests with
bulls starting on test at 6 to 10 months of age. The pretest environmental effect is likely to be less important, or
possible unimportant, if bulls are started on test at an
early age (Preston and Willis, 1970). Lewis and Allen
(1974) reported that the starting age for calves in performance tests in mainland Europe had been reduced to 50 days
in the Federal Republic of Germany, 45 days in Denmark, and
30 days in Sweden. In all these cases, this reduction was
achieved because the bulls were derived from dual-purpose
herds and artificially reared.

Given the continued popularity of central bull performance tests in many countries around the world, and particularly North America, it would appear prudent that more research of the type reported here be undertaken to evaluate
their real effectiveness.

REFERENCES

Baker, R. L. and A. H. Carter. 1975. Progeny testing Angus and Hereford bulls for growth performance. Proc. N. Z. Society of Animal Production 35:102-111.

Baker, R. L., A. H. Carter and P. R. Beatson. 1976. The value of on farm performance selection of Angus and Hereford bulls. Proc. N. Z. Society of Animal Production 36: 216-221.

Baker, R. L., A. H. Carter and J. C. Hunter. 1980. Preliminary results of selection for yearling or 18-month weight in Angus and Hereford cattle. Proc. N. Z. Society of Animal Production 40:304-311.

Carter, A. H. 1971. Effectiveness of growth performance selection in cattle. Proc. N. Z. Society of Animal Production 31:151-163.

Carter, A. H. and E. H. Cox. 1973. Observations on yearling mating of beef cattle. Proc. N. Z. Society of Animal Production 33:94-113.

Carter, A. H., R. L. Baker and J. C. Hunter. 1980. Beef cattle improvement - growth rate versus cow performance. Proc. Ruakura Farmers' Conference 141-153.

Cheong, W. K. 1977. Inbreeding and population structure studies in the New Zealand Angus breed. M. Agric. Sci. Thesis, Massey University.

Dalton, D. C. and R. L. Baker. 1980. Selection experiments with beef cattle and sheep. In - Selection Experiments in Laboratory and Domestic Animals, A. Robertson (ed.), Commonwealth Agricultural Bureaux, Slough, UK, pp 131-143.

Koch, R. M., R. E. Gregory and L. V. Cundiff. 1982. Statistical analysis of selection methods and experiments in beef cattle and consequences upon selection programs applied. Proc. 2nd World Congress on Genetics Applied to Livestock Production, Madrid (in press).

Lewis, W. H. E. and D. M. Allen. 1974. Performance testing for beef characteristics. Proc. 1st World Congress on Genetics Applied to Livestock Production, Madrid, 1:671-679.

Morris, C. A. 1980a. A review of relationships between aspects of reproduction in beef heifers and their lifetime production. 1. Associations with fertility in the first joining season and with age at first joining. Animal Breeding Abstracts 48:655-676.

Morris, C. A. 1980b. A review of relationships between aspects of reproduction in beef heifers and their lifetime production. 2. Associations with relative calving date and with dystocia. Animal Breeding. Abstracts 48:753-767.

Morris, C. A. and R. L. Baker. 1978. Genetic differences in herds and bulls. Comparison by progeny testing. Proceedings of Ruakura Farmers' Conference.

Morris, C. A., L. P. Jones, and I. R. Hopkins. 1980. Relative efficiency of individual selection and reference sire progeny test schemes for beef production. Australian Journal of Agricultural Research 31:601-613.

Nicoll, G. B. 1978. Beef cattle performance in the Angus breeding program. Proc. Ruakura Farmers' Conference.

Preston, T. R. and M. B. Willis. 1970. Intensive Beef Production, Pergamon Press, Oxford, 1st edn. 544 pp.

Seifert, G. W. 1975. Effectiveness of selection for growth rate in Zebu X British crossbred cattle. I. Pre-weaning group. Australian Journal of Agricultural Research 26:393-406.

Smeaton, D. C. and G. W. Winn. 1981. Nutrition of weaner beef heifers; growth, puberty and yearling mating o hill country. Proc. N. Z. Society of Animal Production 41:267-272.

Smith, J. F. 1977. Oestrus synchronization in beef cattle: A review. Proc. N. Z. Society of Animal Production 37:120-127.

Smith J. F. 1978. Techniques of administering progestagens for oestrus synchronization in cattle. Proc. N. Z. Society of Animal Production 38:145-146.

Willham, R. L. 1982. Predicting breeding values of beef cattle. Proc. 1st World congress on Sheep and Beef Cattle Breeding, Palmerston North, New Zealand, 1980. (in press).

30

SYNTHESIZING NEW SEEDSTOCK STRAINS

John Stewart-Smith

BACKGROUND

For the past 10 years the beef industry has been competing for the consumer's protein dollar and has felt increasing pressure from other meats and protein products. In a free-enterprise open-market system everyone has the opportunity to serve the consumer more cheaply and/or better. This poses a real challenge to us beef producers. Pork and poultry producers have used many techniques, including the use of genetic potential in their species, to increase their productivity. Can we do likewise?

In the beef industry, we still use a subjective, antiquated breeding system with pedigrees providing the overall discipline. If beef producers are to utilize the genetic potential in cattle, they must adopt a more up-to-date system for genetic improvement based on objective measurements and functional aims with prescribed selection criteria. First we need to examine some genetic principles.

SELECTION

The genetic approach to increased beef production can only come through the enhancement of the capacity of cattle to reproduce and grow. A consistent selection program must be followed to accomplish this. Selection is defined as a process in which certain individuals in a population are preferred to others for the production of the next generation. Selection is the only directional force available to the cattle breeder in developing functionally superior seedstock. New genes are not created by selection. Under selection pressure the frequency of the more desirable genes is increased whereas the frequency of the undesirable genes is reduced. Thus the main effect of selection is to change gene frequencies.

The response the breeder gets from selection depends on the accuracy with which he identifies his superior bulls, how few he uses, and the variation within the herd from which he is selecting.

Dr. D. D. Kress from Montana State University selected for a number of different traits by measuring genetic progress per generation based on using the top 4% of males and the top 50% of females (table 1).

Table 1 **GENETIC PROGRESS PER GENERATION IN LBS.**

Trait Selected	Birth Weight	Weaning Weight	Post Weaning Gain	Yearling Weight	Mature Weight
Birth Weight	5.4	15	15	29	40
Weaning Weight	2.6	21	12	29	28
Post Weaning Gain	3.2	15	28	41	23
Yearling Weight	3.7	21	24	48	42
Mature Weight	3.8	15	10	32	69

As might be expected, the greatest progress is made in the trait to which selection pressure is being applied, but there is at the same time a response in all the other traits. Mature weight increases less when selecting for the weaning-weight trait rather than for the yearling-weight trait. Mature weight increases least when postweaning gain is the selection trait. Any improvement in all other traits increases birth weights.

Seedstock cattle should be selected in and for the nutritional plane and climatic conditions under which their offspring are expected to perform. If selection is done in an ideal environment, the performance of the offspring could prove to be inconsistent and disappointing.

Economic Traits

Economic traits that concern commercial cattlemen are listed below by their level of heritability.

Dr. R. L. Willham of Iowa State University has outlined the opportunity in the genetic field relative to beef cattle breeding. He compiled table 2 by grouping the economic traits into three main categories along with their relative economic values.

ECONOMIC TRAITS

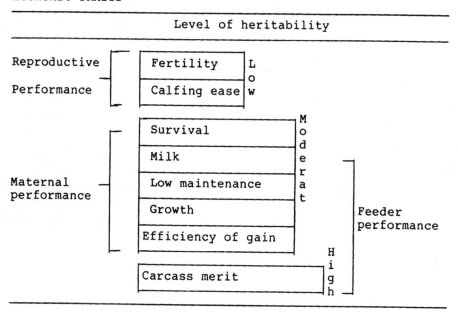

Table 2. **WILLHAM'S TABLE**

Class of traits	Genetic values			Economic values (relation)	
	Breed types differences* (%)	Heterosis+ (percentage increase)	Heritability‡ (percentage of variation)	Breed-ing herds§	Market feedlots
Reproduction	20	10	10	5	0
Production	50	5	40	1	2
Product	10	0	50	0	1

Table 2 shows that the most effective strategy for genetic utilization must be based on (1) exploiting heterosis in the cow by crossbreeding to improve reproduction traits that are of low heritability and (2) utilizing selection to improve the feeder traits, with higher heritability, such as rate of gain, feed conversion efficiency and carcass merit. Also the variation between breeds and within breeds for different traits must be used. The research data and genetic principles cited above indicate that selection combined with heterosis can increase productivity for beef producers.

TYPES OF SEEDSTOCK

To get this increase, two types of seedstock are required because the maternal traits required in brood cows are antagonistic to the terminal traits required in growthy feeder calves. In addition, there are several basic systems of crossbreeding available to cattlemen. An examination of these systems shows that at least five different strains of seedstock cattle are necessary to implement them.

To avoid confusion four terms are commonly used when talking about cattle breeding.

Purebred--pedigree registered, "pure blooded," of unmixed descent.

Crossbred--an animal resulting from the intermixture of two or more breeds.

Composite--an animal resulting from the intermixture of breeds in specific proportions.

Synthetic--an animal resulting from the admixture of different breeds in differing proportions.

Why Synthetics?

I have three reasons for advocating the development of synthetic strains to fulfill our need for functionally efficient seedstock.

First, Dr. Roy Berg of the University of Alberta has demonstrated that the synthetic combination of three breeds to form a new 'gene pool' resulted in a spectacular increase in productivity compared to straightbreds and that thereafter the rate of improvement by selection was faster in the synthetic population.

Second, it has been shown many times that faster progress can be made by incorporating desirable genes into a population by infusion from another source rather than by straightbred selection from within that population.

Third, faster progress can be made by utilizing all the genes that are available, instead of constraining oneself to the confines of any traditional breed.

Prerequisites

First, as shown in table 2, the production of a live calf every year is the factor most important to a commercial cattleman. Since a live calf is the essence of beef production, fertility and calving ease are vital traits in the brood cow. To ensure that these two traits are intensified, three prerequisites should be met by all seedstock:
- The seedstock must be born unassisted.
- Females must calve every year once they have been exposed to breed.
- The breeding period should be limited to 50 days.

Maternal Strains

The function of the maternal strains is to produce brood cows that will raise a superior calf every year in the environment in which they will be expected to perform. In Western Canada, winter feeding is our main expense. In this environment the dam's ability to put on flesh in the fall after her calf is weaned is conducive to thrifty maintenance during the winter. Further, among the maternal strains, which we will call M1, M2, M3, M4, other characteristics are required. The M1 and M2 strains are simply maternal, whereas the M3 and M4 strains have a specialized function in addition to their maternal role.

M1 and M2 strains. The prime objective measurement for selection in the M1 and M2 strains is 'weaning weight.' Weaning weight is the single best measurement of cow productivity. It is considered to be moderately heritable and repeatable and therefore can be improved by selection. Lifetime productivity of dams can be expressed by calculating their MPPA (most probably producing ability).

M3 strain. The M3 strain, although maternal, has another specialized function. M3 sires are used to mate to yearling heifers of any breed or crossbred combination. Hence the prime objective measurement is a restricted birth weight; thereafter yearling weight is used for selection. Yearling weight is used because these cattle can either go back into herds as replacement heifers or the total calf crop can be fed for slaughter.

M4 strain. The M4 strain is also maternal in type. This strain complements the M1 and M2 strains and provides an alternative to the more specialized M3 and Tx strains in some breeding plans. The major selection criteria for the M4 strain is yearling weight, a measure of many economically desirable traits. To retain the capacity to use heterosis in commercial crossbreeding plans, another provision needs to be made in synthesizing these strains. The restriction

is that these strains be kept genetically separated. Therefore, any 'breed' of cattle cannot be infused into more than
one of these strains.

The process of synthesizing a new strain consists of:
- Stipulating its function.
- Choosing the base.
- Deciding which other breeds can be infused into
  the base to enhance the characteristics being
  sought.
- Establishing selection criteria and procedures.

Selection Criteria

M1 strain. The prime function of the M1 is maternal
and is defined as being the capacity to produce and raise a
superior calf each year. Angus is a popular, numerous breed
and makes a suitable base for a maternal strain. Within the
constraints of fertility and calving ease, the two characteristics most desired in a maternal strain are milking
ability and growth. To improve milking ability, one would
obviously turn to a recognized dairy breed such as Ayrshire,
Holstein, Welsh Black, and Tarantaise. For growth one might
use Limousin or Saler. A dual-purpose breed also might be
suitable.

M2 strain. As the base breed one would select the
Hereford because it is numerous and well-adapted to most
parts of North America, although it has one or two imperfec-
tions. Simultaneously with improving the milking ability
and udder of this strain, it is desirable to introduce some
darker pigmentation. The dairy breed most apt for this purpose is Brown Swiss. For growth and milk production, the
Simmental is suitable.

M3 strain. Here light birth weight is the objective.
Probably one would start form a Hereford, Angus, or
Hereford-Angus base. The two breeds that would be the most
suitable to introduce to this base would be Jersey and Longhorn. Jersey would give light birth weights as well as good
milk production. The Longhorn is also an easy-calving breed
and fairly well-adapted to most of the environment in North
America. Other breeds that may have a place in the development of this strain would be the Murray Grey, Red Poll, and
Shorthorn.

M4 strain. All-purpose cattle are the objective in
this strain. The best base to work from would likely be
Gelbvieh or Gelvieh crosses. Again Limousin would be suitable in this strain. To get more growth, and yet keep the
dark pigmentation, two of the Italian breeds, Marchigiana
and Romagnola, might have a role. To get wider genetic
variation, the Beefmaster, a well-recognized, performance-
selected breed, has a place.

Terminal Strain

Tx strain. The function of the Terminal strain is to produce sires that, when mated to crossbred cows, will result in progeny that excell in rate and efficiency of growth and yield a desirable carcass. The progeny of this strain are raised solely for slaughter. The prime measure for the Terminal strain is rate of growth of lean meat after weaning. Leanness is emphasized because half the make-up of the end cross comes from the maternal strains that should have a propensity to fatten.

All of these various breeds could be infused into the base herds or 'gene' pools using either AI or natural service, which is most convenient.

ON-GOING POLICIES

Other factors to be considered:
- Most of the bull power used for the development of these different strains should come from top performing sires produced within the particular strain itself.
- It is desirable that multiple-sire breeding be used whenever possible. This is done to put indirect selection pressure on the mating ability and libido of the bulls.
- None of the strains should ever be closed completely. Semen should be used from any bull that can intensify the traits being sought within that particular strain.
- Color is not an attribute of meat or eating quality and so should be disregarded.
- Polledness should not be a consideration. This is a feature that can be easily obtained by physical dehorning; it is unnecessary to retard genetic progress by introducing this factor into our selection criteria.
- A cow that has to be milked to get her calf sucking automatically should be culled from the herd.
- Cattle with defective feet also should be culled regularly. Seedstock breeders should not indulge in hoof trimming.
- Disposition should also be considered. Any animal that is not manageable or a danger to those who handle it should be removed from the herd.

Finally, I emphasize again that seedstock should be produced under commercial conditions and that selection should be made in the environment under which the cattle are expected to perform--thus selection for genetic variance will be dependable.

31

INNOVATIVE METHODS OF TESTING
AND MARKETING SEEDSTOCK

John Stewart-Smith

INTRODUCTION

In another Stockmen's School paper I outlined some genetic principles and the method used to develop new synthetic strains. I suggested that a minimum of five strains was
necessary and that four of these strains had maternal characteristics and one had terminal characteristics. Further,
I suggested that two of the maternal strains should have additional specialized functions.

Having created our different strains, we are now faced
with the problem of making selections. The crux of cattle
breeding is to be able to identify the superior individuals
for particular traits as quickly and as accurately as possible. This requires keeping detailed records and testing the
animals from conception to cosumption.

TESTING

Types of Tests

There are two types of testing--Performance Testing and
Progeny Testing. Performance Testing is testing the performance of the individual himself and Progeny Testing is the
testing of an individual's progeny.

These tests are also carried out over different periods
of time. Weaning is the critical division in the productive
life of an animal. Prior to weaning the animal's performance is greatly influenced by its dam. After weaning, the
performance of the animal is dependent upon its environment. Therefore, it is necessary to record and test performance over both periods.

Preweaning tests start with the ability of the dam to
conceive. The next performance to be recorded is the birth
of the calf and at this time it is possible to take the
first weight measurement.

The next measurement is taken at weaning, usually 5 to
7 months after birth and consists of weighing the calves
under consistent conditions. The weaning weight can then

be corrected for age, for age of dam, and for the sex of the calf. These data provide much valuable information both on the calf and on its dam.

At weaning, the fist selections are made. This usually concerns selecting the bull calves that will qualify and go on test. These qualifications differ from strain to strain.

Qualifications to Enter Tests

In the M1 or M2 strains, the dam's outstanding performance is the most important attribute for a bull calf to have to qualify for testing. Formulas such as Estimated Breeding Value (EBV) or Most Probable Producing Ability (MPPA), as well as the calf's own weaning weight, will be used to make this initial selection.

In the case of the M3 strain, a limit is established for birth weight. Only those bull calves qualify that are below that birth limit. Among this qualifying group, those with the poorest weaning weights are discarded.

In the MS strain, the actual weaning weight of the calf is as good a measurement as any on which to make this initial selection.

In the Terminal strain, none of the preweaning data is very reliable for predicting postweaning performance. Therefore it is desirable to test as many terminal bull calves as possible.

Having determined the qualifications of the calves to be tested, the next problem is to decide how to test each of the four different types of cattle that are being developed.

Specific Tests

Performance testing. The M1 and M2 bulls have been selected based mostly on their dam's performance and their own weaning weight. Therefore their postweaning performance test is really a culling test with a two-fold purpose: 1) to identify those bull calves that are inferior for postweaning gain and 2) to identify those calves that have poor cold tolerance. Thus, a 140-day test covering the coldest winter months on a low-energy ration is the most effective for this purpose.

The M3 and M4 tests are identical, even though the qualifications for entering these tests are very different, because they are both selecting for yearling weight. The most effective test for these two strains is to test them until the average age of the group is 365 days. With the advent of computers this is easily calculated. Further, because these bulls are required to have some feeder characteristics, they are fed a medium energy ration so that an expression of this characteristic can be obtained during the test.

The purpose of the Terminal strain test is to ascertain the postweaning performance of these animals; consequently, it is desirable they be fed a high energy ration for as long as possible, usually in excess of 200 days. The long test has several advantages.

- It minimizes the effect of compensatory gain and evens out the fluctuations in rate of gain that occur during the test period.
- It is possible to identify those animals that fatten too rapidly.
- Those animals subject to founder will break down when fed for an extended period of time on a high energy ration.

My contention is that the standard 140 day test generally used in the industry needs to be modified to correctly test various types of cattle.

Progeny testing. Progeny testing is desirable in both the M3 and Tx strains.

M3 strain--The purpose of the progeny test of the M3 strain is to ascertain the degree of calving ease of the bulls chosen for testing. This test is difficult to conduct because to be valid the test has to be carried out using a random sample of heifers from different herds.

Tx strain--The progeny test is useful to prove the merit of Terminal sires as exhibited by the performance of their progeny that is measured by:

- Pounds of primal cut produced per day of age.
- Postweaning feed conversion ratio.
- Percentage natural births.

The progeny test can serve another purpose, which is perhaps even more important than determining the merit of the Terminal sires; it can monitor the progress of a particular genetic improvement program.

If the progeny test is carried out on a herd that is representative of the crossbred females produced by the program, and if this herd is fully prformance recorded with data kept on the progeny up to the time of slaughter, then the genetic progress made by the program can be analyzed and assessed.

Breeding Soundness Evaluation

An integral part of the postweaning test for all bulls is a comprehensive breeding, soundness-evaluation program. This program takes into account puberty, scrotal circumference, testicular tone and consistency, semen quality, libido, and mating ability.

Presently, there is research under way to determine how well these measurements predict the actual conception ability under multiple-sire breeding and range conditions.

As a final check, the bulls are visually appraised for overall structural soundness.

MARKETING

Those cattlemen who derive the major portion of their icome from beef production require seedstock that is functionally efficient and genetically superior. They also need consistency in the product and a continuous supply. They are the customers for synthetic seedstock.

Breeding Plan Design

The individual cattleman can be assisted in designing a breeding plan that takes the following points into consideration:
- The size of his breeding herd.
- The number of pastures available for different breeding groups.
- The amount and quality of feed available.
- The labor and facilities required.
- The method of marketing the cattle to be produced.

A breeding plan is a long-term endeavor; the rancher should know what bulls he will need for at least the next ten years and where to obtain them on a dependable basis. A guaranteed supply is the first marketing problem. The Show and Sale method presently used in the industry does not fulfill this requirement; thus an alternative method of allocating supply and determining price is needed.

Tom Lasater of Colorado was the first to use a Formula price and Lottery system for marketing his bulls. Beefbooster has modified the Lasater system.

Formula Pricing

In the Beefbooster system the price is set each November for the bulls to be marketed as yearlings in the following spring. The formula is based on: 1) the current market value of feeder calves in the fall as reported by CANFAX. (Canadian Cattlemen's Market Information Service). and 2) the estimated breeding value of each bull. The bulls are offered for sale in three categories. They are placed in the different categories according to the estimate that has been made of their breeding value for a particular strain. A premium is added to the price of the bulls in the top category and a discount is made on the price of the bulls in the lower category.

Allocation

A Lottery system is used to allocate the bulls to the different customers. A draw is made on 'Selection Day' to establish the customers' "selection priority" for each bull he wishes to purchase. He makes his selection from the bulls available in any of the three categories.

Warranty

As is customary in the industry, all bulls are guaranteed to be as represented. Should a bull be a nonbreeder, the bull will be replaced with an animal of similar quality or Beefbooster will authorize the sale of the bull for slaughter and will refund the customer with the difference between the slaughter value and the purchase price.

Customer Service

Every year a representative visits each client to review his breeding plan, update his bull-purchase schedule, and obtain an order for his requirements for the coming season. This enables the supplier to keep track of the requirements of all the customers and to forecast these requirements for several years ahead. In turn, this ensures that the customers will get the breeding bulls they want and at the time they need them to carry out their plans successfully.

This is an overview of a seedstock development and marketing system, which I foresee will become more and more prevalent in the industry in the years ahead. Specific strains of functionally efficient seedstock cattle will be produced and adapted to a particular climate and/or production system.

Essential Components

In summary, the essential components of a genetic improvement program are:
- The development of various distinct strains of cattle required for crossbreeding using indigenous cattle as foundations.
- A test station where the potential sires from each strain can be further evaluated.
- A herd for progeny testing the top performing Terminal sires on a representative sample of F1 females being produced from the maternal strains.
  Also this herd can be used to monitor the genetic progress of the program.
- A facility where measurements can be made of the post weaning performance of the calves from the progeny test herd.
- Computer capability sufficient to process the data collected to aid faster and more accurate selection.
- A method of allocating the supply of seedstock and determining their price.

Time Frame

Increasing the efficiency of beef production by genetic manipulation of cattle populations is a slow process.
- It takes 2 to 4 years before the first calves are ready for slaughter.
- It takes 6 to 8 years to change the genetic makeup of a cow herd.
- It takes 12 to 15 years for the effects of selection to be significant in a seedstock herd.

Though genetic improvement in beef cattle is slow, it is permanent, cumulative from year to year, and transmitted to future generations--thus worthwhile.

In order to compete viably with other meat-producing species and with vegetable proteins, the beef industry must fully utilize the genetic potential of cattle. The future development and extent of beef production will depend on the rate and efficiency with which cattle can be made to reproduce and grow.

Part 6

GENETICS AND SELECTION:
U.S. PERSPECTIVES

32

THE DESIGN
OF CREATIVE BREEDING PROGRAMS

Richard L. Willham,
Bret K. Middleton

"The breeders of the time of Bakewell suspected him of possessing and concealing special principles of breeding. It is often believed today that successful breeders have some mysterious method of which others are ignorant. Instead, the principles of the successful breeder have been exceedingly simple... The difficulty is not so much in knowing the principles as in applying them."
S. Wright (1920)

A breeding program is a complete system of management that is designed to bring about genetic change in a group of livestock. Such a program is the procedure by which the principles of animal breeding are applied in practice by stockmen. The design and conduct of a creative breeding program involves developing a complete management system and making it work in a practical situation. The purpose of this paper is to use the principles of breeding to design a creative breeding program. Then the stockman has the opportunity to design a program and conduct it using the COMPUTER COW GAME.

BREEDING PRINCIPLES

The many hereditary units or genes that influence traits of economic importance in beef cattle occur in pairs --one member of each pair having come from the sire and the other from the dam. The problem in breeding is that genes have their effect on traits in pairs, but are transmitted singly from one generation to the next. A parent gives to each offspring a sample half (one gene of each pair) of his genes. This is of concern in selection.

Selection

Selection or the choice of parents is the common element of all breeding programs. Selection is the only directional force available to the breeder to bring about genetic change. The basis of selection is the resemblance between parent and offspring. When resemblance is high, selection of superior parents results in above average offspring. When low selection of superiority results in average progeny. This resemblance results from parents giving each offspring a sample half of his genes. The degree of resemblance for a particular trait depends on the importance of gene effects (not gene pair effects) in producing differences among individuals. Researchers in beef breeding have used relative groups, since they have genes in common, to measure the fraction of the variation in traits that is produced by gene effects. This fraction is termed the heritability of the trait. Heritability can be used to predict selection response and thus is of value in the design of optimum breeding programs.

Suppose that the breeding value of each prospective parent could be estimated using the available performance records. If accurately evaluated, these breeding values could be used as the criteria of selection. The accuracy of breeding value estimation depends on the heritability of the trait to be selected. When high (50% to 70%) the performance of the individual is a good estimate of its breeding value. If the yearling weight of a bull is 50 pounds above his contemporaries and heritability is 50%, his estimated breeding value is 25 pounds or 50% of 50 pounds, on the average. The accuracy of this estimate is .71 when accuracy goes from 0 to 1. But when low (10% to 20%) own performance alone is not sufficiently accurate. For a heritability of 20%, the accuracy of own performance is .43. Then performance records of other related individuals, such as paternal and maternal half sibs and progeny, can be used to increase the accuracy.

In design of breeding programs, the most useful formula for predicting selection response and consequently useful in maximizing genetic change per unit of time is as follows:

$$\text{selection response} = \frac{\text{accuracy} \times \text{intensity} \times \text{variation}}{\text{generation interval}}$$

The selection response is per year and is in the units of the trait being selected. The accuracy is the correlation between the true breeding value of the selected parents and the estimated value. The intensity is a standardized form of the superiority of the selected parents over the group from which they came. The variation is a measure of differences among individuals that can be used by selection. The generation interval is the age in years of the parents when the offspring destined to replace them are born.

Antagonisms exist among the components. Usually the more accuracy achieved (progeny test versus performance test), the longer the generation interval will be. So compromise is essential to maximize genetic change. All three (accuracy, intensity, and variation) must be greater than zero and generation interval must be short for a selection response. These are important factors in breeding program design.

Mating

But genes do have their effects in pairs. Usually the average performance of the parents predicts the offspring performance, especially when heritability is high. But for numerous traits of beef cattle when the parents are of different breeds, the performance of the cross is superior to the average of the parental breeds. This well-known phenomenon is called heterosis and can be utilized in many breeding programs. Heterosis is produced by the fact that the dominant gene of a pair is usually more favorable than its recessive partner. When the genetic groups differ in the frequency of genes they have and dominance exists, then heterosis will be produced. The selection of individuals exhibiting heterosis results in little selection gain because the desirable pairs are broken up at segregation before being transmitted. But selection can be a powerful force when used among the genetic groups and among the group crosses. In this way heterosis can be selected.

The production of heterosis is not the only reason for considering crossbreeding in a breeding program. Desirable genes can be incorporated into a group faster by crossing them in rather than by selection of the desirable genes existing within the group. Also several desirable traits from different groups can be incorporated into one individual by group crossbreeding.

The reverse of crossing genetic groups is the mating of related individuals together. This is inbreeding. Inbreeding increases the chance that genes of a pair are alike because they are copies of the same gene from a common ancestor. This is fine if the gene is desirable. The breeder has no control over this chance process. An inbreeding or linebreeding program must be accompanied by intense selection to eliminate the inferior individuals produced. But mild inbreeding (pedigree isolation) is what produced the breed differences now being exploited by crossbreeding. Commonly, inbreeding depression (the reverse of heterosis) occurs because of the less desirable recessive genes occurring together in more individuals. Inbreeding does produce prepotency since more gene pairs are alike and segregation is less apparent. Making inbred lines or closely related families does in a sense produce genetic variance for selection, since the line differences are genetic. But inbreeding in cattle is slow at best and leaves little opportunity for constructive selection since it counters the inbreeding depression.

CURRENT BEEF INDUSTRY VALUES

The principles of breeding need to be synthesized into practical breeding programs in the context of the dynamic beef industry. This tradition-ladened, exploitation-minded, highly segmented queen of the livestock industry has been slow to adopt new practices but now is in transition. Before putting breeding principles in a breeding program context, a look at the relevant beef industry values is in order.

Table 1 gives the current beef industry values. The economically important traits of beef production are divided into three classes of traits. These classes are reproduction, production, and product. Calf crop percentage, average daily gain in the feedlot, and retail yield percentage in the carcass are examples of traits in the three classes. Obviously there is overlap. Based on common commercial production conditions, the first column gives the relative economic value of the three classes of traits. Reproduction is at least 5 times as important as production in terms of net return at the commercial level. A unit increase in daily gain is useless without a live, healthy calf and a cow that can be rebred on schedule. Today, production is twice as important as product improvement. This stems from current buying procedures and could change rapidly with economic shifts.

TABLE 1. CURRENT BEEF INDUSTRY VALUES

Class of Traits	Relative economic values	% Selection variation	% Heterosis increase
Reproduction	10	10	10
Production	2	40	5
Product	1	50	0

The second column gives the average heritabilities for the three classes of traits. These values reflect the percentage of variation among animals treated alike that is due to variation available for selective change. When a large fraction, selection for increased performance will be effective. Numerous evolutionary problems and the sheer complexity of the traits contribute to making the heritability of reproductive traits low (10%). Little selection response can be expected from even intense culling on the reproductive traits. But production and product traits have moderate to high heritability. Selection for increased yearling weight (production) and retail yield will be effective.

The third column gives the average percentage increase in performance derived by crossing breeds of cattle or heterosis. The figures are probably low but are still relative to each other for the trait classes. Reproductive traits show the biggest improvement from heterosis, especially when a crossbred cow is considered in commercial production. Production traits exhibit some heterosis while product traits on the average show little heterosis. The average of the parental performance usually predicts the response.

The column giving relative economic value in commercial production is negatively related to the column relating available variation for selection but is positively related to the heterosis column. Thus, it appears that some sort of commercial crossbreeding including the crossbred cow will become the rule to utilize the reproductive heterosis and the small amount of heterosis in production. The breeding stock utilized by commercial producers can be easily improved by selection for improved production and product traits. This selected superiority will be passed on directly to the commercial producer. Any heterosis for reproduction exploited by the breeders of seedstock would appear to be detrimental at the commercial level.

This table of beef industry values suggests a commercial crossbreeding system designed to maximize reproductive heterosis and put together a desirable combination of traits in the market animal. The table also indicates that selection programs developed to improve production and product traits in the breeding stock herds would benefit the beef industry through direct transfer of the germ plasm of superior parents to their commercial progeny.

BREEDING PROGRAMS

Consider first the basics in the design of a creative breeding program. The general breeding program is to select and combine germ plasm that results in stock superior to that previously produced. What can be combined is circumscribed by what can be incorporated biologically. The definition of the problem indicates that the only directional force available to the breeder is selection. The relation between the problem and the concept of breeding value or the value of germ plasm as parents is evident.

Again, a breeding program is a complete management system that is designed and conducted to bring about directional genetic change. The program must include the specific production system under which a subpopulation of bovines is changed genetically. Depending on the type of program, the production systems to which the product offered for sale is to be relevant must be studied and defined as part of the direction choice.

The form of most breeding programs is cyclic, with one round of genetic change being layered on the previous round. The cycle involves the production of a set of offspring, their evaluation, and the selection of parents to produce the next offspring set. The order in beef systems for any one year is the calving of a set, the yearling evaluation of the previous set, the selection of the next parents, the breeding of these selected individuals for the next set, and the weaning evaluation of the current set. This cycle is repeated yearly over time. Genetic change in the subpopulation is the cumulative differences between the adjacent sets of calves. This assumes no environmental fluctuations, so other methods of evaluating genetic change must be used.

In the total beef production system, there are essentially two types of breeding programs. They differ in the product offered for sale and in the forces available to the breeder to make genetic change. The breeding herd program sells breeding value or the value of his product (parents) in the herds of his buyers, whereas the commercial beef producer sells numbers and pounds however these can be obtained. This leads to a difference in the forces available. The breeding herd program can only utilize genetic differences that contribute to the variation usable by selection; the commercial program can incorporate genetic differences existing among cross combinations of subpopulations or breeds. Selection among groups can thus increase heterosis advantage and incorporate real performance differences among animals in the breeding herd and those destined for market. Further commercial producers can select representatives of these groups as parents that are superior to their group. That is, the market animal of the commercial program does not undergo segregation and subsequent recombination in its offspring as does the breeding stock before its value is determined.

Because selection is the only directional force available to breeders to change the genetic composition of their biological populations (herds), there are really few choices open to the breeder in his design of a breeding program. These choices are three dimensional. They are direction, differences, and decisions. Direction is the paramount choice and the first that must be made since it influences the others. The word direction is used rather than goal since goal implies a fixed object of reference. Really a breeder is making directional change in the mean performance of a biological population in time and space. There is no fixed object at the end; there is no end in adaption of livestock to systems of production that benefit man.

The second choice of the breeder involves the evaluation of each set or calf crop and the parents. The measures chosen relate to the direction. The word difference implies that only differences among animals treated alike are relevant. Variation or difference is the raw material of the selection process; there is no other.

The third choice of the breeder concerns selection decisions to be made based on the differences to move his subpopulation in the chosen direction. This choice depends first on the direction decision. Certain traits are of major importance for optimizing a breeder's production or for the breeder to optimize the production relevant to the herds of his buyers. These traits really define the selection scheme that may include a mating design of the selected animals.

With the three choices defined, each in turn needs examination to fit creative breeding programs in the context of the total beef production system and specific systems of production. Attention will focus on breeding stock programs.

Direction

Direction, especially for the breeder selling breeding stock to the commercial producer, is the paramount and first choice necessary in the initial design stages of developing a truly creative breeding program. First, there is no substitute for putting things on paper. Talk and dreams are cheap. Only when breeders begin to define their direction clearly do they realize how nebulous their ideas have been. The same is true for all involved in the industry. Probably the most important result of the recent work on beef systems is to lay bare the areas of importance in which nothing is really known.

The key to good direction choice by breeders is to gather all the available facts on which direction depends and then integrate these into a working definition. Such facts, once developed, are likely to show the following things:

1. That reproductive performance in production systems that create new wealth (calves) will be primary in economic value. We already know that selection is relatively ineffective, but crossing breeds in the production of brood cows produces real economic heterosis in their reproductive performance. Therefore, breeders that are merchandizing breeding stock to commercial producers must incorporate into their programs the use of crossing by their buyers. Selling breeding value for reproductive performance in commercial bulls to be used in crossing to produce commercial heifers will require new evaluation procedures.

2. That production traits involving growth and development and milk production may well be important economically but may be optimum values rather than the traditional extremes. If this is true, the show ring as a means of evaluating size is

completely out, because animals can only be lined up from small to large or large to small. The genetic trends observed over the first 14 years for yearling weight in the Angus and Hereford breeds may need to be slowed because the results from the integrated systems approach will put traits that respond easily to selection in their proper perspective. Further, milk production levels and their effect on reproduction when evaluated in systems with different resource constraints will surely be a primary issue.

3. That the future price relationships concerned with carcass weight, sex, fat cover, and marbling will have genuine repercussions in many beef production systems. Optimum cow weight and optimum crossbreeding systems may need re-evaluation by the industry. Competition in the market place among meats may have an impact; at the least the frills of the industry can be eliminated.

Of major concern to the breeders in their choice of direction is the future and its correct prediction. Making compounded genetic change over time can be a slow process at best. Therefore, producing breeding stock that will meet anticipated industry needs is necessary. As always, profit is being first. When all adopt a given technology, it becomes survival.

Now to the key aspect of design of breeding stock programs. The market for the product offered for sale (breeding value) is the production systems being used by the buyers of the product. If a general set of economic values can be established, specific breeders can serve a larger market than just one specific system needs.

There is no question that the breeding herd merchandizes breeding value of his stock. Specification of product offered for sale is becoming standard practice in industries such as beef production even when the product offered is biological and will always be subject to the variation inherent in the system. Selling on measures of breeding value is a reality in the beef industry. The problem today concerns the relative economic importance of the traits contributing to production or how the breeding values need to be combined in an index of net merit. Also, not all the necessary breeding values are being calculated currently.

Differences

The second choice of the breeder in the design of a program concerns the measurement and evaluation of differences among his stock for the traits he has determined are important in his chosen direction. Now these traits and their relative economic importance relate to the production systems of his buyers, the commercial beef producers, yet the breeding herd has to have a production system that minimizes the costs of production. For many breeders the available performance program of their breed, with its BIF specified set of tests, has determined their production system. Serious study needs to be given to this problem. We in BIF have assumed that the standard performance program fits all consequently have said that really only one production function is relevant to the gigantic beef industry sprawled all over our continent. Breeders need to seriously consider the needs of their customers and incorporate evaluation of differences among their stock for traits concerning these needs.

The determination of exactly what production system to use by a breeder is a matter of question. The safe bet today is to use the common system being employed by the customer.

Determination of the production system in which differences among breeding stock are evaluated is one of the primary choices of the breeder. Commercial producers have fewer considerations since they are developing a program to improve their production. Their buyer is concerned with numbers and pounds. If their germ plasm suppliers are performing well, fewer records may actually be needed.

Decisions

Decisions constitute the final choice open to the designer of a breeding program. The breeder must design a selection scheme that utilizes the differences evaluated to move his herd in his chosen direction. Selection concerns the choice among groups, such as breeds, as well as the selection of individuals of a breed for parents. The scheme must also be concerned with how the selections are mated to generate the next genetic sample or calf crop. The breeder of breeding stock traditionally has developed a program only after the breed was chosen. Today, breeders are becoming involved with more than one breed with the purpose of serving their commercial producers who use both breeds in a cross combination. Other breeders are in the process of combining germ plasm from breeds they believe can complement each other, and in combination the mix is being developed through selection of parents into synthetics or new breeds. The hope is that there will be retention of a fraction of the heterosis as well as the complementarity.

Commercial producers have the task of developing a systematic crossing system using several breeds or the new synthetics that will provide stock that matches in genetic potential the resources available for production in their system. This is no easy task because the comparative breed evaluations must be studied in depth and then the sources of such germ plasm must be researched.

Thus, the first development in the decision-making choice is the selection of the initial germ plasm and its use in systematic crossing schemes if the producer is commercial. Often this decision accounts for the biggest single jump in the efficiency of production.

Next comes the design of the cyclic selection scheme, especially for the breeder of breeding stock. Selection schemes to maximize the possible rate of genetic change can be devised. The given restrictions of the chosen production system for the herd provide the bounds of the problem.

The first concern is the use of the available differences to develop the best predictions of breeding value for all the possible parents. With the advent of within-herd, mixed-model predictions that use the relationships between every animal with every other animal in the herd, much better predictions will be available. Thus, as these results can be tied across herd by the use of progeny from sires used extensively in the breed, the door will open for breeders to be able to fairly compare all bulls, including theirs, as if all the bulls had progeny in their herd. The intensity of selection on the sire side can be dramatically increased in herds tied to the breed sire evaluation by the use of evaluated sires.

A breeder can be conducting his own breeding program without real use of outside sires, yet the use of an outside sire with 10 to 20 progeny each calf crop, in fair comparison with the sires of his program, will become paramount in general decision-making in the breeding program. When the herd is in comparison to the others for the traits involved in the direction of the breeder, this will help in deciding whether to begin a linebreeding program to concentrate the blood in the herd or whether to continue incorporating superior germ plasm wherever it is found in the breed.

Selection exists two times in the conduct of a breeding program. The first time involves primarily female selection and takes place at the weaning of a calf crop. The second time is at the return of the yearling information on the bulls and possibly the heifers. The heifers may be evaluated for yearling performance at 18 months and then this information can be used in the first selection time. The second selection time involves a bull evaluation and the development of a mating scheme using the selected sires.

The design of the decision-making process or the selection scheme to adopt requires knowledge of the basic principles of animal breeding. Much too much time is spent by breeders in planning each mating. Many times a result of this effort is to bias the needed sire comparisons made in subsequent calf crops because dam differences are included in the expected progeny differences of the sires being compared. Young sires have a difficult time beating their sire when he is mated to highly selected dams. Sire selection and consequently sire evaluation is the paramount issue in beef breeding.

PRACTICE USING THE COMPUTER COW GAME

The computer cow game is a fun exercise designed to let the stockman design and conduct a breeding program and see the response of the computer animals to his program. There is no hay to pitch or calves to pull, but market value is minimal. The stockman receives a cow herd that has produced a calf crop. Without much loss of reality, the calves and their parents are ready to be selected and mated to produce the second calf crop. Birth, weaning, and yearling weight along with hip height at a year are available for the herd.

The objective is for the stockman to design and conduct a breeding program that will make genetic change in his herd. The computer cattle really respond like real ones to selection. This is what makes the game fun. The breeder must select his sires for the next crop from among the males. Any combination of yearling or old bulls may be used. Then the breeder must select his dams for the next crop from the heifers and cow. This process is cyclic and is repeated for each calf crop.

After a few crops, the breeders playing the game will be allowed to use some outside sires artificially. If the breeder uses these sires according to the design required, he can see the bulls he has selected and used in his own herd actually fairly ranked with similar sires used in other herds. That is, the actual sire-evaluation analysis used by breeds on real beef data is used to conduct an over-herd sire-evaluation analysis. The breeder can experience the rewards of his earlier efforts by seeing his sires on the top, middle, or bottom of the sire listing. Then he can utilize this information to further refine his breeding program design and conduct.

SUMMARY

The design and conduct of creative breeding program is discussed. First the basic breeding principles of selection and mating are reviewed followed by a description of the current beef industry values. Then these two are put to work in the design of breeding programs. The choices of the breeder are direction, differences, and decisions. Each choice is detailed, and then the stockman is introduced to the computer cow game that gives him the opportunity to conduct the breeding program he has designed. Also he can observe national sire evaluation in action over herds. And the difficulty will still lie not so much in knowing the principles (which are simple) as in applying them consistently over time!

33

THE NEW AMPHITHEATER WHERE THE BATTLES WILL BE FOUGHT: SIRE EVALUATION

Richard L. Willham

> "You who have heard day after day and night after night the applause of splendid audiences as the final proofs of man's mastery of the mysteries of animal procreation and development have been presented in the great AMPHITHEATER; you who have adjourned from the ringside to the taproom of the Inn, or sought the cozy corners of the Club to discuss the wonders of the shows; you who toil daily within the yards -- may appreciate fully the privileges you enjoy, and again you may not."
>
> Alvin Howard Sanders (1915)

Yes, Alvin, some still discuss the wonders of the shows in the taprooms, but a growing number of progressive stockmen have adjourned from the ringside to present their stock in another, bigger AMPHITHEATER, which involves the evaluation of the performance of the sires of entire breeds. The computer age has unfolded opportunities undreamed of when you sought a cozy corner of the club to reflect, but we do appreciate fully the privileges we have to build upon such a heritage. We witnessed the last show in the AMPHITHEATER in 1975; now we have constructed another that encompasses whole breeds in which to display final proofs to splendid audiences. Breed-wide national sire evaluation is a reality of today in the breef industry. Sires of a breed are now fairly compared on their Breeding Values predicted from the performance of their progeny produced in numerous, different contemporary groups.

The purpose of this paper is to examine the basic breeding principles involved in sire evaluation so that the new AMPHITHEATER of sire evaluation can be better utilized for real breed improvement by beef breeders.

SIRE EVALUATION

The future belongs to beef breeders who are willing to use breeding technology in a creative breeding program designed to make rapid genetic change in economic merit. Sire selection and consequently sire evaluation are of paramount importance in all beef-breeding programs. Because of the higher reproductive rate sires can be more intensely selected, more accurately evaluated, and the superior more easily exploited than dams. Therefore, to design breeding programs today requires the use of sire evaluation.

Genetic Problem and Solution

The primary genetic problem, of both breeding and commercial herds, is to find sires that, when mated to the cow herd, produce progeny that are superior in economic merit to progeny of the current sires. The problem has a solution using the concept of breeding value. Breeding value is simply the value of an individual as a parent. The basic issue is how the progeny of a sire performs in the herd of the buyer. The concept of breeding value is intimately involved in the practice of selection since the prediction of selection response is a function of the breeding values of the parents. Therefore, to solve the genetic problem, it is logical to evaluate potential sires in fair comparisons with each other on their value as a parent or on breeding value. It is the breeding bull that is the issue.

Effective Sire Evaluation

The purpose of national sire evaluation is to enhance the effectiveness of sire selection in the breeding programs of all breeders. The goal is the expansion of the number of sires that can be fairly compared on breeding values from all sources of performance information. To enhance the effectiveness of sire selection, above what can be done within a single herd, requires that sires be evaluated so that fair comparisons can be made. Current sire evaluation programs utilize reference sires as a common base. Given comparable evaluation, the effectiveness of sire selection is the product of the accuracy and intensity of selection and the variation available; all divided by the generation interval. This puts the rate of genetic change on a per year basis. Accuracy is the correlation between the true and estimated breeding value. It is a function of heritability and the number of animals involved in the estimation of breeding value along with their relationship to the individual. The intensity is a function of the fraction saved as parents. The measure of variation among breeding values puts the rate in the units of measure, while generation interval (age of parents in years when the offspring are born) puts the rate on a per year basis.

Increasing the number of sires that can be fairly compared will increase the intensity of selection and the variation among breeding values. Improvement in the accuracy of selection in the minimum of evaluation time increases the rate. To date, evaluating sires on their own performance when sires are evaluated in different performance tests does not give a comparable evaluation of sires. This is the reason BIF developed guidelines for national sire evaluation based on a progeny evaluation and the use of reference sire progeny. However, as more is learned about breed structure through national sire evaluation, all information (including own-performance of sires) can become more useful in breeding value estimation. When fair comparisons are made among sires on their own performance, the generation interval would be two years shorter, but the accuracy would be less than that obtained using progeny. For a 40% heritable trait, own-performance has an accuracy of .63, while evaluation of 20 progeny has an accuracy of .83. For comparison, assume that the generation interval is 2 years for own-performance evaluation and 4 years for the progeny test; that intensity is 2.0 for both, or that the top 5% are selected; that the female generation interval is six years; and that the measure of breeding value variation is 60 pounds. Using own-performance evaluation yields 9.5 pounds in genetic change per year, whereas using a progeny test of 20 produces 10.0 pounds. Usually, intensity is much less for the progeny test. Even so, the resulting rate is surprisingly similar. The net result is that national sire evaluation using progeny and reference sires is to develop fair comparisons -- not to increase the rate of genetic change by a higher accuracy of selection.

National Sire Evaluation

A National Sire Evaluation Program is designed and conducted by an organization having no direct interest in the test bulls. The purpose of such a program is to enhance the effectiveness of sire selection in the breeding programs of breeders. The foundation stones of such a sire evaluation program are the many creative breeding programs being conducted. All that is needed to tie a breeding program to a breed sire evaluation program is to use reference sires in the test to provide comparison with all bulls of the breed so tested. This gives the participating breeder many more bulls from which to accurately select, and thus enhances the effectiveness of his sire selection.

The basic problem in sire evaluation reduces to one of comparison. Since the world is comparative, the issue becomes that of what sires should be compared? The BIF Guidelines for National Sire Evaluation Programs have incorporated the experience of dairy sire evaluation and the realities of the beef industry into a system using reference sires as the base of comparison.

The expected progeny difference of a sire for a particular trait is an estimate from the existing progeny data of half of the breeding value of a sire, or what he is expected to transmit to his offspring. It is an estimate of how future progeny of the sire are expected to perform relative to the progeny performance of the other sires, when both are mated to comparable cows and the resulting progeny are treated alike. The important aspect is to predict future progeny performance from the sample of progeny performance currently available. Therefore, the sire progeny differences are regressed toward the average expected progeny difference, which is zero, depending on the number and distribution of progeny involved in the difference and on the heritability of the particular trait. The expected progeny difference is reported in the units of measure of the trait or in ratios.

Periodically, the organizations conducting the program publish a sire summary that includes information on all of the sires evaluated regardless of their merit. The purpose of such a sire summary is to describe the germ plasm available for the traits considered of major economic importance to the breed. Selection of sires from among those described is the prerogative of the breeder.

Current Sire Evaluation

Beef national sire evaluation has as its goal expansion of the number of sires that can be fairly compared on breeding value differences obtained from all sources of information. Performance programs were sold as within-herd improvement procedures, not useful in over-herd comparisons where the major genetic decisions are usually made. No wonder the show ring has maintained its brilliance. At this time, the breeding-stock segment is becoming aware that progeny tests that involve progeny from a common set of sires yields breeding value differences that can be fairly compared. As more is learned about the beef population through national sire evaluation, all sources of information on breeding value can become more useful. That is, national sire evaluation is a means to an end, not an end in itself. To let the beef industry become married to the progeny test as the only way to estimate the breeding value of sires would be tragic, since only performance and relative information can be used effectively to minimize the generation interval. With the primary trait being sex limited, the dairy industry has few other opportunities, while the beef industry has many, especially when using sequential selection schemes.

Breed associations that have a national sire evaluation program and a sound performance program for their breeders are realizing that a primary reason for existence today is for the collection, processing, and holding of performance records. The rapid increase in volume of records processed, even with major declines in registrations and transfers, has emphasized this point. Short-term direction by national sire evaluation programs reflects this realization by the breed organizations.

FIELD DATA SIRE EVALUATION

Three major British breeds (AAA, AHA, and APHA) have now consolidated their sire evaluation into one report on sires issued annually. The existing designed data has been combined with the field data generated in the respective performance programs--AHIR, TPR and Guidelines. These sire reports list bulls meeting a given level of prediction accuracy, but all sires having progeny records are included in the sire evaluation analyses. The Simmental and Limousin breeds and others have simular sire reports. Details of each need to be obtained from the respective breeds. A general description of the evaluations follows next.

The Expected Progeny Difference (EPD) of each sire for birth, weaning, and yearling weight, if available, is listed. The EPD is exactly what it says it is. It is the best Predictor of the performance of future progeny of a sire in comparison with progeny of similar sires. The prediction is based on all the current progeny data available. The EPD of any sire is directly comparable with the EPD of any other sire, since each EPD is adjusted for progeny number and the heritability of the trait (.3 for birth and weaning weight and .4 for yearling weight).

Each EPD has an Accuracy (ACC) value. The ACC is a relative measure of the possible variation of the EPD. It is a function of the number and distribution of the available progeny. Large numbers of progeny distributed in many herds and contemporary groups gives values of ACC close to 1.00, while few numbers give low numbers closer to zero. Since the EPD is already adjusted for progeny number, attention should be given to ACC when being inaccurate would be a disaster to the breeding program.

Also included is the Maternal Breeding Value Ratio and its accuracy for each sire. These ratios are currently the best method available to predict the ability of sires to transmit maternal performance to their future daughters, as reflected in weaning weight.

The criteria for listing a sire are at least an accuracy of .70 progeny in at least two herds, and sires belonging to the group of sires either directly or indirectly tied by having comparable progeny within contemporary groups. The accuracy of the prediction rather than the value of the EPD constitutes the criteria.

The current listing of sires resulted from a sire evaluation analysis of some 257,000 weights in the Angus data, some 245,000 weights in the Hereford data, and some 249,000 weights in the Polled Hereford data. Approximately 10,000 to 11,000 sires had progeny for the control weight; weaning weight in the polled data and yearling weight in the

other two. Contemporary groups included progeny that had
been raised together and treated alike. Some 500 to 700
sires of each breed have been listed out of the total. Almost 90% of the sires in each breed were tied together;
either directly by having progeny in the same contemporary
group or indirectly through sires used over the breed in
many different herds. Fewer data were available for birth
weight than for the other two weights.

The problem of progeny selection at weaning was considered by doing the analysis on all available weaning
weights. Then the weaning weight was subtracted from each
yearling weight to give a gain figure. This total gain was
then analyzed as yearling weight. The EPD for gain was added to the EPD for weaning weight to give the reported values
for yearling weight EPD. This procedure is comparable to
the method of yearling weight ratios adjusted for selection
weaning recommended in the BIF guidelines.

Records from the designed sire evaluation were treated
differently than those from the field data. Since cows were
randomized to sires and all progeny were given equal treatment within contemporary group, a procedure was used that
gave sires credit for all their progeny. In the field data
no such assumption was made. The possible extra correlation
among calves by the same sire in the same contemporary group
was considered by giving a sire credit for up to 5, 7, or 10
calves in any one contemporary group for yearling gain,
weaning, or birth weight, respectively. This procedure ensures that the reported sires will have calves distributed
over several contemporary groups, making the evaluation more
accurate--or less subject to improper progeny testing than
might be done in particular contemporary groups.

All the sires were placed in birth-year groups for each
trait. The sire equations for a group were summed to give
birth-year group equations and these were solved simultaneously with the sire equations. This procedure regressed the
expected progeny difference of a sire backtoward his group
average rather than the average of all sires. The EPD of a
sire is his group effect plus his effect within one group.
This procedure is used to account for genetic trend. The
group effect expressed the average level of genetic merit of
all bulls introduced in the breed at a specific point in
time. Thus sons of bulls will be in different groups than
their sire. Each son is ranked relative to other bulls introduced to the breed at the same time and group effects
allow one to compare the level of merit of bulls recently
introduced to the breed with those introduced earlier.
Table 1 gives the sire group effects for sire birth year.
The trends represent the response of breeders to the call of
the commercial industry in the mid-1960s to increase the
growth ability of the product offered for sale.

Table 2 condenses the figures of table 1 down to the
genetic trend per year and the accumulation of this trend
over years. The increase in weaning weight and yearling
weight for the bulls of an entire breed is excellent. Note

that the birth weight trend is very slightly negative. This simply indicates that attention has been paid to the birth weight of sires selected in later years, when in earlier years little attention was given to birth weight in the selection of sires. It does not indicate that the positive correlation between birth and yearling weight does not exist. It does exist, and the average birth weight is increasing for the breed.

TABLE 1. GROUP EFFECTS[1] FOR SIRE BIRTH-YEAR GROUPS: THE GENETIC TREND IN THE BREEDS

		Angus			Hereford		Polled Hereford		
Year	Birth	Wean[2]	Year[2]	Wean[2]	Year[2]	Birth	Wean	Year
1964	+ .94	-1.8	-11					
1965	+ .86	-5.7	-16	- 7.4	-17			
1966	+ .46	-4.4	-10	- 6.7	-14			
1967	+ .49	-3.2	-14	-10.0	-18	+.61	-15.5	-18
1968	+ .51	-3.0	- 9	- 6.1	-11	+.17	-13.1	-21
1969	+ .04	-3.4	- 8	- 4.2	- 8	+.27	-10.2	-17
1970	+ .30	-0.9	- 2	- 3.8	- 7	-.27	-10.4	-15
1971	+ .04	-0.5	- 3	- 2.2	- 4	-.38	- 6.4	-10
1972	- .49	+0.4	+ 2	+ 1.0	+ 1	-.03	- 4.2	- 7
1973	- .22	+0.6	+ 3	+ 2.8	+ 5	-.52	- 1.3	- 3
1974	- .27	+2.2	+ 7	+ 5.6	+10	-.06	- .7	+ 1
1975	- .44	+4.0	+11	+ 7.5	+14	-.02	+ 6.2	+ 9
1976	- .37	+3.6	+13	+ 9.6	+19	+.32	+ 6.5	+10
1977	- .65	+5.8	+16	+ 7.9	+17	+.24	+ 9.1	+12
1978	-1.13	+6.4	+20			-.26	+12.5	+20
1979						-.80	+11.4	+16

[1]Pounds
[2]These effects do not consider the selction bias due to culling at weaning, as does the Polled Hereford data, so are not comprable with it.

TABLE 2. THE GENETIC TREND OF THE BREEDS, REGRESSION (POUNDS/YEAR)

Breed	Birth	Wean[2]	Year[2]
Angus	-.12	+ .79	+2.48
Hereford		+1.64	+3.27
Polled Hereford	-.03	+2.45	+3.52

[2]See (2) of Table 1.

The EPDs reported on sires are a better evaluation of one-half of their breeding value than is the Breeding Value Ratio (BVR) for the same weight on the performance pedigree of the sires. This results because the competition of a sire is considered in the sire evaluation analysis. The competition is the progeny of other sires in the contemporary groups. Ratios are averaged in the estimation of BVRs. Thus, a sire with a high BVR may never have been compared with superior sires or one with a low value may have had really stiff competition. Even though the EPD considers progeny only, it is a much more accurate evaluation of the sire than is the BVR. However, the Maternal Breeding Value Ratio is still the best method of evaluating transmitting ability for maternal performance as reflected in weaning weight.

Consider getting young sires listed as soon as possible. To do this in one breeding season using field data with two contemporary groups per herd (male and female), table 3 gives some alternatives in round numbers. This gives an accuracy of .70. It is the young sires that make a breed.

TABLE 3. NUMBER AND DISTRIBUTION OF PROGENY FOR AN ACCURACY OF .70 IN ONE SEASON

Calves/Sire CG[1]	Calves/CG	Herds	Total Progeny
2	10	9	36
5	15	6	60
10	30	5	100

[1]CG = contemporary group

In evaluations to come, we will build on what has been learned will be used and new features will be added that will make future evaluations more accurate than the previous. The results are the first step in utilization of standard performance records for sire evaluation. Treat them as such. In the reports to follow, particular sires will change rank since no EPD has perfect accuracy.

Use of Sire Evaluation

When the breeder has animals involved in his breed's sire evaluation program that rank low on the traits considered important by the breeder, he must acquire sires to bring about change. And the joy within an otherwise grim situation is that the breeder knows the herds to which he can go to get such germ plasm, because the comparisons among sires are fair. That is, genetic differences are compared, not management induced differences.

If the breeder's animals are high-ranking sires then and only then should a breeder even consider intensifying this superiority through line-breeding to a line or a particular superior sire. That is, the breeder can consolidate his superiority and make the product he offers for sale more prepotent for the superiority. Only when a breeder cannot find superior sires to his own sires should such a program be designed. Breeders have no control over the inbreeding process; only through selection can the breeder determine his direction.

Now consider just how best to exploit the results gained by participation in a national sire evaluation program. Breeders not participating can still benefit, but it involves a guess rather than the use of fair comparisons. Suppose a breeder, after studying the sires belonging to a particular breeder, decides such performance is what he wants. There are several avenues open to introduce new germ plasm in his herd.

Now a breeder can use artificial insemination to sample sires he wishes to use in his herd. Better yet, if it can be ascertained that a particular breeder has a creative breeding program, then selection of one or more sons of the evaluated sire provides a way to move a generation ahead. Sons should be selected using their own performance, plus the progeny test of the sire used as a sib test on the son. If the sire evaluated is older, it may be possible to acquire grandsons from the breeder. If the breeder is using a son in his herd, the grandson is a good bet. Breeders doing this are simply reaping extra benefits from the breeding program of the breeder.

Using sons or grandsons of a superior sire brings up some problems that still exist in pedigree breeding. Promotion of individual sires has been the rule. Much time has been wasted by breeders in recovering promotion dollars put on an inferior sire. Many herds have risen to prominence on a particular sire and dwindled from the failure to find a superior son worthy of promotion. What must be promoted and sold from a breeding program with a rapid rate of genetic change is simply the output of the program, which incidently is a group of sires adequately evaluated for the use to be made of them. They may be sons, grandsons, or great grandsons of a particular sire, but this is really incidental. Promotion dollars must be spent publicizing the creative program, not a particular sire. A good way to see if a breeder truly has a superior program that has involved selection over time is to study the ratios of own performance on the ancestors in the performance pedigree put out by the association.

Embodied in yearling sires is the next generation of genetic change. On the average, these yearlings should out-perform the sires from which they came. Of course, this depends on the merit of the breeding programs that produced them. When the available information is coupled with the average performance of collateral relatives, the accuracy of selection can be increased. This is one of the best ways to utilize the progeny test of a sire. His progeny are all half-sibs of the sons he produces. Although only half as accurate as when used as progeny, the sib average is very useful for selecting among the sons of sires on national sire evaluation. Now the progeny evaluation of a sire coupled with the previous information starts to differentiate among breeding bulls and other bulls.

Clearly, elite breeders will be classified as such based on their rate of genetic change. One must be ahead to lead! Making rapid rates of genetic change at the top of the pyramid requires the accurate evaluation and use of young sires and their replacement by even better sons with all deliberate speed. To reduce risk, due to the lower accuracy of the performance test when compared with the progeny test, requires the use of several sons to ensure that the best is used. Provided this is done, the product offered for sale will be sons of sons and so on, so the promotion must be based on the product of the breeding program rather than on the close relationship to an old sire that was highly promoted through the shows or whose progeny did well in national sire evaluation several years ago. Sons of the current sires become the critical issue in buying germ plasm. The power of the performance test to isolate superior sires cannot be ignored. Elite breeders will be using their own yearling sires on their best dams, which should be their heifers. Such breeders will use older sires only if their breeding values are superior to those of the yearlings. Then linebreeding to this sire is a possibility.

SUMMARY

The movement from purely subjective evaluation of breeding stock to more objective means of predicting the breeding value of animals for economic performance has progressed at a surprisingly rapid rate in the last five years, with the involvement of the breed associations in real performance evaluation. The records belong to the associations. They are proud of them and are willing to promote and use them for breed improvement. Adaption by breeders of breeding technology, specifically breeding value use and participation in national sire evaluation with the produce of his creative breeding programs, will produce breeds that can become even more relevant to commercial beef production. Sire evaluation is still the key to beef improvement, even as it was when the International Grand Champion bull was slapped in the AMPHITHEATER at the Chicago stockyards. Now the criteria of choice is the value of the sire as a parent as predicted from the available progeny records of the breed associations.

34

SWITCHING ON THE LIGHT
IN THE GENETIC DARKROOM

Henry Gardiner

It is generally agreed that the main purpose of pure-bred cattle in the U.S. is to improve the commercial cattle herd in the U.S. Registered cattle have been bred and used for this purpose in this country for almost the last 100 years. However, about the only thing our breeds have retained with consistency is their color pattern.

With the use of computers and some complex mathematical formulas, some of the worlds' best animal geneticists have developed ways in which we as animal breeders can finally become much more consistent in improving various traits of economic importance in our cattle.

With the aid of extremely intelligent and dedicated scholars, such as Dr. Richard Willham of Iowa State University, some of the secrets of animal breeding have been unlocked. For 100 years, we have floundered in the dark groping for genes that would make us more efficient and more profitable producers, but more often than not we have produced cattle that were only average, or less than average for their breed.

Now when we go to the genetic supermarket to select the genes for the next generation of calves, we can shop where the lights are turned on and the traits have been progeny tested and accurate numerical values have been given to economically important traits. But do you realize that a large number of breeders are selecting their genes without bothering to shop where the light is turned on? What would your meals be like if your wife bought your food in a pitch dark store? The breeding system we have had until now is similar to a grocery store that has had all the labels removed from its cans and boxes. We thought we could tell what was in the can by its size and shape but if we found one with a purple ribbon on it we knew we had hit a genetic jackpot. The fact is that this type of breeding program has worked very poorly in the past and there is no reason to think it should work any better in the future. It is about time that we realize we cannot choose our genes by looking.

I am reminded of two bulls in the Angus breed, Canadian Colossal and Shearbrook Shoshone. Canadian Colossal weighed 2,500 pounds and was about 58 inches tall. Shearbrook

Shoshone weighed 2,200 pounds and was barely 54 inches tall. To eyeball the two, Colossal would have easily been the better bull. Yet in the sire summary Colossal's calves weighed 7.6 pounds less than the average of all the yearling calves in the summary. Shoshone's calves weighed 56 pounds over the average of the summary. As yearlings, the two sire groups had over a 63 pound difference in favor of the smaller bull. The smaller bull also had calves that were 3.4 pounds heavier at birth on the average, and his daughters gave him a maternal breeding value of 109 compared to 102 for Colossal. I think this is an illustration that you just cannot tell by looking. If you select your genes in a container without a label on it, you may not get what you expect.

Do not get the idea that I am against livestock shows. I am not. They are the best way to promote cattle, meet people, and create enthusiasm for the business, especially among young people. But livestock shows are a very, very poor place to make our genetic decisions. I know, I have tried that way with very little success. I am convinced that for consistent genetic improvement we must use bulls that have enough tested progeny on the ground to give their traits an accuracy of near or above .90. At the age we show cattle there is no way we can have that information available on our show animals.

Since 1970, our ranch has been progeny testing bulls. Our first progeny tests were in the Certified Meat Sire program of PRI and our more recent testing has been in the sire evaluation program of the American Angus Association. We have tested 40 some bulls by getting at least 30 or more progeny from each bull. Our payment for doing this testing was usually taken in the form of semen to use on our registered herd. This might be semen from our own bull we were testing or semen from a more proven bull that was accessible by the herd we were testing.

In the last 12 years we have used a lot of well-known bulls of the breed. My wife, who is my most severe critic, has been complaining for several years that we were not making much progress in our breeding program. I explained to her that genetic change is very slow and that she would just have to have more patience. When the last sire evaluation report came out, there were 23 bulls listed in it that we had used in our breeding program over the last 12 years. Included among these bulls were major show winners, sale-topping bulls at well-known auctions, and sires of show-winning cattle. Quite a few of the major Angus herds in the U.S. were represented. The average genetic value of these 23 bulls is summarized in the table below.

AVERAGE GENETIC VALUE OF 23 BULLS USED OVER THE LAST 12 YEARS

Avg birth weight	Avg weaning weight	Avg yearling weight	Avg maternal breeding value
+.1#	+3#	+9#	99.5

It is a little embarrassing to show anybody these figures, but I think they are probably very typical of a lot of purebred operations struggling to breed better cattle. I showed these figures to my wife and said, "Honey, here is why we have been making such slow progress." She looked at them and said, "Don't you think maybe you ought to let one of the boys manage this operation?"

With a little panic in my voice I pleaded for a little more time. I then showed her the average genetic values of the four bulls used in our herd after I based my selection of bulls on the Angus sire-evaluation report.

AVERAGE GENETIC VALUE OF 4 BULLS USED IN LAST YEAR'S BREEDING PROGRAM

Avg birth weight	Avg weaning weight	Avg yearling weight	Avg maternal breeding value
+.9#	+19#	+51#	106

After looking at my second set of figures she could see that we should make about as much progress on yearling weight in one year as we had before in 6 years and instead of losing maternal value by one-half of a percentage point each year, we should be gaining by 6% in one year. Her comment then was, "I hope your last sets of figures are right. You know you don't have much time left to get something accomplished." She might well have been speaking to all the purebred industry. With our slow generation turnover and a tarnished reputation for breed improvement there aren't any of us who have much time left.

What about my wife's question about the sire evaluation data being an accurate measure of a bull's genetic ability? After progeny testing bulls for 12 years I am convinced that the values that have high accuracies are amazingly accurate. However, I have seen data with accuracies below .80 that can change more than one would expect. In all the bulls we tested, there was not one that did not sire some good calves. The really high ranking bulls have a much higher consistency than the poorer sires. The good ones sire very few calves that will rank below average.

If we study the data available in sire evaluation, it becomes apparent why genetic improvement has been so difficult. When I first became aware of breeding values, it seemed even the best breeding values were very low. A value of 105 is quite good and a value of 110 is about as high a trait value as there is in the breed. If an animal has a trait breeding value of a respectable 104, this would mean that for this trait that particular animal is 4% better than the animals with which it is compared. This isn't impressive until we realize that the progeny of this animal will receive one-half of this value since one parent contributes only half of the genes of an offspring. Thus, on the average, the offspring would only be 2% better for the trait because of the genes received from the parent with 104 breeding value.

It is important to remember that an animal doesn't have just one breeding value but a breeding value for each trait measured. The most common breeding values are given for weaning, yearling, and maternal traits.

I used to think there should be animals in the breed with breeding values of 120 to 130. When Dr. Willham turned the spotlight on the genetic ability of our cattle, such values as this did not exist. We do have a spread of about 20% in breeding values from a low of 90 to a high of 110. This is certainly a significant difference that should allow breeders to make remarkable genetic change if they utilize the superior genes available generation after generation. Until now we have just used an occasional good bull. The odds were very much against using two or three superior bulls in a row. Now, however, we can decide which traits are important in each of our breeding programs and then select sires of known superiority for these traits. The use of top bulls on other tops bulls' daughters should start us on the way to four generation pedigrees with every animal on that pedigree a superior animal. When this occurs we will make dramatic genetic progress. The breeders of such animals will find a ready market for these cattle that will consistently and dramatically out-produce other cattle that have been bred in a hit and miss type program. Maybe then we might see breeding values approach 120. I think it is possible.

Under the present system of computing estimated breeding values for young animals without progeny, it has been my experience that those animals with very high estimated breeding values will usually have those values drop as they get progeny. For example, we just finished a progeny test on eight young sires. The average estimated breeding value for weaning weight for those young sires when they were yearlings was 107.5. The weaning weight breeding value for those eight bulls in our herd after we had tested about 20 progeny from each bull was 99.5. Theoretically about one-half of the values should go down and one-half of the values should go up. In this case, all eight breeding values dropped. They dropped an average of eight points. I repeat—breeding values with low accuracies are not very accurate.

This relatively new way to evaluate cattle and their breeding abilities has a number of new terms that one needs to be familiar with to communicate accurately. Such words and phrases as "estimated breeding value," "ratios," "possible change," "accuracy," "expected progeny difference," "maternal breeding value" should be rather well understood if you are going to communicate and draw conclusions in this new way of breeding cattle evaluation.

I am reminded of the time that I spent in the army in Germany in the 1950s. Several of us had been in a night class learning to speak a bit of German. We went to a local gasthaus to order a meal. Most of the waiters spoke good

English but my friend wanted to order in German so he said "Herr Ober, Ich mogen eine heiss Hund bitte." The waiter gave him a very strange look. If you would translate what he said word for word it would be, "Mr. waiter, I would like a hot dog, please." However what he said in German was, "Mr. waiter, I would like a dog in heat, please."

In a recent ad in the Angus Journal underneath the photograph of a fine looking bull the caption said, "This bull's ability to transmit size and continued growth is unequaled in the breed." Underneath this they gave his yearling estimated breeding value; with a .91 accuracy, the bull's value was 100. In journalists' language this ad described this bull as the greatest. In performance language the ad revealed he is just an average bull for transmitting size and growth. With a .91 accuracy, he is not likely to get much better. It was a poor ad.

Another lesson in animal breeding can be learned by studying a sire-evaluation sire summary to determine the frequency of superior bulls. In the 1981 Angus sire-evaluation report, 673 bulls were evaluated. In yearling weight, expected progeny difference they varied from a low of -46 pounds to a high score of +77 pounds. If you were to go through these 673 bulls and pick out just the bulls that had at least a +40 pound EPD, you would narrow the list to 53 bulls. But I do not think we can select for just one trait in cattle breeding. If you picked all the bulls listed that were +40 or higher for yearling and 102 or higher for maternal, you would have a list of 22 bulls. I believe the birth weight should be limited so that it is not too heavy. If you would not use a bull whose progeny averaged over a +4 pounds at birth, then your list would only include 16 bulls. And if you really wanted to make some improvement on the maternal ability of your cow herd, 105 would be better than 102. This would leave you with only 4 select bulls from the original 673.

These 673 bulls were not just a gate-cut selection. Each one of these bulls was good enough to be used artificially in a number of registered herds, resulting in a large number of progeny, or he would not have been on the list. The oldest bull on the list was born in 1960 and had a registration number of just over 3 million. The youngest bulls were born in 1978 and had a registration number of just over 9 million. About 35% of Angus registrations are bulls so these 673 bulls born over an 18 year period would be the very best of 2 million bulls. With relatively modest performance requirements, those 2 million bulls were culled down to 4 bulls.

It thus becomes evident why genetic progress has been so slow or, in some cases, has gone backwards. If the really great breeding bulls are this rare, are great breeding cows any more frequent? Probably not. Without a large number of progeny, I mean 20 or more from each cow, those highly superior cows are going to be hard to find. That is something to think about when considering embryo transplants.

I guess most good things in life are rare. I recently spent several hours waiting for my flight at O'Hare Field in Chicago. The airport was very busy with hundreds of people walking by where I was seated. I decided I would run a survey of the frequency of pretty girls. On that particular day a pretty girl walked by on the average of only once for each 674 people surveyed. Something certainly should be done to improve this frequency.

As we become aware of how scarce superior bulls are, we may want to reexamine some of our breeding customs. We have heard in the past few years about rolling over the generations. We may want to cull more deeply and roll generations less frequently. Some breeders only use a bull for a couple of years and then go to one of his sons. It appears that many of these top bulls do not produce a son that is better than he is. The really good ones are very rare.

What about narrowing our genetic base too much by widely using just a few bulls to sire a large percentage of our national purebred herd? If those bulls are really superior, I think it would be a good thing. The Holstein breed has been outstanding in the way they have used genetic material available to them to dramatically increase milk production. Last year one-third of all calves registered in that breed were sired by one bull. I think there are more problems caused in all breeds right now by using sorry bulls than there are from too narrow a genetic base. If a bull is average or below in all traits he does not have the genes to broaden the genetic base. If a breed has only 53 bulls instead of 673 bulls that can contribute to its genetic base, they need to realize that fact and proceed to use what will give them some actual improvement in the direction they want to go.

If there are undesirable genes in some of these widely used bulls, it will be very quickly discovered and they can be quickly discarded, if necessary. This is less of a hazard to a breed than the old way of knocking them in the head and hiding them for a generation. Of course, the original breeder many times would be unaware of the problem if he had not inbred the animal very much.

As yet, the breeders of various breeds that I know about are not utilizing the performance data available to them as much as they should. Thus there is a great opportunity for a dedicated breeder who believes in breeding good cattle to get started on a program that should pay big dividends in 10 years. Of course, some breeders do not want to stay in business that long. (The average length of time that a registered Angus breeder stays active as a breeder is 7 years!) The following tables indicate the selection criteria that breeders are using for their herd bulls.

TOP 5 ANGUS BULLS LISTED IN ORDER OF THE NUMBER OF CALVES
REGISTERED IN 1981 WITH THEIR BREEDING VALUES OR EXPECTED
PROGENY DIFFERENCE

Bull	Birth wt EPD	Yr wt EPD	Maternal breeding value
1	+1.8#	+39.6#	101
2	+1.2#	+17.4#	101
3	+5.6#	+56#	109
4	+1.1	+65.6#	105
5	Young show bull no data available yet		

Bulls 1, 2, 4, and 5 have been used because they were
successful show bulls and have produced top show cattle.
The cows bred to these bulls for 1981 calves were bred before this performance data was published. Bull 4's data
would certainly attract breeders looking for performance
after they had seen it. Bull 3 is probably the only bull on
the list that was selected because of his performance. He
has one of the top maternal breeding values in the breed.
When looking at this table remember that the top bull for
yearling weight EPD was a +77.

In the Simmental breed their trait leader for yearling
weight had an EPD of +53 pounds.

TOP 5 SIMMENTAL BULLS LISTED IN ORDER OF THE NUMBER OF
CALVES REGISTERED

Yr weight EPD

Bull 1	+20#	
2	- 7#	(used because of good daughters)
3	+ 9#	
4	+26#	
5	+ 4#	(used because of calving ease)

The Limousin breed is shying away from growth and
muscling. Their most widely used bull has a -22# EPD for
yearling weight.

In summary we now have genetic data available that a
dedicated purebred breeder can use to breed cattle far
superior to any that have ever been produced before. To do
this the superior animal of tomorrow will have to have high
breeding values in almost every animal in its pedigree for
at least three or four generations. To breed a herd of
these super beef builders, there will be no room for experimenting with young unproven bulls no matter how great they
look. There are a very few bulls in each breed that are
capable of making relatively rapid genetic progress. The
consistent and sole use of these kinds of bulls by a large
number of breeders will result in new generations of cows
and bulls that, after rigid culling and selecting, will produce even more superior animals. The dairy people have already shown that such improvement is possible. All we need

to do is set our goals and then be persistent in our pursuit of such goals.

Hopefully we may soon have a breeding value measure for fertility. But whether we do or not, we must maintain a high level of fertility in our herds while we improve other traits.

This is the most exciting time that beef cattle breeders have ever seen. The opportunity is there. I invite you to join me in this challenging future. Maybe, just maybe, we might hold on to our jobs of managing these beef herds a little longer than some of our critics thought we could.

35

IMPROVING SELECTION METHODS
WITH PERFORMANCE RECORDS

Craig Ludwig

INTRODUCTION

Systematic record-keeping of performance is fundamental to genetic improvement in beef cattle. A clear understanding of genetic principles is essential for developing the most effective selection programs using such performance records; these principals are based on results of research and practical application in beef cattle.

Although the beef cattle industry is composed of several segments, it is totally dependent on the genetic engineering of the seedstock or purebred producer for total industry and improvement. The selection and breeding decisions made today in the purebred breeder's herd will require from 4 to 6 years to benefit the commercial industry. Thus, the selection and decisions made by a registered breeder should be based on sound research data to avoid mistakes in this own program and to avoid leading the commercial industry in the wrong direction.

RESPONSIBILITIES OF THE PUREBRED BREEDER

Increased efficiency of production is necessary if beef is to maintain and improve its position among other high-quality protein foods. The opportunity for genetically improving productive efficiency and product desirability rests in the hands of the purebred breeders who produce the bulk of the bulls going to the commercial herds.

Selection and other principles of good animal breeding are the primary tools available to purebred breeders for making genetic improvement. Most of the opportunity for influencing the selection in beef cattle is through the bulls. To fulfill this responsibility to all segments of the beef cattle industry (the commercial producer, the feeder, and the packer, and consumer), the purebred breeder must have a working knowledge of genetics, or the science of heredity, along with an appreciation of all traits economically important to the beef industry. The purebred breeder should also understand the correct procedures for measuring or evaluating differences in these important traits and be able to develop effective breeding practices for making genetic improvement.

The primary goal of all producers, regardless of the area of the country he operates in is, **to maximize profits optimizing production from the available resources.** Optimum production from the available resources does not always mean maximum production. Beef cattle must be adapted to the area and region where produced and must be able to utilize the available feedstuffs to produce the most lean beef possible.

Seedstock producers, as well as the commercial producers, must realize that real genetic progress or genetic improvement is a slow painstaking process that comes as a result of careful planning and selection practiced over a long period of time. Real genetic improvement will not come in a straitght line relationship for any of the traits selected. Regardless of the trait selected there will be some ups and downs in the performance of those traits from year to year, but the overall, long-term effects will be positive.

FACTORS AFFECTING RATE OF IMPROVEMENT FROM SELECTION

Factors that affect rate of improvement from selection include 1) heritability, 2) selection differential, 3) genetic association among traits, and 4) generation interval.

Heritability

Every living thing is a product of environment and genetics. It is the genetic part of each animal that is inherited from its parents, and it is that part of the animal's make-up that must be measured.

Heritability is the portion of the measured or observed differences between animals, that are transmitted to the offspring. Thus, heritability is the proportion of the total variation caused by additive gene effects. The higher the heritability for any trait, the greater the rate of genetic improvement, thus effective selection programs select for those traits that are highly heritable. When traits of equal economic value are being considered, the trait with the higher heritability should receive the most attention.

If heritability is to be meaningful, all animals from a group from which selections are made should receive the same treatment or be exposed to the same environment. When animals of the same sex receive equal environmental treatment they are called contemporary groups.

When contemporary groups are accurately defined and equally treated, a larger proportion of the measured performance differences between the animals in the group will be a result of genetics and the effectiveness of selection will increase.

The heritability of any trait can be expected to vary slightly in different herds, depending on the genetic variability present and the uniformity of environment. Heritability estimates indicate that selection for most of the economically important performance traits should be reasonably effective.

Selection Differential

Selection differential is the difference between the selected animals (females and males) and the average of all the animals from which they were selected. Selection differential is determined by the proportion or percent of the animals needed for replacements and the number of traits considered in selection, and the differences that exist among the animals in the herd. If the average yearling weight of all the bulls in a herd is 1,050 lb and an individual bull retained for breeding weighs 1,150 lb, the selection differential is 100 lb.

In beef cattle, selection differentials are limited for some of the traits of economic importance. The relatively low reproductive rate of beef cattle usually necessitates keeping about 30 to 40 of the heifers for replacements to maintain the herd. Thus, because such a large percentage of the heifers must be saved, the selection differential cannot be very great between those saved vs those that are not.

Most of the opportunity for selection is among the bulls, because a smaller percentage of the bulls must be saved for replacements. If artificial insemination is used, even more opportunity for selection exists because fewer bulls can be selected from a wider source of genetic variation.

As a general rule, herd sires or bulls should be selected from the top 5% to 10% of the available animals. If such herd sires are used, maximum genetic progress can be expected for the economically highly heritable traits.

When too many traits are included in a selection program, the opportunity to increase the performance in any one trait is reduced signficantly because the selection differential is reduced for any one trait. Every effort should be made to obtain the maximum selection differentials for the traits of greatest economic importance with the highest heritability.

Genetic Association Among Traits

A genetic correlation among traits is the result of genes favorable for one trait causing either a favorable or unfavorable expression of another trait. Such genetic correlations can be positive or negative. When the association is favorable between or among traits on which selection is based, the rate of total merit is increased. When the genetic association is unfavorable among traits, the rate of total merit is decreased.

Available information strongly indicates a favorable association between rate and efficiency of gain during growth between 7 and 18 months of age. Likewise, a favorable association exists between yearling and weaning weights and between scrotal circumference and half-sib sister's age at puberty. An unfavorable genetic association reported in beef cattle is the association between birth weight, weaning weight, postweaning gain, and yearling or long-yearling weights.

Generation Interval

The generation interval is the average age of all the parents when their progeny are born. Generation intervals for beef cattle average from 4 1/2 to 6 years in most herds (table 1).

TABLE 1. AVERAGE GENERATION INTERVALS

Species	Males	Females
Beef cattle	3-4 years	4.5-6.0 years
Sheep	2-3 years	4.0-4.5 years
Swine	1.5-2.0 years	1.5-2.0 years
Chickens	1.0-1.5 years	1.0-1.5 years

Taking into account the four factors affecting the rate of improvement include 1) heritability, 2) selection differential, 3) genetic association among traits, and 4) generation interval, how much genetic progress can be made in a trait per year? The yearly genetic progress that can be made in any trait is calculated by the following formula:

Annual progress for a trait = heritability x selection differential ÷ generation interval.

It is generally agreed that the heritability for any particular trait and the potential selection differential within different herds is about the same in most cases. Assuming that this is true, the generation interval is the controlling factor as to how much genetic progress can be made for any selected trait. Progress can be greater when the generation interval is shortened, which can be accomplished by culling cows and herd sires on the basis of the production records.

The opportunity to shorten the generation interval usually comes from selection of young, genetically superior herd sires.

To illustrate how the length of generation interval affects the amount of progress made in selection, let us compare two TPR (Total Performance Records) herds enrolled in the American Hereford Association's performance program (figure 1).

In herd 2, young bulls and young cows are being used for breeding purposes and the generation interval is one-half the generation interval in Herd 1 where older bulls and cows are being used. In a period of 7 years there is the opportunity to produce two generations in Herd 2 as compared to one generation in Herd 1.

Let's assume that both Herds 1 and 2 use yearling weight as their selection criteria. We will also assume that the heritability for yearling weight is 60% in each herd and the selection differential for yearling weight in each herd is 75 lb.

Thus, the annual amount of progress in yearling weight in Herd 1 can be calculated as follows:

Yearling weight improvement
= (.60 x 75 lb) ÷ 6.76 years
= 45 lb ÷ 6.76 years
= 6.66 lb per year

Under ideal conditions, Herd 1 could improve the average yearling weight of its cattle by 6.66 lb per year, or approximately 66 lb in 10 years.

In Herd 2, where the generation interval is 3.40 years, the annual improvement in yearling weight is calculated as follows:

Yearling weight improvement
= (.60 x 75 lg) ÷ 3.40 years
= 45 lb ÷ 3.40 years
= 13.24 lb per year

Under the same conditions as Herd 1, Herd 2 would be able to improve yearling weight by 13.24 lb per year. Over a 10 year period, Herd 2 could improve yearling weight by approximately 133 lb, which is very typical of what the better U.S. Hereford breeders have accomplished.

The above principles, as described for yearling weight improvement, hold true for any of the economically important traits considered in a selection program. It is obvious that the heritability of a trait and the selection

Herd 1

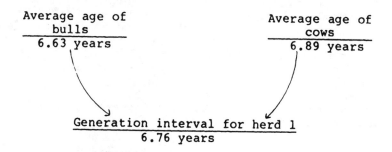

Average age of
bulls
6.63 years

Average age of
cows
6.89 years

Generation interval for herd 1
6.76 years

Herd 1

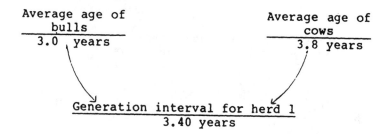

Average age of
bulls
3.0 years

Average age of
cows
3.8 years

Generation interval for herd 1
3.40 years

Figure 1

differential are about the same for all breeders. Thus, breeders who use generation interval to their advantage can make the most real genetic progress.

Over the past 10 years the top beef producers routinely have used superior young bulls in an effort to shorten their herd's generation interval and have, as a result, improved weaning weights, maternal breeding values, yearling weights, efficiency of gain, production efficiency and product desirability.

While breeding, selecting, and culling cattle, one must continually remind himself that the amount of annual progress made for any selected trait is controlled by the following formula and that the generation interval is really an important controlling factor in the formula.

Annual progress for a trait = (heritability   x selection differential) ÷ generation interval.

Beef cattle producers must remember that total performance records provide a basis for selecting the genetically superior animals. If genetically superior animals are selected and used as breeding animals, improved performance and improved efficiency of production will take place over a period of time.

TRAITS FOR CATTLE BREEDER DECISION MAKING

As in deciding upon other production goals, today's cattleman or seedstock producer must decide which traits he will use to select and cull his herd. He should write down the traits considered to be the most economically important. By writing the traits down each year when making selection and culling decisions, the seedstock producer can remind himself that herd priorities selection emphasis should be based on traits that:
- are economically important to production
- are highly heritable
- are easily measured
- can strengthen the herd's overall production level

Some producers of seedstock breeding cattle place too much emphasis on traits of very little economic value and as a result find their breeding program falling behind those breeders who select for economically important traits.

Relative Economic Value and Consideration of Performance Traits

All traits of economic value should be considered in selecting beef cattle. These traits can be divided into the following groups:
1) Reproduction traits, 2) growth traits, and 3) carcass traits.

Reproduction Traits. a high level of fertility or reproductive performance is basic to an efficient beef cattle industry because the percentage of the total beef cattle population composed of cows is high, requiring a major portion of the resources used in beef production. No single factor in commercial cow-calf operations has greater bearing on production costs that does percentage calf crop. Also, a high level of reproduction is fundamental for genetic improvement. A large calf crop decreases the percentage that must be saved for replacement and this increases the selection differential possible for other traits. Both the male and female should be considered in selecting for reproduction because reduced fertility of either the bulls or heifers can cause reduced calf crops.

Fertility is a complex trait. A live calf at weaning is the product of a long sequence of successful events from the time a cow is turned with a bull until her calf is weaned. The bull must have a high degree of libido and be physically capable of mating and producing enough viable sperm to maximize fertilization. The female must reach puberty as a heifer, or must have a sufficiently early calf and short postpartum as a cow so as to exhibit a fertile estrus early in the breeding season. Ovulation, implantation, embryonic and fetal development, and parturition must occur without failure. The calf must consume colostrum and milk vital for early survival and ward off other hazards before weaning. With such a long chain of events involving interrelationships between the sire, dam, and offspring and their environment, the probability that any one event can succeed can be high--yet there may be a low probability that the product of all events will succeed (percentage calf crop weaned). A breakdown at any point in the sequence can be devastating.

Variation in reproduction is large. For example, percentage calf crop can easily range from 75 to 90 percent. Results indicate that heritability of such traits as calf crop, pregnancy rate, and calving interval is low to moderate (10% to 30%). Therefore, most of the variation may not be caused by additive genetic differences between herds but rather by differences in management, nutrition, herd health, and other environmental factors. There are indications that certain components of fertility such as age at puberty and first service conception rate in heifers are more highly heritable than is calf-crop percentage and are related to sire or half-sib scrotal circumference.

When heritability of fertility is low to moderate, detailed records should be kept on reproductive traits. Even with low to moderate heritability rigid culling to remove open cows or problem breeders can help provide a more profitable reproductive pattern in a herd. Selection pressure on reproduction is not wasted, because reproduction is of overwhelming economic importance. The return from culling open and other problem cows and keeping pregnant cows increases the number of calves weaned relative to cost of production and provides greater opportunity to select for other economically important traits because of a larger total number of offspring available.

In a study at Colorado state, the estimates of the genetic correlation between scrotal circumference in yearling bulls and age at puberty in half-sib heifer mates was .71, which indicates a strong, favorable relationship. As scrotal circumference increases, sisters (half-sib heifers) reach puberty at younger ages. Thus, these results indicate that young bulls with above average scrotal circumference and better than average semen should produce heifer offspring with earlier inherent ages at puberty. Further studies at Colorado State indicate that early age at puberty in females is favorably related to other measures of fertility and productivity. Research results indicate that selection for earlier age at puberty should result in earlier conception dates during the breeding season.

Other research on scrotal circumference indicates that the measure is a very important part of a bull's breeding soundness exam. Recent research indicates that scrotal circumference is favorably related to semen quality and volume in young bulls, thus young bulls with larger scrotal circumference measures should produce more viable semen and, consequently, be able to serve more cows.

Scrotal measurement tapes are a very cheap tool that can be used by almost anyone to improve the reproductive performance of a herd. The procedures for measuring scrotal circumference are simple and there are apparently large dividends for use of the measurements.

Growth traits. Growth traits measured through birth weights, weaning weights, yearling weights, weight ratios, and growth breeding value ratios are all indicators of the genetic merit of an animal for growth. Animals with favorable data and above-average ratios within any of these measures of performance should be considered for herd replacements, while those with low measures of performance can be culled from the herd at a relatively young age.

Growth traits are regarded as being moderate to highly (20%-50%) heritable. Data from the American Hereford Association's performance records indicate that there is a significant amount of variation between the performance of animals within and between herds for all the measures of growth.

Birth weight, while considered a growth trait, is generally considered as a reproductive trait by most cattlemen. Birth weight is both highly heritable and highly correlated with all the other major criteria for measuring growth.

A high degree of calving difficulty cannot be tolerated for profitable beef production. Not only is the expense of labor for assistance at calving a prohibitive factor, but calving difficulty can cut deeply into calf crop weaned by reducing calf survival and the rebreeding or conception rate of the cows that had calving difficulty. Research shows that calf mortality is four times greater in calves experiencing difficult births than in those not experiencing difficult births. Conception rates in the next breeding season are from 15% to 20% lower in cows that had difficult births as compared to cows in the same herd that had no difficulty.

If birth weights are a problem, then selection for lower birth weights appears to be the most effective criteria for improving calving ease.

Growth traits such as yearling weight, weaning weight, 18-month weight, feedlot gain, and final weight are all accurate indicators of growth, are easily measured, and are highly heritable. Because these growth traits are so highly heritable and easily measured, breeders can readily improve productive performance in their herds by selecting for the growth traits.

Growth rate is important because of its high association with economy of gain and its relation to fixed costs. Growth rate is generally measured over a time-constant period and results indicate that differences in growth rate can be appraised accurately in this manner.

Ratios for adjusted weaning and yearling or long-yearling weights have been the most common criteria for beef producers to select herd replacements and herd sire prospects. Until 1975, U.S. breeders used ratios for their selection and culling decisions. Now, because of modern technology, U.S. Hereford breeders are able to more accurately select their herd replacements and herd sire prospects because of the calculated growth breeding values for weaning and yearling weights. Breeding values for growth indicate the value of an individual as a parent.

The most accurate estimate of growth breeding value is a progeny test of a number of sires where each sire has a large number of progeny, but this is costly and slow. There are alternatives to use of a progeny test in predicting breeding values for growth. The first alternative is the individual's own performance for the growth traits.

An individual's own record of performance, if available, is always the best single record that can be used to estimate the individual's breeding value because the genetic relationship is as high as possible at 1.0. Thus, for traits that are highly heritable, say 60%, the estimated breeding value and the animal's real breeding value are fairly accurate. If the individual's own performance is not known, 10 progeny records would be required to yield comparable accuracy.

Another alternative is to use relatives' records in conjunction with the animal's own record to estimate breeding values. The primary relatives in beef records are paternal and maternal half-sibs and progeny. Thus, the records of relatives can be used in conjunction with the animal's own record of performance to help estimate growth breeding values of individuals.

Information from all sources including the individual, sire, dam, paternal half-sibs, maternal half-sibs, and progeny, when available, can be combined into a single estimate of breeding value for each animal in a herd.

When selecting herd sire prospects for the growth traits, one generally tries to select those cattle with ratios above 100. The truly superior breeding animals are identified by data analysis of performance measured at birth, at weaning, and as yearlings in performance programs, such as that sponsored by the American Hereford Associations.

Within the Association's Sire Evaluation Listing for 1981, growth breeding values in the form of Expected Progeny Differences (EPD), values are listed of 516 Hereford herd sires. The range in expected progeny diffences between the poorest bull and the best bull in the listing is 56 lb for weaning weight, 102 lb for yearling weight, and 9.0 lb for birth weight. With this amount of selection differential available and the high heritability estimates of the growth traits, today's Hereford breeder in the U.S. has turned to the Sire Listing to improve the economically important growth traits in his cattle. Using bulls with superior traits for growth and reproductive efficiency is the easiest method available to increase beef production from the available feed resources.

For six years now, U.S. Hereford breeders and the American Hereford Association have been calculating the maternal breeding value for all performance-tested bull calves, heifer calves, mother cows and herd sires, although this is not a growth breeding value.

Growth breeding values predict an animal's ability to grow. Maternal breeding values have to do with maternal ability and in particular with milk production potential. Maternal breeding values for young bulls and herd sires predict the bull's potential to sire daughters that have mothering ability to milk in the right amounts to wean a heavy calf yet rebreed for the next calf.

Maternal breeding values, like the growth breeding values, are an important part of the American Hereford Association's Sire Evaluation Listing 1981. Maternal breeding values are listed for all 516 Hereford herd sires listed in the boooklet and, like the growth rate predictors in the same booklet, have proven to be most useful to U.S. cattlemen and their desire to produce beef more efficiently. Unprecedented improvement in milk production from sound-uddered, efficient cows has taken place in the U.S. since the incorporation of maternal breeding values in most U.S. Hereford breeding herds six years ago.

Carcass traits. As known by cattlemen worldwide, carcass traits are highly heritable traits, which, if selected for, can be altered or changed. The relative economic value of carcass traits compared to reproduction or the growth traits is not nearly as important to the cattleman. The trend worldwide is to produce beef carcasses that have a high proportion of edible red meat as compared to fat or waste.

Variation in composition of carcasses (ratio of lean to fat) is the major factor influencing differences in carcass value.

Yearling weight and frame score are both highly related to growth and percentage of retail red meat in carcasses. Research data from numerous sources within the U.S. confirm the importance of yearling weight and frame score as indicators of carcass desirability. Research suggests that cattle of all sizes will produce desirable lean-meat carcasses if slaughtered at a physiological point based on the animal's frame size. The American Hereford Association initiated this research in 1969 and has since confirmed its findings in three separate studies with three other universities. Research personnel all agree that growth and frame size are the two most important items to consider when carcass lean meat is the desired end product.

Scientific and practical knowledge of the U.S. cattleman has shown that a biological equilibrium of fertility, growth, and carcass traits has provided for a more efficient total beef industry. Performance-proven sires, regardless of breed, under U.S. range conditions have proven to be the most profitable kind of bulls for the utilization of the available resources and the production of beef.

SUMMARY

Herd, and breed improvement, and increased efficiency of beef production, depends greatly upon the utilization of genetic selection principles and the use of superior breeding stock. Sire selection and use is the best method to improve the overall efficiency of a beef producing operation. The cattleman's problem is to find and use a new herd bull each year that when used on his cows will sire calves that are genetically superior to those produced the year before.

The selection of herd sires is the only real method available to the producer for herd improvement. Herd improvement results from:

1. Replacing a sire with his superior son.
2. Artificial insemination to a superior sire with proven performance records.
3. Selecting and using a superior herd sire prospect available from a fellow breeder.

The genetic opportunities are there if the breeder chooses to use them. Outstanding sires enhance every cattleman's effectiveness of selection and rate of genetic improvement in the breed or within a herd. Real opportunity exists for all breeders, and especially the purebred breeder, to enhance the effectiveness of his sire selection. The basis of selection is the resemblance between parent and offspring; if the resemblance is high for a trait, selection of superior breeding cattle results in above average offspring. The breeding value expressed in the record of his offspring provides a means of correlation between parent and offspring.

Sire selection and sire evaluation are basic to all beef breeding programs. High-performing herd sires are worth more and present a no-gamble situation for the breeders who use them. Use of superior bull with high values for growth and maternal breeding is an investment in the future for both you and the cow herd. Payoff for the investment is when the calves are born. Today, cattlemen have learned that true herd and genetic improvement is a result of long-range objectives, using cattle that have been selected and modeled to fit the needs of the entire beef industry.

Cattlemen with a clear understanding of the genetic principles--such as heritability, selection differential, and generation interval--can use these principles, along with use of breeding values, in making selection and culling decisions. Breeders who use such principles, plus sound judgement, will have a competitive edge in maximizing profits through optimum production from the available resources.

The shortest route available to the breeder to model and fit superior performance cattle into his breeding program is through proper identification and bull selection every year.

36

IMPROVING CATTLE AS A MARKET FOR MIDWEST FARM RESOURCES

J. David Nichols

We cowboys, ranchers, and farmers consider ourselves to be experts in a lot of different fields. One of them is football...we attend the Iowa State games every year and watch Oklahoma come up and demolish our spirits as well as our football team.

Something I've noticed about the farmers sitting in the stands is that they consistently yell advice to the coach. The day after the game they also have lots of advice as to what our team should have done on the third down and whether it should have punted or passed. I got to thinking about those games and decided that farmers and ranchers would really make lousy football coaches.

Now I'm going to draw a few parallels on how herd management would work if a rancher coached a football team the way he manages his cow herd. Imagine, if you can, a rancher football coach when certain players don't show up for practice. He says, "That's all right, Tommy. You don't need to show up for practice because you're a good guy and I'll give you another chance."

When players don't even show up for the games, the farmer coach says, "Oh, that's all right that you didn't show up for the game. Maybe you'll show up for the game next time."

These are the excuses ranchers give for their cows that don't have calves. So, if ranchers were coaches, they'd make similar excuses for the players who don't show up.

HOW WE'D PICK OUR TEAM

Now let's figure out how we ranchers would select our team members. The first thing we'd do is select them by visual appearance.

We'd line up all our squad. We'd walk out and say, "Hey, you look like a guard, so you play guard. You look like a tackle; you're wide in the hips and shoulders and obviously function follows form, so you play tackle. Oh, here's a long-legged tall one that has a lot of frame--we know that long-legged tall ones run fast and catch passes-- so you can be the end."

Here comes the important part--where we select the quarterback for the team. Now keep in mind that the quarterback for a football team is somewhat similar to the bulls used in a herd, so let's select a quarterback the same way we select our bulls.

We'd say, "Son, you're going to be a quarterback because your mother was an Olympic swimmer, your grandmother won a foot race at the Podunk Center County Fair, and your father played football for the Chicago Bears.

Therefore, you have a wonderful pedigree to be a quarterback. You have slim hips, wide hands, and a square jutting jaw. Therefore, you're the quarterback."

Now I think you all would agree that we wouldn't have much of a football team if this was the way we selected the players. But, in a sense, isn't this really the way we're running our cowherd?

Some of us aren't pregnancy-checking the cows; we let them calve year-around so they can calve when they want to and not have a breeding season. Some purebred breeders' selection programs are based on the fact that "We have got to keep this cow. She's a daughter of Snowflake, and Ole Snowflake is the grand old cow whose grandmother's half-sister won first at the county fair. She's been in the family for years. So what if Snowflake's daughter misses a calf. That's not heritable anyway."

QUESTIONABLE SELECTION METHOD

What we really want to do is breed Snowflake's daughter to Ole Blue Nut. Everyone knows that the Ole Snowflake-Blue Nut combination is the best thing going. If we can get enough cattle of this particular combination, we're going to have something really good. This is the mentality that pervades our industry when it comes to selecting cattle for reproductive traits.

Cow calving is a confused issue. Why are commercial producers so confused? First, when the veterinarian comes out on Easter Sunday morning and pulls a dead calf, the vet says, "My gosh, fella, you gotta quit breeding to that bull. You've got to get a small Angus bull to get rid of your calving problems."

The next night the producer goes to an extension meeting. The extension specialist says, "The thing that's going to make you money is to crossbreed your cows to a large, performance-tested bull." So our producer says, "By golly, that'll solve my problems. I'll breed my heifers to Maine-Anjou next year." So he goes to a breeder who lives across the road who says, "Hey, Hey! Cattle breeding is a super-complicated thing. Here we've got this pen of bulls. This bull is 48 inches high and will cost you $1,000; that bull is only 46 inches. He's only $300."

The producer says, "My gosh, the shorter bull has a better yearling weight." The breeder says, "Yeah, but the tall ones are worth more money."

WHY CONFUSION EXISTS

So what does the producer do? He goes to the bull test station and he finds out that if he doesn't have a bull that gains four pounds a day he's not going to be socially acceptable. And what will the cows think if they aren't bred to a bull that gains four pounds a day? Within days our producer hears he needs a smaller bull, a larger bull, a taller bull, or a 4-pounds-a-day-gaining bull. And you wonder why this guy is confused! We all have a tendency to accentuate our own biases or to merchandise whatever we have to sell. The producer hears what each is selling.

NO NEED TO WORRY

When we started and had only 100 cows, it wasn't too difficult to go check pastures several times a day. When we got up to about 600 cows, we decided that that was the end of the pasture checking.

So here's what we do. Cows at our place calve by themselves unassisted or die trying. I mean they die trying.

It took about two years before we could stop worrying about them. We check our cows once a day, tag and weigh the calves that were born in the last 24 hours, and we come back 24 hours later.

We check heifers three times a day. I used to check the heifers at night and there would be three heifers calving. I'd stand around awhile, go watch Johnny Carson awhile, then go back out and watch the heifers again. Soon I'd go get my wife Phyllis out of bed and say, "Phyllis, we've got to pull this calf." She'd change from her short-sleeved night gown into more appropriate clothing and away to the barn we'd go.

About 2:30 in the morning, we'd get back in after pulling a live calf. The next morning I'd be lying there in bed and the milk-truck driver would knock and say, "Hey, you've got three heifers calving." I'd jump out of bed, go down there, and first thing I knew I was spending all my time calving heifers.

The solution to that is to check them three times a day—morning, noon, and night. When I come in at night and put my feet on that footstool, any heifer that's going to calve that night is going to have to do it herself. This can and must be done.

CAN'T AFFORD TO PULL CALVES

I don't think Iowa farmers can afford to crawl off a tractor and go pull calves. The economics aren't there. With $.80 and $.90 calves, I can't crawl off a corn planter or a soybean planter to mess with calves. I think we have to breed cattle that can calve by themselves. We're running

about 1100 mother cows now. We have found in our operation
that when we check heifers three times a day, we'll have to
pull some calves.   However, when a heifer has to endure a
hard pull, she has only a 50-50 chance of getting re-bred.
This shouldn't be perpetuated.

How do we handle our bulls? We use bulls, plus some
AI. We have a 60-day breeding period once a year. Since we
start calving the first of April, the bulls are with the
cows from June 23 to August 23. As soon as the calves are
weaned, the cows are pregnancy checked. Any cow that's open
goes to market. We may not be making great genetic strides,
but we aren't letting freeloader cows stay around.

We have a few employees, and if one employee doesn't
show up for work, he's not paid. He doesn't expect to be
paid. But cattlemen will let an open cow stay around for a
whole year. It takes the profit from five other cows to pay
the feed bill on that one old cow he's keeping an extra
year.

"After you discover Ole Maude is open, if you love her
so much and want her to spend another Christmas with you,
you're going to have to carry feed to five other cows for a
year just so Maude can spend another Christmas with you."
Now, I don't have any cattle I love that much! You have to
get rid of the open cows.

WHAT THE COWS COST

Another thing that has to do with calf-crop percentage
is problem cows: cows with bad udders, cows that don't
claim their calves, or cows that can't take care of their
calves.

Here's what we do. We figure it takes four hours to
get a cow in. It takes four guys, one hour; two guys, two
hours; or one guy, four hours.

Our employees get between $8 and $10 per hour. Four
hours is $32 to $40. That's the profit on that cow. If
we've got to get a cow in because she isn't claiming her
calf or it can't get started nursing her, we have kept that
cow for one year just for the joy of having that cow.

So what happens? You have a live cow and live calf.
What are you going to do with her? By the time you get that
calf started on that old pendulous-uddered cow and she kicks
you a few times and swishes you in the face, you're so mad
at her that you turn her back out without getting her properly identified, and then you go home and give your wife
hell about supper. A better solution is to get the cow in
and get the the calf started. The last day before we turn
her out, we walk around to the front of this old cow--we're
mad at her anyway--take out the pocket knife and cut off her
right ear.

This helps you feel better by getting even with the
cow. Now, when we're pregnancy-checking the cows, we can't
be standing out there flipping through papers looking up in-

formation on every old cow. So when a bob-eared cow comes down the chute, we save ourselves a dollar on the pre-checking bill. We just say to the vet, "Let her go through." She may cause us trouble one year, but she won't cause us trouble next year. (Also, we know we don't have any of her daughters or sons.)

Another thing we feel strongly about are cows that lose their calves whether it's their fault or not. This has to do with maximizing our resources.

Now, I know ranchers don't like to be called farmers. I live East of the Missouri, so I can still talk about farmers. But, by golly, West of the Missouri, cattlemen are't farmers, They're ranchers!

But, we're all really farmers. We just merchandise what lies in the perimeter of our fences, that's all any of us are really doing. So any cow that loses her calf at our place is sold.

DISASTERS ARE AVOIDED

One spring, C. K. Allen and I were looking at calves. There was a little calf lying in the grass. C. K. was supposed to be looking out for him, and I was supposed to be driving. We ran over the calf and killed it. That calf's mother went to slaughter. Not because we're trying to raise intelligent calves who will get out of the way of a pickup, but simply because there's no point in keeping that cow when I can replace her with two bred heifers for the same feed bill.

Another thing, we used to cry, "Oh, my gosh!" when a first-calf heifer lost her calf. We'd wring our hands and say, "Look what that heifer has cost us. We weighed her at weaning and at a year. We AId her to a select sire, "super bull" and here she goes and she loses this calf. What a disaster!" It's no disaster! The country is full of people who are backgrounding cattle on grass to get them ready for the feedlot. The heifer that loses her calf weighs twice as much as she did at weaning. We haven't lost any money on her. Put her in the feedlot. It doesn't make any difference whether it's her fault, our fault, or whose fault it was. When a heifer loses a calf, we turn it into money.

The gain that heifer puts on from weaning until the time she has her calf is under the same price structure as backgrounded cattle. If you can't take the increase in weight on that heifer and make it wash on her feed cost, you're probably in big trouble anyway.

We calve for 60 days. Are we successful? I think we are. This year, 23 days after we started calving our total cow inventory, 85% of our cows had live calves at side.

This makes the 6th year in a row that we've been above 85%. I think the reason is that we have been selecting and culling for this. We do the best job we can as far as management and nutrition of these cattle are concerned.

Raising cattle is a lot like raising corn. It isn't any one thing you do right, but you get a 2% or 3% advantage from this and 2% or 3% from that. And, I think this is why we're up to 85%.

THE POLLED HEREFORD BUSINESS

We got in the Polled Hereford business 10 years ago. One of the really, really, good moves we've made is to get involved with this breed. We started out with the idea that the best way to get involved in the Polled Hereford business was to buy cattle that were the best performers we could find anywhere that had been subjected to the same kind of selection pressure that we use.

We bought heifer calves from six herds at weaning, brought them home, threw them in with our Angus heifers, and raised them up in the same way.

This was in 1972. We gave $800 for those heifers at weaning in Canada. We breed our heifers for 45 days--20 days AI and 23 days with the clean-up bull.

HIGH-PRICED HEIFERS SELL TOO

We vowed even with these high-priced heifers any heifer that was open was going to the feedlot. Now if you want to see the eybrows on your banker dance, just go in and tell him about your $800 heifers in the feedlot when cattle are worth about $.32 in 1973. But that is the cheapest way to do it!

The Hereford heifers from two herds were the same as our Angus as far as percentage open, percentage that calved, weaning weight, etc. Another herd was slightly below. Two herds were absolute failures. We looked back to see why these two groups of heifers were such absolute failures in getting bred and in calving. In the one herd, a super performance herd, the guy calved in February. He loved them! When you're calving in January and February, you've got to be out there with them and love them. He also liked big calves at birth.

The other herd was from a different place in Canada, but this guy loved his cows. Everyone of them had a name. Using my Minnesota computer, a big Chief tablet and a wooden pencil, I figured out there weren't nearly enough calves on the place for the number of cows bred. Out of one herd, with that guy's heifers, we got 28 heifers. After three breedings, we had four left. We didn't cull any of them for anything except whether they had a calf or not. "LOVE" has its drawbacks.

The other herd, the strong performance herd, where the guy calved early and loved them, we had five left out of 31. But, there were differences between those two groups of cows. The one group of cows that the guy just loved were lazy cattle; fat lazy rips that would stand there and eat while they were trying to have a calf.

LOOK AT THOSE BIRTH WEIGHTS

Another problem in one of the herds was mostly related to birth weight. These cows had huge calves and were all bred to the same bulls. This group of heifers would breed the first time but would either lose the calf calving or come open the second time around.

I think this is something purebred breeders and bull studs have to get serious about--birth weights. Now I know it's fun to stand around and talk about the shape of the calf being more important than the weight of the calf, but that bologna! We have got to get serious about birth weights.

Back in the 1950s we started performance testing. In 1957, commercial men were coming in and saying, "what in the world am I going to do about these calves? I've been buying these high-priced bulls and every time I sell my calves they're 20 pounds lighter than they were the year before. Why are you guys pawning all these little, long-haired toads off on us? I buy bulls at a sale, bring them home, their feet grow out, their hair goes down, and they weigh 1,000 pounds and are a little lighter than that at maturity. We have to have some growth."

Now one thing I'm an expert on is selling bulls. We sell over 400 yearling bulls a year, and they're all sold by private treaty. We have a damn big problem with calving and the commercial man knows it. But when he's down at the lo-cal coffee shop, he's not going to talk about it. We have birth weights available for all our bull buyers and last year was the first time I can ever remember that commercial buyers in general were more interested in birth weights than in yearling weight. Unfortunately, many of these are commercial buyers are the ones who have been spending the most money for their bulls that were selected for yearling weight.

Birth weight is the one thing I really think affects reproduction.

Now, some breeders say, "Well heck, my cows have 120 lb calves with no problem. Birth weights are nothing to worry about." I believe the seedstock operators when they say their cows have 120 lb calves without any difficulty. But, I'll tell you one thing, our cows have problems having 120 lb calves.

The average commercial cow in Adair County, Iowa, cannot have a 120 lb calf. Sure, I can tell you about the old cow that lays down and has a big calf in five minutes, but I'll never live long enough to change my whole herd so they can all have 120 lb calves without problems.

The way the commercial cattle industry works is that people lose the identity of the cattle as far as which calves are sired by which bulls. I don't believe we can ever sell bulls that weigh over 100 lb at birth to our commercial customers and be doing them a service.

Now granted, if they'd stick around and continue using those 100 pound-plus-birth-weight bulls long enough, their cows would probably be able to have 100 lb calves. But, if they go broke in the meantime, I'm going to have a hard time selling them on the idea of going broke so they can have a set of cows that can have calves that weigh 100 lb plus at birth.

I think we have to take cows as they are and genetically engineer the high-performing, fast-growing animal that has birth weights acceptable for the cows in the population today.

37

THE FRUSTRATIONS
OF SELLING PERFORMANCE

J. D. Morrow

Let me assure everyone that I am speaking as an optimist and not a pessimist; that I intend to be constructive and not destructive. I'm not an "outsider" looking in--rather I'm an experienced "insider," who very strongly believes in the successful future of the beef cattle industry. I believe that the cattle business, as with all successful businesses, should be operated for two basic reasons--production and profit. Every cattle operation should stand on its own--with properly established goals following standard business practices. The procedures of a cattle operation should be predicated as much as possible on fact and not on opinions. Comprehensive production records should be maintained, analyzed, and utilized to obtain maximum production from the most efficient cattle available to the producer.

Furthermore, the cattle industry isn't in difficulty just because of high interest rates, over-production, and a slow economy; other basic problems include traditionalism, production inefficiency, and very inadequate promotion of the most nutritious product in the food chain--beef.

Private enterprise and the success of the "American way of life" are based on the production of a product of greater quality, with equal or greater efficiency than anyone else. This incentive is especially strong in the livestock industry, because its market fluctuates strictly on supply and demand with product values determined mainly by the amount someone is willing to pay for that specific product that day. Tax advantages have been generous for depreciation, investment credit, and capital gain income. Inflation of land values has created sufficient borrowing power for the livestock man. Coupled with these economic assets, the American consumer has enjoyed the highest standard of living in the world. So, why--during the last thirty years--has our industry had so many ups and downs? The problems of the present economy make it almost impossible for the most efficient to produce a profit.

Many predict that during the next ten years more operations will sell out than in any decade since the

depression of the 30s. So let's first be honest with ourselves and admit where the deficiencies really lie:

- Many of the present generation, endowed with the
  inheritance from the pioneer breeders, feel that
  the industry owes them a successful existence,
  and that since they own land and cattle, some
  miracle created by God or government will save
  them.
- The present commercial cattle industry is deep-
  rooted in traditionalism and is a parasite
  feeding off of appreciated land values.
- Of the operations that are not inherited, most
  are owned by part-time cattlemen or professional
  people who, for the most part, have very little
  cattle knowledge and operate without goals or
  accepted business methods.
- Lending agencies have been too lenient in subsidizing, without demanding facts on the people or
  the cattle they finance.
- Cattle are selected and propagated on the basis
  of breed and not on profitable efficiency. The
  result is that many operations are sentenced to
  financial losses before they even turn their
  bulls out.
- We have been slow to change constructively but
  sometimes too quick to swallow a gimmick or a
  fad. We have often accepted guidance from unproven leaders who based their judgements on
  opinion and not fact. We've let glamour and
  shiny things cloud our vision of the real
  truths.
- Highly progressive commercial cattle operations
  receive very little premium for proven-perfor-
  mance cattle from the feedlots and none from the
  packers.
- Cattlemen have a proven record of being the
  least successful promoters of their product;
  they let the excuse of wanting independence
  relegate them to taking what someone will give
  for their product.

The basic truth is that the cattle business is primarily motivated by love and tax advantages—neither of
which have any relationship to the acquisition of fact or
the generation of profit. Owning a ranch and cattle is a
status symbol, with the image of agriculture disguising the
complexities involved in successful cattle breeding. Purebred breeders, especially those who are inexperienced in the
cattle business, eternally seek a detailed blueprint that
they can folllow in breeding showring winners and high
sellers at public auction. Suggesting to these people that
the solution is collecting comprehensive data on a cowherd
and utilizing it in a constructive manner bothers their
pride and removes the glamour and mystery that they want and
believe exists.

The purebred industry is responsible for furnishing superior genetics to commercial cattle. The evaluation as to which cattle are superior needs to be done within herds by honest cattlemen emphasizing the important traits that are relative to profit. Maternal traits that emphasize fertility and function must be evident in the selected breeding cattle that produce progeny that weight "heavy" at weaning and exhibit the proper growth patterns until optimum maturity. These facts can only be identified by competent recording.

Breed associations have established programs with computer printouts of herd data. If analyzed and utilized with good judgement this data provides the producer with the best insurance policy any cattle operation can have. These programs are so misunderstood and so seldom used that the conscientious breed representatives become frustrated. With this in mind let me share some of my frustrations, which are probably felt by most of those who are involved in breed assocition work.

- Initially, the breeders and management organizations who merchandize their cattle lack knowledge and interest in the well-designed plans that have been formulated by The Beef Improvement Federation. The purebred breed associations make these plans available competently and inexpensively.
- In the selection of a breed of cattle that can contribute efficiently to the beef industry, recognize their attributes and shortcomings before deciding that they will be all things to all people.
- Assume that cattle that win in the showring are basically superior cattle, but are not guaranteed to reproduce their kind.
- Ignore advertisements and sale catalogs that include personal quotes and show winnings, without the inclusion of production and performance information.
- Design a management program that profitably produces superior cattle.
- Realize that your first priority is to produce superior genetics for the commercial cattleman and that you are mandated to furnish him with as many facts as you possibly can about the cattle you offer for sale.
- Realize that production records are the means of attaining a goal in your breeding program and not something you will start after you have reached a certain plateau.
- Understand that the value of performance information is directly relted to the person who records it and the conditions under which it was obtained--and is not to be valued only for promotion.

- Accept that the first criterion for keeping an animal in a breeding herd should be its level of fertility and functional ability--not its sire or cost.
- Maintain comprehensive records on every animal in the herd; compare each in as large a contemporary group as possible; base the final selections on ratios, kind, and structural soundness.
- Agree that weaning and yearling weights must be adequate to produce a profit on the open market, but superior cattle within a herd can be identified only by how they perform in relationship to their contemporaries.
- Remember that high concentrate feeding programs can produce extreme weaning and yearling weights that may be to the detriment of an animal's reproductive future.
- Collect the data yourself, study it, and utilize it in culling and in selecting your herd bulls and replacement females.
- Learn that the great thrills in breeding cattle come from being able to produce outstanding production records on a cowherd that furnishes the herd bulls for a successful industry. Nothing else can ensure your success or more completely satisfy your pride.

The beef cattle industry <u>can</u> thrive and be profitable if all of us who are involved accept our responsibilities in a dedicated and businesslike manner. Let's substitute excuses with goals and believe that breeding programs founded on fact and directed with prudence will survive in the business world of tomorrow.

38

MEETING THE PUREBRED CHALLENGE

William Eaton

First of all, why are you here or reading this? Why in the world would any of you want to raise beef cattle and especially purebred or registered cattle? Raising cattle of any kind is certainly not the easy way out, and registered cattle are even more work.

For some reason, cattle have had a strange and strong attraction to men for thousands of years--often over and above their utility or profit potential.

I suspect that most of you are voluntary captives of this attraction. I'm also sure that as purebreeders, you especially enjoy the physical and intellectual challenges of molding new generations that are superior to their parent generation. You welcome the opportunity to leave your herds or farms a little better than when you started--for the benefit of your family, the industry, and society as a whole.

Raising registered purebred seedstock is a fascinating mix of challenges, rewards, methods (and probably a little madness), spiced with schemes and dreams and a fair share of disappointments.

My title is vague enough to allow us to explore several directions while looking at various aspects of "Meeting the Purebred Challenge" so we'll wander a bit among these interesting topics.

THE BIG CHALLENGE: SURVIVAL--SO WE CAN DO IT AGAIN NEXT YEAR

Whether we're selling shoes, cars, cattle, or just farming there eventually has to be a little more income than outgo or the guys in the suits will come and say "You can't do that anymore," no matter how much we enjoy it. This means we have to produce and merchandise seedstock cattle for which there is enough demand and total price, or have other income or enterprises that we can justify to subsidize the purebreds, hopefully only until they pay their own way.

THE CHAIN OF DEMAND IS OUR CHAIN OF COMMAND

As long as beef is consumed at cost-plus prices, the demand and profit incentive will feed back through the packer, feedlot, cow-calf, and purebred segments of the industry.

As we all know, beef consumption has declined. The effects of price relative to other meat or nonmeat competition, health and nutrition concerns, the economy and less disposable income, and lifestyle changes have all been widely discussed.

At present, we may be faced with marketing what is perceived as a luxury food item rather than a necessity. Selling luxury in quantity can be a challenge.

The cattle production chain is faced with the need to constantly increase efficiency in all segments of the industry to reduce the obstacle of price differentials with competing foods.

Furthermore, we must do this without significantly decreasing the satisfying characteristics of our beef product. A sirloin steak at $1.95 or a $.59 Big Whopper that "somehow, doesn't satisfy" won't work either.

MORE EFFICIENCY: PLEASE DON'T SQUEEZE THE COWBOY, SQUEEZE HIS GENES

Average operations or efficiency is now a losing or nearly losing proposition. Nowadays the "loss incentive" may be even stronger than the good old "profit incentive" in getting us to change or improve what is needed.

Improved and superior genetics is probably the fastest and cheapest way to increased efficiencies in all phases of agriculture. Whether we raise corn, tomatoes, alfalfa, or cattle, the intelligent use of genetics is one of the major tools we have. It is generally possible to fix desired traits that can be predictably passed to the next generation, and to do it at reasonable costs.

THE COMMERCIAL MAN, PENDULUM GENETICS, CROSSBREEDING, AND THE GOLDILOCKS FACTORS

We purebred seedstock folks often say that our real challenge or purpose is supplying the needs of the commercial cattle industry.

Now, I know, lots of us may sometimes find this a little boring. Most commercial men can't read pedigrees; they don't know who the show champions are and don't care; and most sinful of all, often want to steal our bulls at pound prices. They surely don't get auction fever very often.

Some commercial livestock producers just go out and buy seedstock by the looks or price; they don't read the ratios and breeding values you may have sweated over providing, although more and more are paying attention to them.

We all realize that there would be no real market for purebreds without the commercial segment. We have reason to base our breeding programs upon supplying the traits these herds need. In fact, if you can beat the competition in doing this, the commercial man will generally pay a reasonable premium, especially in "cattle country."

The commercial cowman's demand for larger, faster-growing cattle is really what brought us out of the "pony type" era of British breeds -- especially with the available competition from large Continental breeds. All breeds still generally emphasize growth rate and size as the industry continues to escape from that era.

"Pendulum genetics" is a term we could use to describe how we've changed types in the last few decades in America. The pendulum of desired beef type has swung from selecting for short and blocky to breeding for big and growthy.

Making changes by pendulum genetics within a breed can be beneficial if the right goals are chosen and attainable. It can be a massive exercise in population genetics if a few heritable traits are accurately selected. The problem is that it is generally a very slow process to obtain overall improvement, even within single herds with severe and accurate selection pressure.

The commercial industry has increasingly discovered the potential of crossbreeding and its ability to change their calf crops remarkably in one year, and also to gain the benefits of heterosis in the calf and in the dam. In addition, lowly heritable traits like fertility and reproduction are improved most efficiently.

The rapidity of significant change is very apparent, especially with the use of Continental breeds combined with British breeds. This single factor has tremendously accelerated the transition from our too-small cattle left over from our previous "pendulum." Meanwhile, as we now can change size so quickly, many cattlemen see that laws of diminishing returns can soon take effect in regard to genetic traits such as mature size, milking potential, carcass weight, and birth weight.

We are talking more and more of looking for optimum genetic and environmental inputs and expressions. "Optimum" indicates that traits of interest may not need to be maximum or even minimum in expression but a proper blending for best overall fitness, and for maximum profit.

This we could call the "Goldilocks Factor". Like Goldilocks, we try to get everything "not too hot (big), not too cold (little), but jussst right. "

We must also realize that "optimum" combinations of traits and adaptability are not constant between areas and environments. There are even different optimum conditions within the very same herd, such as between groups of first calf or mature females. "Jussst right" is an elusive target at best. All commerical men need a combination of optimum function for growth, reproduction, birth weight, efficiency, and maternal ability. The combinations just won't be the same in all conditions. And sometimes the current maximum potential expression of these functions could result in too much of a good thing or "undesirable excessives" or even "lethal excessives."

Furthermore, I want to emphasize that optimum function does not necessarily have to be entirely embodied and totally expressed individually by any single sire or breed. I'm using "optimum" primarily as expressed by the progeny in our customers' herds. Seedstock responsiblity is to be able to contribute to optimum function offspring. So, superior seedstock need not themselves necessarily have an optimum phenotypic expression of a trait, but must be able to genetically transmit the appropriate contribution to progeny. Obviously, progeny testing within a given situation will determine the merit and predictability of any parent with the most accuracy, regardless of his or her individuality.

More and more, commercial cattlemen are experimenting and combining and blending breeds to quickly get optimum genetic efficiency for function within their conditions and environment. Some will have shipwrecks, but the potential efficiencies "makes them an offer they can't refuse." Careful planning will be required as they proceed.

Crossbreeding used to be, and still may be, a dirty word in some purebred circles. However, it's not going to go away and actually opens many opportunities to purebreeders.

A major overall challenge for purebred seedstock sources will continue to be to supply or package the traits desired with high predictability for transmission.

Breeds and breeders will have to increasingly emphasize their strong points as contributors to a system, rather than being the entire system. Commercial men will increasingly avoid previous breed loyalty as a major input in their decisions. We will be marketing not only breed traits but also traits of individuals within breeds.

We will also have to demonstrate that our purebreds are needed for continuous infusion into breeding programs and will have predictably more beneficial results, as opposed to simply buying or raising unidentified crossbreds to mate with other unidentified crossbreds. This may take some salesmanship, and the proof will be in the pudding.

We'll also have to show that superior purebreds are needed, rather than just any purebred, and that they have the added value to pay us for our efforts.

We're really talking in terms of specification or tailor-made breeding systems where the magnitude of specific genetic abilities and blending potential identifiable and predictable.

Cowmen will increasingly think of breed traits or individual genetic potential as ingredients for the recipe that they feel will work best in their "kitchen."

To maintain their identity and provide specificity, breeds or lines will need continued maintenance or improvement of their strong points. Line-breeding within breeds could become better utilized as further sources of trait predictability within a breed. Independent culling levels for important traits will probably be more widely used in the pursuit of optimum inputs and will help in the prevention of undesirable excesses in some traits.

Currently, it seems as if most breeds are trying to be all things to all people, and we may need to modify this approach and concentrate on breed or herd strong points. Unfortunately, some desired traits are antagonistic to other desired traits, making it difficult to put everything in one package.

For example, the ability of some breeds or individuals to add desirable growth rate and frame may be accompanied by undesirable birth weights or by unnecessary mature size with inefficient maintenance needs and late puberty. Ability to add high milking potential may lead to excessive nutrition needs for rebreeding on time in some environments. Also, fertile and maternal moderate-sized breeds may still not have the extended growth and efficiency patterns desired. The "excess baggage" that can often accompany desired traits will be a continuing challenge to both purebred and commercial cattlemen.

Genetic-environment interactions or predictable adaptability to certain environments or conditions will be a very important criteria when breeds are chosen for blending into various programs. The same package of genetics may not be the optimum package in all situations. This breed adaptabilty can be a major reason for purebreds or the systematic maintenance of optimum heterosis by planned infusion of true purebreds.

Making "Goldilocks Genetics" work is not a simple task, but breeds and breeders that can meet the needs of the industry will survive and prosper as the industry does. Others will find themselves immaterial, on the periphery, or ignored. Competition among breeders will increase and, appropriately, it will be "survival of the fittest."

SELECTING AND PREDICTING: COMPARED TO WHAT?

Obviously, one of our challenges is in choosing seed-stock predictably superior in important hereditary traits. It requires the ability to sort them out from the general population, either for seedstock producers or commercial men. We need to collect the traits we want in our herds, package them as needed, and send them out to our customers-
-well-identified and predictably available. The practices of recording data and performance-testing for econmically important traits has given us much information. Many data are available from in-herd tests, test stations, etc. Breed associations have made major steps with their sophisticated data collection such as print-outs, performance programs and pedigrees, estimated breeding and values.

The more recent advances in National Sire Evaluation programs are enormously helpful. With these Reference Sire systems, any bull can potentially be compared to any other in the breed. These are remarkable developments. We owe a great deal of gratitude to the consulting geneticists, like Dr. Richard Willham of Iowa State, and the breed officials who kept pushing until this was accomplished. They have made much simpler the challenge of accurate comparisons of individuals, lines, or herds that interest all progressive seedstockers.

There are a few general criteria that we still need to think of when we utilize these data, at least for our best use of them. "Compared to what?" is a question that all informed seedstock breeders continually ask themselves. First, the programs all necessarily assume that all parent animals to be compared are randomly assigned within the available group of mates. Second, all direct progeny or individual performance comparisons are assumed to be conducted under the identical conditions of nutrition and management. The importance of any observed difference depends upon the numbers of animals compared.

Although we all hope for ideal conditions, common sense tells us they may not always occur. intentionally or not, some biased data may be entered into various programs with possible misleading results. When there are many animals, herds, and progeny involved, much of this bias can even out.

In some cases, however, a little extra thought or investigation on our part may yield key information for one of our very major decisions--the correct choice of our seedstock. A few examples follow.

Sire Comparison for Growth Traits: The Possible Effect of Bull Value

These comparisons have high potential for accuracy because of the fairly high heredity and the large numbers that can be generated. Structured and large one-herd sire tests, randomized and overseen by association personnel, offer very good data.

However, my mind sometimes wonders about the source of other data in some cases. For example, purchasers of high priced semen or bulls may quite understandably use either or both on cows they feel are superior, leaving other bulls to cows of less merit.

Obviously, this could bias the data if they can identify superior cows. it can certainly influence a sire's estimated breeding values and those of his relatives. Sometimes, I compare expensive bulls together and cheap bulls together. If I find a $20.00 bull performing better than or equal to a $200.00 bull, I, like most people, am more impressed with the underpriced sire.

Sire Comparisons for Maternal Traits

Usually, we use weaning weights of calves to evaluate milking ability in females. However, the calves of daughters from various sires may be useful for comparing the sires' ability to sire high producing daughters. Remember also that the growth impulse is much involved in any calf's weaning weight in combination with its mother's milking ability.

"Compared to what?" is a question we sometimes need to ask in this test, even though everything is randomized and treated the same. What kind of real or identified competition did the sire have? Most associations don't use a Reference Sire system in this comparison. Maternal Breeding Values are derived from raw within-herd ratios. Therefore, a breed average Sire A compared to sires of poor females in one herd may achieve a fine record, perhaps an average ratio of 106. Now, if instead he had been compared to sires of superior females, he might have signed in at the 100 level. Again, these data will affect the Breeding Values of his sons and daughters. You can see the possible evaluation problems unless he is compared to a sire with a wide spread or accurate identity.

If a sire has lots of daughters in lots of herds with lots of calves, the possibility of this bias is greatly diminished as circumstances "even out." However, if you really need accurate information on a sire's maternal value, you might want to keep in mind the above possibilities and investigate further if possible.

Females Compared for Maternal Traits

Even if cows are assigned randomly in cases of two or more bulls being compared, their own individual maternal evaluation data could be biased in proportion to the real growth differences between the sires involved. Obviously, the best comparisons between females for their milking and growth contributions occur when all cows are bred to the same bull to remove sire-of-calf differences. This seldom occurs nowadays in purebred herds.

If we really need the best information on a certain female or females, and if there are major sire/progeny differences present, we should consider evaluating the cow or cows within her group of cows assigned to a particular sire.

For example, Cow A mated to an inferior sire may have a calf with a 101 ratio across the herd but a 108 ratio within her mating group. Meanwhile, Cow B mated to a superior sire may have a calf with a 107 ratio across the herd but a 99 ratio within her mating group. To me, in this kind of situation, Cow A has made the best of her opportunity but would suffer in the standard across-the-herd comparisons. It's a case where a 101 ratio cow across-the-herd may really have demonstrated potential superiority to a 107 ratio cow, but would go unnoticed without this extra detective work.

Unfortunately, this kind of bias, usually unintentional, will affect her own Maternal Breeding Values as well as her progeny's. Likewise, groups of cows may be identified for culling or selection in this kind of situation and their true merit may be disguised by the major sire differences.

It's ironical that the required and correct randomizing for sire comparisons sets up the possible situation where cow comparisons may be biased, all in proportion to the degree of sire differences. Major sire differences usually may not exist but they are very possible. Also, over a long lifetime, these mating biases may even out within a herd; however, we often select or cull much earlier. This especially applies to first- and second-calf females.

Herd Differences and Their Effects Upon Comparisons

Here's another situation where we may need to dig a little deeper into available data to get the "best" information for our selection decisions. As we all know, ratios are best used within herds, as are the breeding Values derived from them. However, Breeding Values (BV) may be derived from several herds' data, all using within-herd ratios. Sometimes we may bias some data because of very real herd differences when animals are compared across herds.

In a high-performance-selected herd where competition is much greater and actually near breed best, a superior animal might ratio 115 and likewise actually be near breed best, possibly 125 on a breed basis. Meanwhile, in an average or below average herd, an animal with a 115 ratio may really only be a little over breed average. These identical ratios for animals of great difference in true merit can create biases, obviously, when they are plugged into Breeding Values.

Let me illustrate this point with an interesting analogy. Let's say your home high school basketball team has an all star who averages 25 points a game. Within his "herd" he is outstanding. Meanwhile, Dr. J. Julius Erving is an NBA all star who also averages 25 points per game. Within his "herd" he is similarly outstanding. Now, through experience and knowing that the NBA "herd" is more highly selected, with much more competition, we recognize there are "herd" differences. We know that even though these two players have equal within-herd "ratios," there are some very probable differences in their absolute real abilities. But, if we ranked the whole population of all players in the nation from high school through NBA, just on their scoring average, we might assume these two players were apparently equal, if we did not know the data sources.

In cattle, we know these herd differences exist and the interested breeder will try to determine and evaluate the source herds, especially when only a few herds are involved. Once again, the place to turn is often the marvelous National Sire Evaluation data involving Reference

Sires. In addition, personal information from your own or other familiar herd comparisons of animals from different herds is exteremly valuable. Again, in all the examples above, lots of progeny in lots of herds even out opportunity, remove many biases, and make BV's very accurate for large groups. Often, however, this takes more time than we wish and many animals are just not subjected to wide tests.

These examples may also just be my personal peccadilloes. Please don't interpret them as a criticism of any breed or group performance programs or of their derived Breeding Values. The information they all gather and present for us is of immense value in our identification and selection of superior seedstock.

Because I'm a little more fascinated with the challenge and importance of selecting the very best available seedstock, I've tried to show how we can perhaps gain important advantages by extra investigations over and above what are practical to include in association programs. Remember, without their data collection as the starting point, we'd run into blank walls very quickly.

Remember that on a large population basis with deliberate or de facto randomization, BV's can be our best information and extremely revealing in illuminating bloodline, breeder, and sire differnces.

Merchandising: Getting Them Sold for More Than They're Worth

With the tremendous opportunities of open AI and the information from various Breed Performance Programs, it is now relatively much easier for informed breeders to predictably produce the kind of cattle they feel are desirable. The rub often comes in obtaining prices that you feel justify our time, efforts, and investment. We all soon find out that many of our registered cattle command only commercial prices. Generally, only a portion of our herd brings a premium. This total premimum has to be spread over the entire herd and still reward us for our total added effort.

What are purebreds worth? They are worth pound prices plus whatever value customers place on their herd-improving or aesthetic merits. This is our challenge of management, salesmanship, and merchandising. Our goals here are attainable with a combination of good planning, hard work, and, perhaps most important, being in the right place at the right time.

Remember that promotion, recognition, and anticipation are big factors in the value that breeders place on your cattle. Anticipation should be singled out. No matter its basis, anticipation is a primary motivation for any seedstock purchase. In addition, strong promotion and advertising programs need to be considered. If we are producing the best cattle possible, but never prove it or tell others about it, the number of people knocking on our doors will often not justify our efforts. We all realize that our merchandising potential also will vary with the general geo-economy of the area we live in. There are obviously different marketing potentials depending upon whether we are in range cattle country, the corn belt, or diversified-farming areas, and we need to plan our programs accordingly.

In all areas, a major challenge is selling seedstock to other purebred breeders. We all get an extra boost to our ego and our bank account when we can accomplish this. Nowadays, to compete for purebred breeders as customers, we almost have to be selling cattle featuring highly recognized pedigrees or sires. This is on either a show or performance basis, or a combination of the two. The ease of open AI has made top sires available to all of us and made for a much higher level of competition. Furthermore, the herd bull market has declined with this increased availability of super-sires (real or imagined) and is compounded by the decrease of small-breeder herds in many breeds. There is a limited market for the top end of young bulls to go into purebred herds, but times have changed in this market. More and more breeders are raising their own replacement or clean-up bulls. Many now turn to major or primary seedstock herds for young sire purchases because of image and promotion and because many of these herds have put together extremely sound, recognized programs.

Being a primary seedstock source appeals to and motivates all of us. However, the number of these herds that the industry supports is relatively small. Of the 30,000 or so listed breeders in my breed, perhaps only 20 or fewer would be considered primary herds. An analogy might be to compare the proportion of high school football stars that eventually make it to the NFL.

Experience and observation indicate that it is generally difficult to support a seedstock herd with income only from commercial sources. For most of us, our best "mix" is a combination of a strong commercial market and a good outlet for our top end purebreds to other purebreeders. We must remember to emphasize the needs or wants of our best potential customers. Attaining this is achievable to many of us. It requires a well-analyzed and well-planned use of all our resources, including genetic, geographical, and personal ones. it is a very competitive and stimulating situation.

In closing, we've investigated several key areas that present challenges to us as purebred breeders. I hope you can see the intent to tie together industry's trends or needs with selecting the animals to most fit those needs. And finally we've talked a litle about selling our product. There are many other related challenges to all of us, remembering again that we are only as viable as the industry as a whole. Hopefully, as purebred breeders, we can make significant contributions to ensure this viability and a way of life we all enjoy.

39

SIRE SELECTION IN A BEEF CATTLE CROSSBREEDING PROGRAM

Charles W. Nichols

The value of a carefully planned and implemented cross-breeding program to the commercial producer is now well documented. Among the most important considerations in implementing a sound crossbreeding program is the selection of breeds and individual sires from those breeds. No two commercial cattle operations are exactly alike--each operation must select breeds, and bulls within those breeds, to fit their specific program.

SIRE SELECTION

Sire selection in a crossbreeding program can be more complicated than in a straight breeding program because there may be as many as three distinctly different types of sires used. These three types and the most important traits to consider in selecting each are:

First-calf heifer bulls
1. Disposition
2. Calving ease
3. Structural soundness
4. Breeding soundness
5. Growth rate

Sires of replacement heifers
1. Disposition
2. Growth rate and efficiency
3. Evidence of maternal ability from records
4. Structural soundness
5. Breeding soundness
6. Calving ease
7. Carcass merit
   -Fatness
   -Muscle expression

Terminal cross sires
1. Disposition
2. Calving ease

3. Growth rate and efficiency
4. Structural soundness
5. Breeding soundness
6. Carcass merit
   -Ability to grade
   -Muscle expression

Of the various traits listed for the three types of bulls, disposition and structural soundness are completely subjective and must be evaluated by visual appraisal.

I have listed disposition in first place for all types of sires because in our operation we do not tolerate a bull that is wild or mean. Life is too short to put up with wild or mean bulls and their equally crazy offspring--regardless of their other good points. Cattle with wild dispositions require more labor and time to work and both labor and time are in short supply.

Structural soundness is not generally a problem in the bulls we see today, but a good visual examination can help avoid feet and leg problems in particular.

Breeding soundness can best be determined by a physical examination and a semen check performed by a veterinarian or other person qualified to perform such an examination. The evaluation of breeding soundness is somewhat subjective but does detect bulls with obvious physical problems and with very low semen quality. Because of our heavy shinnery oak problem, we try to select bulls with very tight sheaths and bulls that keep their prepuce retracted. Penis and prepuce injuries can take a very heavy toll of bulls under some conditions.

All the other traits listed can be evaluated with varying degrees of confidence from complete performance records. Unfortunately, data on all these traits may not be available from many breeders. In our operation, we prefer to buy bulls with as much performance record background as possible.

First-calf Heifer Sires

First-calf-heifer sires must be selected according to the age and size of the first-calf heifers. Many operations such as our own calve rather small, short-age, two-year-old heifers. In this case, calf weight or size is of great importance. Smaller calves can be expected from Longhorn or Jersey bulls. We use Longhorn bulls because we like the combination of small birth weight and good growth rate in the Longhorn-sired calves. The Longhorn-cross heifers may make good cows for some conditions.

Producers that calve well-grown-out heifers at 2 or 3 years of age may choose bulls from one of their maternal breeds that have been selected for easy calving.

Terminal Cross Sires

The selection of terminal-cross sires is also relatively simple as replacement heifers will not be saved and beef production is the main objective. Terminal-cross sires are primarily selected for a maximum growth rate consistent with an acceptable amount of calving difficulty. This is not always easy because calving difficulty can be a serious problem when using the larger breeds. In selecting bulls that will not give us too much calving trouble, we must depend on the efforts of the seedstock producer to reduce birth weights and calving difficulty in his own herd. The Simmental Breed Association has done an excelllent job of encouraging breeders to use birth weights and calving difficulty for evaluating sires. This information is very helpful to the commercial producer when purchasing bulls.

Growth rate from birth to about 12 or 18 months of age is an important consideration for most commercial producers. Mature size should never be used as a measure of growth rate. Large mature size in either cows or bulls is basically undesirable. Early growth rate has great economic importance and so it is unfortunate that it has a positive relationship with larger mature size.

Since all terminal-cross calves are sold for beef, carcass merit should receive some consideration. Most of the large, growthy breeds have excellent muscling and thin fat cover. Some of these breeds have difficulty reaching the choice carcass grade at a reasonable age and size. There appears to be considerable variation within breeds in their ability to deposit marbling, but the larger breeds in general have more difficulty in grading. I tend to be rather pessimistic about selecting for carcass merit. It is very difficult for the commercial producer to be paid for improvements in carcass merit in his cattle.

In our operation we use Charolais bulls and, to a lesser extent, Simmental bulls as our terminal cross. Several terminal-cross breeds are acceptable, depending on the specific needs of the producer.

Sires of Replacement Heifers

The selection of sires from which replacement heifers will be saved is the most difficult task in a crossbreeding program because the breed selection is so important. I have changed my mind on this matter to some degree in the last few months. I now believe that one should use only those purebreeds in the cow herd that have good performance records under your environmental and management conditions. Perhaps this is too severe a view, but I do not like the lower reproductive performance I see in some crosses that I previously thought might be successful under range conditions.

Complete herd performance records are a great aid in the selection of sires of replacement heifers.

Growth rate should be evaluated from records available on prospective sires. Weaning weight is a complex measure of growth rate of the nursing calf to about 17 months of age. Weaning weight measures not only the genetic ability of the calf to grow but the maternal ability for milk production of its dam. Heavy weaning weights are desirable under all conditions. The optimum level of milk production in the cow herd is that level at which the cows can maintain a 12-month calving interval. This level varies greatly-- from an adequate nutritional level on a lush farm to possibly inadequate nutrition under range conditions during the breeding season. Much the same argument can be used for cow size. If a cow can maintain a 12-month calving interval under a given set of conditions, she is not too large for those conditions.

Yearling weight is an excellent measure of growth rate in bulls as it covers the growth period of most importance to the commercial producer. Yearling weight is a combination of weaning weight and postweaning weight gain. If I had to depend on one measure of growth rate I would use yearling weight. Mature size should never be used as a measure of growth rate.

Maternal ability in a cow herd is very complex and covers all those traits that affect the ability of a cow to reproduce on schedule and to raise a heavy calf. A high level of reproduction in the cow herd is of great economic importance to the commercial cowman. Crossbreeding improves reproductive performance in the herd, but this does not mean that reproduction can be ignored in selecting sires. I prefer to buy bulls from herds that maintain a high reproductive level. Dams of bulls that you buy should have been regular calvers. Fortunately, we seldom see low-fertility cows kept in purebred herds today just because they were show winners or have a great pedigree. I certainly would not like to buy a bull whose sire had become infertile at a young age.

Calving ease is not as serious a problem in the breeds we normally use in our cow herd as it is in the terminal-cross breeds. However, we should avoid buying bulls from those herds that do not attempt to limit birth weights and calving difficulty in their breeding program.

Carcass merit should be considered for the general good in the beef industry. However, it is very difficult for the commercial producer to profit from improved-carcass merit. In buying yearling bulls that have recently finished a postweaning gain test, I attempt to select bulls that have less than 1/2 in. of back fat. I also prefer bulls that show a moderate amount of muscling.

I have not listed price as a factor in bull selection, although it is an important consideration. The price you pay for bulls must be evaluated relative to the possible improvement in profitability a particular bull can bring to your herd. Commercial producers can usually buy genetically superior bulls for less money by being careful to select directly for those traits that have economic importance in the herd. The commercial bull buyer should always stick to the basics and avoid pedigree fads in the purebred industry.

40

WHAT A COMMERCIAL COWMAN WISHES PUREBRED BREEDERS WOULD KNOW AND DO: A SELLERS' AND BUYERS' GUIDE

John L. Merrill

To evaluate comments that might possibly influence one's method of operation, it is helpful to know the background of the individual making them so that you can thresh out the chaff of possible (or probable) preference and prejudice to find any kernels of knowledge or philosophy that might be productive. So I'll provide a brief background. I have owned and owed for cattle approximately 40 years during which time my operation has been divided about half and half between registered and commercial operations. There was some interspersion and overlap of the two operations but always in an all-out attempt to make money because I have never had any to lose. For several of those years, registered cattle were my sole source of income; I have served as president of a regional purebred association. Given this sort of combined experience my problem becomes that of organizing a number of thoughts from a long period of time into some reasonably succinct form and logical order, thus, I have divided my comments into the three basic categories of selection and breeding, production, and marketing.

SELECTION AND BREEDING

While ego and satisfaction are common reasons for owning registered cattle, the basic reason is economic and the possibility of more income than from a commercial herd, as well as to provide bulls that will meet the needs of commercial breeders.

Registered cattle should be selected and culled for commercially significant economic traits, namely soundness and fertility; frame, size, and rate of gain appropriate to the breed; desirable beef conformation; uniformity; and other particular characteristics of the breed in that order of economic importance. An absolute rule is not to pay more for an animal than you are willing to sacrifice if its offspring indicate it should be culled. The longer you keep a poor choice, the more the problem and loss will be multiplied.

The registered breeder must know these principles and how to apply them. There is just as much difference between a cattle owner and a breeder as there is between an airplane owner and a pilot, which explains a lot of crashes in both fields of endeavor.

The only difference between registered and commercial cattle is their known ancestry. Absolute integrity, extreme care in breeding management and record keeping, and fence maintenance are essential.

In many cases, animals in good commercial herds are superior to animals in registered herds because they have been culled ruthlessly and consistently for economically significant characteristics--without the owner's psychological problems associated with their being a son or daughter of old Acey Ducey Domino. I have sat through many association meetings with purebred owners who complained that they could not sell their bulls well. Their problem was that they did not know that their bulls were not as good as the steer calves in the stocker-feeder sale.

Over the years university meats-trained judges in the show ring have typically had no regard for functional soundness of feet and legs and sexual characteristics related to reproduction and production. Most commercial bull buyers at ringside will select for those traits that are feedlot efficient rather than for the more extreme selections of the university judges.

Do not follow fads; you never can catch up. If you buy a son of a major show champion and breed him to your cows, there will be three or four more champions from the same show before you have offspring to sell; the original bull likely will be long forgotton unless, in the meantime, he has proven to be a real breeding bull.

Optimize and standardize the environment for fertility and good reproduction. To the extent you are successful, fertility can be heritable. Select and cull ruthlessly for the most important economic characteristics.

Longevity receives little attention from most breeders, but it is a highly desirable characteristic because of its economic impact when the producer buys costly replacements that must be retained and developed.

To protect your own investment and your customers', any herd bull that is to be used extensively should be bred to 32 of his own daughters for a progeny test to check for all genetic abnormalities, simultaneously. Any animal that produces genetic abnormalities and all its offspring should be removed from the purebred herd. Failure to do so can be disastrous. Cryptochidism (retained testicle/s) and inguinal hernia in bulls, shy quarters in the udders of cows, and predisposition to eye cancer in both sexes are heritable and should receive special culling attention.

A continuing source of amazement to me is the use of artificial insemination from bulls the breeder never sees. If he did, he would never, ever consider them in natural service. There is no magic genetic improvement in the collection and freezing of semen.

My observation is that the purebred herds that have prospered longest did so by using a number of good bulls simultaneously and continuously rather than putting all their eggs in one basket at a time. Another key to prosperous purebred herds is to consistently add herd bulls of the same type but from different bloodlines so as to maintain predictability without losing vigor or adding problems of inbreeding. This breeding plan is absolutely critical for retaining repeat business of commercial buyers who sell their feedlot animals by weight and cannot afford inbreeding from the same purebred breeders' program year after year.

Producers of the American breeds must recognize that genetically they are crossbred animals. The diversity of genes that provides many of their advantages also results in more variations in the offspring than in European breeds that have become stabilized over several hundred years. A higher rate of culling almost always will be necessary in purebred herds until the desired breed characteristics become more fixed. Very often the off color or otherwise atypical animals in a purebred herd make very desirable commercial animals if they are sound otherwise.

PRODUCTION

For true genetic progress, performance, and economy to purebred producers and commercial bull buyers, registered cattle should be raised and selected under commercial conditions as nearly as possible. By doing this to reduce costs, the registered breeder can offer a sounder animal for less money with greater profit. All breeders need to recognize that size and weight due to feeding will not be transmitted to offspring, nor will size and bone developed in cooler, drier areas be passed on to offspring grown in warmer, wetter conditions.

The wise purebred breeder will avoid underfeeding or overfeeding his animals. When bulls are cheap, some owners cut back on their feed until they are sale age and then feed them into condition. There is no ovefeeding problem with growing bulls, but extremely insufficient nutrition of young bulls may permanently impair their fertility.

Conversely, overfattening bulls reduces activity and longevity, and the insulating layer of fat in the scrotum prevents heat dissipation and causes infertility. Liver damage and founder resulting from overfeeding in gain-testing bulls on high grain rations will reduce longevity and functional soundness. Growing bulls fed on a ration of 1% to 1.5% of their body weight in grain will not gain as much per day as if full fed grain, but the expression of genetic-potential-for-gain, based on the ranking of the bulls in the test, will be essentially the same at less cost. They will have enough condition to sell and breed well, and enough rumen capacity and roughage-digesting rumen microflora to do well on pasture when turned with cows for breeding.

MARKETING

Registered breeders should recognize that a strong private-treaty sales program is the backbone of their continuing prosperity. "Silk stocking" auction sales may be used to showcase outstanding cattle, advertise the herd, and bring prospective buyers on the place but are not usually very profitable, do not move large numbers of cattle, and may scare away large-volume commercial buyers. Association or joint consignment sales are a less-expensive form for advertising superior cattle, but if these sales are used as a dumping ground for inferior cattle, the breeders image and sale prices will suffer. All-bull sales timed to coincide with commercial bull-buying season may move large numbers of bulls but they don't meet the needs of commercial buyers who should buy at private treaty with animals in groups of uniform breeding and physical characteristics rather than taking potluck by bidding at an auction.

To be pleased, a buyer must be able to make money from his bull purchase. A rule of thumb as to the price to pay for service-age bulls should include the value of the five calves he sires or the value of three cows he is to breed. Knowing heritability factors for performance traits will allow calculating relative breeding value of bulls with various performance records.

The purebred breeder, when pricing his bulls for sale should know what each young bull is costing him at each stage of its development under existing economic conditions. There may be less cost and more profit for both purebred seller and commercial buyer by trading at a younger age.

The very best form of advertisement is satisfied buyers who have purchased cattle that are genetically and functionally sound, in good health, and have predictable performance at a competitive price. Buyer confidence and satisfaction can be reinforced by providing accurate records of ancestry, performance, health, and feeding regime. Remember that overselling is like overfattening in that the buyer will be disappointed. Stand behind what you sell even when it hurts.

This list is far from comprehensive, but it may stimulate some additions and corrections which altogether may provide direction for increased profitabililily, satisfaction, and longevity to breeders and buyers of worthwhile purebred cattle.

Part 7

PHYSIOLOGY AND REPRODUCTION

41

HORMONAL REGULATION
OF THE ESTROUS CYCLE

Roy L. Ax

INTRODUCTION

All reproductive events are regulated by hormones. In simple terms, if the organs of reproduction correspond to the "plumbing," the endocrine system can be called the "wiring." A delicate balance exists between the nervous system and the endocrine system. We are entering an era in which artificial control of the estrous cycle with hormones promises to become more commonplace. Thus, hormones can be considered valuable management tools. If producers are to use the tools effectively, we must develop a better understanding of the complex hormonal interrelationships between the hypothalamus, pituitary, and ovary.

HORMONES DEFINED

A classical definition for a hormone is that it is a substance produced in one tissue that is transported to another tissue to exert a specific effect. (Some of the confusion about hormone actions should be clarified in the next section). Hormones have many chemical classifications; some of the most common reproductive hormones are briefly described here. Gonadotropin-releasing hormone (GnRH) is composed of amino acids and is thus a polypeptide in nature. The follicle-stimulating hormone (FSH) and luteinizing hormone (LH) are glycoproteins. This means they are composed mostly of protein, with some carbohydrate attached to the protein. Estrogen and progesterone are steroids that are synthesized from cholesterol. Prostaglandins are produced from a fatty acid--arachidonic acid. The diversity of the composition of hormones leads to the variation in their biological functions. Most hormone concentrations are in billionths or trillionths of a gram per milliliter of blood.

HOW DO HORMONES WORK?

The fact that a hormone is produced by a tissue does not necessarily imply that it will exert a physiological effect somewhere else. The ability of one tissue to respond to a particular hormone rests in whether that tissue possesses a <u>receptor</u> to the particular hormone. A receptor functions as the lock, and the hormone functions as the key that fits the lock. Therefore, as an example, if a tissue is going to respond to estrogen, its cells must possess estrogen receptors. After a hormone is bound to its receptor, a cellular response is initiated in the target tissue. A target tissue may possess receptors for several different hormones, and exposure to the various hormones can modulate the final response.

THE HYPOTHALAMUS

The hypothalamus is located at the base of the brain. It contains nerve endings to integrate sensory information and sorts out hormonal signals as well. The major reproductive hormone of the hypothalamus is gonadotropin-releasing hormone (GnRH) that is sometimes called luteinizing-hormone-releasing hormone (LHRH). For purposes of this discussion we will use GnRH nomenclature. GnRH is transported in blood vessels to the pituitary gland to regulate secretion of FSH and LH from the pituitary.

THE PITUITARY

The pituitary is positioned underneath the hypothalamus directly above the roof of the mouth. The major reproductive hormones produced in the anterior lobe of the pituitary are called gonadotropins, which means to stimulate the gonads. Follicle-stimulating hormone (FSH) and luteinizing hormone (LH) are the two gonadotropins that regulate the ovary. They are secreted by the pituitary and transported in the circulation to the ovary where they interact with their respective receptors to affect ovarian functions. The main action of FSH is to initiate growth of follicles on the ovary. Continued follicle growth depends on the presence of both FSH and LH. The major effect of LH is to promote ovulation, but there is increasing evidence that FSH can exert a major influence to facilitate ovulation.

THE OVARY

The ovary has two biological functions: (1) to provide the eggs (ova) for the female genetic contribution to the next generation, and (2) to produce hormones to coordinate behavioral changes with ovulation and prepare the reproductive tract for pregnancy.

Estrogen is the hormone produced by follicles as they develop on the ovary. As the predominant follicle or follicles approach ovulatory size, the increased amounts of estrogen are transported to the hypothalamus to cause behavioral heat. The pituitary also responds to the elevated estrogen by releasing a surge of LH which leads to ovulation. Thus, estrogen coordinates behavioral acceptance of a male when the egg will be released into the female tract. This is Mother Nature's attempt to ensure that the probability of fertilization occurring is maximized.

After ovulation has occurred, the tissue that a moment ago was a follicle starts a dramatic change into becoming a corpus luteum (yellow body). The corpus luteum produces progesterone to prepare the female tract for a possible ensuing pregnancy. The corpus luteum forms, regardless of whether or not mating occurs in farm animals. If the corpus luteum remains functional due to pregnancy occurring, the sustained production of progesterone prevents cyclicity. Until the corpus luteum regresses, the typical pattern of cyclic hormonal changes is absent.

THE FOLLICLE

In a simple sense, the follicle is the dwelling of the ovum. Ovulation of a follicle is the exception rather than the rule because over 99% of all potential oocytes are never shed from the ovary. This loss is called atresia. Atresia can occur at any time during follicle growth. When animals are injected with gonadotropins to induce superovulation, some follicles that would have undergone atresia are rescued. This supports the hypothesis that continued follicle growth is dependent upon continued exposure to gonadotropins and the presence of gonadotropin receptors in the follicle.

OVULATION

Once the LH surge has been elicited from the pituitary, the follicle starts to undergo a series of changes to prepare for impending ovulation. The cells lining the inside of the follicle begin to luteinize and secrete progesterone as the major steroid rather than estrogen. The oocyte commences its maturational steps to get it in the proper meiotic configuration for chromosome pairing with the meiotic contribution from a sperm.

Enzymes are activated to degrade the follicle wall and permit the egg to pass into the oviduct. Biochemical studies have pointed to FSH being responsible for stimulating production of those enzymes. FSH also promotes the spreading apart of cells that are tightly surrounding the egg, which then leads to some of the subsequent maturational changes in the egg. Prostaglandins are required for normal

ovulation. Substances known to inhibit prostaglandin formation prevent ovulation. Due to the enzyme, steroid, and
prostaglandin effects, a hypothesis was formulated comparing
the ovulatory process to an inflammatory reaction.

THE CORPUS LUTEUM

The scar tissue remaining after ovulation becomes the
corpus luteum; this endocrine tissue has been studied extensively. Low amounts of LH from the pituitary are essential
for establishment and continued function of the corpus
luteum. In all livestock species, the corpus luteum functions to maintain early pregnancy by secreting progesterone. Progesterone prevents cyclicity. The placenta of the
developing fetus eventually sustains the pregnancy by producing progesterone in the bovine, equine, and ovine. In the
porcine, the corpus luteum is required to support the entire
gestational period.

If cyclicity is to resume, the corpus luteum must regress and cease progesterone production. In pregnant animals, an embryonic signal leads to maintenance of the
corpus luteum, (discussed in the next section). In nonpregnant animals, the uterus recognizes the absence of an embryo
and secretes prostaglandins. Those prostaglandins are transported to the corpus luteum and cause it to regress. Thus,
the inhibitory effects of progesterone are removed, new follicles start to develop, and heat occurs in a few days.

PREGNANCY RECOGNITION

The corpus luteum continues to function to provide a
signal from the embryo to the dam. These signals are hormones identified in the human as chorionic gonadotropin
(hCG) and in the mare as pregnant mare serum gonadotropin
(PMSG). These two gonadotropins can be used to regulate the
estrous cycle of animals. PMSG is biologically similar to
FSH, and hCG exerts an action similar to LH. Thus, PMSG can
be used to induce follicle growth and hCG will promote ovulation of these follicles. However, both PMSG and hCG are
recognized as foreign proteins by livestock, and the animals
build up antibody resistance to them, if they are used too
frequently.

An ideal pregnancy test would be the identification of
the embryonic signal in the bovine, ovine, and porcine. Experiments have shown that embryonic extracts will maintain a
corpus luteum in an animal that has not been bred. However,
even with sophisticated biochemical tests, specific signals
from embryos in the dam's circulation have not been detected. The livestock industry could benefit significantly from
pregnancy tests of these types if they are ever developed
successfully. Since an embryonic signal would have to be
apparent to prevent a subsequent heat in the dam, a preg-

nancy test would also pinpoint which animals would be returning to heat within a few days.

HORMONAL REGULATION OF THE CYCLE

The preceding sections indicate that follicle growth, ovulation, and corpus luteum formation are a dynamic process of sequential steps in an intricate balance. Administration of a hormone to mimic the effect of that hormone in the animal can be used to regulate the cycle. If hormone administration is to produce the desired result, it must be given at a time that is physiologically compatible with the cycle. The common hormones that have been used experimentally or commercially are progesterone-like drugs (progestins), GnRH, and prostaglandins.

Progestins

These compounds were the first to be experimentally employed to regulate the cycle. Progestins have been injected, fed, implanted, or administered via vaginal sponges. Regardless of what stage of the cycle an animal is in when the progestin commences, cyclic fluctuations in other hormones are arrested. As long as progestin is administered, cyclicity ceases. Removal of the source of the progestin results in renewed follicle growth, and estrus, within a few days. Field trial data suggest that fertility at the first estrus after progestin withdrawal is lowered. Thus, it is usually recommended that breeding be done at the second estrus, since the cycles of the animals will still be in close synchrony.

Prostaglandins

These compounds have largely replaced the use of progestins because (1) only one or two injections are required, and (2) fertility is not affected by use of prostaglandins. Prostaglandins are only effective if the animal possesses a functional corpus luteum. Contrary to some opinions, prostaglandins are not a heat-inducing drug. Rather, they cause a corpus luteum to regress, and the animal secretes her own gonadotropins to regulate the ovary and cause a physiological heat. Success has occurred regularly by breeding animals at a predetermined time after prostaglandin injection. Greater success in conception rates can occur if animals are watched for estrual behavior after receiving prostaglandins and are bred in relation to standing heat. Care must be exercised with prostaglandins, because injections into an animal with a functional corpus luteum sustaining a pregnancy could induce an abortion.

GnRH

GnRH is composed of 10 amino acids. It can now be chemically synthesized in a laboratory, and this has permitted chemists to develop some powerful analogs. There are no noticeable ill effects from administering GnRH; its action is to promote a release of gonadotropins from the pituitary. Maximum gonadotropin output occurs approximately 2 to 4 hours after GnRH injection.

A common use for GnRH is to initiate cyclicity in animals with anestrous. GnRH is the most widely used therapy for treating cystic ovarian degeneration. Cystic ovaries usually result from inadequate gonadotropin production. Thus, GnRH triggers release of gonadotropins to restore ovarian function.

A new use for GnRH is for injection after prostaglandin administration; the interval to the gonadotropin surge, and hence, ovulation, can be coordinated more closely. This reduces the variation in time between prostaglandin injection and standing heat that is ordinarily seen among animals.

We have an ongoing study at the University of Wisconsin to evaluate the efficacy of GnRH injections at the time of insemination in dairy cattle. The heifers receiving GnRH have shown no advantages over heifers receiving the saline control. In lactating cows, administration of GnRH 14 days postpartum or at the first artificial insemination has improved first-service conception rates by 15% to 19%. In cows presented for third service (and thus classified as "repeat" breeders in commercial herds) conception rates were about 30% higher for cows that received the GnRH. The physiological effect elicited by GnRH has yet to be experimentally established. We have postulated that gonadotropins produced in response to GnRH cause a corpus luteum to form that may have otherwise been deficient and led to early embryonic death. GnRH could also promote what would have been a delayed ovulation to occur sooner or have a direct effect on the ovary. The lactational stress imposed on a dairy cow may make her unique to respond to GnRh in this manner. Experiments with other farm animals are needed to determine if similar effects result.

SUMMARY

The reproductive cycle is regulated by fluctuations in different hormones. The cycle can be regulated by administering hormones to mimic the effect that would occur in the animal. Therefore, producers have endocrine tools to assist them in managing their animals. For maximum success the producers must understand how the hormones work biologically and realize that they are powerful drugs. We will see an increasing frequency of producers regulating the reproductive cycle to maximize reproductive efficiency.

REFERENCES

Britt, J. H. 1979. Prospects for controlling reproductive processes in cattle, sheep, and swine from recent findings in reproduction. J. Dairy Sci. 62:651-665.

Britt, J. H., N. M. Cox and J. S. Stevenson. 1981. Advances in reproduction in dairy cattle. J. Dairy Sci. 64:1378-1402.

Foote, R. H. 1978. General principles and basic techniques involved in synchronization of estrus in cattle. Proc. 7th Tech. Conf. on Artif. Insem. and Reprod., Nat'l Assoc. Anim. Breeders, pp 74-86.

Hansel, W. and S. E. Echternkamp. 1972. Control of ovarian function in domestic animals. Amer. Zool. 12:225-243.

Jones, R. E. (Ed.) 1978. The Vertebrate Ovary. Comparative Biology and Evolution. Plenum Press, New York.

Lee, C. N., R. L. Ax, J. A. Pennington, W. F. Hoffman and M. D. Brown. 1981. Reproductive parameters of cows and heifers injected with GnRH. 76th Ann. Mtng. of the Amer. Dairy Sci. Assoc. Abstract P228.

Maurice, E., R. L. Ax and M. D. Brown. 1982 Gonadotropin releasing hormone leads to improved fertility in "repeat breeder" cows. 77th Ann. Mtng. of the Amer. Dairy Sci. Assoc. Abstract P233.

Nalbandov, A. V. 1976. Reproductive Physiology of Mammals and Birds. W. H. Freeman and Co., San Francisco.

42

CALF LOSSES:
WHAT WE CAN DO ABOUT THEM

R. A. Bellows

A recent study of data covering 14 years at the Miles City Station (LARRS) indicated that death loss of calves at calving time or shortly after averaged 6.7% and 2.9% from day 4 to weaning. These percentages don't sound high until one looks at the number of calves lost over the entire 14-year study. The 6.7% loss translates to 821 calves and the 2.9% loss to 372 calves, for a total of 1,193 dead calves.

These percentages are similar to those reported in the industry. Agricultural Statistics report that there are approximately 42.9 million beef females on U.S. farms; thus the percentage would represent a potential loss of 2.9 million calves at calving time and 1.2 million calves from birth to weaning, for a total of 4.1 million dead calves. Determining the causes and preventing these losses is imperative.

The calf losses at LARRS have been summarized by D. J. Patterson for a Masters thesis. His study included loss records on 893 calves and gave some interesting insights into the causes of these losses.

Table 1 summarizes losses according to age of dam.

TABLE 1. CALF LOSSES BY AGE OF DAM

Dam Age At Calving	Number Calving	Calves Lost	Percent of Loss
First-Calf, 2-Year-Old	2,257	245	10.8
First-Calf, 3-Year Old	1,394	121	8.7
Second-Calf, 3-Year-Old	1,461	60	4.1
Second-Calf, 4-Year-Old	1,262	105	8.3
Third-Calf, 4-Year-Old	1,032	50	4.8
Five-Year-Olds	1,760	76	4.3
Mature Cows (6-8 Years Old)	3,209	173	5.4
Aged Cows (9-13 Years Old)	921	63	6.8
Totals	13,296	893	

As would be expected, the greatest losses occurred in the first-calf, 2-year-olds. When these dams calved with their second or third calf, the losses dropped to 4.1 and 4.8% respectively. But, it is interesting to note the losses from dams that were bred to calve first at 3 years of age. The loss with the first calf was 8.7%; the second calf--8.3%; and the third calf--4.3%. We do not know why the loss from the second-calf, 4-year-old dams was more than twice the loss in the second-calf, 3-year-old dams.

In table 2, calf losses are listed by days following calving and by greatest-loss category. A total of 513 calves were lost within the first 24 hours following calving; of these, 357 (70%) were lost due to factors associated with or attributable to calving difficulty. Losses caused by problems encountered at calving continued to be the factor causing the majority of calf losses through the fourth day after calving and resulted in the death of 393 calves. Deaths due to scours and pneumonia became the largest causes of calf losses by day 7 and continued as the largest cause

TABLE 2. CALF LOSSES, BY DAYS FOLLOWING BIRTH

Day of Death After Birth	All Losses		Greatest Loss Category[a]		
	No. Calves	% of Losses	No. Calves	Category	% Losses Within Day
1	513	57	357	Calving Difficulty	70
2	48	5	19	Calving Difficulty	40
3	26	3	8	Calving Difficulty	31
4	27	3	9	Calving Difficulty	33
5	18	2	4	Destroyed	22
6	11	1	6	Accidental Death	54
7	13	2	6	Scours	46
8	14	2	6	Scours	43
9	15	2	4	Pneumonia, Scours, Starvation	27
10	13	2	5	Pneumonia	38
11	10	1	4	Pneumonia	40
12-42	87	10	35	Pneumonia	40
43-102	51	6	25	Missing-- Not Found	49
103-Weaning	47	5	23	Missing or Found Dead	49

[a] Single factor responsible for the most losses.

through day 42. Those diseases accounted for 60 deaths during this time. The majority of calves lost after day 43 were classed as "missing" and "not found," accounting for 48 calves lost during the study.

One of the interesting findings of this study was that age of dam had little effect on time of calf loss. The majority of all losses occurred the first 24 hours after birth regardless of age of dam. Additionally, age of dam was little related to losses from scours and pneumonia, which occurred from days 7 through 42.

Since losses at calving (first 24 hours) was the largest loss group, let's look at them more closely. Losses are summarized by sex and lung status of calf in table 3. A total of 58% of the dead calves were males (losses of 81 more bull calves than heifers). Perhaps one of the most significant findings was that of the calves lost at birth, 193 (39%) had functional lungs. This indicates that only 61% of these calves were actually dead at birth or "stillborn."

TABLE 3. CALF LOSSES AT BIRTH (FIRST 24 HOURS) BY SEX AND LUNG STATUS OF THE CALF

Calf Sex	Lung Status					
	Nonfunctional		Functional		Totals	
	No.	%	No.	%	No.	%
Bull	173	60	113	40	286	58
Heifer	125	61	80	39	205	42
Totals	298	61	193	39	491[a]	

a   Totals will differ among tables due to some calves/data not available for a given classification.

Detailed autopsy information was available on most of the calves. These data included not only functional status of the lungs of the calf, but also noted skeletal or visceral abnormalities. Table 4 is a summary by gross anatomical classification and lung status on calves lost at birth. Note that 77% of the calves were classed as anatomically normal. Lung status of the two groups was almost identical, with approximately 40% of both groups being born alive. Further study indicated that the majority of both normal (61%) and abnormal (64%) calves died during the first 24 hours after calving. Of calves classed as abnormal at autopsy (performed at any time from birth to weaning) average birth weight was 63 lb compared to 76 lb for calves classed as anatomically normal.

TABLE 4. LUNG AND ANATOMICAL STATUS OF CALVES LOST AT BIRTH (FIRST 24 HOURS)

| Anatomical Class | Lung Status | | | | | |
| | Nonfunctional | | Functional | | Totals | |
	No.	%	No.	%	No.	%
Normal	231	61	150	39	381	77
Abnormal	70	60	46	40	116	23
Totals	301	60	196	40	497[a]	

[a] See footnote, table 3.

Data on lost calves also were studied as to presentation at birth, i.e., normal, breech, head back, etc. These data for both normal and abnormal calves and the accompanying average birth weights are summarized in table 5. The majority (67%) of the calves were presented in the normal

TABLE 5. PRESENTATION AT BIRTH AND AVERAGE BIRTH WEIGHTS FOR NORMAL AND ABNORMAL CALVES

| | Normal Calves | | | Abnormal Calves | | |
	No.	%	Birth Weight (Lb)	No.	%	Birth Weight (Lb)
Presentation						
Normal	252[a]	67	79	18	54	62
Backward	38	10	76	6	18	69
Breech	37	10	76	5	15	68
Hiplock	16	4	86	0	--	--
Leg Back	20	5	85	1	3	83
Head Back	4	1	77	1	3	57
Other	7	2	77	2	6	66
Calf Sex						
Male	224	60	82	15	45	72
Female	150	40	74	18	54	60

[a] See footnote, table 3.

position at calving, but 20% of the dead calves were presented backward or breech. Birth weights were not markedly different, but calves lost from hiplock or leg back had the highest average birth weights. During the period covered by the study (1963 to 1977), a total of 215 backward or breech deliveries (live or dead calf) were recorded. The frequency was 2.3% of all first-calf, 2-year-old calvings; and 5.6% of all first-calf, 3-year-old calvings. Calf survival rates were 71% for those presented backward and 33% for breach presentations.

This shows clearly that the best way to reduce calf losses is to improve management. Management at calving time should be directed toward preventing losses from dystocia in first-calf heifers. Management to prevent and control disease becomes of key importance in calves during the first 6 weeks after birth. Prevention of losses from birth to weaning should be directed toward close observation of calves to reduce losses attributed to "missing" calves.

43

REPRODUCTIVE EFFICIENCY:
A COUNTY EXTENSION PROGRAM
IN CALVING MANAGEMENT

Ed Duren

The Idaho Total Beef program relies heavily upon the advice and guidance of the industry leaders including producers, agricultural lenders, practicing veterinarians, and agribusiness--namely feed and animal health suppliers. Frequently, the input of these key people reflects the immediate concerns of ranchers during a specific season of activity. A typical example of this approach deals with calving management and reduction of calf death losses within the Idaho counties of Bannock, Oneida, Owyhee, Nez Perce, Caribou, Bear Lake, and Franklin.

As a result of combined planning efforts and mutual interests, county cooperative extension service agents have conducted five calving-management workshops as a lead to implementing county Total Beef Programs. These workshops use the experience and expertise of University of Idaho staff, with teaching support from practicing veterinarians and selected experienced producers using information generated from the Idaho Pegram Project.

These workshops provide a "hands-on" experience for the young beef producer. Approximately 36 to 40 hours of intensive instruction are offered via formal lecture (15 to 20 hours), audiovisuals (both videotapes and slide-tape cassettes), field technique demonstrations, calving equiment, and ranch tours.

Generally, the workshops scheduled during the onset of calving in the area last for 1 week. On occasion, for efficient use of time, travel, and expense, two workshops in adjacent counties may be conducted simultaneously with extension agents coordinating and scheduling the instructors to service both workshops.

The normal producer charge ranges from $35 to $50 per person, depending upon costs involved. Each participant receives a calving management notebook containing information presented by the instructor.

The successes of the workships have been attested to by both ranch owners and ranch managers. In the words of one father-owner, "My son learned more about calving cows at the workshop than I have in a lifetime." The workshops are

University of Idaho

College of Agriculture
Cooperative Extension Service

COW-CALF MANAGEMENT CHECKLIST

_____ 1) Are your cows tagged?

_____ 2) What are your costs per cow unit?

_____ 3) Performance records (BCI).

_____ 4) Herd Health.

_____ 5) Recognize disease problems.

_____ 6) Pounds of calf weaned per cow?

_____ 7) Preg test cows?

_____ 8) Length of calving season?

_____ 9) Have you checked the need for mineral supplements?

_____10) Do you have an implant program?

_____11) Forage analysis.

_____12) Do you do a breeding soundness evaluation on your bulls?

_____13) Do you have a long term plan to increase weaning weights?

      _____ a) Crossbreeding

      _____ b) Better bulls

_____14) Do you have a range improvement plan?

_____15) What is your marketing plan?

      _____ a) Have you considered alternatives?

      _____ b) Do you do any market charting?

_____16) Heifer weights at breeding time.

_____17) What are your objectives?

      _____ a) This year?

      _____ b) Five years from now?

Developed at the "Make It Happen" workshop 1982. Sponsored by the Idaho Cattlemen's Association, Idaho Cattle Feeders Association in cooperation with the University of Idaho, College of Agriculture.

RALGRO®
International Minerals & Chemical Corporation

COW-CALF MANAGEMENT CHECKLIST REFERENCES

The following reference material will help you put your operation in a positive relationship to the questions on the reverse side.

1. There are many advantages in beef cow management if cows are tagged with a number for reference. Cow-Calf Management Guide Cattleman's Library - CL 730.

2. Without knowing what your costs are per cow unit it is almost an impossibility to become more efficient. Idaho Livestock Budgets Misc. Series - #71, University of Idaho, College of Agriculture.

3. Performance records tell us where the opportunities for improved management are and are monitoring our present management. University of Idaho Beef Cattle Improvement program - See your County Extension Agent.

4. A good, sound herd health program is basic to making all other management work successfully. Cow-Calf Management Guide Cattleman's Library - CL 600 - 695.

5. Recognition of disease problems is helped immensely when good health records are kept. Cow-Calf Management Guide Cattleman's Library - CL 695.

6. Pounds of calf weaned per cow is one of the keys to profitable beef production. Idaho Beef Cattle Improvement Program - See your County Extension Agent.

7. Pregnancy testing tells you more than whether a cow is bred or not. Cow-Calf Management Guide Cattleman's Library - CL 410, 412.

8. Length of the calving is one of the more critical factors in profitability. Idaho Beef Cattle Improvement Program - See your County Extension Agent.

9. Mineral Supplements are probably the most neglected area of beef cattle nutrition. Cow-Calf Management Guide Cattleman's Library - CL 300.

10. Growth implants in calves pays ten to one. Cow-Calf Management Guide Cattleman's Library - CL 755.

11. Unless you know what the feed analysis of your feed is you do not know if you are meeting the needs of the cattle. Cow-Calf Management Guide Cattleman's Library - CL 300, 310, 312.

12. Sound fertile bulls are a must in getting a high percentage of the cows bred in a short time. Cow-Calf Management Guide Cattleman's Library - CL 425.

13. Breeding plans are a long-time program and require dedication. Cow-Calf Management Guide Cattleman's Library - CL 1000, 1020, 1035.

14. Range and pasture improvement and maintenance should be a major part of the overall management program. Cow-Calf Management Guide Cattleman's Library - CL 500, Idaho Forage Handbook.

15. Marketing is usually an afterthought in beef cattle management. Cow-Calf Management Guide Cattleman's Library - CL 800, 805, 815, 825.

16. Proper development of heifers is one of the best investments in time and money. Cow-Calf Management Guide Cattleman's Library - CL 345, 415.

17. Without some clearly identified goals and objectives you will never know where you have been or where you are going. Cow-Calf Management Guide Cattleman's Library - 900, 910, 915, 920, 925.

helpful to veterinarians establishing their professional
services and capabilities with their clients. Agricultural
lenders have indicted that participating ranchers have saved
more calves and thus have increased their cash flow.

The teaching materials and audiovisuals are selected
and documented from the Idaho Cow-Calf Management Guide and
Cattleman's Library published by the University of Idaho
College of Agriculture.

Agricultural lenders, during the loaning process,
assist producers in analyzing the status of herd management. This is accomplished very easily by simply reviewing
the Cow-Calf Management Checklist available to all Idaho
agriculture lenders. The Cow-Calf Management Checklist
serves to identify and prioritize areas of management that
may assist the reproductive efficiency through appropriate
technology and allocation of financial resources.

The checklist also offers the county beef program planning committees insight into educational needs of beef
clientele.

TYPICAL CALVING MANAGEMENT TOPICS--DISCUSSION AND DEMONSTRA-
TION

1. Clean & prepare a cow for examining genital
   tract or calving
2. Technique of epidural anesthesia
3. Closing vulva to prevent prolapse
   a. Placement of prolapse pins
   b. Suture of vulva
4. Techniques of calf delivery
   a. Normal presentation
   b. Abnormal presentation and extraction
   c. Caesarean births
5. Postpartum cow and calf care
   a. Cow
      (1) Oxtocin (P.O.P.)
      (2) Antibiotics for uterine infection
      (3) Retained afterbirth
   b. Calf
      (1) Calf respirator
      (2) Liquid nutrient therapy
      (3) Treatment and medication
      (4) Neonatal diseases
6. Drugs and route of administration
   a. Instrument care and sterilization
7. Diagnostic laboratory services
   a. Selection and handling specimens
   b. Tests
8. Routing veterinary emergency care
9. Reproductive disease, abortion, and vaccination
10. Calving records and planning herd health
11. Cow management during the last trimester

INSTRUCTORS

Instructors participating in the Bannock and Oneida Counties' Calving Workshop are typical of both interest and cooperation anticipated in delivering workshop information.

University of Idaho Personnel -
Tom Ritter - Extension Agricultural Agent, Pocatello, Idaho.
Rauhn Panting - Extension Agricultural Agent, Malad City, Idaho.
J. D. Mankin - Extension Livestock Specialist, Caldwell, Idaho.
Ed Duren - Extension Livestock Specialist, Soda Springs, Idaho.

Practicing Veterinarians -
Mark Ipsen, DVM, Malad City, Idaho.
Bob Miller, DVM, Preston, Idaho.
Stan Hull, DVM, Grace, Idaho.
Chuck Merrill, DVM, Montpelier, Idaho.

Consultants -
Dave Nash, DVM, Western Stockman's Supply; Caldwell, Idaho.
Joe Magrath, DVM, The Magrath Company; McCook, Nebraska.
George Roma, Arizona Feeds; Tuscon, Arizona

Extension Beef Planning Committee -

Bannock County	Oneida County
Glade Davis	Mike Broadhead
Thayne Thompson	Oren Jones
Wayne Wheatley	Bert Smith

COMMITTEE INPUT

In the planning of educational activities, planning with people is more effective than planning for people. The input of a committee must be recognized in successful Cooperative Extension Service programs to meet the beef industry needs for educational material.

The role of each committee member is to represent the producer, professional observations, problems, and priorities as they may influence financial and social welfare of a community, county, or state. Obviously, these people serve as a civic duty without compensation. Frequently, they may become disillusioned with their role, duties, or responsibilities if not allowed to become deeply involved. Then, the question may arise, "What's in it for me?"

Financial conditions are important to those who derive their living and income from an industry. Current information helps to stabilize an industry and make it more efficient. This is true with the beef cattle industry in addition to the other agriculture commodities.

Committee members should personally answer a series of questions if they are committed to growth and prosperity of an industry. These questions include, but are not limited to, the following:

1. How do you expect the beef industry in your area to benefit because of the committee input?

2. What are the goals of your ranch or business?

3. What do you personally want to achieve as a rancher and/or professional?

4. What does it take to reach your goals?

Once these questions have been addressed personally, the committee will blossom and function effectively.

IMPACT AND DIRECTION

Calving Management Workshop participants respond more enthusiastically to the hands-on educational process of the workshop than a typical "road-show" one-day meeting of cafeteria-style information. The workshop provides an educational foundation upon which to build the educational interest of the producer in a problem-solving mode where he shares his experiences with the instructor who, in turn, shares his technical information.

After the workshop, participants begin to implement the technologies on their own ranches. Following one to two calving seasons, professionals in calving observe measureable changes in the herd management. Change does not cease with the ending of the workshop. However, through the workshop, producer participants become aware of the need to integrate appropriate support from various sources in a team effort. Consequently, a team composed of an extension agricultural agent, veterinarian, banker, and producer becomes more involved in dealing with the respective ranch. As problems are attacked, the team rapidly integrates by developing and implementing the Idaho Total Beef Program for a specific ranch or a group of ranches within the community. The Idaho Total Beef Program, however, is never fully completed because it is open-ended and can apply to all levels of management and technological capabilities.

Therefore, the Idaho Total Beef Program continues to grow; the problem-solving mode moves from the simple to the complex. With selected key indicators the results are measurable and results have been both satisfying and stimulating.

44

USE OF LUTALYSE®
STERILE SOLUTION (PGF2a) TO ASSIST
IN MANAGING THE BREEDING
OF BEEF AND DAIRY CATTLE

James W. Lauderdale

WHY REGULATE ESTRUS?

Artificial insemination (AI) allows use of semen from progeny-tested sires to provide selection capability that can increase genetic gain and boost productivity of both meat and milk. In addition, use of AI controls venereal disease more easily than does use of natural service. Even though AI has advantages, only about 50% to 60% of dairy cows are inseminated artificially in the U.S., Canada, and England. A smaller percentage of beef cows (less than 5%) and dairy heifers (about 20% or less) are inseminated artificially in the U.S.

The additional management time and skills needed for estrus detection have been the factors in the failure of herd managers to use AI to a greater degree in both dairy and beef cows. "Cows" as used here refers to all breeding age females unless otherwise indicated. The cost of estrus detection is justified only when AI results in pregnancy. Based on our current knowledge, estrus must be detected by man if either AI or hand-mating programs are to be employed to breed cows. To effectively use AI, dairy cows must be observed at least twice daily for signs of estrus during most of the year; beef cows must be observed at least twice daily during 40 to 60 or 90 day intervals once each year.

The problem of estrus detection can be (1) reduced if females can be inseminated and become pregnant during a 3- or 4-day interval as compared to 20 to 24 days, or (2) further reduced if they can be inseminated with reasonable probability of conception at a fixed time without estrus detection for at least one estrous cycle.

MANAGEMENT OF ESTRUS WITH LUTALYSE (PGF2a)

Early experiments in cows revealed PGF2a to be luteolytic (yellow body regression) between days 5 to 18 of the estrous cycle; most animals in this range of their estrous cycles returned to estrus within 2 to 4 days after the administration of a suficient dose of PGF2a. Cows more than day 18 into their estrous cycle are not considered a problem because this population of cows in a herd of PGF2a-treated cows would return to estrus and would be, coincidentally, fairly well synchronized with the cows responding to PGF2a treatment. However, the nonresponding animals, which are less than day 5 into their estrous cycle, are a major problem because their cycles would be from 15 to 20 days out of synchrony with the rest of the PGF2a treated animals.

During the past 10 years, numerous synchronizations of estrus management systems have been studied that utilize only PGF2a as the synchronizing agent. Most of these management systems have attempted to manage the females from day 0 to 5 in their estrous cycles in a way to efficiently include them in a synchronized artificial inseminatin program. These management systems can be categorized in one of the following systems or as combinations or variations of one or more of these systems (figure 1).

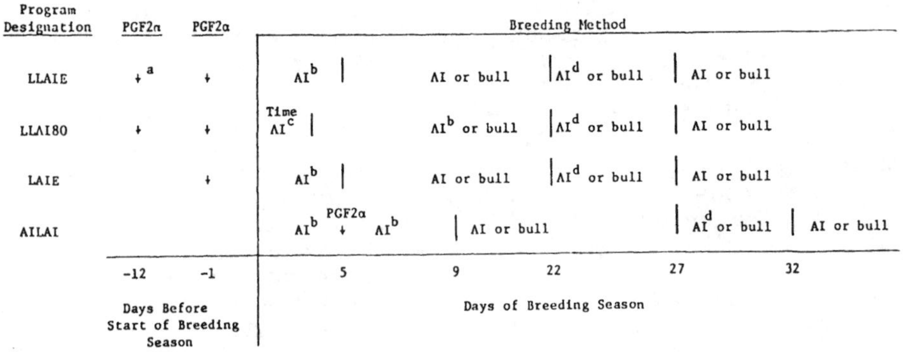

Program Designation	PGF2α	PGF2α	Breeding Method				
LLAIE	+[a]	+	AI[b]	AI or bull	AI[d] or bull	AI or bull	
LLAI80	+	+	Time AI[c]	AI[b] or bull	AI[d] or bull	AI or bull	
LAIE		+	AI[b]	AI or bull	AI[d] or bull	AI or bull	
AILAI			AI[b] PGF2α + AI[b]	AI or bull		AI[d] or bull	AI or bull
	-12	-1	5	9	22	27	32
	Days Before Start of Breeding Season		Days of Breeding Season				

[a] + = 5 ml Lutalyse (25 mg PGF2α) intramuscular

[b] AI = inseminate upon detected estrus

[c] Timed AI at about 77 to 80 hr after the second injection of Lutalyse.

[d] Intervals during which cattle that did not become pregnant to AI during the first synchronized interval would be expected to return to estrus.

Figure 1. Cattle breeding management with lutalyse

- Give two injections of PGF2a 11 days apart, then breed after the second injection (a) according to detection of estrus (LLAIE) or (b) without reference to estrus at a preset time of 75 to 80 hr (LLAI80).

- Cows are detected for estrus and inseminated for at least 4 days and on the morning of the 5th day all cows not detected in estrus or previously inseminated are injected with a single injection of PGF2a. Breeding continues according to detection of estrus, i.e., a 9 to 10 day AI interval (AILAI).
- Inject PGF2a, then inseminate for 5 days according to detection of estrus; this system accepts that females in the first five days of their estrous cycle will not be synchronized (LAIE).

For each program, breeding at estrus subsequent to the first estrus can be with AI, bulls, or a combination of AI and bulls, depending on the goals of the breeding program (figure 1).

The single injection program (LAIE) will be more effective relative to the number of injected females detected in heat within 5 days (1) if either stage of the estrous cycle is known and they are injected on day 6 or later or, (2) if ovaries of the females are palpated and a corpus luteum (CL) is present at time of injection. Effectiveness in the latter case will be dependent on the skill of the palpator to correctly identify a functional CL. Generally, the stage of the estrous cycle will be unknown for individual cows and palpation of ovaries will not be practical for most herds. However, the single injection program has utility in numerous reproductive management programs.

Many factors contribute to success of reproduction management; these factors are important also when time of breeding is to be regulated with Lutalyse. Some of these factors are:
- Physical facilities must be adequate to allow cattle handling without being detrimental to the animal.
- Nutritional status must be adequate prior to and during the breeding season. Nutrition has a direct effect on the age at initiation of estrus in heifers, on the postpartum interval to estrus following calving, and on conception.
- Females must be ready to breed--they must be estrous cycling and must be healthy.
- Estrus must be detected accurately if timed AI is not employed.
- Semen of high fertility must be used.
- Semen must be inseminated properly.

Lutalyse is effective only if cows have a corpus luteum. Therefore, prepuberal and truly anestrous cattle will not respond to Lutalyse.

DEFINITION OF MEASUREMENTS OF EFFECTIVENESS

- Estrus synchronization: no. detected in estrus within 5 days after Lutalyse x 100 ÷ no. injected.
- Conception rate: no. pregnant x 100 ÷ no. detected in estrus and AI. This is a measurement of fertility.
- Pregnancy rate: no. pregnant x 100 ÷ no. injected. This is a measurement of reproductive performance and reflects estrus cycling percentage and conception rate.

For example, in a herd of 100 cows, if 60 were estrous cycling and were detected in estrus in one estrous cycle interval (24 days of AI), and if first service conception rate was 60%, then pregnancy rate would be 36%, i.e., 36 of 100 cows would be pregnant in 24 days in that natural mating or AI program. If the cows were synchronized with Lutalyse, and if 60 of 100 cows were cycling before the start of the regular breeding season, then 60 cows would be detected in estrus within the first 5 days of AI and with a 60% first service conception rate 36% of the herd would be pregnant in 5 days. Lutalyse per se will neither increase the percentage of the herd cycling nor alter conception rates. However, if large numbers of cattle receive AI in a day, or within a few hours, by either an inexperienced or "out-of-condition" inseminator, conception rates can be decreased--sometimes severely.

BEEF CATTLE

The data reported here were derived from a research program with commercial and purebred beef farms and ranches. Although this field research had several management factors imposed that would not ordinarily be factors in beef reproduction management, the data were generated under "real life situations" and I believe the data are predictive of what can be expected with commercial use of Lutalyse. This belief has been borne out during the past 3 years of marketing Lutalyse during which we have not encountered failures of Lutalyse efficacy that were not predicted from our field research.

In each of the investigations to be reported, cows were assigned randomly in replicates of control and Lutalyse experimental groups. Attempts were made to balance age and semen source (bulls) between the experimental groups within the herd so that no bias would be introduced in favor of either control or experimental groups. However, the presence of large numbers of Lutalyse-treated cows in heat in a 5-day interval during the beginning of the breeding season stimulated an unusually increased percentage of control cows to cycle early and become pregnant to AI at that estrus.

DOUBLE INJECTION SYSTEMS OR LLAIE AND LLAI80 (FIGURE 1)

About 3,800 beef cattle in about 50 herds were used to study the efficacy of the double injectin program on a within-herd contemporary comparison. Females assigned to both LLAIE and LLAI80 were injected with Lutalyse intramuscularly in the hip twice at an 11 (10 to 12) day interval at a rate of 25 mg PGF2a per injection. Females of both control and LLAIE groups were observed for estrus and artificially inseminated according to the normal procedures within herd at each detected estrus during the interval of the investigation. Generally, the cows/heifers were observed for estrus twice daily and inseminated about 12 hours after the first observation of estrus. Animals of the LLAI80 group were inseminated between 75 and 80 hours after the second injection of Lutalyse. Under the experimental conditions of this study, LLAI80 females were not rebred for at least five days after the 80-hour AI even if they were detected in estrus; cattle of the LLAI80 group were then inseminated at each subsequent estrus.

Dates of injections of Lutalyse were established such that the second injection would be administered the day prior to initiation of the normal breeding season within herd.

Significantly greater percentages of suckled cows and beef heifers were detected in estrus during days two to five of the AI season for LLAIE (47% cows, 66% heifers) compared to control groups (11% cows, 13% heifers). Similar percentages of control (66% cows, 81% heifers) and LLAIE (70% cows, 84% heifers) animals were detected in estrus at least once during the first 24 days (one estrus cycle) of the AI season. So, not all control animals were exhibiting estrus, even after the first 24 days of the breeding season.

First service conception rates were similar between control (about 60%) and LLAIE (about 60%) females for both days 2 to 5 and 1 to 24. These data reinforce previously reported data that conception rate was not altered following use of Lutalyse.

Pregnancy rate reflects both estrus synchronization and conception rate. During the synchronized interval, days 2 to 5 after second injection of Lutalyse, pregnancy rates were similar between LLAIE (36%) and LLAI80 (36%), but were greater for both of these groups when compared to controls (10%).

Pregnancy rates were similar among control, LLAIE, and LLAI80 groups for days 1 to 18 (45%), days 1 to 24 (50%), and days 1 to 28 (55%). Fourteen to 16 days of AI were required to achieve pregnancy rates in controls similar to a single timed AI.

The data presented in the preceding paragraphs are averages of all herds. Examples of the variety of results achieved among herds are presented in the following paragraphs.

In one herd, pregnancy rate was low (30% in 24 days) due to a conception rate problem. The records indicated 97% estrus detection for days 2 to 5 for Lutalyse-treated animals and 24 days for controls. But, low conception rates (30% first service conception rate) resulted in low pregnancy rates. If conception rate in the herd is low, a high percentage of cows won't be pregnant following use of Lutalyse.

Another beef herd, with 90 females per treatment group, had something more than 50% pregnancy rate to timed AI and more than 85% of the herd pregnancy by the 28th day of breeding. Obviously, the cattle were estrous cycling and conception rate was high. In two beef cow herds, pregnancy rate was more than 60% to timed AI and was more than 80% to 90% by day 28 of the breeding season. These examples demonstrate that good results can be obtained if the cows are estrous cycling and conception rates are high.

We should recognize that with timed AI (LLAI80), unless all cows are cycling, more services per conception (more semen) will be required per pregnancy since all animals are inseminated at 75 to 80 hr whether or not they are cycling. Average services per pregnancy in the field study were 1.7 for controls, 1.8 for LLAIE, and 2.5 for LLAI80.

These data were reinforced by additional studies of LLAIE in about 1,800 cattle in about 24 herds and of LLAI80 in about 1,600 cattle in about 11 herds.

SINGLE INJECTION SYSTEMS

Single Injection Plus Estrus Observation or the LAIE System

The LAIE cattle management system (figure 1) is intramuscular injection of females with 25 mg PGF2a (5 ml Lutalyse) on the day before initiation of the breeding season, followed by observation of cattle for estrus and breeding for 5 days. Breeding for the remainder of the breeding season is with AI, bulls, or some combination of AI and bulls. If cattle ovaries are palpated accurately for the presence of a corpus luteum (CL) and only cattle with a CL are injected, the single injection would be expected to be similar in effectiveness to the double injection (LLAIE). In the absence of ovarian palpation, cows of the LLAIE system compared to cows of the LAIE system should have about a 20% to 25% greater first estrous detection rate and 20% to 25% greater pregnancy rate for breeding the first 5 days. Based on the data, the theoretical calculation is that: (1) PGF2a is ineffective or less effective during days 1 to 4 or 5 after estrus as compared to days 6 to 18 after estrus and (2) cows usually have an 18 to 24 ($\bar{x}$ = 21) day estrous cycle; therefore, a single PGF2a injection per cow in a herd of randomly estrous cycling animals should be about 75% to 80% as effective as the LLAIE system. These computed differences were confirmed in studies of 6 herds with about 1,400 beef cows.

The field study of LAIE was completed in 17 herds with about 2,400 beef cattle.

The percentage of cows and heifers detected in estrus the first time during days 1 through 5 was greater for LAIE vs control for both cows (57% vs 31%) and heifers 52% vs 28%). The percentage of females detected in estrus the first time during days 1 through 24 was similar between LAIE and controls for cows (76% vs 68%) and heifers (82% vs 82%).

The percentages of cows and heifers detected in estrus during the first 24 days of AI were 68% and 82% for control cows and heifers. This should be an over-estimate of the percentage of the herd having estrous cycles on the day of PGF2a injection, because they had 24 more days either after calving or to reach puberty.

Calculation of the predicted estrus detection rates would be as follows for single injections: 75% effective of 68% of estrous cycling cows equals 51% (actual was 57% in LAIE cows) and 75% effective of 82% of estrous cycling heifers equals 62% (actual was 52% for LAIE heifers). Thus, the theoretical estrus detection rates and the actual estrus detection rates were similar, which reinforces the conclusion that a single injection of PGF2a yielded the predicted response.

First service conception rate would be expected to be similar between LAIE and control cattle. Conception rates for days 1 through 5 for LAIE and control females were 54% and 49% for cows and 52% and 47% for heifers. Conception rates for days 1 through 24 for LAIE and control females were 63% and 53% for cows and were 57% and 53% for heifers.

Pregnancy rates for days 1 through 5 were greater for LAIE compared to control cows (30% vs 14%) and heifers (28% vs 12%). Pregnancy rates for days 1 through 24 were similar for LAIE and control cows (60% vs 56%) and heifers (55% vs 49%). Pregnancy rates for days 1 through 28 tended to be greater for LAIE than control cows (66% vs 60%) and heifers (57% vs 52%).

These data on enhanced pregnancy rates after 5 days of AI with LAIE program are consistent with information published previously in three reports. The pregnancy rates for 5 days of breeding in the LAIE management system demonstrated that system to be effective.

Single Injection on Day 5 of Breeding or the AILAI System

The AILAI cattle management system is observation of cattle for estrus and AI for the first four days, inject females not detected in estrus with 25 mg PGF2a (5 ml Lutalyse) intramuscular on the morning of day 5 and continue to observe them for estrus and inseminate accordingly on days 5 through 9 or 10, i.e., a 9 or 10 day AI season. Breeding for the remainder of the breeding season is with AI, bulls, or some combination of AI and bulls (figure 1).

This system takes into account that PGF2a is ineffective during the first 4 to 5 days after ovulation. The program does this through the four (4) day breeding program prior to Lutalyse injection. The four day duration is the minimum for effective control of the nonresponsive period after ovulation. Intervals of 5, 6, or 7 days are effective and are recommended by some AI companies. A second purpose of the 4 to 7 day preinjection breeding interval is to allow for an assessment of estrous cycling in the herd. If all cows and heifers are estrous cycling, about 4% to 5% of the herd should be detected in heat each day. If less than 3% of the herd is detected in heat per day (or a total of about 12% after 4 days of AI), then the females probably are not estrous cycling at a rate high enough to warrant use of Lutalyse at that time. Another option, if a low percentage of the herd was detected in estrus, would be to have an accurate palpator identify those females that had a CL, but had not been bred, and inject only those with a CL with Lutalyse.

The AILAI program allows for about 17 more days postpartum to the time of Lutalyse injection, compared to the postpartum interval to the first injection of Lutalyse in the LLAIE or LLAI80 programs. This extra 17 days is very important in many beef cow herds to allow time for a greater percentage of the herd to initiate estrous cycles following calving or to reach puberty.

This program also allows the inseminator to be reeducated in the art of AI during the first 4 days when only a relatively few cows will be inseminated each day. During this interval the facilities will have minimal stress placed on them and if repairs need to be made, they can be made prior to the days after Lutalyse when relatively more females will be inseminated.

AILAI also appears to have advantages for people just starting an AI program and for herds in which there is question about the percentage of the herd estrous cycling 10 to 12 days prior to initiation of the breeding season. The percentage of cows detected in estrus the first time during days 1 through 9 was greater for AILAI than for controls for cows (54% vs 38%) and heifers (64% vs 38%). Estrus detection rates for the first 24 days of breeding were similar between AILAI and control for both cows (70% vs 73%) and heifers (77% vs 78%).

First service conception rates were concluded to be no different between cows and heifers assigned to AILAI and control groups for days 1 through 9 (58% vs 64% for cows and 53% vs 56% for heifers) for days 1 through 24 (59% vs 63% for cows and 57% vs 59% for heifers).

Pregnancy rates were greater for AILAI than for control animals for days 1 through 9 for cows (39% vs 26%) and heifers (45% vs 24%) and for days 1 through 24 for cows (56% vs 54%). Pregnancy rates for days 1 through 28 tended to be greater for AILAI than for control cows (63% vs 59%) and heifers (63% vs 59%).

The percentages of cattle detected in estrus the first time, first service conception rates, and pregnancy rates should be similar between AILAI and control cattle for days 1 through 5, because the AILAI cattle would not have been injected with PGF2a. That was the case for beef heifers. In contrast, percentage of cows detected in estrus and pregnancy rate were each elevated for control cows for days 1 through 5. The basis for that difference is unknown since the cows were assigned randomly in replicates to the experimental groups, so these differences are assumed to be chance observations. However, this observation reinforces the conclusion that a 4-day estrus observation interval is not always as accurate a predictor as we would like to have of the percentage of a herd cycling. Usually, if the estimate is incorrect, it is incorrect by overestimating the percentage of the herd cycling.

The data on enhanced pregnancy rates after 9 days of AI with the AILAI management system are consistent with data published previously in five reports.

The greater pregnancy rate in the AILAI group for days 1 through 9 demonstrated the effectiveness of the use of PGF2a in that system of cattle management. The trend for more pregnancies in the AILAI group after 28 days of AI reinforces the conclusion that the AILAI management system was effective as measured by percentage of herd pregnant. Since the cows and heifers injected with Lutalyse that do not become pregnant to breeding at the first synchronized estrus (days 6 to 9 of the breeding season) would be expected to return to estrus during days 27 to 32, our data for 28-day pregnancy rate did not allow for the maximum difference between control and AILAI animals. Data published by Dr. Ed Moody support the finding that the difference between control and AILAI cattle would be greater after 32 days of breeding. In Dr. Moody's study of about 1,800 beef cows, pregnancy rate after 32 days of AI was greater in cattle of the AILAI group than those of control groups (72% vs 61%).

DAIRY CATTLE - HEIFERS

Although 50% to 60% of dairy cows are inseminated, only about 20% of the dairy heifers are inseminated. The difference in percentage of AI is due primarily to reduced time available to observe heifers for estrus. Estrus can be managed with the dairy heifer using any of the Lutalyse management schemes outlined for beef cattle. An objection posed to synchronizing dairy heifers is that large numbers of heifers entering the milking string at once is not desirable. If that is the case, then estrus and AI can be managed with Lutalyse only in that portion of the heifer population desired to enter the milking string at any one time. Thus, use of Lutalyse will allow for efficient AI of heifers, and AI in heifers with semen from appropriate PD bulls will allow for more effective increases in milk production in the herd.

DAIRY CATTLE - COWS

Is there need to manage estrus in dairy cows? The answer is an unequivocal "yes." The yes answer is based on the fact that about one-third (33.3%) to one-half (50%) of dairy cows, whether they reside in large or small herds, have calving intervals greater than 13 months. Recent information documents that milk production decreases as calving intervals increase beyond about 13 months (table 1). Thus, milk production within a herd can be increased for that portion of the herd that has a calving interval greater than 13 mo, if the cattle become pregnant by 110 days after calving. To achieve a 12 to 13 mo calving interval, cows must be pregnant by 80 to 110 days postpartum. If average number of services per pregnancy is on the order of 1.8 to 2.0, then cows need to be detected in estrus and artificially inseminated the first time by 60 to 80 days after calving. Several studies have indicated that up to 50% of the cows had not been detected in estrus by 60 days postpartum.

TABLE 1. EFFECT OF CALVING INTERVAL ON MILK YIELD[a]

Calving interval (mo)	Approximate herd average milk yield (lb)
11	15,375
12	16,200
13	16,275
14	15,625
15	14,500

[a]Adapted from C. L. Pelissier, Animal Nutrition and Health (1982).

USE OF LUTALYSE IN LACTATING DAIRY COWS

The objective of this investigation was to evaluate further the effectiveness of Lutalyse sterile solution (PGF2a) for treatment of dairy cows with unobserved (silent) estrus--but with ovarian structures deemed to be corpora lutea, based on digital palpation of the ovaries per rectum. Cows were assigned to control and PGF2a experimental groups in 20 herds with a total of 146 control cows and 167 PGF2a cows. If a cow had a corpus luteum on the ovary, she would be indicated as having been in estrus, even though she had not been detected in estrus.

Control cows were palpated but were not injected. Cows assigned to the PGF2a experimental group were injected with 25 mg PFG2a (5 ml Lutalyse sterile solution) intramuscularly. Herdsmen were instructed to inseminate cows at each detected estrus after injection. If a PGF2a injected cow had not been detected in estrus by 75 to 80 hours post injection, she was to be inseminated at that time. Any cows observed in estrus after the 80 hour AI were to be inseminated as per normal estrus detection and AI procedures at that location. Control cows were to be inseminated at each observed estrus.

Within each herd, artificial insemination procedures were similar between cows assigned to control and Lutalyse groups and were similar between cows on this study and the rest of the herd; i.e., semen storage, bulls, straws, and ampules, AI technicians, and AI techniques were similar.

The uterus of each PGF2a and control cow was palpated per rectum between 35 and 70 days after AI to determine pregnancy status. The date of conception was established based on AI records and the estimated stage of pregnancy by the examiner.

The cows assigned to the study had average postpartum intervals of 119.6 days for controls and 128.5 days for PGF2a groups, i.e., estrus had not been detected for about 4 months after calving at the time of treatment. These cows represented a sample of unobserved estrus problem cows. Cow body weight averaged 1,302 and 1,334 pounds for control and PGF2a cows. The range in body weights was 820 to 1,700 pounds.

The percentage of the group of cows pregnant is the most critical measurement of efficacy. Pregnancy rate is defined as number of cows pregnant x 100 divided by the number of cows assigned to the group. Cumulative pregnancy rates for cows assigned to control and PGF2a groups were 0 and 24.2% by 80 hr, 2.6 and 28.4% by 5 days, 27.1 and 32.7% by 24 days, 46.5 and 52.9% by 60 days, and 55.0 and 57.6% for the duration of breeding after treatment. Thus, PGF2a-treated cows were inseminated and became pregnant at a rate greater than did control cows during both the first 80 hours and 5 days after treatment, a response similar to that reported for estrus cycling cows. Also, the percentage of PFG2a cows pregnant by 5 days after PGF2a (28.4%) was similar to the percentage of control cows pregnant by 24 days (27.1%). In other words, use of Lutalyse allowed cows to be managed so that breeding of individual cows could be completed in 5 rather than 24 days.

Inseminations per pregnancy were 1.4 and 1.8 for control and PGF2a cows. The average interval between initiation of treatment and pregnancy was not different between control and PGF2a cows in this study, but clinical experience has demonstrated that "control" groups become more reproductively efficient by working within a herd. Thus, the increased reproductive efficiency, 27.1% pregnant by 24 days whereas pregnancy rate was zero during the preceding

120 days in these control cows, reflects their assignment to the study and an increased intensity of management to become pregnant.

The pregnancy rate data reinforce the conclusion that silent (unobserved) estrus dairy cows that have a corpus luteum are a management problem rather than a physiological/ endocrinological problem. However, the 60-day pregnancy rates of 46.5% and 52.9% for cows of the control and PGF2a groups, i.e., about 190 days after calving, reinforces the conclusion that cows assigned to this study represented a sample of problem cows. The data support the conclusion that prostaglandin F2a (Lutalyse S.S.) was effective as a treatment or management tool for rebreeding these cows. The data implies the use of Lutalyse to be an effective management aid for ensuring that cows with a corpus luteum would be bred the first time between 60 and 80 days after calving. This management procedure will increase the probability of having a greater percentage of the lactating dairy cow herd calving in a 12 to 13 mo interval.

GENERAL COMMENTS

Both beef cow's estrus synchronization and the dairy cow's estrus-control agent must be effective, have no harmful side, carryover, or aftereffects, be simple to administer and relatively foolproof, and cost must be relative to benefits. The dose of Lutalyse recommended for cows has been investigated extensively and no side effects of consequence have been detected with either the recommended dose or with doses of at least ten times greater. Intramuscular injections of Lutalyse are simple to administer and the cost relative to benefits appears to be acceptable.

Our overall conclusions are that Lutalyse offers a new approach to control of estrous cycles of both beef and dairy cows.

Lutalyse isn't a cure-all. Cows must be estrus cycling and have good fertility in order for Lutalyse to be an effective adjunct to reproduction management.

Under those types of conditions, Lutalyse would be expected to be extremely useful. The effectiveness of the LLAIE, LLAI80, LAIE, and AILAI programs and various combinations of these program supports the conclusion that Lutalyse should be useful in a variety of reproductive management situations with flexibility in AI programs.

Although extensive field investigations have not been completed on bull breeding of synchronized cows, Dr. Ed Pexton, Colorado State University, has reported on the effective breeding of cows with bulls at the synchronized estrus after Lutalyse. In a series of five studies with about 445 beef cows, single sires were introduced to groups of about 20 females, each for a 48-hour interval. Bulls were introduced 48 hours after the second of two Lutalyse injections given 10 days apart. The conception rate (percentage of cows pregnant of those serviced) was 66% in the 48-hour breeding period. As always, females that were not estrous cycling did not come into heat and, therefore, were not serviced by the bull. In addition, the bull serviced about 76% of those females observed to be in heat. Thus, pregnancy rate of cows observed in heat in 48 hours was about 50%. Pregnancy rate for all cows was about 38% (75% in heat, serviced 76% of 75%; 66% conception rate), a value similar to that reported for either timed AI (LLAI80) or 5 days of AI (LLAIE).

A note of caution is in order about breeding a synchronized cow herd with bulls. Although females expressing estrus following Lutalyse are receptive to breeding by a bull, using bulls to breed large numbers of cows and heifers in heat following Lutalyse will require proper management of bulls and females. Other than single-sire mating with about 20 females per group in a small corral or hand-mating, we do not, in my opinion, know how to effectively manage bull breeding of synchronized females.

And finally, a successful AI program can employ Lutalyse effectively, but a poor AI program will continue to be poor when Lutalyse is employed unless other management deficiencies are remedied first. An important factor in the successful use of Lutalyse is an understanding of what Lutalyse can and cannot accomplish. Expectations of success should be realistic relative to the cattle reproduction management each time Lutalyse is used since the cow herd may be different each year depending on herd health, nutrition, semen, etc.

45

PRODUCING VIGOROUS, HEALTHY NEWBORN CALVES

Richard O. Parker

INTRODUCTION

To ensure the survival of their offspring, many animals produce vast numbers of offspring. Unfortunately, cattlemen rely on a species that produces one offspring and promotes its survival through maternal nurturing. Hence, cattlemen must be certain to take steps to ensure that calves are vigorous and healthy from birth onward.

Producing vigorous, healthy calves begins in the uterus of a vigorous, healthy cow. At conception, the genetic material (DNA) from the cow and the bull provides the plans for the formation of calves. From conception onward, the work of producing a calf is rather automatic and proceeds rather quietly, guided by the genetic code. Ideally, at the end of gestation, the uterine environment should have provided for optimal growth and development to meet the adaptation of life outside the uterus. Along the way, however, conditions may arise that affect the developing embryo or fetus, thereby placing the newborn calf at a disadvantage to adjust to life outside the uterus.

All factors governing embryonic and fetal growth are a mystery, but certain factors are recognized as being capable of altering normal processes. Regardless of the factor, the importance of the placenta--the supply line--must be recognized. Aside from genetics, nutrition, hormones, maternal influences, placental size and function, and temperature, all may influence the progress of embryonic and fetal growth and development--and ultimately the successful adaptation as a newborn.

An awareness of the factors affecting the rapid and dramatic changes during development and growth of the embryo and fetus will aid cattlemen in identifying management practices that need changes or improvements.

EMBRYONIC-FETAL DEVELOPMENT AND GROWTH

Development (differentiation) is the formation of major tissues, organs, systems, and the major external features.

Growth is defined as an increase in the number of cells and/or enlargement of existing cells.

Development and growth within the uterus of the cow occur during two stages: the embryonic period extends from conception to about day 45. This period is characterized by rapid growth and development. The fetal period extends from day 45 until birth. This period is characterized primarily by growth and some changes in external form.

Absolute growth--the total increase in weight--increases exponentially, reaching a maximum during the last of gestation (figure 1). Relative growth--the percentage increase in weight per unit of time--is most rapid in the earlier stages and declines as gestation advances. For example, on day 45, the fetal calf weighs about 6 g and by day 72 it weighs about 72 g--an increase of over 1000%. Between day 240 and day 270 weight increases only about 60%, from 17.7 kg to 28.6 kg. Furthermore, organs develop at differing growth rates that seem to be related to functional necessity. For example, the central nervous system (brain and spinal cord), heart, and liver develop first, while muscle and fat develop last. Therefore, factors capable of altering fetal growth and development induce changes depending on the stage of development.

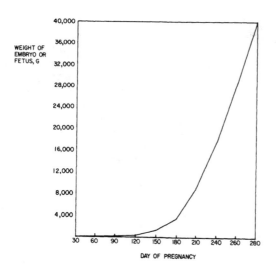

Figure 1. The absolute growth--the total increase in weight--for calves from conception to birth

Scientists often employ measurements other than actual weight and length measurements to determine the progress of intrauterine development. Two common measure are (1) the estimation of cell numbers and (2) the estimation of cell size. The reasoning behind using these two measures is that a reduction in cell numbers or size may limit the ability of the calf in achieving its genetic potential.

DNA (deoxyribonucleic acid) is the chemical that makes genes and is contained in the chromosomes within the cell nucleus. The DNA content per cell (except sperm and ova) within an animal's body is the same, i.e., it does not vary between organs. Thus, total DNA content in a tissue estimates the number of cells. Then in the same tissue, the amount of protein per unit of DNA estimates the size of the cells in the organs or tissues—the larger the number representing the ratio of protein/DNA, the larger the cells.

PLACENTA—THE SUPPLY LINE

The placenta is a versatile, important organ. It substitutes for the fetal digestive tract, kidneys, liver, and to some extent endocrine glands (hormone secreting). Also, it separates the cow from the fetal calf, thus ensuring the separate development of the fetus. Most often, there is a direct relationship between the size of the placenta and the size of the fetus. Placental growth occurs early in gestation and its growth rate slows during the last trimester. Hence, the rapid fetal growth of the last trimester (figure 1) must be supported by a placenta whose growth rate has peaked.

While the blood of the fetus and the mother never come into direct contact, they are close enough for oxygen, nutrients, and waste products to transfer across the placenta. Transfer occurs primarily via two processes: simple diffusion and active transport—an energy requiring process. Electrolytes (Na, K, Cl), water, and carbon dioxide are transferred by diffusion. Amino acids (building blocks of protein), sugars (energy) and most water-soluble vitamins, iron, calcium, and iodine are transferred by active transport systems. Fat-soluble vitamins are impeded by the placenta. In general, substances capable of traversing the placenta vary according to molecular size, charge, and placental type, which differs between species.

In combination, the fetus and placenta form the hormones progesterone and estrogen, which are important for maintaining later pregnancy and preparing for parturition.

FACTORS AFFECTING EMBRYONIC-FETAL GROWTH AND DEVELOPMENT

Much of the embryonic-fetal growth and development proceeds normally and automatically, without fanfare. But it is a complex process and like any complex process, disruption is possible. Science has pinpointed some factors that will influence fetal growth and development, but growth and development within the uterus do not "play" by the same rules as those, in effect, outside the uterus. For example, decapitated fetuses from various species will continue to grow, some quite normally.

Much of the information relative to embryonic-fetal growth and development is derived from laboratory animals rather than actual farm animals, and information for cattle is limited primarily to birth weight, other gross measurements, and some proximate analyses. Nevertheless, the following items can be discussed as factors that will affect embryonic-fetal growth and development: nutrition, hormones, maternal influences, placental size and function, and temperature.

Nutrition

Of all the factors contributing to embryonic-fetal growth and development, nutrition is the most studied in the most species. A variety of studies have clearly shown that maternal feed restriction (protein and/or energy) limits fetal growth, thereby the birth weight of the newborn calves. This is especially true when the restriction is imposed during the last trimester (third)--the time when fetal demands for protein and energy are the greatest (table 1) since fetal weight increases so rapidly. Abundant evidence indicates that smallness per se is a distinct disadvantage for newborn survival, and calves from undernourished cows are less likely to survive.

TABLE 1. DAILY GAIN OF ENERGY AND PROTEIN FOR CALF FETUSES

Day of gestation	Energy (kcal/day)	Protein (g/day)
100	10	1.3
160	90	10.0
220	410	45.6
250	645	80.6
280	780	123.8

Also, some studies indicate a higher morbidity (higher incidence of calf scours) in calves born to severely undernourished dams, and there is some evidence for a correlation between the "weak-calf syndrome" and malnutrition of the dam.

Table 2 demonstrates the effects of a severe feed restriction in pregnant rats. Fetuses in this study were limited to 50% of normal weight, but more importantly the small fetuses possessed less DNA--indicating fewer cells-- and a smaller protein/DNA ratio--suggesting the cells also were smaller. Fewer cells and smaller cells may drastically limit the genetic potential of such individuals. Fortunately, most cases of undernutrition are not so severe.

TABLE 2. SEVERE FEED RESTRICTION DURING THE LAST HALF OF GESTATION AND ITS EFFECTS ON FETAL RAT DEVELOPMENT

	Treatment	
Item	Full-fed	10-day fast
Fetus weight	4.3 g	2.3 g
Total DNA	13.6 mg	9.9 mg
Protein/DNA	25.0	17.0

Table 3 provides some of the results from a study in which ewes were subjected to a 50% reduction in energy and protein intake, compared to well-fed ewes. Underfed ewes

TABLE 3. WHOLE BODY AND ORGAN WEIGHTS OF FETUSES FROM HIGH AND LOW FED EWES ON DAY 144 OF GESTATION[1]

	Ration	
Item	High	Low
Fetus weight (kg)	6.50	5.30
Brain		
Weight (g)	58.00	60.30
Percentage of body	.89	1.14
DNA (mg)	57.30	73.90
Protein/DNA	53.50	57.20
Liver		
Weight (g)	151.70	105.40
Percentage of body	2.30	2.00
DNA (mg)	1283.50	1092.00
Protein/DNA	12.60	10.3
Femur		
Length (cm)	9.70	8.90
Percentage of body	0.15	0.17

1 Ewes were fed two levels of the same ration beginning on day 90 of gestation.

produced smaller fetuses with smaller organs, with the exception of the brain. But none of the organs contained less DNA or had smaller cells. Furthermore, many studies have shown that the brain, and possibly the skeletal system, are protected from nutritional insults. The results in table 3 would support this.

A number of studies clearly indicate that under-nutrition of the cow during the last one third of gestation limits calf birth weight. Several of these are listed in table 4. Most information on fetal calf growth is re-stricted to birth weights, linear measurements, fetal weights, and some data on proximate composition (carbohy-drate, fat, and protein). Very little data are available on cellular growth (DNA, protein/DNA, and RNA) of the fetal calf during undernutrition of the cow. Those data available only describe cellular growth in fetal calves.

TABLE 4. CALF BIRTH WEIGHTS FROM COWS SUBJECTED TO DIFFERING NUTRITIONAL TREATMENTS

Researcher and year	Birth weight	
	High nutrition	Low nutrition[1]
Bellows et al., 1978	32.7	28.6
Clemente, 1978	37.0	36.1
Corah, 1974	37.8	34.7
Dunn, 1964	30.6	28.6
Hight, 1968	33.0	29.4
Wiltbank et al., 1962	26.3	20.4
	35.4	30.4

[1] This division is for simplicity. Actual nutritional treatments of the cow were more sophisticated.

Most nutritional studies deal with undernutrition dur-ing the embryonic period can retard embryo development, possibly affecting subsequent intrauterine growth and devel-opment. This should be considered a possibility; it is an area of research that should be pursued.

While considering nutrition in general, specific nutrient deficiencies should not be overlooked. Several minerals and vitamins affect fetal development. A defi-ciency of calcium leads to increased stillbirths. Vitamin D may also be involved. An iron deficiency of the mother may predispose the newborn to anemia. An iodine deficiency or ingestion of goitrogens causes fetal death, abortion, goiter, lowered birth weights, and debility of the newborn. Manganese and zinc deficiencies are associated with reduced birth weights, fetal resorption, and stillbirth. Copper deficiencies may cause the birth of weak ataxia offspring. In cattle, this can be aggravated by excess dietary molyb-denum. Vitamin A deficiencies can result in the production of calves that, at birth, are weak, malformed, or dead.

Hormones

Fetal endocrine glands and the placenta secrete a variety of hormones during gestation. These hormones induce metabolic changes in the fetus and prepare the fetus for

412

extrauterine adaptation. Any factor hindering or altering the release of the hormones shown in table 5 may affect the vigor of newborn calves. Such factors could be nutrition of the cow or some stress such as heat.

TABLE 5. HORMONES INFLUENCING FETAL AND NEONATAL GROWTH AND DEVELOPMENT

Hormone	General Action
Epinephrine	Glucose mobilization
Estrogens	Behavior of newborn
Glucocorticoids	Activation of a variety of enzymes
Insulin	Glucose utilization
Placental lactogen	Nutrient supply in blood of cow; tissue growth
Progesterone	Behavior of newborn; nutrient supply in blood of cow
Thyroid hormones	Activation of a variety of enzymes

Maternal Influences

Faster embryonic and fetal growth is directly related to maternal size. An early experiment involving reciprocal crosses between large South Devon and small Dexter breeds of cattle demonstrated that the crossbred calves from the large mothers were heavier than those from the small mothers.

The age of the cow also affects the size of the fetus. Older cows have heavier calves. Perhaps heifers continue to grow through their first pregnancy, thus competing with the fetus for available nutrients. Also, the degree of vascularity (blood vessels) may increase with subsequent pregnancies. This may increase blood flow to the placenta.

Whatever the cause, individuals of the same breed vary widely in maternal influence. For example, even in a whole group of cows or heifers subjected to the same degree of undernutrition, some dams will produce calves of normal birth weight, while others will produce calves with extremely low birth weights.

Placental Size and Function

Early in gestation, the placenta grows rapidly and is large in comparison to the fetus. During the last trimester, placental growth slows, fetal growth peaks, and the amount of placenta in comparison to the fetus becomes smaller as time progresses. Thus, the capability of the placenta to provide for fetal growth the last trimester (figure 1), may be burdened, and its functional adequacy depends on early placental development.

Function of the placenta relies upon the blood delivered to it. Anything reducing placental blood flow is apt to limit placental transfer of nutrients. In laboratory animals, restricted feeding of the dam slows the transfer of amino acids across the placenta. It is unclear whether this is due to cardiovascular changes in the dam or the action of nutrition per se on the placenta.

Temperature

Heat stress--82° to 100° F--can limit fetal growth in sheep, resulting in lambs smaller at birth. This effect is similar too, but independent of, the effects of under-nutrition in sheep. Besides being smaller at birth, lambs from heat-stressed ewes are less viable at birth. Likely, high temperatures have a similar effect on cattle--at least in certain breeds not adapted to extreme high temperatures.

How heat stress exerts its effects on the growing fetus is uncertain. Two good possibilities seem to be reduced placental development or decreased blood flow to the uterus and placenta.

PUTTING IT TOGETHER IN A SYSTEM

Awareness is the key. Cattlemen need to be aware of the process of embryonic and fetal growth, realizing the many factors and interactions that ultimately produce vigorous, healthy calves. Through an overall awareness, cattlemen can view their total program; decide to produce vigorous, healthy calves and then based upon awareness and experience, exercise that valuable commodity--judgment.

46

NEW ADVANCEMENTS IN BOVINE EMBRYO TRANSFER TECHNOLOGY

Brent Perry

The world of bioengineering is a fascinating place in which to live. Something new is an everyday event and one is hard pressed to keep up with the outpouring of new data and technology.

The more specialized discipline of bovine embryo transfer holds special fascination for those of us who make our livelihood in the cattle industry. The new developments in embryo transfer can best be discussed in four categories: new techniques, increased efficiency, problem donor cows, and new trade developments.

NEW TECHNIQUES

In new technology, the two most significant advancements to burst upon the scene in 1982 have to be the Rio Vista "One-Step"™ freezing process and the embryo splitting or cloning procedures.

The Rio Vista "One-Step" freezing process is perhaps the cattle industry's biggest breakthrough since artificial insemination. It is a totally new method of freezing and thawing embryos in a single straw container. This process enables embryos to be implanted in a recipient cow in much the same manner as breeding a cow by artificial insemination.

The "One-Step" process further reduces the thawing of embryos from the laborious process requiring the skills of an embryologist with a microscope to one requiring no more than a person skilled in nonsurgical transfer. The older processes also called for repeated handling of the thawed embryos under a microscope during a multistep rehydration procedure. It was not uncommon to actually lose embryos during this procedure.

The "One-Step" process allows a more efficient and economical handling of recipient cows. Embryos can be thawed and implanted as recipients become available, thus doing away with the necessity of synchronization; or you can synchronize and then "clean-up" the stragglers as they become available. The "One-Step" also works well as an adjunct to a .fresh embryo program. Should you have recipients left over due to a lack of fresh embryo production, you can simply fill in with frozen embryos.

Embryo splitting is the other big innovation in embryo transfer to become commercially available in 1982. If cloning is defined as asexual reproduction accomplished by replicating genetic material, then embryo splitting is indeed a form of cloning.

The embryo splitting technique is exactly what it says. After recovery from the donor cows at approximately seven days following insemination, the embryos are actually cut in half and each half is ultimately implanted into a recipient cow or both halves can be implanted in the same cow to produce twins from one recipient. The process is of course accomplished under a microscope using micromanipulation equipment and microsurgical instruments.

Although there is not a great amount of data from commercial application available at this writing, it would appear that embryo splitting, done competently, will yield approximately 50% more than normally expected from embryo transfer. A good technician can expect a 55% to 65% pregnancy rate from nonsurgical transfer. Therefore, from 100 embryos you would expect 55 to 65 pregnancies. If you split those 100 embryos, you would get 200 halves. From the 200 halves, you can reasonably expect 100 pregnancies from that same skilled technician. Therefore, you have gone from 55 to 65 pregnancies per 100 embryos to 100 pregnancies per 100 embryos. Some of us who use only cowboy math would say that equaled 100%! The bottom line is that if the average donor in transplant last year produced an average of 20 pregnancies per year, she can now produce 30 pregnancies per year--a 50% increase in production.

Embryo splitting has significant research implications as well as increasing production. Since splitting produces identical twins, scientists can now accelerate studies involving questions of heredity versus environment. One twin can now serve as the perfect control while response to drugs, hormones, nutrition, etc., can be observed in the other twin. Embryo splitting removes one of the problems of calving twins in the bovine. Since twins from split embryos are identical, you do not have to contend with the possibility of free-martins. A free-martin is the reproductively sterile heifer twin of a bull.

Other technology that should be forthcoming in the immediate future is embryo sexing and the production of chimeric or allophenic calves.

Monoclonal antibody techniques show great promise for the commercial sexing of embryos. Several of these techniques are now undergoing field trials and results should be in by early 1983. Sexing would be most beneficial if the embryos could be sexed at conception with sexed semen. Sexing after the fact could be expensive as it has the effect of cutting production in half if only one sex is desired.

Sexed semen should be a reality for the future when adequate separation techniques are developed.

The production of chimeric or allophenic calves is another exciting possibility. These calves will be produced by actually combining embryos. For example, you can take a Hereford embryo and an Angus embryo, use proper microsurgery techniques, and produce an Fl black baldy calf. This calf would be a true Fl and have four parents. I might mention that you would have a calf produced in nine months that would take three and one-half to four years to produce by conventional means. I have heard it said that you cannot speed up generation intervals. Don't you believe it!

EFFICIENCY

Efficiency of production is being constantly addressed by the embryo transfer industry. Constant improvement of the older collection and transfer methods is leading to lower-cost services and, in some cases, increased production.

While some embryo transfer firms rely upon standard hormone dosages, it has become quite apparent to me that best results are obtained by adjusting hormone doses to specific donor cows. I also believe in minimal doses of hormones to keep a donor producing embryos for a longer period of time.

New collection techniques are also leading to the more efficient retrieval of embryos. Many embryos have been and still are left in the donor cow or are literally poured down the drain due to relatively inefficient technique.

More knowledge has been gained in the use of synchronizing drugs. We can now bring cows into heat much more reliably than we could even a few months ago, using a more judicious dose regimen and schedule of the synchronizing compounds.

THE "PROBLEM" COW

One area of special interest to the industry is the "problem" cows. Traditionally, it has been found that approximately one-third of all prospective donors fail to work adequately, or even at all, in embryo transfer programs. Our company, among others, is placing special emphasis on this group of cows. We are in the process of dividing these problem cows into smaller, more identifiable groups. There are those that do not respond at all to hormone therapy; still others give multitudes of eggs, but they are infertile or degenerate; while others tend not to respond to the synchronizing drugs. We also see the group of cows that have known reproductive or anatomical problems. Cows with calves at side present special problems, especially in certain breeds.

There is no solution for some of the problem cows--that is, often there simply is nothing that can be done to correct their problems. Thus, when we discover these kinds of problems we can all quit worrying about this particular group of cows and get on with worrying about those things that we can do something about. Research tells us that some of the problem cows have abnormal hormone levels. Can we offset this by injection of other hormones? Some cows develop antibodies to semen and even to their own eggs, making fertilization impossible. Can we correct these conditions? Can we convince owners that they would be better off to put calves on nurse cows and send their donors in as dry cows? The answers to these questions are being sought daily. Cows are being bled at frequent intervals to measure their hormone levels. Immunologists are working around the clock to unravel the antibody problems. Consultants are working daily with clients to convince them that optimum embryo production is not magic but very dependent upon direct management of the donor and recipient cows. Yet other researchers are working on removal of ovaries, recovery and maturation of eggs, and in-vitro or test tube fertilization. By the time this paper is published, perhaps we will have answered some of these questions. Maybe in a few months or a few years, we can remove the ovaries from a cow, recover 400 or 500 eggs, use what we want, freeze the rest, and send the cow to the packer. Wouldn't that solve some problems and create a few more?! As I said in the beginning, the world of bioengineering is a fascinating place to live.

MARKETING AND TRADE POSSIBILITIES

Perhaps the most interesting advancements in embryo transfer to you are the enhanced marketing and trade possibilities. After all, we must keep in mind that what we are all about is profit potential--the bottom line. The biggest boost to marketing of bovine genetics has been the frozen embryos. We now have a simple and reliable method of storing and shipping bovine embryos to any place in the world. The next boost will be sexed embryos. Then we can ship exactly what a buyer wants to exactly where he wants it, and the calves can be born exactly when he wants them to be born. And don't forget, the calves will be born as native calves having all the natural and acquired immunities of the recipient cows native to that region.

I might point out in closing that the use of embryo transfer has become a standard tool of the purebred cattle industry. New advancements have added new dimensions to the uses of embryo transfer and just over the horizon are miracles untold and unbelievable to us today. Embryo transfer is also available to you today at a much lower cost than in the past. I do not know of many things in the world today that are getting cheaper, but embryo transfer certainly is. The new technologies have allowed us in the industry to lower our prices significantly. If you realistically consider inflation, the costs of embryo transfer are even less than the raw figures indicate. I am pleased to bring you the good news of better services at lower costs.

47

IDAHO TOTAL BEEF PROGRAM:
A SYSTEMS APPROACH
TO REPRODUCTIVE EFFICIENCY

Ed Duren,
J. D. Mankin

The Idaho Total Beef Program is an educational concept based upon cattle management by objectives that provides tangible results such as improved profits, shorter calving season, more rapid growth, and increased calf survival. Intangible effects may reflect pride in the ranch, self-satisfaction, peer recognition, improved employee morale, and strengthened team effort. These factors all have a direct and indirect benefit to the ranches participating and cooperating in the Idaho Total Beef Program which was developed by the University of Idaho College of Agriculture. It is an action program designed for producer benefit and is in harmony with the priorities of the Idaho beef cattle industry.

Let's examine what makes the Idaho TBP unique.

PROBLEMS IDENTIFIED

Beef cattle will respond in a given environment to the limit of their genetic capability or to the limits set by the environment. The environment is everything that interacts with that beef animal. These interactions form two broad systems. The abiotic system is made up of such factors as heat, cold, wind, solar radiation, etc. The biotic system interactors are those forms of life that impact the animal such as disease organisms, plant life, other animals, and of course, man. This means to me that in beef production we have cows, bulls, and calves with certain genetic potential. They have to exist and produce in an environment that consists of two broad systems. The production that is generated from this environment is limited more by the environment itself than by the genetic potential.

Nationally, the beef cattle industry has identified low reproductive efficiency as a high-priority problem. An educational effort to improve reproductive efficiency can be intitiated on the ranch with the identification of a problem or problems associated with the lower reproductive efficiency of a specific herd. A ranch or a community of ranches may participate in the Idaho Total Beef Program at whatever point the problem is perceived in the biological cycle of the herd. However, the problem must be identified and key indicators selected to measure improvement in efficiency of reproduction from base data collected on a ranch or a community of ranches.

Defined simply, beef production at the commercial cow-calf level is "growth management." A beef animal starts to perform at conception and continues to perform from that point until it dies. However, growth in itself is meaningless unless set in a time frame; the amount of growth achieved in a given time period must be measured through **costs** and **returns**.

Growth management can be broken into two segments: (1) the management of genetic material to maximize growth, and (2) the management of the environment to maximize growth.

If we agree that we must manage the environmental system to get the most growth economically possible, then we must learn how to analyze that system.

ESTABLISHED BENCH MARKS

Currently, the educational thrust of the Idaho Total Beef Program is to improve the reproductive efficiency of the cow herd. This direction and emphasis resulted from the collection of base data with producer and professional cooperators.

In Idaho last fall (1981), the Extension Livestock Specialists' 841 Planning Unit conducted a survey for open cows. Six veterinary clinics across southern Idaho co-operated in the survey, with thirteen practicing veterinarians reporting their pregnancy examinations of client beef cow herds. The following table summarizes the overall results:

```
Total cows and heifer................17,378
Percent age open..........11% (range 0-25%)
Percent age calving in first 40 days....46%
Percent age calving in second 40 days...28%
Percent age calving in 80+ days.........15%
```

The seriousness of the open-cow problem is emphasized when you combine those that are calving 80+ days into the calving season with the 11% open. You are looking at roughly 26% of the cows as reproductive failures--certainly failures in terms of paying bills with those light calves.

The summary table shows that fewer than 50% of the cows will calve in 40 days. The practical implications of this finding are that for each day one calf is younger than another it will be 1.5 lb to 2 lb lighter in weight. Thus, calving later by one heat period represents a loss of 30 to 40 lb. When translated into dollars, a long calving season becomes the biggest "robber" in the beef industry. Improving reproductive efficiency of beef cow herds is a key to ranch survival in Idaho during the 80s.

MEASURABLE INDICATORS

There are four key indicators of the level of production in a seasonal calving commercial herd.

Open cows. The indicator that has the largest economic impact. If it takes $350 plus to keep a cow for a year, you can quickly figure out that 14% to 20% open cows will knock a large hole in the budget. In spite of what you may have read in some popular press reports you can't tolerate open cows and you don't need a sophisticated computer program to deal with those candidates for the golden arches.

Length of the calving season. In those herds that do not have a seasonal calving system, days from the last calving or calving interval is the key indicator. I repeat "A long calving season becomes the biggest 'robber' in the beef industry. This cannot be overemphasized.

Calf death loss. These losses range from 2% to 30%; with the wide range reflecting actual variations from year to year and the lack of accuracy in reporting. We are beginning to get a handle on this in Idaho. Annual calf losses will vary somewhat by season, but are consistent with ranch management. Preliminary indications are that annual baby calf losses vary from 12% to 40% or more. Nationally the USDA statistics suggest annual losses in a range of 12% to 15%. It is most important that this key indicator of production be determined on a herd, county, and state basis.

Growth. Within the operation there are two systems that can impact upon growth. Since growth is the major product we sell, it is important that we realize that the total growth of the herd is a composite growth of the individuals and of the total growth produced by the herd. The individual's growth is a product of the breeding system, and the total herd growth is a product of the management system. It is the management system that sets the limits of tolerance for open cows, length of the calving season, death loss, and minimums of growth.

KEY CAUSES

If you analyze the production system from these four key indicators, there are four key causes of variation in levels of production. Remember, there are only four key causes to open cows.

Nutrition. Nutrition is, without doubt, the most common problem in getting cows to breed. We used to call it "hollow belly," but we are more educated now with more ruminant nutritionists around. It now has a scientific name, "lackus feedosis."

Disease. Some diseases are implicated in the open cow problem; however most of the health problems in reproduction can be dealt with very successfully.

The bulls. The problem of open cows also is associated with bulls that are not as fertile as they could be, nor as young, nor as sound as they once were. Perhaps there are not enough bulls for the terrain.

Management. This cause can be called environmental manipulation or gross management: Were the bulls out? Did the cows have enough feed? Were the females cycling?

ANALYZING THE SYSTEM

In analyzing the system let's take one of the possible key causes, disease, and follow it through some steps of the analysis. There are two things we can do to determine the possible cause of open cows in relation to disease. We can give the cows a physical exam to determine their fitness and give them a clinical exam to determine if disease is implicated. If for example, there is no physical problem but the clinical exam reveals that Vibrio is probably the key problem, the next step is to recognize that a change needs to be made in the management system. The key change would be a vaccination program for Vibrio.

We should be able to measure this against the base data for some key results. In this case, it should show up as a decrease in the number of open cows as a key indicator of production level. If management changes do not impact upon the key indicators, they may not be important contributions to the production efficiency of a cow herd.

EDUCATIONAL SUPPORT

The Idaho Total Beef Program is a University of Idaho College of Agriculture educational thrust for improving the reproductive efficiency of the state's beef cow herd. It is a total college effort to provide technical information and enthusiastic leadership to cattlemen managing the cow herd, using an integrated approach by extension, research, and teaching through interdepartmental cooperation within the College of Agriculture.

The technical information available in western beef cattle production has been assembled by beef cattle specialists from the states of Idaho, Colorado, Montana, Oregon, Utah, Washington, and Wyoming into a ready, usable, handy reference recognized in Idaho as the Idaho Cow-Calf Management Guide and Cattleman's Library. This reference is published by the University of Idaho College of Agriculture and distributed by the Idaho Cowbelles in cooperation with the County Cooperative Extension Service Offices throughout the state of Idaho for a fee of $35.00. This reference presents a management system based upon the biological cycle of the beef cow.

The Idaho TBP Cow-Calf Management Guide and Cattleman's Library provide reference information that is assembled and arranged according to the biological cycle of the cow. This reference service is supported by appropriate audiovisual aids, i.e., twenty-four videotape and slide-tape cassette lessons utilized in various extension teaching methods. The technical information offered in the Idaho Total Beef Program Guide, Cattleman's Library, and audiovisual aids are continually updated with practical and reliable information gathered from cooperating ranches throughout various Idaho Exten-Search type products.

The handbook does not constitute the Idaho Total Beef Program, but is a reference of technical information.

Performance records in commercial herds provide producers with information for maximizing growth management of their herds, as well as for individual animals within their herds. These records can be used to pinpoint factors in the environmental system that can be changed so that growth potential will be maximized and potential profit increased.

The Idaho Beef Herd Improvement, a computer analysis of herd reproduction and growth records, is a vital tool with a management input to system analysis.

IMPACT PROJECTS

The system analysis approach of the Idaho Total Beef Program is practical and sound. This has been successfully executed and demonstrated on two completed projects, with several other projects now in the planning and development stages.

Pegram. The completed Pegram project (3 herds, 1600 cows) in southeast Idaho reduced baby calf death loss (a key indicator) from 16% to 2% in a 3-year period. The key objectives of the Pegram Exten-Search Project were to:
- Prioritize and implement management changes to minimize death loss in young calves.
- Develop an effective economical disease control program to minimize death loss in young calves.
The problem of calf death loss has been identified with a breakdown in reproductive management, a high priority of the Industry Extension Beef Resource Committee. Calf survival is a recognizable problem of the nation's cow herds at all levels of the beef cattle industry.

The Idaho Beef Council, which helped in the funding of this project, has directed that a system be developed for saving calves with any technique that can be profitably used by a cow-calf producer. The Idaho calf-survival problem, readily identified by the Idaho Beef Council, became an industry crusade and a University of Idaho challenge.

Boise River. The other complete project is the Boise River Project (A Method & Result Demonstration) where the problem was open cows--at least that was the rancher's worry when 23% of 350 cows were diagnosed as open. Fortunately, this rancher had 2 years of herd records that showed, in addition to 23% open cows, a 140-day calving season, 7% death loss in calves, and 352-pound weaner calves at about 250 days on irrigated pastures. By using the four key indicators for analysis, we discovered that the major causes could be lack of selenium and copper. The key management change we made was a mineral supplementation program. This past year his cows had 93% conception, a 60-day calving season, and 1.5% death loss. Weaner heifers were 87 pounds heavier and steers 60 pounds heavier than were those of the previous year. (For more complete detail, see flow chart on page 428.)

The Idaho Total Beef Program, in contrast, is in line of action to improved management executed by key people requiring the integration of disciplines; of multifunctional units of extension, research, and teaching within the University of Idaho College of Agriculture (the Idaho land grant system); of the U.S. Department of Agriculture in harmony with agribusiness; and of professionals associated with agriculture lending, beef cattle health, and nutrition in cooperation with beef cattle producers.

PEOPLE INVOLVEMENT

The Idaho TBP is not just reference material, computer usage, or meetings—it is a line of action that involves the cooperation of people in the educational process to speed up the flow of information from scientist to producer. The rate information is adopted depends on the motivation and participation of the producers.

The impact of the Pegram Project is an example of the Idaho Total Beef Program--a concept in Integrated Reproductive Management to improve beef cattle reproductive efficiency in Idaho. This program began with the problem--the survivability of neonatal calves. The Pegram Project, conducted on working ranches in southeastern Idaho, represented a combined interest of the Idaho cattle industry, agribusiness and the University of Idaho College of Agriculture.

The Exten-Search type project derived its description from the fact that it didn't fit the typical characteristic of either a research project or a county extension program. From the standpoint of research, the project lacked environmental and administrative control experienced by animal and veterinary scientists. The project did not totally reflect extension work simply because of the need to investigate the unknown and apply the known in a compatible system.

Ranchers and their families in the Pegram Project resonded with full cooperation and commitment of cattle, time, labor, and other ranch resources. These people were toally dedicated and willingly responded to planning and implementation of the project.

The three technicians lived and worked in the ranch community during prime activity periods--collecting field data, assisting ranches with imposed treatments, sharing technical information, teaching skills, routinely observing cattle, and coordinating activity with the project leader. Technicians were selected from animal science majors recommended by the teaching staff in the departments of Animal and Veterinary Science. The technicians employed were senior level or B.S. graduates with ranch experience and skills.

The project leaders were held accountable for "why it happened." The field worker assembled academic information from the scientific community within the land grant college system and the allied agribusiness community that could be applied to the project. The administrative responsibility directed the application of funds, assured the flow of materials and services, and protected the project from harassment within the system.

The allied professions contributed expertise in the respective fields of financial analysis, health, and nutrition. This involvement came from agricultural banks, practicing veterinarians, feed stores, and four national animal health and nutrition companies.

Communications were paramount. They ranged from one-on-one contact to periodic semiformal committee meetings and formal brain-storming type conferences. The dialogue among people built upon confidence and commitment remained open and honest.

KEY PEOPLE

A state or county Total Beef Program will be motivated by knowledgeable key people willingly contributing to a data base. Field data, i.e., open cows, dead calves, etc., are essential to problem identification, priority determination, and program direction by beef industry leaders.

Implementing a Total Beef Program is one of integration. The delivery and application of technical information useful to total ranch management must be supported by professionals. Each individual associated with a ranch or county program has a role to play in developing and implementing a Total Beef Program. For example, in the Pegram Project and the Boise River Project, these contributors communicated information in basically three areas, namely, data collection, information analysis, and application of technology that resulted in the integration of both technology and people.

Key people can be categorized based upon their previous training and experience. Each group will play a significant role in the integrating process contributing to the problem-solving mode if their efforts are organized and properly stimulated.

The research scientist is technically trained to structure problem-solving activities in terms of statistical treatments, replicas, and controls that are the parameters of a research project. They tend to build projects requiring continued study and refinement. Frequently, a research scientist will experience total frustration in an attempt to study field problems on a ranch as opposed to a research project on an experiment station.

The extension scientist is relatively unstructured in the sense that no two recipients of the same technology are the same, nor do the results have to be the same. This experience with people combined with technology can bridge the gap between the experiment station and ranch environment.

The practicing veterinarian associated with crisis management reacts successfully in the face of an outbreak and often becomes less interested in preventing the crisis. (In other words, don't fix it if it isn't broken!)

An agricultural lender is basically trained to analyze monetary repayment ability with little regard to the producer or ranch management. However, bankers recently demonstrated a renewed interest in management and can be extremely vital contributors in an integrated program.

The producer is a recipient of technology and ultimately responsible for the integration resulting in ranch management change. Frequently the rate of adoption of technology delivered to the producer is influenced by tradition, peer pressure, and restricted to family structure.

The industry representatives of animal health and nutrition products operate within the confines of product sales to the parent company. In these attempts to meet the sales quota the representative will fail to recognize the management level of the ranch. Most products result in a higher return in proportion to the management applied and must be considered accordingly for the dollar invested.

In beef cattle production, man is the organism that has the greatest impact. The decisions that men can make influence all of the effects of an integrated cow-calf management sytem affecting reproductive efficiency.

SUMMARY

In looking ahead to management in the 80s, the day is gone when we can operate a cow outfit out of our hip pocket. We must remember we are the largest impactor on the production of our ranches because we manipulate the environmental system through our system of management. We must have a system of analysis on the key indicators of level of production. We must be willing to change. All of the technology in the world is useless until it is incorporated into the management system. Seldom will we be successful unless we use an interdisciplinary approach. In Idaho we call it the Total Beef Program; nationally it is called

Integrated Reproductive Management. You who are producing cattle need to obtain all of the technology available and incorporate as much as you economically can into your management. Those of us in education need to be innovative in our efforts to speed the rate of technology adoption. We have read that from time of discovery of technology to general adoption is 7 to 15 years, with 12 years as average. This is too long in these fast, exciting, changing times of the 80s and 90s.

Flow Chart of TBP Approach

BOISE RIVER EXTENSION PROJECT
(A RESULT & METHOD DEMONSTRATION)
300 HEAD COW HERD

DATA BASE	KEY INDICATORS	KEY CAUSES	KEY APPROACH TO PROBLEM ISOLATION	KEY PROBLEMS IDENTIFIED	KEY MANAGEMENT CHANGES	KEY RESULTS
PERFORMANCE	OPEN COWS (23%)	DISEASE	PHYSICAL EVALUATION CLINICAL TESTS	NO PROBLEMS / NO PROBLEMS		
		BULLS	PHYSICAL EVALUATION CLINICAL TESTS	1 GIMPY / NO PROBLEMS	SOLD 1,GIMPY	
		NUTRITION	FEED ANALYSIS CLINICAL TESTS	LOW Se, CU, CA, P / LOW Se, COPPER	SUPPLEMENT Se,CU, P	93% CONCEPTION
		MANAGEMENT	REVIEW MGT. PROGRAM PARASITE CONTROL RECORDS-SUMMER PASTURE	HIGH FLY POPULATION	ECTRIN EAR TAGS	LESS FLIES
	CALVING SEASON (140 DAYS)	DISEASE	PHYSICAL EVALUATION CLINICAL TESTS	NO PROBLEMS	VACCINATION PROGRAM	60-DAY CALVING SEASON
		BULLS	PHYSICAL EVALUATION CLINICAL TESTS	NO PROBLEMS		
		NUTRITION	FEED SAMPLES CLINICAL TESTS	LOW Se, CU		
		HEIFER DEVELOPMENT	CHECK FOR SIZE, AGE, CHECK PERF. RECORDS	50% LARGE ENOUGH TO BREED	PREG TEST -CUT OFF LATE ONLY HEIFERS OF SUFFICIENT SIZE KEPT & BRED. BULLS KEPT IN HEAT THEN PREG TEST	
	CALF DEATH LOSS (7%)	DISEASE	NO CLINICAL SIGNS OF DISEASE		COMPLETE VACCINATION PROGRAM-BOSE	REDUCTION OF 5.5%
RECORDS		NUTRITION	NURSING	SUPPLEMENT	SUPPLEMENT	
		MANAGEMENT	SAME AS COWS			
	GROWTH 352 LB W/W, 232 DAYS/AGE	NUTRITION	FEED ANALYSIS LEVEL OF FEEDING	LOW Se LOW COPPER	SUPPLEMENT Se, CU	HEIFERS 87 LB HEAVIER THAN 1981
		PARASITES INTERNAL & EXTERNAL	EXAMINE FOR EXT. INT. PARASITES	HIGH INT. POPULATION. SOME EXT.	WORM CALVES POUR ON	STEERS 60 LB HEAVIER
		DISEASE	CHECK FOR CALF DISEASES	SOME BVD, PI3 IBR SUSPECT	VACC. PROGRAM FOR CALVES	
		GENETICS	EXAMINE PERFORMANCE RECORDS OF COW HERD & HERD BULLS	VARIATION IN PERFORMANCE	COW EVALUATION & CULLING	

KEY ACTIVITIES TO ADOPTION

INFORMATIONAL (MEETING)

A. DISCIPLINES REPORTED WHAT EACH CONTRIBUTED TO PROJECT RESULTS
B. PROCESS OF TBP-APPROACH EXPLAINED
C. LIST OF PRODUCERS PRESENT OBTAINED- (70 PRODUCERS PRESENT)
D. NEWS MEDIA PRESENT-RADIO, TV, PRESS
E. NOTICES OF MEETING SENT TO LIVESTOCK PRODUCERS ON THE BOISE RIVER
F. PUBLISH COMPLETE REPORT OF ACTION ON PROJECT.

FOLLOW-UP (SURVEY OF PRODUCERS)

EXTENSION AGENTS SURVEYING FOR BASE DATA FROM EACH PRODUCER ATTENDING MEETING.

(ANALYSIS OF DATA COLLECTING)

1. IDENTIFY AREAS OF OPPORTUNITY AND IMPACT KEY PRODUCTION INDICATORS.
2. IDENTIFY POSSIBLE AREAS OF RESEARCH.
3. IDENTIFY PRODUCERS THAT HAVE SIMILAR DATA WILLING TO COOPERATE IN FURTHER DEMONSTRATIONS.

Part 8

ANIMAL MANAGEMENT

48

TROPHY WHITE-TAILED DEER MANAGEMENT

Robert L. Cook

INTRODUCTION

Management of white-tailed deer herds for production of trophy-antlered buck deer is a controversial and often misunderstood issue. The purpose of this paper is to discuss the practical rationale, requirements, "dos and don'ts," and results of such a management plan implemented on a private ranch in central Texas.

The land manager or the deer manager should understand the terms "trophy-deer management" and "high-quality-deer management." First, what constitutes a trophy deer? It is appropriate to rearrange an old saying, "The trophy is in the eyes of the beholder." What constitutes a trophy buck for one hunter or manager may be considered only an average deer by others.

Second, controversy exists between deer managers and between officials of various state and federal agencies as to which is better--trophy- (or high-quality) deer management or management of deer herds for the most deer (or quantity management). Quantity versus quality. It is generally believed that the majority of hunters just want the opportunity to harvest a deer--it doesn't have to be a trophy. On the other hand, there are those hunters who do not want to harvest just any deer and who, in fact, will not harvest a deer unless it is, in their judgment, a "trophy." These trophy hunters are often willing and financially able to pay large sums of money for the privilege to hunt for and harvest trophy deer on private lands. Today's prices for harvesting a trophy buck often range from $1,000 to $2,000 per buck and it is common to hear prices quoted from $2,500 to $3,000 per buck in the "brush country" of south Texas.

It is popular for the deer manager to be referred to as a "trophy manager." Certainly this title carries more esteem and prestige than "mass production manager" of deer. Economically it appears to be beneficial to a land manager to be referred to as one who produces trophy deer, even if he doesn't actually live up to this title. Therefore, titles are often misapplied and misused.

However, true trophy-deer management and management for maximum number of deer are actually the same and would be better referred to as quality deer management. Quality deer management is optimum utilization of natural resources without waste or depletion. The most important factor in quality deer management is nutrition. The deer herd and each individual within the herd must have an abundant supply of a variety of preferred native forage plants from which to fulfill its daily nutritional requirements. Adequate nutrition eliminates die-offs and waste due to starvation and ensures optimum production and survival that are essential to both quality and quantity management. Factors such as genetics, mineral supplements, and age structure are important to long-range deer-management plans but are far overshadowed by nutrition.

The manager actually cannot manage for quality without managing for quantity at the same time. Production of quality deer results in the maximum production and survival of deer; therefore, proper harvest becomes the second most important factor in quality deer management and in production of preferred deer foods.

GOAL

Following acquisition of the South Fork Ranch in August 1979, Robert R. Shelton initiated long-range plans for its wise and efficient management. These plans included protection, conservation, and management of natural resources, rangelands, wildlife populations, and wildlife habitat. The objective of these plans on South Fork Ranch was the sustained production of high-quality white-tailed deer and Santa Gertrudis cattle.

Location and Description

South Fork Ranch contains 35,112 acres of rangeland and is located on the western edge of Kerr County in the Edwards Plateau Ecological Region of central Texas. The topography is rolling to moderately rough, with steep slopes and canyon walls associated with drainages. Dominant vegetation includes live oak, shin oak, post oak, Spanish oak, ash juniper, and bluestem grasses.

Background

During the previous ownership, the ranch was grazed by crossbred cattle, Angora and Spanish goats, horses, sheep, native deer, and a variety of exotic animals. Exotics, including Axis deer, blackbuck antelope, Fallow deer, and Sika deer were introduced to the ranch over 20 years ago. These species have been hunted sparingly and are well established on the ranch.

Limited hunting of white-tailed deer was permitted by previous owners. Overpopulation, starvation, and periodic die-offs were common to the ranch's deer herd. In 1975, 1976, and 1977, a commercial hunting group harvested 365, 254, and 175 white-tailed bucks, respectively, from South Fork Ranch. Data collected from these bucks indicated that they were typical, mature, Hill Country bucks. Eyewitnesses rated 30 to 40 of this total of 794 bucks as exceptional Hill Country deer. If the manager was interested in maintaining a harvest of primarily mature bucks, this was considered an overharvest, as exhibited by the drastically declining harvest rate from 1975 to 1977. The harvest during these years was also unsound because almost no antlerless or doe deer were harvested.

DESCRIPTION OF DATA COLLECTION PROCEDURES

Detailed data collection and analysis were initiated during the fall of 1979, using the following procedures.

Deer census.

Spotlight deer surveys: Four permanent, spotlighted, deer-survey transects were established on South Fork Ranch during September 1979. Each spotlighted transect was counted twice during the month of September. Data from the spotlighted surveys were the primary source used to compute deer density, buck-doe ratio, and fawn production.

Incidental observations: During August and September, biologists observed, identified, and recorded as many deer as possible during the performance of other duties. These observations provided supplemental data concerning sex-ratio and fawn production.

Helicopter surveys: Permanent helicopter survey transects were established east-west through the ranch at one-half mile intervals during September 1979. Each transect was counted twice (once in the morning and evening) during the last week of September and the first week of October. Deer were observed and recorded by two observers and the pilot from a Bell 47 G4A helicopter flying at an altitude of 50 feet and a ground speed of 35 miles per hour. All deer observed within 100 yards of each side of the flight line were counted. Helicopter survey data have been used as supportive data for deer density and herd composition estimates resulting from ground survey techniques.

Deer harvest.
　　Total deer harvest: A record was maintained of
the total number of deer harvested or removed by
paying hunters, guests, employees, scientific
permit, and culling.
　　Physical and morphologic harvest data: The following data were collected from a sample of deer
harvested on South Fork Ranch:
- Date of kill.
- Name and address of hunter and number of days
  hunted.
- Time of kill.
- Caliber of firearm used by successful hunters.
- Sex of deer.
- Field-dressed weight of deer.
- Age of deer as determined by tooth replacement and wear.
- Name of pasture and quadrangle in which deer
  was killed.
- Main beam antler spread measured at widest
  inside main beam spread and widest outside
  main beam spread.
- Main beam antler circumference of one antler
  measured immediately above antler burr.
- Number of points on each antler. An antler
  point must extend at least one-half inch past
  antler surface.
- Length of one main antler beam from base of
  burr along and around outside curve of antler
  to tip of antler.
- Length of longest tine on each antler.
- General physical condition of each deer harvested was rated either excellent, good,
  fair, or poor depending upon criteria established.
- Whether or not female deer were lactating.
- Number of tags on each doe hunter's license
  at check-in.

Range and Weather Conditions

　　Range conditions, rainfall, and special weather conditions on South Fork were recorded monthly. Deer condition
and notes on fawn crop and antler development were recorded
monthly.

RESULTS

　　Initial census efforts in 1979 indicated the need to
reduce the total number of deer on South Fork Ranch (table
1). Census data collected in 1979 indicated almost 6,000

deer or one deer per 6 A. Previous research on the nearby Kerr Wildlife Management Area had indicated that when combined with reasonable numbers of domestic livestock in deferred-rotation grazing systems, this area could safely carry a deer density of about one deer per 9 A to 11 A. Therefore, South Fork Ranch should carry a deer herd of about 3,200 to 3,900 deer. As a result of this information, South Fork was stocked with Santa Gertrudis cattle in deferred-rotation grazing systems at about one animal unit (mother cow and calf) per 30 A and the deer herd was reduced by heavy harvests of antlerless deer. A total of 1,241, 417, and 406 antlerless deer were legally harvested and removed from South Fork Ranch in 1979, 1980, and 1981, respectively. The majority of this harvest was accomplished during 3-day anterless deer hunts held on the ranch during the last 2 weeks of the season each year. During these hunts paying hunters were permitted to harvest only antlerless white-tailed deer. This has proven to be a successful method of controlling deer herd numbers and harvesting antlerless deer.

TABLE 1. SOUTH FORK RANCH WHITE-TAILED DEER POPULATION COMPOSITION SPOTLIGHT DEER CENSUS 1979, 1980 AND 1981

Deer per 1,000 acres	1979	1980	1981
Bucks per 1,000 acres	54.8	59.1	40.5
Does per 1,000 acres	70.7	35.8	33.5
Fawns per 1,000 acres	44.0	12.5	39.2
Total deer/1,000 acres	169.5	107.4	113.2
Estimated deer population	5,952	3,771	3,976
Acres per deer	5.9	9.3	8.8
Does per buck	1.35	0.93	0.68
Fawns per doe	0.47	0.14	1.10

During this same period it was essential to repair or rebuild the ranch's perimeter deer-proof fence. This was not done to keep "our deer" in but to keep "our neighbor's deer" out as our range and forage conditions rapidly improved because of reduced deer and cattle numbers on South Fork.

A total of 157, 118, and 180 antlered males were harvested on South Fork during this same period. Not all of these bucks were considered trophies. Trophy buck hunters harvested 71, 93, and 127 trophy bucks during 1979, 1980, and 1981, respectively. The majority of the other antlered bucks harvested or removed were considered culls or died of accidents or unknown causes.

As a result of the heavy harvest of antlerless or doe deer and the regulated harvest of antlered bucks, the deer herd composition has improved dramatically (table 1). For example, in 1981 there were 68 does per 100 antlered bucks and 110 fawns per 100 doe deer—both of which are considered excellent. The main criteria by which to judge the results of this quality-deer-management program, however, are field-dressed weights and antler development. Deer quality was poor when the program was initiated in 1979. For example data collected in 1979 indicated that adult bucks 4.5 years old and older averaged field-dressed weights of only 85.7 lb (table 2). Buck fawns field-dressed weights averaged 23.7 lb and yearling bucks 1.5 years old averaged field-dressed weights of 42.3 lb in 1979. Substantial increases in field-dressed weights were recorded in all age classes of male deer harvested on South Fork from 1979 to 1981. Adult buck weights, for example, increased from an average of 85.7 lb in 1979 to 98.5 lb in 1980 and to 100.0 lb in 1981. Likewise, average weights of buck fawns increased from 23.7 lb in 1979 to 36.0 lb in 1980, and to 38.4 lb in 1981. Antler developments also improved during the first 3 years of the program on South Fork (tables 3, 4, 5 and 6).

Production of quality deer does not occur only in the adult age classes. A buck does not have to be 5 years old or older to be a high-quality or trophy deer. In fact, the risk of losing bucks to natural causes is so great that it is probably a mistake to manage for bucks in the older age classes of 6 years old or older. Effective, efficient deer management dictates that the majority of a quality- or trophy-buck herd be harvested at the ages of 3.5, 4.5, and 5.5 years of age to prevent excessive waste through losses to natural causes. In fact, it is the author's opinion that no more than 30% to 35% of the total trophy-buck harvest should be 5.5 years old or older. As demonstrated by the bucks harvested in 1981 on South Fork Ranch by "trophy hunters" paying $1,150 per buck, bucks in the 3.5 year old age

TABLE 2.   SOUTH FORK RANCH TRENDS IN DRESSED WEIGHTS (LB) BY AGE CLASS MALE DEER 1979, 1980, AND 1981

Age	1979	1980	1981
0.5 *	23.7	36.0	38.4
1.5 *	42.3	53.6	58.3
2.5	65.0	82.0	89.3
3.5	78.3	88.0	90.6
4.5	87.7	95.5	96.7
5.5	86.9	99.3	104.0
6.5	84.8	102.8	100.0
7.5	92.8	97.0	87.0
8 +	77.1	98.0	–
Adult (4.5 yr+)	85.7	98.5	100.0

*Harvested as anterless deer by doe hunters.

TABLE 3.   SOUTH FORK RANCH TRENDS IN BEAM CIRCUMFERENCE
           (IN.) BY AGE CLASS 1979, 1980, AND 1981

Age	1979	1980	1981
2.5	2.8	2.8	3.3
3.5	3.3	3.3	3.5
4.5	3.7	3.5	3.8
5.5	3.8	3.7	4.1
6.5	3.9	3.7	4.1
7.5	4.0	4.1	3.7
8 +	3.8	3.9	-
Adult (4.5 yr+)	3.8	3.7	4.0

TABLE 4.   SOUTH FORK RANCH TRENDS IN INSIDE MAIN BEAM
           ANTLER SPREAD (IN.) BY AGE CLASS 1979, 1980,
           AND 1981

Age	1979	1980	1981
2.5	10.0	10.6	12.4
3.5	14.3	13.0	13.6
4.5	13.9	14.8	14.6
5.5	15.2	14.6	15.4
6.5	14.9	15.3	14.8
7.5	14.2	15.7	14.6
8 +	14.3	14.0	-
Adult (4.5 yr+)	14.7	14.9	15.1

TABLE 5.   SOUTH FORK RANCH TRENDS IN OUTSIDE MAIN BEAM
           ANTLER SPREAD (IN.) BY AGE CLASS 1979, 1980,
           AND 1981

Age	1979	1980	1981
2.5	11.0	11.8	13.7
3.5	15.7	14.2	14.5
4.5	15.5	16.1	16.1
5.5	16.7	16.1	17.0
6.5	16.4	16.7	17.5
7.5	15.7	17.2	16.5
8 +	15.7	15.5	-
Adult (4.5 yr+)	16.2	16.3	16.8

TABLE 6. SOUTH FORK RANCH TRENDS IN OUTSIDE MAIN BEAM
LENGTH (IN.) BY AGE CLASS 1979, 1980, AND 1981

Age	1979	1980	1981
2.5	12.0	12.5	16.1
3.5	17.8	14.6	16.5
4.5	17.8	16.7	17.6
5.5	18.2	17.5	18.9
6.5	18.3	17.9	19.4
7.5	18.4	19.1	20.5
8 +	18.1	16.7	-
Adult (4.5 yr+)	18.2	17.6	18.5

class were very acceptable as trophy deer (table 7). In
fact, with long-range good management and improvement of
forage and range conditions, it is believed that the quality
of deer will continue to improve to the point that an even
larger percentage of the trophy-buck harvest will be composed of 2.5- and 3.5-year-old bucks. However, it is recommended that the manager interested in the long-range production and harvest of trophy-antlered buck deer and optimum
economic return maintain at least 40% to 50% of his trophy-
buck harvest in the 4.5 year-old-and-older age classes.
     In addition to the harvest of trophy or high-quality
bucks, cull bucks should be harvested from the well-managed
deer herd on an annual basis. It is estimated that cull
bucks might comprise 10% of the average antlered buck herd.

DISCUSSION AND RECOMMENDATIONS

     Trophy-deer production is a valid enterprise if ap-
proached as "quality deer management." Adequate quantities
of preferred native forage (forbs and browse) are the key
ingredients to quality deer management. Harvest of only
adult-age class bucks is not trophy management; in fact, it
is mismanagement. All age classes and both sexes of deer
must be provided adequate nutrition. Does should be healthy
and productive. Does should be liberally harvested to prevent overpopulation and overutilization of preferred forage
plants.
     Fawns should have high-survival rates and should
exhibit adult-deer characteristics early in life, such as a
few doe fawns should breed their first winter and a few buck
fawns should produce hardened antler nubs their first fall
season. Bucks should develop antlers acceptable to many
trophy hunters by the time they are 3.5 years of age. Bucks
should exhibit a distinct rutting season that is characteristic of a healthy deer herd. Field-dressed weights will
vary between areas within the white-tail's range, but in
Texas adult bucks (4.5 years and older) should normally
field-dress an average of 110 to 150 lb.

TABLE 7. SOUTH FORK RANCH AVERAGE FIELD-DRESSED WEIGHTS AND ANTLER MEASUREMENTS WHITE-TAILED DEER MALES 1981

Age class	Number in sample	Percent of harvest	Average field-dressed weights (lb)	Average number antler points	Average inside beam spread (in.)	Average outside beam spread (in.)	Average main beam length (in.)	Average beam basal circum (in.)	Average length longest tine (in.)
2.5	3	2.2%	89.3	8.7	12.4	13.7	16.1	3.3	6.6
3.5	33	23.9%	90.6	8.7	13.6	14.5	16.5	3.5	7.1
4.5	42	30.4%	96.7	9.3	14.6	16.1	17.6	3.8	7.5
5.5	38	27.5%	104.0	9.1	15.4	17.0	18.9	4.1	8.8
6.5	21	15.2%	99.4	9.9	15.8	17.5	19.4	4.1	8.4
7.5	1	0.7%	87.0	8.0	14.6	16.5	20.5	3.7	11.8

The deer manager should seriously consider such factors as genetics and mineral supplementation only when he has done everything possible to provide for the nutritional requirements of the deer herd through production of high-quality native forage. There is currently a considerable amount of fanfare about genetically inferior and genetically superior deer in Texas. The genetics of a deer herd are important and should not be ignored in a long-range management effort. However, some managers erroneously look at genetic manipulations as the panacea or quick and easy way out of their deer-herd-management problems. There is no quick or easy way to good deer management. Years of mismanagement require years of corrective efforts. Nutrition and habitat are the key ingredients to good deer management. A wider variety and increased abundance of preferred deer foods and less competition between deer and livestock will produce bigger, healthier deer, larger antlers, and better survival of fawns. Research in Pennsylvania, Louisiana, Michigan, and Texas has proven that antler development and some antler characteristics are hereditary and are also indicative of other physiological parameters such as body size. For example, it has been clearly demonstrated that bucks that are spike antlered as yearlings (1.5 years old) are inferior in antler development and body size to forked-antlered bucks of the same age on the same nutritional level. In addition, the spike-antlered bucks remain inferior in all physical characteristics (antlers and body size) to the forked-antlered bucks of the same age throughout their entire lives. Research has also proven that the male offspring of bucks that were spikes as yearlings are more likely to be spikes as yearlings than are the offspring of bucks that were forked antlered as yearlings. Therefore, it would normally be desirable to exert a considerable amount of hunting pressure on spike-antlered bucks. Under almost no circumstances should spike bucks be protected while the harvest of other bucks is permitted. Spikes are apparently inferior deer--whether this is a result of genetics or poor nutrition or a combination of several factors is not known. Spikes should not be protected, especially when associated with a long-range, quality-deer-management program. Although the data is not conclusive at this time, it is believed that this same criteria applies to bucks that have no brow tines, or that have 5 antler points or fewer. They probably should not be protected from harvest. In fact, students of the European system of culling Red Deer believe that bucks with no brow tines or bucks with otherwise inferior antlers of any age should be culled from the deer herd as soon as the deficiency is recognized. Certainly the deer manager in Texas and throughout the white-tail's range would need to know that the deer herd's nutritional requirements were satisfied before embarking upon such a culling program on an extensive basis. In addition, the manager should clearly recognize from the start of such a program the difficulty of acquiring hunters capable of judging inferior or "cull" deer in sufficient numbers to legally and adequately remove undesirable bucks from his herd.

The overall annual buck harvest should not exceed 15% to 25% of the total antlered-buck herd if the manager is interested in the production of trophy-age-class bucks.

Deer managers frequently search for a mineral supplement that will produce larger deer and larger antlers. Again, there is no easy answer. A successful, cost-effective mineral program for deer would involve extensive research and chemical analysis of forage and soil samples over a long period of time. A "shot-in-the-dark" approach to mineral supplementation is likely to be expensive and of no recognizable value.

Quality deer management requires quality habitat and range management. Trophy or quality deer cannot be produced efficiently without proper attention to the entire ecosystem. Efforts have been initiated and quality deer management is being attempted on South Fork Ranch. it will be several years before the benefits of the program are fully realized.

49

INITIAL STEPS IN DEER MANAGEMENT FOR THE FARMER OR RANCHER

Robert L. Cook

INTRODUCTION

This paper discusses management practices that singly or in combination will benefit the production and survival of white-tailed deer. Hopefully, the reader will be stimulated to study this fascinating creature further and to learn more about deer requirements and habits. In no way should this paper be assumed to be a complete treatment. The farmer or rancher who is sincerely interested in deer management must study the subject in depth and sift through many opinions--some worthwhile, some worthless.

BACKGROUND AND HISTORY

The white-tailed deer of North America carries the scientific title <u>Odocoileus virginianus</u>. Although there are several different subspecies, this separation is often only geographical, and quite frankly, there is practically no difference in these deer genetically, except that one may occur on one side or the other of the Mason-Dixon line. In other words, the white-tailed deer in Texas is, for all practical purposes, the same deer that occurs in the north woods of Maine or the pine forests of the Carolinas.

White-tailed deer were very important to our ancestors and to the native Indians of North America as food and clothing. However, the white-tail was almost wiped out in many areas during the period 1850 to 1880 because of indiscriminate slaughter by commercial meat and hide hunters; near the end of the century, they were nearly exterminated because of ignorance of the deer's habitat requirements. However, thanks to laws enacted by the states' legislators and enforced by dedicated game wardens, the white-tail "caught its breath" and began to repopulate most of its former range during the 1930s and 1940s. In addition, throughout the eastern and southern United States, thousands of family farms that had been cleared and farmed during the 1800s and early 1900s were abandoned and deserted as the nation became more industralized and urbanized. These thousands of acres of farmlands have slowly but steadily become reforested and again serve as home for the white-tailed deer. Likewise, forestry practices in southern pine forests have generally been beneficial to the production of white-tailed deer. The white-tailed deer is now the most numerous big game animal in Texas and in the U.S. Current census estimates indicate that there are over 3 million white-tails in Texas alone. Therefore, many farmers and ranchers in Texas and North America are somewhat familiar with white-tailed deer. Many of these land managers desire to know basic information about the white-tail so they can better manage this resource and their habitat. Thus, some of the initial steps in white-tailed management are outlined below.

HABITAT

Food

White-tailed deer occur, to some extent, in almost every habitat type present in the eastern two-thirds of the U.S. Therefore, it is difficult, if not impossible, to intelligently generalize about their habitat requirements. Yet, habitat is the single most important factor affecting white-tailed deer herds. The production of an abundant supply of nutritious preferred deer forage is the key to producing the maximum number of high-quality deer. A wide variety of preferred plants is extremely important. Deer primarily prefer to eat forbs (what most people call weeds) and browse (leaves, twigs, and tender shoots of woody plants and vines). Deer also prefer to eat what most people call brush. Brush not only provides escape cover but also is often a very nutritious food source. Forbs such as buttercups, filaree, clover, vetch, verbenia, oxalis, wild lettuce, plantain, pigweed or carelessweed, partridge pea, silda, wild onion, and lespedezas are preferred by deer and are readily eaten when available. Browse or woody plants preferred by deer include oak leaves, twigs and acorns, yaupon, greenbrier, hackberry, mulberry, rattan or supplejack, sumac, mesquite beans and dried leaves, hawthorns, beautyberry, wild cherry, plum, wild grape, honeysuckle, dogwood, elm, blackberry, dewberry, gum elastic (chittum), acacias (catclaws), walnut, guayacan, prickly pear leaves and fruit, wild chinaberry, kidneywood, and condalias. Deer eat very little grass throughout most of the year but will consume limited amounts of grass during the early spring when the grasses are fresh and tender.

The farmer/rancher should be able to recognize preferred deer foods if he wants to improve his land for deer. Competition for the preferred plants should be minimized--sheep and goats compete directly with deer for preferred foods. Large numbers or concentrations of sheep and/or goats will drastically reduce the available foods for white-tailed deer and will result in fewer and poorer deer. Continuous grazing of rangeland by cattle is detrimental to deer and deer habitat. Even though cattle are primarily grass eaters they will consume available deer food if forced to by continuous heavy grazing. However, if the manager will stock cattle in reasonable numbers and will utilize deferred-rotation grazing systems, it has been proven that deer and cattle will successfully occupy and utilize the same range. In fact, grazing a reasonable number of cattle on a deferred-rotation grazing system may actually help stimulate production of preferred forbs by harvesting dense, tall stands of mature grasses. Improved grass pastures such as coastal Bermuda or bufflegrass are extremely poor deer range and, in general, do not provide food or cover for white-tailed deer.

Deer, of course, will feed on most winter grains such as oats, rye, and wheat, but these do not provide forage for long periods of time. In addition, deer are notorious for feeding on vegetable gardens, yard shrubbery, and truck farms. However, these areas do not provide cover for deer and usually are available only during short periods of the year. There is no substitute for native forage (browse and forbs) when discussing deer food and this is where the manager should put his efforts.

Cover

Cover requirements appear to be less critical than food requirements for the white-tail. Although deer occasionally occur in grassland prairie habitats where there is practically no woody overstory, it is certainly deemed preferable to provide a deer herd with sufficient cover for escape from hunters and predators and for protection from extreme cold or hot temperatures. It is important for the manager to remember that much of this cover can and should be understory within about 5 feet of ground level to provide food as well as cover. In general, it is recommended that at least 20% to 30% of the ground surface be covered with woody vegetation of suitable height and thickness for deer to hide in. In some areas, woody cover is so important that to remove even the smallest quantity of brush will result in a decrease in the quality of deer range. Finally, the oak cover typical of the Texas Hill Country and the brush of the South Texas Plains are so valuable as food and cover for the extensive and economically valuable deer herd in these areas that the control of any substantial amounts of this brush is questionable.

HUNTING AND HARVEST

Almost any established deer herd can and should be hunted. Reasonable annual harvest rates of 10% to 25% of both sexes of the deer herd will not deplete most deer populations in North America. Obviously, the quality of the habitat, effect of predation, production, and survival rates will influence the number of deer available for harvest from any given piece of property. Land managers in areas of extensive deer herds must clearly understand the necessity to harvest antlerless deer to control deer numbers to prevent overpopulation, starvation, and range depletion. Recent research also has proven that spike bucks should not be protected from harvest. Many land managers worry that they do not have enough bucks to harvest or to breed their does. This is particularly true in areas where doe hunting is illegal or is practically nonexistent and the manager observes many more does than bucks. However, there is not a documented case of an area that didn't have enough bucks to breed the does. The deer herd should be harvested and managed so that there are about 2 to 3 does per buck going into the breeding season. Generally, a good rule of thumb would be to harvest as many female deer as male deer.

DEER CENSUS, SEX RATIOS, AND POPULATION ESTIMATES

Many managers are baffled by questions dealing with deer numbers and sex ratios. Management of deer herds in some ways is very similar to management of cattle. True, the manager never knows the exact number of deer he is dealing with, but he must not let this prevent him from implementing a good deer-management program. There are currently several deer census methods available that have been tried and proven, including the spotlight census and the helicopter census. There are numerous agencies and individuals ready and willing to assist the private land manager in using these techniques to better manage his deer herd. The deer manager must have a reliable annual deer population estimate to properly harvest and manage his deer herd. Population estimates collected annually will permit the manager to evaluate the trend of the deer herd, either increasing, decreasing, or steady, as he implements various harvest- and habitat-management techniques.

The spotlight deer census and the helicopter deer census are both good techniques. The manager may find one preferable to the other, depending upon his situation and his habitat; or he may prefer to use a combination of these techniques. Data from spotlight surveys can be used to estimate deer density, buck/doe ratios, and fawn production. The spotlight deer-census technique has been used extensively by the Texas Parks and Wildlife Department in some areas of the state for almost 20 years. The first widespread use of the spotlight census by the game department was in the northwestern portion of the Edwards Plateau in the vicinity of San Angelo, Sweetwater, and Abilene. This technique has been used as the primary deer census method on the Kerr Wildlife Management Area near Kerrville since about 1965. The technique was further refined, developed, and tested throughout Texas by the Texas Parks and Wildlife Department during the period 1975-1978. The technique has recently been implemented in several southeastern states and is being tested there.

The basic description and procedures for conducting spotlight deer-census surveys are as follows:

- Spotlight deer census lines should be continuous (not segmented) and may vary in length from 5 to 15 miles.
- Each spotlight-census count should be started about 45 minutes after official sunset.
- Visibility readings (perpendicular distance from the survey route that an observer could see a deer) are estimated by spotter personnel every 1/10 mile and sample acreage calculated by mile and total. The driver-recorder notes visibility on each side of the line at each station, with a reading taken at the start and finish of the line. Maximum visibility readings should not exceed 250 years and are taken at night during the deer survey.
- A map should be prepared and maintained showing the exact location of the permanent spotlight lines with starting and ending points clearly designated.
- Every effort should be made to prevent the possibility of confusing the census effort with illegal night hunting activities. Survey personnel should notify game wardens and other concerned persons prior to conducting the census. These persons should be advised of the time and date that the census is to be conducted.
- The standard work force consists of three personnel (two spotters equipped with spotlights and a driver to record visibility and number of deer seen). Spotlights (2) are 100-watt, 12-volt, 4-inch sealed-beam aircraft landing lights. These lights are easily mounted in plastic bottles for protection and ease of handling. Spotter personnel operate the lights from immediately behind the pickup cab and high enough above the cab so it does not interfere with their line of vision. The driver does not assist spotters in sighting deer.
- Surveys are driven at a speed of 5 to 8 miles per hour. In addition to recording total deer seen, observers record the identity (buck, doe, fawn) of observed deer when possible. When a group (two or more deer) is observed, the entire group are classified as unknown unless all deer in the group are identified. Binoculars are used only to identify observed bucks, does, and fawns. Deer observations are recorded on the basis of one-mile intervals.
- Each spotlight line should be counted twice during September and/or October.
- Weather data are recorded on the survey sheet.

Various types of helicopter deer census methods have been used in Texas for several years. Many managers prefer to utilize what is referred to as "total count" by observers from a low-flying helicopter. Using this method, the pilot and observers fly adjoining consecutive strips of approximately 300 to 400 yards in width throughout an entire ranch, counting and identifying all deer seen. Controversy exists, however, over the observers' ability to 1) distinguish deer previously observed which ran over from the adjoining transect, 2) to dependably maintain proper spacing between transects for ensuring full coverage, and 3) to see all or most of the deer present on the ranch because of canopy cover.

It is recommended, therefore, that the manager use the helicopter census like the spotlight survey and "sample" only strips of his ranch. Permanent helicopter survey transects should be established east-west through the ranch at one-quarter or one-half mile intervals. Each transect should be counted twice during September or October. Each transect should be counted once in the morning and once in the evening. Deer are counted by two observers and the pilot from a Bell 47 G4A helicopter flying at an altitude of about 50 ft and a ground speed of about 35 miles per hour. All deer observed within 100 yards of each side of the flight line are counted. Helicopter survey data can be used as supportive data for deer-density and herd-composition estimates resulting from ground survey techniques and as trend estimates of the population level and density.

Deer managers should observe, identify, and record as many white-tailed deer as possible during the performance of other duties in the months of August and September. Individual members of groups of deer should be recorded only if all individual deer in the group are identified. This information should be used to compare and to supplement spotlight or helicopter sex ratio and fawn production data. Incidental observations have no direct application in estimating deer density but are extremely useful to the manager as estimators of sex ratios and production.

DEER MOVEMENT BETWEEN LANDOWNERSHIPS

Many managers believe that deer management is a waste of time because their deer move between ownerships and are often harvested when they cross the fence onto the neighbors' land. This is true to some extent on small acreages (less than 2,000 A to 3,000 A), especially where habitat is limited and deer are drawn out of their normally small home range to food plots or "baited" hunting areas. The only absolute answer to control this movement is construction of deer-proof fences that usually are cost prohibitive. Construction labor and materials for a 7-ft, deer-proof fence currently costs $6,000 to $12,000 per mile depending upon materials and terrain. Deer movement can be minimized by the land manager who, through grazing practices and deer-herd-population control, keeps his range and deer habitat in good condition.

PREDATION AND OTHER DECIMATING FACTORS

Although predation certainly has some effect on some deer herds, it is not a serious limiting factor in most areas. The "Brush Country" of the South Texas Plains probably has the highest concentration of coyotes and bobcats anywhere within the range of the white-tailed deer in North America. Studies have proven that predators kill and eat large numbers of fawns from the deer herds in this area during the first few weeks of the fawns' lives. However, the deer herds in this area are extensive and are generally very healthy. In fact, South Texas has become quite famous as a favorite area for trophy-deer hunters.

White-tailed deer are plagued by almost every disease and parasite known to man or beast. However, in most cases, deer herds are quite capable of coping with these problems if their habitat is in good condition and the area does not become overpopulated and overgrazed. Diseases and parasites combined with starvation and inclement weather have caused extensive, massive die-offs in many deer herds.

LEASING AND ECONOMICS

In today's world of increasing interest rates and high taxes, the land manager must optimize production and manage his resources efficiently. Lease revenues from hunting can be sizable and in some areas are beginning to approach or even exceed revenues from livestock production. Season deer-hunting leases in Central and South Texas currently bring approximately $2.00 to $4.00 per acre. Deer-hunting revenues on ranches where the hunter is provided lodging, guides, transportation on the ranch, and the opportunity to harvest a quality or trophy deer commonly return $4.00 to $8.00 per acre.

COMMON MISCONCEPTIONS

New bloodlines - new genes. You cannot significantly improve your deer herd by bringing in new bloodlines or better genes from Michigan, Wisconsin, South Texas, or anywhere. That is simply not a viable, practical answer. You can improve your deer herd by 1) improving what they eat; 2) providing more nutritious forage by decreasing competition with domestic cattle, sheep, and goats; 3) implementing deferred-rotation grazing systems with your cattle; and 4) maintaining your deer herd within the carrying capacity of your range by harvesting both sexes of deer to control deer numbers while keeping sex ratios balanced. These practices will benefit your range, your cattle, and your wildlife. New bloodlines (good or bad) are quickly absorbed and dispersed within your deer herd.

Supplemental feed and minerals. Do not attempt to significantly increase deer size and maintain larger deer by feeding them shelled corn, pellets, or mineral supplement out of a sack. It is simply too expensive to be practical. Again, the key to more deer and/or larger deer is an abundant supply of native, nutritious, preferred forage.

Brush control. The word "brush," like the word "weed," implies that the plant is undesirable. This is not always true. You cannot destroy deer habitat or deer food and expect to maintain your deer herd. If you chemically or mechanically control extensive areas of good deer cover and food such as hardwood creek bottoms or the brush of South Texas, you will have fewer and poorer deer. Remember coastal Bermuda and bufflegrass fields are not deer habitat--and they never will be. Not all brush is undesirable. Brush control should be done only after careful evaluation of its cost and effect on your entire ranching enterprise.

Stockpiling deer. Do not attempt to "stockpile" deer. They won't "pile." Deer will not necessarily live until the next year, or to the next, just because you don't allow hunters to harvest them. Proper range and deer management requires that deer numbers be kept at or below the carrying capacity of the range. When you see the first poor deer, it is too late for correction.

Judging deer age and condition. There is only one way to know the age of a deer, and that is to study tooth replacement and wear on the lower jaw. No one can tell the age of a deer running across a pasture. Likewise, there is only one way to know if your deer are in good condition from year to year and that is to weigh, age, and measure harvested deer. Do not attempt to evaluate the effects of your range and deer management practices by looking at deer on the hoof. You should census your deer herd and collect data annually from harvested deer. Likewise, don't trust your memory to recall facts about your deer herd--write it down.

Dry does and buck fawns. Many managers encourage their hunters to harvest "dry, barren does" and they go to great lengths to prevent the harvest of buck fawns. As a results, their deer herds are not properly harvested or managed. Do not attempt to harvest only "dry, barren does." Research has proven that a "barren" doe rarely exists. Almost all doe deer are capable of conceiving, giving birth to, and raising fawns every year. Do not worry about harvesting a doe just because she is obviously still being followed by a fawn. When the hunting season starts in November, most fawns in most areas are 100 to 120 days old and for all practical purposes are not dependent upon the mother's milk or guidance for survival. In areas where the hunting starts earlier, doe deer should not be harvested until late November of December. During the harvest of antlerless or doe deer, some buck fawns will be harvested. Do not worry about the harvest of buck fawns by doe hunters. It would be a mistake to prevent an adequate harvest of doe deer simply because a small percentage of the buck-fawn crop will accidentally be harvested.

SOURCES OF INFORMATION AND ASSISTANCE

If you are interested in trying to improve your deer herd, and you would like to manage your herd more efficiently, there are many experts willing to give advice. You can find numerous world experts on white-tailed deer in almost every cafe and coffee shop south of the Canadian border. Domino parlors and billiard halls are the favorite habitat of these experts in the South, although quite a few deer experts are found in "fillin' stations" and drug stores in Texas and adjoining states. However, the following sources are recommended once the manager has exhausted all others and is still wondering if it is worth the trouble or not.

1. State game departments--Almost every state within the range of the white-tailed deer has a wildlife division within its game department with professional biologists ready and willing to help you with your deer-management program. Usually their main office is located in the vicinity of the state capital. All it will cost you is a telephone call to get their assistance. In Texas, call the Texas Parks and Wildlife Department Wildlife Division in Austin--(512) 479-4978.

2. State and federal conservation agencies--This is an extensive, diversified group and includes the following: the USDA Soil Conservation Service, which has a fine staff of regional wildlife biologists; the United States Forest Service; the United States Fish and Wildlife Service; and the Texas Agricultural Extension Service, which has a keen interest in wildlife and an excellent field staff at your service.

3. Universities--In recent years many universities throughout the United States have developed extensive wildlife science departments and they are deeply involved in research and management of white-tailed deer across the nation. Again, simply call the university and get the number of their wildlife department. Some of the university wildlife departments in the south include: Texas A&M, Texas Tech, Louisiana State University,

Stephen F. Austin, Sul Ross University, Mississippi State University, Auburn University, Oklahoma State University, Virginia Polytech Univer-
sity, Clemson University, and Texas A&I Univer-
sity.

50

A RANCH PROGRAM THAT WEANS 720-POUND STRAIGHTBRED CALVES FROM TWO-YEAR-OLD DAMS

Henry Gardiner

On July 22, 1982, we weaned 56 straightbred Angus steer calves from their two-year-old Angus mothers: their average weight was 723 lb. This was the first time in over 40 years that our ranch has been weaning calves that we have averaged over 700 lb. They had not been on creep feed. These calves are part of a progeny test for the American Angus Association Sire Evaluation Program and are sired by five different bulls. Since they are on a progeny test, we have a lot of data on their performance.

At the time of weaning, their average age was 310 days or 10 months and 10 days. Their average birth date was September 15. Their adjusted 205 day weight was 536 lb, taken May 4. They gained 2.14 lb per day from May 4 until weaned on July 22. They were on native grass during this time and were nursing their mothers. From November 1 to May 4, they were on wheat pasture when it was not covered by snow.

In checking back through our records, we found that these calves are much heavier than any other set of calves handled in a similar way. In fact, if we look at a 10-year period of our ranch records from 1964 to 1973, the average weight of the steer calves sold during this time was 523 lb, or exactly 200 lb less than these 1982 steers. The 1964-73 steer groups varied from a low of 477 lb average in 1966 to high averages of 550 lb in 1965 and 563 lb in 1971. Incidently, the price per pound received for these steers varied wildly. The low price per pound was $.23 in 1964 and the high was in 1973 at $.65. (The price in 1974 was $.325.) There is actually more than 200 lb difference in these steers because the 1964-73 calves were all weighed a month later than the 1982 steers.

What has caused this difference? I would have to say it was a combination of a number of different management changes. As a purebred breeder, I would like to say that the only factor was genetic--but it was not. We might have gotten a 30 lb increase from genetics, but not much more. We have tried very hard to get genetic improvement but have not been effective until now. I believe that, if animal breeders will utilize new genetic information available to

them in sire-evaluation summaries, in three or four generations we may have bulls capable of adding another 100 lb to
calves this age available to the commercial men. We have
not used any crossbreeding. I think most herds would benefit from crossbreeding. We have been trying to breed a better purebred that, when used in a crossbreeding program,
will give even better results.

In this paper, I would like to take a detailed look at
our program and maybe you can use some of these methods to
get 200 lb more for each calf from your cows.

OUR KANSAS RANCH

First, a little background information on our ranch.
It is located in southwest Kansas, just north of the east
end of the Oklahoma Panhandle and 50 miles south of Dodge
City. Our annual rainfall is about 20 inches. Out temperatures range from a -15° F to a +110° F. We have about
12,000 A in the ranch that was started by my father, and
4,700 A are in cultivation. There are 3,700 A in wheat, 700
A in sudan, and 300 A in alfalfa. The cow herd consists of
450 commercial Angus cows and 200 registered Angus cows. My
wife, three sons, and two full-time men help me run this
ranch. We do not have irrigation (thank heavens) but the
alfalfa is subirrigated and produces about 6 tons to the
acre.

BREEDING AND CALVING PROGRAM

The length of breeding season varies for different
groups of cows. The replacement heifers are AI for 45 days
and anything that does not settle is culled. A clean-up
bull is not used. The commercial cows are bred naturally
with home-raised yearling and two-year-old bulls for a 60-
day breeding season. The registered cows are bred artificially for a 90-day breeding period. A clean-up bull is
used sparingly. The registered cows that settle in the last
part of the breeding season will be sold in our sale in the
spring if they are not too old or too sorry--those two kinds
of cows will be sent to the packer.

The cows will calve in September and October, with the
registered cows having a few November calves. If wheat pasture is available, we turn the cows on it in November; if
not, we will start feeding a lot of alfalfa. Our breeding
season is December and January. As the wheat becomes short
and frozen down, we feed more alfalfa. In 1972, we started
growing and feeding alfalfa. Before that, we had tried to
get by on wheat pasture and sorghum hay. In the fall of
1973, we had 100 more calves on the ground by November 1
than we had had the previous year. At that time, we had a
120-day breeding season. The cows settled much quicker when
fed better.

As previously mentioned, we use AI as the only way to breed our replacement heifers and the main way to settle our registered cows. The use of heat synchronization works extremely well in our AI program. We use two different methods of heat synchronization: MGA for the heifers and prostaglandin shots for the cows. This management tool probably added about 70 lb to the 1982 steers over the 1964-73 steers. The average birth date of the 1982 steers was September 15. The dams of the 1982 steers were bred so that they were supposed to start calving September 3. Thus, you can see we got a lot of calves on the ground quickly. We don't have any average birth date records on the 1964-73 steers, but with a 120-day breeding season, I would guess that they were about 30 days younger than the 1982 steers out of the two-year-old cows. When these calves were weaned in July, they had an average weight per day of age of 2.33 lb--30 days times 2.33 lb equals 70 lb more weight.

Use of MGA

We synchronize the replacement heifers with MGA, which is a feedlot additive commonly used on heifers to keep them from coming into heat. We buy this in a commercial protein mix. We prefer to get it in Purina Receiving Chow (it needs to be in a palatable feed so that every heifer will eat her share every day). Each heifer is supposed to get .5 mg for 14 days. The amount that we buy only gives each heifer .42 mg for only 12 days, but it still works.

The heifers are brought into the dry lot about 6 weeks before we want to start breeding them. We feed them in a fenceline bunk. This is the same lot where they were weaned and broken to eat at the bunk at weaning time. The first 2 weeks we feed them the same product without the MGA in it so that the heifers are eager to consume it. Then 30 days before we want to start AI, we feed them the MGA for 12 days. Three days after we stop feeding the MGA, the heifers start coming into heat--but since they don't settle very well, this heat period is skipped and they are bred the next period. About 85% of them will be in heat within 10 days. Chart 1 shows the pattern in which they came into heat. This allows us to breed them through three heat periods in 45 days. It is quite a bit cheaper than prostaglandin. Each heifer can be synchronized for $.30 per head, or $54 for 180 head. One shot of prostaglandin costs a little over $800 for 180 head, or $4.44 per head. MGA will not work if the heifers are not cycling when the MGA is fed. Since it is fed 1 month ahead of breeding, it will not work very well on cows because they won't all be cycling at that time.

The heifers are all identified with a number branded on their hip with a hot iron 6 weeks before they are bred. At the time of breeding, this brand is starting to peel. Repeated riding when the heifer is in heat will make the brand bright red for a few hours during and after the time

Mon. Nov. 23	Tue 24	Wed 25	Thu 26	Fri 27	Sat 28	Sun 29	Mon 30	Tue Dec 1	Wed 2	Thu 3	Fri 4	Sat 5	Sun 6	Mon 7	Tue 8	Wed 9	Thu 10	Fri 11	Sat 12	Sun 13
110	680	200	20	480	70	140	450	80	400	01	014	058	250	060		230		062		
10	070	420	40	840	90	370	750	180	03	004		073	490	065						
170	081	09	100	880	520	460	011	530	047	021		090	08							
240	086	071	130	001	702	950	019	07	059	016										
300	088	075	330	012	730	02	037	015	079											
610	04	092	340	013	05	059	039	017	094											
630	020	097	360	016	010	038	040	018	011											
9L		014	390	022	020	064	041	025												
041		09	570	024	087	093	044	030												
031			640	026	099	02	043	034												
061			790	033	017	06	015	054												
077			890	042	024		021	05												
085			1690	046			022													
089			000	049			023													
01			027	056																
019			032	052																
			035	053																
			036	057																
			043	063																
			048	074																
			051	071																
			055	080																
			067	083																
			068	084																
			069	091																
			072	093																
			082	095																
			096	03																
			025	07																
				012																
				018																

Chart 1. Breeding and Heat Detection Chart, Registered and Commercial Yearling Heifers, November 23, 1981. The numbers listed below each date represent the heifers bred on each day. These 180 heifers were synchronized by feeding MGA for 12 days.

they are in heat. This helps us find the heifers that have been in heat during the night and out of heat by the next morning. We use only visual observation, and we will detect about 95% of the heifers or cows being artificially inseminated.

We will usually AI from 400 to 450 females. I do most of the inseminating but have help detecting heat. The most I have ever inseminated was 96 in one day. If your arm is already used to that job, this is not too many.

Use of Prostaglandin

I like to use prostaglandin on cows. We AI for 5 days before we give the prostaglandin shot to any cow that has not been inseminated. When you use prostaglandin, you get a lot more calves on the ground sooner. It has the added advantage of aborting the neighbor bulls' calves. Which I want it to do. I am not too thrilled when one of my better registered cows has a Longhorn calf.

Last year in working with 234 cows, 44 were bred the first 5 days, then we gave a shot of prostaglandin to 190 cows (chart 2). These cows were located in three different corral systems. After giving this shot, we just left the cows in these corrals for the next 5 days. This made a lot less work and hassle when it came to getting the cows in heat in and sorted off. The cows were bred only on observed heat. After 10 days, 197 cows had been bred or 84% of the herd that we were working with, and we had identified the 37 cows that apparently were not cycling. These were mainly 2- and 3-year-old cows. This group was sorted off and given extra feed. The calves were removed from these 37 cows for 36 hours. This calf removal triggered about one-third of these cows to come into heat in the next 7 days and most of the other cows cycled during the next 3 weeks. Five weeks into our breeding period we had only two cows that were not bred. When we removed the calves for 36 hours we did not have any sickness in the calves, but it sure was noisy.

There was no difference in conception rate between the cows that were given a shot before they came into heat and then were bred from the cows that were bred the first five days of the breeding season without a shot. Both groups of cows had a 61% conception rate.

If you count the replacement heifers plus the cows that were bred on our two heaviest breeding days, 155 head were bred. The first reaction to this is dismay at the number that may be calving on the same day. It really does not happen that way. If we get 60% conception, we will have about 90 cows calving from these two days of breeding. Some of these will calve 10 days early and some will calve 7 or 8 days late so that this group will calve over a 16- to 18-days period. There will be a bulge in the middle of this period, but this just lets you be more efficient with your time as you check a larger group as they calve.

Chart 2. Registered and Commercial Cow Breeding and Heat Detection Chart, Nov. 23 - Dec. 13, 1982. This breeding chart illustrates the way this group of 234 cows came into heat. There were no cows detected in heat on November 23. This was probably a case of poor heat detection. On November 27, all cows (190 head) who had not been detected in heat and bred the first 5 days of the breeding season were given a shot of prostaglandin. This caused 125 cows to come into heat Nov. 30 and Dec. 1. There were 21 cows that came in November 28 and 29 that were given the shot but came into heat before the shot took effect.

Mon Nov. 23	Tue 24	Wed 25	Thu 26	Fri 27	Sat 28	Sun 29	Mon 30	Tue Dec 1	Wed 2	Thu 3	Fri 4	Sat 5	Sun 6	Mon 7	Tue 8	Wed 9	Thu 10	Fri 11	Sat 12	Sun 13
	1638	92	742	631	1591	352	270 / 48	779	1603	34	35	569			126	609		339	165	
	288	935	215	233	37	752	131 / 58	81	165		920	921			419	659		679	833	
	117	146	5A	218	28	1A	221 / 138	291	385		941				829			418		
	698	436	9	528	78	586	322 / 148	531	356									8804		
	833	488	76	808	498	257	732 / 208	173	366											
	862	938	267	918	966	170	658 / 295	278	358											
		259	617	838		338	353 / 348	363	3											
		737	637	94		428	463 / 518	433	879											
			118	880		438	503 / 568	255												
			508			430	24 / 708	505												
			558			828	54 / 718	19												
			728			369	344 / 758	516												
			309			907	434 / 778	184												
			499			776	504 / 818	326												
			649			706	534 / 49	646												
			902				664 / 591	686												
			965				854 / 269	746												
			982				15 / 746	87												
			988				85 / 389	97												
			890				235 / 926	137												
			786				245 / 912	247												
							345 / 933	327												
							375 / 940	337												
							465 / 964	467												
							484 / 969	18												
							545 / 971	98												
							605 / 985	245												
							23 / 998	288												
							34 / 95	328												
							92 / 83	408												
							306 / 818	538												
							456 / 320	248												
							516 / 321	29												
							606 / 845	79												
							726 / 849	124												
							17 / 719	209												
							217 / 749	299												
							229 / 675	479												
							447 / 802	729												
							517 / 883	539												
							369	90												

Additional values under Tue Dec 1 (second sub-column): 99, 975, 877, 330

Calving Ease

Our replacement heifers are used to progeny test some of the larger Angus bulls in the breed. We never select a bull for calving ease alone. Although calving ease is important, what most people want measured in a progeny test is growth. We have never had a wreck in calving these heifers out, but we do see quite a difference in calving ease. We pulled only 4% of the calves from the easiest calving bull we have ever used. We pulled 60% of the calves from the hardest calving bull we have had. We have never had a caesarian. Our death loss on calves at birth runs from 2% to 4%. We have been calving two-year-old heifers this way for 18 years.

These heifers are calved out in the same dry lot where they were bred. This is located just 50 yards from our house so that one of us can check on them frequently. One thing we do that makes it easier to calve these heifers is to regulate their feeding time. We feed only in the evening. Heifers are fed about one-half of their hay at 7:30 p.m. and the last half of their daily ration is fed at 10:30 p.m. I do not know why, but when fed this way almost all calves will be born from 8:00 in the morning until midnight. We have a group of heifers that had less than 2% of their calves from midnight to 8:00 a.m. even though this represents 33% of a 24-hour day. Since most respectable people prefer to sleep during these hours, this lets us get a bit more sleep. I have seen as many as five heifers calving at one time during the day, but this does not cause problems.

EAR TAGS FOR FLY CONTROL

We have used the Ectrin ear tags for fly control and find they do an excellent job. Our fly control is much better than when we used to spray every 30 days. This should add at least 10 lb more weight to our calves. The recommendation is to put 2 tags on per animal, but we have found in our area that one tag on just half of the animals gives about as good control and costs only 25% as much. We do put one tag on each bull. We also implant the steers with Ralgro twice while they are on their mothers. Research indicates this should add another 40 lb to their weight.

FEEDING BULLS FOR SLAUGHTER

Recently there has been quite a bit of talk about feeding bulls for slaughter instead of steers. It has been our experience that steers will gain faster than bulls while they are on their mothers. Before bulls are weaned, they will start chasing anything in heat, even cows that are not in heat, when they are about 4 months old. In 1981, 74 bull calves gained 1.53 lb per day over a 77 day period from May 1 to July 17 when they were taken from their mothers. Over the same time period, 75 progeny-test steers, whose mothers were mainly 2-year-olds, gained 2.20 lb per day from May 1 to July 16. The steers were gaining .67 lb/day more than were the bull calves when they were both on the same kind of pasture. In 1982, 77 bull calves gained 1.76 lb per day from May 4 until July 11, a 71-day period or .38 lb/day more than the bulls. If you adjust the bulls' weight by adding 14 lb to compensate for the bulls being weaned 11 days earlier, the steers were still 39 lb heavier when they came off the cows. Producing bull calves for slaughter may not be as profitable for the cow-calf men as steers. If the cowman is going to do it, he will need to maintain ownership until slaughter. After the bulls are weaned, they probably will gain faster than steers in the feedlot.

We started this dissertation discussing steers that were over 200 lb heavier than steers produced from this same ranch a few years earlier. To summarize, I would guess this weight came in the following ways:

- 30 days older on the average =   70 lb
- Genetic improvement of parents =   30 lb
- Two Ralgro implants =   40 lb
- Ear tags to prevent flys =   10 lb
- Extremely good wheat pasture year = <u>70 lb</u>
                                       220 lb

FEEDLOT MANAGEMENT

Another management change that is helping us is maintaining ownership of these calves after they come off the cows. We make more money than by selling them as feeders. A word of caution: be careful what feedlot you use. I am a member of a seven-person feeding club that feeds about 10 pens of cattle a year. We have been a business since 1972. Some feedlots will feed cattle for a much cheaper cost of gain than other feedlots. It is not uncommon to see a $.10/lb difference in cost of gain. This is not all the fault of the feedlots, but a lot of it is.

Last year, the cost of gain on our 75 head of progeny-test steers fed at Brookover feedlot at Garden City, Kansas, was $43.75 a hundred. At the same time, Kansas State Extension Service published a summary of 11,237 steers coming out of six different Kansas feedlots. Their average cost of gain was $52.26, and they had a range from $49.10 to $57.75. Our cost of gain was $5.35 a hundred cheaper than the best pen of that 11,000 head, and it was $14 a hundred cheaper than the worst pen. Not all of this is because of the feedlot. For some reason, younger cattle will put on gain cheaper. Our calves were slaughtered at 14 months of age, which is about the age that many feeder calves go into the lot, then to be slaughtered 5 months later.

Even if the younger cattle weigh the same as the older cattle, they will gain cheaper. I have seen it happen several times. If I were going to guess, I would say that older cattle should gain cheaper because they should have more compensatory gain--but it does not work that way unless the older cattle have really been starved.

Our pen of progeny-test steers gained 3.24 lb/day and were on feed 138 days. We sold them grade and yield to get the carcass data on them. We were paid $60.68 a hundred on a live-weight basis. The 14-month-old calves weighed 1108 pounds after being shrunk the standard 4%. They were sold December 7, 1981, at a time when most fats were losing money for the feeder. Because of their low cost of gain and high gain per day, they netted us $67.06 more than they would if we had sold them as feeders.

These are some of the things we are doing to keep our ranch in business. I hope that we are good enough at it that 50 years from now my three sons will still be in the ranching business.

51

USING THE O'CONNOR MANAGEMENT SYSTEM TO IMPROVE PRODUCTIVITY

James N. Wiltbank,
Roy Anderson

HERD NONPRODUCERS

The O'Connor Management System was devised to increase the economic return in a beef cow herd. To produce calves economically, most cows must wean a heavy calf. Most beef herds contain many nonproducers, such as dry cows, replacement heifers, bulls, and light calves. For example, in a beef herd containing 100 cows there might be 15 replacement heifers and 5 bulls. If 90 cows weaned a calf, this herd would have 30 nonproducing animals.

Nonproducers in a 100 Cow Herd

No. calves weaned	Dry cows	Replacement heifers	Bulls	Nonproducers		Cost per calf[a]
				No.	%	
90	10	15	5	30	25	$333
80	20	15	5	40	33	375
70	30	15	5	50	42	428

[a] $250 per animal carrying cost.

The nonproducers must be reduced to make production of calves economically feasible. The cost of keeping nonproducers is as great or greater than the cost of keeping producers.

Weaning Weight and Net Return

Weaning weight	Gross return at $.70	Cost of keeping cows	Net return
500	$350	$250	$100
450	315	250	65
400	280	250	30
350	245	250	- 5
300	210	250	-40

Calves that are light at weaning will not pay the cost of keeping the cow. As an example, consider calves weaning at different weights. It does not take a mathematician to calculate the value of the heavy calf.

Calves wean light because they are born late in the calving season or do not grow, or both. As an example, look at the following table.

Weaning Weight As Influenced by Time of Birth and Average Daily Gain

Day of calving[a]	Average age weaning	Average daily gain birth to weaning (lb)		
		2.25	2.0	1.75
0- 20	220	565	510	455
21- 41	200	520	470	420
41- 60	180	475	430	385
61- 80	160	430	390	350
81-100	140	385	350	315
101-120	120	340	310	280
121-140	100	295	270	245

[a] 0 = Optimum date of calving.

The weaning weights in this herd varied from 565 lb to 245 lb. The calves dropped late were light weight even when they gained 2.25 lb a day. You cannot leave a calf on the cow, wean it later, and expect the calf to continue gaining. Calves stop growing when grass dries up and the cow's milk production stops; therefore, to wean heavy calves--they must be born early and have the genetic ability and the necessary nutrients to grow. A cow must wean at least 350 lb of calf to pay her own costs and to help cover costs of nonproducers.

To get the complete picture, consider the concept of heavy calves and nonproducers together.

Influence of Nonproducers and Weaning Weight on Pounds of Calf Weaned and Net Return in 100 Cow herd

Calves weaned in in 100 cow herd	Total animals in herd	Non-producers	Pounds of calf weaned per animal			Net return per animal[b]		
			500[a]	400	300	500[a]	400	300
90	120	30	375	300	225	$ 12	$-40	$-92
80	120	40	333	267	200	-17	-63	-110
70	120	50	292	233	175	-46	-87	-128

[a] Average weaning weight per calf.
[b] Calves at $.70 and $250 carrying cost.

To make money, the number of nonproducers must be kept low and the average weaning weight must be high. In cows, weaning calves averaging 500 lb, the pounds of calf weaned per animal in the herd varied from 375 lb to 292 lb. The pounds of calf weaned must be averaged out over a lot of nonproducers. Most of the figures on net return are negative. Only in herds with cows weaning 500 lb of calf and only 30 nonproducers are the results positive. Now look back to the last table and see how many calves weighed 500 lb or more. Only those calves born early and gaining 2 lb or more a day weighed over 500 lb.

THE O'CONNOR MANAGEMENT SYSTEM

The O'Connor method was devised to encourage early calving and thus optimize the pounds of calf weaned per animal in a cow herd and to increase the net return. This management system was first put into practice by Mr. Tom O'Connor, Victoria, Texas. The reproductive performance in a small group of cows was found to be exceptionally high.

Reproductive Performance in a Herd at O'Connor's

	% Pregnant after breeding		
21 days	42 days	63 days	84 days
80	87	87	93

A large proportion of the cows in the O'Connor herd became pregnant in a short period because:
- All cows in this group calved at least 30 days prior to the breeding season.
- Cows were in moderate or good body condition at calving time.
- Cows were gaining weight for 3 weeks prior to and during the first 3 weeks of the breeding season.
- Calves were removed from cows for 48 hours at the beginning of the breeding season.
- Cows were bred to fertile bulls.

The number of cows involved was small; therefore, an experiment was designed at Brigham Young University to further test the concepts of this management system and to compare pounds of calf weaned with a control group. The work was done cooperatively on a ranch managed by Dale Jolley at Elberta, Utah. Two hundred and thirty cows were checked for pregnancy in October. An attempt was made to group the cows by stage of pregnancy. The cows had been exposed to bulls for 5 months but some cows were only 35 to 40 days pregnant at the time of examination. Cows selected to be in the O'Connor management group were all early calvers (having calved 30 days before the start of the breeding season) while cows in the control group were expected to

calve for the 150-day period. The controls contained the
same percentage of early-calving cows as was found in the
original group. Cows were scored for body condition and
were allotted so that each group was similar. Most cows in
both groups were in moderate or good body condition at calving time. Cows in the O'Connor group were full-fed corn
silage starting 2 weeks before breeding and were continued
on this ration for the first 3 weeks of breeding. Calving
started in the last of January, and bulls were turned with
cows on April 22. All bulls were evaluated for fertility 4
weeks before the start of the breeding season. All bulls
turned into the O'Connor group had testicles larger than 32
cm in circumference and had more than 70% normal sperm. For
the 48 hours that the bulls were placed with the cows, the
calves were removed.

Thirty-three of the 85 cows in the O'Connor management
group showed heat within 48 hours after calf removal.
Twenty-one days after the start of the breeding season, 93%
had been bred. This increased to 97% after 42 days of
breeding.

Reproductive Performance at Elberta Using O'Connor System

	Number of cows	Bred after 21 days %	Bred after 42 days %	Conceived 1st service %	Pregnant after 21 days[a] %	Pregnant after 42 days %
Control	83	53	69	50	27	52
O'Connor manage- ment	85	93	97	81	75	93

[a] Estimate made from pregnancy exam giving number present
after 11 days of breeding.

Conception rate at first service was high in the
O'Connor group (81%). Seventy-five percent of the cows in
the O'Connor group were shown to be pregnant after 21 days
of breeding. At the time of pregnancy exam, only cows bred
in the first 11 days of breeding season could be checked for
pregnancy: 54 cows (64%) of the 85 cows were pregnant. It
was estimated from heat dates and conception rate that 10
more cows would be pregnant in the first 21 days of breeding. Thus a 75% pregnancy rate was estimated after 21 days
of breeding. Application of the 5 principles used by
O'Connor resulted in many cows becoming pregnant in a short
period of time. The following programs must be developed to
improve fertility in cow herds.

1. A 60-day breeding season.
2. Nutrition program to ensure that all cows are in
   at least moderate body condition at calving.

3. Nutrition program to make certain that cows are gaining weight for a 3-week period prior to breeding and during the first 3 weeks of breeding.
4. A method of removing calves for a 48-hour period at the start of the breeding season.
5. A program for evaluating bulls for potential fertility each year.

The importance of each of these will be mentioned and some methods for implementing them discussed.

A Sixty-Day Breeding Season

The length of the breeding season is an important factor in determining pregnancy rate. Late-calving cows have smaller calf crops than early-calving cows. As an example, in cows calving from November 15 to May 21, pregnancy was 88% in early-calving cows compared to 60% in late-calving cows.

Calving Time and Pregnancy

Time of calving	Feb. 10 to April 11: 60 days	Feb. 10 to June 11: 120 days	Feb. 10 to August 9: 180 days
Nov. 15 to Feb. 10 (%)	70	85	88
Feb. 11 to May 21 (%)	36	57	60

Similar results have been noted in an 80-day breeding season; pregnancy rate decreased from 88% in early-calving cows to 60% in late-calving cows. Cows calving early have more time to show heat before start of breeding, thus more will become pregnant.

The only reliable method for making sure cows calve early in the calving season is to have a short breeding season. Our results indicate that the breeding season should not last more than 60 days.

Shortening the breeding season from 150 days, or even from 90 days, to a 60-day season may present a cash-flow problem. The first year the breeding season is shortened there could be fewer calves for sale. The first step to avoiding the cash-flow problem of the short breeding season is to estimate how many calves were dropped in each of the weeks of the previous calving season. Relate this present breeding season to ascertain when cows will be bred. Next, an estimate of the amount and quality of forage available in different months of the year should be made. A chart that shows the nutrient requirements of cows should be obtained and a breeding season should be selected so that the maximum nutrient requirements of cows match as nearly as possible the best available forage supply. The present calving pattern should be compared with the desired calving pattern and changes made objectively. There are two methods for shortening the breeding season with only small losses in calf numbers. First, a plan is developed in which the breeding season is shortened 2 to 4 weeks per year. A heifer development program whereby heifers are bred only 45 days is an important part of this program and must be implemented or the plan will not work. Second, a plan can be developed in which cows are bred in a fall and spring program. Forage supply must be carefully evaluated in this type of program --calf numbers may actually be increased.

Table 1 indicates a procedure in which the breeding season might be shortened from 150 days to 60 days by adding 30 replacements per 100 cows each year for 3 years. To get these 30 replacement heifers to calf in a 45-day period, 35 heifers are bred and open heifers culled. The cost per animal in the herd is increased from $250 to $270. The net return is changed from $-39 to $+45. This is assuming a 90% calf crop each year. Generally, with a long calving season, the calf crop is lower.

This particular method increases revenues--but it requires a place to carry an extra 25 heifers each year for 3 years, and thus might not be feasible. This method could be implemented by checking cows for pregnancy and culling open and late-calving cows. Using this system, the number of cows replaced would be determined by the number of pregnant replacement heifers available to be placed in the herd.

Nutrition Program to Maintain Moderate Body Condition

Body condition is important in determining the proportion of cows showing heat and becoming pregnant. Many cows in thin body condition do not become pregnant. In one study, the proportion open varied from 77% in very thin cows to 5% in cows in good body condition.

Relationship Between Body Condition and Pregnancy Rate in Florida

	Very thin	Thin	Slightly thin	Moderate	Good
No. of cows	115	545	564	344	234
% open	77	49	27	14	5
Early calvers (%)	5	15	19	40	56

Only 5% of the thin cows will calve early compared to 56% of the cows in good body condition.

Thin cows do not become pregnant, or calve late, because many are delayed in showing heat. The table below indicates that the proportions of cows that showed heat by 60 days after calving were: good body condition (91%), moderate (61%), and thin (46%).

Body Condition at Calving and Days to Heat After Calving

Body condition at calving	No. of cows	Days to heat after calving				
		40 %	60 %	80 %	100 %	120 %
Thin	272	19	46	62	70	77
Moderate	364	21	61	88	100	100
Good	50	31	91	98	100	100

By 100 days after calving, only 70% of the cows in thin body condition had shown heat.

TABLE 1. CHANGING LENGTH OF CALVING SEASON

Expected day of calving	No. of cows and heifers bred				
	1st year	2nd year	3rd year	4th year	5th year
1- 20	10	30	50	70	75
21- 40	10	20	20	25	20
41- 60	10	10	20	5	5
61- 80	20	20	10		
81-100	20	20			
101-120	10				
121-140	5				
141-150	5				
Total no. pregnant	100	100	100	100	100
No. replacement saved	35	35	35	12	12
Pregnant replacements placed in herd	10	30	30	30	10
Cost per animal	$250	$270	$270	$270	$250
Calf crop weaned	90	90	90	90	90
Animals per 100 calves	127	135	135	135	127
Lb calf weaned per animal	281	308	336	352	393
Net return	$-39	$-39	$-18	$-6	$+45

There are two approaches to maintaining cows in moderate body condition. First, the cows should be carefully observed 1 or 2 months before calves are scheduled to be weaned. If cows are thin, then calves should be weaned right away. This will give cows a few months of good feed before the quality of the forage declines. Calves are probably growing at a slow rate because of low-quality feed available.

The second approach is to sort cows by body condition at weaning time. Cows should be scored for body condition from 1 (thinnest) to 9 (fattest) as shown below (Spitzer, 1977). Decisions on feeding should then be made. The amount of weight gain needed to change body condition must be kept in mind; the following table can be used for such calculations.

Recommended scores for evaluating body condition in beef cattle:

Thin.
1. Poor, starving, bordering on inhumane, survival questioned during stress.
2. Very thin, poor milk production, chances for rebreeding slim to none.
3. Thin, lowered milk production, poor reproduction, fat along backbone and slight amount of fat cover over ribs.

Moderate.
4. Borderline, reproduction bordering on inadequate, some fat cover over ribs.
5. Moderate, minimum necessary for efficient good overall appearance, fat cover over ribs feels spongy.
6. Moderate to good, milk production and rebreeding very acceptable, spongy fat cover over ribs and fat beginning to be palpable around tailhead.

Good.
7. Good, fleshy, maximum condition needed for efficient reproduction.
8. Fat, very fleshy, unnecessary, no advantage in rebreeding from having cows in this condition; cow has large fat deposits over ribs, around tailhead, and below vulva.
9. Extremely fat, extremely wasty and patchy, may cause calving problems, cow extremely over-conditioned.

The body condition desired at calving is a 5. A cow that scores a 5 at weaning must gain 100 lb to calve with a body condition of 5. This 100 lb represents the weight of the calf, fluid, and membranes. Thus, even a cow with ideal body condition at weaning must gain nearly .8 lb a day to calve in ideal condition. A cow that scores only a 3 at

Body condition at weaning	Body condition at calving	Weight gain			Days weaning to calving	ADG
		Calf fluid and membrane	Fat or muscle	Total		
5 (moderate)	5	100	0	100	130	0.77
3	5	100	160	260	130	2.00
3	5	100	160	260	200	1.30
3	5	100	160	260	100	2.60
2	5	100	240	340	130	2.60
7	5	100	-160	-60	130	-0.46

weaning time must gain 2.0 lb a day when there are 130 days from weaning to calving. If calves are weaned earlier so that there are 200 days between weaning and calving, she only has to gain 1.3 lb. However, when calves are weaned late and there are only 100 days from weaning to calving, a cow scoring a 3 at weaning must gain 2.6 lb a day to score a 5 at calving time. To change a cow from one body condition to the next requires the cow to gain or lose approximately 80 lb of fat or muscle.

Each year is different. Cows are different. You must assess the body condition of your cows, the forage available, and then put together a plan so cows will score a 5 or 6 at calving time. The problem is that thin cows will either be open or calve late.

Flushing and Forty-eight-hour Calf Removal

Flushing and 48-hour calf removal can be helpful in improving reproductive performance. Neither practice alone is as beneficial as the combination of the two. A study conducted at Howell's in South Texas with first-calf cows that were slightly thin (scored at 4) at calving time demonstrates this principle.

Pregnancy Rates Following Calf Removal and Flushing

	Control	Fl[a]	Cr[b]	Fl + Cr
No. of cows	18	21	21	21
Pregnant (%)				
21 days	28	14	38	57
24 days	56	52	62	72
63 days	72	76	62	86

[a] Flushed 10 lb corn for 2 weeks before breeding and first 3 weeks of breeding.
[b] Calf removal for 48 hours at start of breeding.

Pregnancy rate was increased only in the group which used both flushing and calf removal. Flushing cows for 3 weeks before breeding did not increase pregnancy rate.

Feeding thin cows (a score of 3 or less) for short periods of time after calving to get them to show heat is not an effective practice. The principle is illustrated in the following table.

Body condition score		Weight gain needed	Days calving to breeding	ADG
At calving	Needed at start of breeding			
3	5	160 lb	80	2.0
3	5	160 lb	60	2.7

A minimum of 2 lb a day must be gained by the cow scoring a 3 at calving if she is to have sufficient body condition to show heat early in the breeding season. If, in addition to scoring 3 she has only 60 days from calving to breeding, she must gain 2.7 lb per day. As soon as you increase her food level, she will increase her milk production; thus only a small amount of the nutrients fed will contribute to weight gain. It is difficult, if not impossible, to get her to gain 2 lb a day while nursing a calf. Therefore, the cow should be in good condition before calving.

Cows that score a 4 or greater will respond beautifully to a little extra feed for at least 3 weeks prior to breeding, if the calves are removed for 48 hours when the bulls are placed in the breeding pasture.

So, how do cows gain weight prior to breeding? You could feed grain or put them on good pasture with added dry matter. However, don't expect a cow to gain weight on short green grass; such grass is 90% water. Use good hay, grain, or a pasture that has some good growth to avoid disappointment.

Removing calves for 48 hours can be a problem in some situations. The best way to accomplish this without extra labor is to remove calves for 24 hours prior to the day when they are to be worked. Work the calves and then turn them back to their mothers at the end of the 48 hour period. Calves must not nurse for 48 hours to get maximum results.

Fertile Bulls

Fertile bulls must (1) produce adequate amounts of sperm, (2) have a large proportion of normal sperm, (3) and have the desire and ability to deposit the sperm in the cow. A good measure of semen production is scrotal circumference. It can be measured quickly and easily with a tape.

Available data indicates that bulls with a scrotal circumference of less than 30 cm have reduced fertility. Ten to 15% of the bulls in most breeds have little or no desire to breed. Simple reliable tests for identifying these bulls have not been developed, although reliable tests have been developed for bulls that have been handled regularly.

The effect of selecting bulls for semen quality was demonstrated recently at the King Ranch. Semen from 79 bulls was collected and evaluated. Twenty-seven of these bulls were selected and placed with 675 cows. These 27 bulls had 80% or more normal sperm. Another 26 bulls were placed with 655 cows. These bulls were a representative sample of the original group of 79 bulls (52% of the original group had 80% or more normal sperm). In the control group of bulls, 14 of the 26 or 54% had 80% or more normal sperm. Like in the original group, 16% of the control group had less than 40% normal sperm. The pregnancy rates after 120 days of breeding was 93% in the selected group and 87% in the control; a study the second year showed a 5% or 6% improvement in pregnancy rates.

Bulls Selected for Semen Quality at King Ranch

	Multiple sire-1980 (Four bulls per 100 cows)		Multiple sire-1981		
	Control	80% or over	Control	80%+	70%+
Number exposed	572	656	1,179	522	769
Pregnant (%)	87%	93%	85%	90%	91%

Bulls should be evaluated each year. Semen quality will improve in certain bulls from the first semen collection to the second. If a bull has poor semen, collect a second time immediately. Evaluate, and if semen is still poor, collect semen from the bull 3 or 4 weeks later. Then make a decision, but don't use a bull with poor semen.

SUMMARY

More calves can be produced in your herd by decreasing the nonproducers. This means reducing the number of dry cows and replacement heifers. The number of dry cows can be reduced and more calves will calve early if you:
1. Have a 60-day breeding and calving season.
2. Have cows in good to moderate body condition at calving time.
3. Flush cows for 5 to 6 week period near breeding time.
4. Remove calves for 48 hours at the start of the breeding season.

5.   Breed to fertile bulls.

This system requires a plan that uses all five principles. Use only one or two of the principles and you will be disappointed. It takes a complete plan to make the system work for you.

REFERENCE

Source: Spitzer, J. C. 1977. Body condition and rebreeding in the cow. Texas Animal Agricultural Conference.

52

DEVELOPING REPLACEMENT HEIFERS

James N. Wiltbank

Proper development of replacement heifers leads to (1) higher pregnancy rate first breeding season, (2) less calving difficulty, (3) higher pregnancy rate second breeding season, and , (4) consequently, higher returns.

The economic importance of developing heifers can be seen by comparing Brahman crossbred heifers fed to weigh either 600 lb or 700 lb at the start of the breeding season. In the group fed to weigh 700 lb (TW2) 19% more calves were weaned the first year than in the group fed to weight 600 lb (TW1). In the second breeding season, 28% more cows became pregnant in the TW2 group than did in the TW1 group. This should lead to 28% more calves at weaning time for the TW2 group in the second year (table 1). The first calves born to TW2 cows were 30 lb heavier at weaning than were calves born to TW1 cows. A 16 lb advantage for second calves also was estimated for TW2 cows. TW2 heifers weaned 21,512 lb more calf for the first two calves than did TW1 heifers, or 215 lb per heifer exposed. This difference in pounds of calf weaned was obtained for approximately 500 lb of concentrate per heifer. This means each pound of concentrate fed produced 2.3 lb more calf. With calves selling for $0.65, each pound of concentrate was worth $1.50 for the first two calves. Other data would indicate this trend of early calving would continue throughout the lifetime of these cows.

More cows weaned calves in the TW2 group in the first year because more cows became pregnant early in the breeding season in this group; this trend also was apparent the second year (table 2). This difference in pregnancy rate occurred because heifers were in heat and bred early in the breeding season both years.

These data show the advantages of feeding the Brahman-cross heifer to weight 700 lbs. However, 700 lb is not the magic number for heifers of all breeds and crosses; target weight differs by breed of heifer.

Information available indicates that the number of heifers showing heat and becoming pregnant early in the breeding season is dependent on age and weight at puberty (and breed or cross) of the heifer. Table 3 shows the pro-

TABLE 1. VALUE OF DEVELOPING BRAHMAN CROSS HEIFERS TO TWO
WEIGHTS

	600(TW_1)	700(TW_2)	Difference
No. of heifers	100	100	--
Calves weaned			
1st year	58	77	19
2nd year[a]	40	68	28
Total	98	145	47
Weaning weight			
Average			
1st year	356	386	30
2nd year[b]	408	424	16
Total lb	37,020	58,532	21,512
Per heifer exposed			
1st year	206	297	91
2nd year	163	288	125
Total lb	370	585	215
Costs of development per heifer			
Hay (lb)	1978	1788	-190
Concentrate (lb)	924	1416	492
$for feed	100	122	22
Pregnant cows			
2nd breeding season	40	68	

[a]Estimate from cows pregnant.
[b]Estimate from time of conception 2nd breeding season.

portion of heifers that showed heat at different ages and
weights. At 12 months of age, only 15% of the Hereford
heifers weighing 600 lb had shown heat, compared to 40% in
the Angus heifers, and crossbred heifers. The percentage of
heifers weighing 600 lb that had shown heat by 14 mo. of age
increased to 70% in Angus heifers, to 82% in A x H heifers,
but was only 30% in Hereford heifers. However, 90% of the
700 lb Hereford heifers had shown heat at 14 to 15 mo of
age. These data indicate that age, weight, and breed can
affect time of puberty. Most of the heifers in these two
breeds will show heat by 14 to 15 mo of age IF they have
sufficient weight. The weight needed to reach puberty
varies according to the breed of the heifer.

Puberty will be delayed in heifers until they attain
sufficient weight. Table 4 shows the weight needed for

TABLE 2.   REPRODUCTIVE PATTERN IN HEIFERS DEVELOPED TO TWO
WEIGHTS

| | 600(TW_1) | 700(TW_2) | |
	110	111	Difference
No.			
1st breeding sea on			
% showing estrus by			
20 days	33	63	
40 days	56	80	
60 days	71	92	
90 days	97	100	
Conceived 1st service (%)	46	63	
% pregnant by			
20 days	9	39	
40 days	27	57	
60 days	47	74	
93 days	66	82	
2nd breeding season			
no. of cows exposed	65	88	
% showing estrus by			
20 days	12	24	
40 days	48	70	
Conceived 1st service (%)	69	81	
% pregnant by			
20 days	8	20	
40 days	33	57	
60 days	59	79	
90 days	68	85	

heifers of different breeds to reach puberty at 14 to 15 mo
of age.

As an example, 50% of the Hereford heifers 14 to 15 mo
of age would be expected to be in heat at 600 lb - this is
the average weight at puberty. If you want 85% to 90% of
Hereford heifers to show heat, each should weigh 700 lb before being bred.   This doesn't mean that the group of
heifers should average 700 lb.   It means each heifer should
weigh 700 lb.   This can be done by sorting heifers and feeding the light ones to make more gain and the heavy heifers
to make less gain.   Results presented are similar for
heifers of other breeds.

Use a scale to determine the desired weight not the
"eyeball" technique.   Heifers must be weighed or heart girth
measured monthly.   If discrepancies are noted, rations
should be adjusted so that heifers reach the desired breeding weights.

On the Tom O'Connor Ranch, heifers were divided into three groups by weight at the start of breeding: (less than 550 lb, 550 to 600 lb, and over 600 lb). Only 65% of the heifers weighing less than 550 lb became pregnant in a 60-day breeding season as compared to 90% in heifers weighing over 600 lb (table 5). Only 40% of the heifers weighing

TABLE 3. PROPORTION OF HEIFERS IN HEAT AT VARIOUS WEIGHTS AND AGES

Weight (lb)	Age in months			
	12	13	14	15
Hereford				
500 (%)	0	0	0	0
600 (%)	15	20	30	37
700 (%)	--	65	90	90
Angus				
500 (%)	0	33	57	77
600 (%)	40	65	70	80
700 (%)	--	80	100	100
A x H				
500 (%)	27	36	73	91
600 (%)	40	75	82	96
700 (%)	--	78	96	100

TABLE 4. WEIGHT AT WHICH 14 TO 15 MONTH OLD HEIFERS SHOW 1st HEAT

Proportion desired in heat	Weight (lb) needed by						
	Angus	Hereford	Charolais	AxH	SxE	LxE	BRxE
50%	550	600	700	550	650	650	650
65-70%	600	650	725	600	700	700	700
85-90%	650	700	750	650	750	750	750

A = Angus
E = English
H = Hereford

L = Limousin
S = Simmental
BR = Brahman

TABLE 5. REPRODUCTIVE PERFORMANCE IN HEREFORD HEIFERS AS
INFLUENCED BY WEIGHT AT START OF BREEDING
(TOM O'CONNOR)

	Less than 550 lb	551 to 600 lb	Over 600 lb
No. heifers	40	166	45
Pregnant 60 days (%)	65	77	90
Calves weaned (%)	40	71	86
Losses, pregnancy diagnosis to weaning (%)	25	6	4
Wet cows pregnant 2nd year (%)	18	57	69

less than 550 lb weaned calves, compared to 71% and 86% in
the other two groups. Losses from pregnancy to weaning were
25% in light heifers, compared to 6% and 4% in the other two
groups. In the light heifers, only 18% of the cows suckling
calves became pregnant with a second calf, compared to 69%
in the heifers that weighed over 600 lb at the start of
breeding.

Other data indicate that heifers on higher levels of
feed have larger pelvic openings near calving. Brahman
crossbred heifers fed to weigh 700 lb were 9 cm larger than
were heifers fed to weigh 600 lb at the start of breeding
and at similar levels of feed thereafter (table 6). In
1981, Bellows reported a 12 cm difference in pelvic area between heifers of high or low levels of feed during the de-
velopmental period. A difference of 10% in calving difficulty was noted in Bellow's study.

TABLE 6. EFFECT OF HEIFER DEVELOPMENT ON PELVIC OPENING
NEAR CALVING

	Target weight at breeding		
	600	700	Difference
No. heifers	69	89	
Pelvic area (cm^2)[a]	249	258	9

	Feed level during winter after weaning[b]	
	Low	High
No.	30	30
Precalving pelvic area (cm^2)	240	252
Calving difficulty	46	36

[a]50 days prior to start of calving season.
[b]Bellows (1981).

Many producers underfeed their replacement heifers. in outlining a breeding program the amount of gain should first be determined according to the breeding weight needed for that breed of heifer (table 4). Each heifer should be individually weighed and the amount of gain calculated to reach a target (breeding) weight (table 7). For example,

TABLE 7. GAIN AND DAYS TO REACH TARGET WEIGHT OF 700 LB

Initial weight	Total gain	Days to reach target weight when ADG is:		
		1.0	2.0	2.5
300	400	400	200	160[a]
400	300	300	150	120
500	200	200	100	80

[a]This gain is difficult to achieve for this weight of heifer.

if 700 lb is chosen as the target weight, a heifer weighing only 300 lb must gain 400 lb or gain 2 lb/day to reach the target weight in 200 days. Contrast this to heifers weighing 400 lb that must gain only 300 lb and can reach target weight by gaining 1.5 lb/day for 200 days. Target weight is even easier to reach for 500 lb heifers.

Table 8 shows the approximate amount of corn needed to achieve different weight gains. To gain 2 lb/day, from 6 to 8 lb of grain per head per day are needed for a 300 lb heifer, whereas, a 400 lb heifer needs 4 to 5 lb/day of

TABLE 8. POUNDS OF CORN NEEDED TO MAKE ADG FROM INITIAL WEIGHT TO TARGET WEIGHT OF 700 LB[a]

Initial weight (lb)	ADG		
	1.0	2.0	2.5
300	2	8	12[b]
400	2	7	10
500	0	6	9

[a]All heifers assumed to have full feed alfalfa hay.
[b]This gain is difficult to achieve because of grain consumption needed.

corn and a full feed of hay to gain 1.5 lb/day. Thus, costs are considerably greater in lighter animals.

Feed intake of light heifers is extremely limited (table 9). As an example, a 300 lb heifer will eat only 9

TABLE 9. AMOUNT OF HAY AND GRAIN[a] HEIFERS WILL EAT IN GAIN-
ING FROM INITIAL WEIGHT TO TARGET WEIGHT OF 700 LB

Initial weight lb	Initial intake lb	Intake at halfway lb	Intake at target weight of 700 lb lb
300	9	15 (500)[b]	21
400	12	17 (580)	21
500	15	18 (600)	21

[a]Silage is about 40% dry matter so heifer will consume about
3 lb silage for each pound of dry feed.
[b]( ) weight at halfway.

lb of hay and grain per day. So if you are striving to
reach a target weight of 700 lb with a 300 lb heifer, a gain
of 2.5 lb/day will be difficult--because a light heifer can
eat only about 2.5% to 3.0% of its body weight. Remember,
many heifers are light because they are young. Trying to
feed them to target weight may cause them not to cycle, if
they are only 11 or 12 months of age.

The heart of a good reproductive program is a heifer
replacement program that has heifers old enough and heavy
enough to breed early in the breeding season.

Lightweight heifers at the start of breeding cause
three problems: (1) pregnancy rate at first breeding is
low; (2) losses are high at first calving; (3) cows suckling
calves do not breed for the second calf. Heifers that are
bred at target weight will have a better reproductive performance and net income will increase.

53

CULLING COWS AND BEEF PRODUCTION

R. A. Bellows

Many studies have been reported on the importance of developing good replacement heifers. This phase of beef cattle management is one of the keys to successful beef production. But, we often overlook what the probabilities are that a cow will be culled and what we can do with the cull cow. These topics are the subject of this presentation.

CULLING PROBABILITIES AND HERD LIFE

A study was made of cow records from 1943 through 1976 at the Livestock and Range Research Station, Miles City, Montana (Greer et al., 1980). The records identified age at culling and reason for culling in the entire herd. Animals included were predominantly Hereford, including several inbred lines. Data from 1958 on included Angus, Charolais, Brown Swiss, and all reciprocal crossbreds. The breeding season varied from 45 to 60 days and included heifer groups bred to produce their first calf at either 2 or 3 years of age.

Herd management is typical of ranches in the Northern Great Plains, and level of culling is approximately representative of the area. The level of culling for individual performance records may be higher, however. In addition, cows are culled for age at 10 years after their calves are weaned.

Specific reasons for culling were physical impairments and management criteria. Physical impairment categories included:
- Dead or missing
- Bad udder
- Lump jaw
- Eye cancer
- Vaginal or uterine prolapse
- Bad feet
- Other injury or illness
Management criteria categories included:
- Nonpregnant - based on fall pregnancy test.
- Fertility - failed to produce a calf in 2 consecutive years; freemartins.

- Performance - slow growth of individual or calves.
- Conformation - undesirable physical traits.

Tables 1 and 2 show the percentages of cows culled, summarized by reason and cow age. These percentages are estimates of the chance (probability) that a cow of a given age will be culled for a specific reason. The values listed are percent culled; all reasons are estimated of the chance that a cow of a given age will be culled.

Tables 1 and 2 show some differences in culling percentages for heifers bred to calve first as a 2- or 3-year-old. These figures were pooled and give some useful values shown in table 3 and table 3A.

TABLE 1. ESTIMATED PROBABILITIES OF FEMALES BEING CULLED EACH YEAR, BY AGE AND REASON. HEIFERS BRED TO CALVE FOR THE FIRST TIME AS 3-YEAR-OLDS (TO CONVERT PROBABILITIES TO PERCENTAGES, MOVE THE DECIMAL POINT 2 DIGITS TO THE RIGHT)

Reason for Culling	Age in Years							
	3	4	5	6	7	8	9	10
Dead or missing	.0135	.0113	.0073	.0146	.0164	.0153	.0166	.0104
Bad udder	.0000	.0010	.0020	.0034	.0066	.0104	.0157	.0131
Lump jaw	.0028	.0025	.0020	.0017	.0041	.0028	.0018	.0091
Eye cancer	.0016	.0050	.0120	.0146	.0246	.0389	.0406	.0666
Prolapse	.0195	.0081	.0147	.0137	.0128	.0090	.0101	.0104
Bad feet	.0007	.0028	.0030	.0043	.0062	.0056	.0083	.0052
Other injury or illness	.0030	.0040	.0030	.0026	.0087	.0069	.0157	.0026
Management decision criteria	.0958	.1322	.1288	.1017	.0921	.1028	.1456	.3525
Proportion culled, all reasons	.1369	.1670	.1729	.1566	.1714	.1918	.2544	.4700

TABLE 2. ESTIMATED PROBABILITIES OF FEMALES BEING CULLED EACH YEAR, BY AGE AND REASON. HEIFERS BRED TO CALVE FOR THE FIRST TIME AS 2-YEAR-OLDS (TO CONVERT PROBABILITIES TO PERCENTAGES, MOVE THE DECIMAL POINT 2 DIGITS TO THE RIGHT)

Reason for Culling	Age in Years								
	2	3	4	5	6	7	8	9	10
Dead or missing	.0165	.0179	.0100	.0186	.0114	.0042	.0144	.0000	.0000
Bad udder	.0000	.0012	.0000	.0029	.0023	.0000	.0144	.0563	.0067
Lump jaw	.0000	.0006	.0010	.0000	.0000	.0000	.0000	.0000	.0000
Eye cancer	.0004	.0012	.0010	.0057	.0160	.0127	.0000	.0563	.1333
Prolapse	.0032	.0025	.0000	.0014	.0023	.0000	.0072	.0000	.0000
Bad feet	.0000	.0006	.0010	.0000	.0023	.0042	.0000	.0000	.0000
Other injury or illness	.0012	.0018	.0020	.0014	.0046	.0000	.0072	.0000	.0000
Management decision criteria	.1717	.1694	.1170	.1461	.1461	.1941	.2302	.5070	.5333
Proportion culled all reasons	.1930	.1953	.1320	.1762	.1849	.2152	.2734	.6197	.7333

TABLE 3. PROBABILITIES OF COWS OF A GIVEN AGE GROUP BEING CULLED OR RETAINED IN THE HERD: POOLED SAMPLE ESTIMATES. (TO CONVERT PROBABILITIES TO PERCENTAGES, MOVE THE DECIMAL POINT 2 POINTS TO THE RIGHT.)

Row	Item	Cow Age in Years									
		2	3	4	5	6	7	8	9	10	10B[a]
A	Probability of cows of an age being culled	.1930	.1528	.1599	.1735	.1611	.1761	.1990	.2768	.4750	1.0000
B	Probability of cows of an age remaining in the herd	.8070	.8472	.8401	.8265	.8389	.8239	.8010	.7232	.5250	.0000
C	Probability of a cow that enters the herd at age 2 remaining in the herd to a given age	1.0000	.8070	.6837	.5744	.4747	.3982	.3281	.2628	.1901	.0998

[a] Age 10B is used here to designate the proportion of cows that meet or surpass management criteria and are not suffering a physical impairment at age 10, but which are removed from the herd as a result of the general policy to cull all cows at the end of age 10. The decision is made at or shortly after weaning in the fall.

TABLE 3A. AGE DISTRIBUTION OF COWS CULLED IN AN AVERAGE YEAR AND COWS IN THE HERD IN AN AVERAGE YEAR (PERCENTAGES)

Row	Item	Cow Age in Years									
		2	3	4	5	6	7	8	9	10	10B[a]
A	Cows culled	19.30	12.33	10.93	09.97	07.65	07.01	06.54	07.27	09.03	09.98
B	Cows in the herd	21.19	17.10	14.49	12.17	10.06	08.44	06.95	05.57	04.03	00.00

[a] Age 10B is used here to designate the proportion of cows that meet or surpass management criteria and are not suffering a physical impairment at age 10, but which are removed from the herd as a result of the general policy to cull all cows at the end of age 10. The decision is made at or shortly after weaning in the fall.

Row A in table 3 indicates the probability of a cow being culled at a certain age, given that she remains in the herd to that age. Row B in table 3 is the complement of Row A, i.e., the probability that the cow will remain in the herd. Row C in table 3 indicates the probability of a cow entering the herd as a 2-year-old and remaining in the herd to each of the ages shown. Row A in table 3A is the age distribution of the culled cows; and Row B in table 3A is the proportion of cows in the herd of each age--or, simply, the age distribution of the herd.

It can be seen that .2119 or 21.19% of the females in the herd were 2-year-olds. This represents the replacement rate in the Miles City herd.

The proportions of cows culled for various criteria are summarized in table 4. It can be seen that the number culled for death or missing did not change greatly over the cow-age categories. However, marked changes did occur in culling for physical impairments and management criteria.

TABLE 4. PERCENTAGE OF COWS OF AN AGE REMOVED FOR PHYSICAL IMPAIRMENT AND MANAGEMENT CRITERIA REASONS: POOLED SAMPLE ESTIMATES

Cow Age, Years	Total Percentage Culled	Proportion Culled For:		
		Dead or Missing	Other Physical Impairment	Management Criteria
2	19.30	01.65	00.48	17.17
3	15.28	01.47	02.22	11.59
4	15.99	01.11	09.16	12.92
5	17.35	00.95	03.19	13.21
6	16.11	01.41	03.83	10.87
7	17.61	01.51	05.79	10.31
8	19.90	01.52	06.98	11.40
9	27.68	01.56	09.34	16.78
10	47.50	01.02	10.89	35.60

Expected herd life is important in the prediction of the likelihood of future performance. Values are summarized in table 5. These values show that if a cow remains in the herd until 3 years of age, her expected herd life is an additional 3.73 years for a predicted culling age of 6.73 years. If a cow in the herd is 6 years of age, her expected herd life is 2.69 years or 8.69 years. The average age of cows culled was 5.72 years, and the average age of cows in the herd was 4.78 years.

TABLE 5. EXPECTED HERD LIFE FOR COWS AT EACH AGE

Cow Age, Years	Expected Herd Life, Years
2	3.82
3	3.73
4	3.41
5	3.05
6	2.69
7	2.21
8	1.68
9	1.10
10	.53

This study indicates that longevity of animals in a range beef herd is not great. Production of replacement females is an expensive process, and this cost ·must be charged against the income from only three or four calves that she is expected (based on these probabilities) to produce.

BEEF PRODUCTION FROM CULL COWS

This study by Greer et al. (1980) indicated a yearly replacement rate of 21.19%. This means that over 21% of the cow herd will be marketed within a given year. Cows culled from breeding herds produce a large amount of beef consumed in the United States. Many of these cows are in average to poor condition and are slaughtered soon after culling. These animals represent a savings in feed grains since they are marketed as grass fed with the meat usually marketed as ground beef. Increasing weight gains and carcass quality of these animals were studied recently at Miles City by Bellows et al. (1979).

Two experiments involving a total of 118 crossbred cows were conducted to determine effects of spaying, a 36-mg zeranol (Ralgro®) implant and a vaginal device on weight gains and carcass characteristics of mature, cull cows grazing range forage.

Experiment 1 involved 59 cows ranging in age from 5 to 9 years. All animals were crossbred with varying percentages of Hereford, Angus, and Charolais breeding. All animals had been culled as nonpregnant in the fall preceding the study. Early the following spring, cows were assigned to a study involving intact or spayed cows with or without zeranol implants.

Thirty cows were spayed by surgically removing both ovaries. Zeranol was administered as a 36-mg implant placed just under the skin of the ear. Surgical and implant procedures were routine.

Cows were placed on the study on April 6 and slaughtered on July 6. During the 91-day grazing period, all cows were pastured on the same native range pasture. A mineral mix of salt and dicalcium phosphate was fed free choice throughout the study.

Animals were slaughtered by a commercial packing plant and routine carcass information was obtained.

Experiment 2 was conducted one year later than Experiment 1 and involved 59 cows ranging in age from 5 to 10 years. Breeding and reasons for culling were the same as for Experiment 1. This study involved intact and spayed cows, with or without a 36-mg zeranol implant and with or without a vaginal device.

Spaying and implanting procedures were the same as described for Experiment 1. Vaginal devices were inserted as specified by the manufacturer. Implant and vaginal devices were administered an average of 36 days after surgery. Animals were then placed on native range pasture until June 24, giving a grazing period of 70 days.

Animals were slaughtered at a commercial packing plant, and data collection was the same as described for Experiment 1. Results are summarized in table 6.

Spaying resulted in a nonsignificant depression of weight gains in both experiments. Zeranol increased weight gains by 10.3% in Experiment 1 and 17.3% in Experiment 2. Zeranol affected carcass quality measures but the effects were not consistent between the two studies. Weight gains of the spayed cows with zeranol implants were equal to those of intact, control cows. The effects of the vaginal device on weight gains were nonsignificant.

These studies indicate that weight gains of cull cows can be improved with zeranol implants. However, spaying and use of vaginal devices did not improve beef production. Producers should consider implanting their cull cows with zeranol and grazing the animals for 60 to 90 days before marketing.

TABLE 6. LEAST SQUARES MEANS FOR WEIGHTS AND CARCASS DATA

Experiment and treatment	No. Head	Initial body weight lbs	Total weight gain lbs	Hot carcass weight, lbs	Kidney, pelvic and heart fat, %	Area in.2	Fat thickness in.
Experiment 1							
Intact	29	1107	136	653	1.71	11.6	.4
No implant	14	1098	133	639	1.68	11.6	.4
Implant	15	1116	140	664	1.73	11.6	.4
Spayed	30	1082	138	637	1.37	11.4	.4
No implant	15	1076	128	626	1.53	10.8	.4
Implant	15	1089	148	648	1.21	12.0	.3
Experiment 2							
Intact	30	1074	156	646	1.69	12.0	.3
No implant	15	1074	147	666	1.74	12.4	.3
No device	7	1076	157	659	1.86	11.5	.3
Device	8	1074	136	670	1.62	13.3	.2
Implant	15	1030	164	622	1.64	11.5	.2
No device	7	1038	169	626	1.71	11.8	.2
Device	8	1019	160	615	1.56	11.2	.3
Spayed	29	1091	153	659	1.72	12.0	.2
No implant	14	1087	137	662	1.68	11.7	.2
No device	7	1041	140	670	1.78	11.6	.2
Device	7	1133	135	653	1.57	11.9	.2
Implant	15	1098	169	648	1.77	12.3	.3
No device	8	1122	149	675	1.75	12.6	.3
Device	7	1069	190	624	1.78	12.1	.3

REFERENCES

Bellows, R. A., R. B. Staigmiller, J. B. Carr, and R. E.
    Short.   1979.   Beef production from mature cows on
    range forage. J. Anim. Sci. 49:654-663.

Greer, R. C., R. W. Whitman, and R. R. Woodward.   1980.
    Estimation of probability of beef cows being culled and
    calculation of expected herd life.   J. Anim. Sci.
    51:10-19.

54

THE PEGRAM PROJECT:
INTEGRATED CALF MANAGEMENT

Ed Duren

The death rate of newborn beef calves has steadily increased from 4% to 12% in Idaho over the past ten years. During a three-year period from 1976 to 1978, the annual statewide death loss of newborn calves has ranged from 10% to 12%.

Neonatal calf losses in Idaho beef herds have been associated with various management practices, environmental stresses, and pathogenic agents. Ranchers identify these death losses with clinical observations ranging from instant death to diverse scour conditions. The complex condition has been labeled as the weak-calf syndrome in specific areas of Idaho.

The Pegram Project

The Pegram Project, funded by the Idaho Beef Council, was developed to attack the death-loss problem. Three ranches located near Pegram, Idaho, a ranching community in Bear Lake County near the Idaho-Wyoming border, cooperated in the 3-year study.

The three cooperative ranches had several similarities in their normal operations: all three grazed on similar summer range and all wintered their cattle at Pegram within a ten-mile area. The combined three ranches managed 1,300 to 1,600 mother cows plus replacement heifers and summer yearlings.

Previous ranch records from the spring of 1974 through the spring of 1976 reflected a calf death loss of 16% to 22%.

Typically, the Pegram ranchers calved their range beef cows in open meadows, depending upon sagebrush, willow, hills, and other natural windbreaks to provide protection for newborn calves against environmental stresses such as wind, rain, snow, and cold.

There were many reasons for this high percentage of newborn calf deaths, but, frequently, newborn calves became chilled, weakened, and died as a result of prolonged exposure to cold and extreme temperature variations. Calf scours were invariably observed in many sick calves. Calves that survived the scours were usually set back in growth and did not fully recover their thriftiness. Typical clinical symptoms of diarrhea, fever, and dehydration were observed in most calves that died. In most cases, the medication was futile and added financial costs.

OBJECTIVES AND RESULTS

The primary objectives of the Pegram Project were to:
- Prioritize and implement management changes to minimize death loss in young calves.
- Develop an effective, economical disease-control program to minimize death loss in young calves.

During the 3-year Pegram Project study, death loss was reduced to 2.8% during the calving period from birth to turnout on summer range--from approximately February 15 to June 1, or about 100 days. In previous years, the loss had been as much as 22%.

The increased survival rate of calves born during the Pegram study can be attributed to a number of management changes and calving techniques that have been applied by ranchers, veterinarians, and University of Idaho staff associated with the Pegram Project.

SICK CALF TREATMENT

The calf treatment system is designed to correct common stress factors influencing neonatal calf mortality. These stress factors include the following:
- Pathogenic agents both bacterial and viral
- Chill and body heat
- Starvation
- Dehydration
- Acidosis
- Electrolyte losses and imbalances.

Calf mortality is related to a complex interaction of stress factors but is most commonly associated with calf scours. Nearly everyone involved in the cattle industry has developed his own "sure fire" remedy for curing calf scours. The scientific community gave limited help because of fragmented and incomplete research data that required continued study.

The calf treatment system devised on the Pegram Project has been successful for the Pegram ranches. The response has been dramatic, but is not limited to a single approach. The managment combinations dealing with the stress factors under field conditions include:
- Complete diagnostic and problem analysis
- Nonspecific colostrum antibody protection
- Specific immunologic calf response
- Nutrient fluid therapy
- Selective medication

It is recognized that this refined herd-managment program may not reduce all calf mortality on all Idaho ranches, but when applied judiciously it should increase calf survival.

CALVING MANAGMENT

The University of Idaho, Department of Animal Sciences, employed a field technician on the Pegram Project. The three technicians employed (one each year) were animal science majors at their senior level, or graduates with a bachelor of science degree. The technicians lived on the Project in a home provided by the ranches. The normal employment period extended for approximately six months during the spring calving season. A pickup truck and two-way radio became standard equipment for hauling supplies, instruments, and field equipment.

The technicians were responsible for collecting field data, conducting the study, and assisting ranchers with treatments and calving management by routine daily ranch visits. During the calving season, technicians were on call 24 hours per day and developed daily work schedules compatible to both the field study and ranch management.

The expertise and knowledge of the technicians were shared with the ranchers. As each new practice, technique, treatment, or diagnosis was introduced, a complete explanation or demonstration followed with the ranchers either as a group or on a one-to-one basis. As a result, the cowboys working on the ranch improved or acquired new skills associated with calving and cattle management. The following techniques were demonstrated:

- Calf bleeding
- IV administration of fluids
- Epidural anesthetics
- Cleaning and sterilization
- Heifer spaying
- Suturing vaginal prolapse
- Rumen fluid therapy
- Caesarean surgery
- Calf extraction and delivery
- POP (oxytocin) administration
- Postmortem examination procedure
- Vaccine selection and routes of administration
- Field calving and medication records and their use
- Medication treatment systems

Ranchers were trained to employ these techniques and improved their skills of implementation so that the practices now have become routine on all three Pegram ranches.

Technicians were instructed to employ the maximum sanitation possible under field conditions. After each ranch visit, technicians washed their boots in soapy water. Field instruments were routinely scrubbed in soapy water and disinfected properly. All equipment, including the pickup and the technician, remained clean at all times. As a result of this effort, ranchers have become extremely conscientious about sanitation and cleaning of equipment.

Technicians assembled some 50 pieces of practical field equipment essential to calving management and a complete inventory list was maintained. Ranchers were trained by the technicians to maintain and properly use each piece of equipment and are now acquiring duplicate equipment for their individual ranch use.

HERD HEALTH PROGRAM

All death losses occurring in various ages of cattle have been confirmed by necropsy and laboratory diagnosis. Routine examination performed by cooperating laboratories included gross pathology, blood serology, bacterial cultures and virus isolation.

The following diseases have been identifed:
- Clostridium
- IBR
- BVD
- Chlamydial infection (Epizootic bovine abortion)
- Pasturella
- TEM (Thromboembolic Meningoencephalitis)
- Mycotic gastro-enteritis
- Polyarthritis
- Congenital malformation
- Rota-virus
- Weak calf syndrome

To combat these diseases and others, a disease prevention program has been designed for and adapted by the Pegram ranchers. The herd health program includes a complete breakdown of all recommended vaccine combinations, their estimated price, and dosages, according to the the season and to the age of the cattle. In addition to the herd health program, forms have been developed for maintaining records regarding drugs and vaccine supplies, immunization records, and treatment records. The program has expanded the disease prevention by adding immunization for existing disease organisms.

The program of preventive medicine initiated with the Pegram herds has improved the herd health and has reduced the annual death loss of cattle in all stages of maturity. In addition, to be effective, a vaccination program also should be economical and practical to administer.

As the need may arise, ranchers may elect to revise an adopted program of preventive medicine based on herd history, health records, and veterinary service supported by laboratory diagnosis. The introduction of new vaccines or the combination of vaccines also may alter the vaccination program.

WINTER COW NUTRITION

Samples of all harvested winter forages were tested for protein and phosphorus from each of the ranch units in the Pegram Project. The wide range in the crude protein content of these harvested forages provides an opportunity to match the forage quality with the nutrient needs of the animals being fed. The crude protein requirements for dry, pregnant, and mature cows is about 6% of the diet. The crude protein requirement will range up to approximately 11% during lactation. It is obvious from this analysis that the lower-quality or lower-protein hay could be fed in early fall as the cows approach the last third of pregnancy. Hay with crude-protein levels greater than 10% could be fed through the last third of pregnancy and during lactation. This would simplify the supplementation program that would be needed. Phosphorus levels would still be inadequate and a phosphorus supplement should be provided to all cows during the winter and early spring. A mixture of 75% trace mineral salt and 25% monosodium phosphate offered free-choice in covered mineral feeders would serve this purpose.

Because of the great variability in crude protein and phosphorus content of these forages, a supplement was formulated to be fed in late gestation and during lactation to those cows receiving hay low in crude protein and phosphorus. This supplement was formulated to be fed free-choice with consumption at approximately 2 lb per head per day.

PHYSICAL COW EVALUATION

The annual physical evaluation of the cows allows the producer to keep an accurate count of his cattle. From this he can develop income and profit information related to the number of animal units he has on his acreage. The records, when kept over a period of years, will afford him a reasonable estimate of how long a cow will remain productive under his management scheme. Each year he will have a complete record of the physical condition of his herd on an individual basis. From these records, he can plan his culling for the following summer after calving and project the number of replacements he will require to maintain the herd size. Annually, the cows with reproductive disorders or disease can be removed from the herd. Knowing the status of the herd allows the producer to project ahead for the cow herd and to look back to find the reasons cows are culled.

Annual physical evaluations are very beneficial in planning for nutritional needs. Cows of various ages and physical conditions require different nutritional care to ensure maximum production. After evaluation, the herd can be separated to allow group feeding for the most efficient utilization of feedstuffs. Eliminating nonpregnant cows and late calvers at the earliest practical date will ease the nutritional burden of the herd and make the shorter calving period more efficient and manageable. Reduced death loss at calving and during the first few weeks of life can be anticipated. A more uniform calf crop can also be weaned if the herd calves during a shorter interval. All things considered, physical examinations to include pregnancy diagnosis will simplify calving management, increase the calf crop, lower nutritional expenses, and produce a more readily marketable calf crop.

CALVING FACILITIES

Few Idaho ranches have adequate calving facilities to provide ample protection for newborn calves and nursing cows. Because a calving facility suggests confinement many ranchers fear it may cause the spread of infectious disease organisms. However, on ranches without calving facilities, cows are frequently confined to relatively small fenced-in areas for calving. Neonatal calves without adequate protection and facilities in the confined area are prone to death losses caused by environmental stress and invading infectious pathogenic organisms. Thus, the scours associated with poor management is the cause of a high death loss.

During the first year of the Pegram study, it was generally agreed by ranchers and university staff that improved calving sheds and facilities would contribute substantially to a reduction of newborn calf losses. The University of Idaho developed plans for a 30ft x 105ft type calving barn, which is both practical and economical in any ranch operation. This well-designed calving barn provides adequate protection and ensures proper lighting and suitable ventilation. The practical arrangement of the barn reduces labor and facilitates ease of handling of both the cows and newborn calves. Three basic areas of the calving barn include holding, working, and warming areas.

PUBERTY INDUCTION IN HEIFERS

Previous evaluation of calf losses at the three Pegram ranches suggested the calving season should be advanced so that calving would occur during more favorable weather conditions. Advancement of the calving season would more likely occur by breeding heifers earlier, rather than by working with the cow herd.

Breeding heifers earlier poses problems. Many heifers born late in the calving season will probably be light in weight and be prepuberal at the next desired breeding time. For this reason, it is desirable to induce puberty.

BEEF CATTLE IMPROVEMENT (BCI)

At the beginning of the Pegram Project, all mature cows were individually identified with a numbered tag. Each calf born was also identified with a numbered tag, and complete cow-calf records were maintained on the Idaho BCI program. Complete weaning-weight records were only obtained in 1977. Four factors have been discovered through the BCI records:
- Calving seasons on some ranches over 145 days.
- High incidence of open cows.
- Small light-weight heifers being returned to the cow herd.
- Light calf weaning weights.

Calves not born because of delayed calving contributes to calf loss even though the calves may not have died. Hopefully, the BCI records emphasize the importance of growth and development of replacement heifers and the reduction of open cows through improved management.

HERD RECORDS

The printing and distribution of ranch record forms for herd management would include:
- Physical cow evaluation.
- Treatment record for sick animals.
- Immunization record for sick animals.
- Drug and vaccine supply records.

Currently, these record forms are in use by the Pegram ranches. They have also been requeted for use by other ranches in southeastern Idaho.

RANCHER SKILLS

Calving-management workshops have been conducted by the University of Idaho extension agricultural agents in cooperation with the Pegram ranches, practicing veterinarians, and extension livestock specialist in the southeastern Idaho area.

The practical application of the Pegram information has been used on a pilot basis to test the acceptance by ranchers. The potential training program involves cooperative efforts of extension staff and practicing veterinarians utilizing the support materials, audiovisual aids, and live demonstrations of various techniques.

The data collected and management systems established provided support information for development of the calf-survival section to the Idaho Total Beef Program. Idaho Total Beef Program is a Cooperative extension thrust to improve reproductive management in Idaho's beef cow herds.

The widespread distribution of this information is to be correlated with introduction of TBP calf-survival symposium. Further distribution will be stimulated by county extension beef program planning and development committees. Several Idaho county extension programs requested release and use of this information for county use in early 1980.

The Pegram Project is described and published in more detail in the final report to the Idaho Beef Council. the report entitled, "Management Systems to Minimize Death Loss in Young Calves (Pegram Project)," is available by contacting the Idaho Beef Council, 2120 Airport Way, Boise, Idaho 83705.

Part 9

FEEDS AND NUTRITION

55

PROTEIN REQUIREMENTS OF BEEF CATTLE

Allen D. Tillman

It has been known for more than 150 years that animals require protein for survival and growth. However, the concept that amino acids are essential for growth of non-ruminant animals and that these can be categorized as dietary essential and dietary unessential amino acids has developed in the present century. With the discovery that threonine is a dietary essential amino acid, there came rapid developments that made it possible to feed amino acids to nonruminants on a quantitative basis. However, the feeding of protein to ruminant animals is more complex because of the intermediary effects of microbes in the reticulo-rumen. For this reason, the protein and amino acid requirements of ruminants remain much more vague than in the case of nonruminants. This paper discusses some of the factors that affect the protein requirements of ruminants and makes suggestions for possible improvements in the feeding standards for protein for ruminants.

IMPORTANCE OF PROTEIN

The primary purpose of dietary protein is to supply amino acids to the metabolic pool in the animal's body. Since amino acids are the basic building units for all animal cells and tissues, including blood, vital organs, brain, nervous tissue, muscles, skin and hair, there must be a ready supply of these in the body if new cells and tissues are to be formed. In addition, all protein secretion in the body, including mucin, hormones, enzymes, milk and others, requires a special assortment of amino acids for its synthesis.

Young animals require high levels of dietary protein because muscle and other high-protein tissues are being formed at a fast rate. Likewise, lactating cows are producing milk, that contains 27% protein in the milk solids; thus their protein requirements, too, are high.

FACTORS AFFECTING REQUIREMENT OF RUMINANTS

The digestive trace of ruminants differs markedly from that of nonruminants (figure 1); the main differences are in relative sizes and makeup of the stomach. In comparison to a pig weighing 190 kg and whose digestive tract has a capacity of 27 liters, a cow weighing 575 kg has a capacity of 260 liters. The cow weighs three times as much, but has a digestive tract capacity ten times that of the pig.

The stomach of the ruminant animal is made up of four compartments--the rumen, reticulum, omasum, and abomasum, the true stomach. In brief, the reticulorumen is a large and an effective "fermentation vat" that is situated at the head of the digestive tract. The fluid contains billions of microorganisms, primarily bacteria, which have prior access to all nutrients entering this tract. Therefore their fermentations take place before dietary nutrients and nutrients produced by the microbes can undergo enzymic digestion as a result of enzymes produced by the ruminant's body in the abomasum and small intestine. The microbes have the abilities to degrade dietary nutrients and then to synthesize new nutrients required for the building of the bodies of new microbes. For example, they can degrade dietary crude fiber to produce the simple sugars and then to utilize these sugars as an energy source or for metabolic essentials. In turn, they make available to ruminants the end-products of their carbohydrate digestion and metabolism--the volatile fatty acids (VFA)--that the host ruminant can use either as an energy source or as metabolic essentials for synthesizing body constituents. Also, the microbes can degrade proteins to peptides and then to amino acids, which are used to build microbial proteins. However, most the resultant amino acids are further deaminated by the microbes producing ammonia and branched-chain fatty acids. Some of the ammonia may be used to synthesize new amino acids while the remainder is absorbed into the body (figure 2). Also, the microbes can utilize nonprotein nitrogen (NPN) compounds as an ammonia source and combine these with carbon skeletons to form amino acids that, in turn, are used to synthesize microbial protein.

The omasum is a main absorptive area in the ruminant's stomach for small molecular products, which are produced by microbial actions. Afterward, the food particles and dead microbes move from the omasum to enter the abomasum, which has functions quite similar to those in the stomach of nonruminant animals. All digestion in the abomasum and small intestine is mediated by body enzymes in both ruminants and nonruminants; the small intestine in both is a major absorptive site for amino acids that were freed by digestion in the stomach and small intestine. Therefore, the ruminant animal has two major sites for the synthesis of proteins--the reticulorumen and body tissues, while non-ruminants have only one, the tissues.

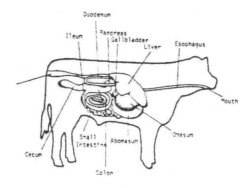

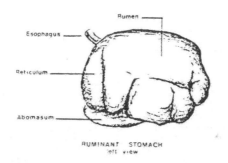

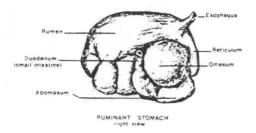

Figure 1. The digestive tract of cattle (from Ensminger and Olentine, 1978).

502

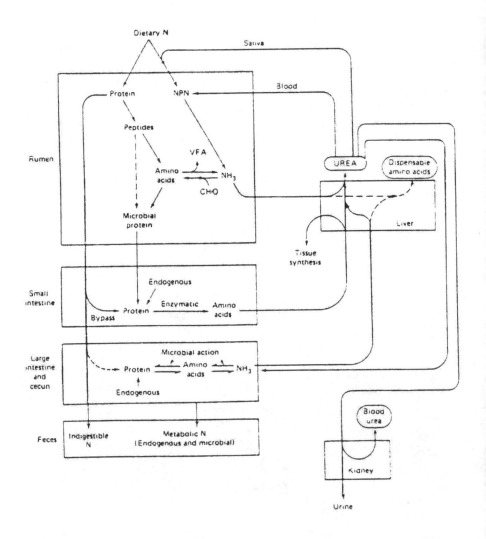

Figure 2. Pathways of digestion, absorption and metabolism in the ruminant. (From Maynard, Loosli, Hintz and Warner, 1979.)

MICROBIAL REQUIREMENTS

The reticulorumen microbes require ammonia, energy, carbon skeletons (from carbohydrates and amino acids) and minerals for their multiplication or growth. A deficiency of any of the above essentials will limit their growth. Ammonia can be supplied as dietary true protein, body protein (via saliva), dietary NPN, and body NPN (via saliva or blood); NH_3 is transferred across the reticulorumen wall. Therefore, an ammonia deficiency can result because of either a true protein deficiency or from the feeding of dietary true proteins, which are poorly degraded by the microbes; these are called "high bypass" proteins. A deficiency of ammonia in the reticulorumen reduces the rate of cellulose (fiber) degradation and causes reduced feed intake in the host animal, probably because the reduced rate of digestion in the reticulorumen.

There seems to be no general agreement on the optimum level of ammonia in the reticulorumen fluid. However, Roffler and Satter (1975) reported that 5 mg of ammonia-nitrogen per 100 of reticulorumen fluid will promote a satisfactory growth rate in the microbes.

Bryant and Robinson (1963) found that most bacteria in the reticulorumen will grow if ammonia is the only source of nitrogen. However, added protein stimulated microbial growth, presumably through the incorporation of peptides and amino acids into microbial protein. It is possible that added protein provided a source of branched-chain fatty acids, which also will stimulate microbial growth when ammonia is the source of nitrogen (Bryant, 1973). Many workers have found that sheep, beef cattle, and dairy cattle can grow and produce on purified diets in which urea is the sole nitrogen source. However, growth rates and level of performances are poor. Urea is the principal source of NPN for practical ruminant feeding. Successful utilization of this compound requires a knowledge of urea utilization and of ration factors affecting it. Steps in urea utilization appear to be the following (NRC-1976):

1. Urea $\xrightarrow[\text{urease}]{\text{microbial}}$ $NH_3 + CO_2$

2. Carbohydrate $\xrightarrow[\text{enzymes}]{\text{microbial}}$ VFAs + keto acids

3. NH_3 + keto acids $\xrightarrow[\text{enzymes}]{\text{microbial}}$ amino acids

4. Amino acid $\xrightarrow[\text{enzymes}]{\text{microbial}}$ microbial protein

5. Microbial protein $\xrightarrow[\text{and small intestine}]{\text{body enzyme in abomasum}}$ amino acids

6. Amino acids $\xrightarrow[\text{enzymes}]{\text{tissues}}$ body proteins

Step #1 proceeds rapidly, and if there are not sufficient carbon skeletons present, the reticuloruminal ammonia level rises and it is rapidly absorbed in the body, possibly causing ammonia toxicity if the level of urea is excessive. Toxicity can be avoided and good utilization can be obtained if good management practices are used: added urea must be well-mixed into the other ration ingredients. Rations should not contain more than 1% urea, dry matter basis. Another safeguard is to use urea at a level to supply no more than one-third of the ration protein. In general, these precautions are easy to follow. After many years of research, the Iowa workers (Trenkle, 1982) have proposed a system that provides excellent guidelines on the amount of urea one may feed with success in ruminant rations as follows:

$$\text{UFP (g/kg feed DM)} = \frac{10\% \text{ TDN per kg feed DM} - \text{g protein degraded per kg feed DM}}{2.8}$$

Wisconsin scientists (Satter, 1982) have developed a system that is based upon rations that supply an ammonia level to provide 5 mg NH_3-N per 100 ml of reticulorumen fluid; their calculations are shown in table 1. Data were obtained by use of a multiple regression equation:

$$NH_3\text{-N (mg/100 ml)} = 38.73 - 3.04\% \text{ CP} + 0.171 \text{ CP}^2 - 0.49\% \text{ TDN} + 0.0024 \text{ TDN}^2$$

The data in table 1 represents the percent of dietary crude protein at which reticuloruminal NH_3-N levels reach 5 mg/100 ml of fluid. The rations contain variable levels of TDN and added NPN. It is obvious that the addition of NPN to any ration containing less than 60% TDN would be useless. Even with high-concentrate diets, those containing about 75% TDN, the permissible level of urea for dairy cows, would be around 1%

When NPN is used in ruminant rations, one must evaluate the ration especially for its phosphorus, potassium, and sulfur contents. Protein sources that are replaced by urea and grain are usually excellent sources of all of these essential minerals--essential for the microbes as well as for the host animal.

TABLE 1. UPPER LIMITS FOR NONPROTEIN NITROGEN UTILIZATION[a]

% CP in ration DM before NPN is added	TDN (DDM)	TDN and digestible dry matter (DDM) (expressed as percent of dry matter)[a]					
		55-60 (59-63)	60-65 (63-68)	65-70 (68-72)	70-75 (72-76)	75-80 (76-81)	80-85 (81-85)
		(Percent crude protein after NPN addition)					
8		NO[b]	10.0	10.5	10.9	11.2	11.4
9		NO[b]	10.4	10.9	11.3	11.6	11.8
10		NO[b]	10.8	11.3	11.7	12.0	12.2
11		NO[b]	11.2	11.7	12.1	12.4	12.6
12		NO[b]	NO[b]	12.1	12.5	12.8	13.0
—		—	—	11.4[c]	12.2[c]	12.8[c]	13.6

[a]From Satter (1980) TDN values from NRC (1978). DDM values were calculated.

[b]No benefit would be obtained from NPN addition.

[c]Dietary crude protein where ruminal NH_3 begins to accumulate when only plant proteins are in the diet.

Any discussion of reticuloruminal ammonia and its sources must go hand-in-hand with a discussion of bypass protein sources. In general, one can classify protein feeds as having low, medium, and high bypass dietary protein. Low bypass protein feeds are those in which 60% to 80% of the protein is degraded (20% to 40% bypass) and include soybean and peanut meals; medium bypass protein feeds (40% to 60% bypass) include brewers dried grains, corn grain, cottonseed meal, and dehydrated alfalfa meal; high bypass feeds (60% to 80% bypass) include blood meal, corn gluten meal, fish meal, and meat meal. These are only estimates as the extent of bypass in each is affected by processing conditions (Thomas et al., 1982), heating (Sherrod and Tillman, 1964), dietary variables (Chalupa, 1982), level of feed intake (Chalupa, 1982; Owens and Gill, 1982), reticulorumen fluid conditions, and microbial variables (Allison, 1982). Unfortunately, these factors are not yet fully understood; therefore, the use of bypass data in the future protein requirement tables will require much additional definitive research.

Solubility of feed proteins in reticulorumen fluid or other solubles has been proposed as an index for the amount of bypass protein. However, this system is not yet completely reliable for all ration conditions and protein systems (Satter, 1982).

When high bypass proteins dominate the ration, reticulorumen microbes usually need supplemental ammonia nitrogen.

It appears that the digestibility and biological value of microbial protein is lower than that of bypassed true protein. Therefore, theoretically, one could make a case for developing a practical feeding system in which high bypass protein would be fed to meet most of the tissue demands for amino acids and then to use a NPN source to be fed at a level to meet the NH₃-N requirements of the microbes. Such a system would theoretically permit the feeder to have the best of "two worlds." Unfortunately, in practice there are many constraints on the system as follows: Some treatments that are used to increase bypass protein also reduce protein digestibility in the abomasum and small intestine, thus there is much waste protein in the feces. Also, it has been shown that there is differential degradation of dietary proteins; therefore, the bypassed protein has a different assortment of amino acids than the original feedstuff. Much research is needed before such an ideal can be put into practice, particularly as the level of concentrates, reticuloruminal fluid pH level of feed intake, and other factors affect the bypass phenomenon.

POSTRUMINAL RUMINAL PROTEIN SUPPLY

The proteins found in the abomasum and the small intestine represent the sum of bypass feed protein plus microbial protein that is synthesized in the reticulorumen microbial growth; thus the amount of microbial protein is affected by specific conditions in the reticulorumen, including pH, dilution rate, limiting nutrients such as ammonia-N, phosphorus, potassium, sulfur, energy, or organic matter, which is digested in the reticulorumen and others. The presence of digestible organic matter is probably one of the most important of the above factors. Microbial crude protein synthesis in the reticulorumen is estimated to range from 77g to 270g per kg of organic matter fermented. The higher values usually are found in the high-roughage rations, while lower values are found in high-concentrate rations, probably due to lower growth of the microbes. Also, grain fermentation causes a lower pH in the reticuloruminal fluid, which promotes greater bypass of the dietary proteins.

Digestibility of bypassed protein varies with feedstuff, processing methods, and heat treatment. It is estimated that approximately two-thirds or more of both microbial and bypass protein are digested in the small intestine. Heat damage is particularly damaging to the digestibility of both forages and silages. The true digestibility of feed protein can be approximated by two techniques--pepsin and the acid detergent fiber digestion. Acid detergent insoluble nitrogen (ADIN) values can be determined as a part of the crude fiber partitioning methods of Van Soest et al. (1970). If heat damage is severe, one may need to increase the ration protein above the stated requirement figures to compensate for the reduced digestibility.

PROTEIN REQUIREMENTS

Protein is so important both as an essential metabolite and as a "high cost" economical factor in the practical feeding of beef cattle that it behooves producers to feed enough dietary protein so as to obtain expected performance and yet maximize production profits by feeding only the levels of protein recommended in feeding standards. At the same time, it behooves animal scientists, who are responsible for establishing and publishing feeding standards, to publish the most accurate data possible. As indicated earlier, the situation is much more complex in ruminants than in nonruminants because of the degradation and synthetic actions that take place in the fermentation vat, the reticulorumen of ruminant animals, and the effects of these actions upon the utilization of dietary protein and NPN compounds.

In general, two primary methods are used by scientists to establish protein feeding standards—the factorial method and the use of feeding trial data. This paper examines some of the present methods of setting protein feeding standards and considers relative strengths and weaknesses of these in regard to how well each takes into account the effects caused by reticulorumen microbes and by other factors.

The NRC (1976) used the factorial method and published beef cattle protein requirement data that are used for various purposes (growing-finishing steer calves and yearlings, growing-finishing heifer calves and yearlings, and the breeding herd). Requirements were estimated for optimum production; the factorial method was based on the endogenous urinary nitrogen ($gN = 0.12 \ W^{0.75}$), N losses through the hair and skin cell ($gN = 0.02 \ W^{75}$), metabolic fecal N ($4g$ N/Kg feed DM), and composition of gain. Steers weighing 100 kg were considered to contain 18% protein in their gains; those weighing 50 kg, 9%. Corresponding values for heifers were 18% and 7%. There were, of course, intermediate values.

Nitrogen values were converted to digestible protein values by (1) multiplying digestible N by the factor of 6.25, and (2) by dividing it by a biological value of 0.775. Total protein requirement (X) was expressed as a function of digestible protein (Y) by the equation: $Y = 0.877X - 2.64$, or more simply as:

$$X = \frac{Y + 2.64}{0.877}$$

Because the protein requirements for dry, pregnant cows and for cows nursing calves appeared to be too high when this procedure was used, the authorities used the NRC (1970) (revised edition) figures. The 1976 standards made provisions for using urea and other NPN sources as protein substitutes. It is of interest to quote one section from the NRC (1976) edition: "Protein and amino acid requirements of beef cattle and the influence of protein solubility on these requirements are being investigated currently. The results

of this research may influence future recommendations."
Another NRC edition of "The Nutritive Requirements of Beef
Cattle" is slated to appear in late 1982 or early 1983, and
the new standards in that edition will be a great improvement over those published in 1976.

The Agricultural Research Council (ARC) of Great
Britain has revised its earlier edition of "The Nutritive
Requirements of Ruminant Livestock," and these standards
were published in 1980. They are new standards and do take
into account the effects of reticulorumen fermentations upon
the total protein requirements. In general, the ARC
approach considers the components affecting total N requirements of animals. Tissue amino acid N requirements are calculated factorily using an approach similar to that of the
NRC (1976). However, as we saw earlier, reticulorumen
actions on dietary protein and NPN compounds can modify the
total N requirement. Because of this, the total N requirement in their calculations is the sum of the reticulorumen
degradable N (RDN) and the reticulorumen undegradable N
(UDN) requirement. The derivation of the equations takes
into consideration the metabolizable energy content of the
diet, but this will not be considered here. It is a good
system. However, their feeds and ration conditions differ
widely from those used in the U.S., thus I feel that it
would be of more value to consider several different
American models for determining protein requirements of
ruminants.

Preston (1982) suggested that it would be desirable to
develop a single mathematical expression that could consider
all the factors affecting protein requirements of
ruminants. This becomes a difficult task because of wide
variations in animals, feeds, ration conditions, production
performances, etc., found in the U.S. However, there are
steps in this direction. For example, the first step in the
measure of utilization of dietary crude protein is to determine its digestibility. For this reason, digestibility
trials became an important part of ruminant research, and
the use of digestible protein in feeding beef cattle became
common in most feeding standards. However, we seem to have
forgotten that DP is greatly affected by the metabolic fecal
nitrogen level (Blaxter and Mitchell, 1948) and the later
work by Glover et al. (1957) which demonstrates that DP in
reality is a function of the level of crude protein in the
ration. Therefore, if DP is used in expressing protein
requirements, it would be more accurate to calculate DP,
using the following equation (Glover et al., 1957):

$$COD = 70 \log (\text{percent CP}) - 15$$
$$\text{where}$$
$$COD = \text{coefficient of protein digestibility}$$

SOME AMERICAN MODELS

Trenkle (1982) described the Iowa State University model, which is based upon metabolizable protein. Details of equations used in the model were described in earlier Iowa publications (Burroughs et al., 1974) and will not be discussed here. In general, they consider that microbial protein is related to synthesis in the reticulorumen that , in turn, is related to the level of ration TDN (10.44% of total TDN is converted to microbial protein) and the microbial protein is estimated to have a true digestibility of 80%; the model is shown in figure 3. It will be noted

RETICULORUMEN

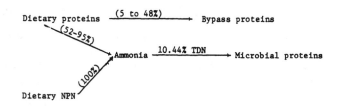

SMALL INTESTINE

Bypass proteins —(90%)→ Amino acids

Microbial proteins —(80%)→ Amino acids

METABOLISM

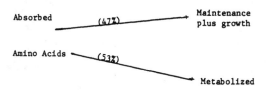

Figure 3. The Iowa State University Metabolizable Protein Model (Trenkle, 1982).

that the upper part of the model relates to the urea fermentation potential (UFP) that was discussed earlier. UFP, of course, expresses the balance between ammonia utilization and microbial protein production in terms of urea; other sources of ammonia-N can fit into the equation. UFP values can be positive or negative. If negative, the addition of ammonia-N will not be useful because the N requirements of the microbes have already been met.

The variables that account for different metabolizable protein (MP) and UFP values are levels of TDN and protein, extent of degradation of feed protein, and the digestibility of the bypass and of the microbial protein. The equations are:

1. MP (g/kg feed DM) = P$_1$ x 0.9) + (P$_2$ - 15 x 0.8)

2. UFP (g/kg feed DM) = $\dfrac{10.44\%\ \text{TDN/kg feed DM} - \text{g protein degraded kg feed/DM}}{2.8}$

where

P$_1$ = g undegraded (bypass) feed protein/kg feed DM

P$_2$ = g microbial protein/kg feed DM, postreticulorumen

15 = g post reticulorumen protein required to supply the metabolic fecal N

0.9 = true digestibility (%) of bypass protein

0.8 = true digestibility (%) of microbial protein

The net protein requirements were estimated factorially using a method similar to the NRC (1976). The figures in the model estimate that about 53% of the metabolizable protein is metabolized, while 47% is available for maintenance and growth.

The greatest advantage of the Iowa system is the fact that the metabolizable protein portion of the standard permits the determination of the most economical protein supplements for feedlot cattle. it also presents a means for determining whether urea will be useful in beef cattle rations; this is another step forward. The standard has been quite accurate for measuring the value of treating proteins for less extensive degrading in the reticulorumen. These high proteins, of course are usually more efficient in supplying amino acids for tissue synthesis.

The Wisconsin workers (Satter, 1982) have developed a metabolizable protein system for dairy cattle. Since this system is keyed to reticulorumen ammonia concentration, it sidesteps the need for determining protein degradability of low protein feeds. Some feel that the Wisconsin system might be more reliable than other methods until there is better information on factors that affect the degradation of feed protein and subsequent reticulorumen synthesis of microbial protein.

The conceptual basis of the system can be illustrated by figure 4 which shows reticuloruminal actions on dietary protein and NPN sources. As regards the concept, it does

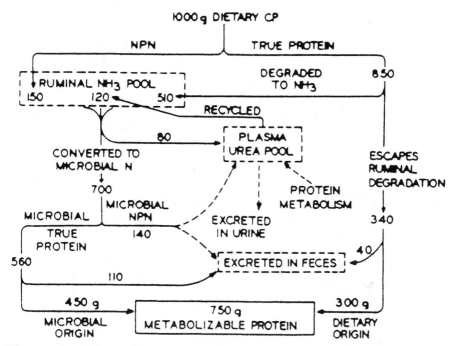

Figure 4. Schematic summmary of metabolizable protein calculations. (From Satter, 1982.)

not really matter to the ruminant whether dietary protein is degraded as long as the reticulorumen microbes are able to utilize all of the ammonia produced. In either manner, dietary or recycled N ends up as protein that is presented to the small intestine. However, this is not true if ammonia production exceeds the ability of the reticulorumen microbes to convert ammonia to microbial protein. When this happens, if complete, only dietary true protein that escapes degradation will contribute to the amino acids in the small intestine. Also, in the above situation, ammonia from NPn will have no value. Therefore, the method concentrates on determining the point of "ammonia overflow," and the Wisconsin workers (Roffler and Satter, 1975a and 1975b) assumed that a level of 5 mg NH_3-N/100 ml of reticulorumen fluid will support maximum yields of microbial protein. Then they developed a multiple regression equation to predict reticuloruminal NH_3-N concentration as follows:

$$NH_3\text{-}N \ (mg/100 \ ml) = 38.73 - 3.04\% \ CP + 0.171 \ CP^2 - 0.49\%$$

$$TDN + 0.0024\% \ TDN^2$$

When part of the ration is replaced by NPN compounds, the amount of true protein escaping degradation in the reticulorumen is reduced; the fraction going to produce ammonia is decreased, adding to ammonia overflow. As the level of NPN increases in the ration, the point of zero utilization of NPN is reached at a corresponding lower protein content of the ration.

Table 1 shows the integration of the effects caused by dietary contents of crude protein and TDN, as well as NPN substitutions for true protein on ruminal ammonia. As indicated earlier, the data represents the percent of ration CP at which reticuloruminal ammonia reaches 5 mg/100 ml in rations containing variable amounts of TDN (or digestible DM) and added NPN. These values represent the upper limits to which the total ration CP (DM basis) can be raised by adding NPN. In other words, NPN additions above these CP levels will result in poor or no utilization of the added NPN.

A schematic summary of metabolizable protein (MP) calculation in the Wisconsin study is shown in figure 4. Because this system is primarily for dairy cows in production, further discussion on derivation of the various equations and concepts is not indicated. It is obvious that the MP values could be converted by use of appropriate factors to net protein values for milk production. In fact, Chalupa (1982) of the University of Pennsylvania has proposed the following factors for an MP and NP scheme, as shown in table 2.

Chalupa (1982) proposed that many of the discrepancies concerning optimum concentrations of dietary protein can be reconciled on the basis of protein degradability and energy supply. Using a basal diet containing 1.7 Mcal/kg net energy, 10% crude protein that had a protein degradability of 40%, he calculated (table 3) the effects of varying protein levels (12.5, 15.0, 17.5, 20.0, 22.5, and 25.0%) and different protein degradabilities (50, 60, 70) upon milk production. When dietary protein was 40% to 50% degradable, increasing the CP level to 15% resulted in large increases in milk yields, but beyond 15% CP, additional responses were prevented by a deficiency of energy. When dietary protein was 60% to 70% degradable, milk production was predicted to respond to higher levels of dietary protein.

The Cornell University model (Van Soest et al., 1982, and Fox et al., 1982) considers rumen dynamics as well as the net protein needs of ruminants. Their submodel, which considers only the reticulorumen, dynamically models protein digestion and is based upon rates of passage and digestion. The actual operation of the model is dependent upon more precise definition of the nitrogen fractions in feedstuffs than can be obtained from only solubility measurements. If these added values were in hand, the model offers a flexible means for predicting ruminant responses to specific diets.

TABLE 2. DIGESTIVE AND METABOLIC PARTITIONS OF DIETARY
CRUDE PROTEIN[a]

Fraction	Estimation
Crude protein	N x 6.25
Undegradable protein	Variable
Degradable protein	Variable
Microbial crude protein	NE(MCal) x 44
Wasted degradable protein	Degraded-microbial crude
Microbial true protein[b]	Microbial crude x 0.80
Intestional true protein	Undegraded plus microbial true
Metabolizable protein[c]	Intestinal x 0.75
Net protein[d]	MP x 0.70

[a]From Chalupa (1982).
[b]Nucleic acids constitute about 20% of microbial crude.
[c]Assumes that 75% of MP is absorbed.
[d]Assumes that a wastage of 30% of MP in its conversion
to maintenance and production.

TABLE 3. ESTIMATED EFFECTS OF PROTEIN DEGRADABILITY ON
PRODUCTION RESPONSES TO DIETARY PROTEIN[a]

Dietary Crude Protein (%)	Protein Degradability (%)			
	40	50	60	70
	Daily Milk Production			
10.0	19.5[c,d]	18.8[c,d]	18.2[c,d]	17.6[c,d]
12.5	26.8[c,d]	26.0[c,d]	24.9[c]	20.9[d]
15.0	34.1[c]	32.8[d]	28.1[d]	23.3[d]
17.5	35.0[e]	35.0[e]	31.3[d]	25.7[d]
20.0	35.0[e]	35.0[e]	34.4[d]	28.1[d]
22.5	35.0[e]	35.0[e]	35.0[e]	30.5[d]
25.0	35.0[e]	35.0[e]	35.0[e]	32.8[d]

[a]From Chalupa (1982), 600 kg cow fed 20 kg/day of a diet
containing 1.7 MCal/kg net energy.
[b]Production calculated using protein requirements of Chalupa
(1980) and energy requirements of the NRC (1978).
[c]Production limited by supply of degradable protein[c],
undegradable protein[d], and net energy[e].

The method is also dependent upon specific dietary and feed composition data.

The conceptual model of the Cornell scheme is shown in figure 5.

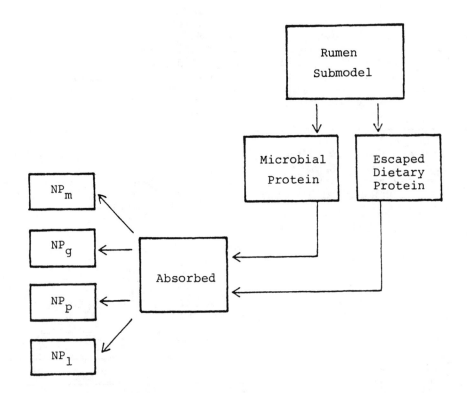

Figure 5. Net protein system for ration formulation: conceptual model. (From Fox, Sniffen and Van Soest, 1982.)

The submodel is used to estimate the amount of micro-bial synthesis; the net protein requirements are calculated using the following equation:

$$NP = (0.235 - 0.00026W)(DG)$$

where

DG = expected daily gain
W = equivalent weight in kg

Equivalent weight (Fox and Black, 1977) represents the animal's weight multiplied by an adjustment factor that takes into account the effect of sex (male, female, castrates) and frame sizes upon expected daily gains. The

Cornell workers state that the use of equivalent weight eliminates the need for separate equations for each cattle type.

Feed protein utilization factors used in the Cornell model do not differ widely from those used by other researchers and are shown in table 4.

TABLE 4. FEED PROTEIN UTILIZATION FACTORS[a]

Fraction	Estimation
Microbial true protein	Microbial crude x 0.80
Metabolizable protein	Microbial true x 0.80
Metabolizable protein	Microbial crude x 0.64
Metabolizable protein	Bypass protein x 0.90
Net protein	Metabolizable x 0.60

[a]From Fox et al. (1982).

Fox et al. (1982) state that further refinement could be made if one would adjust for the effect of rate of gain upon composition of gain. However, the system, as presently used, does adjust for some major errors: environment, use of growth stimulants, frame size, sex, and stage of growth. The system offers much promise of developing into one that would permit one to maximize production and still permit judicious feeding of dietary protein. As with all new systems, it needs further research and testing for improvement. Sensitivity analyses (Fox et al., 1982) have shown that the varying of protein fraction portions has a large predicted impact upon microbial protein yields and protein bypass, both of which affect the net protein yields and protein bypass, both of which affect the net protein vale of the ration.

The Nebraska Growth System and use of the data generated are based upon several points (Klopfenstein, 1982), as follows:
- The NRC (1976) protein requirements for beef cattle are based, primarily, upon rations containing corn silage, hay, grain, and soybean meal.
- The proteins in all of the above feeds are extensively degraded, i.e., low bypass feeds.
- Most ranchers produce their own feeds for beef cattle feeding, except for protein supplements; therefore, their choices of purchased feeds are limited to protein supplements.
- The bypass of feed proteins is a critical factor in protein feeding.

For the above reasons, the Nebraska system relates the value of a particular protein feed to that of soybean meal, which is given a value of 100. It ignores finishing cattle

on high grain rations because NPN can meet the supplemental
N needs of these animals.

The general procedure for protein evaluation in the
Nebraska system is as follows: Basal diets for each test
protein are formulated to contain 61% TDN and 11.5% of crude
protein (N x 6.25).   Sixty percent of the supplemental N is
supplied as natural protein, and 40% by urea.   These diets
were combined with the urea control (100% of supplemental N
from urea) in such way that the supplemental N supplied by
natural protein was 20, 30, 40, 50, or 60% of the total
nitrogen supplement.   The test rations contained 60% corn
silage, 30% corn cobs, and 10% supplement.   Animals fed the
urea control diet were fed ad libitum while the others were
essentially limited to the feed intake of the urea control
animals.   The rations were balanced for protein using the
NRC (1976) standards.   All animals received the same level
of feed above maintenance.
Protein efficiency (PE) was defined as the daily gain
obtained above the gains on the urea control per unit of
natural protein in the supplement.   Regression lines were
calculated in which added gain and amount of natural protein
above the urea control were the factors.   A segmented model
was used that related daily gain above the urea control to
increments of natural protein.

$$\text{Daily gain above urea control} = BX \ (XX_O)$$

where

B = slope of the regression line

X = amount of natural protein fed

X_O = the amount of protein that yields maximum gain

The system is demonstrated in figure 6.   Because both
100% soybean meal and 40 or 50% blood meal furnished more
protein than was needed by the steers, these points were not
used; therefore the soybean plots were for levels of 20, 40,
60, and 80%, and for blood meal only points for 10, 20, and
30 units were used.   Inspection of figure 6 shows that   the
slope of the regression lines is equal to the PE value of
each protein (PE = gain on natural protein - gain on urea
divided by amount of natural protein fed/day).   As SBM was
assigned a value of 100% (slope = 0.635), blood meal would
have a relative value of 313% (1.985 ÷ 0.635).   Using an
average of the 80% and 100% and the 30% and 40% BM, a value
of 0.158 kg of gain above the urea control appeared to be
the maximum achievable gain for these steers when their protein requirements were met.
The same approach was used to test other protein
systems, and the results are shown in tables 5 and 6.   It
will be noted that all of the high bypass proteins (blood
meal, meat meal, dehydrated alfalfa meal, and corn gluten

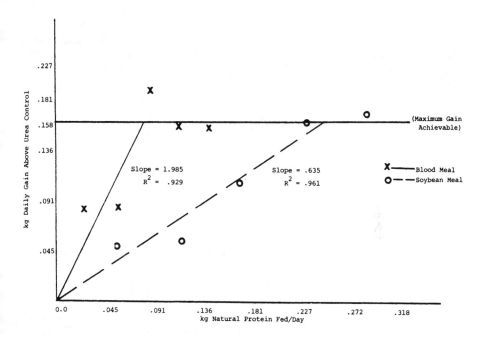

Figure 6. Natural protein fed/day vs. daily gain above urea control. (From Klopfenstein, 1982.)

meal) greatly improved gains of steers fed the high roughage diets. The results can be explained on the basis of previous discussions on bypass protein; therefore, no further discussion on this point is indicated except to note the synergistic effect of combining corn gluten meal with dehydrated alfalfa meal. It seems likely that the amino acid pattern available to the animal tissue for body protein synthesis was better balanced than when either was fed alone. Using this system, the Nebraska workers have found that distillers' dried grains (DDG), DDG with solubles, and brewers' dried grains with urea have relative values of 200, 180, and 188, respectively. Since daily feed intake exerts a tremendous effect upon the amount of bypass protein reaching the abomasum, the actual values reported by the Nebraska workers for the high bypass protein feedstuffs could be too high: Zinn and Owens (1982) varied daily feed intakes of one concentrate diet at levels of 1.2, 1.5, 1.8, and 2.1% of the body weights of the cattle and found that the percents

TABLE 5. EFFECT OF PROTEIN SUPPLEMENTS ON GAINS OF GROWING STEERS[a]

		Protein		Source
Item	Urea	Soybean Meal	Blood Meal	Meat Meal
Daily gains, lb	1.58	1.80	2.00	1.88
Daily feed, lb	13.63	14.02	13.87	13.34
Feed/grain	8.63	7.79	6.94	7.10
Gain/protein	–	0.67	1.43	1.38
Relative value (SBM = 100)	–	100.00	213.00	204.00

[a]From Klopfenstein (1982).

TABLE 6. EFFECT OF PROTEIN SUPPLEMENTS ON GAINS OF GROWING STEERS[a]

			Protein Sources		
Item	Urea	Soybean Meal	Dehy[b]	CGM[c]	Dehy + DGM
Daily gain, lb	0.90	1.02	1.18	1.12	1.20
Daily feed, lb	12.80	12.70	13.30	12.80	13.30
Feed/gain	14.20	12.50	11.30	11.40	11.10
Gain/protein	--	0.38	0.79	0.72	0.92
Relative value (SBM = 100)	--	100.00	208.00	189.00	242.00

[a]From Klopfenstein (1982).
[b]Dehydrated alfalfa meal.
[b]Corn gluten meal.

TABLE 7. THE ECONOMIC VALUE OF HIGH BYPASS PROTEIN IN RATIONS OF GROWING BEEF CATTLE[a]

Source	Protein (%)	Protein efficiency value (%)	Protein source (lb)	Corn (lb)	Urea (lb)	Cost/ton SBM equivalent[c]
Soybean meal	45	100	2000	-	-	$261
Corn gluten meal	62	200[b]	643	1218	139	$181
Brewers' grains	28	190	1619	229	152	$147
Distillers' grains	28	173	1853	12	135	$167
Distillers' grains plus solubles	28	137	2343	-	87	$202
Dehy alfalfa	17	190	2788	-	152	$191
Blood meal, ring dried	85	250	365	1473	162	$182
Blood meal, conventional	85	200	458	1405	137	$179
Meat meal	50	185	885	984	131	$210

aFrom Klopfenstein (1982).
bWhen fed with a high quality protein.
cBased upon the following prices/ton:

SBM = $261   BMRD = $420
CGM = $290   BMC = $350
BDG = $145   MM = $311
DDG = $165   Corn = $121
Dehy = $126   Urea = $200

of bypass protein in the abomasal flow were 43.8, 48.0, 61.9, and 70.6, respectively. In other words, the protein system used in the Oklahoma workers' trial was a medium bypass protein (43.8%) as a daily intake of 1.2% but became a high bypass protein when the daily feed intake was increased to 2.1% of the animal's body weight.

It is suggested that feeding trial data, if properly conducted, yields data which can be used to calculate least-cost rations using the computer. The Nebraska data are excellent for this purpose, especially if similar rations were used. In fact, the Nebraska workers have published (table 7) the calculated value of protein supplements when SBM is given a relative protein efficiency value of 100% and costs $261.00 per ton (2,000 lb). It is obvious that all supplements found in this table are high bypass proteins and that all are cheaper to feed than was soybean meal. Relative price changes could alter this situation quite rapidly.

SUMMARY

The purposes of this paper were to discuss some of the factors which affect the protein requirements of ruminant animals. The primary factor is the makeup of the intestinal tract of ruminants. Because the reticulorumen is large and is an ideal fermentation vat for microorganisms, primarily bacteria, and is located ahead of the major adsorptive sites, omasum and small intestine, dietary proteins can be fermented to form peptides and amino acids, both of which may be incorporated into microbial protein or the peptides can be further degraded to amino acids. The amino acids in excess of the needs of the microbes are deaminated and converted to fatty acids and ammonia, both of which may be used to form new amino acids, if needed. Ammonia in excess is absorbed into the body and converted by the liver to urea, which may be excreted via urine or recycled back into the reticulorumen via saliva or blood; urea is able to pass from blood through the walls of the reticulorumen. In general, if a protein is highly soluble in rumen fluid or other solvents, it is a low bypass protein and most of it may be degraded and converted if needed to microbial protein. Some proteins are resistant to microbial digestion; therefore are high bypass proteins.

Because microbial protein contains a high level of nucleic acid and usually has a lower biological value than bypass protein, it might appear desirable in many feeding conditions to have a high level of bypass protein in the diet. If this were true, it would be desirable if, in establishing protein feeding standards for beef cattle, we had means for predicting the amount of dietary protein which would be bypassed under any particular feeding conditions. Unfortunately, the situation is complex because many factors affect the degradability of dietary protein and the subsequent incorporation of the released ammonia into microbial

protein. The two most important factors are levels of protein and energy (TDN, digestible organic matter, metabolizable energy, or net energy has been used). Other factors
affect the solubility of dietary protein: reticulorumen pH,
dilution rates, rate of feed passage, daily feed intake,
limiting nutrients, and others. Differential degradation of
different fractions of the total protein in feeds also could
make general statements on relative protein qualities or
biological values of microbial vs bypass protein useless.
Moreover, it is also known that some bypass proteins are so
severely damaged by heat and other factors that they are
poorly digested in the abomasom and small intestine of
ruminants. In spite of these reservations, there is no
doubt that reticulorumen dynamics are important and should
be considered in setting protein requirements of ruminants.
New methodology can overcome the difficulties.

Various methods of determining the protein requirements
of beef cattle were discussed, including those of the United
States National Research Council (NRC) and the Agricultural
Research Council (ARC) of Great Britain. In addition, some
American models for the determination of metabolizable protein and net protein levels in feeds were discussed, including that of Iowa State University and those of the Universities of Wisconsin, Cornell, and Nebraska. The merit of
these new approaches were discussed.

522

REFERENCES

Agricultural Research Council (ARC). 1980. The Nutritive Requirements of Ruminant Livestock. Commonwealth Agric. Bureaux, Farnham Royal, England.

Allison, M. F. 1982. Nitrogen Requirements of Ruminal Bacteria. In Protein Requirements for Cattle, F. N. Owens, Editor. Oklahoma State University, Stillwater, Oklahoma.

Blaxter, K. L. and H. H. Mitchell. 1948. The Factorization of the Protein Requirements of Ruminants and of the Protein Values of Feeds, With Particular Reference to the Significance of Metabolic Fecal Nitrogen. J. Animal Sci. 7:351.

Bryant, M. P. and I. Robinson. 1963. Apparent Incorporation of Ammonia and Amino Acids Carbon During Growth of Selected Species of Ruminal Bacteria. J. Dairy Sci. 46:150.

Burroughs, W., A. Trenkle, and R. Vetter. 1974. A System of Protein Evaluation for Cattle and Sheep Involving Metabolizable Protein (Amino Acids) and Urea Fermentation Potential of Feedstuffs. Vet. Med. and Small Animal Clin. 69:713.

Chalupa, W. 1982. Protein Nutrition of Dairy Cattle. Proc. Distillers Feed Conference. 37:101.

Ensminger, M. E. and C. G. Olentine. 1978. Feeds and Nutrition, 1st Edition. Ensminger Publishing Co., Clovis, CA.

Fox, D. G. and J. R. Black. 1977. A system for predicting performance of growing and finishing cattle. Mich. Agr. Expt. Sta. Res. Rpt. 328, pp. 141-162.

Fox, D. G., C. J. Sniffen, and P. J. Van Soest. 1982. A Net Protein System for Cattle: Meeting Protein Requirements of Catle. In Protein Requirements for Cattle, F. N. Owens, Editor. Oklahoma State University, Stillwater, Oklahoma.

Glover, J., D. W. Duthie, and M. H. French. 1957. J. Agri. Sci. 48:373.

Klopfenstein, T. 1982. Nebraska Growth System. In Protein Requirements for Cattle, F. N. Owens, Editor. Oklahoma State University, Stillwater, Oklahoma.

National Research Council (NRC). 1976. Nutrient Requirements of Beef Cattle, 5th Rev. Ed., National Acad. of Sciences, Washington, D.C.

Owens, F. N. and D. R. Gill. 1982. Influence of feed intake on site and extent of digestion. Proc. Distillers Feed conference.

Preston, R. L. 1982. Empirical Value of Crude Protein Systems for Feedlot Cattle. In Protein Requirements for Cattle, F. N. Owens, Editor. Oklahoma State University, Stillwater, Oklahoma.

Roffler, R. E. and L. D. Satter. 1975a. Relationship Between Ruminal Ammonia and Non-protein Nitrogen Utilization. I: Development of a Model for Predicting Non-protein Nitrogen Utilization by Cattle. J. Dairy Sci. 58:1880.

Roffler, R. E. and L. D. Satter. 1975b. Relationship Between Ruminal Ammonia and Non-protein Utilization by Cattle. II. Application of Published Evidence to the Development of a Theoretical Model for Predicting Non-protein Nitrogen Utilization. J. Dairy Sci. 58:1889.

Satter, L. D. 1982. A Metabolizable Protein System Keyed to Ruminal Ammonia Concentration, the Wisconsin System. In Protein Requirement for Cattle. F. N. Owens, Editor. Oklahoma State University, Stillwater, Oklahoma.

Sherrod, L. B. and A. D. Tillman. 1964. Further Studies on the Effects of Different Processing Temperatures on the Utilization of Solvent-extracted Cottonseed Meal Protein by Sheep. J. Animal Sci. 23:510.

Thomas et al, 1982. Estimations of Protein Damage. In Protein Requirements for Cattle, F.N. Owens, Editor. Oklahoma State University, Stillwater, Oklahoma.

Trenkle, Allen. 1982. The Metabolizable Protein Feeding Standard. In Protein Requirements for Cattle, F. N. Owens, Editor. Oklahoma State University, Stillwater, Oklahoma.

Van Soest, J. 1970. Basis for the Relationship Between Nutritive Value and Chemically Identifiable Fractions. National Conf. on Forage Quality Evaluations and Utilization. U1-U19, Lincoln, Nebraska.

Van Soest, et al., 1982. A Net Protein System for Cattle: The Rumen Submodel for Nitrogen. In Protein Requirements for Cattle, F. N. Owens, Editor. Oklahoma State University, Stillwater, Oklahoma.

Zinn, R. A. and F. N. Owens. 1982. Influence feed intake on site of digestion in steers fed a high concentrate ration. J. Animal Sci. (In press)

56

TRACE MINERALS FOR BEEF CATTLE:
METABOLIC FUNCTIONS AND REQUIREMENTS

Allen D. Tillman

The practical role of trace minerals in the nutrition of beef cattle has become more important since many unsolved health problems have been shown to be related to these nutrients. In some cases, problems arise when specialized feeding programs, which result from innovations in mechanization, are put into practice in larger units. Examples are widespread. Likewise, any change in the numerous agronomic practices, in feed processing, and in the use of new or modified ingredients, can alter the mineral nutrition of beef cattle. This paper considers some of the functions of individual trace minerals in the body and lists reasonable requirement figures for those that are needed. Special attention will be given to those that should be added to the diets of beef cattle under practical feeding conditions.

ESSENTIAL TRACE MINERALS

Mertz (1981) has defined an essential element as one required for maintenance of life; its absence results in death of the animal. Deficiencies of a trace mineral sufficient to produce death are difficult to produce, therefore, a broader definition of essential trace minerals has been adopted. An element is considered essential when a deficient intake consistently results in an impairment of a function from optimal to suboptimum, and when supplementation with physiological levels of this element prevents or cures the impairment. When more than one investigator has demonstrated the above on more than one animal species, the element is considered essential. Using diets, techniques, and environments that are not practical with animals as large as beef cattle, scientists have identified the following to be essential minerals: silicon, vanadium, chromium, manganese, iron, cobalt, nickel, copper, zinc, arsenic, selenium, molybdenum, and iodine. In addition, there are reports that make one suspect that fluorine and tin are essential, but they are not yet generally accepted as such.

For beef cattle, the author will consider only cobalt, copper, iodine, iron, manganese, molybdenum, selenium, and zinc to be essential. The fluorides present a special case as the element (Fl) is not discussed as an essential dietary essential for beef cattle rations. Rather, it is the toxic properties that are of interest to those beef cattle producers in specific areas.

METABOLIC FUNCTIONS, REQUIREMENTS, AND SIGNS OF A DEFICIENCY

Cobalt

In reality, beef cattle do not require dietary cobalt (Co) per se: the reticuloruminal microbes require dietary Co (Ammerman, 1981) in order to synthesize vitamin B_{12}, which the beef cow's tissues need for the metabolism of propionic acid, one of the important volatile fatty acids (VFA) produced in the reticulorumen by microbial fermentation. The primary physiological manifestation of a vitamin B_{12} deficiency is an impairment of propionic acid metabolism. In biochemical terms, vitamin B_{12} is a constituent of the enzyme methylmalonyl-CoA isomerase, which is essential for the conversion of methylmalonyl-CoA to succinyl-CoA. Vitamin B_{12} is also a constituent of an enzyme essential in the recycling of the amino acid methionine from homocysteine. It is also required in normal folic acid metabolism in the liver.

The National Research Council (NRC) (1976) estimated the Co requirements of beef cattle to be between 0.05 and 0.10 mg/kg (0.05 to 0.10 ppm) of ration dry matter (DM), and it appears that any diet containing 0.10 mg of Co/kg DM is adequate for beef cattle. However, Underwood (1977) reported Co requirements ranging from 0.07 to 0.11 mg/kg DM. Calves appear to be more sensitive to a Co deficiency than adult cattle. It has been reported that cattle that graze phalaris tuberosa suffer a condition known as phalaris staggers and that this condition may be prevented by added Co. The physiological reasons for the apparent cue are not known. It has been reported that injected vitamin B_{12} is not a preventive factor.

Soils in many areas of the world are deficient in cobalt. Kubota (1968) has studied the distribution of cobalt in soils in forages found in the United States, and McDowell (1976) found that about 43% of the forages analyzed from South America were deficient. Underwood (1977) reported on the Co status of soils and forages in other regions of the world. Also, diets affect Co requirements. For example, Raun et al. (1968) found that an all-barley diet for fattening beef cattle was improved by Co supplementation.

Signs of Co deficiencies are related to its biochemical function--the deficient animal loses appetite and, after a while, has the appearance of a starved animal. The haircoat is rough, and the animal may be anemic. Recovery is rapid after supplementation with Co is initiated.

The question of toxicity of Co to beef cattle is appropriate since Co supplementation of the diet of these animals is fairly general in the United States and elsewhere. The NRC (1980) has reviewed the relevant research work on this subject and has set the maximum tolerable dietary level for cattle at 10 mg/kg of DM (10 ppm), or a tolerance:requirement ratio (10 mg ÷ 0.10) of about 100. With this wide tolerance ratio, there is little likelihood to Co toxicity under practical feeding conditions, unless mixing errors are encountered.

The sources of Co supplements have been studied by many workers (Ammerman and Miller, 1972), and it has been proposed that the carbonate, chloride, sulfate, and oxide forms of Co are satisfactory dietary sources.

Copper and Molybdenum

The requirement and tolerance of beef cattle for Copper (Cu) are so dependent upon the level of dietary molybdenum (Mo) that these two elements are discussed together. For example, additional dietary Cu will, if sufficient, alleviate the toxic effects of high levels of dietary Mo. Also, requirements for dietary Cu increases with increasing levels of dietary Mo.

Copper is necessary for hemoglobin formation, for iron absorption, and for the mobilization of iron from body tissue stores. The oxidation of iron, permitting it to combine with transferrin, the iron transport protein, is a function of Cu. The crosslinking of polypeptides in connective tissues and collagen is carried out by a copper-containing enzyme, lysyl oxidase. In addition, Cu is found in many other body enzymes that catalyze a variety of functions in the body.

There are indications that Cu requirement for cattle in the United States is from 5 to 8 mg/kg DM. However, the NRC (1980) has set Cu requirements that range from 1.2 mg/kg ration in the preruminant calf up to 20 mg in adult animals. The British scientists also list requirements for cattle kept for various productive functions, and there is variation from one group to another. The NRC (1980) lists requirements for dairy cows as varying from 4 to 10 mg/kg diet, indicating that 4 mg/kg diet will meet requirements under some conditions but that 10 mg/kg diet will meet requirements under some conditions but that 10 mg/kg diet would be a more practical level for the wide variety of conditions found in the United States. This is a reasonable requirement for Cu because requirements vary greatly, depending upon Mo levels in the ration. For example, Goodrich and Tillman (1968) found that high levels of both Mo and inorganic sulfur decreased the storage of Cu in the livers of sheep. Because of the high Mo levels in European forages, the Cu requirements of their cattle are, in general, higher than U.S. cattle.

The deficiency symptoms (most are nonspecific—usually associated with Cu) have been reported by Allcroft and Lewis (1957), Becker et al. (1953), and Underwood (1971), including reduced growth, loss in body weight, and unthrift-iness. If the deficiency is severe, there is severe diar-rhea, rapid weight loss, rough hair coat (it could be bleached, graying, or dirty yellow), a change in hair tex-ture, swelling at the ends of the long bones (especially above the pasterns), fragile bones that often break, stiff joints, delayed estrus in cows, birth of calves with congen-ital rickets, falling disease, or sudden death due to heart failure. The animals have anemia, and performance is sub-normal.

Copper toxicity can occur in cattle, but it is not as common as in sheep. Maximum tolerable levels (NRC, 1980) are 115 mg/kg diet for cattle. If 10 mg/kg diet is the requirement, the tolerance requirement is 11.5.

Grains are generally lower in Cu than forages, and most forages supply more Cu than beef cattle require. There are geological regions where the soils are deficient in Cu. If soil Mo is high in such regions, Cu supplementation is essential for health and performance of beef cattle. The inorganic salts of Cu are available to beef cattle--$CuCo_3$, $Cu(NO_3)_2$, $CuSo_4$, $CuCl_2$, and others.

Molybdenum is an indispensable component of xanthine oxidase, which is important in the tissues of beef cattle, thus is an essential element. However, a deficiency has never been reported in cattle; thus we are more concerned with Mo toxicity than with requirements.

Cattle are less tolerant of high levels of Mo than are other farm animals (Ward, 1978). Molybdenum toxicity is of little importance to many sections of the world (Underwood, 1977). Forages are likely to have high levels of Mo when grown on poorly drained soils, especially on granite allu-viums, the black shales that are high organic soils, such as the peats and mucks. Also, soils with high pH increases Mo availability to the forages.

As indicated earlier, Mo and Cu are autogenetic to each other within the body. If dietary Cu is low, a smaller amount of Mo is toxic. The major symptoms of Mo toxicity are the same as those of a severe Cu deficiency--scours or severe diarrhea. The names "teart" pastures or "peat scours" stand out in the world's literature on Mo toxicity. The Mo content of teart pastures runs from 20 to 100 mg/kg DM, with 3 to 5 mg/kg being considered normal (Underwood, 1977).

Molybdenum toxicity is modified by other dietary compo-nents, but these effects are not yet well-defined and are complex. In general, high levels of the sulfates help to protect against dietary Mo by increasing Mo excretion from the body. The maximum level of Mo that can be normally con-sumed without toxicity is about 6 mg/kg DM in the diet. However, levels of dietary Cu can change this upward.

The requirement of cattle for Mo is not yet established but is probably less than 2 mg/kg diet DM. As in the case of Cu, an exact estimate of the Mo requirement is impossible because both Cu and sulfate alter Mo metabolism. Molybdenum levels of a 5 to 6 mg/kg diet inhibit Cu storage (NRC, 1980) and, of course, elevate levels of Cu alleviates Mo toxicity.

Iodine

Iodine (I) is an important element in beef cattle nutrition. It is found in every cell and is stored in the thyroid gland, which takes up inorganic iodine from the blood and uses it for the synthesis of the thyroid hormones, primarily thyroxine, which play active roles in thermoregulation, intermediary metabolism, reproduction, growth, circulation, development, and muscular functions (Underwood, 1977).

About 80% of the stored I in the body is found in the thyroid gland. Also, muscles and kidneys contain fairly high concentrations. Reproducing females have relatively high concentrations of I in the ovaries, presumably because of its importance in reproduction.

Deficiencies of I in the U.S. and elsewhere appear to be area problems. For example, soils in the Northwest and in the Great Lakes regions of the U.S. are low in I, producing plants and grains deficient in I. A deficiency of I results in lowered thyroxine production and one manifestation of the deficiency is an enlarged thyroid gland called an endemic goiter. Failure to reproduce by female animals is another important deficiency symptom. In severe deficiency states, the young may be born hairless and with goiter. Retarded growth, development, and slow maturity are also common deficiency symptoms. In some feeding situations, the feeds used may contain sufficient levels of I but at the same time contain goitrogens, which interfere with I uptake by the thyroid gland. Some of the feeds that are goitrogenic (those containing goitrogenic compounds) are kale, cabbage, rape, soybeans, linseed, peanuts, and the peas or pulses. Cassava, which is an important energy source in the tropics, contains a cyanogenic glycoside. In addition, thiourea and other goitrogenic substances are found in feeds and all reduce the uptake of I, thereby reducing thyroxine formation. The detrimental effect of these can be overcome in most cases by supplemental I.

The requirements for I have not been extensively studied. The NRC (1976) estimated the requirements for pregnant and lactating beef cows to be from 400 to 800 micrograms/day. They recommended that the requirements could be met by feeding salt to cattle in mixtures containing 0.007 of I; I must be provided in a stabilized form. If the salt mixture is fed at 0.10% of the diet, the level of I would be 0.8 mg/kg diet DM. Therefore, the American work indicates that the daily allowances of about 1 mg of stabilized I would meet the requirements of adult cattle. The NRC (1980) suggested that the level of I should be increased to about 2 mg/kg diet DM when substantial goitrogens are present. Otherwise, the requirements are about 0.5 mg/kg DM.

This is reasonable. Also, the NRC (1978) set the requirements for dairy cattle at 0.5 mg/kg diet.

Supplemental I can be furnished in a number of forms
and may include calcium iodate, ethylenediamine dihydroi-
odide (EDDI), sodium iodate, sodium iodide, and others. The
most common way of supplying iodine is through the use of
salt blocks; this method is satisfactory if stabilized I is
used and the blocks are not exposed to the rain and sun
weathering.

Iodine toxicity in cattle has been studied (Underwood,
1977); however, the results have been variable. Newton et
al. (1974) found that an excess of 50 mg/kg diet depressed
growth of calves and that diets having 50 to 100 mg/kg diet
are excessive. The Agricultural Research Center (ARC)
(1980) suggests that the maximum safe limits lie between 8
to 20 mg/kg diet DM.

Iron

Although the body contains only about 0.004% iron (Fe),
this element is important in life processes. Iron is a constituent of the respiratory pigment, hemoglobin, which is
essential for carrying oxygen to cells and removal of CO_2
from these. Iron is found in the iron-porphyrin nucleus,
therefore, it functions in hemoglobin as well as in the
iron-containing enzymes including cytochrome C, peroxidase,
catalase, and others, and plays a central role in respiration and in the oxidizing enzymes. In cattle, most of the
body Fe is in the hemoglobin; the lower levels found in the
muscles are myoglobin. Iron is also found in cytochrome.

The absorption of Fe from the intestinal tract is complicated by many factors (Underwood, 1977). Factors favoring absorption include vitamin C, many amino acids, sugars,
citric acid, and certain chelating agents; some chelating
compounds reduce the ferric form of Fe to the ferrous form.
Factors reducing absorption include dietary phytates,
oxalates, and high levels of phosphorus. Many of these may
decrease the absorption of Fe in nonruminants, but their
effects appear to be less important in ruminants.

Iron requirements of cattle in general are poorly established. However, Fe nutrition is more critical in very
young animals that subsist on milk. The NRC (1978) suggests
a calf requirement for Fe is about 100 mg/kg diet; the adult
requirement is about 50 mg/kg diet. The ARC (1980) suggests
that a diet containing 30 mg/Fe/kg diet DM should be adequate for adult cattle.

Deficiency symptoms are not likely to occur in adult
cattle, but could occur in young calves subsisting on milk
diets. Symptoms include anemia, listlessness, pale mucous
membranes, reduced food intake, and reduced gains.

Iron toxicity has been studied, and the maximum safe
levels appear to be 1000 mg/kg diet DM (NRC, 1980). Therefore, there is not much danger of deficiencies or toxicities
in adult ruminants. The Fe sources were ranked in decreas-

ing order of their availabilities: ferrous sulfate, ferrous carbonate, ferric chloride, and ferric oxide (Ammerman et al., 1967).

Manganese

Manganese (Mn) is found primarily in the animal's liver, but it is also found in appreciable quantities in other organs and in the skin, muscles, and bone. Manganese is the preferred divalent metal cofactor for many enzymes that are involved in carbohydrate metabolism and in muco-polysaccharide synthesis; the list of enzymes and their specific functions in carbohydrate metabolism is long and will not be discussed. However, in addition to carbohydrate metabolism, Mn is a cofactor in two amino acid synthetases, arginine and glutamine. Some enzymes that require magnesium (Mg) will permit the substitution of Mn for Mg. In addition, divalent Mn can replace divalent zinc (Zn) in several Zn-dependent enzymes, but their catalytic activities are lowered. Tissue levels of Mn in cattle are relatively constant over a wide range of Mn intakes, thus the animal appears to have an effective homeostatic control over intestinal absorption of Mn. Many elements ar antagonistic to Mn metabolism in the body--high levels of phosphorus, calcium, iron, zinc, copper, magnesium, and molybdenum appear to reduce its absorption and metabolism (Underwood, 1977).

Manganese requirements of beef cattle are complicated because of the effects of other minerals upon its absorption and metabolism. The NRC (1976) estimated the requirements for growth to be 1 to 10 mg/kg diet. However, the NRC (1976) cautioned that elevated levels of dietary calcium and phosphorus will increase requirements. The NRC (1978) estimated the requirements of dairy cattle to be 40 mg/kg diet; these figures were based upon the experiments of Hawkins, et al. (1955) and Vagg and Payne (1971). Underwood (1977) states that 20 mg/kg diet is likely to be sufficient for growing finishing cattle. The ARC (1980) estimates the requirements for growth to be 10 mg/kg diet DM but that 20 to 25 mg/kg diet DM is required to permit optimum skeletal development. Requirements for reproduction were also set at 20 to 25 mg/kg diet DM.

Deficiency symptoms reflect the many functions that Mn plays in the different enzyme systems: deficiencies lead to degenerative reproduction failures in both bulls and cows. Bone malformations and severe crippling are noted in all animals. Animals develop ataxia, depigmentation, and deterioration of the central nervous system. Reproductive disorders in female adults are an early deficiency symptom, including delayed estrus, reduced fertility, abortions, and deformed young. Calves from Mn-deficient diets exhibit deformed legs (enlarged joints, stiffness, twisted legs, "over-knuckling," weak and shortened bones, and poor growth).

Toxicity studies on cattle are limited but those completed indicate that cattle can tolerate high levels of dietary Mn without obvious adverse effects (Underwood, 1977). The NRC (1980) set the maximum tolerable level for cattle to be 1000 mg/kg of diet. This level is much higher than the Mn content of feedstuffs.

Selenium

The main interest in dietary selenium (Se) for many years concerned the problems of toxicity in animals (Underwood, 1977). However, toxicity is an area problem and is confined to several relatively small areas in the U.S. In contrast, Se deficiencies are observed over a much wider area in the U.S. and the world, thus Se deficiency is a much more practical problem for farm livestock. Selenium deficiencies are much more likely to occur when feeds are grown on acid soils.

Selenium is an integral part of glutathione peroxidase (Rotruck et al., 1973) and the activity of this enzyme is useful as a biochemical measure of Se status (Ammerman and Miller, 1975). While the role of Se in glutathione peraxidase is its best defined function, the element does appear to have other roles: it is a component of muscle protein, which has properties similar to those of cytochrome. It is also postulated that Se is a component of the liver microsomal system. The element is also involved in reproduction and growth. The sulfur-containing amino acids serve as precursors for glutathione, thus are necessary for proper glutathione peroxidase function (Underwood, 1977). It is postulated that glutathione peroxidase prevents membrane damage due to its antioxidant properties (Hoekstra, 1974). Because of these antioxidant properties, it has been postulated that Se and vitamin E could become mutually replaceable in cattle diet. Research work shows that though the functions of Se and vitamin E are intertwined and exert mutually sparing effects, Se cannot be fully replaced by vitamin E (Hoekstra, 1974; Buchanan-Smith et al. 1969a and 1969b; Underwood, 1977). Selenium compounds will reduce the toxicities of cadmium and mercury.

Deficiency symptoms of Se have been described by Underwood (1977), Ammerman and Miller (1975), and many others. A major sign of a pronounce Se deficiency is nutritional muscular dystrophy, commonly known as "white muscle disease (WMD)," which usually occurs in young calves. Animals with WMD have chalky white striations, degeneration, and necrosis in skeletal and cardiac muscles. There is paralysis of the hind legs and heart failure is common. The animals show elevated serum glutamic oxaloacetic transminase (SGOT) values. There is growth depression and a progressive loss of condition, with diarrhea occurring in some animals. Fertility is affected in females, and some reports (Julian et al., 1976) indicate that the incidence of retained placenta is reduced with Se supplementation.

Cattle requirements for Se are not well-defined and appear to depend upon the vitamin E content of the diet. The NRC (1978) estimated the requirements to be about 0.1 mg/kg diet DM while a range of 0.1 to 0.3 was suggested by the NRC (1980). The maximum tolerable level for all animals was set at 2 mg/kg diet DM (NRC, 1980). Se may be provided to animals as a drench, as a subcutaneous injection, as a sublingual injection, in fertilizers to the soil, and as a feed additive.

Selenium is a toxicant in several areas of the world, particularly in South Dakota, and has been called alkali disease by many workers (Underwood, 1977). Acute Se poisoning causes dullness, slight ataxia, a rapid weak pulse, a characteristic posture, labored respiration, diarrhea, lethargy, and death by respiratory failure. Signs of chronic Se toxicity may include lameness, loss of vitality, loss of appetite, emaciation, sore feet, deformed, cracked, and elongated hoofs, loss of hair from the tail, liver cirrhosis, and nephritis. High protein diets appear to offer some protection.

Zinc

Zinc (Zn) has been known to be an essential element for almost 50 years, based upon its functions as an enzyme activator and as a constituent of enzymes. The element is largely involved in nucleic acid metabolism, protein synthesis, and carbohydrate metabolism. The enzymes involved include several dehydrogenases, peptidases, and phosphatases (Underwood, 1977).

Zinc is found throughout the body, but in high concentrations in the pancreas, liver, kidneys, pituitary gland, and the adrenals. It is also found in the bone, teeth, hair, skin, and the eye. The testicles and male accessory glands have high levels of Zn, probably because of its role in reproduction. Normal blood plasma levels range from 80 to 120 mcg/100 ml (Underwood, 1977).

The Zn requirement of beef cattle is not known with any degree of accuracy. The NRC (1976) recommended that requirement be set at a level of 20 to 30 mg/kg diet DM. In reality, a severe Zn deficiency does not appear to be a practical problem in beef cattle except in certain areas. The NRC (1980) concluded that cattle diets providing 30 mg/1 of Zn/kg diet DM will meet Zn requirements. Field trial data on Zn requirements yield indefinite conclusions, probably because many dietary factors affect Zn absorption from the intestinal tract, for example, dietary levels of cadmium, calcium, iron, magnesium, manganese, molybdenum, and selenium. Zinc requirements vary due to age, growth rate, and other functions. It is known that the efficiency of Zn absorption reduces with increasing age (Stake et al., 1973). The work of Underwood and Somers (1969) indicates that the requirements for reproduction in sheep was 32.4 mg/kg diet DM while 17.4 mg/kg diet DM was sufficient for growth. Therefore, it appears that levels of 20 to 40 mg/kg diet DM should meet cattle requirements under a variety of conditions.

Deficiency symptoms of Zn in calves is characterized by decreased gains, lower feed consumption and feed efficiency, decreased testicular growth, listlessness, swollen feet, scaly lesions on feet and legs, loss of hair and general dermatitis. The dermatitis is more severe on legs, neck, head, and around the nostrils (Miller et al., 1962; Ott et al., 1965). In addition, parakeratotic lesions are found on all animals. When lactating cows were fed diets containing only 6 mg/kg diet DM, they developed symptoms almost identical to those in calves (Schwarz and Kirchgessner, 1975). One feature of a Zn deficiency is the failure of wounds to heal normally (Miller et al., 1965). Since milk contains a high level of zinc, high-producing cows will develop deficiency symptoms quite rapidly.

The toxicity threshold for Zn is influenced by other dietary variables. Levels of 900 mg/kg diet DM in growing cattle fed corncob-concentrate diets produced subnormal gains and efficiencies (Ott et al., 1966). When the level was increased to 1700 mg/kg diet DM and above, there were depraved appetites. Cattle fed toxic levels of Zn have higher than normal levels of Zn in the blood plasma, kidney, pancreas, and liver. Blood hemoglobin and hematocrit are decreased while liver Cu is decreased and liver Fe is increased.

Plants vary in their levels of Zn; legumes contain higher levels than do grasses. The grains levels of Zn are considerably lower than those of hays and silages (10 to 30 vs about 60 mg/kg DM, respectively). Plant protein supplements contain about 50, while animal protein sources contain about 90 mg/kg DM. The NRC (1980) report that the inorganic salts of Zn are excellent sources, including acetate, carbonate, chloride, and sulfate. In addition, metallic Zn is a highly available source.

Fluorine

Studies with laboratory animals tend to indicate fluorine (F) to be an essential mineral, but it is still not accepted (Mertz, 1981). There is no known beneficial metabolic role of F in the animal body. However, F does prevent tooth decay when levels of about 1 part per million (ppm) are used in drinking water.

The body supply of F is found primarily in the bones and teeth, and normal bone levels are from 500 to 1000 mg/kg bone DM (Underwood, 1977). Soft bone tissue has a higher concentration of F than compact bone (Shup et al., 1963a, b).

The F requirement, if any, for cattle is unknown. In reality, F is important in beef cattle nutrition because of its toxicity, which resulted primarily after the introduction of rock phosphates as a phosphorus supplement. At the present time, the major sources of F for cattle are the nondefluorinated rock phosphates, which may contain 3 to 4% F, and contaminated forages near industrial plants. In some areas, waters from springs may contain excessive levels.

Usually acute, F toxicity develops in cattle suffering from chronic fluorosis, which develops slowly over a long period of time. In such cases, the excess F accumulates in bones and teeth, and this induces changes. Developing teeth in young cattle are quite sensitive to excess F, even if fed over a short period of time (NRC, 1974). Changes in the teeth included mottling, staining, 'excessive wear, erosion, or pitting of the enamel, depending upon the degree of fluorosis. Excessive levels of F ingestion causes bone lesions including enlarged, white, and chalky areas. The animals become lame and stiff. Appetites are reduced, general health declines, and performance is lowered. Maximum tolerable levels are 100 mg/kg diet DM for finishing animals on feed for only several months. However, for breeding animals, 50 mg/kg diet DM is the maximum tolerable level (NRC, 1980). Dental changes are the most easily recognized symptoms because of the softening of the enamel, resulting in mottling and staining (NRC, 1974).

Most food contains only trace levels of F, as plants neither require F nor accumulate it in their tissues. Phosphates are the primary source of dietary F and this danger is removed by effective defluorination. Bone meal from animals receiving high levels of dietary F may contain high levels of the element.

A summary table listing beef cattle requirements, as well as the maximum tolerable levels for each trace element, is shown in table 1.

The utilization of any mineral may be affected by many dietary factors. In addition, there are many mineral interactions that may affect individual mineral utilization and requirements. Some of these are shown in summary figure 1.

TABLE 1. ESTIMATED BEEF CATTLE REQUIREMENTS AND MAXIMUM TOLERABLE DIETARY LEVELS FOR THE ESSENTIAL TRACE MINERALS AND FLUORINE[a]

Mineral	Requirements mg/kg ration DM (ppm)	Maximum tolerable level mg/kg ration DM (ppm)
Cobalt	0.10	10
Copper	10.0	115
Molybdenum	?	?
Iodine	0.1	8 to 20
Iron	5-100	1,000
Manganese	20-25	2
Zinc	20-40	500
Fluorine	–	50-100

[a] Tabulated from a review of the literature

536

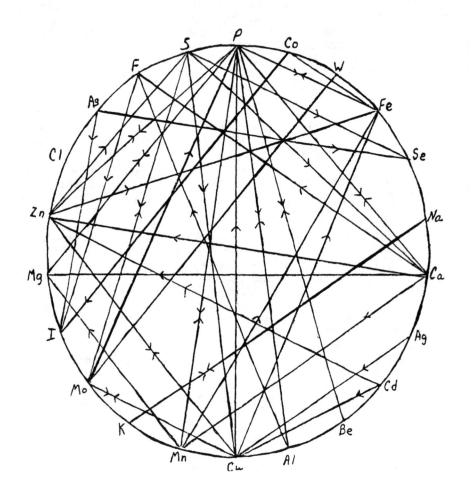

Figure 1. Mineral interrelations

Explanation: When the line connects two minerals on the circle, it indicates that there is an interrelationship between these. If the arrow on the line points in the direction of the mineral, i.e., calcium——>——zinc, the utilization of zinc is interfered with by an excess of calcium. If the arrow points in both directions, i.e., calcium __>__<__mg, it indicates that these minerals are each antagonistic to the other, if excess levels of either exist.

Figure 1 (continued)

Phosphorus: (a) Ca, Mg, Mn, Zn, Fe, Al, and Be interfere with the phosphorus absorption and vice versa due to the formation of insoluble phosphates. (b) Low Cu-high Mo intakes increase the loss of body P (Cu is required for phospholipid synthesis).

Calcium: (a) High Ca levels in the diet reduce the absorption of the other minerals due to the formation of insoluble tricalcium phosphate. (c) Large intakes of either Ca or Mg increase the urinary excretion of the other minerals, but both Ca and P prevent the absorption of excess Mg. (d) SO_4 increases the excretion of Ca.

Copper: (a) Cu is required for the proper metabolism of Fe. (b) Cd and Ag increase the severity of Cu deficiency. (c) High dietary Zn reduces liver stores of Fe and Cu while low Zn favors excess storage of Fe and Cu. Excess Cu causes low storage of Zn. (d) Mo limits Cu storage in the presence of adequate sulfate. Sheep with high liver cu have low Mo levels and Cu toxicity may develop with low Mo intakes.

Sulfur: (a) Sulfate - S limits Cu and Ca storage and protects against Se toxicity. (b) High Zn increases fecal S. (c) SO_4 decreases live Mo.

Molybdenum: Tungstate (W) increases urinary excretion of Mo.

Cobalt: (a) Cobalt increases the urinary excretion of I. (b) Fe accumulates during Co deficiency (Co needed for Fe metabolisms).

Iodine: As and F have goitrogenic activity.

Fluorine: Ca and Al salts protect against F toxicity.

Selenium: SO_4 and As reduce Se toxicity.

Sodium-Potassium: Deficiency symptoms of either of these elements are aggravated by an excess of the other. The high K content of forage may explain the high salt requirement of Herbivora.

Manganese: (a) High Mn interferes with Fe utilization. (b) High Mn lowers serum Mg.

Zinc: (a) High Zn interferes with Fe utilization. (b) High Zn interferes with Cu utilization.

REFERENCES

General

ARC.   1980.   The nutrient requirements of ruminant livestock.   The Agricultural Research Council, Commonwealth
Bureaux, Farnhan Royal, England.

Ensminger, M. E. and C. G. Olentine.   1978.   Feeds and
Nutrition (1st ed.).   Ensminger Publ. Co., Clovis, CA.

Maynard, L. A., J. K. Loosli, H. F. Hintz, and R. G.
Warner.   1979.   Animal Nutrition (7th ed.).   McGraw-
Hill, Inc., New York.

Mertz, Walter.   1981.   The essential trace elements.
Science 213:18.

Miller, W. J.   1981.   Mineral and vitamin nutrition of dairy
cattle.   J. Dairy Sci. 64:1196.

NRC.   1976.   Nutrient Requirements of Beef Cattle (5th rev.
ed.).   National Academy of Sciences - National Research
Council.   Washington, D.C.

NRC.   1980.   Mineral Tolerances of Domestic Animals.
National Academy of Sciences - National Research Council, Washington, D.C.

Underwood, E. J.   1977.   Trace Elements in Human and Animal
Nutrition (4th ed.).   Academic Press, New York.

Cobalt

Ammerman, C. B.   1981.   Cobalt, ANH Mineral Series, Animal
Nutrition and Health, October 1981.

Ammerman, C. B. and S. M. Miller.   1972.   Biological availability of minor mineral ions: A review.   J. Anim.
Sci.   35:681.

Kubota, J.   1968.   Distribution of cobalt deficiency in
grazing animals in relation to soils and forage plants
in the United States.   Soil Sci.   106:122.

McDowell, L. R.   1976.   Mineral Deficiencies and Toxicities
and Their Effects on Beef Production in Developing
Countries.   Symposium on Beef Cattle Production in
Developing Countries.   U. of Edinburgh, Scotland.

NRC.   1976.   (See general references.)

NRC.   1980.   (See general references.)

Raun, N. S., G. L. Stables, L. S. Pope, O. F. Harper, G. R. Waller, R. Rensbarger, and A. D. Tillman. 1968. Trace mineral additions to all-barley rations. J. Anim. Sci. 17:695.

Underwood. (See general references.)

Copper and Molybdenum

Allcroft, R. and G. Lewis. 1957. Copper nutrition in ruminants. Disorders associated with copper-molybdenum content of feedstuffs. J. Food Agric. 8 (Suppl.):96.

ARC. 1980. (See general references.)

Becker, R. B., P. T. D. Arnold, W. G. Kirk, M. G. Davis, and R. W. Kidder. 1953. Minerals for dairy and beef cattle. Fla. Agr. Exp. Stn. Bull. 513.

Goodrich, R. D. and A. D. Tillman. 1966. Copper, sulfate, and molybdenum interrelationships in sheep. J. Nutr. 90:76.

NRC. 1970. (See general references.)

NRC. 1980. (See general references.)

Underwood. (See general references.)

Ward, G. M. 1978. Molybdenum toxicity and hypocuprosis in ruminants. A review. J. Anim. Sci. 46:1078.

Iodine

ARC. 1980. (See general references.)

Newton, G. L., E. R. Barrick, R. W. Harvey, and M. B. Wise. 1974. Iodine toxicity. Physiological effects of elevated dietary iodine on calves. J. Anim. Sci. 38:449.

NRC. 1976. (See general references.)

NRC. 1978. Nutrient Requirements of Dairy Cattle. (5th Rev. Ed.). National Academy of Sciences - National Research Council, Washington, D. C.

NRC. 1980. (See general references.)

Underwood, E. J. 1977. (See general references.)

540

Iron

Ammerman, C. B., J. M. Wing, B. G. Dunavant, W. K. Robertson, J. P. Feaster, and L. R. Arrington. 1967. Utilization of inorganic iron by ruminants as influenced by form of iron and iron status of the animal. J. Anim. Sci. 26:404.

ARC. 1980. (See general references.)

NRC. 1976. (See general references.)

NRC. 1978. (See general references.)

NRC. 1980. (See general references.)

Underwood, E. J. 1977. (See general references.)

Selenium

Ammerman, C. B. and S. M. Miller. 1975. Selenium in ruminant rations. A review. J. Dairy Sci. 58:1561.

Buchanan-Smith, J. G., E. C. Nelson, and A. D. Tillman. 1969a. Effects of vitamin E and selenium deficiencies on lysommal and cytoplasmic enzymes in sheep tissues. J. Nutr. 99:387.

Buchanan-Smith, J. G., E. C. Nelson, B. T. Osburn, and A. D. Tillman. 1969b. Effect of vitamin E and selenium deficiencies in sheep fed a purified diet during growth and reproduction. J. Anim. Sci. 29:808.

Hoekstra, W. G. 1974. Biochemical role of selenium in trace elements metabolism in animals by W. G. Hoekstra, J. W. Suttie, H. E. Ganther, and W. Mertz. Univ. Park, Maryland.

Julian. 1976.

NRC. 1976. (See general references.)

NRC. 1980. (See general references.)

Rotruck, J. T., A. L. Pope, H. E. Ganther, A. B. Swanson, D. G. Hafeman, and W. G. Hoekstra. 1973. Selenium: biochemical role as a compound of glutathione peroxidase. Science 179:588.

Underwood, E. J. 1977. (See general references.)

Manganese

ARC. 1980. (See general references.)

Hawkins, G. E., G. H. Wise, G. Matrone, and R. K. Waugh. 1955. Manganese in the nutrition of young dairy cattle fed different levels of calcium and phosphorus. J. Dairy Sci. 38:536.

NRC. 1976. (See general references.)

NRC. 1978. Nutrient Requirements of Dairy Cattle. (5th Rev. Ed.) National Academy of Sciences - National Research Council, Washington, D.C.

NRC. 1980. (See general references.)

Underwood, E. J. 1977. (See general references.)

Vagg, M. J. and J. M. Payne. 1971. Effect on raised dietary calcium on the gastrointestinal absorption and rate of excretion of manganese by dairy cows. In: Mineral Studies with Isotopes in Domestic Animals, p. 121. International Atomic Energy Agency, Vienna.

Zinc

ARC. 1980. (See general references.)

Miller, J. K. and W. J. Miller. 1962. Experimental zinc deficiency and recovery of calves. J. Nutr. 76:467.

Miller, W. J., J. D. Morton, W. J. Pitts, and C. M. Clifton. 1975. Proc. Soc. Exp. Biol. and. Med. 118:427.

NRC. 1976. (See general references.)

NRC. 1978. (See general references.)

NRC. 1980. (See general references.)

Ott, E. A., W. H. Smith, M. Stob, H. E. Parker, and W. M. Beeson. 1965. Zinc deficiency symptoms in the young calf. J. Anim. Sci. 24:735.

Ott, E. A., W. H. Smith, R. B. Harrington, and W. M. Beeson. 1966. Zinc toxicity in ruminants. J. Anim. Sci. 25:419.

Schwarz, W. A. and M. Kirchgessner. 195. Experimental zinc deficiency in lactating dairy cows. Vet. Med. Rev. No. 1/2, p 19.

Stake, P. E., W. J. Miller, M. W. Neathery, and R. P. Gentry. 1975. Zinc-65 absorption and tissue distribution in two-and six-month old calves and lactating cows. J. Dairy Sci. 58:78.

Underwood, E. J.   1977.   (See general references.)

Fluorine

NRC.   1974.   National Research Council.   Effects of Fluorides in Animals.   National Academy of Sciences, Washington, D.C.

NRC.   1976.   (See general references.)

NRC.   1980.   (See general references.)

Shupe, J. L., L. E. Harris, D. A. Greenwood, J. E. Butcher, and H. M. Wilson.   1963a.   The Effect of Fluorine on Dairy Cattle.   V. Fluorine in the Urine as an Estimator of Fluorine Intake.   Amer. J. Vet. Res. 23:777.

Shupe, J. L., M. L. Miner, D. A. Greenwood, L. E. Harris, and G. E. Stollard.   1963b.   The effect of fluorine on dairy cattle.   II.   Clinical and pathological effects. Amer. J. Vet. Res. 24:964.

Underwood, E. J.   1977.   (See general references.)

57

EFFECTS OF RUMENSIN®
ON REPLACEMENT HEIFER PERFORMANCE

Herb Brown

INTRODUCTION

Rumensin® has been fed to feedlot cattle since 1975 and to slaughter, feeder, and stocker cattle on pasture since 1978. Rumensin affects rumen fermentation by causing a shift in the ratio of volatile fatty acids (VFA). Fewer acetic and butyric acids are produced, but more propionic acid. Propionic acid has been calculated to be a more efficient source of energy to the animal compared to acetic or butyric acid. Therefore, approximately 10% more energy is available from the same amount of feed. This VFA shift improves (1) feed utilization in cattle that are fed high grain rations and (2) both feed utilization and rate of gain for cattle fed high roughage rations or grazing on pasture.

Rumensin improves feed utilization of replacement heifers resulting in improved average daily gains. Thus, heifers being developed for herd replacements should benefit from an improved physiological state, theoretically, should exhibit improved reproductive efficiency.

MATERIALS AND METHODS

Five hundred and fifteen heifers were used to study the effects of Rumensin on weight gain and reproduction. Five individual trials were conducted using different types of cattle in several geographic areas (table 1). Three trials were with beef-type heifers and two trials were conducted with dairy-type heifers. Basal rations were the same within each trial for both treatment groups. Rumensin-fed heifers received 200 mg of Rumensin per head daily for the duration of the experiment. All heifers were prepubertal at the initiation of each trial.

Average daily gain was monitored at approximately 28 day intervals and reproductive activity was observed daily to determine the date of first estrus for each heifer.

TABLE 1. SUMMARY OF MAJOR EXPERIMENTAL VARIABLES OF PREPUBERTAL HEIFERS
SUPPLEMENTED WITH AND WITHOUT RUMENSIN

Trial No.	Location	Breed	Trial length (days)	No. of heifers	Initial weight	Type of ration
1	Texas	Brangus	316	90	535	Alfalfa hay, micronized milo supp.
2	New York	Holstein	441	56	515	Corn silage, alfalfa hay protein supp.
3	Penn.	Holstein	448	60	430	Corn silage, alfalfa hay, protein supp.
4	Kansas	A & A X H	278	168	370	Complete mixed ration of sorghum silage, soybean meal, milo
5	Montana	A X H	203	141	470	Chopped crested wheat hay concentrate

RESULTS

A summary of overall average daily gain for each trial
and for the five trials combined is shown in Table 2. The
overall average daily gain from the five trials combined was
1.35 lb per head per day for the control heifers and 1.45 lb
per head per day for heifers given Rumensin. Feeding Rumensin in trial 1 did not result in the typical improvement in
average daily gain; however, the time to first estrus was
shortened from 198 to 174 days in this trial.

TABLE 2. SUMMARY OF AVERAGE DAILY GAIN (LB) OF HEIFERS
SUPPLEMENTED WITH RUMENSIN

Trial No.	Rumensin, mg/hd/day	
	0	200
1	1.43	1.32
2	1.50	1.59
3	1.33	1.52
4	1.40	1.48
5	1.08	1.45
5-Trial average	1.35	1.45

The feeding of Rumensin dramatically reduced the number
of days to first estrus in four of the five trials. In the
combined five trials, the time to first estrus was reduced
from 152 to 139 days by the addition of 200 mg Rumensin per

head daily. There was no apparent estrous response for Rumensin in trial 3; however, average daily gain was increased from 1.33 to 1.52 lb. All heifers were bred as they exhibited estrus and no treatment differences were noted between control and Rumensin-fed heifers in subsequent reproductive performance.

Randel et al., in a series of studies at the Texas A&M Center at Overton, has confirmed that feeding 200 mg of Rumensin daily to beef heifers hastens puberty. The study demonstrated that changes in rumen fermentation are correlated with earlier onset of estrus and may be mediated through the hypothalmic-pituitary axis to affect ovarian function. They concluded that judicious use of some nutritional parameters may be a method of improving management systems for developing replacement heifers.

TABLE 3. AVERAGE DAYS TO FIRST ESTRUS FOR HEIFERS SUPPLE-MENTED WITH RUMENSIN--SUPPLEMENTED AND FOR NONSUPPLEMENTED HEIFERS

Trial No.	Rumensin, mg/hd/day	
	0	200
1	198	174
2	136	113
3	150	153
4	90	79
5	178	163
5-Trial average	152	139

SUMMARY

The feeding of Rumensin (200 mg/hd/day) in replacement heifers has been shown to increase the rate of weight gain and to decrease the age of puberty by nearly two weeks. The mode of action may be through alteration of rumen fermentation that, in turn, may affect hypothalmic-pituitary axis and ovarian function.

REFERENCE

Nutrition influence on reproductive development of replacement heifers. Technical Report No. 82-1. Texas A&M, University Agricultural Research and Extension Center, Overton, Texas.

58

DEVELOPING MAXIMUM PROFIT
FEEDLOT DIETS

Donald R. Gill

Formulation of finishing rations for beef cattle is a complex scientific process for combining feed ingredients and other variables to maximize feedlot profits. This means that balanced or even least-cost balanced rations are not necessarily going to lead to profitable feeding programs. The degree to which this concept is understood is highly correlated to the success of any given feedlot operation.

Balancing a ration for maximum profit requires additional steps beyond what is normally considered in nutritional balance. New formats are developing to pinpoint nutritional requirements of feedlot cattle. These formats make it possible to balance rations for least-cost production. These requirements recognize the fact that requirements are based on at least 3 factors: (1) the sex and size (weight) of the animal, (2) the level of production (daily gain), and (3) nutrient intake.

It is usually wasteful to feed more of any nutrient than is necessary for utilization of the most limiting nutrient in the diet. Because of these relationships, it is no longer possible to state that the ideal ration will have a certain percentage of nutrients, since intakes differ. It instead appears more desirable to state the nutrient requirements in terms of daily requirement based on the sex, size, (weight of the animal) and the level of gain desired.

Table 1 shows the nutrient requirements in terms of energy and protein; Table 2 lists minerals and vitamins. The protein and energy tables are based on the factors listed above.

WHY CHANGE RATION FORMULATING SYSTEMS?

The reason for considering change in formulating rations can best be demonstrated by using examples of two

TABLE 1. PROTEIN AND ENERGY REQUIREMENTS OF FEEDLOT CATTLE

Daily requirements for maintenance

Body weight = 400 lb			Body weight = 600 lb		
Daily gain lb	NE$_m$ 3.85 Mcal Steers NE$_g$	Heifers NE$_g$	Crude prot.	Daily Gain lb	NE$_m$ 5.21 Mcal Steers NE$_g$	Heifers NE$_g$	Crude prot.
1.0	1.25	1.39	1.00	1.0	1.70	1.88	1.21
1.2	1.52	1.69	1.07	1.2	2.06	2.30	1.29
1.4	1.79	2.01	1.14	1.4	2.43	2.73	1.36
1.6	2.07	2.34	1.22	1.6	2.81	3.17	1.44
1.8	2.35	2.68	1.30	1.8	3.19	3.63	1.51
2.0	2.64	3.03	1.37	2.0	3.58	4.11	1.59
2.2	2.94	3.39	1.45	2.2	3.98	4.59	1.67
2.4	3.24	3.76	1.52	2.4	4.39	5.09	1.74
2.6	3.54	4.14	1.60	2.6	4.80	5.61	1.82
2.8	3.86	4.53	1.68	2.8	5.22	6.14	1.89
3.0	4.17	4.93	1.75	3.0	5.65	6.68	1.97
3.2	4.49	5.34	1.83	3.2	6.09	7.23	2.05
3.4	4.82	5.76	1.90	3.4	6.54	7.80	2.12

Body weight = 500 lb			Body weight = 700 lb		
Daily Gain lb	NE$_m$ 4.55 Mcal Steers NE$_g$	Heifers NE$_g$	Crude prot.	Daily Gain lb	NE$_m$ 5.85 Mcal Steers NE$_g$	Heifers NE$_g$	Crude prot.
1.0	1.48	1.64	1.10	1.0	1.91	2.11	1.31
1.2	1.80	2.00	1.18	1.2	2.31	2.58	1.39
1.4	2.12	2.38	1.25	1.4	2.73	3.06	1.46
1.6	2.45	2.77	1.33	1.6	3.15	3.56	1.54
1.8	2.78	3.17	1.41	1.8	3.58	4.08	1.62
2.0	3.12	3.58	1.48	2.0	4.02	4.61	1.69
2.2	3.47	4.01	1.56	2.2	4.47	5.16	1.77
2.4	3.83	4.44	1.64	2.4	4.93	5.72	1.84
2.6	4.19	4.89	1.71	2.6	5.39	6.30	1.92
2.8	4.56	5.35	1.79	2.8	5.87	6.89	2.00
3.0	4.93	5.82	1.86	3.0	6.35	7.50	2.07
3.2	5.31	6.31	1.94	3.2	6.84	8.12	2.15
3.4	5.70	6.81	2.02	3.4	7.34	8.76	2.22

TABLE 1.　(continued)

| | Daily requirements for maintenance | | | | | | |

Body weight = 800 lb				Body weight = 1.000 lb			
Daily gain lb	NE_m 6.47 Mcal Steers NE_g	Heifers NE_g	Crude prot.	Daily Gain lb	NE_m 7.65 Mcal Steers NE_g	Heifers NE_g	Crude prot.
1.0	2.11	2.33	1.41	1.0	2.49	2.76	1.60
1.2	2.56	2.85	1.49	1.2	3.02	3.37	1.67
1.4	3.01	3.39	1.56	1.4	3.56	4.00	1.75
1.6	3.48	3.94	1.64	1.6	4.12	4.66	1.83
1.8	3.96	4.51	1.71	1.8	4.68	5.33	1.90
2.0	4.45	5.09	1.79	2.0	5.26	6.02	1.98
2.2	4.94	5.70	1.87	2.2	5.84	6.74	2.05
2.4	5.45	6.32	1.94	2.4	6.44	7.47	2.13
2.6	5.96	6.96	2.02	2.6	7.05	8.23	2.21
2.8	6.48	7.61	2.09	2.8	7.67	9.00	2.28
3.0	7.02	8.29	2.17	3.0	8.30	9.80	2.36
3.2	7.56	8.98	2.25	3.2	8.94	10.61	2.43
3.4	8.11	9.68	2.32	3.4	9.59	11.45	2.51

Body weight = 900 lb				Body weight = 1,100 lb			
Daily gain lb	NE_m 7.06 Mcal Steers NE_g	Heifers NE_g	Crude prot.	Daily gain lb	NE_m 8.21 Mcal Steers NE_g	Heifers NE_g	Crude prot.
1.0	2.30	2.55	1.51	1.0	2.68	--	1.69
1.2	2.79	3.11	1.58	1.2	3.25	--	1.76
1.4	3.29	3.70	1.66	1.4	3.83	--	1.84
1.6	3.80	4.30	1.73	1.6	4.42	--	1.92
1.8	4.33	4.92	1.81	1.8	5.03	--	1.99
2.0	4.86	5.57	1.89	2.0	5.65	--	2.07
2.2	5.40	6.23	1.96	2.2	6.28	--	2.14
2.4	5.95	6.90	2.04	2.4	6.92	--	2.22
2.6	6.51	7.60	2.11	2.6	7.57	--	2.30
2.8	7.08	8.32	2.19	2.8	8.23	--	2.37
3.0	7.67	9.05	2.27	3.0	8.91	--	2.45
3.2	8.26	9.81	2.34	3.2	9.60	--	2.52
3.4	8.86	10.58	2.42	3.4	10.30	--	2.60

well-established feeding programs. The first is based on all corn silage plus a minimum amount of protein supplement. The second is based on a 90% concentrate ration. Numerous tests and claims have been made for the merits of both of these programs; however, an analysis of the projected gains and economics may be made using Lofgreen's and Garrett's net energy system for feedlot cattle.

Tables 3 and 4 show the two typical high silage and high grain programs, respectively, with fairly typical costs.

TABLE 2. ESTIMATE OF OPTIMUM MINERAL LEVELS FOR CATTLE

% in 100% DM Ration		PPM in 100% DM Ration	
Salt	0.2 - 0.3	Copper	10 PPM
Calcium	0.4 - 0.6	Iron	100 PPM
Phosphorus	0.3 - 0.45	Manganese	30 PPM
Magnesium	0.1	Zinc	60 PPM
Potassium	0.4 - 0.7	Cobalt	0.05 - 0.2 PPM
Sulfur	0.15	Iodine	0.2 PPM

Vitamin A: 20,000 to 25,000 IU/head/day

TABLE 3. HIGH SILAGE PROGRAM

	Ration DM (%)	As fed dry matter (%)	Lb to make 100 lb DM	As fed composition (%)	Purchase basis (% DM)	Cost
Corn silage	94.90	32.00	296.56	98.13	32.00	1.50
44% supl.	5.00	90.00	5.55	1.84	90.00	9.00
Additives	0.10	90.00	.1111	0.04	90.00	16.00
Total			302.21	100.00		

Calculated dry matter content of ration 33.09%.
Cost of DM ration = $4.97/cwt.
Cost of as-fed ration = $1.64/cwt.

Tables 5 and 6 show the projected gains based on energy alone for 800 lb steers at various levels of feed intake.

Considerable experience has demonstrated the accuracy of these projections. Projected gains will be obtained only if all other nutrients other than energy are adequate for level of gain indicated.

In the case of the high silage ration, an 800 lb steer must eat 22 lb of dry matter to gain 2.55 lb. Table 1 shows that for an 800 lb steer to gain 2.6 lb, it would require 2.02 lb of protein daily. Using these tables, it is possible to conclude that the high silage ration should contain

2.02/22 lb of feed = 9.18% protein to allow full utilization of the energy in the ration. In the case of the high concentrate ration, the 800 lb steer would only need to eat 16 lb to gain approximately 2.6 lb. Again, the animal's daily protein requirements are the same at 2.02 lb per day, but the ration should contain 2.02/16 lb of feed, or 12.65% protein.

TABLE 4. HIGH GRAIN PROGRAM

	Ration DM (%)	As fed dry matter (%)	Lb to make 100 lb DM	As fed com- position (%)	Pur- chase basis (% DM)	Cost
Corn	85.00	84.50	100.59	74.41	84.50	5.71
Soymeal	5.00	90.00	5.55	4.11	90.00	10.00
Additives	0.10	90.00	.11	0.08	90.00	11.00
Corn silage	8.90	32.00	27.81	20.57	32.00	1.50
Mineral	1.00	90.00	1.11	0.82	90.00	4.00
Total			135.17	100.00		

Calculated dry matter content of ration 73.97%.
Cost of DM ration = $6.76/cwt.
Cost of as-fed ration = $5.00/cwt.
Note that these rations are formulated on a dry matter basis to avoid confusion.

Protein and energy make up the bulk of the costs of nutrients in cattle-finishing rations, with energy being by far the most costly. In terms of total pounds, either of the two rations needed to supply in excess of 10 lb of usable energy components, about 2 lb of crude protein and less than 1/2 lb daily of all minerals and other nutrients to meet the nutrient requirements of 800 lb steers.

Thus, ration-balancing systems designed for maximum profit must tie all other nutrients to energy, the largest component. Then all other nutrients should be adjusted to make maximum utilization of energy. Developing a ration for finishing cattle, therefore, becomes primarily a function of energy and fixed-cost economics.

Tables 5 and 6 provide an interesting comparison of all corn silage and the 90% corn rations referred to previously. The gains possible on the high silage programs are lower than on the high grain ration because the animals will only be able to eat between 18 and 26 lb of dry matter per day. At 22 lb intake (66.5 lb as fed with 68% moisture silage), gains of about 2.55 lb should result on this weight cattle. Feed cost per pound of gain should run about 43 cents using the prices shown in table 3. Similar animals on the high grain ration (table 6) would probably eat about 20 lb of dry matter (27 lb as fed feed--table 4) and would gain about 3.54 lb. Feed cost per pound of gain would run about 38 cents.

TABLE 5. ESTIMATED DAILY GAIN FOR FEED CONTAINING 74 MCALS/CWT OF ENERGY FOR MAINTENANCE AND 44 MCAL/CWT OF ENERGY FOR PRODUCTION WITH A COST OF $.45 PER CWT

Body weight	Daily feed	Feed req. maint.	Steers				Heifers				Feed cost /day	Total cost /day
			Daily gain	Feed conv.	$ Cost /lb	Total cost gain	Daily gain	Feed conv.	$ Cost /lb	Total cost gain		
800	17	8.74	1.66	10.22	0.51	0.78	1.47	11.56	0.57	0.88	0.84	1.29
800	18	8.74	1.85	9.74	0.48	0.73	1.63	11.07	0.55	0.83	0.89	1.34
800	19	8.74	2.03	9.37	0.47	0.69	1.78	10.68	0.53	0.78	0.94	1.39
800	20	8.74	2.20	9.07	0.45	0.66	1.93	10.38	0.52	0.75	0.99	1.44
800	21	8.74	2.38	8.83	0.44	0.63	2.07	10.13	0.50	0.72	1.04	1.49
800	22	8.74	2.55	8.63	0.43	0.61	2.21	9.93	0.49	0.70	1.09	1.54
800	23	8.74	2.72	8.46	0.42	0.59	2.35	9.77	0.49	0.68	1.14	1.59
800	24	8.74	2.89	8.32	0.41	0.57	2.49	9.63	0.48	0.66	1.19	1.64
800	25	8.74	3.05	8.20	0.41	0.56	2.63	9.52	0.47	0.64	1.24	1.69
800	26	8.74	3.21	8.10	0.40	0.54	2.76	9.43	0.47	0.63	1.29	1.74

TABLE 6. ESTIMATED DAILY GAIN FOR FEED CONTAINING 98 MCALS/CWT OF ENERGY FOR MAINTENANCE AND 63.5 MCAL/CWT FOR GAIN WITH A FEED COST OF $6.76 PER CWT. NONFEED FIXED COST $.45

Body weight	Daily feed	Feed req. maint.	Steers				Heifers				Feed cost /day	Total cost /day
			Daily gain	Feed conv.	$ Cost /lb	Total cost gain	Daily gain	Feed conv.	$ Cost /lb	Total cost gain		
800	15	6.60	2.36	6.37	0.43	0.62	2.05	7.31	0.49	0.71	1.01	1.46
800	16	6.60	2.60	6.15	0.42	0.59	2.26	7.09	0.48	0.68	1.08	1.53
800	17	6.60	2.84	5.98	0.40	0.56	2.46	6.92	0.47	0.65	1.15	1.60
800	18	6.60	3.08	5.84	0.39	0.54	2.65	6.79	0.46	0.63	1.22	1.67
800	19	6.60	3.31	5.74	0.39	0.52	2.84	6.69	0.45	0.61	1.28	1.73
800	20	6.60	3.54	5.65	0.38	0.51	3.03	6.61	0.45	0.60	1.35	1.80
800	21	6.60	3.76	5.58	0.38	0.50	3.21	6.55	0.44	0.58	1.42	1.87
800	22	6.60	3.98	5.52	0.37	0.49	3.38	6.50	0.44	0.57	1.49	1.94
800	23	6.60	4.20	5.48	0.37	0.48	3.56	6.47	0.44	0.56	1.55	2.00
800	24	6.60	4.41	5.44	0.37	0.47	3.72	6.44	0.44	0.56	1.62	2.07

A feeder, using the price relationships in tables 3 and 4 and feed cost as the only consideration in ration formulation, would use the 90% concentrate program.

Other factors that should be considered in ration formulation: research indicates that cost of ownership of 800 lb cattle, including interest, labor, equipment, drugs, veterinary expenses, death losses, and taxes is about 45 cents per head per day. If this 45 cents overhead cost per day is charged the cost per pound of gain on the silage program then becomes 61 cents compared to 51 cents on the high grain program.

If rations are formulated to provide adequate protein, minerals, and vitamins to allow full utilization of the energy in the ration, then energy density in the ration becomes very important. Obviously, the reason better gains could not be obtained on the high corn silage ration illustrated in tables 3 and 5 is simply that the animals seldom can eat enough to reach their full gain potential. Most feeders think of energy density in terms of roughage concentrate ration. If nutritionally adequate rations were formulated at various energy levels ranging from very low (level of wheat straw) to very high (level of shelled corn), the relationship in figure 1-A would become apparent.

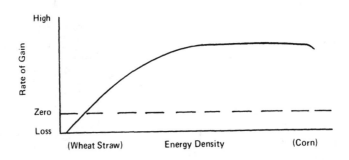

Figure 1-A. Effect of energy density on rate of gain

When energy density is very low, animals simply cannot eat enough feed to make gains; as energy density increases, gain increases. Maximum gain usually occurs on rations of about 50% to 70% percent concentrate, if the roughage portion of the ration is equivalent to good quality alfalfa hay. Higher or lower energy roughages will tend to move the relationship in the appropriate direction. A slight depression in gains is frequently observed on very high concentrate rations.

Figure 1-B shows feed conversion (pounds of feed dry matter per pound of gain) superimposed on the diagram shown in figure 1-A. Note that as energy density increases that the conversion ration will similarly show an improvement in balanced rations.

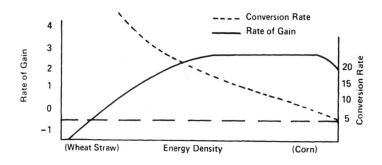

Figure 1-B. Effect of energy densities on rate and efficiency of gain

Rate of gain and feed conversion ratio are very important factors in ration formulation because of their affect on cost of gain. However, neither maximum rate of gain nor the most efficient conversion ratio ensures economical gains. Feed prices are important in determining the cost of gain.

Figure 1-C illustrates a hypothetical cost-of-gain curve superimposed on figure 1-B.

The object in the formulation of beef cattle finishing rations is to feed at the low point on the cost-of-gain curve. **It is probably true that more feeders who experience high cost gains do so from being away from optimum on the cost curve rather than ration imbalance itself.**

The cost-of-gain curve is affected by the factors shown on both the rate and efficiency curve; however, feed price and overhead costs can move this curve either to the right or left. If, for example, a feeder had good quality alfalfa at a price of $10/ton delivered to the cattle, then he may achieve maximum profit on a total ration made up of alfalfa, providing he could achieve very low nonfeed fixed costs. Rate of gain and feed efficiency for cattle on this ration would most likely be less than maximum, yet they would produce maximum net profit. For most feeders, concentrates such as milo, corn, or barley may actually cost less per pound than any available roughage. For them, a high concentrate ration will usually lead to lower cost gains.

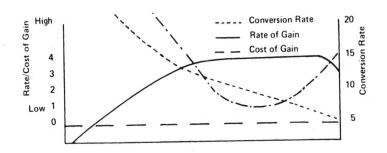

Figure 1-C. Effect of energy density on rate, efficiency, and cost of gain

All cattle feeders are affected by the relationships that affect the cost-of-gain curve. In areas where both corn and corn silage are available, the cost relationship of silage of grain frequently changes so that one year the high silage program may be better than the grain program and in the next year this can be reversed.

The steps necessary to approximate the low point on the cost-of-gain curve (figure 1-C) are easy to follow and should be an essential part of balancing any feedlot ration.

Using the net energy values of common feeds (table 7) and requirement tables (table 1), it is possible to project the cost-of-gain on any ration. By formulating and projecting higher or lower energy rations, the feeder can test to see if an improvement may be made in terms of cost of gain.

The examples shown in table 8 are intended to illustrate extremes. The protein supplement is assumed to contain 0.60 Mcal NEm and 0.40 Mcal NEg per pound. Feeds are priced in the first ration and remain the same in all rations.

Procedure for calculating projected gains and costs for a particular ration:

Step 1: Calculate net energy for maintenance and net energy gain of the ration.

Step 2: Estimate feed consumption (an 800 lb steer would probably eat about 24 lb of DM of this ration.)

TABLE 7. NET ENERGY AND PROTEIN CONTENT OF COMMON FEEDS[1]

	NEm Mcal per lb	NEg Mcal per lb	Crude prot. %
ROUGHAGES:			
Alfalfa hay, excellent	0.71	0.40	22.0
Alfalfa hay, good	0.57	0.27	20.6
Alfalfa hay, fair	0.51	0.18	17.1
Alfalfa hay, poor	0.47	0.09	13.7
Barley straw	0.46	0.07	4.1
Bermuda hay	0.48	0.13	7.9
Bermuda hay, good	0.54	0.24	10.2
Corn stover	0.56	0.26	6.5
Cottonseed hulls	0.47	0.10	4.3
Milo stover	0.41	0.13	3.6
Oat hay	0.52	0.18	9.9
Prairie hay, early cut	0.52	0.18	8.6
Prairie hay, average	0.49	0.13	8.1
Prairie hay, mature	0.49	0.11	5.5
Wheat straw	0.16	0.07	4.3
SILAGES:			
Alfalfa silage, good	0.60	0.30	17.5
Alfalfa silage, average	0.54	0.24	17.7
Corn silage, excellent	0.73	0.46	8.4
Corn silage, average	0.73	0.43	8.0
Corn silage, fair	0.66	0.40	7.2
Sorghum silage	0.67	0.30	7.9
BY-PRODUCT FEEDS:			
Alfalfa dehy 15%	.59	.27	16.4
Alfalfa dehy 17%	.60	.31	19.2
Beet pulp	.73	.47	9.6
Beet pulp M.D.	.92	.61	9.9
Beet molasses	.92	.62	8.5
Cane molasses	1.03	.67	3.8
Cottonseed meal, O.P.	.82	.54	44.7
Cottonseed meal, sol.	.77	.50	45.6
Cottonseed, whole	.91	.54	25.6

	NEm Mcal per lb	NEg Mcal per lb	Crude prot. %
BY-PRODUCT FEEDS (cont.):			
Fat, animal	2.08	1.27	--
Guar meal	.83	.56	41.1
Hominy feed	1.04	.68	11.8
Meat meal	.71	.48	55.2
Rice bran	.75	.50	14.2
Rice mill feed	.38	.20	6.4
Soybean meal, solvent	.87	.59	48.9
Soybeans, whole	1.08	.70	42.1
Wheat middlings	.87	.49	17.2
Barley 48-50#	.97	.64	13.0
Barley 44-46#	.80	.53	13.3
Barley 38-42#	.73	.46	11.7
Corn, dent no. 2	1.02	.67	10.0
Milo, minimum process[a]	.80	.53	10.0
Milo, semiprocess[b]	.90	.60	10.0
Milo, extreme process[c]	.97	.64	10.0
Oats 38# BU test	.83	.57	12.2
Oats 36# BU test	.79	.52	13.5
Oats 32# BU test	.74	.47	13.8
Oats 28# BU test	.66	.44	13.8
Wheat	1.00	.65	14.3
MINERALS:			
Ammonium chloride	--	--	163.0
Bone meal	--	--	5.5
Dicalcium phosphate	--	--	--
Ground limestone	--	--	--
Monosodium phosphate	--	--	--
Oyster shell flour	--	--	--
Potassium chloride	--	--	--
Rock phosphate	--	--	--
Sodium tripoly phosphate	--	--	--
Urea	--	--	281.0

a Coarse, ground, or rolled

b Very fine grind, excellent dry roll, high volume steam flaker, less than optimum high moisture harvest or reconstitution.

c Excellent quality milo with good job of steam flaking (24-26# BU test weight), optimum high moisture harvest.

1 All feeds expressed on a dry matter basis.

TABLE 8

Ration	Commodity	Percent	Cost per cwt
1	Alfalfa hay, fair Corn dent. no. 2 Supplement	90% 5% 5%	$4.00 6.30 8.00 Cost $4.32
2	Alfalfa hay, fair Corn dent. no. 2 Supplement	50% 45% 5%	$4.00 6.30 8.00 Cost $5.24
3	Alfalfa hay, 28% fiber Corn dent, no. 2 Supplement	10% 85% 5%	$4.00 6.30 8.00 Cost $6.16

Step 3: Calculate gain from net energy requirement tables and Step 1.
 a. NE required for maintenance = 6.47 (table 1).
 b. Feed required for maintenance = 6.47/0.54 = 11.98 lb.
 c. Feed available for gain = 24 - 11.98 lb = 12 lb.
 d. NE_g available for gain = 12 x 0.216 = 2.59
 e. Expected gain (table 1) = 1.20 lb

Step 4: Cost projection
 a. Ration cost per cwt = 4.32.
 b. Daily feed cost = 4.32 x 24 = $1.04
 c. Feed only cost per pound of gain = $1.04/1.02 lb/day=$.8667 per pound.
 d. Total cost of gain = feed cost $1.04 + overhead $.45/gain = $1.24/lb.

It is apparent that this ration will not give economical gains because of its low energy density.

The formulator should follow the steps to project the performance of Ration 2. In this case, a feed intake of about 22 lb appears reasonable.

The calculated NE_m of the ration is 0.744 Mcal/lb and the NE_g is 0.41 Mcal per lb. With an 800 lb steer consuming 22 lb a day, 8.7 lb will be used for maintenance, leaving 15.3 lb for gain. 15.3 lb x 0.41 Mcal/lb NE_g gives 6.27 Mcal NE_g daily for gain. Using table 1, this level of energy would allow gains of about 2.7 lb per day. Feed cost would run $1.26/2.7 or 46.6 cents per pound. However, total cost would run approximately 63.33 cents per pound.

The third ration is a high nutrient density ration. Expected feed intake on 800 lb steers would be about 20 lb per day. Using the procedure, 20 lb of this ration would give gains between 3.3 and 3.4 lb per day. Feed-only cost of gain is 36.78 cents. When 45 cents per day overhead is added, total gain costs would be 50.20 cents per pound of gain.

Nutritionists should test rations with different energy densities around the 90% concentrate level to see if even more reduction in feed cost might be afforded. After selecting a ration, a comparison should be made of estimated feed consumption and actual feed consumption. It may be necessary to make additional adjustments because of errors made in estimating feed intake or the abilities of the cattle.

Because of the relationships illustrated in the three ration examples, it should be apparent that feed recommendations cannot be made in terms of optimum energy levels. Price relationships differ greatly among individual cattle feeding operations and within areas of the U.S. so that an optimum ration for one feeder is frequently impractical for another. One fact is readily apparent--regardless of area and feed prices, gains of very low orders are impractical because of the high fixed overhead costs in cattle feeding.

High interest rates, high labor costs, increased cost of death loss resulting from high-priced feeders have, in many cases, doubled the fixed nonfeed costs associated with the ownership of cattle. These costs will add 40 to 50 cents a pound to the cost of low-order gains, i.e., $0.75 to $1.25 lb per day. Even with gains at 3 lb per head per day, few of the most efficient cattle feeders have overhead costs per pound of gain of less than 15 or 16 cents.

Corn silage has in the past consistently proven to be the best roughage available for finishing cattle. Feeders who do not have good corn silage should be cautious not to conclude that the high roughage levels shown in tables 3 and 5 are always desirable. For example, alfalfa hay (fair) has 39% of the gain potential per pound of dry pound of dry matter after maintenance as that of good corn silage. Substitution of "fair" alfalfa hay for corn silage in the ration shown in table 3 would result in gains of less than 1.00 lb per day, and if the hay cost $60/ton, feed cost alone would run 72 cents per pound of gain, and the total cost would be approximately $1.17 /lb.

The net energy system developed by Lofgreen and Garrett (1968) was a major breakthrough in beef cattle nutrition. By using this system, progressive cattle feeders can calculate rations that can be formulated to give least-cost gains. Computer programs are already developed (Oklahoma State University "Feedmix.") that can formulate any number of least-cost rations, project these rations in terms of performance and cost, and then allow the feeder or nutritionist to select the ration most likely to produce least-cost gains. Tables 3 through 6 are examples of these outputs.

It is apparent that feedlot cattle are capable of handling extremely wide ranges of energy levels, as illustrated by the rations in tables 3 and 4. It is up to those men responsible for cattle feeding to select the optimum program for his situation.

Nutritional Recommendations

Energy Levels. Optimal varies with feed costs, cattle cost, overhead, and cattle type.

Protein Levels. Beef cattle nutritionists now recognize the relationships between protein and energy. It is suggested that the tentative protein allowance tables (table 2) be used.

Feeding of excessive levels of protein is usually costly. Protein requirements are much higher for young cattle and animals whose growth consists mostly of muscle tissue. A general review of recent research on protein requirements indicates a trend among research groups to recommend higher protein levels than were believed necessary in recent years, especially on high concentrate rations.

Urea. The use of urea in finishing rations for feedlot cattle is well accepted. Excessive levels usually reduce rate and efficiency of gains. Including urea into rations at a rate of 0% to 0.45% of the ration has given excellent results with most but not all types of rations. If natural proteins are expensive, sometime it may be more economical to go to levels up to .80% of the ration as urea. In most cases gains and efficiency will be reduced somewhat from the more conservative levels. In formulating rations, feeders should note that some natural protein supplements are much lower in energy than are others. Substitution of urea plus grain for cottonseed meal may be much more desirable than the same substitution for soybean oil, because soybean oil meal is much higher in energy for gain than cottonseed meal.

Phosphorus. Levels of phosphorus in the ration dry matter between 0.30% and 0.50% (higher level with very light cattle) seem to be accepted.

Calcium. Levels of 0.40% to 0.60% in the ration dry matter are apparently adequate. Calcium phosphorus ratios of 1:1 to 2:1 seem to be indicated; little problem is encountered, however, if higher amounts of roughages such as alfalfa hay. The addition of calcium sources such as calcium carbonate at levels in excess of 1% of the ration dry matter do not seem to be beneficial and frequently depress performance.

Potassium. Potassium level is of little concern in high to moderate roughage rations. In high concentrate rations, potassium level probably should be brought up to at least 0.4% but little evidence supports going higher than 0.7% of the ration dry matter. When feed intakes are very low, higher levels may improve performance.

Salt. If cattle feeders wish to sell or use feedlot manure as fertilizer, they should give serious consideration to reducing salt levels of finishing rations to 0.2% to 0.3% of the ration. These levels are adequate and do not contribute to the salt pollution problem to the extent as do the generally accepted (but excessive) levels.

Trace Minerals. It has been very difficult to demonstrate trace mineral deficiencies in feedlot rations. Optimum levels of the minor elements are suggested in table 3; in most rations, these are met with no additional elements. Where careful chemical analysis does indicate a shortage of a trace element, care should be taken to add the element in an available form. Many trace minerals today are added to rations in poorly available forms, generally oxides, which tend to support university research showing that these elements are usually unnecessary. Many dollars are wasted nationally each year on unnecessary trace minerals, and a number of deficiencies are produced by the unskilled addition of trace elements to rations that actually interfere with the availability of other elements.

Roughage--Roughage Factor. It appears that the ruminant animal requires a minimum amount of roughage factor for normal rumen function. All concentrate rations show a response to small amounts of roughage factor. As little as 1 lb per day of some low-quality roughage such as alfalfa stems, rice hulls, or sorghum straw has given improvements in efficiency ranging up to 15% on all-concentrate rations. Five pounds of "as fed" corn silage daily is an excellent roughage. Most nutritionists would agree that it is hard to beat the consistency of performance on a "high roughage-high energy ration," which is a ration containing a generous amount of roughage yet having a high nutrient density. This would imply a high energy roughage or possibly the addition of fat to offset a limited amount of a low-energy roughage.

General recommendations would state that finishing rations for cattle should contain at least 7% "high roughage factor" roughages for optimum results.

Vitamins. Only the addition of vitamin A at a rate of 20,000 to 30,000 IU per head per day is universally accepted for cattle finished in open lots. There is little evidence to support the need for other vitamin fortification of rations.

All cattle feeders should periodically review their rations and calculate how much they are spending on unnecessary feed nutrients and additives; energy cost will be the major cost in feeding cattle.

The other nutrients and proven feed additives should be adjusted to levels that allow maximum use of the energy in the ration. Once this is accomplished, little more may be gained.

59

FEEDING CATTLE TO OBTAIN
OPTIMUM CARCASS AND OFFAL VALUES

Donald R. Gill

Cattle feeders have known for years that as cattle approach or pass normal marketing weights, three things happen: 1) daily gain decreases, 2) feed per unit of gain increases, and 3) cost of gain increases. The net energy system proposed by Lofgreen and Garrett (1968) can be used to demonstrate that energy requirements (feed) do in fact increase with each unit increase in weight. It is not surprising that some have proposed shortening the period of time that cattle are fed.

The factors that should determine the length of feed should be: 1) final product quality control, and 2) optimization of the cattle value within the biological limits imposed by the cattle, feed, etc.

The biological-economic factors that interact while cattle are on feed are fairly well understood on a live weight basis, but are often not understood at all on a carcass basis. The data in figure 1 can be used to illustrate the effect of time on feed on the economic parameters normally considered by cattlemen. This figure represents a simulation of a 700 lb black baldy steer of medium-frame size. Cattle, feed, lot, and money costs are typical for High Plains feedlots about the first of March 1982. Before drawing any conclusions from this figure, carefully study all the input parameters; a change in any of these may cause a significant change in the simulation. The net energy equations can be used to estimate live weight gains on cattle, either on a daily basis or over a period of time. This is illustrated under the "Gain" column below the heading "Days Lb." Note that the gain for day 0-180 given at 20 day increments ranges from a high of 3.55 lb to a low of 2.14 lb on the last day.

Feed conversion on a daily basis reaches a low 5.64 lb of feed per lb of gain on day 20 and progressively increases to 9.17 on the last day. However, it should be noted that the pay-to-pay feed conversion improved through the first 100 days due to dilution of fixed costs inherent in cattle feeding.

Days	Animal weight	Days feed	Days feed cost	Gain		Conversion		Cost days total	Gain PY_W total	Break even $/cwt
				Days lb	PY_W lb	Days lb	PY_W lb			
0	665	0.00	0.00	0.00	0.00	0.00	0.00	0.00	0.00	70.75
20	724	19.73	1.28	3.50	1.21	5.64	14.39	0.45	1.43	69.07
40	797	20.44	1.34	3.55	2.42	5.76	7.76	0.47	0.69	66.86
60	866	20.87	1.37	3.36	2.76	6.21	7.02	0.51	0.61	65.41
80	931	21.05	1.38	3.16	2.89	6.66	6.86	0.55	0.59	64.52
100	992	21.02	1.38	2.95	2.92	7.12	6.87	0.59	0.58	64.05
120	1,049	20.83	1.37	2.74	2.91	7.59	6.95	0.63	0.59	63.90
140	1,102	20.50	1.35	2.54	2.87	8.08	7.06	0.68	0.60	63.97
160	1,150	20.08	1.32	2.23	2.81	8.61	7.20	0.73	0.61	64.24
180	1,195	19.59	1.29	2.14	2.75	9.17	7.36	0.79	0.62	64.68

Input parameters:
Steers
Purchase weight	700
Purchase cost/cwt	$66.50
Intake factor	5.4
Medical cost/head	5.0
Shrinkage %	5.0
Equity per head	0.0
Interest rate %	16.50
Yardage per day	.05
Death 1%	.40
Day loop 1	15
Death 2%	.40
Day loop 2	100

Rations used:

Number	Days fed	$/cwt	Nem	Neg
1	10	6.26*	86	55
2	10	6.48	91	59
3	160	6.56	95	62

* $2.65/bu. corn, $70/ton alfalfa; $25/ton corn silage; $200/ton soy-meal; ration mark-up = $20/DM/ton

Figure 1. Simulation of a 700 lb feedlot steer on March 5, 1982

In the cattle business, there is little information on how to relate cattle and financial interactions to consumer desires and the final value of the retail product.

The data shown in figure 1 are typical of feedlot experience. The major factor is that as cattle become heavier more feed is required per pound of live weight gain. While the industry has talked about live weight gains for years, the value of the feedlot steer is determined in the carcass.

How the beef feeding segment works becomes clearer when the simulation in figure 1 is expanded to show the current market value of cattle as they progress through the feeding period. It is very important to recognize that daily changes do occur on the production and on the market side, and while the conclusions developed from the simulation in figure 2 are accurate for the conditions set forth, there will likely never be another day that will give the same numbers.

As any cattleman knows, the total beef market is large and there seems to be a market for all types of carcasses. On any day, carcass beef sells from 60 lb veal carcass to 900 lb Y-5s. If there are too few of a type or class, these bring a premium price; conversely, too many bring a discount price.

The beef price structure of Friday, March 5, 1982, is shown in figure 3. The column "Max live bid, $" is calculated on the basis of what these cattle would have brought on a beef market where a packer would hold a constant $30 kill cost, including profit margin. The value is determined by adding the offal credit minus the kill cost to the wholesale carcass value.

Usually cattle increase in value as the feeding period progresses. Cattle on feed are usually worth far less in the beef trade than feeder cattle sell for during the first few weeks on feed. In the data in figure 2, these cattle increased in live value (maximum live bid based on total carcass and offal value less $30) from a low of $52.99 to a high of $70.61/cwt over the 180-day projection. Note that the feeder animal in this example was valued at $66.50/cwt; if the feeder cattle had cost less than $52.99, you could expect this type cattle to start going directly to slaughter. Most of the time, cattle are "cheapened back" in the feedlot.

The exceptions to this have occurred a few weeks out of the last decade when feeder cattle sold for far less than fed cattle. When feeder cattle are high in relation to fed cattle, i.e., 1979, the increase in the "Max live bid, $" value is very steep as cattle progress on feed. If for some reason this did not occur, then feeder cattle would need to sell for much less per pound than did fed cattle. Note the sharp increase in the value of the carcasses in figure 2,

Days	Animal weight	Cost days total	Gain PY W $	Break even $/cwt	Est. dress %	% Ch.	% Y-4 plus	Car. value $	Max live bid $	F.L. P&L $	Increase in value per day	Total cost /day
0	665	0.00	0.00	70.75	52.00			352.4	52.99	-118.00	--	
20	724	0.45	1.43	69.07	53.50						--	1.67
40	797	0.47	0.70	66.86	55.00			452.2	56.74	-80.70	2.50	1.68
60	866	0.51	0.61	65.41	57.50	20		520.6	60.11	-45.90	3.42	1.71
80	931	0.56	0.59	64.52	60.50	30		595.7	63.99	-32.50	3.76	1.71
100	992	0.60	0.59	64.05	61.50	40		652.9	65.82	-6.99	2.86	1.71
120	1,049	0.64	0.59	63.90	62.50	60	2	709.2	67.60	38.84	2.81	1.72
140	1,102	0.69	0.60	63.97	63.50	70	5	755.9	68.70	50.97	2.34	1.72
160	1,150	0.74	0.62	64.24	64.50	80	10	803.9	69.90	65.17	2.40	1.71
180	1,195	0.80	0.63	64.68	65.50	85	25	843.9	70.61	70.96	2.00	1.71

Price structure: Friday March 5, 1982
Under 600 lb--$2.00
Good Y-2 or better $1.02
Good Y-3 or better $1.01
Choice Y-2 $1.07
Choice Y-3 $1.05
Choice Y-4 $ .98
Choice Y-5 $ .93

Offal credit = $5.50 per cwt live
Kill cost = $30 per head

Figure 2. The value of cattle slaughtered on March 5, 1982

which were increasing at over $3.00 a day in the early feed-
ing period. Even in the last 20 days, the value went up
$2.14 per day. This was still more than the cattle feeder's
total cost, which averaged about $1.71 per day during the
last 20 days.

The major factor that affects the "Max live bid, $"
value of cattle is dressing percentage. Grade of cattle is
like the tail on the dog--it has an effect but only on the
value per pound. Yield grade works the same way. It will
take much larger premiums than have typically been paid to
improve the profitability of the shorter-day cattle in
figure 2 under the conditions that existed at that time.

In the simulation in figure 2, the dressing percentages
were estimated based on our experiences with our feedlot
research cattle. Dressing percentages increased from 52% as
the cattle were started on feed to over 65% with a long
feeding period. With heavy-muscled cattle, higher dressing
percentages can be achieved with fewer days and fat cover--
with thinner-muscled cattle, the reverse is true. Walters
and Hintz (1981) have developed preliminary equations to
predict yield grade and dressing percentages on feedlot
cattle. Better equations are under development for yearling
cattle. The beef industry needs to develop more data on
these types of relationships to be able to plan and manage
cattle to their maximum economic potential.

In figure 3, carcasses were quoted from $100 to $103.94
per cwt; thus, to break even, it would have been necessary
to sell them at $94.18 for the longest-fed cattle up to
$134.16 for the feeder cattle. On this market, it is quite
apparent that it would not have paid to short-feed cattle.
The relationships between live "Break even" and "Max live
bid, $" are very important to any cattle feeder.

If the consumer or retailer finds, for example, that
lower-grading cattle with less fat cover are worth more to
them, then they could possibly bid more for short-day
cattle. Using the example in figure 2, had the trade bid
$11.05 more per hundred on a carcass basis at 100 days, it
would have yielded the same money to the cattle feeder as
the 160 day cattle.

Knowledge of how the market works, and of how to produce the kind of product that the consumer wants to buy, is
the real challenge for the beef industry. The use of the
computer will make it possible for today's cattleman to
study the impact of costs and the biological potential of
cattle to obtain an optimum balance between consumer desires
and beef production costs. The cattle business is too complex for any one man or group of persons to know how to
optimize cattle feeding for maximum profit without taking
into account the many factors that affect profitability. A
computer can be a useful tool to this end. In the past and
in the future dressing percentage has been and will likely
remain the most important factor in determining value of

Days on feed	Steer weight lb	Break even live	Est. dress %	Carc. weight lb	Carc. & net offal value $	Max live bid $	Feeder P&L $	Packer carc. break even $/cwt	Carc. price to break even feeder $/cwt live
0	665	70.75	52.00	345.80	352.38	52.99	-118.11	100.00	134.16
20	724	69.07	53.50	387.34	452.19	56.74	-80.68	100.00	118.41
40	797	66.86	55.00	438.35	520.56	60.11	-45.89	101.00	110.22
60	866	65.41	57.50	497.95					
80	931	64.52	60.50	563.26	595.73	63.99	-4.95	102.00	102.88
100	992	64.05	61.50	610.08	652.94	65.82	17.56	103.00	100.12
120	1,049	63.90	62.50	655.65	709.15	67.60	38.84	103.94	98.02
140	1,102	63.97	63.50	699.77	755.92	68.70	50.97	103.65	96.37
160	1,150	64.24	64.50	741.75	803.93	69.90	65.17	103.90	95.11
180	1,195	64.68	65.50	782.73	843.89	70.61	70.96	103.25	94.18

Beef price structure: March 5, 1982
Carcass under 600 lb--$2.00

Good Y-2 or better	$1.02
Good Y-3 or better	$1.01
Choice Y-2 or better	$1.07
Choice Y-3 or better	$1.05
Choice Y-4	$ .98
Choice Y-5	$ .93

Offal credit = $5.50 per cwt live
Kill cost = $30 per head

Figure 3. Study of cattle value market of March 5, 1982

cattle at time of slaughter. The cattleman of the future as well as the experienced packer buyer will know the economic impact of this factor and the other factors that affect carcass value at any point in the feeding period on traditional finished cattle.

REFERENCES

Lofgreen, G. P. and W. N. Garrett. 1968. A system for expressing net energy requirements and feed values for growing and finishing beef cattle. J. Anim. Sci. 27:793.

Walters, L. E. and R. L. Hintz. 1981. Preliminary development of yield grade and dressing percentage equation for beef steers. Okla. Agr. Exp. Sta. Res. Rep. MP 108, p 49-51.

60

USING A COMPUTER TO SIMULATE FEEDLOT CATTLE FOR BOTH NUTRITIONAL AND FINANCIAL PURPOSES

Donald R. Gill

THE DECISIONS IN FEEDING CATTLE

A key step in the decision to feed or not feed cattle is the process of pricing cattle and then figuring out or estimating a break-even price for the cattle at marketing time. If the feeder figures a break-even below what he thinks cattle will bring, he may buy the cattle and place them on feed. To aid cattle feeders in this process, hundreds of tables, calculators, nomographs, and fudge factors have been developed over the years. While many of these served a purpose at the time of its development, they aren't worth much today because something about them is out of range.

The factors that affect cost of gain have not changed over the years. They are:
1. Cattle price (laid in)
   a. Commission
   b. Freight
2. Shrink
3. Death Loss
4. Medical expenses
5. Interest rates
6. Yardage changes, and taxes
7. Feed cost per unit of gain

Not many years ago, most cattle feeders did not pay much attention to items 3, 4, and 5--as they didn't really make much difference in the final break-even figure. Today it would be interesting to rate these seven items in order of importance.

The dollar amount of each of these items can be generated using the Oklahoma State University (OSU) computer gain simulator and the following assumptions:
1. Cattle price--$62.00 cwt laid in
2. Shrink--5%
3. Death loss--1.86%
4. Medical expense--$5.75
5. Interest rate--16.5%
6. Yardage--$.05/head/day

7. Feedlot costs @ $3.16/bu. corn
   Ration (includes feedlot mark-up to generate $.14 to $.15¢ per day):
   1. $7.20 per cwt DM
   2. $7.30 per cwt DM
   3. $7.60 per cwt DM
   4. $7.70 per cwt DM
   5. $7.80 per cwt DM
   6. $7.75 per cwt DM

Table 1 is a projection of the cattle performance for progressive ten days periods to show costs etc. Cattle cost as affected by shrink is apparent on day 0 where we would have $66.32/cwt in the cattle at feedlot weights and with medical costs added. At this point this represents $362.11 for a 546 pound in weight. In table 2, we can find the dollar amount for each of our input items. They are as follows:

1. Cattle cost at the feedlot--$356.50
2. Shrink changed a $62.00 steer into a $66.32 steer, or about $18.00 was spent to feed the animal back to its original purchase weight
3. Death loss--1.86% cost $7.61 per head
4. Medical expenses--$5.75 per head
5. Interest--$39.02 per head
6. Yardage and taxes--$8.10
7. Feed cost/unit of gain
   a. pay-to-pay all costs include $63.52 cwt
   b. feedlot (when the appropriate figures are added up, it will take $655.63 or $62.68/cwt to break even on a 1046 pound steer with these costs), feed, medicine, and yardage $50.55 cwt death loss subtracted from in weight
   c. total feed bill--$238.48, feed only cost, less feedlot markup and shrink--$207.61

A percentage breakdown on the total costs shows the following:

Original cattle costs	54.37%
Death loss	1.16%
Medical expenses	0.88%
Interest	5.95%
Yardage	1.31%
Feed including mark up	36.37%

As can be seen in tables 1 and 2, the steer in the simulation is a good one; will it be profitable to feed with an expected break-even of $62.68 in the latter part of June? If you are a rancher/feeder and these are the only cattle you have to consider, you may decide either to sell feeders or handle feedlot cattle. The professional cattle

TABLE 1. OKLAHOMA STATE UNIVERSITY BEEF GAIN SIMULATOR

Date	Day	Fat wt	Days feed	Days feed cost	Gain			Conversion			Cost of Gain			Break even $/cwt
					Days lb	Average IN W	Average PY W	Days lb	Average IN W	Average PY W	Days Total	Average IN W	Average PY W	
1/01/83	0	546	0.00	0.00	0.00	0.00	0.00	0.00	0.00	0.00	0.00	0.00	0.00	66.32
1/10/83	10	564	11.12	0.80	1.75	1.75	-0.00	6.36	6.29	-0.00	0.69	0.81	-0.00	66.37
1/20/83	20	596	16.42	1.25	3.37	2.49	1.06	4.88	5.45	12.86	0.50	0.53	1.59	65.46
1/30/83	30	630	16.96	1.32	3.44	2.80	1.84	4.93	5.23	7.97	0.51	0.48	0.92	64.65
2/09/83	40	664	17.45	1.36	3.39	2.95	2.23	5.15	5.18	6.85	0.48	0.46	0.75	63.73
2/19/83	50	698	17.86	1.39	3.33	3.03	2.46	5.37	5.29	6.42	0.50	0.45	0.68	63.00
3/01/83	60	731	18.21	1.42	3.26	3.07	2.60	5.58	5.25	6.22	0.52	0.45	0.64	62.45
3/11/83	70	763	18.50	1.44	3.19	3.10	2.69	5.80	5.32	6.13	0.54	0.45	0.62	62.05
3/21/83	80	794	18.73	1.46	3.11	3.10	2.74	6.02	5.39	6.10	0.56	0.45	0.61	61.77
3/31/83	90	825	18.90	1.46	3.07	3.09	2.78	6.15	5.47	6.10	0.57	0.46	0.61	61.59
4/10/83	100	855	19.03	1.47	2.99	3.09	2.80	6.37	5.55	6.12	0.59	0.46	0.60	61.48
4/20/83	110	885	19.10	1.48	2.90	3.08	2.82	6.59	5.63	6.16	0.62	0.47	0.60	61.45
4/30/83	120	913	19.14	1.48	3.81	3.06	2.82	6.82	5.72	6.20	0.64	0.47	0.61	61.50
5/10/83	130	941	19.13	1.48	2.71	3.04	2.81	7.05	5.80	6.26	0.67	0.48	0.61	61.62
5/20/83	140	967	19.08	1.48	2.62	3.01	2.80	7.28	5.89	6.32	0.69	0.48	0.61	61.79
5/30/83	150	993	19.01	1.47	2.53	2.98	2.79	7.52	5.98	6.39	0.72	0.49	0.62	62.02
6/09/83	160	1018	18.90	1.47	2.43	2.95	2.77	7.77	6.07	6.46	0.75	0.50	0.63	62.30
6/19/83	170	1042	18.77	1.45	2.34	2.91	2.75	8.02	6.15	6.53	0.78	0.50	0.63	62.62
6/21/83	172	1046	18.75	1.45	2.32	2.91	2.74	8.07	6.17	6.55	0.78	0.51	0.64	62.68

feeder on the other hand may decide on a performance projection on a number of weights and classes of both steers and heifers to see what kind of cattle offer him the best profit potential.

TABLE 2.

Input Parameters		Recap at Sale Weight (6/21/83	
Sex (1=S, 0=H)	1	In weight	546
Purchase weight	575	Out weight	1046
Purchase cost/cwt	62.00	Days fed	172
Starting factor	0.7	Gain/head	471
Feeder grade	4.3	Feedlot gain/head	500
Medical cost/head	5.75	Avg daily gain	2.74
Shrinkage %	5.00	Feedlot avg daily gain	2.91
Selling weight	1045	Conversion	6.55
Equity/head	0.00	Feedlot conversion	6.17
Interest rate %	16.25	Total cost of gain/cwt	63.52
Overhead/head/day	0.05	Feedlot cost of gain/cwt	50.55
FRT + Comm/head	0.00	Breakeven/cwt	62.68
In date	1/01/83	Total interest	39.02
Death 1	1.50	Total overhead	8.60
Day 1	30	Death loss cost	7.61
Death 2	0.50	Profit	45.16
Day 2	200		
Print increment	10		

Ration	Days	$/cwt	NE_m	NE_g	Days fed	Pounds	Cost
1	12	7.20	86	55	12	141	10.22
2	5	7.30	89	57	5	80	5.89
3	5	7.60	90	60	5	82	6.24
4	5	7.70	92	61	5	83	6.42
5	60	7.80	94	62	60	1082	84.40
6	300	7.75	94		85	1616	125.28
Totals		7.73			172	3087	238.48

The ability to evaluate the potential breakeven on many weights and classes of cattle in a short time with a reasonable degree of accuracy is important to any feeder. Fortunately, the ability to run breakevens, or simulations similar to the ones illustrated in tables 1 and 2, is within the financial reach of anyone who can afford to buy cattle. The OSU version developed in 1974 and shown in the table has now been adapted to the TI-59 programmable handheld calculator and the inexpensive microcomputer.

CATTLE GAIN SIMULATIONS

The essential mathematical formulas to do cattle gain simulations are listed below. Most of the better feedlot cattle simulations use the net energy system of predicting gain. The simulator must also simulate feed intake. This is the most difficult part of developing these programs. The program we use has an input factor called feeder grade. The original idea was to tie feed intake to the body-type score developed by the University of Wisconsin. We use a numerical input from 1 to 10. This number does two things. First, it determines how much feed the simulator feeds the animal and, second, the shape of the feed-intake curve. The feed-intake equations are influenced by other factors: (1) the initial weight of the animal as it enters the feedlot (remember that a 600 pound steer with a 10% shrink weighs only 540 when it starts), (2) the weight of the animal on the day fed, and (3) the feeder grade. Feeder grade is the only factor that the user can control. Of the three factors, the weight at which the cattle start on feed is the most powerful. Two formulas are used to generate feed intake.

$$K = .21 \times \frac{\text{in wt} - 300}{1000} + FG_{100}$$

$$\text{Daily Feed} = K \times \frac{W^{.75}}{2.2} - \frac{W-500^2}{200} \times .90 \times DM$$

K = internal constant
FG = feeder grade
in wt = feedlot in weight (in pounds)
W = weight on any day (in pounds)
DM = ration dry matter content (expressed as a decimal)

When designing these equations, we tried to tie how much feed an animal would consume to his body-type score, i.e., 1 = the smallest frame type cattle and 5 = a large frame type Hereford. On this scale a 2.5 was average. This concept works with calves except that the average body type received at the feedlots is about a 3.5 today (with very young cattle). Yearling cattle, especially those from wheat pastures, often eat much more feed, and feeder grades of 4.5 to 7.5 may be used with these 650 to 800 pound cattle. For yearlings, start with the following body-type score: 600 lb = 4.2; 650 lb = 4.8; 700 lb = 5.5; 750 lb = 6.0.

Correct feeder grade is essential for an accurate simulation of cattle, and the user should test every run to determine the average intake. This figure should be compared to similar closed pens and to current feedlot daily management reports to assure a reasonable simulation.

Each unit added or subtracted to the feeder grade will change feed intake for the entire feeding period by about 1 pound of dry matter per day.

EQUIVALENT SIZE

An added feature that can really increase the usefulness and accuracy of simulations is the capability to do an equivalent size adjustment on the net energy equations. We have called this an Efficiency Factor. We accomplished this by multiplying body weight by the formula:

$$1 - [(\text{Efficiency Factor} - 5) \times .05]$$

The current weight of the animal is multiplied by this factor before using both the net energy formulas for maintenance and for gain.

Efficiency Factor

The normal efficiency factor is 5, and when this value is used the original net energy equations are used without modification. If all cattle in a feedlot were averaged for a period of several years, this value would give an estimate of feedlot performance. Some cattle would beat the simulation and some would fall behind. Within a year, cattle placed in the late spring and early summer would possibly do better than those placed in the late fall and early winter.

The efficiency factor may be used for two purposes: (1) some cattle, because of breeding, background, age, and flesh condition, will be more efficient than normal. (2) Some cattle come to the feedlot over-aged and over-fleshed and, they can be expected to perform more poorly than average.

Experience has shown that, for average cattle in the panhandle lots, a factor of 6 works for the best feeding months and a factor of 4 works for the worst. The user should carefully compare his results with the cattle and history.

Starting with 5 as normal, each digit up or down will increase/decrease gain and feed efficiency about 5%; it will change gain about .15 to .16 lb and conversion about .25 to .32 lb. Each unit of efficiency factor changes equivalent body size in the equation 5%.

Table 3 shows the power of efficiency factor when applied to a 650 pound steer fed to 1050 lb on a typical ration sequence.

Net Energy Equations

To solve for gain, knowing ration NE_m and NE_g, use the following equations and examples:

A Megacalories for maintenance $MMC = .043W^{.75}$

B Pounds for maintenance MP x $\dfrac{MMC}{NEm}$

C Pounds for gain GP = C - MP
D Mcal for gain GMC = GP x NE_g

E Gain for steers

$$\dfrac{.00017474 + \dfrac{.003112 \text{ x } GMC}{W^{.75}} - .00322}{.001556}$$

F Gain for heifers

$$\dfrac{.00019740 + \dfrac{.005760 \text{ x } GMC}{W^{.75}} - .01405}{.002880}$$

TABLE 3. EFFECT OF EFFICIENCY FACTOR ON RATE AND EFFICIENCY OF GAIN IN A SIMULATION (FEEDER GRADE = 4.8)

Efficiency Factor	ADG	Conversion
9	3.55	5.35
8	3.35	5.69
7	3.19	5.97
6	3.03	6.32
5	2.86	6.69
4	2.74	6.98
3	2.60	7.37
2	2.47	7.76
1	2.35	8.17
.001	2.27	8.48

Example 650 lb steer, feed 16 lb NE_m = 0.90 NE_g 0.60 per lb

A 650 $.^{75}$ = 128.73155
   128.7315 x 0.042 = 5.5354 Mcal/day

B MP = $\dfrac{5.5354}{.90}$ = 6.1505 lb for feed for maintenance

C GP = 16 - 6.1505 = 9.8494 lb of feed for gain
D GMC = 9.84949 x 0.60 = 5.9096 Mcal NE_g/day

E Gain =

$$\cfrac{.00017474 + \cfrac{.003112 \times 5.9096954}{650^{.75}} - 01322}{.001556}$$

$$= \cfrac{.00017474 + \cfrac{.0183909721}{128.731556} - .01322}{.001556}$$

$$= \cfrac{.00017474 + .000142863 - .01322}{.001556}$$

$$= \cfrac{.0178214188 - .01322}{.001556}$$

$$= 2.9572$$

Gain = 2.96

Repeating the same example for the 650 pound heifer consuming 16 lb of a ration that contains 0.90 NE_m and .060 NE_g per pound.

A MMC = $650^{.75} \times .043$ (128.7315 × .043 = 5.5354)

B MP = $\dfrac{5.534}{0.90}$ = 6.1105

C GP = 16 − 6.1505 = 9.8494
D GMC = 9.8494 × 0.60 = 5.9096

E Gain =

$$\cfrac{.0001974 + \cfrac{.00576 \times 5.90965}{650^{.75}} - .01405}{.002880}$$

Gain = 2.583358217 or 2.58

Most errors in these equations occur when the mathematical operations are done in the wrong order. Thus, when solving the gain equation, first multiply GMC x .003112; divide by $W^{.75}$; add .00017474; take the square root, and subtract .01322; finally divide by .001556. I prefer to get $W^{.75}$ in the maintenance equation and store the value for use in the gain equation. If you have a machine that cannot do exponentials but has the square root function, $W^{.75}$ is the same as W x W x W and then taking the square root twice.

Many cattle feeders have stated that they want a "hedge" program--a program that would pinpoint hedging opportunities as cattle progress on feed. This may be accomplished by changing the generated information in a gain simulation with actual feed intake and ration energy values. Many feedlots have taken the net energy equations and from actual daily feed intake have calculated or estimated daily gain. These daily estimates are added to the previous days to give a calculated weight. These calculated weights are generally as accurate as most feedlot scales.

Developing Your Program

You are limited only by your imagination in developing cattle management program. We can tell little difference in programs that simulate cattle on a daily basis vs those calculated every 5 or 10 days providing that all rations used in the actual cattle feeding are evaluated in a 5- or 10-day program. Good simulations are a very valuable management tool.

The data shown in table 4 are typical of feedlot experience. The major factor is that as cattle become heavier more feed is required per pound of liveweight gain. While the industry has talked about liveweight gains for years the value of the feedlot steer is determined in the carcass.

How the beef feeding segment works becomes clearer when the simulation in table 4 is expanded to show the current market value of cattle as they progress through the feeding period. It is very important to recognize that daily changes do occur on the production and on the market side, and while the conclusions developed from the simulation in table 5 are accurate for the conditions set forth, that there will likely never be another day that will give the same numbers.

TABLE 4. A LIVE WEIGHT AND ECONOMIC SIMULATION AS FEEDLOT CATTLE PROGRESS ON FEED (MARCH 5, 1982 MARKET PRICES)

Days	Anim wt	Days feed	Days feed cost	Gain days lb	Gain py_w lb	Conversion days lb	Conversion py_w lb	Cost days tot.	Gain py_w $	break even $/CWT
0	665	0.00	0.00	0.00	0.00	0.00	0.00	0.00	0.0	70.75
20	724	19.73	1.28	3.50	1.21	5.64	14.39	0.45	1.43	69.07
40	797	20.44	1.34	3.55	2.42	5.76	7.76	0.47	0.69	66.86
60	866	20.87	1.37	3.36	2.76	6.21	7.02	0.51	0.61	65.41
80	931	21.05	1.38	3.16	2.89	6.66	6.86	0.55	0.59	64.52
100	992	21.02	1.38	2.95	2.92	7.12	6.87	0.59	0.58	64.05
120	1049	20.83	1.37	2.74	2.91	7.59	6.95	0.63	0.59	63.90
140	1102	20.50	1.35	2.54	2.87	8.08	7.06	0.68	0.60	63.97
160	1150	20.08	1.32	2.23	2.81	8.61	7.20	0.73	0.61	64.24
180	1195	19.59	1.29	2.14	2.75	9.17	7.36	0.79	0.62	64.68

Input parameters

Steers
Purchase weight	700
Purchase cost/cwt	$66.50
Intake factor	5.4
Medical cost/head	$5.0
Shrinkage %	5.0
Equity per head	0.0
Interest rate %	16.50
Yardage per day	.05
Death 1 %	.40
Day loop 1	15
Death 2 %	.40
Day loop 2	100

Rations used:
Number	Days fed	$CWT	NEM	NEG
1	10	6.26*	86	55
2	10	6.48	91	59
3	160	6.56	95	62

*$2.65/bu corn, $70/ton alfalfa; $25/ton corn silage; $200/ton soy-meal; ration mark-up = $20/ton/DM

TABLE 5. A DETAILED BREAKDOWN OF CARCASS AND ECONOMIC TRAITS AS CATTLE PROGRESS ON FEED (MARCH 5, 1982 MARKET PRICES)

Days	Anim wt	Cost days tot.	Gain PY W $	Break Even $/CWT	Est dress %	CH. %	Y-4 plus %	Car. value $	Max live bid $	F.L. P&L $	Increase in value per day	Total cost /day
0	665	0.00	0.00	70.75	52.00	—	—	352.4	52.99	-118	——	
20	724	0.45	1.43	69.07	53.50						——	
40	797	0.47	0.70	66.86	55.00			452.2	56.74	-80.7	2.50	1.67
60	866	0.51	0.61	65.41	57.50	20		520.6	60.11	-45.9	3.42	1.68
80	931	0.56	0.59	64.52	60.50	30		595.7	63.99	-32.5	3.76	1.71
100	992	0.60	0.59	64.05	61.50	40		652.9	65.82	- 6.99	2.86	1.71
120	1049	0.64	0.59	63.90	62.50	60	2	709.2	67.60	38.84	2.81	1.72
140	1102	0.69	0.60	63.97	63.50	70	5	755.9	68.70	50.97	2.34	1.72
160	1150	0.74	0.62	64.24	64.50	80	10	803.9	69.90	65.17	2.40	1.71
180	1195	0.80	0.63	64.68	65.50	85	25	843.9	70.61	70.96	2.00	1.71

Price structure: Friday, March 5, 1982
Under 600 lb -$2.00

Good Y-2 of better $1.02
Good Y-3 or better $1.01
Choice Y-2 $1.07
Choice Y-3 $1.05
Choice Y-4 $ .98
Choice Y-5 $ .93
Offal credit = $5.50/CWT live
Kill cost = $30/head

TABLE 6. A DETAILED BREAKDOWN OF CARCASS AND ECONOMIC TRAITS AS CATTLE PROGRESS ON FEED (MARCH 5, 1982 MARKET PRICES)

Days on feed	Steer weight lb	Break even live $	Est dress %	Carc. weight lb	Carc. & net offal value $	Max live bid $	Feeder P&L $	Packer carc. break even $/CWT	Carc. price to break feeder even $/CWT live
0	665	70.75	52.00	345.80	352.38	52.99	-118.11	100.00	134.16
20	724	69.07	53.50	387.34					
40	797	66.86	55.00	438.35	452.19	56.74	- 80.68	100.00	118.41
60	866	65.41	57.50	497.95	520.56	60.11	- 45.89	101.00	110.22
80	931	64.52	60.50	563.26	595.73	63.99	- 4.95	102.00	102.88
100	992	64.05	61.50	610.08	652.94	65.82	17.56	103.00	100.12
120	1049	63.90	62.50	655.63	709.15	67.60	38.84	103.94	98.02
140	1102	63.97	63.50	699.77	755.92	68.70	50.97	103.65	96.37
160	1150	64.24	64.50	741.75	803.93	69.90	65.17	103.90	95.11
180	1195	64.68	65.50	782.73	843.89	70.61	70.96	103.25	94.18

Beef price structure: March 5, 1982
Carcass under 600 pounds = -$2.00

Good Y-2 or better $1.02
Good Y-3 or better $1.01
Choice Y-2 or better $1.07
Choice Y-3 or better $1.05
Choice Y-4 $ .98
Choice Y-5 $ .93

Offal credit = $5.50/CWT live
Kill cost = $30/head

61

COMPUDOSE®: A CONTROLLED-RELEASE LONG-ACTING ANABOLIC IMPLANT FOR BEEF CATTLE

Herb Brown

The precise means by which anabolic effects of estrogens are achieved are not fully understood; however, Trenkle (1970) reported a significant increase in plasma growth hormone concentration following estrogen treatment of sheep and cattle. It also has been established that elevated growth hormone levels increase nitrogen retention, resulting in an increase in the production of lean meat with no adverse effect on carcass quality. The hypothesis is that estrogens exert their anabolic effect on the hypothalamus or pituitary directly to stimulate increased pituitary size and secretion of growth hormone.

Compudose is a new drug-delivery concept for improving both growth rate and feed efficiency in steers. Compudose is a long-acting, controlled-release implant containing the natural steriod, estradiol 17β. The properties of Compudose are such that it provides the advantages attained by other anabolic drugs on the market--with the added benefit of continuous delivery for over 200 days.

The silicone rubber implant is the drug-delivery mechanism controlling the daily amount of estradiol released to the animal (figure 1).

The implant is a 3 cm (1.18 inch) long cylinder, 4.767 mm (3/16 inch) in diameter and provides 4.84 cm^2 of surface area. Estradiol microcrystals are impregnated in the outer 250 microns of the implant. The implant contains 24 mg of estradiol 17β and is designed to provide a continuous release of this hormone for at least 200 days. Note in table 1 the surface areas of the different length implants. By measuring the amount of estradiol lost from each used and recovered implant it was relatively simple to calculate the average estradiol released each day for the different-sized implants and the amazingly consistent payout based on surface area of the implant.

FIGURE 1.

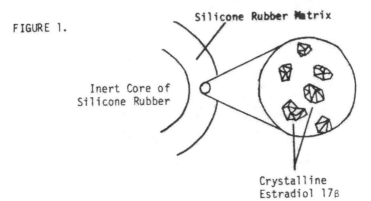

Silicone Rubber Matrix

Inert Core of
Silicone Rubber

Crystalline
Estradiol 17β

Implant Dimensions

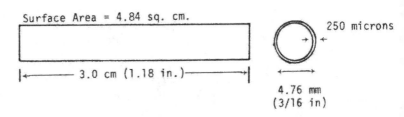

Surface Area = 4.84 sq. cm.

|←———— 3.0 cm (1.18 in.)————→|

250 microns

4.76 mm
(3/16 in)

After Compudose is placed subcutaneously in the middle of the back of the animal's ear, estradiol begins to diffuse from the implant into the body fluids surrounding the implant (figure 2). The estradiol microcrystals dissolve slowly and are released in an active form from the silicone rubber at a rate dependent on surface area of the implant.

TABLE 1. ESTRADIOL 17β DELIVERED FROM COMPUDOSE IMPLANTS

Implant length inches (cm) area	Implant surface area cm^2	Mean implant weight loss[a]mg	Estradiol payout	
			Average mcg/day	mcg/day/cm^2 surface
Control	4.15	0.15	--	--
0.25 (0.64)	1.31	3.56	17.0	13.0
0.50 (1.27)	2.25	5.85	28.0	12.4
1.00 (2.54)	4.15	10.49	50.2	12.1
1.18 (3.00)[b]	4.84	12.85	61.5	12.7
1.50 (3.81)	6.06	17.01	81.4	13.4

[a]Each mean represents 24-39 implants - average 209 days' duration.
[b]Values estimated based on average value of 12.7 mcg/day/cm^2.

FIGURE 2.

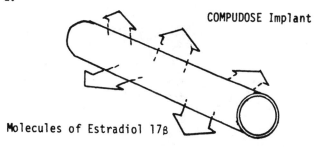

COMPUDOSE Implant

Molecules of Estradiol 17β

    The initial Compudose implant studies were conducted with 0.25, 0.50, 1.0, and 1.5 inch implants. Weight gain and feed efficiency data from the three finishing trials with 253 steers show that the minimal dose required for maximal anabolic response plateaued with the 1.0 in. (2.54 cm) implant.

FINISHING STEERS

Two hundred fifty-three steers were used in three trials to evaluate Compudose in finishing cattle. Average initial weight was 560 lb. The average time to slaughter-weight termination was 188 days. Barley was fed in two trials while milo was fed in the third trial. The rations were approximately 75% grain.

Average daily gains and feed conversion data are listed by trial and treatment in table 2. All Compudose implants produced improved, average daily gains and feed efficiency. A 17.4% improvement in average daily gain was obtained with 1.0 and 1.5 in. implants. This increase in average daily gain over the 188-day, average trial length results in 71.4 lb of total weight gain advantage per steer.

All sizes of implants improved feed efficiency with maximum improvement from the 1.0 and 1.5 in. implants.

The payout of the marketed implant is plotted in figure 3, which shows the estimated range of estradiol 17$_h$ required to obtain the maximum anabolic response as indicated by the dose titration studies for feedlot cattle.

TABLE 2.   SUMMARY OF AVERAGE DAILY GAIN AND FEED CONVERSION OF FINISH-
ING STEERS IMPLANTED WITH COMPUDOSE—3-TRIAL SUMMARY

Length of implant in.	Average daily gain (lb)/feed efficiency							
	Trial no.						Average adjusted mean	
	503		502		501			
	ADG	F/G	ADG	F/G	ADG	F/G	ADG	F/G
Control	1.99	9.64	2.21	7.74	2.61	9.30	2.18	9.19
0.25	2.47	9.03	2.22	7.99	—	—	2.45	8.97
0.50	2.51	8.70	2.32	7.70	—	—	2.51	8.68
1.00	2.54	8.72	2.42	7.24	3.00	8.59	2.56	8.51
1.50	2.50	8.69	2.32	7.51	3.11	8.52	2.56	8.52

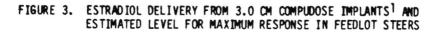

FIGURE 3. ESTRADIOL DELIVERY FROM 3.0 CM COMPUDOSE IMPLANTS[1] AND ESTIMATED LEVEL FOR MAXIMUM RESPONSE IN FEEDLOT STEERS

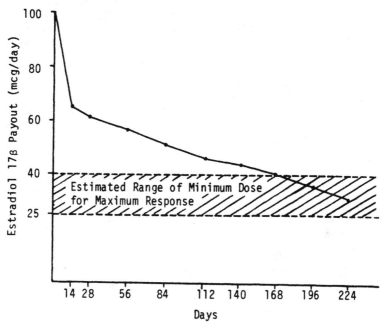

[1]Values adjusted to 3.0 cm implant, based on 2 manufactured lots of 1.0 inch (2.54 cm) implants (2450-55-1 and 24945).

SUCKLING CALVES

Six trials involving 400 suckling steer calves were conducted under normal pasture conditions in various geographic locations to determine the minimum size of implant required to provide maximum growth response in suckling steer calves. All of these trials were with calves dropped by spring-calving cows. The average initial calf weight at implanting was 188 lb, and the trial period, from implant until weaning, was 148 days.

TABLE 3. SUMMARY OF AVERGE DAILY GAINS OF SUCKLING STEER CALVES
IMPLANTED WITH COMPUDOSE

Trial number	Average daily gain (lb) Length of implant (in.)				
	Control	0.25	0.50	1.00	1.50
601	1.94	2.01	1.94	2.00	2.04
602	1.74	1.64	1.82	1.73	1.70
606	2.23	2.38	2.30	2.44	2.36
608	1.78	1.88	1.89	1.87	1.89
701	1.98	2.22	2.07	2.09	2.00
703	1.85	1.86	1.92	1.99	2.01
Average adjusted mean	1.92	2.02	1.99	2.02	2.00

Average daily gains for the individual trials are shown
in table 3. All lengths of the Compudose implant were
effective in increasing the average weaning weight of suckling steer calves. Adjusted means for different implant
lengths from all six trials did not differ, however, all
lengths of implant resulted in increased rate of gain of 4%
to 5% when compared to the control group.

GROWING STEERS

Nine trials involving 778 yearling steers were conducted to determine the minimum amount of Compudose required
to provide the maximum growth response in growing yearling
steers. The average initial weight of these steers was 488
lb and the average length of trial was 130 days. Several
different types of pasture forage were represented in the
nine trials.

Average daily gain for the individual trials is shown
in table 4. All levels of Compudose increased average daily
gain with the 1 in. implant resulting in the maximum
improvement of 8.6% compared to nonimplanted steers.

TABLE 4. SUMMARY OF AVERGE DAILY GAINS OF GROWING STEER CALVES
IMPLANTED WITH COMPUDOSE

Trial no.	Average daily gain (lb) Length of implant (in.)				
	Control	0.25	0.50	1.00	1.50
603	1.68	1.68	1.68	1.71	1.55
604	1.38	1.44	1.63	1.51	1.60
605	2.00	1.99	1.97	2.06	1.97
607	0.99	1.04	1.05	1.09	1.06
614	1.37	1.61	1.60	1.69	1.59
615	1.34	1.34	1.39	1.52	1.55
702	0.98	0.98	1.06	0.99	1.02
704	0.61	0.67	0.84	0.66	0.36
706	0.11	0.14	0.06	0.11	0.20
Average adjusted mean	1.16	1.21	1.25	1.26	1.22

COMPUDOSE AND RUMENSIN® FOR GRAZING STEERS

Five trials involving 512 steers weighing an average of
547 lb were used in these experiments. The treatments were:
(1) no supplement, (2) supplement, (3) supplement plus
Rumensin, and (4, 5, 6) each of the three treatments with
all of the steers implanted with Compudose. Improved and
native pastures were used in different trials.

A summary of average daily gain for all five experiments is shown in table 5. Unimplanted control cattle
gained 1.22 lb per day for the 124 day experimental period.
Two lb of an energy supplement produced increased weight
gains to 1.35 lb per day. Rumensin added to the supplement
produced increased daily gain to 1.45 lb and Compudose
implanted cattle gained more than each comparable treatment
without Compudose. Weight gains obtained when cattle were
implanted with Compudose were additive to both energy
supplementation and the feeding of Rumensin.

TABLE 5.   EFFECT OF COMPUDOSE AND RUMENSIN ON AVERAGE DAILY GAIN OF
YEARLING STEERS—SUMMARY OF 5 GRAZING TRIALS

Main plot treatment group	Average daily gain (lb)	
	Control	Compudose
Control	1.22	1.39
2 lb energy suppl./ hd/day	1.35	1.56
2 lb energy suppl. + 200 mg Rumensin/hd/day	1.45	1.72

These data leave little to doubt that all steers
receiving Compudose should also receive Rumensin. Although
this Rumensin-plus-Compudose information was generated only
with growing steers on pasture, it can be concluded that the
benefits of Rumensin and Compudose would be completely additive for finishing steer use.

THE PROCESS OF IMPLANTING WITH COMPUDOSE

An implanting tool with a spare needle is provided with
each 500 implants.   To properly administer Compudose
implants the following procedures should be used:
1.   Properly restrain the animal.
2.   Cleanse the implant site to reduce the pos-
     sibility of infection.   Ears contaminated
     with manure should be washed with an anti-
     septic solution prior to being implanted.
3.   With the Compudose implant properly loaded,
     grasp the ear with one hand and position the
     needle flat on the back side near the tip of
     the ear.
4.   Insert the entire needle under the skin using
     the correct angle to stay between the skin
     and the cartilage.
5.   As the needle is withdrawn, push the plunger
     to deposit the implant as far as possible
     from the site of insertion.
6.   After the needle has been withdrawn from the
     ear, grasp the ear with thumb and forefinger
     between the site of insertion and the
     implant.   The implant should rest approxi-
     mately in the middle of the back side of the
     ear.
7.   Disinfect the needle between uses to help
     avoid infection and the spread of disease.
     Poor implant technique that mutilates the ear
     or introduces infection into the implant
     insertion site or anything else that en-
     courages excessive bleeding or infection may

result in some implants being expelled from the ear.

A successful implant is one that has been positioned properly in the ear and remains there without infection for over 200 days, releasing the minimum amount of estradiol daily for maximum growth response.

SAFETY

Estradiol and its major metabolite, estrone, are present in the normal physiological environment of both males and females, as well as in the human food supply. In addition, all bovine tissues (bulls, steers, and heifers) contain estradiol.

The occurrence of estradiol and estrone in milk from the pregnant and nonpregnant cow is shown in table 6.

The data in table 9 indicate that tissue levels of estradiol in Compudose implanted steers are minimal when compared to the estrogen present in tissues and milk derived from normal pregnant females (tables 6, 7, and 8).

SUMMARY

1. Compudose provides long-term payout of activity over a period of 200 days.
2. Field trials have resulted in an overall rate of gain improvement of 5% in suckling calves, 9-16% in pastured growing cattle and 17% in finishing cattle when compared to nontreated animals. Feed efficiency could only be evaluated in finishing cattle and was improved 7.4%.
3. Compudose is an efficient new drug-delivery system that provides maximum response with a minimum dose.
4. The silicone rubber composition of Compudose provides continuous structural integrity. It cannot be crushed or damaged during or after implantation to effect drug payout.
5. Estradiol 17β, the active ingredient in Compudose, is a natural steriod and very safe when used as direct.
6. Compudose may be used with steers fed Rumensin with benefits of both compounds being completely additive.

TABLE 6. CONCENTRATION OF ESTRADIOL AND ESTRONE IN MILK OF PREGNANT AND NONPREGNANT COW (pg/ml)[a]

Reproductive stage	Number of observations	Estradiol 17β	Estrone
Nonpregnant	61	27	43
Pregnant			
1st trimester	5	85	57
2d trimester	4	52	35
3d trimester	4		
6 days precalving	9	1461[b]	

[a]Adapted from Monk, et al., J. Dairy Science 58:34-40 (1975).
[b]This mean represents total of both estradiol 17β and estrone with values ranging from 643 to 2142 pg/ml.

TABLE 7. CONCENTRATION OF ESTRADIOL AND ESTRONE IN KIDNEY FAT OF CYCLING AND PREGNANT HEIFERS

Heifer reproductive stage	Number of observations	Estradiol 17β		Estrone	
		pg/g	Range	pg/g	Range
Cycling	14	10	(5-28)	44	(5-176)
Pregnant					
1st trimester	3	5	(5-6)	40	(11-100)
2d trimester	4	22	(5-50)	964	(380-1920)
3d trimester	5	163	(22-440)	3870	(750-8200)

TABLE 8.   ESTRONE CONCENTRATION IN TISSUE OF COWS AND BULLS (pg/g)[a]

Sex	Reproductive stage	Fat	Liver	Kidney	Muscle
Cows	Luteal	26	14	13	19
	Follicular	28	23	15	40
	1st trimeter	16	24	8	12
	2d trimester	474	122	314	153
	3d trimester	3687	258	550	208
Bulls		35	12	2.7	15

[a]Dr. Donald M. Henricks, Clemson University, personal communication (1981).

TABLE 9.   ESTRADIOL AND ESTRONE LEVELS IN TISSUES OF COMPUDOSE TREATED AND UNTREATED STEERS (pg/g)

	Treatment	No. steers	Lean meat	Kidney fat	Kidney	Liver
Estradiol 17β	Compudose	10	3.4	15.0	21.0	10.0
	Control	20	5.8	6.8[a]	5.7	4.0
	Treatment difference		-2.4	8.2	14.3	6.0
Estrone	Compudose	10	10.4	35.0	19.0	9.0
	Control	20	4.8	10.5[a]	7.0	6.5
	Treatment difference		5.6	24.5	11.1	2.5

[a]Concentrations of estradiol 17β and estrone in kidney fat from an additional 306 untreated steers wre 3.6 pg/g and 6/5 pg/g, respectively.

62

ENERGETICS IN CATTLE PRODUCTION: FEED AND ENVIRONMENT

Stanley E. Curtis

Knowledge in the areas of nutritional and environmental energetics generated in the last decade make it possible to be more precise in managing energy in cattle production nowadays. Feed and the environment are closely related in terms of their impact on the animal's energy budget.

Energy metabolism in cattle--the phenomenon Max Kleiber poetically called "the fire of life"--is the driver of all the physiological processes exploited in cattle production. The source of that energy is the feed.

Cattle cannot use all of the chemical-bond energy in their rations. That part of the dietary energy consumed (intake energy) that is not digested, together with the energy in certain compounds of metabolic origin, appear as fecal energy. The difference between intake energy and fecal energy is called digestible energy. The digestibility of feedstuffs (digestible energy/intake energy) varies mainly with the chemical nature of the material. Other energy losses show up in the digestive gases and the urine. What is left after fecal, gas, and urine energy losses have been accounted for is called the metabolizable energy portion of the diet.

Cattle use metabolizable energy to drive maintenance and productive processes. The metabolizable energy eventually becomes the energy of some product (such as meat, milk, or fetus) or else it is lost to the environment as heat. This heat originates in the maintenance processes themselves and in the heat increment of feeding. The heat increment, in turn, is the heat produced during processing of the feed by the body. It thus reflects energetic inefficiency, unless the animal can use the heat to maintain its body temperature in a cold environment.

The idea that the digestibility of dietary energy is affected by the thermal surroundings of the animal ingesting the ration is a relatively new one. Available evidence supports the conclusion that dietary-energy digestibility is related directly with environmental temperature. This effect appears to be apart from other influences such as feed-intake and digesta-passage rates.

David Ames has suggested adjusting the nutrient value of dietary components to account for thermal effects on digestibility according to the formula:

$$A = B + [C (T - 20)]$$

where A is the adjusted nutrient value, B is the standard nutrient value (representing thermoneutral conditions), C is a specific thermal correction factor, and T is effective environmental temperature (°C). The specific correction factor for dry matter is .0016; energy components, .001; acid-detergent fiber, .0037; and crude protein, .0011.

Cattle are warm-blooded animals: over a wide range of environmental temperature, their body temperature remains roughly the same. This phenomenon is called homeothermy. As environmental temperature becomes progressively colder, insulative mechanisms--such as constriction of blood vessels in the skin and subcutaneous fat, erection of hairs, and huddling and other postural adjustments--are called more and more into play. But eventually the point is reached where these insulative responses become inadequate to cope with the coldness of the environment, and heat-production rate must be increased to make up for the fast rate of heat loss from the body to the environment.

Similarly, the physiological, anatomical, and behavioral reactions cattle use to combat heat stress require the expenditure of energy.

Estimates of the energy costs of themoregulatory processes indicate that these reactions can amount to significant inefficiencies. In essence, cattle under cold stress or heat stress have altered maintenance-energy needs, because stress responses are maintenance processes. In the first place, acclimatization status alters the animal's maintenance requirement for net energy. This is due to hormonal and metabolic changes that occur as the animal becomes accustomed to one environment or another. An estimate of maintenance-net-energy requirement of cattle (in megacalories per day) is:

$$a \, W^{.75}$$

where a is a constant of proportionality that varies depending on the animal's acclimatization status, W is the animal's body weight (in kilograms), and $W^{.75}$ is the animal's metabolic body size. The constant of proportionality, a, is equal to .070 when cattle are acclimatized to an average daily temperature around 30°C; .077 at 20°C; .084 at 10°C; and .091 at 0°C.

Second, the maintenance energy requirement is affected directly by heat stress and cold stress. For heat stress, there is an approximately 7% increase over the thermoneutral level when the animal is in first-phase panting, and an 11% to 25% increase during second-phase panting.

On the cold side, the lower critical temperature--the environmental temperature below which the animal must increase heat-production rate to maintain normal body temperature--of a 6-month-old calf gaining .5 kg of body weight daily is around -17°C in a dry, still environment, and 10°C in wet, windy conditions. Comparable values for a yearling gaining around 1 kg daily are -34°C and -10°C, and for a dry cow at midpregnancy, -25°C and -7°C. The increased dietary energy requirement for the calf would be around .1 Mcal of metabolizable energy daily per Celsius degree of coldness in the dry, still environment, and around .2 Mcal in the wet, windy surroundings. Comparable values are .2 Mcal and .3 Mcal for the yearling, and slightly higher, respectively, for the cow.

Ames has provided as an example a 500 kg, dry, mature Hereford cow on a diet of brome hay in the last third of pregnancy. He evaluates the cow's daily maintenance requirement in four environmental conditions:

- a) thermoneutrality;
- b) a hot, dry environment where air temperature averages 30°C, but with short daily excursions to 35°C resulting in first-phase panting;
- c) a cold, dry environment where air temperature averages -25°C and there is a dry-bedded area with an effective windbreak; and
- d) a cold, dry environment where air temperature usually averages -15°C each day, but estimates are required for a period of several days of winter storm with air temperature averaging -25°C, high winds, drifting snow, and inadequate bedding and shelter. Results of the analysis are tabulated below:

| | Environmental condition as described above | | | |
	a	b	c	d
Maintenance energy requirement (Mcal of metabolizable energy/day)				
Adjusted for acclimatization	16.4	14.9	23.1	21.6
Adjusted for direct thermal effects	0.0	1.0	0.0	5.9
Total	16.4	15.9	23.1	27.5
Brome hay (dry matter basis) energy content adjusted for environmental conditions (Mcal ME/kg)	1.87	1.90	1.79	1.79
Cow's feed requirement for maintenance (kg/day)	8.8	8.4	12.9*	15.4*

*Brome hay's bulkiness might limit intake at these levels.

It is obvious from this example that environmental temperature affects not only the nutritive value of a given feed, but also cattle's need for feed. Most feed allowance recommendations are based on cattle being kept in an environment having a temperature range between 15°C and 25°C-- this is assumed to be the thermoneutral range of environmental temperature, the range in which feed intake is at the basal level.

The range from 25°C to 35°C might result in a 3% to 10% depression in voluntary feed intake, while temperature above 35°C commonly results in a 10% to 35% reduction in intake in full-fed cattle, and 5% to 20% reduction in cattle on a maintenance ration.

In the other temperature direction, feed intake is typically stimulated by 2% to 5% in the 5°C to 15°C range of environmental temperature, and by up to 8% in the -5°C to 5°C range. Intake might be 10% above basal at -15°C, and up to 25% above in even colder surroundings.

Part 10

HEALTH, DISEASE, AND PARASITES

63
INFECTIOUS LIVESTOCK DISEASES: THEIR WORLDWIDE TOLL

Harry C. Mussman

By good fortune, plus nearly a century's worth of cooperation between stockmen and government, the United States has managed to wipe out or keep out the most serious of the world's livestock diseases.

Contagious bovine pleuropneumonia, introduced in the mid-1800s, was the first major disease we battled together and eradicated. Hog cholera was the last to be wiped out--just four years ago. In between these two efforts were nine outbreaks of foot-and-mouth disease, which has not been back since the last remains were buried in 1929.

These diseases and others still bring major losses and reduce livestock productivity in much of the world. This is particularly true in the developing nations of Africa, Asia, and Latin America. The popularly held notion that Third World livestock are infected with all of the most dreaded diseases is not true, but many of these countries do live with at least a few of them, often superimposed one on the other. Losses due to animal disease in some developing countries are estimated at 30% to 40% annually--twice as great as losses recorded by most industrialized nations.

Looking at meat and dairy productivity, developing nations produce one-fifth the beef and veal per animal that developed countries do, one-half the pork, one-half the eggs, and one-eighth the milk. North America and Europe are roughly four times more efficient in mutton, lamb, and goat meat production than is South America, and 25% more efficient than Africa.

The so-called Third World is hungry already, and is projected to host 90% of the global growth in human population expected by the year 2000. Obviously, greater and more reliable sources of protein must be found and developed, or many of these people will not survive. A diseased animal that may not survive can hardly be counted a productive resource. A virus or other disease agent can wipe out entire herds in a very short time--or, in a chronic state, can leave them debilitated, more expensive to feed, and capable of producing far less milk or meat.

Of all the factors involved with production, animal diseases have by far the most dramatic impact. Unfortunately, disease control is not improving rapidly enough in the Third World. And many of the most feared diseases--because of the tremendous economic toll they can take--are appearing in locations where they never existed before or reappearing in places where they had been eradicated years ago.

Some of these diseases are listed next.

Foot-and-Mouth Disease. Foot-and-mouth disease (FMD) is the most feared of all the animal diseases worldwide because of its ease and speed of spread and the susceptibility of all clovenhooved animals. It has long been established in livestock populations of the Mideast causing abortions, deaths, weight loss, mastitis, and lowered milk production. Europe has managed to keep these Asian strains at bay in Turkey, however.

Prior to World War II, FMD swept through Europe every 6 to 8 years, severely reducing livestock productivity. Finally in the 1950s, better quality vaccine was introduced and stemmed the spread of the epizootics. Outbreaks that occurred in the early 1960s, and again in 1973, were halted by teams from the United Nations working with the affected countries and regions. Then, in March 1981, FMD was confirmed in France, in Brittany, and near Cherbourg. Within a few days it broke out on the Isle of Jersey and the Isle of Wight in Great Britain. Just prior to that positive diagnosis came in from the south of France, from Portugal, and from Spain--outbreaks that most likely originated in Spain. Austria, free of FMD for six years, was infected during the winter of 1981 from imported Asian buffalo meat.

After 10 years of freedom from the disease and despite a vigilant quarantine program, Denmark discovered an outbreak of FMD on the Isle of Fyn just last March. The origin is unknown, but suspicions are that it came from one of the Eastern bloc countries where severe FMD outbreaks were reportedly occurring just prior to Denmark's infection. The loss Denmark has suffered from those outbreaks is estimated at almost a quarter of a billion dollars. Most of the loss has come from the cessation of fresh pork sales to the U.S., Canada, and Japan, although sixteen other countries were also compelled to embargo Danish meats and animals. The cost of eradication alone was also significant.

U.S. policy prohibits imports of fresh, frozen, or chilled meat products, or swine, or ruminants for a full year from any country that experiences an FMD outbreak. We had to refuse United Kingdom exports for a year following their 1981 outbreaks. We also refused farm products from across the Canadian border in 1952 following an outbreak there--which also curtailed big game hunting for U.S. hunters up north. (Certified "clean" animals, however, may be brought in from FMD-infected countries through the Harry S. Truman Animal Import Center on Fleming Key, Florida, after three months of high-security quarantine and testing at the importer's expense; or they may be shipped directly to an approved zoo.)

Some call this "politics," aimed at protecting the home meat industry. U.S. meat producers do gain a competitive edge from such an embargo, certainly, but that is not the aim. The productivity of more than 200 million fully sus-ceptible cattle, swine, sheep, and goats is at stake. The collective worth of that stock is estimated at $24 billion --if they stay healthy. Healthy animals with their high-quality protein will keep America healthy, help alleviate worldwide hunger, and help keep the U.S. trade balance healthy.

If FMD were to invade the U.S. and enter a major mar-ket, it could spread to over a dozen states within 24 hours. By the end of the first year, direct losses would be an estimated $3.6 billion--indirect losses to allied indus-tries and curtailed exports could climb to $10 billion.

Rinderpest and Peste de Petits Ruminants. Another serious, almost 100% fatal among cattle and resurging as a serious problem in Africa, is rinderpest. As with FMD, the United States (by law) cannot import cattle or other rumi-nants from rinderpest-infected countries (unless, as with FMD, they are brought in through the Truman Center on Flem-ing Key or are destined for an approved zoo.)

Rinderpest was essentially eliminated from Africa in 1975, following a decade of intensive and widespread vacci-nation. A team of international cooperators trudged from West to East Africa, vaccinating some 80 million head of cattle at a cost of $30 million. Unfortunately, though, the people became complacent once the disease had been sup-pressed. As the livestock populations were built up again, the young were not all vaccinated and were allowed contact with remaining carriers and wild animals.

So, the incidence is climbing again, especially in West Africa, and the virus is spreading east. Egypt is affected now, and the Arab Gulf States are reporting such frequent outbreaks that it appears the disease has become enzootic.

Peste de petits ruminants is a virus quite similar to rinderpest and readily infects goats and sheep in West Africa. It is actually classified as a strain of rinderpest that has lost its ability to infect cattle under natural conditions.

Rift Valley Fever. Another disease that has moved up from Kenya through the Sudan into the Sinai is Rift Valley fever. RVF primarily affects sheep but may also affect goats. It is also a serious human health problem. An epi-demic in Egypt a few years ago killed nearly 600 people, along with a large number of livestock. The disease can spread rather easily. A human body incubating the virus could carry it anywhere in the world or a mosquito could carry it while hitchhiking on an airplane.

Israel has been vaccinating its animals against RVF for self-protection as well as to help curb further spread of the disease around the Eastern Mediterranean Basin.

Because of grave concern over Rift Valley fever, the U.S. Department of Agriculture has developed diagnostic capability of its own and is making preliminary vaccine studies. Work on possible domestic insect vectors is also planned.

African Swine Fever. Over the past 20 years, swine-producing nations have faced a rising risk of African swine fever infection. ASF has been established for many years in wart hog, bush hog, and giant forest pig populations in tropical Africa, south of the equator. Other animals may not have the same tolerance to the virus: in 1909 Europeans tried bringing domestic hogs into Africa, and the hogs were dead from the virus in a matter of a few days.

The first appearance of ASF in Europe was in Portugal in 1957 where the disease was at first mistaken for classic hog cholera and finally recognized as ASf. More than 16,000 pigs became infected before the virus was recognized as ASF and finally stamped out. In 1959, the disease was found in Spain, and in 1960 reappeared in Portugal. France was invaded by ASF three times in the 1960s and 1970s, though the French managed to eradicate it each time at an early stage--a tribute to their veterinary surveillance system.

Italy became infected in 1967 through the Rome airport. More than 100,000 swine were slaughtered before the disease was wiped out. Finally, about four years ago, Malta and Sardinia became infected. In Malta, the entire swine population was slaughtered to achieve eradication--at a cost of some $50 million (U.S.). Italy is still working on eradication of the disease on Sardinia.

The Western Hemisphere managed to avert an invasion of the dread ASF until 1971 when the virus hit Cuba. Approximately half a million pigs died or were destroyed before the disease was eradicated. Then, another Cuban outbreak occurred in 1980 and was successfully eradicated in about six weeks. ASF appeared in Brazil 5 years ago and is still there today. Positive diagnosis came from the Dominican Republic in 1978 and in Haiti early in 1979.

The U.S., Canada, and Mexico are particularly and acutely concerned about African swine fever in the Western Hemisphere. All three countries are free of the disease but are working together in the Dominican Republic toward eradication. The program used a hard strategy: they completely depopulated the country's swine farms after which they restocked with healthy, imported pigs. Not a trace of the virus has reappeared since the restocking was completed. The Haitian effort repeated the model program (and first of its kind in the West) completed in the adjacent Dominican Republic last year. Equal success is anticipated in Haiti.

Success is crucial to the three big cooperators. Mexico's Yucatan Peninsula is just a short hop across the

Caribbean from the Dominican Republic; Canada has trade interests all over Latin America and the Caribbean; and the threat of ASF lies right at the doorstep to the United States. The U.S. strictly prohibits swine imports from ASF-infected countries, but the virus could enter through contaminated food waste from ships or planes, through farm soil still clinging to a returning traveler's shoe, or as a result of illegal immigration. The investment in such an eradication program is sizable. Direct costs for the Dominican Republic--shared by the UN's Food and Agriculture Organization, the Inter-American Development Bank, and the United States--amounted to approximately $20 million. The bill for Haiti should run about the same.

The cost of chronically battling a disease like ASF can be far higher, however. Spain, for example, has spent roughly $315 million fighting it over the past 20 years. In addition, exports from Spain (along with Portugal and Brazil) have been restricted because of ASF-infected status.

Heartwater Disease. Latin America was jolted last year by the diagnosis of heartwater disease in a goat on Guadaloupe, an island in the eastern Caribbean. This was its first appearance in the Western Hemisphere and the implications are serious. We are cooperating with other countries in the region to make certain that it does not spread.

Heartwater is native to southern Africa and Madagascar and may be a major cause of animal deaths reported in West Africa. We have no simple and reliable means of diagnosing the disease; a blood test, especially, is badly needed.

Contagious Bovine Pleuropneumonia. Contagious bovine pleuropneumonia (CBPP) recently broke out in France for the first time in years. The disease may have been smouldering in the Basque region in Spain for some time. CBPP was the first major exotic disease to invade U.S. shores back in the 1800s. Cattle ranching had become so productive after the Civil War that a surplus of more than a quarter million head was available for export to England. When pleuropneumonia showed up in the animals, however, Britain closed off the market. That was the spark that set off the first animal disease eradication program in the United States.

Goat Diseases. In the United States evidence is accumulating of two possibly serious diseases of goats: contagious caprine pleuropneumonia and caprine arthritis encephalitis. CCPP is well established in much of the developing world--the regions that depend heavily on goats. The true impact of the disease is clouded as yet, largely because of common and widespread disagreement among diagnosticians. More research is needed to develop definitive diagnostic tests and effective vaccines.

Caprine arthritis encephalitis was only recently recognized in the United States and may exist yet undiscovered in other countries also. Research may give us some answers to this relatively new entity.

Trypanosomiasis. A full third of Africa--taking in most of the wide band of savannah and forest across the middle of the continent--is plagued by infestations of the tsetse fly. A blood parasite of animals as well as people, the tsetse fly transmits trypanosomiasis, or animal sleeping sickness, from animal to animal. If the tsetse-infested areas could be reclaimed for cattle production, it is estimated that annual meat production in Africa could double.

Bacterial Disease, Intestinal Parasites, and Others. The more dramatic virus diseases, such as FMD, ASF, rinderpest, bovine pleuropneumonia, and external parasites that transmit East Coast fever and trypanosomiasis, are the most highly visible of the livestock plagues worldwide.

Much more research remains to be done on bacterial diseases and intestinal parasites, particularly with small ruminants. Also more research is needed to understand the incidence and significance of the clostridial diseases--tetanus, anthrax, blackleg, and malignant edema--and paratuberculosis, and brucellosis. In the U.S., eleven states are free of cattle brucellosis now and another 25 have an extremely low incidence. Eradication of the disease requires only time. The damage done by different strains of the disease worldwide, however, is not entirely known. Since all strains can infect people as well as animals, the public health problem is always important. Anaplasmosis and babesiosis also are economically significant blood-parasite diseases in many tropical areas of the world, but the actual extent of the damage they do is not fully appreciated, much less known.

Another virus disease--one less dramatic in its toll, but gaining more and more visibility--is bluetongue. Bluetongue affects cattle, sheep, goats, and wild ruminants. It is particularly damaging to sheep, with a mortality rate running as high as 50% among affected animals; in cattle and goats, bluetongue lowers reproductive ability. Currently, U.S. cattle cannot be exported directly to Europe because of bluetongue in the South and southwestern states. Canada has a competitive edge over the United States in the European livestock trade--with the exception of the Holstein dairy breed--because the small gnat that carries the disease does not appear that far north.

Like caprine arthritis encephalitis in the United States, the possibility of diseases yet undiscovered still remains--diseases that may have been smouldering since the first goat herds on record were tended some 10,000 years B.C.

THE NEED FOR STRONGER CONTROLS

Proper vaccines, a solid border inspection and quarantine program, and accurate diagnosis could take care of many of the most severe disease problems in the world.

Vaccine

The alarming resurgence of rinderpest in Africa shows what can happen when a successful vaccination program breaks down. On the other hand, a laboratory opened in Mali just 5 years ago has been researching, tracing, and producing vaccine for several diseases rampant in West African livestock --rinderpest and bovine pleuropneumonia. The Central Veterinary Laboratory (CVL), as it is called, is also coordinating regional vaccination programs, and CVL veterinarians and technicians are being welcomed now by Malian herdsmen as they drive their cattle back and forth for the grazing season. The incidence of disease appears to be declining, at least locally.

Quarantine

Contributing to the rinderpest problem in Africa was the failure to keep sick animals from the healthy. Quite simply, quarantine can prevent the spread of many infectious diseases. For example, double fencing, as practiced in some parts of Africa, effectively protects domestic swine from African swine fever.

Border Inspection

Strict import controls over meat, feed, and animal products and a tight border guard are essential to keeping disease out. Years ago when most travel was by boat, a virus disease could probably not survive a transocean trip. Today a few hours on a jet plane are little threat to virus survival.

Customs and Agriculture inspectors at all 82 U.S. ports-of-entry inspect as much luggage and as many carry-on bags and travelers as possible. The goal of 100% inspection is a difficult one to achieve, however. International travel rose 8% in the United States last year, and 6% worldwide. Also more and more cargo is containerized every year.

Diagnosis

Research must proceed and develop simpler and more effective means of diagnosing heartwater disease, contagious caprine pleuropneumonia, and other diseases. Clinical signs of certain diseases are changing--and we must keep up with them. Genetic research must continue so that animal science can continue to breed for resistance to disease. Biological, chemical, ecological, and other means of controlling insect vectors are unfolding through field and laboratory trials.

CONCLUSION

Each country or region of the world will have its own formula for prioritizing the danger that different animal diseases present. Generally speaking, the most dangerous diseases are those that are spread not only by animal-to-animal contact (contagious diseases) but whose virus, mycoplasma, bacteria, or rickettsia may be transmitted by live vectors, inanimate fomites, or meat scraps. The most dangerous diseases do significant economic damage to producers and exporters and carry a human health hazard, such as brucellosis or Rift Valley fever.

Most important before the animal disease situation in the Third World can improve, veterinary services will have to be strengthened. Developing countries have nearly 50% of the world's total livestock population but less than 20% of the world's veterinary forces. A number of developing nations, in fact, have little or no organized veterinary structure at all. Certain nations in Africa, in fact, have only one or two veterinarians working the entire country; and in others they are often young, with little or no field experience. The expertise of technicians, if it can be found, is usually limited--gained through in-service training. Veterinary technicians have no reliable means of communication or transportation--yet national animal health policy decisions are based on their input. Thus, the global data on disease prevalence, productivity losses, and effectiveness of control measures on which many veterinary directorates rely is woefully deficient.

It becomes clear why diseases such as contagious bovine pleuropneumonia, foot-and-mouth, and rinderpest still exist and spread. With existing reliable diagnostic tests for both, as well as effective vaccines, eradication of CBPP and rinderpest would require little more than organization and finance, backed up by the government's strong commitment. With heavier investment in livestock health programs--and greater cooperation among nations--the great toll that diseases take can be curbed. The people can be fed.

64
IMPACT OF ANIMAL DISEASES IN WORLD TRADE

Harry C. Mussman

IMPACT OF ANIMAL DISEASES ON WORLD TRADE

Animal diseases are an important factor inhibiting world trade and hampering the free movement of both live animals and animal-derived products. The foot-and-mouth disease outbreak in Denmark in early 1982 provides a recent example of the impact of animal disease on trade. At one point, Denmark's export trade in meat and dairy exports was suffering badly--$7 million was lost each week because of the outbreak. The U.S.--as well as several other countries--placed Denmark on the list of countries from which animals and animal products cannot be imported until free status is regained. Overall costs of the outbreak will run in excess of one billion dollars.

Another example closer to home is our bluetongue situation. In 1980, the European Economic Community banned animal imports from the U.S. because this cattle and sheep disease was found in a portion of our country.

U.S. COMPETITIVE EXPORT POSITION

Despite the bluetongue situation and other domestic diseases such as brucellosis, tuberculosis, and leukosis, the U.S. export position remains competitive. In 1981, our animal and animal-product exports had a market value of $3.24 billion; and, in the same year, we had a $307 million trade surplus in these commodities, the first since 1977.

A U.S. animal export health certificate enjoys high credibility. One reason is that we have eradicated 12 major animal diseases that still plague many other countries of the world--diseases such as foot-and-mouth, rinderpest, hog cholera, and contagious bovine pleuropneumonia. Only a half-dozen or so other countries can claim freedom from these diseases. Largely because of our strict animal import health requirements and procedures, we are successful in keeping animal diseases out of our country.

IMPORT PROCEDURES

Livestock destined for this country must first be examined, tested, and certified by government veterinary officials as being healthy and meeting U.S. requirements in the country of origin. The foreign exporter must obtain health certification papers from his government and, in a majority of cases, an import permit in advance from us. At the U.S. port of entry the livestock must be examined again, this time by our veterinarians. This examination sometimes includes further testing and port-of-entry isolation, depending upon the kind of animal and the country it came from.

Although we are strict in our import controls, we try to meet the needs of American importers, particularly U.S. livestock breeders needing new bloodlines and exotic breeds of cattle. We recently opened a specialized import-quarantine facility--the Harry S. Truman Animal Import Center, at Fleming Key, Florida. Imported cattle are carefully tested and held there in quarantine for 3 months. They are mingled with a select "sentinel" group of susceptible U.S. cattle and swine to make sure they will present no disease threat to U.S. livestock.

Because some swine diseases, such as African swine fever, could devastate our swine herds, we do not accept swine imports from most of the world. Imports of sheep and goats are also severely limited because of the threat of scrapie, which has an incubation period of up to four years or more.

Restrictions on horse imports are generally less stringent than those required for swine, cattle, sheep, and goat. Still, incoming horses are tested for such diseases as dourine, glanders, equine infectious anemia (EIA), equine piroplasmosis, and contagious equine metritis (CEM).

Because of Venezuelan equine encephalomyelitis, horses from all Western Hemisphere countries--except Canada and Mexico--are quarantined for at least a week. Horses from countries known to have African horse sickness are quarantined for two months.

Contagious equine metritis is a recently identified disease of breeding horses. It has been found in parts of Europe, Japan, and Australia. As a result, horses cannot be freely imported from these countries.

Stallions and mares can be imported only after extensive treatment and negative culturing of the genitalia, both in the country of origin and again in the U.S. while under quarantine.

PROCEDURES FOR IMPORTS FROM CANADA AND MEXICO

The entry procedures for animals from Mexico and Canada are generally less strict than those for animals from overseas nations because Mexico and Canada have animal disease situations relatively similar to ours. The exception is hog cholera in Mexico, and for that reason we do not import swine from that country. In the case of other animals, however, entry quarantine and advance import permits are not required for either Mexico or Canada.

The U.S. has 16 crossing points on the Mexican border and 43 on the Canadian border where APHIS veterinarians examine and process animal imports. Entry procedures vary. For example, cattle from Mexico must be dipped in a pesticide solution as a precaution against cattle fever ticks and scabies.

ANIMAL SEMEN, EMBRYOS

Since animal semen is as much a potential disease threat as live animals, it must be subjected to strict standards for collection, handling, and shipping. These standards are spelled out in agreements between USDA and the foreign countries involved. While the technology for testing animal semen is well researched and established, the same cannot be said for embryos, the newest practical method for exporting animals. More research is needed to determine the diseases to which embryos are immune.

Our restrictions on animal imports may seem extreme to some. But we have a major responsibility for maintaining the health of our livestock--for example, a $9 billion swine industry, a $35 billion cattle industry, and a $10 billion poultry industry. Considering what is at stake, our restrictions are reasonable.

EXPORT PROCEDURES

Just as we insist that only healthy livestock be imported into this country, we have an obligation to see that only healthy animals are exported to other countries.

We ensure the health status of our exported animals in two ways. First, we establish our own health rules for exports; second, we cooperate fully in meeting the import rules of receiving nations.

All animals we export are subjected to special testing and certification requirements to indicate freedom from certain diseases found in the United States: bluetongue for cattle, sheep, and goats; brucellosis for cattle and goats; anaplasmosis for cattle; equine infectious anemia for horses; and pseudorabies for swine.

Our own export rules are necessary because we do have some health problems, i.e., EEC bluetongue of concern to foreign importers. We want to avoid damaging our position in the world market by making sure the animals and animal products we export are free of disease.

The animal health requirements of foreign countries vary, reflecting the particular animal disease problems and danger they face in their part of the world. Testing and certification are performed by private veterinarians accredited by us (usually at the shipper's expense) who normally conduct their tests on the farm or cattle ranch. Adult dairy and breeding cattle and goats must be tested and found free of brucellosis and tuberculosis within specified time limits before shipment. An animal health certificate is endorsed by the APHIS area veterinarian after the private accredited veterinarian completes his testing. The endorsement certifies that the private veterinarian is qualified to conduct the examination and tests.

Once the certificate is endorsed, livestock can move to a port of embarkation. There they must rest for 5 hours at a USDA-approved facility while the animals and paperwork are given a final check by an APHIS veterinarian. If the animals are healthy, if they are properly identified, and if the health certificate is in order, they are loaded on a ship or aircraft under APHIS supervision for export.

COMPLEXITY OF FOREIGN REQUIREMENTS

The health requirements imposed by foreign governments can be quite complex, so whenever possible we work out agreements with these countries. We do all we can to negotiate standard requirements, but we have over 150 agreements with some 70 foreign governments, and they can be changed on short notice. It is nearly impossible for an exporter or an examining veterinarian to keep track of the different rules for every overseas livestock shipment. To avoid delay and frustration, the exporter and the accredited veterinarian are urged to contact the APHIS Veterinary Services office in their state before attempting the process livestock or animal products.

Each APHIS Veterinary Services office keeps a current file of foreign animal import health requirements. Each area office of APHIS Veterinary Services also has a veterinarian assigned to work with exporters and their veterinarians. His job is to check the export health tests and certifications and place the final endorsements on the health papers before the livestock can leave this country.

CANADIAN, MEXICAN EXPORTS

As with our imports, our exports to Canada and Mexico are handled more simply and quickly than those to overseas nations. Exported livestock do not need an APHIS veterinary examination at the port of export. Once the health tests, certification, and APHIS endorsements are completed, the animals move directly to the border where they are examined by the Canadian or Mexican officials. Canada and Mexico are by far our biggest customers for exported livestock (and poultry as well), so it is important to devote some special attention to health matters on shipments across our borders.

Exports to Mexico move with few special problems. However, Canada has some requirements that exceed our own export rules. The Canadians are particularly concerned about chemical residues in cattle shipped to slaughter, so feed additives and antibiotics should be withdrawn from livestock within the recommended time limits. This is the responsibility of the exporter.

CLOSE COORDINATION REQUIRED

Successful U.S. exports, particularly those to overseas nations, require close coordination between the exporter, the private veterinarian, APHIS officials, the forwarder, the broker, the insurance underwriter, and the carrier. Among the most common causes of costly delay is the failure to conduct all required tests and failure to allow enough time for completion of the tests at the diagnostic laboratory.

Even if a plane or ship is waiting, animals cannot move to the port of embarkation unless APHIS endorses the health papers. Therefore, exporters should be aware of all the requirements and plan to allow sufficient time for testing when making plans to ship animals to another country.

INTERNATIONAL INVOLVEMENT

The U.S., along with its major partners, is actively involved in international organizations such as the Food and Agriculture Organization (FAO), the Office of International Epizootics (OIE), and the General Agreement on Tariff and Trade (GATT) in dealing with animal diseases worldwide and taking steps to assure the expeditious movement of healthy livestock and livestock products. We are actively involved in international health programs because the more we can do to reduce diseases worldwide, the more freely our own animals will move in international markets--an advantage of the U.S. exporter.

CONCLUSION

APHIS does all it can to help the stockman with his exports. We safeguard his markets by making sure no diseased animal gets out; and we make sure no diseased animal gets in to infect his livestock. We are against the unduly restrictive animal health import requirement that functions as a nontariff barrier and, more often than not, serves to protect a country's livestock industry--more from foreign competition than from foreign animal diseases.

The U.S. strongly supports "free trade" and endeavors, whenever possible, to make it a dominant principle in world

trade, which includes trade in live animals and animal products. This is compatible with our free-enterprise tradition. If we adhere to this principle and keep our disease
defenses strong, our livestock should remain the healthiest
in the world. And export opportunities for U.S. stockmen
should expand as world trade in animals expands.

65

ANIMAL PARASITES: PROFIT REDUCERS

Harry C. Mussman

With livestock prices at today's levels, the U.S. stockman must cut production costs to maintain even a relatively moderate to slim profit margin. Such cost cutting could begin with a review of his program for keeping his herds free of animal parasites--the great profit reducers.

The adverse economic effects of parasites on livestock production vary, depending on the kind of animal involved, its general state of health, and factors independent of the animal such as geography, season, weather, and condition of pasture. Given their pervasiveness and ubiquity in nature, parasites (both internal and external) can cause significant economic loss if the stockman fails to safeguard his herd from infestation. Basically, two types of pests affect livestock: 1) arthropods--external parasites such as flies, ticks, and mites; and 2) helminths--worms living in the gastrointestinal tract, liver, lungs, and other body organs of the animal.

EXTERNAL PARASITES: ARTHROPODS

Among the external parasites, arthropod (insect) pests causing major losses in cattle production in the U.S. are: horn, stable, and face flies; mosquitoes; cattle grubs; lice; ticks; and the scabies and mange mites. Arthropod pests in the U.S. are responsible for livestock production losses estimated to be more than $2.4 billion a year: $2.2 billion of the losses occur in cattle production (from flies, ticks, mites, and lice); $231.6 million in swine production (from lice and mange mites); and $54.4 million in sheep production (from keds and bots).

Arthropod Pests of Cattle

Flies. Flies are among the major parasites of beef and dairy cattle in the U.S. Bloodsucking flies, often called "biting" flies, reduce weight gains by interrupting the animal's feeding activities, in addition to causing loss of blood. They also can spread diseases such as anaplasmosis, anthrax, tularemia, and bluetongue. The most economically damaging of the "biting" flies is the horn fly, which accounts for an estimated $730 million in cattle production losses in the U.S. annually. Studies have shown that horn flies cause a 12 lb/calf weight loss, a 10 lb/head weight loss in stocker cattle, a 4.6% weight loss in slaughter cattle, and a 1% reduction in milk flow during the 6-month horn fly season.

The stable fly, another bloodsucking fly, has been shown to reduce milk production by as much as 40% to 60%. Stable fly losses for cattle in the U.S. have been estimated at $400 million a year.

The face fly does not suck blood; nevertheless, it causes sizable annual losses (an estimated $53.2 million a year). Attracted to the moist areas of the mouth and eyes of cattle, the face fly transmits pinkeye, which has been shown to cause a significant decrease in weight gain, especially in young cattle. In one test, 10% of the calves with pinkeye lost 33 lb each.

The extreme annoyance to cattle caused by other nonbiting flies, such as the house fly that flourishes in large numbers in barn and feedlot areas, also can cause significantly less weight gain.

Flies, biting and nonbiting, transmit cattle diseases as well. Among flies producing myiasis in cattle are the warble fly and the screwworm fly. Happily, the screwworm fly has been eradicated in the U.S., and a joint U.S.-Mexican screwworm program in Mexico is pushing this pest southward, farther away from the U.S.

Mosquitoes. Not to be overlooked as a great profit reducer is the mosquito that causes an estimated loss to cattle production each year of $39 million. In a series of tests in Louisiana, mosquitoes reduced the weight gain of feedlot steers by 0.05 kg/animal/day.

Cattle grubs. Cattle grubs have been shown to cause a 12 lb/head weight loss in calves in feedlots, a 41 lb/head weight loss in older cattle. Further, a $1.00/animal loss is experienced in meat trim of 40% of the cattle slaughtered when they have cattle grubs in their backs, and there is a $1.00/hide loss for hides that have more than five holes. In addition, a 10% decrease in milk flow can occur for 3 mo of the year when cattle grubs are in the backs of cattle. The annual loss caused by cattle grubs in the United States is estimated at over $600 million.

Lice. Cattle in most sections of the U.S. are attacked by three species of sucking lice and one species of biting lice. Lice spend most of their entire short life on the host, and when found in large numbers can be serious pests of beef and dairy cattle. Infestations of the shortnosed cattle louse can cause anemia and increased winter-weight loss of up to 81 lb/heifer and 55 lb/bull. Cattle losses due to lice amount to an estimated $126 million a year. Lice caused an average 68 lb/head loss in 12% of the cattle slaughtered during one reported test.

Ticks. Cattle fever ticks can spread bovine piroplasmosis, a severe and often fatal disease of cattle. The cattle fever tick has been largely eliminated in the U.S. At present the slow process of eradicating this tick is in progress in Mexico. It should be remembered that the U.S. required 37 years to eradicate this pest.

To prevent cattle fever tick reinfestation in the U.S., APHIS Veterinary Services has "tick" patrols along the U.S.--Mexican border to make sure no stray infested cattle cross over into U.S. territory. A reintroduction of the cattle fever tick into the U.S. would prove costly to the stockman. Cattle fever is capable of killing 90% of the cattle in infested herds. Even when the ticks do not transmit the disease, infested animals need extra feed to meet the demands of the parasite. As a result, growth is retarded and cattle become emaciated and unprofitable. In dairy cattle, milk production drops.

Before our eradication program began in 1906 (when cattle were selling for $.02 to $.03/lb), cattle fever ticks were costing the U.S. cattle industry losses of $40 million annually. At today's cattle prices, losses could be catastrophic. Other ticks still cause considerable cattle production losses in the U.S.--an estimated $275 million a year.

Scabies. Scabies, a contagious skin disease of cattle, is produced when tiny parasitic mites pierce the animal's skin to feed. Discharge from the mite wound oozes onto the surface of the skin and forms scabs or crusts. The affected areas often become infected with bacteria. Cattle with scabies invariably lick, rub, and scratch themselves to relieve intense itching. As a result they gain less weight but require more feed, resulting in higher production costs. Tests have shown that 100 additional days of feeding are required to bring untreated, infested cattle to marketable condition.

One study reports scabies mites caused a loss of $45/head for lack of weight gain for 340,000 cattle found infected in feedlots. Daily feed consumption of feedlot cattle with scabies can decline by 21% and reduce weight gains by 0.5 lb/day. According to the same study, scabies treatment may run from $5.00/beef cow to as much as $6.50 for each stocker heifer and bull over 500 lb. Scabies and mange mites cause a total estimated loss of $30 million a year in cattle production in the U.S. The cost of treatment, quarantine, and market discrimination of diseased cattle was $65 million in one epidemic year in the U.S. During a recent two-year scabies outbreak in the Southwest, a scabies control cost the state, federal, and producer operations $40 million.

Sheep and Goat Pests

Although scabies is still extant among cattle in the U.S., the scabies mite of sheep has been eradicated (1973). But sheep and goats can be attacked by other parasites that live in their hair, suck their blood, and invade their tissues, causing decreased weight gain and loss in value and production of wool and mohair.

Lice and keds. Sheep can become infested with the sheep biting louse, the sucking body louse, and the sucking foot louse. Sheep, and to a lesser extent goats, can be infested with the sheep ked, a wingless bloodsucking fly that is the cause of cockle, a sheepskin defect. It is estimated that sheep keds cause an 8% reduction in total weight gain, a 15% reduction in wool production, and a 30% reduction in the value of sheepskins.

Nose bot, mange mites, ticks, and gnats. Sheep nose bot larvae cause a 4% decrease in total weight gain. Keds and bots together cause an annual estimated loss of $54.4 million in production. Goats also can be infested by mange mites. Both sheep and goats are susceptible to ticks like the spinose ear tick. The biting gnat attacks sheep and--of great importance--transmits bluetongue.

Swine Pests

Swine in the U.S. are attacked by stable flies, mosquitoes, black flies, gnats, horse flies, and other biting flies. In addition, they are parasitized by the hog louse and a mange mite, which cause stunted growth and decreased weight gain through irritation. House flies also can cause extreme irritation and weight loss. Swine production losses in the United States from lice and mange mites alone have been estimated at $231 million a year.

INTERNAL PARASITES: HELMINTHS

The effect of worm infestations on the production of wool in sheep, and on weight gains and deaths of sheep, has been well-documented. However, the adverse effects of gastrointestinal helminths in beef and dairy cattle production is less well-established. In some studies, anthelmintic treatment has had no effect; in others, there have been increased weight gains as high as 73 to 74 lb/calf and from 0.15 to 0.18 lb/ADG in field and feedlot trials. Also, gains of up to 150 lb/calf have resulted from one drenching and subsequent movement to worm-free pastures.

Losses in U.S. cattle production from worms have been estimated at $250 million a year. Yearly losses from other internal parasites have also been extrapolated: $200 million from anaplasmosis, $70 million from coccidiosis.

Internal parasite losses in U.S. sheep and goat production have been estimated at $45 million a year. Helminths cause substantial losses in swine. The large roundworm is found in the small intestine where the female produces a large number of eggs that are passed with the feces. When later ingested, the eggs hatch into larvae that penetrate the intestinal wall and enter the portal circulation and later damage the lungs and liver.

STOCKMAN ROLE IN COMBATING PARASITES

Stockmen should constantly remind themselves of the dangers that parasites pose to their profit margins--especially stockmen whose operations have rarely, or never, experienced severe infestations. Lack of experience with infestations tends to lower vigilance. Coping with parasites is a complex problem requiring year-round attention on the part of the stockman. There are no "magic wand" treatments: some may last only a few hours, others perhaps for weeks, so year-round attention is necessary.

To minimize scabies mite infestations in high-incident areas, livestock should be dipped each time after mingling with new animals or coming off a new pasture. Dipping has to be done twice, two weeks apart, to eliminate any parasite eggs that may have hatched.

Sanitary conditions and pesticide spraying in holding and loafing sheds also help to control infestations. Worming is always important and can be conveniently done whenever cattle are brought together for branding. Many brochures and pamphlets on combating parasites are available from feed companies and other sources at little or no charge. Both literature and advice are available from colleges, universities, and government agencies. Feed firms and pharmaceutical houses usually make extension services available free of charge. Literature is also available from government agencies, both state and federal.

Stockmen should look for signs of internal parasites, which are not readily apparent in the early stages. They should also consult their veterinarians before an infestation becomes serious.

CONCLUSION

Parasites are many and everywhere. They constantly threaten to reduce the profits of stockmen. Government agencies and international organizations try to safeguard a nation's livestock assets from parasites, but it is chiefly the individual stockman, through deliberate management decisions, who can reduce and control animal parasites that eat away at his profits. His individual efforts will become increasingly important as government projects, at the national and international level, are phased out and the job of control is turned over to the private sector.

66

THE STOCKMAN'S ROLE
IN NATIONAL ANIMAL HEALTH PROGRAMS

Harry C. Mussman

STOCKMAN SUPPORT AND INITIATIVE

One of the most important elements necessary for a successful national animal-health program, is the support and participation of stockmen. The initiation of such a program depends on the desires of the livestock sector. The federal agency responsible for a nation's animal health seriously considers a program only when a majority of stockmen and other sectors of the industry can justify its need scientifically, agriculturally, and economically.

In the U.S. a particular animal disease can cause significant damage to the livestock industry that inevitably extends to the consumer's pocketbook through higher food prices. This situation rallies consumers and subsequently the Congress and obviously influences the implementation of a national program to combat the disease.

Bluetongue Example

Bluetongue is an example of a disease problem affecting a small group of our livestock industry and having little impact on the national economy. The bluetongue problem affects only livestock breeders who want to export to disease-free areas such as Europe.

Other factors making a bluetongue program impractical include the lack of both scientific knowledge about the disease and eradication tools, plus the cost of improving each. These factors are considered when determining the feasibility of a bluetongue eradication program.

Legislation and Funding

Specific legislative authority from the U.S. Congress is needed before a bluetongue eradication program can be initiated. Getting such legislative authority requires the support of a majority of livestock producers--if the knowledge and eradication tools are available. This support would extend to obtaining funding for the program.

Research

Livestock industry support also would be needed for research to develop additional scientific knowledge and refine the tools a bluetongue eradication program would require. Support could come later in the form of participation by producers in the research itself.

Voluntary Cooperation

We may have various means of enforcement--laws, regulations, and fines--but voluntary cooperation gets the best results. Therefore, industry cooperation has to be on a voluntary basis. Stockmen usually cooperate voluntarily if they are well-informed about the program and the consequences of not taking action. Each stockman must become aware that everyone suffers if a disease is permitted to reach epidemic proportions.

Outside Support

Every successful program needs the participation of scientists and experts with specialized skills both in and out of government. However, despite available money, it often takes effort and imagination to pinpoint them. As an example, during the exotic Newcastle disease outbreak in California in the early 1970s, we borrowed U.S. military personnel because we lacked civilian personnel to do the job. They functioned as civilians. Other assistance at the time came from a specialist in Japan who was able to take a year's leave of absence to work for us.

I must stress the importance of having the support of the majority of the producers to start a program and keep it going to a successful conclusion.

Brucellosis Example

The brucellosis program is a case in point. In one-fourth of the states where brucellosis incidence is still high there are some doubts about continuing the program. In the other three-fourths of the states the disease has been either eradicated or greatly reduced. That support in the majority of the states justifies keeping the program going and striving for an eventual total eradication of the disease.

VULNERABILITY OF U.S. LIVESTOCK

The effectiveness of our programs must be maintained because we have become more and more vulnerable to penetration by animal diseases from outside the U.S. Our inspection and quarantine defenses are maintained at our major ports of entry, but the volume of animals, plants, food products, other cargo, and people coming into the U.S. increases each year--and so do the chances of disease introduction.

For example, African swine fever (ASF) is pesently at our doorstep in Haiti. Cuba and the Dominican Republic were infected by it within the last two years. Fortunately, we were able to help the Dominican Republic eradicate the disease, and now we are helping Haiti. Cuba, with the help of other countries, successfully eradicated ASF in 1971 and in 1980. The large-scale concentration of thousands of animals packed into feedlots makes us more vulnerable to wide-scale disease today than when we were a nation of small and scattered producers.

We not only have livestock in the multi-millions but multi-thousands of them are on the move daily. Thousands of head of cattle born in Florida soon end up in feedlots in California. As livestock move through today's mass marketing channels, exposure and infection can take place widely and multiply rapidly before detection because most animal diseases have a week to 10 day incubation period. One infected animal in the system can spread a disease like wildfire before we know it.

It has been some time since we had a major cattle disease outbreak of foreign origin. We tend to forget the cost. From 1914 to 1916 we had an outbreak of foot-and-mouth disease in the U.S. and it came close to us again when it was found in Canada in the 1950s and in Mexico in the 1940s. Fortunately, with U.S. help, the disease was stopped in both countries before spreading to U.S. livestock. Experts calculate that such an outbreak today, given the concentration and rapid movement of our livestock, could spread to 48 states, and the cost would run into billions of dollars for both producers and consumers.

SELF-HELP

We have considered what the stockman can do to help animal health officials, now how can the stockman help himself?

Every stockman should:
- Have a good, ongoing sanitation and disease-prevention program designed to deal with the chronic and parasitic diseases, especially those that result in poor weight gain and lower profits (an example of annual animal production losses; from the horn fly, $730.3 million; from the stable fly, $398.9 million; from mosquitoes, $38.7; from the face fly, $53.2 milllion; from the cattle grub, $607.8 million; and from lice, $126.3 million.)
- Consult his veterinarian and ask him for help in setting up a disease prevention program similar to a maintenance program for an automobile or an anticavity program for children's teeth.

- Join and actively support his livestock associations, which blend his single voice with those of his fellow stockmen to forge a mass demand capable of gaining industrywide action, the passage of legislation, and establishment and funding of government programs.

CONCLUSION

The role of the stockman in national animal-health programs is crucial both in terms of his initiative and support and in terms of what he does to maintain the health of his own herd. With his full support, animal health officials can do what is necessary to keep a country's livestock industry healthy and prosperous. The system works in the U.S. and can be made to work wherever livestockmen recognize the role they play in assuring a healthy environment in which to raise their animals.

67

ECONOMICS IN ANIMAL HEALTH

Don Williams

GENETIC ENGINEERING

I am excited and expectant about the coming evolution in new products for animal health care through genetic engineering. In the coming decade, there will be dramatic changes in our approach to disease control and treatment through this new field of science. The scientist in this field will transfer genetic material from infectious viruses to simple bacteria that will then produce our vaccines.

The chemical compounds that comprise the genes in the nucleus of a virus, or other living cells, are referred to as DNA. This DNA determines the function and growth of cells. Each cell, bacteria, or virus expresses its identity partially by the proteins that the DNA has manufactured. The genetic engineer and immunologists are particularly interested in the proteins that are on the surface of a virus because this is how the body recognizes the virus as a foreign body; these surface proteins also stimulate the body to make antibodies. The body's immune system will recognize these surface proteins as foreign, just as it will an isolated protein if protein is attached to the virus. The immune system will respond and make antibodies.

With this knowledge, the genetic engineer identifies this ultramicroscopic portion of the DNA that informs the cell to make this surface protein and then removes this DNA segment from the virus. The isolated DNA is then put into the nucleus of a bacterial plasmid, which has the ability to transfer the isolated DNA into the recipient bacteria so that the bacteria now produces the desired protein (that is the original surface protein of the virus) simply by fermentation. This production by fermentation is very similar to production methods used to produce penicillin and terramycin. The advantages are that large amounts of this surface protein, which will be the vaccine, can be produced very economically and yet do not contain any live virus in the vaccine.

Another procedure which is available to genetic engineers is the hybridoma cell. This cell is formed by the fusion of a cell from mouse lymph tissue with a mouse cancer cell. The new cell grows continuously and forms antibodies against one specific infectious agent. These antibodies can be used as a direct treatment of a disease. A genetic engineering company, Molecular Genetic Inc., hopes to market antibodies against the K-99 E. coli bacteria by the end of this year. This particular E. coli plays a major role in calf diarrhea. The administration of large quantities of these specific antibodies produced by hybridoma cell lines should prove very effective in the treatment of calves exposed and sick with calf diarrhea.

The new procedures now being developed by genetic engineers will give many new producers. Among the products now being developed in the laboratories of this nation are: new foot and mouth vaccines, a pseudo-rabies vaccine, a more effective brucellosis vaccine, a new vaccine for calf diarrhea, a bovine virus diarrhea vaccine, and a bacteria for Pasteurella pneumonia.

Now I would like to direct your attention to some new products that will soon be on the market, and which you will be invited to purchase, we will consider management procedures that you can use today that require only limited expenditures.

HERD HEALTH FOR THE COW HERD

Let's consider the economic cost of four diseases in the cow herd.

Vibriosis. Vibriosis is a venereal infection of cattle that, conservatively, will reduce your calf crop by 5% and delay the calving in half of your cows by at least 30 days. When vibriosis is first contracted, the female will fail to conceive for two or three periods until sufficient immunity is developed to control the infection. The persistence of the infection varies among individuals. It is introduced by adding an infectious individual to the herd. In a commercial herd of 100 cows for one year, the cost of this disease (table 1) on today's market would be a minimum of $1,810.00. Fortunately, we do have effective vaccines for vibriosis. If your labor for vaccinating is $75.00 for the 100 cows, your return on your investment the first year is 17:1 .

Trichomonas. Trichomonas infection is another venereal disease of cattle. A recent survey of 280 bulls being sold in a commercial sale barn in Oklahoma showed an infection rate of about 8%. A similar survey on 109 bulls in Florida revealed an infection rate of over 7%. It is a disease of economic importance in this nation and is spread by introducing an infected animal into your herd. As in the case of vibriosis, no one gives the animals this disease, you actually go out and buy it. In a herd in which the disease is established, your calf crop will be down 5% (table 2), and since it takes longer for cows to develop immunity, half of your cows will be delayed at least 60 days in calving. The total cost for a year would be at least $2,685.00 for 100 cows.

TABLE 1. VIBROSIS COST - 100 COWS

Reduction in calf crop - 5 calves 5 x 400 lb x $.60	$1,200
45 cows delayed 30 days in calving 45 x 30 lb x $.60	610
	$1,810
Vaccine	
100 Doses	30
Labor	75
	$105
Return on Investment $1,810 ÷ 105 =	17:1

TABLE 2. TRICHOMONAS COSTS - 100 COWS

Reduction in calf crop - 5 calves 5 x 400 lb x $.60	$1,200
45 cows delayed 60 days in calving 45 x 55 lb x $.60	1,485
Total	$2,685

Brucellosis. Brucellosis is the most controversial disease in our nation's cow herds. In the state of Oklahoma (table 3) the records show that the average infected herd will lose 14% of the original cows while it will take 5 tests to clean them up. There will be a 5% decrease in the calf crop and you will lose 10 lb per calf due to the testing of your cows. The total cost would be $4,935.00. This is another disease that you probably bought.

TABLE 3. BRUCELLOSIS COSTS - 100 COWS

Loss on 14 cows infected @ $200	$2,800
Labor on 5 herd tests @ $75	375
Weight loss - 2 lb/calf - 5 tests 800 lb @ $.60	480
5% decrease in calf crop 5 x 400 lb x $.60	1,200
Total	$4,935

If it costs you as much as $3.00 per head for calfhood vaccination (table 4) and the labor costs are $50.00, your total investment in calfhood vaccination of 100 cows would be $550.00. This would yield a return on your investment of 9:1, provided that calfhood vaccination would completely prevent the disease.

TABLE 4. RETURN ON CALFHOOD VACCINATIONS

20 heifer calves/year @ $3.00/hd	$ 60
Labor on vaccination	50
	$110
	X 5
	$550
Return on investment $4,935 ÷ 550 =	9:1

Anaplasmosis. Anaplasmosis, of course, is not a disease of the reproductive tract but may occasionally cause abortion if sufficient red blood cells are destroyed so that insufficient oxygen reaches the fetus. The conservative costs (table 5) for a herd of 100 cows that has had anaplasmosis introduced is $1,365.00 per year. Since anaplasmosis is primarily spread by horse flies and ticks, and since the horse fly that is engorged with blood travels only a short distance, this disease does not usually cross the fence unless the cattle cross the fence.

TABLE 5. ANAPLASMOSIS COST - 100 COWS

Death loss - 2 cows @ $525	$1,050
3 monthly injections terramycin	90
Labor	225
Total	$1,365

Now let's see what good management procedure might be worth to you (table 6). If you add no replacements to a clean herd but virgin heifers and bulls that have been adequately tested for brucellosis and anaplasmosis, you may be saving yourself a potential $10,795.00 for a herd of 100 cows. Yes, this added trouble and management will cost something. I will let each of you estimate your own cost, but I am sure that the return on your investment will be excellent and without the purchase of any antibiotics or vaccines--only good management procedures.

TABLE 6. COSTS OF DISEASES - 100 COWS

Brucellosis	$4,935
Vibriosis	1,810
Trichomonas	2,685
Anaplasmosis	1,365
Total	**$10,795**

Prevention: Buy virgin replacement heifers and bulls tested for anaplasmosis and brucellosis.
Return on investment: Excellent!

HERD HEALTH FOR STOCKERS

In stockers or backgrounding operations, we are faced primarily with freshly weaned calves that have been assembled from several different herds and often several states. With a little luck, a person in this business can buy a few calves carrying a variety of diseases until he gets a truck load of calves that represents a museum containing all of the respiratory diseases in the U.S. Most calves are probably carrying resistance to the diseases that were present in the herd in which they were raised, but not to all of the other diseases in the assembled museum. In an attempt to save a reasonable number of these new calves, many programs are developed. Some of these programs are successful and some are even economical.

An example of an unsuccessful program in a commercial backgrounding yard is shown in table 7. The cattle were vaccinated and treated with just about every product on the market. It cost $19.95 for medicine per calf to treat the illnesses and there were still a 12% death loss. These were not sale-barn calves, but all were from one ranch and about half of these calves had been weaned prior to shipment. Many of you could give similar examples. The total cost of this program was $5,105.00 for 100 steer calves without estimating the poor performance of the calves that lived but did not gain well.

In contrast to these results, let us look at another program that receives 8,000 to 10,000 head of sale-barn calves per year. The program has a minimal vaccination program and has sufficient grass traps for the cattle so that they can be put in natural surroundings. The 2% death loss (table 8) and $4.25 per head treatment is only that high on sale-barn calves received in the worst weather. The average death loss for the past few years has been under 1% and treatment costs have been under $1.50/head.

TABLE 7.   STOCKER OR BACKGROUNDING 100 HEAD, POORLY MANAGED

1.	(a)	Vaccinated 7-way, IBR, PI_3, BVD, and Lepto, H. somnus	
	(b)	Injected with Vitamins A and D, L.A. 200 wormer	
	(c)	Pour-on or dipped	$ 600.00
2.		Drugs and treatment	1,995.00
3.		12% death loss 12 x 450 lb x $.65	3,510.00
Total			$5,105.00

TABLE 8.   STOCKER OR BACKGROUNDING 100 HEAD, WELL-MANAGED

1.	(a)	Vaccinated with 2-way IBR-Lepto	
	(b)	Injected with Vitamins A and D and wormer	
	(c)	Pour-on or dipped	$ 300.00
2.		Drugs and treatment	425.00
3.		2% death loss 2 x 450 lb x $.65	585.00
Total			$1,310.00

Evaluation of different programs across the country has led to the development of these cardinal rules for receiving stocker calves:

1. Obtain fresh, moderately fleshed calves with the shortest haul possible.
2. Make them feel at home:
   Natural environment
   Easy access to feed and water
   Shade in the summer, protection in the winter
   Reduced temperature stress
3. Feed them well.
   Adequate energy and protein
   Adequate high quality roughage - nonirritating
4. Overdose them with "tender loving care" by experienced people.
5. Isolate new arrivals for three to four weeks

If these two examples given are representative of the difference between good and bad management, the return for good management is 4 to 1. Actually, to use a well worn cliche, we need to return to the basics. The tender loving care program was the standard in the early 1940s before we began depending on the sulfonamides and antibiotics.

HERD HEALTH FOR THE FEEDLOT

The same principles that have been discussed for stocker calves apply equally well to all feedlot animals since one of the primary problems is the assembling of animals from diverse origins after the stress of shipment. If it takes 10 to 14 days to put a group of cattle together in a pen, it is similar to starting a small fire and adding sufficient kindling until you have a full-grown bonfire. If a few animals are placed in a pen and then others added at intervals of three to four days, the additions are received just when the previous cattle are the most contagious. As each group becomes sick, they have a more severe infection than the previous group. The virologist uses the same method in the diagnostic laboratory to coax a virus to grow under lab conditions. His first attempt to grow a virus may be difficult, but if he transfers the virus from one tub to the next at the appropriate intervals, the virus grows quite well. He refers to this as serial passage.

Of course, this cycle can be prevented by obtaining all of the cattle from one origin so that they have already established resistance to the infections in this group of animals. If cattle must be assembled from different origins, it is often advisable to segregate the shipment for three weeks until they are no longer contagious and over the stress of shipment.

Conclusion

In any well-designed, well-managed animal health program, whether it be cow-calf, stocker operations, or feedlot, the single most rewarding and economic practice is the isolation of all incoming cattle. The returns on the investment in these practices are excellent.

68

HEALTH MAINTENANCE PROGRAMS: WHY THEY FAIL

R. Gene White

No program in the cow-calf or feedlot area will pay the dividends that a health maintenance program will pay. A health maintenance program is very similar to buying insurance on a house. Many of our cow-calf and feedlot operators are working on borrowed capital; it is only prudent that they develop a health maintenance program to protect that investment. In fact, many bankers and lending agencies have insisted that a health maintenance program, or close working relationship on a herd health program with your veterinarian, be maintained as a condition for borrowing money.

Many times producers try to get their health management through a syringe. More than vaccination is involved in a health maintenance program. A good management program requires a close, working relationship with the banker, with the nutritionist or agronomist (who helps produce the feed), with the cowboy whose eye selects good genetic replacement material and feeders, and with the veterinarian. In many cases a good management program involves the tremendous devotion and support of the family of the manager and the cowboys who are involved. Each individual production unit will require a different type of management program that works best for its operation or else it will be doomed to failure.

For most of my discussion, I will be talking about the reasons for health maintenance failure that is basically involved with the health care and the immunization program of the cows, replacement heifers, the calves, stocker feeder, and the bull.

CATTLE DISEASE RESISTANCE

Cattle disease resistance is the interaction of numerous biological activities within the animal's body.

First line of Defense

The skin is the body's first line of defense against disease. When the skin is healthy and not broken mechanically, it provides a wall of protection against numerous disease-producing microorganisms. Examples of mechanical penetration are bite wounds, nails, and surgical procedures that break the continuity of the skin.

The mucous membranes that line the body openings such as the eye, mouth, digestive and reproductive tracts, have a number of ways of fighting infection. They produce enzymes and control the acidity that counteracts many of the infective organisms. Many of the cells making up the mucosal lining have cilia or small whip-like projections that extend into the body openings. For example, in the respiratory tract, these cilia have a whip-like action that moves them forward to prevent foreign material such as viruses, bacteria dust, etc. from entering these openings and causing disease. The secretions of the mucosal cells attract foreign bodies that are expelled in secretions such as the nasal discharge. The mucosal cells also may be involved in what is described as local immunity or the production of antibodies in specific areas of the body that counteract disease-producing organisms. The health and well-being of these cells are highly dependent upon a good nutritional program with an adequate supply of vitamin A. Mucosal lining has numerous microorganisms that live continually on its surface. These are normal microflora and their presence aid in counteracting invasion by disease producing organisms.

Second Line of Defense

The white blood cells are known as the body's second line of defense. This system can be rapidly activated and is highly effective. These cells are located throughout the body, but primarily in the blood, spleen, lymph nodes and bone marrow. Some of these cells can readily move to the area of the infection or wound. Others are static or attached to the lining of spaces within the body organs. These cells become attracted to and attack undesirable or foreign objects within the body such as bacteria, viruses, etc., and engulf them. If all the organisms are engulfed, they are destroyed before they have an opportunity to produce disease.

The Last Line of Defense

Antibodies are in essence the last line of defense from infection. Antibodies are protein fractions in the fluid portion of the blood (serum) and in some other body secretions from the mucosal cells lining the body openings. Antibodies are formed by the antibody-forming cells of the body upon stimulation by antigens and vaccines or disease organisms. The antibody-forming cells are distributed throughout the body in the blood (white blood cells), lymph nodes, spleen, liver, bone marrow, and mucosa. Antibodies are highly specific so that they provide protection only against the specific organism that stimulated their formation. Antibody titers are used to measure the amount of protection possessed by an animal or to demonstrate the presence of a disease. The presence of antibodies for a specific organism indicates that they have been exposed to that organism or vaccinated with the organism or antigen that stimulated antibody protection.

Species immunity. There are several different kinds of immunity. One form of immunity is known as the natural or innate immunity. An example of innate immunity is the occurrence of hog cholera in swine but not in cattle, or of canine distemper in dogs but not in people.

Individual immunity. Another aspect of natural immunity is the difference between individuals within a herd or flock. An outbreak of pasteurella pneumonia in a beef cattle herd will result in some of the cattle dying while others recover, and some will exhibit no signs of illness. The variation in the amount of individual immunity may be due to age, maternal antibodies from the colostrum, inherited difference or unobserved infections that are followed by recovery and immunity.

Active immunity. Antibodies are usually produced when an animal is vaccinated or recovers from disease. This is the active form of immunity or immunity produced within the vaccinated animal.

Passive immunity. Antiserums, antitoxins, or colostral antibodies are known as passive immunity. Passive immunity results from antibodies that are produced in an animal and given to another animal that requires immediate protection. In contrast with the active immunity, passive immunity provides comparatively short protection but is immediately available to the recipient.
Antibodies obtained by the newborn from colostrum as a result of nursing its mother is a good example of passive immunity. Antibodies received in this manner will protect the young animal during early life when it has no other antibody protection. If calves are vaccinated too early in life, or while they still have some of the maternal antibodies, a lasting or effective immunity seldom results. The problems of developing immunity as a result of vaccination can be very frustrating.

Failure to induce antibody production. When an animal is exposed to a vaccine or an antigen, it will recognize this antigen and start producing antibodies against it only if enough vaccine (antigen) mass is present for an extended period of time. The absolute amount of antigen necessary to induce a good immune response has not been determined for most infectious agents. In general, however, the more antigen available, the better the immune response. This is one of the problems that is enountered when using inactivated or killed bacterins. The killed bacterins must be incorporated into an adjuvant that will maintain the presence of vaccine particles over an extended period of time. When live virus vaccines are used, they commonly do not have enough antigen to produce an immune response in the animal, but they reproduce or replicate in the host after vaccination and this growth will provide enough antigenic mass to induce production of antibodies. Therefore, care of the modified live vaccines is very critical in order that they are not accidentally inactivated. Another way to effectively decrease the available antigenic mass of the virus vaccine is the presence of maternal antibodies or passive immunity in young animals. In this case, the maternal antibodies will immediately attach to the virus particles so they cannot infect cells and replicate. When the antigen antibody complex is taken up by the white blood cells, the antigen will be degraded to the point that it is not recognized as an antigen and cannot stimulate antibody production. if the antigenic mass of the vaccine is large enough or the level of passive immunity is low enough, no interference occurs. One way of overcoming passive immunity is the practice of administering live vaccines by the respiratory route as in the internasal vaccination of calves. Since maternal antibodies are secreted into the respiratory tract, this area can be infected with viruses and the production of secretory antibodies (IGA) can be induced.

In cattle, passive immunity against some of the viral diseases may last as long as 7 months. This can interfere with vaccination of replacement heifers, feeder and dairy calves.

The use of modified live vaccines recommended for internasal innoculation is effective in young calves even in the presence of passive immunity, at least so far as the induction of the secretory antibody (IGA) is concerned. In general, live virus vaccines will give long lasting immunity and the killed viruses and bacterial vaccines give a relatively short immunity.

If calves are vaccinated at weaning for IBR and PI3, the calves that are kept for replacements should be revaccinated 30 to 60 days prior to breeding for IBR, PI3 and BVD. Indications are that this regime will provide a lifetime antibody protection against these diseases. Since no vaccine gives 100% immunity, cows can be vaccinated at 3 to 4 year intervals to make sure that immunity has persisted.

Another consideration for immunization failure is the possibility of other viruses causing a similar disease for which there is no vaccine. In the respiratory disease, there is no vaccine against respiratory mycroplasma infection. In most cases, the immunizing agents do a relatively poor job of protecting the respiratory tract. If cattle are vaccinated with pasteurella bacterins, pasteurella organisms can be cultured from many animals that die from respiratory disease to indicate that pasteurella is involved with this pneumonia process.

Abortions. Unless the manufacturer indicates otherwise, live virus vaccines may induce abortion in pregnant cattle. Live-virus vaccines are dependent upon virus replication in the animal to obtain a sufficient antigenic mass to induce a good immune response. Live vaccine virus replicates in a pregnant animal that has not been previously immunized and some virus may infect the fetus. The fetus may be more susceptible to the virus than the dam and may become ill and die. The vaccine virus also may be shed by the vaccinated animal and this virus can infect other non-vaccinated animals in the herd.

Stress. It is difficult to define stress accurately. Stress may be due to malnutrition, shipping, parasitism, diseases, inclement weather, or other related causes. Animals that are under severe stress may not react to the immunizing agents to produce adequate and protective antibody levels. Cordicosteroids depress the immune mechanism, and animals that are stressed have a normal increased output of these substances.

Dangers of Mixed Vaccines. Most live viral products grown on tissue culture use bovine serum as a nutrient. Bovine serums are tested exhaustively to show that no live agents are present (especially BVD virus); however, this serum may contain antibodies against viral agents other than the virus it is helping to grow. For example, IBR viruses may be grown on bovine tissue using a bovine serum that contains no IBR antibody, but may contain either or both antibody to BVD or PI3. If this product is mixed with a BVD or PI3 product, it could possibly contain enough antibody to inactivate the other viral entities.

Another problem is vaccinating animals that are incubating the particular disease. If the infection already has been established, and viral replication has taken place, even though clinical signs are not yet evident, the virulent virus will prevail over the modified live virus and disease may result. Following exposure of animals to a specific antigen, there is a transient period before antibody is produced during which exposure to the specific infectious agent may induce a more severe disease than if the animal had never been vaccinated. This hypersensitivity reaction is more common with killed vaccines but can occur with live viral products. This is often suspected in cases of severe shipping fever following the vaccination with pasteurella bacterins or at times with killed PI3 vaccines.

Care in use of vaccines. Never purchase vaccines from suppliers that do not keep the vaccines continually under refrigeration. Be sure to keep all vaccines out of sunlight as much as possible. When vaccines are taken to the field for use, they should be kept on ice and away from sunlight until administered. You cannot detect a deteriorated vaccine by looking at the bottle or the vaccinated animal.

Usually the first sign of ineffective protection is a disease outbreak. Before using the vaccine, read the labels or description insert and follow instructions as to the route of administration and dosage. Don't mix several vaccines together unless instructions indicate this is permissible.

On making up a vaccine, use sterile syringes and needles. Syringes and needles used for attenuated dessicated viral or bacterial vaccine should not be chemically sterilized but should be boiled in water for several minutes. Chemical sterilization may leave a residue that will destroy the virus and lower the potency of the vaccine. Contaminated vaccines can cause severe disease outbreaks or acute postvaccinational reactions. In preparing a vaccine, always mix thoroughly so that the antigenic material (bacterial or virus) will be administered uniformly to each recipient. Unused portions of the vaccines should never be stored for future use. Be sure all animals to be vaccinated are in a good state of health. The stress of vaccination to an already sick animal, particularly with attenuated vaccines, can lead to disaster. When administering vaccines, every effort should be made to be sure the entire desired dose is obtained by the recipient animals. Needles that are too large, too short, or not inserted deep enough into the muscle may allow some vaccine to run out, thus the animals may not receive a sufficient amount of vaccine to stimulate immunity. It should always be remembered that vaccination done under the very best of conditions will not always provide complete protection. However, when good vaccines are used according to instructions, maximum possible protection will be achieved. This should be sufficient to prevent serious disease outbreaks in a herd or flock in geographic area.

Always consult your veterinarian concerning a vaccination program for your herd.

69

BOVINE PEDIATRICS

R. Gene White

Bovine pediatrics is a term that can vary considerably in its overall connotation. First we must identify the type of animal that we are talking about: for instance, the beef calf reared on the range under clean environmental conditions as opposed to a dairy calf that is taken off its mother at birth, put in a cage or close confinement, and reared under unsanitary conditions. These are the two extremes. There can be tremendous variation between the type of husbandry practiced in different types of operations. In this discussion, we will use the term bovine pediatrics to cover the care of the beef calf from birth until weaning.

ENVIRONMENT

There is probably no better place for a newborn calf than out in the open if the weather happens to be dry and not too hot. A newborn calf can survive extremely cold weather if he is dry, has some form of protection, and plenty of ventilation. Ideally, the best place to calve is outside--with some type of a windbreak or open-front shed, if the weather is cold. There appear to be fewer problems with respiratory disease and diarrhea in the newborn if the environmental temperature and humidity are not extreme in any direction.

COLOSTRUM

In the formation of colostrum, the antibodies in the mother's bloodstream are concentrated in the first milk that the calf receives. The age of the calf when it consumes colostrum is extremely important. As the calf ages, the matter of a few hours can reduce the ability of the intestinal tract to absorb this antibody molecule and allow it to pass into the bloodstream. The older calf gets more exposure to extraneous protein and its absorption mechanism is less efficient. This antibody is the first protection against the many microorganisms that he will be exposed to during his lifetime. If the mother has a high level of antibodies or has been exposed to many of these disease microorganisms, she will pass a better grade of disease resistant-colostrum to her calf. Cows that are purchased and brought into a new environment may not show evidence of disease at calving time, but their calves may have a much higher incidence of disease than calves from cows raised in that particular operation. The production of healthy calves is affected by the immune status of the dam, the timing of the calf's consumption of colostrum, and the amount of exposure to disease-causing organisms.

VITAMIN A

Calves are born with low reserves of vitamin A. Colostrum is an excellent source, providing the dam has adequate reserves in her system or is receiving adequate vitamin A in her diet. (Many producers inject a million units of vitamin A at birth.) This is a well-accepted practice in areas where the winter is long and a vitamin A deficiency might be suspected in the cowherd.

Under normal circumstances, the cow will maintain a vitamin A reserve within the liver for approximately 3 to 6 months. However, if the mother is deficient in vitamin A, the vitamin A in the colostrum will also be deficient.

Deficiency of vitamin A affects the health of the epithelium or the first layer of moist cells that protects the calf. Vitamin A is involved with the mucosa of the respiratory tract, salivary glands, eyes, lacrimal glands, intestinal tract, urethra, kidney, and reproductive tract. The calf that is deficient in vitamin A is more susceptible to infection and organisms that cause diarrhea or pneumonia during the first weeks of life.

In drought areas or during long winters there can be a decrease in the supply of vitamin A. In many areas of the west, the grass is green only until the middle of July so that the cow does not receive an adequate supply of vitamin A from July until the next spring. Under these low vitamin A conditions, and especially during gestation, it is a good idea to supplement the cow either by injection or in the feed.

WHITE MUSCLE DISEASE

Milk is not a good source of vitamin E. The nonsaturated fats may tie up vitamin E and create a deficiency. There is a close interaction between vitamin E and selenium.

White muscle disease occurs in young calves where there is a deficiency or a tie up of vitamin E. This disease is usually characterized by hyline degeneration of the muscle and necrosis of the cardiac and skeletal muscles.

In reports from the Pacific Northwest, vitamin E was shown to be ineffective in preventing white muscle disease; however, the addition of one part per million (.001 ppm) of selenium was effective in precaution. The disease itself can be very complex. There is some indication of a correlation between the use of phosphate and sulfate commercial fertilizer when applied to selenium-deficient soils. This association may increase the incidence of the disease problem and would require a higher dosage of selenium to prevent white muscle disease.

The disease has two forms--cardiac and skeletal. The "cardiac hyperacute" form affects the heart muscle. Normally this results in sudden death, especially during exertion and exercise. In this form, the calf, which was apparently healthy a few hours previously, is found dead by the owner. These cases should be differentiated from other acute diseases such as enterotoxemia, injuries, acute pneumonia, abomasal ulcers and others. Occasionally, the calf appears (clinically) to have pneumonia--the temperature is usually normal, the heart is very erratic, and the lungs are congested with large amounts of fluid. Such calves have labored respiration and usually die within a few hours. In the cardiac form of white muscle disease, the lesions are usually easily recognized by the white areas of degeneration in the muscle fibers of the heart. Some of the heart muscle fibers may be calcified and pulmonary edema is usually evident on necropsy (autopsy).

The "skeletal" form of the disease is the second form in which the muscle necrosis affects the skeletal muscles. This form is recognized by a gradual weakness of the limbs in calves two weeks of age or older. The calf usually takes a sawhorse type of a stance and over a period of a few days is unable to rise unless assisted. Appetite, temperature, pulse, and respiration remain normal. When the calf is in this condition, you will occasionally see the shoulder blade protrude above the backbone in the thoracic area while the animal is standing. These animals will usually respond to vite-selenium injections and recover with no ill effects. In the skeletal form, white or grayish areas of degeneration are seen in groups of skeletal muscles and are usually present on both sides of the animal. The muscles are usually very pale in color, which gives a striated appearance. Histopathological studies of these areas are necessary to confirm the diagnosis.

In the cardiac form, the calves usually are found dead or are observed when near death. Treatment is seldom effective at this stage. However, in the skeletal form of the disease, selenium injections of Bo-Se (R) or Mu-Se (R) (selenium and alpha tocopheral preparations) may be effective. Prevention is by far the best method of deterring the disease.

ABOMASAL ULCERS

Ulceration of the liing of the abomasum ocurs commonly in cattle in association with various kinds of stresses and as secondary complications of other diseases.

In some geographic areas of the U.S. abomasal ulcers are found quite frequently, usually affecting calves from 2 to 8 weeks of age, and the thriftiest calves that are still nursing. Quite frequently these calves are found dead, or they appear to be bloated or have a large abdomen. If the calf is found before death, the abdomen is usually "pear-shaped" with the calf in a lot of distress; it kicks at its abdomen, stands alone, and usually dies quickly.

On necropsy, an extensive peritonitis and a perforated ulcer are found, usually in the greater curvature of the abomasum. These ulcers are usually 1/2 to 3 inches in diameter. When the lining of the abomasum is examined, this may be the only ulcer or erosion that is found. The author has observed beef herds in which up to 4% of the calves are affected by abomasal ulcers. Usually, however, there are isolated instances of one to two calves in a herd. It may show up in one calf the first year and none in subsequent years; or it may show up with an increase in incidence in the following years.

The cause of abomasal ulcers in the young calves is unknown. Perhaps a fungus or mycotic agent may be involved. Clostridium perfringens is usually cultured from these calves; however, this is not unusual because C. perfringens can be cultured from most calves. The significance of this finding is open for debate. E. coli is usually cultured from the abomasal ulcer; however, E. coli usually can be cultured from calves following death, although seldom cultured from so far forward in the intestinal tract.

WEAK CALF SYNDROME

The weak-calf syndrome is an immune-suppressed disease of the newborn calf. It is characterized by abortions, still births, and weak calves. In 1963, a disease of unknown origin was recognized in individual herds of beef calves that was characterized by 1% to 2% abortion rates and considerable death loss in newborn calves up to six weeks of age.

It has been identified in well-managed herds with sound nutrition, selected breeding, and disease control programs. This disease reached epidemic proportions in 1969, causing heavy loss of calves. Ninety-five percent of the cows become immune as they abort or produce a typically weak calf. Many causes of this disease have been incriminated, such as nutritional deficiencies, toxicities, bacterial and viral infections, and stress brought on by weather conditions. Several viral agents have been isolated and are suspected of being the causative agent. As many as four different viruses have been isolated from one calf with weak-calf syndrome. Bovine virus diarrhea, virus, endo viruses type 2, 5, and 7 and some unclassified viruses have been isolated. It is a very complex disease and some research workers have suggested the causative agents may be a combination of one or more viruses. The heaviest losses from the weak-calf syndrome occurred in the first part of the calving season. It often was followed by some type of stress on the cowherd. Abortions occur in the last third of pregnancy; full-term calves are stillborn or weak at birth. Some of the calves die in utero and are delivered 24 to 48 hours afterward. There is an apparent uterine inertia in the cows that have delivered weak calves. There may be difficult delivery, retained placenta, and cows that fail to claim their calves. The cow will usually show an increased temperature, drastic milk reduction, and rapid weight loss. Many calves will die within minutes after normal delivery. Other calves will be very weak and unable to rise unless hand fed and taken into a warm surrounding. Many calves show a reddening of the muzzle, small hemorrhages on the third eyelid with larger diffused hemorrhages on the sclera. Some calves will not nurse and need assistance. An edema of the legs causes a disturbance in circulation that makes the legs, ears, and tail subject to freezing. The fluid from the infected joint will have a strawberry to a dark port-wine color. Calves are depressed, have an increased temperature, and exhibit pain during abdominal examination. The labored respiration is often seen in the infected calves. Many of these calves will die by the third day.

A necropsy examination on full-term and older calves shows hemorrhage and edema around the joints and the extremities with some subcutaneous fluid in some of the muscles. Edema of the gall bladder, bile duct, and attachments of the first segment of the intestine are found in newborn calves up to 3 to 5 days of age. The abomasum usually reveals small petechia hemorrhage in the lining. Calves that survive to older age may show signs of chronic debilitation and loss of body fat. The thymus gland is greatly reduced in size. Histopathology may be required to confirm this disease. At the present time, no known prevention is available. However, the cows that have produced calves with the weak-calf syndrome apparently are immune for quite some time.

BLOATING

A high incidence of bloating in calves from 1 to 4 weeks of age has been observed. The bloating is abnormal with the gas in the intestine instead of the rumen. The treatment is mineral oil, castor oil, or some type of a laxative to help stimulate intestinal movement. The cause of this condition is unknown, but it has been observed in areas of high sandy soil.

DIARRHEA (SCOURS)

Calf diarrhea or scours causes more financial loss to the cow/calf producer than any other disease-related problem. Calf scours is not a disease – it is a clinical sign of a disease that can have many causes. In all diarrheas, the intestines fail to absorb fluids and the secretions into the intestine are increased thus causing diarrhea.

A claf is approximately 70% water at birth. If the calf begins to lose fluids through diarrhea, he dehydrates very rapidly. This dehydration and loss of certain body chemicals (electrolytes) produces a change in body chemistry and severe depression in the calf. Even though infectious agents may be the cause of primary damage to the intestine, death from scours is usually due to the loss of electrolytes resulting in dehydration, acidosis, and hyperbalemia.

Calf survival from scours depends on the age of the calf when the diarrhea begins; the younger the calf, the greater the chance of death.

In addition to the financial loss caused by diarrhea, scours in the herd is very demoralizing for the producer. In the range country, one man could be calving out 300 cows and get about 10 calves on the ground when infectious diarrhea hits. Scours will usually go through the older calves and then affect all of the new calves as they are born. Many calves require treatment three, four, or five times. The producer becomes demoralized by the never-ending job and a lot of the fun and profit is taken out of the cow/calf operation.

The cause of diarrhea can be very complex and involves interactions between many of the microorganisms such as bacteria, viruses, chalymidia, or protozoa. The death loss as a result of a diarrhea outbreak is dependent upon several different factors. The important factors in preventing a severe diarrhea outbreak are: amount of immunity or resistance that the calf has, the number of microorganisms that the calf is exposed to and the virulence of these organisms, the stress to which the calf is exposed, the nature of the diet, and the level of management.

Diarrhea is most common in calves from 2 to 10 days of age. It may occur as early as 12 hours after birth or at 6 weeks of age. In these cases, calves are born without antibodies present to protect them against these infectious agents (see section on Colostrum). Calves from first-calf heifers usually have less colostrum, with fewer antibodies being passed on to the calf. Therefore, the problem of neonatal diarrhea is much higher in calves from first-calf heifers. In our management practice, we have tended to isolate the first-calf heifer from the older cows during calving time. In doing this, she is missing the exposure to the infectious agents tht might be present in that herd and therefore does not have a high level of antibodies in her system to pass on to the calf. Calves that are weak or have a swollen tongue from a prolonged, difficult delivery may not be able to suck for several hours, which causes a marked decrease in ability to absorb colostral antibodies. Excess colostrum from older cows can be frozen in quart containers and stored for an indefinite period of time. This is very valuable to use in calves from cows that may not have colostrum or in weak calves that are unable to nurse.

Management practices affect the incidence of neonatal diarrhea. During periods of inclement weather when cows are brought together in close confinement to get away from the storm, there is an increased incidence of diarrhea spreading from one calf to another. The extra stress of crowding may prevent calves from nursing as soon as they would under normal conditions; this reduces the amount of colostrum consumed.

Sources of Infectious Organisms

The primary source of infection for calves is the feces of infected animals. Calves obtain the organisms from contaminated bedding, other diarrheic calves, overcrowded conditions, and from the skin around the udder of the mothers. The organisms are then spread within the herd through the feces of the infected calves and all of the objects that can be contaminated by the feces: bedding, boots, treatment instruments, clothing, feed, and water supplies. If the cows are overcrowded into calving grounds and heavily used calving pens, this contributes to the dynamics of the population of these organisms. Calving sheds may become so contaminated with these disease-producing organisms that they may have to be abandoned and clean facilities obtained.

The major clinical findings in diarrhea are abnormal feces, dehydration, weakness, and death within one to several days after onset. Based on the clinical findings alone it is difficult to make a definite diagnosis of the cause of the diarrhea. A presumptive diagnosis may be made by considering the history, the age of the animals affected, and the character of the feces. There may be mixed infections that require considerable research and diagnostic expertise to arrive at the actual causative agent or agents.

We will discuss the main causes of diarrhea as separate entities because in many cases they are mixed infections and the cause of the disease may be difficult to determine.

Bacterial Scours or Escherichia coli (Colibacillosis)

E. coli has been incriminated as a major cause of scours. Many times, following routine bacteriological culturing, this is the only organism identified. Certain E. coli can cause diarrhea. These are termed enteropathic or enterotoxigenic strains of E. coli. These organisms have the ability to colonize in the small intestine and produce endotoxins that cause an increase in the net secretion of fluid and electrolytes from the blood stream into the lumen of the intestine where large amounts of fluid containing sodium, chloride, potassium, and bicarbonate become trapped. There may be minimal or no structural changes in the intestinal epithelial cells.

Some sero types of E. coli can cause scours while others do not. Some sero types are always present in the intestinal tract and can cause secondary infection following viral agents or other intestinal irritants.

E. coli scours or colibacillosis is characterized by diarrhea and progressive dehydration. Death may occur in a few hours without diarrhea from the severe enterotoxigenic strains. The color and consistency of the feces are of little value in making a diagnosis. This course of the diarrhea varies from 2 to 4 days. Severity depends upon the age of the calf when the scours started, on the immune status of the calf, and on the particular sero type of E. coli.

Upon necropsy examination, lesions are nonspecific; however, the small intestine may be filled with fluid and the large intestine contain yellowish feces. Diagnosis depends upon an accurate history, clinical signs, and culture of the internal organs for bacteria and sero typing of these organisms. Also, location of the culture from the intestine is important. Control of E. coli scours can be difficult in a severe herd outbreak. All calves should receive colostrum as soon after birth as possible. This should give some immunity to the calf. Early isolation and treatment of the scours helps to prevent new cases. The temperature of the calf is usually normal but may become subnormal as the disease progresses. Mildly to moderately affected calves may be diarrhetic for a few days and recover without treatment.

In cattle, the primary enteropathogenic E. coli appears to possess the K99 pili antigen; however, 977P pili antigen has also been reported to cause an enteric condition in young calves. E. coli immunization used in the K99 antigen should be administered in two doses, 6 and 3 weeks before calving.

Salmonella

There are more than 1,000 sero types of salmonella. All types are potential disease producers. Salmonella produces a potent toxin or an endotoxin (poison) within its own cells. Calves are usually affected from 1 to 6 weeks of age, corresponding very closely to the age of some of the virus infections. The source of salmonella infection in a herd can be from other cattle, birds, cats, rodents, water supply, or human carriers. Clinical signs associated with salmonella infection include diarrhea, blood and fibrin in the feces, depression, and elevated temperature. The disease is more severe in young or debilitated calves. Finding a membrane-like coating in the intestine on necropsy is strong evidence that salmonella may have been involved.

Oral or intravenous fluid therapy may be more success-ful than other forms of treatment in salmonellosis. Feces should be cultured from several different calves and at dif-ferent intervals to obtain a positive diagnosis for salmonellosis. A bacteriological sensitivity test should be performed to determine the antibiotics of choice.

Enterotoxemia

Enterotoxemia can be highly fatal to young calves. Toxins produced by clostridium perfrengens organisms cause enterotoxemia. The disease has a sudden onset. Affected calves become listless, display uneasiness, and strain or kick at their abdomen. Bloody diarrhea may occur.

Enterotoxemia can be associated with a change in weather, change in feed of the cows, or management practices that cause the calf not to nurse for a longer period of time than usual. The hungry calf may over-consume milk, which establishes a condition in the gut that is conducive to the growth and production of toxins by the clostridium organ-isms. In many cases, calves may die without signs being observed.

Necropsy lesions may be a hemorrhagic intestinal tract, thus the common name, "purple gut." In the small intestine there may be large hemorrhagic or bloody, purplish areas where the tissue looks dead.

Enterotoxemia is diagnosed using laboratory methods to find the toxin in the small intestine. This toxin breaks down rather rapidly so that the contents of the intestinal tract must be collected very soon after death and preserved by freezing. Finding hemorrhagic enteritis in a calf that has died suddenly can provide a tentative diagnosis.

The disease is best controlled by vaccinating the cows with toxoid, 30 and 60 days after calving. A single booster dose of the toxoid should be given annually thereafter be-fore calving. If this problem is diagnosed in calves from nonimmunized cows, antitoxins can be given to the calf.

Viral Scours

Rotavirus infections can cause scours in calves within 12 hours of birth; however, when the infection is first introduced into a herd it can affect calves up to 30 days of age or older. Infected calves are severely depressed and may have a slight excess of salivation and profuse watery diarrhea. The feces may vary in color from yellow to green. Calves lose their appetite and the death rate may be up to 50%, depending on the secondary bacteria present and the effectiveness and consistency of treatment.

Diagnosis of rotavirus depends upon an accurate his-tory, clinical signs, and proper specimen collection and submission. The rotavirus infection alone causes no gross lesions in the small intestine, but there is an increased volume of fluid in the small and large intestine (See corona virus). An oral vaccine that is specific for rotavirus is given by mouth to the calf soon after birth and is very effective if that is the only agent causing the scours.

The cow's colostrum can be fortified with antibodies by using a cow vaccine guard given IM 30 days before calving.

Corona Virus

Corona virus usually causes scours in calves over 5 days of age. When the infection first starts in a herd, calves up to 6 weeks of age or older may scour. As the calves are born, calves about 5 days old are infected. These calves are not as depressed as those infected with the rotavirus. Initially the fecal material may have the same appearance as that caused by the rotavirus. However, if the calves continue to scour, the fecal material may contain mucus that resembles the white of an egg. Diarrhea may continue for several days. Mortality from corona virus scours ranges from 1% to 25%.

If gross lesions are observed in the intestine, they are usually the result of secondary bacterial infection. Many herds have been found infected with both the rota and corona viruses.

Diagnosis of the Rota and Corona Virus Scours

For an accurate diagnosis of viral scours, fecal material is collected very early in the infection or a section of the small intestine is tied off and frozen and submitted to a diagnostic laboratory. These samples should be frozen as quickly as possible and delivered to the laboratory in the frozen state.

Microscopic lesions of the rotavirus are seen as a shortening and denuding of the epithelial liing in the posterior part of the small intestine. When the epithelial cells are sloughed off, that leaves an avenue for E. coli to grow and pass from the intestinal tract into the circulation causing evidence of bacteremia or septicemia.

The large intestine is the organ of choice for corona virus diagnosis. Fecal samples or sections of the large intestine should be frozen and submitted to the laboratory in a frozen state. In either rota or corona viruses, fluorescent antibody tests are run on the submitted samples. This is the quickest way to confirm a diagnosis of rota or corona viruses.

Bovine Virus Diarrhea

The virus of bovine virus diarrhea can cause scours and death in young calves. Diarrhea begins about 28 hours to 3 days after exposure and may persist for quite a long time. Ulcers on the tongue, lips, and in the mouth are the usual lesions found in the live calf. The lesions are similar to those found in yearlings and adult animals affected with the same virus. Diagnosis is made by using the animal's history, observing for lesions, and sending samples to a laboratory. Bovine virus diarrhea is controlled by vaccinating all replacement heifers 1 to 2 months before breeding. A killed vaccine for bovine virus diarrhea is now on the market. It can be used in the cowherd if problems arise in pregnant cows. **Caution: Do not vaccinate pregnant heifers or cows with modified live virus. Consult your veterinarian before starting a program of bovine virus diarrhea vaccination.**

Other Enteric Viruses

The author has observed other viruses that can cause neonatal diarrhea that have been identified from calves with scours; one virus is particularly bad in first-calf heifers that have been in a calving facility. The calves will scour within about 4 hours after delivery, showing profuse watery diarrhea and severe dehydration. The calf will be dead in 12 hours if the fluid balance is not maintained.

Coccidiosis

Coccidiosis is caused by one-celled "parasites" that invade the epithelium or cells living in the intestinal tract of animals. Ther are many species of coccidia; however, Eimeria zuerni and Eimeria bovus are usually associated with clinical infections in cattle. Coccidiosis has been observed in calves from about 4 weeks of age and older --usually it follows some stress, poor sanitation, overcrowding, sudden change of feed, or turning the cow and calf on pasture. In some areas, this is common 7 to 14 days after the cow and calves are paired out and put on pasture.

In the young calf, coccidiosis may occur without hemorrhage and with very few coccidia present in the feces. Histopathology may be required for diagnosis. A typical sign of coccidiosis in young calves is diarrhea with fecal material smeared over the rump as far around as the tail will reach. This material may contain blood. Death may occur from secondary complications during the acute period or later.

Morbidity may reach 25% to 50% of the calves during an outbreak. If this has been consistent over the years, perhaps some type of prevention, such as amprolium, may be required prior to putting the calves on pasture. Amprolium is difficult to get into the calves, but when the disease is observed, amprolium should be supplied at the rate of 5 mg/kg of body weight for a period of 21 days to cover the critical time.

Nutritional Scours

Under range conditions, a calf adapts a pattern of nursing that fills its needs. Nutritional scours can be caused by anything that disrupts this pattern. A storm, strong winds, or the mother searching for new grass disrupts the normal nursing pattern. When the calf finally nurses, it is overly hungry and the cow may have more milk than normal, so the calf may overload, which causes nutritional scours (see enterotoxemia). This is usually a white scour caused by undigested milk passing through the intestinal tract. Many of these calves, if they are still active and alert, do not require treatment. If the calf becomes depressed or quits nursing, treatment should be started.

General Considerations for Treatment of Diarrhea

Treatment of calf scours is very similar regardless of cause. Treatment should be directed toward correction of the dehydration, acidosis, electrolyte loss, and hyperbalemia. Isolation of the affected animals should be done immediately, with antibiotic treatment given simultaneously with the treatement for dehydration. Dehydration can be overcome with simple fluids given by mouth early in the course of the disease. If the dehydration is allowed to continue before treatment starts, intravenous fluid may become necessary.

Calves with diarrhea should be identified with an ear tag and an accurate record kept for monitoring the effect of the treatment. Calves with diarrhea should be removed to a separate pen or paddock for treatment. This limits the exposure healthy calves get to the infection. The animal attendant or cowboy should not treat sick animals before caring for the well animals. He should change clothing and disinfect the boots or wear protective clothing before coming in contact with the healthy calves.

In viral and other types of diarrhea, the digestive and absorptive capacity of the intestinal tract are reduced; therefore milk should be restricted from the diet if at all possible. Oral glucose-electrolyte mixtures are commonly used as a source of energy, fulid, and electrolytes during the period of diarrhea. Such mixtures are inexpensive, easy to use, and readily available. An esophageal feeding tube can be used to administer oral fluids to a calf to help prevent dehydration and acidosis.

Antibodies should be used both orally and by injection whenever treating calves for diarrhea. In acute salmonella outbreaks, antibiotics may cause the release of excess endotoxins; therefore, cosideration should be given to using fluid therapy only.

Diarrhetic calves should be eartagged for identification. A daily record kept on the treatment administered aids in evaluating and utilizing follow-up treatment. If an outbreak of scours occurs, persistent treatment and keeping good records are essential.

70

BEEF-COW-HERD HEALTH MAINTENANCE

R. Gene White

THE VETERINARIAN

The need to develop an ongoing herd health maintenance program in the beef-cow production unit is less apparent than that in dairy or feedlot operations.

If a herd health program is to be meaningful, there should be a correlated effort that includes ownership, management, labor, and specialists such as the veterinarian, nutritionist, diagnostic laboratory, accountant, and the banker. In recent times, economic pressures and tax reporting procedures are making for a better coordination of the many facets of the beef industry.

The goal of the health maintenance program is to maintain and improve a profit margin. Health problems are more easily solved if they are recognized quickly. The opportunity to observe early signs of trouble is greatest for the cowboy, secondly for the manager, and lastly for the veterinarian. Thus, for optimum results, the cowboy must be taught to observe and report. This is one of the areas where the "art" of cow-calf management comes into play.

The veterinarian is usually only on a ranch for emergency situations. He should have an opportunity to visit with the owner, manager, and the cowboys to outline the basics of a herd health maintenance program based upon each management situation. The veterinarian should understand the goals and objectives of your operation. He should know what your operation is capable of doing and what you are interested in doing for yourself. With this information, a total herd health management practice can be designed to fit your operation.

The veterinarian is aware of diseases in the area and can recommend a vaccination program that is economically feasible.

The physical examination, necropsy, laboratory analysis, and interpretation of the findings are a veterinarian's responsibilities. Each death and each abortion should be investigated as a possible herd health problem. Do not expect miracles. Sometimes a complete diagnostic process is completed without a definite diagnosis; however, if you miss one infectious disease, it can be very expensive.

The veterinarian should set up a treatment schedule with you and provide information on how you can treat many conditions alone. If a disease process reaches a stage where he feels that he needs to take a look, then have confidence in his abilities to make that decision. In other words, do not go too far before you discuss your problem with your veterinarian. Give him an opportunity to be successful in treating those animals.

A good health-maintenance program involves several different facets that are very closely related and cannot be separated from total management of the cow-calf operation. Nutrition at the proper time during the life cycle of the animals will be discussed by the nutritionist. The management at calving, care of the calf, weaning, raising replacements, getting the replacements into the cow herd, and selection of the genetic material for replacements are all very closely related to the herd health maintenance program. The best advice available should be sought to protect the dollars that are invested in a cow-calf program.

THE BULLS

The average bull/cow ratio is probably about one bull for 25 cows. This ratio can be lowered by careful selection and culling. Introduce only virgin bulls into a beef herd. Don't buy the other man's problems. Bulls must be sound and have an interest in following the cows and be capable of breeding a cow.

Of all the evaluation techniques to consider in the bull, the size of the testicles is probably the most important anatomical structure. Do not buy bulls with testicles smaller than average size for age and breed.

It is preferable to buy bulls that have been subjected to breeding-soundness evaluation. A breeding-soundness evaluation is much more than a semen check. It should take into consideration a general physical examination. The feet, eyes, general attitude, disposition, and confirmation as related to his ability to follow the cows should be considered. A breeding-soundness evaluation involves a collection of semen from the bull and during this period the penis should be extruded and examined for any abnormalities. After the reproductive capability has been assured to the extent possible, the buyer should look for confirmation, pedigree, and price. Usually bulls from a breeder with production records are worth extra money to the producer.

Foot, eye, and traumatic problems from fighting or breeding take their toll on the bull battery. Constant surveillance is necessary, especially during the breeding season. Only sound bulls should be wintered over in the bull battery. Old ranchers have indicated that in multisire units, at least three bulls should be involved--two to fight, and one to breed the cows. The bull should be kept in good condition and should only be left with the cows during the breeding season. Bulls can be pastured together; there will be some fighting, but usually a peck order will be established with few harmful consequences.

Worming programs should be established, especially for young bulls. Where there is a fly problem, the flies are attracted to the bulls; therefore a vaccination program for anaplasmosis should be considered on the bulls. Normal vaccination programs on the bull should be consistent with the cow program.

REPLACEMENT HEIFERS

Development of the replacement heifers to calve at two years old can be profitable under most management practices. The additional cost of developing heifers to calve at two years of age depends upon the cost of feeds in the different areas. Additional labor may be needed with the heifers at calving time.

It is not considered advisable to calve heifers at two years of age that weighed less than 400 lb at weaning. Replacement heifers should be fed enough during the first winter to support an average gain of 3/4 to 1 lb/head/day. At breeding time, they should weigh from 600 to 650 lb and be fed to gain 200 to 250 lb during the summer. She should weigh 750 to 800 lb going into the winter carrying her first calf.

Replacement heifers should be bred to calve at least 30 days ahead of the main cow herd. This will give her an extra 30 days to start recycling and be bred back to be on schedule with the cows. After the heifers are calved they will need some extra feed to get them back in condition to breed with the cow herd.

Heifers to be kept for replacement should be vaccinated for brucellosis between four months and one year of age. They should be vaccinated for infectious bovine rhinotracheitus, vibriofetus, and leptospirosis 30 to 60 days prior to breeding. If Bovine Virus Diarrhea (BVD) vaccine is going to be used in the herd, this is the ideal time to immunize heifers. In my experience, heifers vaccinated with a modified live BVD vaccine 30 to 60 days prior to breeding have not had problems with BVD infection during their reproductive life, unless exposure was extremely high. From this point on, the heifer should have the same vaccination program as the cow herd.

COWS

The herd health maintenance program is usually geared to increasing the reproductive efficiency of the cow herd. This program should work into the total management and be given whenever the cows are being processed or at least twice a year.

In the spring before the bull is turned with the cows, they should be given a booster for vibriosis and leptospirosis. Leptospirosis vaccine only protective for perhaps two or three months to a maximum of a year, depending upon the degree of infection to which they are exposed. Many times the cows should be vaccinated every six months and, in some instances, every three months to effectively control leptospirosis.

Modified live vaccines for the viral disease should not be given to pregnant cows. If you wish to maintain an immunity in the cows, live vaccines need to be given while the cows are open. Intranasal vaccine or killed viral vaccines are safe for pregnant cows.

'Internal parasite control is extremely variable in a cow herd. It has been shown that adult cows develop some immunity against internal parasites. Work with your local veterinarian in establishing a worming program that is suitable for your geographical location and management type.

Establish a firm breeding season. Forty-five days is ideal; this gives two cycles for the cow to get bred. Remember that a cow must be safely settled within 85 days from the time she calves to maintain a 12-month calving interval.

Cows should be pregnancy checked 40 to 60 days after the bulls are removed. this enables you to do your culling before the cows go into the winter season, which is the expensive feeding period. While the cows are in the chute for pregnancy check, the production records, feet, teeth, eyes, and udder should be checked. If the cow is open, she should be culled immediately. If her production record does not meet the goals set for your herd, she should be culled. If the cow has a good production record, is pregnant, and has a broken mouth, then consider whether you can give her the extra feed and care she may require to deliver the calf. Eyes should be checked very closely. Many cows without pigment around the eyes may have a plague or the beginning of a cancerous growth around the eye or on the corneascleral junction. These plagues can be surgically removed and the cows maintained in the herd for several years. If there is a lesion on the eye that has advanced beyond the stage for surgical removal, the cow should be removed from the herd immediately. Cows with big teats or poor feet should be culled. A disposition problem is grounds for culling.

You should plan on replacing approximately 20% of the cows per year. A good rule of thumb in selecting replacement heifers is to select 50% more than you will need. To take advantage of the genetic ability to reproduce, turn the bulls with the heifers for a 45-day breeding period, and select the replacement heifers from the ones that are bred during that 45-day period. This will establish a good genetic base for reproductive performance in the cow herd.

CALVES

The calves should be numbered as they are born, and a permanent record started on those calves. They should be identified to the cow by cross reference and notes made on anything that occurs to that calf. A note should be made of the difficulty of the cow in delivering that calf.

Calves should be castrated and implanted within the first week of life. If conditions are appropriate, the calves also should be dehorned at this point. Dehorning paste works very well if used properly. However, it is the best time to dehorn and castrate the males. Implant all calves that are not to be kept for replacement.

Many times when the calf is about 3 to 4 months of age, there may be an outbreak of respiratory problems. This is about the time that calves are losing immunity that they received through the colostrum against the respiratory viruses, such as IBR, BVD, and PI_3. If an outbreak occurs, you should consider vaccinating the calves with the respiratory vaccine. You may have to wean them to be able to get them to where you can observe them closely for treatment.

Thirty days prior to weaning, calves should be vaccinated for IBR, PI_3, leptospirosis, 4-way Blackleg, and pasteurella; hemophilus should be considered also. The heifers should be vaccinated for brucellosis.

These calves should be put back on the cows to develop immunity prior to the stress of weaning. Regardless of the vaccination program, the chances of respiratory problems at weaning are pretty good.

At weaning, the calves should be revaccinated with pasteurella and hemophilus, if this was included in the preconditioning program. Calves should be wormed especially if they are going on to new pasture. Grub and louse control should also be considered at this time.

If you maintain ownership of the calves, they should be started on some hay at weaning. If they are range calves, you should note whether they are familiar with the water tank or the water facilities that are available. Close observation will eliminate a lot of problems.

A treatment program should be worked out with your veterinarian. If calves are pulled with signs of a respiratory problem, they should be treated immediately. In many cases, you are buying time by early treatment and the veterinarian may not be available to administer all these treatments.

Bovin respiratory diseases cause a greater loss to the beef cattle industry than any other disease. All of the vaccines and management may not prevent an outbreak of respiratory disease from occurring. You are inviting disaster by stressing the calves until immunity is built up.

We have discussed the vaccination programs for the cow herd to increase reproductive efficiency. Discussion of this type would be incomplete without some specifics on the diseases that are involved. Therefore, the following is a brief discussion of the reproductive diseases that we are trying to prevent by vaccinations in order that you might have a better understanding of the vaccination program.

REPRODUCTIVE DISEASES

Brucellosis

The most obvious sign of this disease is abortion. There are also dead calves at term and an increased incidence of retained placentas. Although the incidence of brucellosis has been greatly reduced by the national eradication program, there are some parts of the country where infection is still quite high. diagnosis is made by the use of the blood test.

Brucellosis vaccination should be done when the heifer calves are four months to one year of age. The decreased vaccine dosage has helped to decrease the number of vaccine reactors when they are tested later on.

Brucellosis is still a very costly disease to the beef cattle producer.

Leptospirosis

Leptospirosis is an important infectious disease that affects all domestic animals as well as man and wildlife. The disease is responsible for significant economic losses to the livestock industry primarily due to abortions, milk reduction, reduced weight gains and, secondarily, due to deaths.

There are five serotypes of Leptospira organisms that may be found in cattle in the United States. The L. pomona serotype commonly affects cattle, swine, and occasionally wildlife. L. hardjo is the predominant lepto spiral serotype involving cattle that are the only host apparently other than man. L. grippotyphosa has its reservoir in wildlife, especially the small carnivores. Occasionally, it is responsible for serious infections both in cattle and swine. Rats are usually the reservoir for the serotype for L. icterohaemorrhagiae. L. cannicola primarily affects dogs but also can be found affecting cattle.

Leptospira organisms are shed in the urine and transmitted to other animals by direct contact with the urine droplets or with urine-contaminated ponds or water. Antibodies are usually detected 5 to 10 days after infection. Acute leptospiral infections usually involve the kidney, brain, and uterus, an may become chronic, persisting for extended periods. These organisms usually enter the uterus during the acute infection in pregnant cattle. As a result, an infection of the fetus causes death, abortion, and stillbirths.

Leptospirosis is initially indicated by elevation of temperature that may persist only a few hours to several days. You may see a wine-colored urine, anemia, icterus, muscular weakness, pulmonary congestion, encephalitis, and death in the most severe infections. Abortions occur a few days to several weeks following the initial infection of the fetus. Usually at abortion, the feti have undergone a considerable amount of decomposition that results in killing the organism before the fetus is expelled. This makes it very difficult to isolate the leptospiral organism from the aborted feti. Serology on the cow is about the only method of accurately diagnosing the disease. Two serum samples-- one collected early in the acute stages of disease and another two weeks later--are refined: a two- to four-fold increase in titer is diagnostic. A single positive serological test cannot reliably indicate the course of the disease.

Prevention consists of eliminating exposure to streams, ponds, marshes, etc. that may be contaminated by the leptospiral organism. All new animals should be tested before adding to the herds. Animals can be tested for leptospiral titers at the same time brucellosis testing occurs. There are 5-way vaccines available that contain the five serotypes that are most likely to cause problems in beef cattle. These will usually give a protective antibody level for 6 months to a years. However, in cases of a high degree of infection, the virulent organism may override the immunity and require a vaccination at three month intervals until the disease is under control.

Vibriosis

Bovine vibriosis is a venereal disease of cattle caused by the organism Campylobacter fetus, subspecies fetus. This organism was formerly called vibrio fetus, variety venerealis, thus the name of the disease. It is transmitted from infected to noninfected animals through the act of breeding. It is characterized by temporary infertility of the female. The chief clinical sign of the infection is return to estrus. Usually animals will develop their own immunity following two or three breedings, and if left with the bull, most will be pregnant within nine months. This organism is usually brought into the herd by an infected bull. When he breeds a susceptible cow, the organisms are deposited in the area of the cervix. The organisms multiply and by 7 to 14 days will have invaded the ovaducts and caused the death of the fertilized ovum. The cow usually returns to estrus; instead of the normal 21-day cycle it may be a 28- to 30-day cycle.

Large numbers of the organism usually will be found in the mucus of the vagina for 3 weeks to 3 months following exposure. It appears that carrier animals perpetuate the infection from year to year. The disease is superficial in nature in the bull and he shows no sign of the disease.

A tentative diagnosis of vibriosis can be made by a thorough study of the herd history and application of the knowledge of the pathogenesis of the disease. Classical disease characteristics are repeat breeding, delayed conception, and 20% to 30% pregnancy rates after 60 days of breeding. Diagnostic laboratory work-up is required to confirm a diagnosis of vibriosis in the herd. A bacterial culture of a freshly-aborted fetus or the cervical vaginal mucus are the most positive tests.

This disease can be controlled by the use of artificial insemination after the semen has been properly treated. The most practical means of control in a beef herd is by vaccination of all female cattle. For best results, a single injection of the vaccine should be given 60 to 90 days before breeding. Since both the convalescents and vaccinal immunity decrease with the passage of time, annual revaccination is required to maintain maximum reproductive rates. This vaccine can be given either in the fall to the pregnant cow or before breeding.

Vaccination of bulls usually has not been recommended. Recent research in Australia and Belgium suggests that the carrier state of bulls may be modified by vaccination.

IBR (Infectious Bovine Rhinotracheitis, Rednose)

The IBR virus is a herpes virus and has normally been thought of as the cause of respiratory disease of cattle. However, this virus also causes vulvovaginitis and abortion. Abortions occur about 20 to 35 days after infection. Most cows abort during the last third of gestation. If replacement heifers are vaccinated from 5 to 6 months of age or at weaning and then revaccinated with a modified live virus 30 to 60 days prior to breeding, there has been little problem with IBR abortion. Perhaps vaccination of the entire cow herd should be considered on a 3 to 4 year basis. If new cattle are brought into the herd or cows are in contact with respiratory outbreaks, yearly revaccination would be advisable.

BVD (Bovine Virus Diarrhea)

BVD virus may cause abortion and birth of weak calves or calves with brain damage or deformities. The degree of damage depends upon the time of gestation and age of the fetus when the infection occurs. Diagnosis is by virus isolation or paired serum samples. BVD vaccine should be used very cautiously in a cow-calf operation. In my opinion, it should only be used in the replacement heifer 30 to 60 days prior to breeding. This has provided adequate protection against this virus. Problems could develop, however, if new animals are brought in that are carrying the virus or the cattle were exposed to a hot infection of BVD virus. One caution, do not give modified BVD vaccines to pregnant cows. Should an outbreak develop, a killed BVD vaccine is available that should be safe in pregnant cattle.

71

BIOLOGY AND CONTROL
OF INSECT PESTS OF BEEF CATTLE

R. O. Drummond

Beef cattle raisers continually battle a variety of flies, grubs, lice, ticks, and mites that bite, sting, annoy, injure, give diseases to, and suck blood from their cattle. Annual losses incurred by United States beef cattle raisers because of insect pests reach the hundreds of millions of dollars, despite the use of millions of pounds of insecticides to control these pests on beef cattle and in livestock buildings. Knowledge of the biology, life history, and local abundance of external parasites and of the proper use of safe and effective chemicals and other control practices is the most effective way the beef cattle producer can prevent losses, reduce costs of treatment, and control pests without hazard or danger to the animal, the applicator, the consumer, or the environment. This article contains information about the biology of a number of pests of beef cattle, describes accepted techniques for their control, and lists precautions for the safe use of insecticides on beef cattle.

EXTERNAL PARASITES

Flies

The most common pests of beef cattle are bloodsucking flies (commonly called "biting flies") such as horn flies, stable flies, horse flies, deer flies, and mosquitoes; non-biting flies such as house flies, face flies, blow flies, and screwworm flies; and nonfeeding flies such as heel flies or gadflies.

Biting flies. The horn fly, Haematobia irritans (L.), is found on beef cattle in pastures throughout the U.S. Thousands of these small dark flies may be present on the heads, shoulders, and backs of cattle. Adult flies suck blood as often as 20 times per day. The constant irritation produced by these flies causes infested cattle to bunch together, to shake and toss their heads, and to switch their tails--as a result, they do not graze normally. Female horn flies lay eggs on freshly dropped manure, and maggots live in undisturbed manure pats. The life cycle takes about 2 to 3 weeks and the species has the potential to build up very large populations on cattle in a short time.

The horn fly can be controlled by treating cattle with dips, sprays, dusts, or dousing with insecticides; often these treatments have to be repeated every 2 or 3 weeks during the horn fly season, which lasts from the last frost in the spring to frost in the fall. Insecticide-impregnated ear tags may control horn flies for several months. Back-rubbers or dust bags can be placed in pastures or at openings to water and mineral feeders, so that the cattle can treat themselves with insecticides in oils or in dusts. Also available are salt or mineral blocks or feed additives and supplements that contain insecticides or an insect growth regulator. Cattle consume enough of these materials daily so that the manure is toxic to the horn fly larvae or adults do not emerge from pupae, but adult flies may migrate in from neighboring herds.

The stable fly, Stomoxys calcitrans (L.), another bloodsucking fly, is found around feedlots, dry lots, and other areas where livestock are confined. The adults, about the size of a house fly, usually suck blood from cattle once or twice a day and spend the rest of the time resting on fences, buildings, etc. Female stable flies lay eggs on and maggots develop in moist straw, hay, spilled feed, and other decaying organic matter that is mixed with manure.

The first and most important step in control of the stable fly is to remove all the breeding material, usually manure-contaminated wastes, from the cattleholding area. This material should be spread and dried so it will no longer provide a breeding site for the flies. Also, breeding material can be treated with insecticides. Another method of control is to spray resting surfaces with an insecticide that has a long period of effectiveness to kill flies when they rest on the treated surfaces. Finally, beef cattle may be treated with insecticides to repel or kill stable flies. However, such treatments tend to be relatively ineffective or give protection for only a short time. Insecticide-impregnated ear tags may be used as an aid to control stable flies. An insecticide given daily to cattle as a feed additive or supplement can control stable fly larvae in the manure of treated cattle. Insecticidal baits used to attract and kill house flies around pens, lots, and facilities are not effective against stable flies.

Other biting flies, mosquitoes, horse flies, deer flies, gnats, and black flies are difficult to control. For example, mosquitoes and other flies are best controlled by eliminating or treating the water in which the larvae is found. Repellents applied to beef cattle usually do not provide satisfactory protection, and insecticides that control horn flies and stable flies on beef cattle may be effective for only short periods on other biting flies.

Nonbiting flies. Although the house fly, Musca domestica (L.), does not suck blood from cattle, it can be a nuisance to beef cattle owners and his neighbors. The flies breed in decaying organic matter, and large numbers are often found around barns, feedlots, and other cattle facilities. As with stable flies, the most effective method of control of house flies is removal of breeding material--manure, contaminated feed and hay, and rotting organic matter. Also, breeding material may be treated with insecticides, the surfaces of livestock-holding structures may be sprayed with insecticides, or insecticide-treated baits may be used to attract and kill house flies. Cattle can be treated with sprays or mists of insecticides to control house flies. Also, a feed additive/mineral supplement that contains an insecticide will control house fly larvae in manure of treated cattle.

The face fly, Musca autumnalis (De Geer), a recent introduction into the U.S., is found in all states except Texas and New Mexico. This fly, which is about the size of the house fly, is an important pest of beef cattle because of its habit of feeding on liquids around the eyes and nostrils of cattle. The feeding activity of face flies injures the eye, interferes with grazing, and increases the incidence of eye problems such as pinkeye. Female face flies lay eggs on freshly dropped manure and maggots develop there. Treatment of the head and neck of cattle with insecticides, such as dusts, smears, sprays, and ointments, may give some short-term control of face flies. Certain insecticide-impregnated ear tags will control face flies, while others can be used as an aid to control face flies. An insecticide added to the diet of cattle by feed or by mineral-salt block can control face fly larvae in the manure of cattle; however, the treatment is not very effective in reducing numbers of face flies on cattle because of flies that migrate from neighboring cattle.

The screwworm fly, Cochliomyia hominivora (Coquerel), has been eradicated from the U.S. and is currently the subject of a highly successful eradication campaign in Mexico. Although screwworms have been eradicated from the U.S.; cattle producers in states along the Mexico border should routinely examine their animals for wounds. If the wounds contain maggots, some should be collected and sent to the Screwworm Eradication Program, P.O. Box 969, Mission, Texas 78572, for identification. Wounds should be treated thoroughly with a spray, dust, aerosol, or smear of insecticide to kill maggots and protect the wound from reinfestation.

Other flesh flies and blow flies may be found in wounds on cattle, but generally these flies lay eggs on dead or decaying flesh and do not destroy living flesh as does the primary screwworm. Such larvae should be controlled by treating wounds with insecticide dusts, aerosols, smears, and sprays.

Nonfeeding flies. Beef cattle are infested with larvae
of two species of cattle grubs: Hypoderma lineatum (de
Villers), the common cattle grub, is found throughout the
U.S.; and H. bovis (L.), the northern cattle grub, is found
in the northern two-thirds of the U.S. Cattle grubs are the
larval stages of flies called heel flies or gadflies because
they "chase" cattle to lay their eggs on hairs of cattle.
After 3 days, larvae hatch from eggs, penetrate into the
animal's body, and then migrate during a period of 3 to 6
months to the esophagus (common cattle grub) or to the
spinal column (northern cattle grub). After several weeks,
the larvae migrate to the animal's back where they secrete
an enzyme that dissolves the hide from the inside, thereby
creating a hole; then the grubs are found in warbles or
wolves, which are characteristic swellings, usually along
the middle of the back. Several weeks later, cattle grubs
leave the animal's back and drop to the ground, where they
pupate. The pupal period may last 2 to 6 weeks or more,
depending on temperature. Then the nonfeeding adults emerge
from pupae and mate, and females chase cattle, which usually
respond with typical gadding. The life cycle takes about a
year, so in any given location, the same stage--adult
activity (called heel fly season) or warbles in backs--is
found about the same time each year.

The annual multimillion-dollar losses in the beef
cattle industry due to gadding (decreased grazing and milk
flow) and to the presence of grubs (meat trim and hide
damage) can be prevented by treating cattle with animal
systemic insecticides during the period after the end of
heel fly season and before warbles appear in the animals'
backs. These insecticides, whether they are applied in the
feed, in a salt-mineral supplement, or as a spray, dip,
"pour-on" or "spot-on," travel throughout the blood system
of the cattle and kill the migrating larvae, thereby eliminating the hide value and meat trim losses. However, the
time of treatment of cattle with animal systemic insecticides is very critical, and it is important for each rancher
to know when the heel fly season is over in his region.
Usually this information is available from the local Agricultural Advisor, County Agent, or Extension Official.

Lice

Beef cattle can be infested with two types of lice--
biting lice and sucking lice. All lice have a similar life
cycle in that all stages are found on the host, females
attach eggs (called nits) to hair, and immature forms
(called nymphs) hatch from eggs and molt one or more times
before they become adults.

Biting lice. Beef cattle are infested with one species
of biting louse, the cattle biting louse, Bovicola bovis
(L.). Biting lice live off skin scales, debris, hair, and
other matter on the surface of the animal's body. Usually

biting lice are found in long hair on the shoulders and neck of cattle. Infestations are largest in cooler months when cattle have their heaviest hair coats.

Sucking lice. Beef cattle are infested with 4 species of sucking lice: the longnosed cattle louse, Linognathus vituli (L.); the so-called little blue louse, Solenopotes capillatus (Enderlein); the shortnosed cattle louse, Haematopinus eurysternus (Nitzsch); and (rarely) the cattle tail louse, Haematopinus quadripertusus (Fahrenholz). All these species suck blood from cattle, and massive infestations can cause anemia, loss of condition, a scurfy appearance, and loss of hair; they also may lead to the death of the animal. As with biting lice, infestations of sucking lice are usually greatest in the winter. Certain animals, called louse carriers, have greater infestations than normal and often serve as sources of reinfestations for the remainder of the herd.

In general, lice on beef cattle can be controlled by the application of effective insecticides to the body of the animal. Cattle may be dipped, sprayed, or dusted thoroughly for louse control. Self-application by backrubbers and dust bags may aid in louse control, but these treatments are not as thorough as sprays, dips, or dusts. "Pour-on" or "spot-on" treatments may also provide louse control.

Ticks

Beef cattle are attacked by a variety of ticks. These 8-legged relatives of spiders may cause heavy losses for beef cattle breeders because ticks suck blood, transmit diseases, cause paralysis, and create unthrifty animals because of massive infestations and constant irritation ("tick worry").

Soft ticks. Beef cattle are infested with two species of "soft ticks," a title given to a large group of ticks because of the wrinkled, leathery texture of their "skin." The most common soft tick is the spinose ear tick, Otobius megnini (Duges), which lives deep in the ears of beef cattle and other animals. Adults of this species are free living (they do not feed) and are found in protected places such as cracks and crevices of buildings, fences, under salt troughs, etc., where females are bred and lay eggs. Small 6-legged larvae hatch from eggs, seek cattle, attach deep in the animals' ears, feed for a short period, and molt to the spiny-appearing nymph that may feed for several months. The species is found on cattle in most states but is most common in the southwestern states.

Spinose ear ticks are usually controlled by the thorough treatment of ears of beef cattle with insecticides such as dusts, low-pressure sprays, aerosols, and smears. Effective treatments will provide adequate control for a month or longer. Insecticide-impregnated ear tags will control spinose ear ticks for several months. Attempts to control the adults in the environment are generally unsuccessful.

A second soft tick that is a pest of beef cattle is the pajaroello tick, Ornithodoros coriaceus (Koch), a vector of epizootic bovine abortion (EBA) in California. This tick feeds for a very short time on cattle and thus is very difficult to control. Ranchers may reduce losses to EBA by not exposing first-calf heifers to ticks during early pregnancy.

Hard ticks. Most ticks in the U.S. have a hard covering on all or part of the back and thus are called "hard" ticks. These ticks have 2 types of life cycle. One is the "1-host" cycle in which larvae (or seed ticks) attach to a host, engorge on blood, and molt to the next stage called the nymph, which engorges and then molts to the adult male or female. The adults mate on the host; the females completely fill with blood, detach, drop to the ground, find a secluded spot, and lay eggs from which larvae will hatch. All the molting and engorging take place on a single host. The other cycle is the "3-host" cycle. Larvae of a 3-host tick feed on a host and, when fully fed, drop off the host and molt to nymphs on the ground. The nymphs find another host on which they engorge. Fully engorged nymphs drop off the host and molt to the adult stage on the ground. The adults find a third host and mate; then the females engorge, drop off, and lay eggs.

Each region of the U.S. has a group of hard ticks that attacks beef cattle. The Pacific Coast tick, Dermacentor occidentalis (Marx), is found on the Pacific Coast west of the coastal mountains. The winter tick, D. albipictus (Packard), is generally found throughout the northern tier of states and far south as Texas. The Rocky Mountain wood tick, D. andersoni (Stiles), is generally distributed in the northern Rocky Mountain States. The American dog tick, D. variabilis (Say), is found generally distributed over the eastern half of the U.S. The blacklegged tick, Ixodes scapularis (Say), is found on cattle in southcentral U.S. The lone star tick, Amblyomma americanum (L.), is found throughout the southeastern one-third of the U.S. The Gulf Coast tick, Amblyomma maculatum (Koch), is limited to south Atlantic and Gulf Coast States though large populations are found in eastern Oklahoma and surrounding states.

Because of the variety of tick species, their variable life cycles, and their seasonal abundance, each beef cattle rancher must know which species is attacking his cattle so that he can apply the most effective treatments at the proper time of the year. Generally tick control consists of treating cattle thoroughly with insecticides applied as dips or sprays. Less thorough treatments such as "pour-ons," "spot-ons," dusts, or self-treatment devices usually do not provide adequate control. Insecticide-impregnated ear tags effectively control Gulf Coast ear ticks because they attach to ears of cattle.

In some situations, ticks may be controlled by the application of approved insecticides to the ground to kill larvae, newly molted forms (of 3-host species), and engorged females.

Mites

Beef cattle are infested with several species of itch, mange, or scab mites. These very tiny species live on or in the skin and can cause intense irritation, itching, loss of hair, thickening of skin, and considerable discomfort to infested cattle. One species, the common scab mite, Psoroptes ovis (Hering), is the subject of a national eradication campaign: Cattle infested with this species are subject to quarantine and compulsory treatment to eliminate infestations. Beef cattle may be infested with other species of mange or scab mites that are of less importance, but all should be controlled by thorough treatment such as dipping or whole-body spray. Less thorough, low-volume treatments are only partially effective.

USE OF INSECTICIDES ON BEEF CATTLE

Treatments

Insecticides remain our first line of defense against insect pests of beef cattle. Properly used, approved insecticides can kill the pests and thus prevent or reduce losses due to the infestations of flies, lice, ticks, mites, etc. It is necessary to seek the advice and recommendations of local officials such as county agents and agricultural advisors who should have current information on kinds of insect pests in the area and will know correct techniques of application, recommended times of treatment, and a variety of other facts about local external parasites, their biology, and their control.

Precautions

The insecticides that can be used to kill insect pests of beef cattle also can be toxic to the cattle and the humans who apply them. In addition, these insecticides can create illegal residues in tissues of treated cattle and can be destructive to the environment if not used and handled in a safe manner. The following are a few precautions to follow when choosing and applying insecticides for the control of insect pests of beef cattle:
- Use only those insecticides recommended and approved for use on livestock by a recognized authority, usually a government official, such as an agricultural agent or advisor.

- Use a formulation of the insecticide that is approved and designed for use on cattle. In dipping vats use only those formulations designed specifically for dipping vats.
- Follow the label directions exactly. The label contains all the information on dilution, time of retreatment of animals, antidotes for poisoning, methods of disposing of unused insecticide, and other important facts.
- Avoid treating cattle in cold, stormy weather, and avoid treating stressed, overheated, or sick animals.
- Be sure that spraying equipment is clean, working properly, and provides sufficient agitation to allow for thorough mixing of insecticides.
- Be aware of safe practices when mixing and applying insecticides. Wear protective clothing; do not smoke, drink, or eat while applying insecticides. Do not contaminate feed or water troughs.
- Learn to recognize signs of insecticide poisoning in livestock (and humans) to avoid delaying antidotal measures.
- Store all insecticides in original containers. Do not store insecticides with food or where they can be reached by children, animals, or unauthorized persons.
- Avoid contamination of the environment: dispose correctly of all containers, unused concentrate, and used diluted insecticides.

72

PLANNING A HERD HEALTH PROGRAM

Ed Duren

Many so-called beef herd-health programs are a result of slipshod vaccinations, careless treatment, and indiscriminate medication--then, when all else fails, a veterinarian or beef-cattle specialist may be contacted for assistance.

Sound beef herd-health programs are based upon the interrelationships of breeding, physiology, nutrition, disease, and management. Normally, disease prevention programs are less costly to the rancher than are herd or individual animal treatments. The goal should be to prevent the disease rather than to wait for an outbreak and subsequent treatment. Animals subjected to therapeutic drug treatment over prolonged time periods may also experience a direct loss in body weight and condition. Due to the difficulty in assessing the value of these losses, the cost of animal treatment may not include these costs or may not even be realized by the rancher.

Animal thriftiness is essential to all beef herds if the ranch is to make a sound financial return on the investment. Only a foolhardy rancher believes that an injection of a wonder drug will solve all herd health problems.

DISEASE OUTBREAK

Disease outbreaks in beef herds result from exposure to invading disease organisms, and stressed and unthrifty cattle are frequently most susceptible. The nutritional status of the herd is vitally important to maintaining herd health and the impact of environmental condition should be emphasized in planning herd health programs.

The severity of a disease outbreak is directly related to environmental stress factors, previous herd management, nutritional status, and degree of confinement.

Timing, diagnosis, and treatment are the keys to reducing costs associated with body weight loss and medication. In dealing with a disease outbreak, producers frequently make several mistakes that can be prevented by:

- Early recognition of clinical signs of disease.
- Accurate diagnosis by a practicing veterinarian, which may include a postmortem examination of dead animals.
- Proper treatment, not only for obviously sick animals, but also animals exhibiting fever in early stages. Frequently, medication is limited to individual sick animals and not to animals in early stages of disease.
- Inadequate nutrition to prevent dehydration and physiological rumen disturbances.

A single method of prevention and control that is satisfactory to all ranch situations has not been developed. For this reason a total program of diagnosis, immunization, treatment, and nutrition should be developed specifically for a ranch by a competent team including a veterinarian, herd owner, and beef-cattle specialist.

DISEASE DIAGNOSIS

The success of an accurate diagnosis is consistent with the completeness and accumulation of information. Inadvertently, ranchers tend to withhold information from the veterinarian and (or) beef-cattle specialist. A systemized approach of observing, collecting, and recording appropriate and useful information relative to the herd will be highly beneficial to all participants in a team effort for solving problems related to breeding, nutrition, health, and management.

HERD HISTORY

Ranch records should be routinely maintained relative to the herd condition and management. The herd history should include the following items:

Nutrition Program
a. Amount
b. Quality
c. Type

d. Ration changes or alterations
e. Water supply and availablility
f. Consumption

Disease
a. Previous exposure or outbreaks
b. Treatments (herd or individual)

c. Drugs administered and dosage
d. Immunization and vaccines

Fertility
a. Breeding dates – length of calving season
b. Calving records – death loss
c. Physical evaluation of cows and bulls
d. Open cows
e. Abortions

Clinical Examination (Ranchers should regularly observe the herd daily for any unusual conditions that may suggest a disease condition.):

a. Off feed or water
b. Coughing (difficult breathing)
c. Diarrhea
d. Listlessness – lack of coordination
e. Discharge – nasal, eye, rectal, vaginal, etc.

Veterinarians will examine individual animals closely for:

a. Temperature
b. Pulse and respiration
c. External and (or) internal palpation
d. Coordination, gait, reflexes.

Herd-health programs are established on the basis of accurate disease diagnosis by a competent veterinarian in cooperation with the rancher and a beef-cattle nutritionist. Frequently, confirmation of the diagnosis may require the staff services of a veterinary diagnostic laboratory.

POST MORTEM (NECROPSY)

Upon death, animal carcasses may be posted for observation of internal organs and tissues. Due to rapid degeneration of body tissues, an animal must be posted within a short time following death, preferably when the animal body is still warm.

Veterinarians determine the cause of death or sickness from gross pathological conditions of the tissues. However, the veterinarian also may select tissue samples for submission to a veterinary diagnostic laboratory for a series of laboratory tests and cultures, including:

histopathological
toxicological
parasitological
urological
bacteriological
virological
blood chemistry

A number of tests can be performed by laboratory procedures, but due to the difficulty in performing these tests, the tissues may, if necessary, be removed aseptically and free of other contaminating organisms or drugs.

Accurate diagnosis is more likely when based on an accumulation of data information, observation, examination, and laboratory tests. Veterinarians then can proceed to prescribe an effective treatment and (or) immunization selected for the disease encountered in the herd.

Medication and treatment regimes change as new drugs are developed. Generally, veterinarians prescribe a broad spectrum of drugs (which include specific antibiotics in combination with long acting sulfas) that are administered at therapeutic levels for periods of 5 to 7 days to reduce fever and mixed-disease complications. The treatment must be combined with proper feeding and nutrition to prevent dehydration. Routine injections of antibiotic may have questionable therapeutic value in combating unknown disease outbreaks.

DEAD ANIMAL DISPOSAL

The disposal and removal of dead animals poses a problem on most ranches. It is not uncommon for dead carcasses to remain on a premise for natural degradation. This continues to be a source of contamination to other animals within the herd. Disposition of dead animals in a river, stream, or bone pile is certainly not acceptable. The safest assumption is that all animals that have died on the ranch have some infectious agent present that should be isolated from the rest of the herd.

Rendering trucks, which will pick up dead animals from the ranch, can be a source of contamination unless properly cleaned and disinfected. It is not uncommon for drippings from other dead animals to leak out and spread infection. Therefore, this practice of dead animal removal is questioned by animal-health officials.

The disposal of dead animals with specific diseases is frequently covered by law. However, regulations usually permit cremation by burning or burying in a pit at least six feet deep and covering with lime and then soil.

ENVIRONMENT AND STRESS

The mother cow provides a great deal of security and protection to her calf that might be fearful of being handled. The calf's rumen during the nursing stage is still adjusted to milk as a primary source of nourishment.

Many natural and man-caused environmental conditions create stress factors on a calf during birth or following weaning, which reduces thriftiness and increases the susceptibility to predisposing disease organisms. Practical calf management may eliminate many man-made stress factors and reduce the shock from natural stress factors.

Weather is a natural stress factor over which a cowman may have little control except for anticipated changes. Weather becomes a problem later in the spring and fall with cool nights and warm days compounded by dampness from rain or snow. Calves adjust faster with earlier weaning during moderate temperatures. Calves nursing cows to mid-December or later just drain the cow of vital nutrients and reduce her body condition; if the feed is inadequate she may not recover by spring calving. Generally, October or early November weaning is easier on calves than is weaning in late November or December in the intermountain region.

Handling, minor surgery, disease immunization, and pest control programs can be applied and manipulated by the cowman to reduce stress to a minimum. The recovery rate is more rapid from various medications, insecticide treatments, vaccinations, and surgical procedures if these occur while the calf is still "mothered-up" to the cow. Stress during weaning can be reduced if these procedures are completed 10 to 14 days before weaning and the calves run through the chute one time. It is easier on the calf to do this early while the calf is nursing.

If for some reason these practices cannot be applied and completed prior to weaning, a sequence of these practices should be applied to the calf following the initial shock and recovery of weaning. It takes 10 to 14 days alone for a calf to adjust physiologically and forage on its own. This is a high risk period because the calf is highly susceptible to disease organisms and a disease outbreak can be costly.

It is an advantage to wean calves in small groups of 100 head or less, and preferably in 50-head bunches. This isn't always possible because of existing ranch management, but a producer can provide better care and be able to watch the calves more closely for disease in smaller groups. Weaning takes longer but it could well be more successful.

Normal range management generally requires relocating calves by moving them to meadows or fields with similar range and water conditions and completely out of hearing distance of the mother cows. Availability of similar forage and water reduces the transition stress during the weaning phase. After the adjustment calves can gradually become accustomed to more concentrated rations fed from bunks. If meadows aren't available for weaning, avoid dry, dusty, crowded corrals.

Stress factors include:
- Changes in temperature.
- Dust that irritates the mucus membranes of the respiratory tract.
- Shock caused by hauling and handling.
- Fatigue.
- Dehydration from long periods without feed and water.
- Indigestion from change of feed and water.
- Hormonal influences.
- Fear and anxiety caused by strange surroundings, animals, people, crowding, age.
- Other diseases that may exist at the same time.

Resistance increases with age; good physical condition affords little protection from infection but is important in avoiding complications and losses.

DISEASE PREVENTION

The days are gone since one vaccine, namely a Black-leg-Malignant Edema Combination injection, provided an animal with all the protection needed.

In this area, ranchers expect more production from a cow herd. The herd is also subjected to a wider range of environmental stress factors and greater exposure to transient cattle. Protection for other common diseases is considered essential to good herd management.

Numerous vaccines are available for preventing disease. A crucial question is: Have you checked with your practitioner for his suggestions on a sound vaccination program? And followed his advice?

It is desirable to review the herd vaccination program with a practicing veterinarian who is familiar with existing disease conditions in your herd's area. The diseases listed here are prevalent in the southeastern Idaho area.

The following list of diseases is not complete. However, it does include those diseases for which vaccines are available. It is important for ranchers to become familiar with common and technical names of the various diseases to reduce potential error in the implementation of a vaccination program.

Bacterial Diseases

        Pasteurella
          P. multocida
          P. haemolytica
        Haemophilus
          Haemophilus sommus
        Clostridial
          Blackleg Cl. chauvoei
          Malignant edema Cl. septicum
          Novyi Cl. novyi
          Entertoxemia Cl. perfringens Type C & D
          Sordelli Cl. sordellii
          Red water Cl. haemolyticum
        Lepto
          L. pomona
          L. grippotyphosa
          L. hardjo
          L. canicola
          L. icterohoemorrhagiae
        Vibrio
          V. fetus
        Brucellosis
          Brucella abortus
        E. coli
          Escherichia coli

Virus Disease

IBR
  Infectious Bovine Rhinotracheitis--Red Nose
PI$_3$
  Para-influenza
BVD
  Bovine Virus Diarrhea
Rota Virus
Corona Virus

Immunization and Medication Records

Many vaccines are prepared either singly or in various combinations for injection or nasal administration. Some vaccines may be compatible, while others may react adversely when given in combination. Multiple injections should be administered in different body sites and preferably on opposite sides of the animal for best immune response.

Commercial laboratories are rapidly developing new vaccines and related technical information is distributed to professions associated with the beef industry. Therefore, it is important for a rancher to review this information periodically with a veterinarian to determine if a given product may have application to the herd.

Disease prevention is more complex today than in the past. Cattle require different immunizations to the prevalent disease conditions. Vaccines against some diseases are more effective than others, but no vaccine is 100% effective. Effectiveness depends on age of animal, passive immunity the animal may have when vaccinated, the stress on the animals, disease exposure, and other factors that we don't even understand. When used properly, vaccines are an aid to practical sanitation, management, and nutrition.

Read the label and administer each product as recommended by the manufacturer or veterinarian. They require care in handling and storing. A few dollars spent for a used refrigerator to store animal health products is sound management; tape a record sheet to the storage refrigerator for quick reference. Ranchers should develop a record of disease prevention--a notebook containing product information can be very useful. Record information on each drug or vaccine as follows:

Product:
- Drug or vaccine
- Trade name
- Company
- Product or serial or lot number
- Expiration date
- Date purchased
- Date administered
- Firm of purchase

Cattle:
- Number vaccinated or treated
- Date administered
- Which group, herd, or individual dosage used and how administered

Herd health record forms developed by the University of Idaho College of Agriculture in the Pegram study have proven popular with progressive beef producers in southeastern Idaho.

Vaccination Schedule

Ranchers should select appropriate vaccines and plan an annual vaccination schedule based on seasonal cattle management and handling. Herd health program may be altered as the need arises with consultation and diagnosis by a veterinarian. Dollars spent on a well-designed herd-health program and vaccination schedule will contribute to improved reproductive efficiency in a cow herd. Other contributing factors include management and nutrition. The Idaho Total Beef Program is an integrated approach and a University of Idaho College of Agriculture thrust assisting ranchers in improving reproductive efficiency of their cattle.

Vaccination is only one factor in an integrated herd-management concept. The herd-health program and vaccination schedule must be tailored to a specific ranch. A suitable program is the result of both the expertise of the professionals and the experience of the cowboys associated with the ranch.

PEST CONTROL

Infestations of lice and grubs will reduce the thriftiness of wintering cattle, particularly calves. A management plan should provide control of these two pests.

Grubs. Cattle are infested with grubs as a result of heel fly striking during the summer grazing period. These eggs hatch, then the larvae burrow into the skin and proceed to migrate through the body to complete the life cycle by spring when they drop from the animal's back.

The best time to kill grubs and break the life cycle is in the fall with an application of a systemic insecticide. This is commonly done by pour-on or dip. Two other effective methods include spot-on (low volume) and salt-mineral additive.

Lice. The lice population will begin to build up during the late fall and winter. There are a number of lice species, but they all feed by sucking or biting the animal. If the population on an animal builds up, the animal may become anemic and eventually die. Therefore, an early fall treatment for lice followed by a spring treatment will usually control them. Louse carriers are the exception and may require periodic treatments at 10 to 14 day intervals.

Many effective insecticides are on the market--the key to success is the method and concentration of the application. Insecticides effective for grubs also may reduce lice populations.

Dip and Pour-On Applications

The effectiveness of a pour-on application for lice is dependent upon timing and population. The insecticides listed will control lice, but the application method is critical. A pour-on insecticide primarily kills lice systemically, which means that the louse must consume blood containing the insecticide within a relatively short period of time following application.

An organo-phosphate insecticide applied as a dip has two methods of kill: contact and systemic. This factor, plus less cost, favors dipping over the pour-on application. Having a dip vat is an advantage. Dip vat solution should be checked periodically for correct dilution. At the present time, no cost is required for analysis of dip solutions containing CoRal, Prolate, or Toxaphene. Contact your supplier regarding this service.

SUMMARY

It appears that ranchers are headed for closer surveilance and tighter restrictions on the use of drugs, vaccines, and pesticides. Now is the time to get in the habit of keeping records on proper use and precautions.

An organized cattle health program requires that you:
1.   Keep records.
2.   Follow directions on label.
3.   Consult regularly with the veterinarian.
4.   Use common sense in handling cattle.

APPENDIX MATERIAL
FOR
PLANNING A BEEF HERD HEALTH PROGRAM

ANNUAL BEEF HERD HEALTH AND VACCINATION SCHEDULE
CALVES

Spring

Calving - at birth
   1. Number ear tag calf at birth
      (a) Record birthdate, sex, calf number, cow number
   2. Clostridial disease (8-way)
   3. IBR
   4. BVD
Prior to turnout
   1. Castrate - knife preferred
   2. Dehorn - Barned dehorners preferred
   3. Brand
   4. Implant growth stimulant

Fall

Preweaning (minimum of two weeks prior to weaning)
   1. Clostridial diseases (8-way) - 2nd booster
   2. Pasteurella
   3. IBR-PI$_3$ - 2nd booster
   4. Grubicide
   5. Individual calf weights
Weaning
   1. Pasteurella - 2nd booster
   2. Implant steers and nonreplacement heifers
Postweaning - replacement heifers
   1. Bangs vaccination (third through ninth month of age)--thrifty replacement heifers 2 to 3 weeks after weaning

ANNUAL BEEF HERD HEALTH AND VACCINATION SCHEDULE
YEARLINGS

Spring

Steers and open heifers
1. Clostridial diseases (8-way)
2. IBR-PI3
3. Implant growth stimulant
4. Spay heifers
5. Fly tag
Replacement heifers
1. Clostridial diseases (8-way)
2. Lepto (5-way)
3. IBR-PI3-BVD
4. Vibrio
5. Yearling individual weights
   (a) Breed Hereford heifers at 650 lbs and over
   (b) Breed Angus heifers at 500 lbs and over
6. Fly tags

ANNUAL BEEF HERD HEALTH AND VACCINATION SCHEDULE
COWS

Spring

Open cows - prior to breeding
1. Vibrio
2. Lepto (5-way)
3. Redwater (if needed)
4. Lice treatment
   (a) Systemic insecticide (pour-on application)

Summer

1. Fly control
   (a) Salt
   (b) Dust bag
   (c) Ear tag for flies

Fall

1. Redwater (if needed)
2. Louse - grubicide combination
   (a) Systemic insecticide (pour-on application)
3. Lepto (5- way) if not administered in spring

ANNUAL BEEF HERD HEALTH AND VACCINATION SCHEDULE
BULLS

Spring

Virgin Bulls
1. IBR-BVD-PI$_3$
2. Redwater (if needed)
3. Lepto
4. Lice treatment
5. Clostridium (8-way)
6. Fly tag
Mature Bulls
1. Lepto
2. Redwater (if needed)
3. Lice treatment

Fall

1. Redwater (if needed)
2. Grubicide

Cooperative Extension Service
University of Idaho
College of Agriculture

Drug and Vaccine Supply Record

Brand name	Company	Type of vaccine	Serial number	Lot number	Expiration date	Supplier	Doses	Cost	Doses returned

Appendix B

Cooperative Extension Service
University of Idaho
College of Agriculture

Immunization Record

Disease	Brand	Serial number	Sex group	Number of head	Date	Storage	Reactions

Appendix C

Cooperative Extension Service
University of Idaho
College of Agriculture

Treatment Record — Sick Animals

Disease	Sex group or animal ID no.	Product used	Company	Expiration date	Date of treatment	Dosage	Route of administration	Results

674

Appendix D   DRUG AND VACCINE SUPPLY RECORD

Brand Name	Company	Type Of Vaccine	Serial Number	Lot Number	Expiration Date	Supplies	Doses	Cost	Doses Returned
ViBRIN	Norden Lab	Vibrio Fetus Bacterin				Steve Regan			
BAR Four F80+ Reanose	Anchor St Joseph Mo-Cusdi	Bovine Rhinotracheitis Parainfluenza3	Bovine tissue culture Orgin			Steve Regan			
Seven Way	Jen-Sal Kansas city Mo.(with)	Clostridium Chauvi-Septicum Novyi-Soebelli Perfringens-Type c+d Bacterin				Steve Regan			
IBR-PI-3	Abbott Lab. North Chicago Ill.	Bovine Rhinotracheitis Parainfluenza-3 Live Vaccine	3 Bottle. 67070 BQ37 4" 240 to BQ37		10-9-77 5-14-77				
Electrotox 70(43271?) (2 Bottles)	Richardson Merrill Kan.C. Mochvilfer?	Clostridium septicum-Novyi Sordelli Perfr. C+D	749		3-2-79				
ViBRIO-Fetus Bacterin	Annarillo-Tex 79105 Franklin Lab	Bacterin Vibrio (Bovine IsolAtes)	2-(L)V203B 3-(L)V203A		3-1-78 5-1-78				
ViBRIO Fetus Bacterin	Anchor Lab St Joseph Mo-CWSDz	Bovine strains Alum.Hydro.Adsor. 128 A Vibro 3	12 Bottles 128 A CODA #110.72L		5-26-78				
RAL-620 Implant	IMC Chemical Group Inc Terre-Haute Ind-when?		#70293 (1-240 Dose 30x)						
BVD	ABBOT LAB	BVD sterile Diluent	#62050 Bg32 #60370 Bg11		8-11-78 8-11-78	WALCO Mike G.			
BVD	Anchor Lab		#135-B 2 Bottles		7-31-77	WALCO Mike G.			

Appendix E

IMMUNIZATION RECORD

Disease	Brand	Serial Number	Sex Group	Number Of Head	Date	Storage	Reactions
PI-3 IBR, BVD 4			Replace Hfrs		Spring-1975		Hfrs - Done Twice
Lepto IBR - PI-3 Nasplin - Entero Navi y Sordelli + BVD			Cows Bulls Calves Calves	ALL ALL	Spring-1975 Fall -1975 11-9-1975 Fall-1975		
3-way Lepto Enterotox - 4-way	Cooper - Entero Ven-Sal (4-way)		Replac. Hfrs Calves "	ALL ALL "	Dec-1975 March-3-1976 " " "		
Vibrio Lepto + Anabac 7-way - Bacterol - Anchor War Bex	Norden A.A.B.		Replace-Hfrs Cows Calves	1/5 Head ALL ALL	May 1-1976 Oct 30-1976 Nov 3-1976		
All Calves Doctored and given Tylan			+ Span Bolets Suffa Boen pills - Oct-1976				Six calves died
Pre-tested 3-way Lepto with Rex			Replac. Hfrs "		Oct-1976 "		
All Cows were Brucellis (Bangs) Tested Nov 29,1976. Results were All Negative							
All Heifer Calves were Bangs Vaccinated — Dec 10,1976							
Four Way Booster	Ven-Sal		Calves	ALL 179- 27 76 61 = 312	Feb 9,1977 5-18-77 4-30-77 5-16-77 5-17-77		Clostridium — Novyi. Sordelli Types C&D Perfringens - Bacterin Toxoid
Seven - Way	Electrotox 70 (1 Bottles) 749		Calves	312	Repeat of Above		
Vibrio	Anchor-Lab 1288 A one 2 (L) V203B (50 does) 3 (L) V203A (10 dose)		Cows	122	5-10-77		
Vibrio	Franklin		Replac. Hfrs	5	5-13-77		
Implant	IMC Chemical Lot # 70893		Hfrs Stres	96 102	5-13-77 5-13-77		
Ral - Gro							

Part 11

MEAT AND MEAT PROCESSING

73

CURRENT STATUS OF MEAT IN THE DIET AND ITS RELATIONSHIP TO HUMAN HEALTH

B. C. Breidenstein

Throughout the recorded history of man, and in all likelihood prior to that time, the primary diet/health concerns revolved around the need to obtain sufficient nutrients to maintain life. The priority status associated with consumption of meat and other animal products is well documented in the Old Testament of the Bible. Whether this status can be attributed to the perceived nutrient contribution to the diet or to taste appeal is not clear, but one could well speculate that it might have been a combination of those two reasons.

Nutrient deficiencies have been observed and recognized as such for several hundred years. The use of certain foods in the prevention or correction of these deficiencies has been commonly practiced. For example, scurvy, a symptom of vitamin C deficiency, was a common problem during long ocean voyages of European explorers and was found to be prevented by the consumption of citrus fruits such as limes, which are now known to be an excellent source of vitamin C. Nevertheless, the scientific study of man's nutrient needs and the food sources for satisfying those needs is relatively recent. Those "needs" of the human body are, however, not definable as a single quantifiable entity. Rather, age, sex, level of physical activity, ambient conditions, genetics, and general health factors are among the factors known to influence nutrient requirements. Further complicating the issue is the fact that certain nutrients exert a sparing effect on the need for other nutrients. The bioavailability of nutrients from different sources also varies. For example, the heme iron contained in meat is known to be one of the most bioavailable forms of the mineral. In addition, its presence in the diet makes the less available plant sources of iron more available to humans. On the other hand, consumption of an excess of some nutrients may result in metabolic waste of others.

The human not only has a threshold of minimum requirements, but in many cases, is also very tolerant of an excess beyond those requirements. It is generally recognized, for example, that approximately 200 mg of sodium are required

daily for humans. Yet about 70% to 80% of the U.S. population is believed able to tolerate daily intakes of sodium of
10 to 20 times that amount without adverse effects. The
magnitude of the range between the minimum requirement for
any person and the intake level resulting in adverse effects
is undoubtedly different for different nutrients—and is
probably influenced by the same factors that affect the
minimum requirement levels.

In spite of the continuous challenges to the desirability of its inclusion in the diet, per capita meat consumption in the U.S. has increased steadily during the
current century. It is probably that, during at least the
early part of that time period, its popularity in the diet
was due largely to its taste appeal and secondarily to a
perceived preferential source of nutrients. During the past
several decades, however, meat has become increasingly
recognized as a premiere dietary source of high-quality protein, B vitamins, and minerals. The reputation of meat as a
source of these essential nutrients is rarely, if ever,
seriously challenged.

Red Meat Consumption* Pounds Per Capita/Year

Decade	Beef	Veal	Lamb & mutton	Pork Exc. lard	Total red meat
Last half 1940s	63.5	10.2	5.7	69.5	148.9
Last half 1950s	75.3	8.1	4.3	66.0	153.7
Last half 1960s	98.7	4.8	4.3	63.3	171.1
Last half 1970s	117.2	2.9	2.5	64.1	186.7
1980	105.6	1.8	1.6	75.1	184.1

*Carcass weight disappearance.

Probably because the basal statistics are readily
available, meat consumption figures are frequently expressed
as total carcass weight of livestock slaughtered, divided by
that year's total population. It is true that meat consumption per capita expressed on that basis increased during the
current century, but obscured by such a figure is the change
in carcass composition that has occurred in response to consumer demand. For example, prior to the wide availability
of plant source oils swine were viewed as somewhat dual-
purpose animals. They were prized not only for their
ability to produce appetite-satisfying meat in a wide
variety of products, but also for their ability to convert
plant materials to fat, thereby representing a primary
source of edible fats in the U.S. The rapid increase in the
supply of relatively low-cost vegetable oils since World War
II resulted in a depressed price for lard and the diminished
status of swine as a source of edible fats. In recognition
of this economic reality, the swine industry has, since the
early 1950s, been very actively engaged in highly successful
genetic, nutritional, and management programs designed to

maximize muscle and minimize fat production per animal marketed. As a result of these efforts, lard production per head slaughtered in 1977 has been reduced to about half the level produced in 1950. This has been accomplished simultaneously with ever-closer trimming of fat from the cuts offered to the consumer.

While cattle and lamb fats have been, and remain, an important by-product of their slaughter, the reliance on them as converters of plants to fat has never been viewed as an economic asset as was the case historically for swine. However, consumers undoubtedly purchase all red meats primarily for their muscle content and consider the fat contained as an economic loss. Thus, the consumer pressure for increased leanness of meat products, which expressed itself forcefully in pork consumption in the early 1950s, became fully recognized in the beef and lamb consumption perhaps a decade later. As a result of this awareness on the part of the cattle and sheep industry, their response was similar to that of the swine industry. Even a casual observer of carcass composition must recognize the dramatic changes toward less fat and more lean that have taken place in the meat supply during the past three decades. As a result, if one assumes that little of the fat attached to retail cuts is actually consumed, then the true consumption of _meat_ has increased to a greater extent than indicated by carcass weight disappearance data.

It is obvious from observing the per capita consumption, which has increased by more than 25% over the past three and one-half decades, that red meat is viewed by consumers as a desirable dietary component. Pork consumption has remained relatively constant over that time period while lamb and mutton and veal consumption has declined. Those declines, however, have been more than compensated for by increases of more than 80% of beef consumption.

The diet/health controversy, insofar as red meat is concerned, is not a new issue. During the first two decades of this century, the annual per capita consumption of red meat and lard declined by more than 27 pounds, with beef consumption accounting for nearly 23 pounds of that total. Thomas E. Wilson, president of the Institute of American Meat Packers speaking before the American National Livestock Association on January 12, 1922, said: "One of the outstanding factors operative for the last two decades has been the fostering and development by propaganda of an impression that meat is harmful to the health. In this connection, meat has been misrepresented in a damaging fashion and in a widespread way. The food value of meat has been mis-stated, its place in the diet minimized, and its healthfulness challenged. People are naturally sensitive to any propaganda relating to their health. They are quick to avoid foods said to be harmful. In this way the public, no doubt, has been materially influenced. Almost every other food interest has made invidious comparisons of its products with ours to the disparagement of meat. Many of these comparisons have not reflected the truth from a scientific standpoint." Characteristic of such information is an article from the April 1905 issue of The Ladies Home Journal entitled "Why I Do Not Believe in Much Meat," an exerpt from which is: "Nourishing diet, in the minds of most people, is a meat diet, which soon upsets the digestive organs and makes the person a martyr to rheumatism and gout. Women grow constipated and have cloudy complexions; men red-faced, or very thin, according to their various resisting powers."

Mr. Wilson suggested: "Scientific data wherewith to correct adverse propaganda should be collected, compiled, and disseminated showing the high value of meat in the diet. Such information should be circulated among dietitians, physicians, hospitals, teachers, home demonstrators, household editors, agricultural colleges and others."

In its January 21, 1922, issue, the National Provisioner reported on the creation of the National Live Stock and Meat Board stating that it "shall be created to conduct and direct an adequate educational campaign counteracting the widespread and insidious propaganda against the food value of meat and disseminating through all possible avenues correct information about meat in the diet."

Differences of opinion among scientists is to be expected. One of the valuable attributes of a scientist is the ability to think in independent and original terms. Such traits lead to the expansion of the frontiers of scientific knowledge and, as such, are a valuable asset of the scientific community and indeed to the community at large. It should not be surprising that such characteristics make it more difficult to reach even a consensus opinion, let alone a unanimous one, regarding such a complex issue as diet/health. Some recent examples of truly dedicated, knowledgeable, and well-meaning scientists reaching differing conclusions from observing a common body of knowledge reflect that difficulty. Dr. D. M. Hegsted of the Harvard School of Public Health in a press release on 2/14/77 regarding the second edition of dietary goals said: "The diet of the American people has become increasingly rich--rich in meat, other sources of saturated fat and cholesterol, and in sugar. It should be emphasized that this diet which affluent people generally consume is everywhere associated with a similar disease pattern--high rates of ischemic heart disease, certain forms of cancer, diabetes and obesity. These are the major causes of death and disability in the U.S."

Only some two years later in a 1979 report from the Surgeon General appears the statement that "the population of the United States has never been healthier" and "mortality rate for coronary heart disease has declined 20% during the last 20 years and is currently falling at 2% per year." The following year, 1980, the Food and Nutrition Board of the National Academy of Science said, "The American food supply on the whole is nutritious and provides adequate nutrients to protect essentially all healthy Americans from deficiency diseases" and "the excellent state of health of the American people could not have been achieved unless most people made wise food choices."

It should be recognized that a diet is a massively complex and variable combination of a multitude of nutrients and other components. Ingested by the human, which is an organism of an extremely complex and nonuniform nature, those dietary ingredients are broken down into simpler forms by various biochemical and physical means to exert their ultimate effect on the human organism. The oversimplification of the role of diet or even a single food entity on the health and well-being of the human leads to potentially dangerous conclusions. It is somewhat analogous to characterizing a single tree while not recognizing the existence of the forest of which the single tree is only a small part.

There are those in the scientific community who seem willing to make sweeping dietary recommendations on the premise that such changes involve an acceptable risk level for adverse effects and may possible result in benefits.

It is disturbing that such a segment of the scientific community seems to be gaining in influence, especially in the popular press. It would seem that challenges to the safety of our environment or our food supply whether or not properly founded in fact, are deemed newsworthy and hence receive widespread media distribution. On the other hand, a statement such as made by the Food and Nutrition Board in 1980 that "the excellent state of health of the American people could not have been achieved unless most people made wise food choices," is viewed as not very dramatic and hence receives little attention by the news media.

Although the diet/health controversy regarding meat has not subsided over the eight decades of this century, there has been, and continues to be, an increasing body of scientific knowledge reflecting the positive contributions of meat to the diet of man. Meat has long been recognized as an excellent source of high quality protein. A 3 oz serving of meat supplies more than half the daily human requirement for protein. The fact that it is a high quality protein simply means that it contains all of the essential amino acids in amounts appropriate for best use by the human. It therefore needs no supplementation from other proteins to be effectively utilized.

Meat is an excellent source of the B vitamins and ranks as the principal dietary source of most of them. Pork is an exemplary source of thiamin with a 3 oz serving supplying well over 50% of the U.S. RDA. A 3 oz serving of meat supplies about 15% of the U.S. RDA for riboflavin and more than 15% for niacin. Meat is also recognized as a good source of pyridoxine (B_6) and vitamin B_{12}.

Meat is a good dietary source of minerals, especially of iron. The heme iron in meat is highly available to the human and its presence also aids in the assimilation of the nonheme iron present in the diet from plant sources. Meat also supplies about 15% of the U.S. RDA for the trace mineral zinc.

Because meat contains such significant quantities of these important nutrients in relation to its caloric contribution of about 200 calories, it is described as a food of high nutrient density.

There are many human conditions that the scientific community and/or the popular press have purported to be related to the diet. It is the intent of this discussion to limit such considerations to those having implications for the livestock and meat industry and likewise perceived to be important by the public-at-large. Dietary implications regarding obesity, cardiovascular diseases, hypertension, and cancer are discussed.

Probably obesity is a condition that excites the least controversy. The Food and Nutrition Board contends that obesity, or excess fatness, is the most common form of malnutrition in the Western nations of the world. It has a multiple etiology and is influenced by neurohumoral, endocrine, metabolic, and social factors. It is generally recognized that, in many persons, obesity is associated with significant increases in morbidity and mortality from such diseases as hypertension, diabetes, coronary heart disease, and gall bladder disease, and that mortality from these diseases is reduced with weight reduction. Obesity results from the failure, for whatever reason, to balance energy intake with energy expenditure. Such a balance is difficult to achieve when energy expenditure is low, as is generally true for the adult population of the United States.

Because meat is of high nutrient density, it fits very well into either a weight-reduction diet or a diet low in energy and designed to maintain body weight at the desired level. Dr. Maria Simonson of John Hopkins has compared response to vegetarian vs meat diets in weigth-reduction programs. She reports that those on a meat diet 1) lost weight somewhat more slowly, though not significantly so, 2) were more consistent in weight loss, 3) had fewer drop-outs and less cheating, 4) had no feelings of hunger, 5) had far fewer physical and psychological problems, 6) had no anemias, 7) improved their work productivity and work efficiency (by 13%), and 8) showed excellent weight maintenance for up to one year for those who achieved goal weight. To obtain the necessary nutrient intake in diets of reduced energy level, it is obvious that nutrient-dense foods, such as meat, must play a prominant role.

Arteriosclerosis and its complications, i.e., coronary artery disease, stroke, and peripheral vascular disease are the leading causes of death in the U.S. Cardiovascular disease is the leading health problem accounting for about 50% of the deaths in this country. Mortality rates from cardiovascular disease, as is true for other degenerative diseases, increase sharply with increasing age. Several risk factors for cardiovascular disease have been identified through epidemiological studies. Among these is hypercholesterolemia or high levels of cholesterol in the blood serum. Risk factors are those factors found to be statistically associated with an increased incidence of the disease and cannot, without independent evidence, be considered causative agents of the disease. In any event, the identification of high serum cholesterol levels as a risk factor has led to concern about cholesterol and fat, especially saturated fat, in the diet. Cholesterol is found primarily in foods of animal origin with a 3 oz serving of lean meat containing about 75 mg. The human synthesizes between 800 and 1500 mg of this essential metabolite per day. Dietary cholesterol is poorly absorbed by man, with only 10% to 50% of dietary intake actually absorbed.

Diet modification with respect to level and kind of fat and the amount of cholesterol has been shown to alter serum lipid and lipoprotein concentrations of subjects in metabolic wards under rigid dietary control. A high intake of saturated fat as a percentage of calories is a major factor in elevating serum cholesterol and LDL (low density lipoproteins) levels and the reverse is true for a high intake of polyunsaturated fat. Studies among free-living subjects, however, indicate that such dietary modification is only about 60% as effective in altering serum cholesterol and lipoprotein levels as was the case for the controlled-metabolic-ward studies. Obviously, other factors still to be identified, in addition to diet, influence serum lipid values of free-living persons in an as yet unpredictable manner.

Intervention trials utilizing diet modification to alter the incidence of coronary artery disease and mortality in middle-aged men have generally been negative. It has not been proven that lowering serum cholesterol and LDL levels by dietary intervention will consistently affect the rate of new coronary events. The Food and Nutrition Board "recommends the dietary fat content be adjusted to a level appropriate for the caloric requirements of the individual" and "that sedentary persons attempting to achieve weight control may be well-advised to reduce the caloric density of their diets by reduction of dietary fat." Regarding dietary cholesterol, the Food and Nutrition Board says, "no significant correlation between cholesterol intake and serum cholesterol concentration has been shown in free-living persons in this country" and "the board makes no specific recommendations about dietary cholesterol for the healthy person."

A genetic predisposition for hypertension is believed to exist in the U.S. for about 15% to 20% of the population. While sodium has been implicated in this condition, an association between blood pressure and salt intake has not been demonstrated within selected U.S. populations.

Nevertheless, dietary sodium reduction is a frequently made recommendation for those persons with elevated blood pressure.

Sodium appears to exist in red meats in a reasonably constant relationship to protein. Red meat contains about 3.5 mg of sodium per gram of protein. Thus a 3 oz serving of fresh meat having 16% protein would contain about 48 mg of sodium. It can, therefore, be used very effectively in diets restricted in sodium.

Salt (NaCl) has been widely used for centuries as a preservative for meats in addition to its flavor-enhancement properties. Salt is the primary contributor to the sodium content of processed meats. At this time, the meat industry has no economically viable preservative alternative to the use of sodium chloride. Research is currently in progress to develop alternative compounds, combinations, etc., to permit reduction in the amount of sodium used. It is probable that some reduction in amount will be possible as a result of such research, with potential reductions estimated at about 25%.

Processed meats, such as ham, bacon, and a myriad of different sausages, have provided a multitude of dietary variations and have added greatly to our food enjoyment. They have greatly enhanced the overall palatability of the American diet. It seems that the first requirement of a suitable diet is frequently overlooked; namely that it be palatable and otherwise appealing to the consumer. Processed meats have much to contribute in that regard as well as making a substantial contribution to nutrient requirements.

The issue of diet/cancer was recently addressed by a special committee of the National Research Council reported on June 16, 1982, in a document entitled Diet, Nutrition and Cancer. As acknowledged by that committee, the extent of the relationship between diet and cancer is not precisely understood at this time. That committee further stated: "It is not now possible, and may never be possible, to specify a diet that would protect everyone against all forms of cancer. Nevertheless, the committee believes that it is possible, on the basis of current evidence, to formulate interim guidelines that are both consistent with good nutritional practices and likely to reduce the risk of cancer."

Two of those interim guidelines have potential for direct impact on the red meat industry: 1) the consumption of both saturated and unsaturated fats be reduced in the average U.S. diet from the current level of about 40% of calories to about 30% and 2) the consumption of food preserved by salt curing (including salt pickling) or smoking be minimized.

Regarding dietary fat, it is of interest to note that the per capita consumption of animal fats has declined over the past 70 years by about 7%, but the per capita consumption of vegetable fats has increased by 180%.

As consumers consider reduction in caloric intake, they should be fully aware of the nutrient contributions made by the fat-containing foods they consume. Animal-source foods are often nutrient-dense, providing significant quantities of nutrients in relation to calories contributed. Eliminating or reducing such foods because they contain fat may result in a deficiency of other essential nutrients provided by those foods.

To further illustrate this point, the U.S. Department of Agriculture's Nationwide Food Consumption Survey 1977-78 has shown that women from 20 to 50 years of age did not receive the Recommended Dietary Allowances for the following nutrients: vitamin B_6 (78% of RDA), folacin (71%), iron (77%), and zinc (69%). This short-fall occurred with a dietary pattern in which 42% of the calories originated from fat. It seems prudent to suggest that any change in food consumption aimed at reducing dietary fat should be made only after consideration of its effect on other essential nutrients, especially those already perilously low.

The committee's recommendation that the consumption of salt cured or smoked food be minimized appears to be based primarily on epidemiological data. It is presumed by the committee that such foods contribute to the incidence of esophageal and gastric cancer, especially gastric cancer. This recommendation is made despite the acknowledgement by the committee that gastric cancer in the U.S. is low and is decreasing. It has been described by others as "almost a vanishing disease." That declining incidence has occurred during a time period in which per capita processed-meat consumption has increased by about 50%.

The transfer of associations revealed by epidemiological studies form extremely diverse life styles would seem to be tenuous at best. In this case, populations in China, Japan, and Iceland that consumed diets containing salt-cured or smoked foods were referenced as having high incidences of esophageal and stomach cancer. The body of the committee's report identifies salt-pickled vegetables as being widely consumed in China and states that they are frequently contaminated by fungi. The committee acknowledges that the N-Nitroso compounds (carcinogens) contained in such foods were believed to originate with the fungus contaminants. To suggest that this food consumption pattern and its association with a high incidence of stomach cancer is useful in establishing recommendations for the U.S. population regarding processed-met consumption would seem, at best, questionable.

Red meat continues to be a vital contributor to the satisfaction of the human nutrient requirements in the U.S. The recommendation to consume a balanced and varied diet, including foods from all food groups, remains the most effective approach to good nutrition. The red meat industry takes justifiable pride in its contribution to the current health condition described by the Surgeon General of the United States, "The population of the United States has never been healthier."

Finally, the Food and Nutrition Board of the National Academy of Sciences, since 1941 the nation's recognized authoritative voice regarding nutrition/health, expressed its concern in 1980 over excessive hope and fears in many current attitudes toward food and nutrition. The Board said, "Sound nutrition is not a panacea... Good food that provides appropriate proportions of nutrients should not be regarded as a poison, a medicine, or a talisman, IT SHOULD BE EATEN AND ENJOYED."

74

NEW TECHNOLOGY
IN MEAT PROCESSING AND MARKETING

B. C. Breidenstein

For purposes of this discussion, the term "new" is used to describe innovations that have either evolved over the past 20 years or appear now to be in an emerging evolutionary state. Technological innovations are usually evolutionary in nature rather than revolutionary, and observations over an extended time frame tend to place events in a more appropriate perspective. Some of these innovations are discussed next.

METHODS OF ANIMAL DISPATCH

Methods of livestock dispatch have undergone marked changes during the past two decades. These changes were motivated, not so much for their impact on the resulting meat entity, as for compliance with a desire to render the dispatch process more humane.

In the case of cattle, they had historically been rendered unconscious by a sledgehammer blow to the pate prior to bleeding. That was an effective means of rendering them unconscious but was subject to operator error in aim or to sudden movement of the animal after the swing was begun. Thus, occasionally, the animal was injured but not rendered unconscious by the blow. It was generally recognized that the humanitarian aspects of the process could be improved upon. The method of general preference that has emerged is the use of a gunpowder charge to propel a hammer-like device against the pate. It is activated by a "trigger" that is depressed upon contact with the forehead of the animal. Thus, the "aim" of the operator is greatly improved and the possibility of injury as opposed to rendering unconscious is greatly diminished.

In the case of both hogs and sheep, the method of general choice for rendering the animal unconscious is electrical stunning. In spite of several years of experience and considerable experimentation with regard to voltage, amperage, and duration of electrical energy application, certain adverse effects upon muscle tissue are purported to result from improper stunning procedures. These adverse effects include capillary hemorrhages resulting in blood spots. Such spots may appear on the muscle surface and although possible to remove them during processing, the result is a rather obvious reduction in product value. The spots may, however, occur deep in the muscle tissue so that they are not visible during processing. Such deep hemorrhages must surely have an adverse impact on the consumers' perception of value.

BEEF DRESSING METHODS

Innovations in beef dressing methods have had a significant impact on the industry over the past two decades. Cattle dressing through the late 1950s and into the 1960s was largely accomplished on stationary skinning beds where the job was essentially done on an individual animal basis, rather than on an assembly-line basis that has characterized the hog kill and cut for decades. More recently, on-rail dressing procedures have had a dramatic effect on the area of kill floor required per head, through-put per unit of time, and man-hours required per head killed. Because of these and other factors, individual plants are able to handle a much higher capacity per unit of time. The on-rail dressing approach has led to a totally different perspective regarding the economics of scale and has resulted in appreciably lower kill costs than would have been possible with the old stationary-bed systems.

A phenomenon called "cold-shortening" has been a recognized factor in postmortem conditioning of meat in which tenderness of meat is adversely affected. The term "cold-shortening" is relatively new as an apt description, but the condition itself is not new. It is now recognized as a postmortem shortening of muscle fibers during the onset of rigor mortis and is believed to be energized by the residual glycogen of the muscle. That energy is believed to be released by adenosine triphosphate (ATP). Upon complete disipation of the energy source for the muscle contraction, the muscles relax and bring an end to the shortening phenomenon. In the case of typical postslaughter chilling/handling, that relaxation requires four to seven days for its completion, at which time muscle tenderness is similar to immediate postslaughter muscle.

The use of electrical stimulation shortly after slaughter has been found to induce rapid and rather violent muscle contractions that quickly dissipate the glycogen, thereby eliminating the energy source for further muscle contraction. Consequently, the cold-shortening phenomenon does not occur and muscle tenderness is maintained at the immediate postslaughter state. Typically, electrical stimulation improves tenderness. Biological variances affect both the rate and extent of cold-shortening resulting in increased variability in tenderness among carcasses handled in a similar fashion postslaughter. Electrical stimulation has been found to improve uniformity of tenderness among beef carcasses of similar visible quality and postmortem handling. In addition, electrical stimulation appears, for unknown reasons, to hasten the "set-up" of carcasses. That phenomenon is a recognized requirement for maximized grade achievement. Upon complete set-up, the muscle is firm and marbling is most prominently visible, thereby facilitating grading at an earlier time postmortem than would otherwise have been the case. Lean color uniformity of the rib eye also influences carcass grade. A phenomenon called "heat ring" results in a darker color for the outer (skin side) region of the rib eye. Such a condition in the absence of electrical stimulation is more common in thinly finished carcasses than in those having a more abundant fat cover.

It has been theorized that the darker color results from rapid chilling because of the greater prevalence of this condition in cattle carrying minimal fat as an insulation on the outer surface of the muscle. It has been postulated that the more rapid chill slows down the biochemical reactions in that area and that this reduced reaction rate is expressed in a visibly darker color of lean. In any event, electrical stimulation has been observed to reduce the incidence and(or) the severity of this "heat ring" phenomenon.

The form in which beef was distributed from the kill/chill plant to retail stores began to undergo change beginning in the mid 1960s. Prior to that time, beef was delivered to the individual retail stores primarily in the form of quarters, with limited supplementation of requirements in the form of primal cuts. Thus, each site required the skills and expertise necessary to convert that carcass form to the individual retail cuts in the kinds and amounts required to meet that market's need.

Under those conditions, the back room of the retail market also required workers with sufficient physical strength to lift and transport the ungainly beef quarters weighing 125 to 200 pounds from the storage cooler to the processing area. The difficulty of maintaining the most capable personnel in each individual retail market resulted in some sacrifice of control and adherence to a prescribed merchandising plan. Additionally, the reliance on "cattle" as the beef raw material in each individual store tended to limit merchandising programs that emphasized the anatomy of cattle, rather than the consumers' wishes in that particular market area. Balance in the movement of the product was thus achieved in many cases by reducing the broader market prices on slow-moving cuts to a price level that reflected the lack of demand for them in that narrowly defined market. Conversely, to maintain a profitable operation, the more popular cuts were priced at unduly high prices compared to the price level that reflected their demand in the broader market. Thus, it seems probable that relatively low price levels on slow-moving cuts created a lower-value perception by the retailer for beef carcasses than would have been the case for a broader market outlet, i.e., a nation-wide market. It also is probable that the compensating higher prices ascribed to those cuts in high demand in a narrowly defined market tended to reduce demand for them.

The advent of boxed beef in the industry emerged as a significant factor during the late 1960s. Essentially, it involves the breakdown of carcasses into subprimal cuts that are cut, boned, and trimmed to conform to either a single or very limited number of end uses. For example, rib eyes or loin strips are boxed for broiling steaks and chuck clods and chuck rolls are boxed for boneless pot roasts, etc. With boxed beef as the purchase form, it became possible for the retailer to identify cost much more definitively for each retail cut and to make his purchase pattern conform much more closely to the consumer demand in his defined market area.

In addition to improved merchandising flexibility, boxed beef offers more opportunities for improved value/cost during the distribution process. Since boxed beef is generally vacuum packaged in film having both a high moisture barrier and a high-oxygen barrier, losses through dehydration and trimming of oxidized, dehydrated, and discolored muscle surfaces are minimized. The removal of a significant portion of fat and bone at a central location instead of at an individual retail store represents significant improved value for beef carcasses. The centralized location for the removal of those portions of fat and bone also reduces freight costs per pound of retail cuts sold. Boxed beef with its uniform symmetrical cartons can be mechanically handled to lower costs and reduce the physically demanding part of the job for both the distributing warehouse and the retail store. Because of the film/vacuum packaging, the product is protected against bacterial contamination during the distribution process. Storage space required for boxed beef is about 40% less than that required for a comparable amount of beef stored as quarters. Because each carton is identified as to item and weight, inventory control is simplified.

Because regulatory control of water content of many processed meats is based on a protein multiplier, as well as the processors' desire for finished products of uniform composition, the piece-to-piece variance in the introduction of water-carried ingredients is an important consideration. The more uniformly such solutions are introduced into the meat, the more closely one can set targets to the regulatory limits. Thus, reduced variance leads to improved yield. Significant progress is being made in the pumping machines used to introduce such solutions into the meat.

To conform to consumer desires, it is often necessary to combine two or more pieces of meat in such a way that they give the appearance of a single muscle unit. With the consumer demand for reduction of both subcutaneous and intermuscular fat, it is frequently necessary to "muscle bone" as a means of removal of the intermuscular fat. As a result, the muscle units may be considerably smaller than desired. Contained within the muscle cell is a natural binder or adhesive that is activated by heat. So long as it remains within the cell, however, it cannot function as a binding agent. These intracellular proteins, primarily myosin, are salt-soluble and can be extracted by the application of mechanical energy and (or) by cell membrane rupture. The application of mechanical energy is commonly achieved by use of mixers, tumblers, or stirring devices in vats. The term commonly applied to the process is "massaging" or "tumbling" or sometimes both. Frequently, a combination of cell rupture and massaging are employed. The widespread adoption of such techniques in the meat processing industry occurred during the decade of the 1970s and still continues. Systems are available that combine the pumping, cell rupture, and mechanical agitation within a single totally closed system. Such a system may also combine all these operations under vacuum. The application of vacuum to a meat mass increases its volume, creates a presumably looser muscle structure, and distributes the curing solution in the meat mass more rapidly and uniformly.

The finished product, appearing as a single, solid muscle mass, is formed by placing the meat mass in some kind of a shaping, cooking container and subjecting it to heat. The bonding together of the individual pieces is a heat-induced phenomenon commencing at a temperature of about 145°F. Temperatures above 175° to 180°F do not further enhance the "bind."

The subject of restructured products has received considerable attention in the trade press in recent months. The concept of restructured products is not particularly new--ground meat formed into patties is one form of restructuring. What is new are new applications for existing equipment and new equipment to produce products having a different texture and mouth feel than ground products. The principles of restructuring in this newer sense involve the extraction of intracellular proteins to the meat-mass surface where, upon heating, the particles adhere to each other forming a solid, more nearly muscle-cut-like-structure than is true for ground products.

Restructuring basically involves particle-size reduction, mechanical agitation, usually the addition of salt and phosphates to extract the intracellular proteins, forming or shaping, freezing and portioning. Great variations in texture and/or mouth feel are possible, depending upon raw material used, extent and/or nature of particle-size reduction, amount and time duration of mechanical energy applied, kind and amount of additives used, etc. Thus, a broad spectrum of products can be produced through the restructuring process. Much good technology exists, but the amount of meat surface area exposed during the process makes the product more susceptible to oxidative changes affecting color and flavor than is true for whole muscle cuts. It is, therefore, essential that technology be properly applied if consumer satisfaction is to be maximized.

The use of liquid smoke as an alternative to direct application of wood smoke became much more widespread in the 1970s. The primary stimulus was probably the need to minimize atmosphere discharge from cooking/smoking operations as a means of compliance with the Clean Air Act. Additionally, the advances made in suitability of liquid smoke and the desire for a higher level of process control also were contributing factors. In any event, liquid smoke is now widely and successfully used as a part of meat-processing systems. The use of additives to enhance functional properties has been practiced as long as meat processing has been done. As an example, various forms of phosphates have been used for the purpose of water retention and altering the mouth feel of finished products. Not only do phosphates aid in yield control, but products containing phosphates tend to have less free moisture accumulation in consumer packages. The absence of free moisture in the package greatly improves consumer acceptability. Recent regulatory changes of the USDA now permit the use of phosphates in a broader array of products including some sausage products. Ascorbate is another additive that has been used for some time but has had increased emphasis placed on it in recent years. Its presence in bacon has been shown to inhibit the formation of nitrosamines.

Processing of prerigor meat, as opposed to traditional kill, chill, and heat process is receiving increased interest and attention. As energy costs increase, a frequently asked question is: why start with a carcass at a temperature of about 100°F, expend energy for refrigeration to reduce its temperature to about 30°F or 40°F, then expend energy to heat it back through that temperature range to its ultimate processing temperature? That is obviously an oversimplification, but it has resulted in increased efforts to examine the issue of prerigor processing of those carcass portions that are expected to be processed further prior to sale or distribution.

PACKAGING MATERIALS

Packaging materials for primary containers have undergone dramatic improvement over the past two decades. The barrier properties of films to both moisture and oxygen have been sharply improved as well as the film resistance to the rigors of distribution. Only quite recently have films been used in which the product could be cooked and also serve as the primary packaging container. In conditions of water cook, such as boiled hams, product exposure to the cooking water results in leaching of meat components and curing ingredients. Protein loss thereby represents a loss of yield potential not only of the protein but of the water as well, because of the protein multiplier. Controlling yields thus becomes a matter of estimating cooking losses so that the pump can be adjusted to compensate for it. The new cook-in-

-the-container packaging does not allow exposure of the meat product to the water--thus cooking loss (one of the previously estimated variables) is eliminated.

The film package has recently appeared in retail markets as a replacement of the metal can as a primary container; however, it is too soon to make any judgments regarding its success. Its primary advantages would appear to include lower container cost, product visibility through the transparent film, and less hazard in the home upon opening. The primary question would appear to be container integrity and ability to withstand the rigors of distribution expected of products in metal containers.

RETAIL STORE DELIVERY

A major shift in retail-store delivery of meat and other products has occurred over the past two or so decades. The use of distribution centers by supermarket chains, cooperative groups, etc., has permitted the combination of products for delivery to greatly reduce the number of deliveries to the individual stores. In earlier years, each different meat supplier sent delivery trucks to each store as frequently as required. The distribution center now takes quantity deliveries from each supplier, warehouses them, and combines all suppliers in accord with each store's orders to make a single delivery. Such a system has resulted in significant distribution efficiencies.

During the late 1960s and early 1970s, beef fabrication facilities were often a part of a distribution center. Generally, these were justified on the basis of improved quality control and meeting perceived requirements as compared to the alternative of purchasing boxed beef from a slaughterer /fabricator. Many such operations have been discontinued. One can speculate as to the reasons. Most assuredly, product specification standardization among boxed-beef suppliers has vastly improved since the childhood of boxed beef in the late 1960s. Secondly, the ability of the boxed-beef industry to adapt to a true market requirement has largely accommodated most "unique" retailer requirements. Finally, many of the features of economy that characterize boxed beef packaged at the source of supply were not captured if naked carcass beef was transported to a distant facility to be fabricated.

The use of piggy-back refrigerated trailers once appeared to be a very viable meat-shipping mode. It seemed to offer the economy of rail transport, the flexibility of truck delivery, and the advantages of mechanical refrigeration. Experience in the industry, however, was that it was entirely too unreliable to meet the needs for transport of a valuable, perishable product such as meat. The over-the-road truck/trailer has evolved as the method of choice for the transport and delivery of the vast majority of meat products. Air transport has been championed for meat from time to time but has never accounted for any significant proportion of the volume of meat product delivery. It would seem possible that air transport might be useful in the delivery of ready-to-cook fresh meat items for export. Such a situation could conceivably arise from demand for high-quality American meat products overseas, i.e., to satisfy tourist requirements.

COMPUTER APPLICATIONS

As in almost all elements of life, the computer revolution has had a dramatic impact on the meat industry. The necessity for timely information has always characterized the meat industry. Its perishability and narrow margins per unit of volume make it essential that management information regarding costs, yields, inventories, shipments, etc., be available at frequent intervals, i.e., daily or more frequently. The computer has made such things possible, though probably few companies have fully exploited the computer for such purposes.

Least-cost sausage formulation has been a widely used computer capability for at least two decades. However, the ever increasing computer capacities and capabilities have led to ever more sophisticated models that provide an expanded examination and interpretation of data.

In the early 1970s, the National Livestock and Meat Board assembled industry representatives and assumed the leadership role in the development of Uniform Retail Meat Identity Standards. This activity resulted in the publication of a book by the Board in which those standards are set forth. That effort has been widely accepted as a basis for improving the consumer information contained on the label as well as reducing the confusion resulting from geographic variation in the names by which each retail meat cut is known.

Beyond their original intent, however, the presence of those widely accepted standards has provided the opportunity for the adoption of computers by the meat department of retail stores. Universal product coding of catch-weight packages is now a feasible use of the computer. Optical scanning makes it possible to track meat sales and other operational parameters of the retail meat department much more comprehensively than was previously possible. Numbers have been assigned to the standard cuts described in the Uniform Retail Meat Identity Standards thus greatly facilitating the use of the computer.

To that end, the Food Marketing Institute and the National Livestock and Meat Board have cooperated in sponsoring educational seminars attended not only by the representatives of the meat department but also by his coworker, the computer systems analyst, and/or programmer. Such a joint approach for each to better understand his coworkers' problems and needs is expected to lead to a smoother and more effective transition to the computer age.

The presence in each retail store of minimanufacturing operations has long been recognized as something less than maximally efficient. Decentralized preparation and packaging of retail cuts have been on the horizon for decades. As new packaging and preservation technologies evolve, they have been seized upon as possibly the missing element required for that dream to become a reality.

Freezing, in combination with advances in packaging, has been tried several times with, in most cases, less than satisfactory results in the U.S. Some centralized retail cutting operations appear to be successful in Europe. In the U.S., however, some vacuum-packaged subprimal cuts, prepared and packaged using the boxed-beef concept, have been successfully marketed at retail. Such products are available as beef, pork, and lamb cuts in many areas of the country.

Finally, the productivity of American agriculture is the envy of the world. It is a real challenge for our meat-processing distribution and marketing industries to achieve comparable proficiency. The extent to which we are successful in meeting that challenge, thereby reducing marketing costs and/or improving upon the perceived value of our products, plays a very important role in the economic health of this great industry.

75

PRINCIPLES OF MEAT SCIENCE
AFFFECTING LIVESTOCK PRODUCERS:
MICROSTRUCTURE

Donald M. Kinsman

INTRODUCTION

The adage "We are what we eat" is a truism for the human race as well as it is for the livestock we produce. If we wish a high-protein result, then we consume a high-protein food; if we wish high energy, we consume that source, assuming each is available to us at a cost we can afford. Fortunately, we generally ingest a balanced diet from many sources, and we usually eat those things we like most, such as meat. Meat supplies a complete protein, chiefly from muscle; ample energy, from the accompanying fat; vitamins, especially the B complex; and minerals, most certainly iron. In addition meat can provide palatability which is described as a combination of appealing tenderness, juiciness, and flavor, plus the satiety value or "lasting value" that provides "staying power." The fact that meat is highly digestible (93 - 97%) makes it especially valuable and a highly efficient source of assimilative nutrients. This is all documented, but we do need to remind ourselves, the nation, and the world that it is a nutrient-paced product. As our consumers become more and more remote from the production scene, it becomes increasingly important that we call these facts to their minds.

Furthermore, if we are to continue to produce this commodity, we must have an appreciation of how it is manufactured. Our urban citizens may believe that it grows in cellophane wrappers. We, of course, have watched it grow from "conception to consumption" and know the care and costs involved. But do we really know how it grows? Certainly we know that feed goes in one end, waste products out the other, and the difference is used for maintenance and growth. But what are these growth units? How are they formed? What do they contribute to the end-product?

MUSCLE TISSUE

The chief component of meat is muscle, which the consumer generally refers to as lean. There are three types of muscle:
- Striated (sometimes called voluntary or skeletal) which represents the greatest amount of body muscle.
- Nonstriated (also referred to as smooth, visceral, or involuntary muscle) is found only in the body linings, such as the digestive system.
- Cardiac or striated-involuntary muscle is peculiar to the heart only.

We will confine our remarks to the striated muscle. Without becoming too involved with the ultrastructure of muscle, let us take a broad overview of this tissue. There are more than 600 muscles in the animal body, which vary decidedly in size, shape, activity, and function. They are attached to bone by tendon and, of course, have a blood supply and nerve supply.

Starting with the smallest physical structure, the myofilament is composed of the proteins actin and myosin, which are vitally involved in muscle activity - contraction and relaxation - in the living animal. These, in combination, form myofibrils, which course through the muscle fiber or muscle cell in conjunction with the cell fluid or sarcoplasm, nuclei, and other cellular inclusions (figure 1).

Numerous individual muscle fibers (cells) in turn combine as muscle bundles, and these bundles or fasciculi compound to form the approximately 600 individual muscles attached to the skeleton. Yes, "mighty beasts from little myofibrils grow."

CONNECTIVE TISSUE

Each of these units--muscle fibers, muscle bundles, and muscles proper--have connective tissue associated with them. As the term implies, this tissue serves to connect, envelope, or strengthen and give elasticity to these structures. There are three types of connective tissue:
- Collagen is the most abundant. It is white and transparent and gelatinizes when exposed to moisture and heat. Thus, it will be tenderized when meat is cooked by moist heat.
- Elastin is also prevalent but, unfortunately, does not tenderize easily. It is yellow in color and dense and elastic in consistency. It is best seen in the backstrap or neckleader (ligamentum nuchae) of carcass meats, especially in the neck and chuck.

- Reticulin is composed of small fibers that form
networks around cells. They may be precursors
to collagen and/or elastin.

Although connective tisue is not readily apparent, it
is associated with muscle structure and is a major factor
influencing meat tenderness. Fortunately, conective tissue,
collagen in particular, is an excellent source of protein.

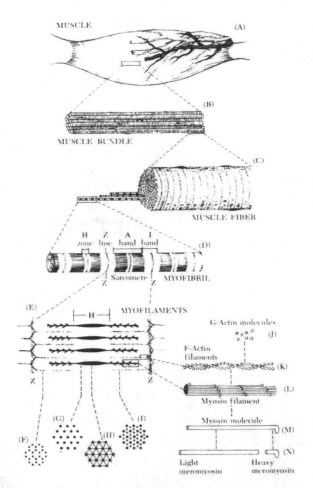

Figure 1.

Source: **Principles of Meat Science** by J. C. Forrest, E. D.
Aberle, H. B. Hedrick, M. D. Judge, and R. A. Merkel. W.
H. Freeman and Co. Copyright © 1975.

ADIPOSE TISSUE

In polite society, we refer to fat as adipose tissue. It is made up of fatty acids of two general types, saturated and unsaturated. Meat animals produce both types of fat, with ruminants incorporating more highly saturated fatty acids such as myristic, palmitic and stearic, whereas the monogastrate produces more unsaturates like oleic, linoleic and linolenic fatty acids. Thus, cattle and sheep have harder, more saturated fats, and hogs produce softer, less saturated fats (table 1). In total, lamb has a ratio of about 58% to 42% saturated to unsaturated fatty acids, while beef runs approximately 55:45, and pork 45:55. These facts all have ramifications relating to quality, consumer acceptance, processing, and shelf-life; however, time does not permit discussion of these topics.

However, we are most concerned with the amount of deposition of fat on and in the carcass and its influence on cutability and quality. Fat is laid down first internally, around the kidneys, heart, and pelvic region. Next, it is deposited subcutaneously to form covering or finish. Simultaneously, but slightly later, it is being infused between the muscles as seam fat (intermuscular fat) and within the muscles as marbling (intramuscular fat).

This sequence causes us much concern if it is necessary to accept all the internal and external fat that appears to be necessary to realize the degree of marbling preferred. Fortunately, once identified, we can select genetically for marbling or grade without producing wasty carasses to obtain these characteristics. Thus, from a quality or palatability standpoint, it is reassuring to know tht we can produce desirably finished animals that will "grade" without being excessively wastey.

HOW DOTH YOUR MUSCLE GROW?

Visualizing now the make-up of muscle, let us consider its growth and development on a unit basis and as manifested in the living animal and its resulting carcass (table 2). Growth is basically an increase in size. It may be accomplished in muscle by either, or both, of the following:
- Hyperplasia is the production of new cells and is largely determined prenatally. The number of muscle fibers (cells) does not appear to increase significantly after birth.
- Hypertrophy is the enlargement of existing cells. The greatest increase in muscle size occurs postnatally as the muscle fibers increase both in diameter and length. The diameter increase is caused by a proliferation of the myofibrils within the muscle fiber (cell), possibly by as much as 10 to 15 times from birth to

TABLE 1. FATTY ACID AND TRIGLYCERIDE COMPOSITION OF SOME ANIMAL FAT DEPOSITS (PERCENTAGE BY WEIGHT)

Component	Chicken	Pig Subcutaneous Outer	Pig Subcutaneous Inner	Pig Perirenal	Cattle Subcutaneous	Cattle Perirenal	Sheep Subcutaneous	Sheep Perirenal
Fatty acids								
Lauric	–	trace	trace	trace	0.1	0.2	0.1	0.1
Myristic	0.1	1.3	0.1	4.0	4.5	2.7	3.2	2.6
Palmitic	25.6	28.3	30.1	28.0	27.4	27.8	28.0	28.0
Stearic	7.0	11.9	16.2	17.0	21.1	23.8	24.8	26.8
Arachidic	–	trace	trace	trace	0.6	0.6	1.6	2.6
Total saturated	32.7	41.5	47.3	49.0	53.7	55.1	57.7	59.5
Palmitoleic	7.0	2.7	2.7	2.0	2.0	2.2	1.3	1.9
Oleic	20.4	47.5	40.9	36.0	41.6	40.1	36.4	34.2
Linoleic	21.3	6.0	7.1	11.8	1.8	1.8	3.5	4.0
Linolenic	–	0.2	0.3	0.2	0.5	0.6	0.5	0.6
Archidonic plus chipandonic	0.6	2.1	1.7	1.0	0.4	0.2	0.6	0.8
Total unsaturated	67.3	58.5	52.7	51.0	46.3	44.9	62.3	41.5

Source: Principles of Meat Science by J. C. Forrest, E. D. Aberle, H. B. Hedrick, M. D. Judge, and R. A. Merkel. W. H. Freeman and Co. Copyright © 1975.

TABLE 2: APPROXIMATE COMPOSITION OF MAMMALIAN SKELETAL MUSCLE
(PERCENTAGE FRESH WEIGHT BASIS)

	Percent		Percent
Water (range 65 to 80)	75.0	Non-Protein Nitrogenous Substances	1.5
Protein (range 16 to 22)	18.5		
Myofibrillar	9.5		
myosin	5.0	Creatine and creatine phosphate	0.5
actin	2.0		
tropomyosm	0.8	Nucleotides	
troponin	0.8		
M protein	0.4	(adenosine triphosphate [ATP],	
C protein	0.2	adenosine diphosphate	
α-actinin	0.2	[ADP] etc.)	0.3
β-actinin	0.1	Free amino acids	0.3
Sarcoplasmic	6.0	Peptides	
		(anserine, carnosine, etc.)	0.3
soluble sacroplasmic and mitochondrial enzymes	5.5	Other nonprotein substances (creatinine, urea, inosine monophosphate [IMP]	
myoglobin	0.3	nicotinamide adenine	
hemoglobin	0.1	dinucleotide	
cytochromes and flavo-proteins	0.1	phosphate [NADP]	0.1
Stroma	3.0	Carbohydrates and non-nitrogenous substances (range 0.5 to 1.5)	1.0
collagen and	1.5	Glycogen (variable	
recticulin elastin	0.1	range 0.5 to 1.3)	0.8
other insoluble proteins	1.4	Glucose	0.1
Lipids (variable range 1.5 to 13.0)	3.0	Intermediates and products of cell metabolism (hexose and triose,	
Neutral lipids (range: 0.5 to 1.5)	1.0	phosphates, lactic acid, citric acid, fumaric acid, succinc acid	
Phospholipids	1.0	acetoacetic acid, etc.)	0.1
Cerebrosides	0.5	Inorganic Constituents	1.0
Cholesterol	0.5	Potassium	0.3
		Total phosphorus (phosphates and inorganic phosphorous	0.2
		Suflur (including sulfate)	0.2
		Chlorine	0.1
		Sodium	0.1
		Others (including magnesium, calcium, iron, cobalt, copper, zinc, nickel, manganese etc.	0.1

Source: Principles of Meat Science by J. C. Forrest, E. D. Aberle, H. B. Hedrick, M. D. Judge, and R. A. Merkel. W. H. Freeman and Co. Copyright © 1975.

maturity. In hogs, the maximum proliferation, and therefore muscle fiber size, is at about 5 months of age. Individual muscles vary in their rates of growth. The larger muscles, hind limb and back generally, have the greatest rate of growth. However, the greatest influence is the number of muscle fibers initially present to be hypertrophied. In turn, the diameter expansion, is influenced by a number of factors such as species, breed sex, age, nutritional level, and exercise. The muscle fibers of sheep are smaller than those of cattle and swine; intact males have muscle fibers larger in diameter than those of castrates, which have larger fibers than do females. Generally, muscle fiber diameter increases with maturity, nutritional plane, and physical activity.

We should be able to control some of these factors to develop maximum muscle while minimizing unnecessary fat; we have an opportunity to produce the most of the best for the least. Like much theory applied to practice, such production may be more easily talked about than done. Nonetheless, using this knowledge, proper records, wise selection, and culling, adaptation of genetics, nutrition, and environmental manipulation, we should be able to put much of this information to work and thereby improve our production of the "right kind" of livestock for the marketplace of the future. The goal then should be, in most basic terms, to increase muscle fiber size and numbers and to decrease adipose tissue cell size and numbers!

The time is upon us for action to feed our nation and much of the world on the most nutritious single foodstuff we have to offer and known to man - red meat!

We should be aiming for Choice quality grade and yield grade 1 to meet the demands of the future. <u>It can be done, it is being done, it must be done</u>. We can refute the saying of "No waste, no taste." Let's maintain the quality we desire, but eliminate the inefficient waste fat that we have tolerated in the past. This challenge I leave you!

76

PRINCIPLES OF MEAT SCIENCE AFFECTING LIVESTOCK PRODUCERS: GROWTH AND DEVELOPMENT

Donald M. Kinsman

INTRODUCTION

In mathematics, "The whole is equal to the sum of its parts." In nature and with all biological materials, everything is interrelated and, under normal conditions, the harmonious blending of all cells, organs, and systems produces the final organism. Understanding these micro and macro structures is important in the production of all animals, especially those we are raising for meat purposes. Certainly we need to be cognizant of the end-product of this chain of events to be certain that we are producing the kind of product that will sell, that the consumer desires and demands, and that can be produced and marketed as efficiently and economically as possible for a reasonable profit.

WHAT'S UNDER THE HIDE?

Let us first examine this end-product by its components, as viewed from a production perspective. In essence, meat is the edible tissue of animals. That which we ingest is principally muscle, but it is accompanied by some form of fat, either intramuscular, subcutaneous, or intermuscular. Connective tissue in the muscle and fat is also consumed. Although we do not normally consume bone, it is the framework on which the muscle and fat develop and is frequently included as part of the roast or steak. How these basic components grow and develop is a function of many interrelated factors--with genetics, nutrition, and environment serving as major influences. Table 1 depicts the heritability estimates of the growth and carcass characteristics of cattle, sheep, and swine. Nutrition certainly plays a major role in the growth and development of our meat animals, and figure 1 portrays the influence of plane of nutrition on the rate of skeleton, muscle, and fat growth in swine. Environmental effects are more difficult to measure, including weather, temperature, climate, and stress, but we have the greatest control over management.

TABLE 1. HERITABILITY ESTIMATES OF GROWTH AND CARCASS
CHARACTERISTICS OF CATTLE, SHEEP, AND PIGS

Characteristic	Species	Heritability estimate[a] approximate average
Weaning weight	Cattle	25
	Sheep	33
	Pigs	17
Postweaning rate of weight gain	Cattle	57
	Sheep	71
	Pigs	29
Feed efficiency in feed lot	Cattle	36
	Sheep	15
	Pigs	31
Carcass grade	Cattle	48
Tenderness	Cattle	61
	Sheep	33
Fat thickness	Cattle	38
	Sheep	20
	Pigs	49
Longissimus muscle cross sectional area	Cattle	70
	Sheep	48
	Pigs	48

Source: Principles of Meat Science by J. C. Forrest, E. D.
Aberle, H. B. Hedrick, M. D. Judge, and R. A.
Merkel. W. H. Freeman and Co. Copyright © 1975.

[a]Estimates were compiled from various sources.

These all combine to yield a growth curve as shown in figure
2. Note that this curve starts slowly, rises gradually,
and then escalates very rapidly at an early stage before
leveling off at its eventual mature state. It is during
this rapid escalation period that we need to capitalize on
the animal's greatest feed and growth efficiency in producing maximum muscle with minimum fat to meet the market
demands.

NOTHING IS STATIC

The term "homeostasis" expresses the maintenance of a
physiologically-balanced internal environment. This is virtually a system of checks and balances that permits the

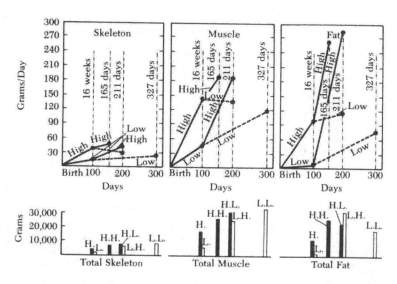

Source: **Principles of Meat Science** by J. C. Forrest, E. D. Aberle, H. B. Hedrick, M. D. Judge, and R. A. Merkel. W. H. Freeman and Co. Copyright © 1975.

Figure 1.   Effect of plane of nutrition on rate of skeleton, muscle, and fat growth in the carcasses of pigs with live weights of up to 91 kilograms. Key: H = high plane of nutrition; L = low plane of nutrition; HL = high plane, then low plane (to 91 kg); HH = high plane (to 91 kg); LH = low plane, followed by high plane (to 91 kg); LL = low plane of nutrition (to 91 kg). [Data from McMeekan, C. P., "Growth and Development in the Pig, with Special Reference to Carcass Quality Charcters; III. Effect of the Plane of Nutrition on the Form and Composition of the Bacon Pig," J. Agr. Sci. 30, 511 (1940).]

animal to cope with the stresses that tend to alter that environment. This mechanism is controlled by the nervous system and the endocrine glands. Fortunately, they usually maintain the harmony and equilibrium that permits adaptability of the animal to its oft-changing environment. However, change is always occurring in any organism. Certainly, amounts and proportions of the basic meat components change in meat animals during both prenatal and postnatal growth and development, and the stage at which these factors change is of importance to all of us as producers of the nation's meat supply. Let us study figure 3 and recognize the changes in the percentage of bone, muscle, and fat in beef carcasses during growth. Bone has the least percentage change, decreasing from about 25% of the carcass in the neonatal animal to about 13% at maturity. Muscle percentage of

708

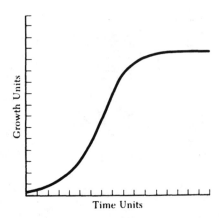

Source: <u>Principles of Meat Science</u> by J. C. Forrest, E. D.
Aberle, H. B. Hedrick, M. D. Judge, and R. A.
Merkel. W. H. Freeman and Co. Copyright © 1975.

Figure 2. A Typical Growth Curve

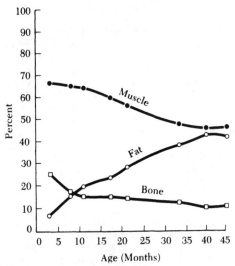

Source: <u>Principles of Meat Science</u> by J. C. Forrest, E. D.
Aberle, H. B. Hedrick, M. D. Judge, and R. A.
Merkel. W. H. Freeman and Co. Copyright © 1975.

Figure 3. Changes in the percentage of bone, muscle, and
fat in beef carcasses during growth. [Data from
Moulton, C. R., P. F. Trowbridge and L. D. Haigh,
"Studes in Animal Nutrition: III. Changes in
Chemical Composition on Different Planes of
Nutrition." Mo. Agr. Exp. Sta. Res. Bul. 55
(1922).]

the carcass decreases as fat increases. Thus, muscle represents approximately 68% of the newborn animal and may decrease to 50% or less as fat approximates that proportion. On an age basis, fat is about 30% (the maximum fat permitted in ground beef) at 20 to 24 months of age in beef cattle. Thereafter, it climbs steadily, approaching 40% to 50% at maturity. This is, of course, dependent upon nutritional planes. Furthermore, we need to concentrate on those stages that affect the final product and then interrupt it at the optimum point of greatest value to producer, packer, processor, and consumer.

Where do you believe that point or segment lies? To carry that thought further, we may wish to relate the percentages of muscle and bone to various percentages of fat in the carcasses of beef and sheep to determine the influence of carcass fat on muscle, in particular, and to bone, secondarily (table 2).

TABLE 2. APPROXIMATE PERCENTAGES OF MUSCLE, AND BONE PLUS TENDON, IN BEEF CATTLE AND SHEEP CARCASSES HAVING VARYING PERCENTAGES OF FAT

Fat	Muscle	Bone plus tendon
8	66	26
12	62	26
16	62	22
21	61	18
26	59	15
32	54	14
37	49	14
42	46	12

Source: Principles of Meat Science by J. C. Forrest, E. D. Aberle, H. B. Hedrick, M. D. Judge, and R. A. Merkel. W. H. Freeman and Co. Copyright © 1975.

WHAT IS THE OPTIMUM "FINISHING' POINT?

At what point is it best to "finish" our livestock? There is no single, uniform answer as we consider efficiency, economics, profits and the multitude and variability of our markets. However, if we are producing for USDA Choice, quality grade and yield grade 1 or 2 then we must harvest these meat animals while young and tender, with sufficient quality to satisfy the goals of palatability and cutability. This generally implies slaughter of beef cattle at 15 to 20 months of age, and of hogs and lambs at 4 to 8 months of age--animals with adequate genetic, nutritional, and environmental background to provide the tenderness, juiciness, and flavor with minimum waste to meet these precise goals.

It <u>can be done</u>, it <u>is being done</u>, and it <u>must be done</u> if we <u>are</u> to continue to provide this highly-nutritious, heavily-demanded, reasonably-priced basic foodstuff for America's dinner table with decreasing energy and costs. <u>Go forth and do likewise.</u>

Part 12

MARKETING, ECONOMICS,
AND FARM BUSINESS

77

CATTLE FUTURES TRADING

Dick Crow

Futures trading in feeder and finished cattle is one of the hottest topics in the industry today. You can walk into any coffee shop where ranchers are known to hang out and hear one, if not several, conversations about futures prices, the Chicago Mercantile Exchange (CME), and their respective effect on cash cattle prices. Many, many cattlemen condemn futures trading, saying it adversely affects cash prices and exerts a downward effect on the amount of money each of you receives for your year's effort in producing cattle.

The problem, as we see it, is that some traders within the CME have used futures trading in a way that is not necessarily in the cattle industry's best interest. Compounding the problem, the CME and the Commodity Futures Trading Commission haven't policed the ranks as effectively as is necessary.

A SURVEY OF LIVESTOCK PRODUCERS

Several editorials and numerous articles have appeared in the pages of Western Livestock Journal concerned with futures trading. In addition, the editorial staff surveyed its readers a year ago to determine their feelings about futures. Our survey, while certainly not comprehensive enough to satisfy many statisticians, reflects the feelings of many livestock producers in the United States. As one respondent said, "In any business today there's room for improvement, and I'm sure there is in the futures market also." Survey results show that 69.9% of replying stockmen have traded cattle futures anywhere from once a year to 500 times a year or more. Of that number, 41.3% have hedged, 10.8% have speculated, and 47.8% have maintained both the hedge and speculative position. Of those stockmen who have hedged, 41.2% were statisfied with their hedging results and 58.8% were not. The chairman of the board of the Texas Cattle Loan Company, and one of the most astute cattlemen in the history of the industry, the late Jay Taylor of Amarillo, wrote: "We own and operate two feedlots and feed all our own raised cattle and buy quite a few in addition to those we raise. We try very hard to hedge everything that goes in the lot. We don't always have an opportunity to hedge, but the moment we do, we hedge. I would hate to have this avenue of insurance taken away." Cattlemen shouldn't confuse hedging and speculating. There is a time and a place for both. In response to the survey question, if cattlemen believe that some firms (that both feed cattle and speculate on the cattle futures) have attempted to manipulate the delivery process at terminal markets in an effort to gain price advantage, 94.4% answered yes while only 5.6% answered no. There's a lot of suspicion in the countryside about futures trading.

In responding to how the cattle futures market has helped manage risks, 36.5% of those responding said yes and 63.5% said no. A cattleman from Texas, who responded no, said the futures market encourages feeding of cattle that are otherwise not economical resulting in a false market and oversupply. On the other hand, a cattleman from Nevada who said the futures market has helped him manage risk, claims that over the past 10 years he has not seen a 12-month period when he could not hedge at a profit.

Another respondent to our survey said he believed the futures market is used by many feeders and stocker operators. He further believed that the majority of the cattlemen do not have enough cattle, the financial backing, or the knowledge of the futures markets to be able to trade in them. Citing specific examples of what he believed are suspicious patterns of cattle futures trading, another respondent said he gave his broker an order to sell all contracts at the market. The market was falling. It took three days to complete the sale. The three-day delay cost him $10,000 in margin money that he had to dig up. Another respondent said he hasn't traded futures in years because they are being manipulated by insiders and that this is totally damaging to legitimate cattlemen.

The survey, which drew replies from 18 states, shows nearly 85% believe block trading by managed accounts beyond the speculative limits occurs at Chicago Mercantile Exchange. Nearly 80% cited examples of what they believe to be suspicious patterns of cattle futures trading.

Views of cattle futures trading vary widely. Offered a California rancher, "I believe that it is unreasonable to have a bunch of speculators more or less setting the price on cattle. Before futures, the price was controlled by supply and demand. Now you never know what a bunch of speculators will guess it is." A Wyoming rancher responded, however, that futures trading is a tool. Learn how to use it... "A hammer is also a tool, but you don't throw it away when you hit your thumb." From a Utah stockman, "Futures are for a few with money and a lot of them never see the cattle or handle them. I hope it comes to an end."

In response to the question of whether a clearing firm had required a regional or local brokerage office to follow house rules on buying or selling, even though it was not in the best interest of that office or the cattlemen's client-trading position, a Colorado stockman said one company made him liquidate his short position on the feeder cattle that he had hedged the first of the delivery; it was either liquidate or put up the full contract value. He said he trades with three companies and the other two don't require this on speculative positions. Responded a South Dakota cattleman, "I'm convinced that there's manipulation in all futures contract trading just like stock, bonds, cash-grain prices, and packer-to-packer transactions of dressed beef. No amount of regulation will stop it completely. Another Wyoming stockman said he has felt many times that he has been taken by the futures market. Why should he feel otherwise when he sees people invest in cattle who will never see a cow? He feels this type of investor does cattlemen no good.

Exactly half of those who traded cattle futures claim that brokers had encouraged them to a close position that they did not want to close out. Most indicated, however, that it was not a regular occurrence. One-fourth of the respondents (26.3%) said their broker arranged for a trade to be designated as a hedge trade when in fact it was speculative.

Nearly half of those responding believe cattle futures are a useful tool for cattle feeders while more than one fourth (26.4%) believe futures are a useful tool for cow/calf operators.

OTHER REACTIONS TO THE FUTURES MARKET

Divergent opinions? You bet. An emotional issue? Beyond a shadow of a doubt. We receive many articulate, well-written letters in addition to the survey form we published in our weekly newspaper. Comments in these letters ranged from absolute belief in futures trading to some calling for immediate abolition of this trading. There's no doubt that many feeders and some cow/calf and stocker operators use futures successfully and find it a valuable financial tool. There's also no doubt that the majority of cattlemen, as shown by our survey, distrust futures trading. Why can some cattlemen successfully use futures while others can't? Is it because those who distrust futures simply haven't taken the time to really learn how the process works and how to apply it to their operations? This is a theory proposed by both cattlemen and brokers. One broker interviewed by WLJ feels cattlemen's distrust of futures stems from misunderstanding. The reason cattlemen have trouble using futures is the fault of cattlemen, not the market, he says.

The broker, Matthew Burns, a commodities and financial futures advisor in the Denver Boettcher and Company office, feels the major problem is one of understanding exactly what futures can and can't accomplish. Another problem, he says, is greed. He feels most cattlemen are not willing to settle for an assured bankable profit but instead want to sell at or near price-rise peaks. By doing this, though, Burns says cattlemen almost always take a price beating because it's impossible to react as quickly as does the market. Burns feels by selectively using futures and picking a competent broker who understands the market, cattlemen can assure themselves of a reasonable profit and avoid risking their very survival by operating with no price protection throughout the year and risking a loss on the cash market.

There are some cattlemen who agree with this thesis. Two respondents to our survey use futures very successfully in their operations. One, a cow/calf and stocker operator from Wilson, Wyoming, says the survey shows a tragic picture of the people in our industry. The results, he says, show a lack of knowledge not only by producers but by the related businesses of banking and extension. This cattleman feels many cattlemen are blaming poor management on the futures board. The other cow/calf, stocker operator, and a feeder from South Dakota, says cow/calf operators are probably the biggest objectors to futures, yet they have the most reason to use them because futures provides the only price protection available. In their letters to us, both stressed that cattlemen should make an honest effort to learn all they can about the futures market. They both claim that futures can work for any producer, but only if each producer will make them work. The Wyoming cattleman said ranchers are actually getting the better end of the deal. He thinks ranchers get more money on the cash market because of the futures board. If futures were done away with, cow/calf and stocker operators would have a price disadvantage because feedlots would only buy cattle at a price at which they could make money. The South Dakota rancher feels futures are by no means perfect, but he says he is a better cash marketer because of his understanding of the futures market.

THE CHICAGO MERCANTILE EXCHANGE (CME), COMMODITY FUTURES TRADING COMMISSION (FTC), AND NATIONAL CATTLEMEN'S ASSOCIATION

Many may interpret my report to this point as an excellent example of the old saying--if you get 10 ranchers together, you'll get 10 different opinions on whatever you want to talk about. Such is the case with futures: the opinions and ideas are as divergent as the industry that uses them. Some cattlemen swear by futures and others swear at them. Regardless of whether an individual uses futures and understands the market or not, there is evidence to suggest that the Chicago Mercantile Exchange has not done an adequate job of regulating its members and ensuring that trading practices are entirely above board. Many feel that when the CME does detect wrongdoing, the penalties handed down in no way compensate for the potential profits made by manipulation.

A recent investigation by the Government Accounting Office (GOA), the investigative arm of Congress, revealed that the Commodity Futures Trading Commission (FTC) needs to work harder to assure that futures trading is reasonably free from abuse. The GAO based its conclusions on a one-year study of the CFTC. The report called for CFTC improvements in its reparations-program audits and financial-surveillance registration and rule enforcement. In terms of exchange-rule enforcement review, GAO said the CFTC has improved marginally since the 1978 review and substantial improvements are still required. The GAO listed, as an example, that a CFTC review of the New York Mercantile Exchange conducted in 1981 overlooked serious problems brought to light in a separate 1981 CFTC enforcement investigation. A report on this appeared in WLJ's March 1, 1982, newspaper. At that time, CFTC had yet to receive key reports on the Chicago Board of Trade and the Chicago Mercantile Exchange. GAO felt the Commodities Futures Trading Commission should improve its market surveillance efforts by upgrading its data processing operation, which includes some 10-year-old software. The government agency also suggested amending the commodity exchange act to follow the CFTC to share routinely with exchanges referral to futures traders who also hold government positions on futures exchanges and make decisions affecting trading.

As a result of the furor arising from rank and file cattlemen over the futures market, the National Cattlemen's Association formed a surveillance committee to review futures trading practices. As we wrote in our Comments Column, published March 8, 1982, the National Cattlemen's Association has done the country a service by bringing to everyone's attention that the trading practices and market surveillance procedures at the Chicago Mercantile Exchange are possibly suspect and that, if this is true, something must be done about it now. The nation's stockmen have entered 1982 looking at a bumpy economic road that some believe has been made all the more difficult by cattle futures trading.

Nevertheless, the ranching and feeding industry has tremendous underlying strengths. The ability to hedge risks offered by cattle futures trading should be one of those strengths. Returning credibility to the Chicago Mercantile Exchange would open major new management opportunities to rank and file stockmen. But it will require possibly revolutionary changes in the ground rules in the CME. We've learned, for example, that the CME's own investigation shows that one firm and its officers violated the speculative limits on live or feeder cattle futures on at least 15 separate occasions. Further, our sources indicated that such violations at the CME went on for nine months or more before any action was taken. When the CME did respond to these serious, repeated violations, fines and suspensions in no way compensated for the potential profit. An isolated incident? Some will argue it is. Many sincere stockmen believe, however, that this is not an isolated incident, but a continuing problem involving many traders. Rank and file cattlemen will likely never know the exact nature, timing, and size of these violations and what impact they may have had on either futures or cash cattle prices.

Another editorial by Glen Richardson in the March 1, 1982, issue, Richardson wrote, "Led by concerned cow/calf producers, the National Cattlemen's Association has been given until the summer of 1982 to review trading practices and internal market procedures of the Chicago Mercantile Exchange and develop methods to prevent market misuse and manipulation in cattle futures trading." NCA Stockmen who pushed the futures directive into a floor resolution at NCA's annual convention in San Antonio, Texas, say the emergency action by the CME on the November 1981 Feeder Contract implies that there exists an opportunity for traders to seriously threaten the liquidity of the cattle contract.

Other stockmen argue that cattle futures have not been reflective of fundamental factors in the cash market. On the other hand, those that favor futures trading point out that futures don't affect the supply of available cattle—only supply and demand consideration should influence a stockman's decision to sell cattle. The point these stockmen and feeders make is that the industry must recognize that the paper market is, many times, a market unto itself and that it does not necessarily relate to the cash market. The idea that the Chicago Mercantile Exchange has not been more responsive to the needs of the average stockman and feeder is serious enough. But the real serious issue is the validity of some cattlemen's contentions that they are being undercut in the market by a few big companies that feed cattle and trade commodities. Richardson continued that the average stockman wants such trading stopped, and if their charges are sustained, they are right in demanding it. They're also entitled to better treatment from the CME.

The NCA surveillance program may be the best method to give cattle futures trading at the CME a second chance to fulfill its public trust. We believe that futures trading and the CME should be given a fair hearing. There appears to be some movement towards cleaning up past problems within the futures trading industry. Did the hue and cry that arose from the grassroots of America's cattle industry cause this movement? I think it's rather presumptuous to suggest that cattlemen were wholly responsible, but I also think that without the impetus from the grassroots cattle farmers, changes would not have occurred as rapidly, nor would the change have considered the needs of the cow/calf and ranch producer.

THE NATIONAL FUTURES ASSOCIATION

The Commodity Futures Trading Commission went through a reauthorization process recently and some responsibilities were realigned. Specifically, the formation of the National Futures Association, I think, helped immensely in the effort to return credibility and usability to the futures industry---credibility that was desperately needed by stockmen. The Commodity Futures Trading Commission's reauthorization was completed in September of 1982. As part of the realignment of regulations within the commodities futures industry, the National Futures Association (NFA) was formed. The NFA was given the authority to regulate the futures industry, relieving the CFTC of this task. NFA is set up as an industry self-regulating organization, with the CFTC acting as an oversight group. It is our hope, which is echoed by many within the cattle industry, that the NFA can return confidence to commodity futures trading. But their task will not be easy, at least within the livestock industry. Indeed, it will be exceptionally difficult to convince rank and file stockmen that regulation of futures trading is sufficient to protect them from market manipulation.

Nonetheless, sufficient regulation to ensure the public that there are no hoodwinked deals going on in the futures business is one of the primary goals of NFA and one that must be quickly achieved. To paraphrase Mark Anthony--I can neither condemn future trading nor praise it. My objective has been to present some different views, relate the general pulse of the industry, and report what has been done to right past wrongs. The role of futures trading in the livestock industry, as our business progresses through the 1980s, is difficult to predict. Moreover, its acceptance by grassroots cattlemen is hard to foretell. It is, however, reasonable to assume that futures trading and feeder and fed cattle will continue. Commodity futures exchanges must realize that they have an obligation to serve not only trading members but also the livestock producer as well. Maybe they have, in the past, overlooked that obligation.

78

USING TECHNICAL PRICE ANALYSIS
METHODS TO AID IN HEDGING DECISIONS

James N. Trapp

Some cattle producers have never "hedged" cattle and never intend to. For others, hedging has become a way of life. To some the futures market and all activities associated with it are a form of gambling. But others regularly use hedging to lock in prices as a form of "price insurance" and risk management. This paper briefly discusses the functions of hedging, focusing on "technical analysis" methods that can be used to make hedging decisions. Technical analysis refers to an analysis of the market's action; it includes use of charts and graphs and various mathematical formulas. Using such methods, rules can be formulated to indicate when one should buy or sell futures contracts. Technical analysis methods have long been a tool of the speculator and testing of several technical methods indicates that they also can improve the success of a hedging program.

HEDGING FUNCTIONS

A straightforward hedge can be described as taking a position in the futures market opposite the one held in the cash market, i.e., a cattle feeder purchasing feeder cattle in January to be finished in June would sell a June contract at the time he purchased the cattle. Thus he would have, in effect, established a price and delivery date at the time he committed to put the cattle in the feedlot. In June, when the cattle were ready to be sold, the producer could either opt to deliver the cattle to fulfill the contract or "buy" a June contract. The latter option would cancel his obligation in the futures market and permit him to sell the cattle on the cash market at any delivery point and day he chose. This is generally the option taken in a hedging program. A similar example could be developed for feeder cattle beginning at the point when calves are born or weaned.

The basic purpose for hedging a commodity is to remove the risk of uncertain future prices. The removal of price risk by hedging or "locking in" a price for the commodity does not ensure higher profits, only more predictable profits. In fact, in a rising market a hedged position will generally lose money relative to an unhedged position. In a falling market, the opposite is true. Technical analysis is designed to take advantage of this point.

A producer is encountering price risk any time he owns a commodity and does not have it hedged, or any time he has a position in the futures market unrelated to his cash commodity holdings. The latter activity is commonly referred to as speculating while the former is simply unhedged production. However, both unhedged production and speculating are subject to the same type of price risk. The role of the futures market is in essence to allow the producer to rid himself of commodity price risk by letting the speculator assume the price risk. Hence it should be recognized that speculators are necessary for hedging to be conducted and that unhedged production involves speculation in the cash market.

USING TECHNICAL PRICE ANALYSIS

Technical analysis involves the study of the market's actions or patterns as opposed to studying the factors that affect the supply and demand for a commodity. The basic assumption of technical analysis is that by studying the statistics and chart patterns generated by the market, useful predictions of future market prices can be deduced. Numerous types of technical analysis procedures can be used, including price chart patterns, trend-following methods, or character-of-market indicators. This paper focuses on a trend-following technique referred to as "moving averages", which is simple but effective.

A moving average price is a progressive average in which the number of prices averaged remains the same, but a new price is added to the front of the series at periodic intervals (e.g., daily) as an old price is dropped from the end of the series. A 10-day moving average, for example, would always be the average of the prices observed in the most recent 10 days.

The moving-average technical strategy involves using two or more moving averages of different lengths. The strategy is based upon the fact that the two different moving averages when plotted will generate "crossing" actions as depicted in figure 1. These crossing actions are used as sell or buy signals. Intuitively one can ascertain that when prices are rising the shorter moving average, made up of more recent prices, will be above the longer moving average. Likewise when prices are falling, the shorter moving average will be below the longer moving average. Hence when the shorter moving average crosses the longer moving average from the top, a sell signal is generated. Likewise when the short average crosses the longer average from the bottom side, a rising market and buy signal are indicated.

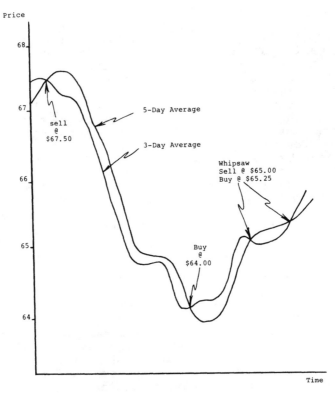

**Figure 1.   Illustration of Moving Averages Crossing Action
Signals to Place and Lift Hedges**

Several modifications to the preceding basic moving-
average strategy are commonly used.   The first modification
is to follow a "penetration rule" that requires the moving
averages to cross by at least a certain amount before the
crossing action is perceived as a valid buy or sell signal.
A second modification is to use three moving averages simul-
taneously--a short, medium, and long average.   A valid signal is not considered to be given unless the short and medi-
um-length averages cross the long average in sequence.   For
a sell signal to be generated, the short average must first
cross the long average, followed by the medium average and
in such a way that the short average has remained below the
medium average.   Still a third modification involves using a
"weighted" moving average for one or more of the averages.
A linear weighting procedure that places the largest weight
on the most recent price is usually used.   Table 1 below indicates the calculation procedure for a five-day, linearly
weighted, average price.   Once calculated, the weighted moving average is used the same as any other moving average.
In all cases the prices used are closing futures-market

TABLE 1. FIVE-DAY, LINEARLY WEIGHTED, AVERAGE PRICE
CALCULATIONS

Day	Price	Weight		Product
June 11	63.25	X	5	= 316.25
June 10	63.00	X	4	= 252.00
June 9	62.42	X	3	= 187.26
June 8	63.27	X	2	= 126.10
June 7	64.10	X	1	= 64.10

946.15

Five-day, linearly weighted, average price = 946.15/15 = 63.08

quotes and signaled trades are placed at the close of the market.

Execution of the signals generated can be done in two ways. The first (and easiest) involves placing a market order on the morning of the day after the closing price causes the moving averages to signal a buy or sell. This method requires only a daily calculation of the moving-average value and use of a basic buy or sell market order. The second method involves anticipating a price that would generate a signal to place a "sell-stop" or "buy-stop" order at this price. It is usually recommended that this order be placed to be acted upon only "at the closing price." Calculating, or anticipating, the price at which an action should be taken requires some algebra. An example of the calculations required for a 3-day, 6-day moving-average combination is worked out below.

Prices for the Last 6 Days

Date	Price
May 1	51.00
May 2	50.00
May 3	49.00
May 4	48.00
May 5	49.00
May 6	50.00
May 7	?

Step 1: Moving Averages for May 6th
3-Day Moving Average
$(48.00 + 49.00 + 50.00)/3 = 49.00$

6-Day Moving Average
$(51.00 + 50.00 + 49.00 + 48.00 + 49.00 + 50.00)/6 = 49.50$

Step 2: Moving Averages for May 7th
3-Day Moving Average
(49.00 + 50.00 + ?)/3 = 3DAVG

6-Day Moving Average
(50.00 + 49.00 + 48.00 + 49.00 + 50.00 +
?)/6 = 6DAVG

Step 3: Calculation of Buy-Signal Price for May
7th

$$3DAVG = 6DAVG$$
$$(99.00 + ?)/3 = (246.00 + ?)/6$$
$$33.00 + .33? = 41.00 + .167?$$
$$.163? = 8.00$$
$$? = 49.01$$

As can be observed from the price series, prices "bottomed"
on May 4th and began to turn up. A buy signal is about to
be given, if the market continues up. The May 6th moving
averages are still in a sell position since the 3-day average is below the 6-day average by $.50, i.e., $49.00 versus
$49.50. Step 2 shows the calculations that will be made for
the May 7th set of moving averages when the May 7th price is
known. The question is what price will cause the May 7th,
3-day average to equal or exceed the May 7th, 6-day average. Step 3 makes these calculations by solving for the unknown price when the two averages are equal to each other.
The solution is $49.01. Hence a buy-stop order at $49.02
should be entered for the close on May 7th. This order will
be executed if the closing price is at or above $49.02.

Countless combinations of moving-average lengths, penetration rules, and weighting schemes could be used. The
success of any moving-average strategy depends upon picking
the proper combination. The shorter the moving average used
the more sensitive it is to changing market prices and the
quicker it will signal an action. However, it is also more
prone to false signals that are quickly reversed in "whip-
saw" actions. Through time, and by use of computer-aided
testing, combinations of moving averages that are superior
to others have been found for various commodities.

USING MOVING AVERAGES IN A HEDGING PROGRAM

The traditional concept of hedging requires an immediate hedge on a commodity at the time it is acquired, which
is typically referred to as a "blind hedge" because it is
done with no consideration of the price or the profit implied. A second, and perhaps more common, approach to hedging is to hedge at a selected price based upon some predetermined selection criteria such as a 15% profit level.
This approach is often termed the "selective hedge"
or "hedging by objective." Once the selective hedge is
placed, it is held until the commodity is sold.

In using moving averages to assist in a hedging program, "multiple hedging" is done, whereby the hedge may be placed and removed several times as the name implies. Such hedging differs from traditional hedging in which the hedge is placed once and held until the commodity is sold.

The concept of multiple hedging requires that the cattle be hedged when there is significant risk of adverse downward price movement. When prices are expected to be moving up, cattle should remain unhedged or any hedges previously in place should be removed. This will allow the producer to achieve the benefits of a rising market instead of having a "locked in" price while prices are rising. In other words, price should be "locked-in" or insured only when there appears to be significant danger of adverse price changes.

The success of a multiple-hedging strategy depends upon being able to know when to place and lift hedges. The moving-average technique is used to do this. Hedges are placed when the technique generates a sell signal. They are removed when a buy signal is given. Note that if a producer were a hedger he would never establish or buy a long position in the cattle market. This would constitute speculation in the futures market. What is being done in the multiple-hedging program is to selectively determine when the producer wants to remain unhedged and, in effect, speculate in the cash market. As such the multiple-hedging program is a middle-of-the-road approach to price-risk exposure. The policy of never hedging cattle totally exposes the producer to market price changes or risks, but with the blind hedge the producer is totally unaffected by price changes. Thus the multiple-hedging strategy allows one to select or manage the amount of price risk wanted.

By using moving averages, subjectivity and emotion are removed from the multiple-hedging process. The moving-average procedure functions mechancially to indicate when to place and to lift hedges. All that is required is the obtaining of closing futures prices and a few minutes of calculations each day to determine at what level closing buy or sell orders should be placed. This is not to say that the strategy will work perfectly every time. A certain amount of intestinal fortitude is required to ojectively continue with the moving-average technique when it gives false signals. Such fortitude and a long-term commitment to the system are, however, required to give the system a chance to work.

HOW WELL DO MULTIPLE HEDGING PROGRAMS WORK?

Multiple hedging was developed because traditional "blind-hedging" programs did not work well. Observation and testing of blind-hedging programs typically indicated that they did stabilize profits, but they also lowered average profits. For many producers, the increased profit stability was not worth the reduction in average profits. Tables 2 and 3 demonstrate this trade-off for feeder cattle and live cattle.

Profit stability is defined in terms of the standard deviation of profits. A standard deviation is defined so that the range obtained when one standard deviation is added and then subtracted from the average contains 66% of the observed values.

The profits and loss levels reported in tables 2 and 3 are not of particular importance since they are dependent upon the accounting process used, i.e., commissions charged and production costs assumed. What is important is that the accounting was done consistently for all strategies so that the results are a fair comparison. In both the feeder-cattle and fat-cattle cases, the blind-hedge strategies showed the lowest average profit and the most stable profits (when a low standard deviation is taken as a measure of stability.) The multiple-hedging strategies using moving averages showed the highest average profit and greater profit instability than did the blind-hedge strategy, but it showed substantially less instability than did the no-hedge strategy.

The multiple-hedging strategy was able to raise profits per head because when a sell hedge is successfully placed, and then lifted, profits from the futures market are generated from selling high and buying back low. Losses also can be generated if false signals are given and hedges are placed in a rising market and then later lifted at higher prices. The moving-average technique is not sufficiently sophisticated to generate only profitable trades. In general, the technique will tend to make as many if not more unprofitable trades than profitable trades. The key to its success is that the profitable trades tend to generate large profits while the unprofitable ones are quickly terminated so that losses are small.

Table 4 gives an indication of the number of hedges placed and lifted over a 140-day feeding period, the number of profitable trades, and their average profitability net of trading commissions. As can be seen, profit per trade, on the average, is not large. The multiple-hedging strategy is not intended to make profits from futures trading, rather its purpose (as with any hedging program) is to reduce price risk and profit instability, with as few costs as possible. The fact that the strategy typically increases profits while helping to stabilize profits can be viewed as a fringe benefit.

For some producers, the fact that the moving-average technique can make profitable futures trades on the average over the long run suggests it should be used independent of ones cash position--i.e., for speculation. If this were done, results of most studies indicate profits could be made but the volatility of such profits would be quite high-- higher than those for unhedged cattle. Hence, risk, as measured by profit volatility, is increased by speculative

TABLE 2. AVERAGE PROFIT PER HEAD AND STANDARD DEVIATION
OF PROFITS FOR FIVE FEEDER CATTLE HEDGING
STRATEGIES (1972-1977)[a]

Strategy[b]	Average profit per head	Standard deviation of profit per head
1. No hedge	13.20	50.13
2. Blind hedge	4.67	13.92
3. 3-10[b]	21.64	20.76
4. 4w-5-10[b]	22.04	23.20
5. 8-4w(.05)[b]	21.67	16.63

[a]Source: John R. Franzmann. Moving Averages as an
Indicator of Price Direction in Hedging Applications (Iowa
State University, 1981).

[b]3-10 denotes a 3 day and 10 day set of moving average;
4w-5-10 denotes a 4-day linearly weighted, 5-day and 10-day
set of moving averages; 8-4w(.05) denotes an eight day and
4-day linearly weighted set of moving averages with a five
cent penetration rule.

TABLE 3. AVERAGE PROFITS PER HEAD AND STANDARD DEVIATION OF
PROFITS FOR THREE FAT-CATTLE-HEDGING STRATEGIES
(1972-1975)[a]

Strategy[b]	Average profit per head	Standard deviation of profit per head
1. No hedge	-19.65	77.54
2. Blind hedge	-28.88	56.39
3. 3-10[b]	10.41	60.79

[a]Source: Wayne D. Purcell. More Effective Approaches to
Hedging (Oklahoma State University, 1976).

[b]3-10 denotes a 3-day and 10-day moving average.

use of moving averages. Only through the use of moving averages as a multiple hedging tool is risk reduced.

TABLE 4. PROFIT NET OF COMMISSIONS FROM SELECTED MOVING AVERAGES FOR FAT CATTLE CONTRACTS, 1975-1979[a]

Length of moving average	Average net profit per short trade	Percent profitable trades	Average number of trades per pen fed
3-4-7w	$52.03	39.5	4.86
1-3-5w(.09)	96.43	47.7	1.93
3-4-6w	16.66	39.7	5.24
3-4-6(.09)	56.40	41.9	3.16
4w-5-15	8.77	37.2	2.74

[a]Source: Mike Shields. Simulated Multiple Hedging Programs Employing Optimized Moving Average Combinations for Use in Continously Operated Feedlots.

AN EXAMPLE OF MULTIPLE HEDGING

A concept of how the multiple-hedging strategy works can be obtained by following an example case. The case considered here is purchasing feeder cattle on November 1, 1981, and selling them on May 1, 1982. A multiple-hedging strategy is followed using the May 1982 feeder cattle contract. In this case a 3-4-6 (.07) moving average is used consisting of a 3-day, 4-day, and 6-day moving average and a 7¢ penetration rule between the 4-day and 6-day moving average. The results of the strategy are reported in table 5 and illustrated in figure 2.

On the purchase date, the moving-average set indicated the cattle should not be hedged. However, on November 17 a sell signal was given and a hedge placed at the closing price for the day of $67.20. The placing of this hedge is indicated in figure 2 by the left most downward-pointing arrow. The market proceeded to decline until late in the month and then rallied on November 30th. The moving averages indicated the hedge should be lifted on December 1. This was done at the closing price of $65.45, which yielded a gross return of $735 on the trade from November 17 to December 1.

In hindsight, it can be seen that the hedge should not have been lifted, because the rally proved to be short-lived. But this was not evident on December 1. The moving averages signaled the hedge should be replaced on December 7th at $62.50, some $3 below the point it was lifted. This hedge proved to be a very protective action because the mar-

Figure 2. Multiple Hedging Points Signaled for the May 82 Feeder Cattle Contract by a 3-4-6(.07) Day Moving Average, November 1, 1981 to May 1, 1982.

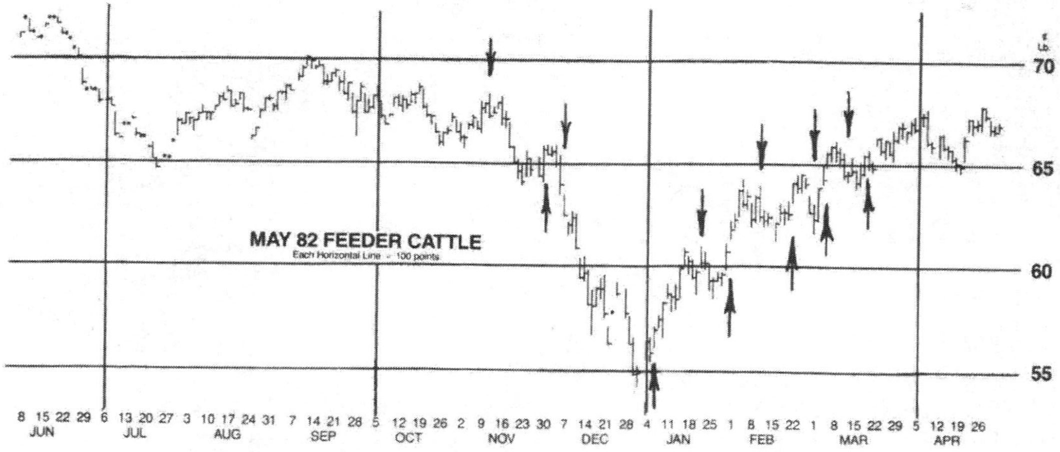

MAY 82 FEEDER CATTLE
Each Horizontal Line = 100 points

ket subsequently declined to below $55. The December 7th hedge was eventually lifted at $55.95 on January 6th, grossing $2331 of futures market profit. In hindsight, the turning point called on January 6th could be seen to be very near the bottom of the market for the duration of time the cattle were owned. It probably would have been best, if this had been known, to have left the cattle unhedged until May 1 when they were sold. However, several reversals were seen in the futures market between January 6th and May 1. On each of these reversals, the moving average signaled that hedging protection should be taken. In each case, the reversal proved to be temporary and the hedge was lifted within a week or so--but the temporariness of the reversals was not known at the time. Hence, hedging protection was taken and losses were encountered on the futures transactions.

TABLE 5. SUMMARY OF MULTIPLE HEDGES SIGNALED BY A
3-4-6(.07) DAY MOVING AVERAGE FOR THE MAY 82
FEEDER CATTLE CONTRACT[a]

Hedge Placed Date	Price	Hedge lifted Date	Price	Gross profit	Cumulative gross profit
11-17-81	$67.20	12-1-81	$65.45	$735.00	$735.00
12-7-81	62.50	1-6-82	56.95	2331.00	3066.00
1-21-81	60.20	2-1-82	61.82	-680.40	2385.60
2-10-82	62.45	2-22-82	64.10	-693.00	1592.60
3-1-82	62.37	3-4-82	65.55	-1335.60	357.00
3-11-82	64.52	3-18-82	65.05	-222.60	134.40

[a]3-4-6(.07) refers to a 3-day, 4-day, and 6-day set of three moving averages with a 7¢ penetration rule between the 4-day and 6-day moving average.

In summary, the first two downturns signaled developed into prolonged trends for which hedges were needed, whereas the last four downturns were only temporary and no hedging protection was needed. At the time, however, this could not have been forseen. Over the period for which the cattle were owned, seven hedges were placed and lifted that resulted in a gross futures market profit of $134.40. If commissions of $70 per contract traded are charged, the net return from the futures market in this case will be a negative $286. This is somewhat atypical of average results; on average, the multiple-hedging strategies for feeder cattle and fat cattle generated a positive rate of return. The case presented here also is slightly atypical in that an above average number of trades were generated.

SUMMARY

Test applications of multiple-hedging systems using moving-average technical indicators to determine when to place and lift hedges have indicated the system can raise average profit levels and reduce the volatility of profits received over time. Blind-hedging strategies that arbitrarily hedge cattle when they are acquired result in more stable profits than multiple-hedging systems, but have the negative effect of lowering average profits relative to unhedged operations. For producers who desire a moderate amount of risk reduction, relative to operating with unhedged production, the multiple-hedging system offers an attractive alternative.

The use of moving averages to conduct the multiple-hedging program makes the execution of the hedging activities systematic and nonsubjective. The producer need not closely follow all the market news and go through long agonizing subjective decisions. A few calculations a day and half-a-dozen or so phone calls to the broker per month are all that are typically required to conduct multiple-hedging activities.

The multiple-hedging program is not for everyone. Producers who need a large degree of price risk protection should consider other more conservative hedging programs. Individuals who psychologically cannot be comfortable with encountering losses (or profits) on the futures market should avoid the program. Those who believe they do not have the discipline to calculate the required averages daily and follow their signals religiously should also give careful consideration to using the multiple-hedging approach. Those hoping to "get rich quick" through the multiple-hedging approach also will be disappointed. The program is recommended to producers who wish to eliminate a considerable amount of price risk from their operations, without reducing their long-run average profits.

REFERENCES

Franzmann, J. R. and J. D. Lehenbauer. 1979. Hedging feeder cattle with the aid of moving averages. Oklahoma Agricultural Experiment Station Bulletin #746. Oklahoma State University.

Franzmann, J. R. and M. E. Shields. 1981. Multiple hedging slaughter cattle using moving averages. Oklahoma Agricultural Experiment Station Bulletin #753. Oklahoma State University.

Franzmann, J. R. and M. E. Shields. 1981. Long-hedging feeder cattle with the aid of moving averages. Oklahoma Agricultural Experiment Station Bulletin #754. Oklahoma State University.

Franzmann, J. R. and M. E. Shields. 1981. Managing feedlot price risks: fed cattle, feeder cattle, and corn. Oklahoma State University Experiment Station Bulletin #759. Oklahoma State University.

Franzmann, J. R. 1981. Moving averages as an indicator of direction in hedging applications. Proceedings, applied Commodity Price Applications and Forecasting Conference. Iowa State University.

Lehenbauer, J. D. 1978. Simulation of short and long feeder cattle hedging strategies and technical price analysis of the feeder cattle futures market. Unpublished M.S. Thesis, Oklahoma State University.

Purcell, W. D. 1976. More effective approaches to hedging. Proceedings, Oklahoma's 12th Annual Cattle Feeder's Seminar. Oklahoma State University.

Rife, D. A. 1976. A simulation analysis of the financial effects of alternative hedging strategies for cattle feeders. Unpublished M.S. Thesis, Oklahoma State University.

Shields, M. E. 1980. Simulated multiple hedging programs employing optimized moving average combinations for use by continously operated feedlots. Unpublished M.S. Thesis, Oklahoma State University.

Teweles, R., Harlow and Stone. 1977. The Commodity futures game: Who wins? Who loses? Why? McGraw Hill.

79

INFORMATION AND DECISION MAKING

Topper Thorpe

> "The cost of knowing is nothing
> compared to the cost of not knowing."

Times have changed. If the conclusion is true that the meat industry is now mature, information will become more important than ever before.

In the past, people in the cattle industry, and to some extent those in the pork industry, have done well financially during the boom periods because of the thrashing they took during the busts. Maturity of an industry suggests that the boom and bust periods no longer exist, or at least not in as big a way as they have in the past. This means that livestock producers will have to be more concerned with maintaining a consistent level of profitability--probably much smaller than during the boom periods, but larger than during the busts. To maintain profitability, cattlemen will have to make sound plans and decisions. And that will require the best possible market information.

The decisions that you and I and others in the industry will be making will be--and should be--based on economics--making use of the information that we have generated on our own operations. We must know where and at what prices we can market our product. We must know just what it costs to produce a calf or a hog. Then we will add information from outside sources--from groups like Cattle-Fax, which I represent. Cattle-Fax devotes all of its efforts to collecting, analyzing information and data, and preparing projections for members' use in making marketing and management decisions.

We all seek help in the form of information to make better decisions. It may come from a conversation with an associate or other knowledgeable person; or in written form from a newspaper, newsletter, or public or private analysis service. Some information is obtained by listening to the radio or TV. Some comes from seminars, short courses, or similar programs.

Decisionmakers generally use information from different time periods. Historical data is useful in detecting previous trends, responses to given changes--and it is useful as a basis for looking ahead. Current information is helpful in establishing a "bench mark" from which to base projections. And, of course, projections of the future--which incorporate the historical and current information--are widely used by decision makers.

With the explosion in the amount of information available, we have to do some sorting. One basis for sorting is to review the characteristics of information we use in making decisions.

TIMELY INFORMATION

Normally, week-old newspapers, year-old research reports, and market quotes that are several hours old are useless to today's decisionmaker. With advances in electronics, information is available to more people more rapidly than ever before. Consider the tick-by-tick commodity quotes that now appear in offices and homes all across the nation, nearly instantaneously relaying changes in the market to those actively involved in the business.

ACCURATE INFORMATION

All of us who collect and offer information seek to have perfect accuracy, and users want basically the same thing. This remains an elusive goal--not as it relates to the current situation but as it relates to future projections. Nonetheless, it is important that information users be able to count on the information being relatively accurate and unbiased.

USABLE INFORMATION

Information must be clear, concise, and presented in a form that the recipient can quickly and easily use to make a decision. The tendency is to add so many "qualifiers" to analyses that a clear conclusion is never reached, limiting its value in decision making.

ALL-ENCOMPASSING INFORMATION

Information should be all-encompassing in the sense that it takes into account the multitude of factors that influence a given price trend or market. We are well beyond the point where we can ignore what happens in foreign countries or at the other end of our continents, for all these factors influence our situation and the decisions we must make. Ideally, the information we use considers all factors but sorts out and isolates the ones that are really important.

We who use information to make decisions come in several categories, which makes it difficult for one information group to fit the needs of all of us. A lot of us are looking for someone who will tell us what to do. Then, if things don't turn out right, we have someone to blame rather than face the real decisionmaker who appears in the mirror.

Some of us go to the other extreme. We want all of the facts, but no analysis, no interpretation, no insight into the strengths and weaknesses of the data or the analysis, to make our decision.

Then there are those of us who want to see the facts and figures and also want to review the analyses and conclusions resulting from the data. We may then accept, reject, or modify the conclusions. In addition, we may seek several opinions to compare analyses and projections.

In many cases, we depend entirely too much on information of the type and from sources I have described. Actually, when evaluating options and alternatives, no information is more important than records on our own business. Most cattle producers, and many other livestock and poultry producers, don't know their real costs of production. They will pay hundreds or thousands of dollars for reams of information on which to base decisions but the information does not fit their operation because of its unique characteristics. Then they wonder why the decision was wrong.

There are many today who say there is too much information available and that the wider availability of information has wrecked the commodity business system. Perhaps greater availability of information has wrecked the system for the relatively few operators who had good information and could operate profitably within the wide market swings because they were a step ahead. But now that information is much more widely available, and those relatively few operators no longer have as great an advantage over others.

Computers and other electronic devices have added a new dimension to the availability of information. A major job now is to sort out the useful information and ignore the rest. We have moved quickly from a period of limited information used by a few to a period of unlimited information available to large numbers of persons. Those who ignore the greater availability of information do so at their peril. If one is to make correct decisions more than 50 percent of the time, he simply must obtain and use available economic and market information.

The current information explosion will require quicker decisions, about more complex matters.

Thus, we say:

"The cost of knowing is nothing
compared to the cost of not knowing."

80

WILL THE PACKER BUY YOUR CATTLE?

Don Williams

With the consumers' greater concern for the safety of their food supply, the livestock industry has high stakes resting on the consumer public's perception of beef as a safe, healthy, and nutritious source of protein. We are fortunate that beef is the preferred meat in the U.S. and that it continues to furnish the highest volume of sales in this nation's supermarkets when compared to all other items. With the stakes so high, it behooves each of us to use extra diligence to keep the American public confident of their domestic beef supply. We can then let the Australians convince us that imported products contain no horsemeat or kangaroo meat. Let us take a few minutes to consider certain areas in which errors can give unfavorable press as to the wholesomeness of beef.

ANTIBIOTICS

Antibiotic treatment has become an important and necessary component of all beef-production operations. With the slim margins and high production costs, a producer must make every effort to keep each animal healthy and productive and cut the death loss to a minimum. This has created two problems: 1) to salvage something from diseased animals, some producers do not follow the labels on the antibiotic bottles that state the time an animal must be held before slaughter after being administered a specific antibiotic; 2) the reliance on antibiotics has resulted in many producers rushing out to give their animal "a shot" and forgetting other good management practices, i.e., adequate shelter, good palatable feed, good fresh water, etc.

Scientists have developed very critical tests to detect minute residues of antibiotics and we must be sure that none of the animals that we sell are detected as having antibiotic residues. The reason for the concern over antibiotic residues is the allegation that consumption of small amounts of antibiotics over time could possible cause the bacteria that are normally in your and my digestive tract to become resistant to that specific antibiotic. Should they become antibiotic resistant, then it is remotely possible for these normal bacteria to transfer this resistance to disease-causing bacteria that may enter the body. You and I may argue about the possibility of this sequence of events occurring, but that is not the point. The point is that the regulations are already in effect that prohibit antibiotic residues in meat, and there are many hungry reporters who would write an article very derogatory to the beef industry if they thought they could make the front page or the evening news. The few dollars a person gets for an animal is not worth the risks to the industry.

INSECTICIDE AND CHEMICAL RESIDUES

With the proliferation of new chemicals in our world, a legitimate concern has developed for the possible introduction of these chemicals into the food chain. This contamination can occur from (1) the failure to observe clearance times for chemicals used in the production practices of beef cattle and (2) from chemicals consumed by animals ingesting contaminated feedstuffs. Some of the agents that have caused problems in the past few years are:

- PCB or polychlorinated biphenyl compounds that have been used in electrical transformers and are known carcinogens, or cancer-producing agents. A leak in a transformer at a feed company resulted in the contamination of a large number of cattle in the upper midwest a few years ago. Many of these animals had to be destroyed. The EPA has held hearings in 1982 to develop improved regulations for the use of PCB in transformers and capacitors to ensure that there will be no further contamination of animal feed from this source.
- Aldrin and mercurial seed treatments. Aldrin is an insecticide used for treatment of seed grain and was the issue in a large lawsuit in the midwest when some grain treated in an elevator was accidentally sold to a feedlot, resulting in the contamination of the animals being fed. The various compounds of mercury that had been used to treat seed grain have been involved in several accidents wherever the treated grain was fed to livestock, which resulted in either the death of the livestock or the contamination of the meat, making it unacceptable for human consumption.
- D.E.S. or diethylstilbestrol is a synthetic female hormone that increases gain and feed efficiency in cattle. It was classified as a carcinogen and its use was permitted for several years provided proper withdrawal times were followed to allow its clearance from the animal's

body. Public pressure finally resulted in its ban from use in animals. The beef industry received another black mark when numerous producers continued to use it after the ban. The beef industry does not need this type of publicity.

- Many of our present drugs used to treat internal and external parasites are labeled with withdrawal times required before the treated animals may be used for food. To date, we have not had any national publicity due to the misuse of these products, but the danger is there. As in the case of antibiotics, I cannot stress too strongly, READ THE LABEL!

UNACCEPTABLE PACKER CONDITIONS

Up to this point in our discussion, our subjects have been those items that might result in the contamination of beef and/or in the packer not taking your cattle because of contamination. I hope that I have impressed upon you the danger and need for a constant vigil to prevent these occurrences. However, there are some conditions that may occur that would make your cattle unacceptable to the packer. Let us consider some of these problems.

"Measles". "Measles" is the common name used to refer to the infested beef carcasses at the intermediate stage of tapeworm of humans, Taenia saginata. The life cycle of this tapeworm includes the passage of tapeworm eggs in the excreta contaminating the feedstuff of cattle. The cattle ingest the contaminated feed and the tapeworm develops to the larval stage and develops a cyst in the muscles of the animals that have ingested it. The cysts are most numerous in cheek and heart muscles. A person who eats the infected meat then develops tapeworms. This tapeworm is quite rare in persons who have lived their lives in the U.S. because our excellent meat inspection services have detected the cysts at slaughter. However, the incidence of this tapeworm is quite high in Mexico and in the immigrants from Mexico who work in our cattle production units of the Southwest. Though the incidence is not high, the danger is great enough that most feedlots in the Southwest carry insurance to indemnify the packer should cattle he buys from the feedlot have "measles." Contamination of feedstuff may be prevented by furnishing adequate toilet facilities to Mexican immigrants. The worst problem that I am aware of in this regard occurred when a discharged employee maliciously contaminated the corn silage of a commercial feedlot with the excreta of Mexican immigrants. Over 30,000 cattle were involved in this debacle.

Grubs. Grubs are the larvae of the heel fly. These larvae hatch from the eggs that are laid by the fly on the hair of the legs. After hatching, the larvae burrow into the skin and migrate through the body for 5 to 6 months before reaching the area under the skin of the back. There they form a cyst that has a hole in the skin so that the larvae can breathe. After 40 to 60 days, the larvae emerge, drop to the ground, and mature to the adult fly. The damage to the carcass, and thus the loss to the packer, is primarily in the back. The larvae secrete enzymes that irritate the tissue under the skin causing this tissue to become watery and discolored. Occasionally the holes in the skin become infected and can lead to abcesses. All of this damage must be trimmed away under the direction of the meat inspection service so that only good wholesome meat is left. However, the mutilation of the carcass from this trimming can result in discounts of $5.00 to $50.00 per carcass. With the small margins that packers operate under, they cannot accept this type of discount. Fortunately, there are several products on the market that do an acceptable job of controlling cattle grubs and packers have come to expect the cattle that they buy to be grub-free. Consequently, if your program does not cover grub treatment, you may find that the packer will decide to buy your neighbor's cattle rather than yours.

"Dark Cutters". "Dark Cutters" is the term used to designate those carcasses from fed animals that do not have a good, bright, cherry-red color. To understand the cause of dark cutters, we need to understand a little about the biochemistry of the muscle cell. Energy is stored in the muscle cell as a carbohydrate called glycogen. Under normal feedlot conditions, the muscle cells of cattle have a high glycogen content. After slaughter the glycogen is converted by enzymes in the muscle cell to lactic acid, and the lactic acid lowers the pH of tissue, acting as a preservative. The low pH maintains the pigment of the muscle, myoglobin, in a chemical state so that when it is exposed to atmospheric oxygen, the myoglobin (and thus the muscle) develops the bright cherry-red color that the consumer associates with fresh desirable beef.

If the animal expends energy just prior to slaughter, the reserve of glycogen is depleted and enzymes do not produce sufficient lactic acid to adequately stabilize the myoglobin. When beef from these animals is exposed to oxygen, the myoglobin does not develop the cherry-red color, but rather it is a dark reddish-brown and is referred to as a "dark cutter." There is nothing unwholesome about this carcass and beef from it is just as nutritious as from a normal appearing carcass. This meat can be utilized in areas in which the meat is sold cooked rather than fresh, but since it will not sell well on a retail meat counter it is discounted throughout the beef industry. Thus, when the packer ends up with "dark cutters" in his cooler, he is upset and

wants to be sure that it doesn't happen again. It then follows that if you are selling fed cattle to a packer, it is
wise to prevent the dark cutters if at all possible. Some
of the causes of dark cutters are:

- Mixing strange cattle together at least 12 hours
  before slaughter. The introduction of new cattle into the social order of an established pen
  of cattle will cause sufficient fighting and
  physical exertion to produce "dark cutters."
  Though this can occur with steers and heifers,
  it is a major problem with bulls. There are so
  many things that can increase the physical
  activity in a set of young bulls that we prefer
  to arrange the shipment of them in a manner so
  that they can be slaughtered in 6 or 8 hours
  after they leave their home pens. Penning bulls
  next to heifers at the packing plant overnight
  will increase the incidence of "dark cutters."
  Placing steers next to a pen of fighting bulls
  will cause enough excitement in the steers to
  produce "dark cutters."
- Feeding a product to heifers in the feedlot that
  prevents estrus and increases gain. When this
  product was first placed on the market, some
  feeders stopped feeding it 4 or 5 days before
  slaughter and many of the heifers exhibited estrus within 18 to 48 hours. With a high percentage of the heifers in estrus at one time, there
  was a tremendous amount of riding and physical
  activity. This depleted the heifers muscle glycogen and caused a very high incidence of "dark
  cutters." The product was said to cause "dark
  cutters" when in reality it was the management
  of the system. Many heifers are finished every
  day on this product, and if it is maintained in
  the feed until just before slaughter, there are
  no problems.
- Chilling that causes cattle to shiver and use
  their muscle glycogen to stay warm. Packers
  will tell you that the highest incidence of dark
  cutters occur in the fall as cattle are becoming
  acclimated to winter weather. After cattle become acclimated, the incidence goes down. In
  March 1982, we shipped a pen of cattle that had
  over 50% "dark cutters" upon slaughter. The
  cattle were shipped 24 hours after we had had a
  freezing rain and light snow; this change of
  weather blew in after 10 unseasonably warm days
  during the latter part of February when temperatures had reached 90°. The change in temperature, the moisture, and the accompanying wind
  produced sufficient chill that the body's metabolism was increased to maintain body heat.
  This increased metabolism burned up a large part

of the muscle glycogen and led to dark-cutting carcasses.

In summary, any management procedure that will reduce physical exertion, anxiety, or chilling will reduce "dark cutters." There is also evidence to support the contention that keeping the animals on a palatable high carbohydrate ration as near to the time slaughtered as possible also will help.

81

WHAT IS A COW WORTH?
IMPROVING CULLING AND REPLACEMENT
DECISIONS USING A COMPUTER MODEL

James N. Trapp

What is a cow worth? The typical economist's answer to that questions would be "It depends...and then again it's always changing." You could never get an economist to say something as specific or simple as four-hundred-seventeen dollars and seven cents. The reason the economist will offer no simple answer is that he perceives this question as basically asking two things. First, how productive is that cow? And second, what value is the market going to place upon what that cow produces? These are not simple questions. But the producer who knows the answer to them can most certainly increase the profitability, i.e., productivity, of his herd.

A key to maintaining a productive cow herd, which is synonymous to having a profitable herd, is to know which cows can earn a positive return. Unfortunately, at times, this boils down to none of them. Of course, a producer needs to strive to practice sound management that results in good calving rates and quality calves. But beyond this, there are some basic decisions to be made using organized rationale as to which cows to keep in the herd and which to add to the herd. This paper will present some basic concepts in answering the question "How much is a cow worth?" and then proceed to use the concepts to develop some general guidelines to culling and replacement strategies for coping with cyclical cattle prices.

EXPECTED COW PRODUCTIVITY BY AGE

The first part of the answer to the question of what a cow is worth depends upon her physical productivity. Although this will vary from herd to herd, a fairly regular pattern of productivity by age exists. Table 1 presents a typical set of productivity rates by age for English breeds of cows. By observing columns 3 and 4 of the table for weaning weights and calving rates by cow age, the biologically most productive years of a cow's life can be ascertained.

TABLE 1:   AGE RELATED PHYSICAL CHANGES IN COW PRODUCTIVITY

Cow age	Calving year	Average[1] weaning weight	Birth[2] rate %	Death rate %	Cow weight	Steer[4] weaning weight	Heifer[4] weaning weight
(1)	(2)	(3)	(4)	(5)	(6)	(7)	(8)
2	1	425	85.5	2.25	821	439.45	410.55
3	2	444	89.0	2.25	905	459.096	428.904
4	3	465	92.7	2.30	986	480.81	449.19
5	4	488	94.5	2.35	1041	504.592	471.408
6	5	488	94.3	2.45	1100	504.592	471.408
7	6	488	93.0	2.8	1100	504.592	471.408
8	7	488	90.8	3.25	1100	504.592	471.408
9	8	488	87.0	3.7	1100	504.592	471.408
10	9	488	82.0	4.35	1100	504.592	471.408
11	10	465	76.6	5.8	1100	480.81	449.19
12	11	465	70.0	6.3	1075	480.81	449.19
13	12	465	63.6	6.5	1050*	480.81	449.19
14	13	465	56.2*	6.6*	1025*	480.81	449.19
15	14	465*	45.0*	6.6*	1000*	480.81*	449.19*
16	15	465*	41.0*	6.6*	1000*	480.81*	449.19*

*Values for these characteristics could not be found in the literature. Estimted values were provided based upon extrapolation of the preceding series and the author's knowledge of the cattle industry.

[1]Bently, Waters, and Shumway
[2]Rogers, 1971.
[3]Kay and Rister.
[4]Average Weaning Weight from Earnest, Shumway and Waters, and relative breakdown of heifer and steer weaning weights from Rogers (1971).

The biological productivity of different-aged cows, as presented in table 1, can be summarized by multiplying the expected weaning weight and calving rate for each age level. This will indicate the expected pounds of calves per cow according to her age. Figure 1 depicts the results of doing this. The figure indicates that cows typically reach their productive peak at 5 years and continue to maintain a high level of productivity until approximately 9 or 10 years old.

Assuming that cows cost approximately the same amount to maintain, irrespective of age, it is obvious that, over a 1-year period, a 5-year-old cow would earn the greatest returns. But her value? The value of a cow is more nearly reflected by the total productivity of her remaining

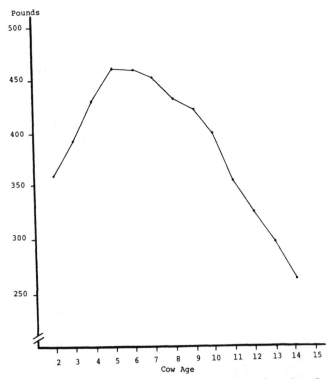

Figure 1. Expected Pounds of Calf Production by Cow Age

expected life span. For example, a 10 year-old-cow in her next calving season is expected to produce more pounds of calf then a 2- or 3-year-old cow. However the remaining life-time productivity of a 10-year-old cow is much less than that of a 2- or 3-year-old cow, hence the 2 or 3-year-old cow is probably worth more than a 10 year old. Of course, a cow also reaches a point in her productive life cycle when she is worth more as a slaughter animal than as a calf producer. In past studies this age has been estimated to be about 8 or 9 years of age. In principle, the age at which a cow should be culled will change depending upon the relative prices of calves, price for slaughter cows, and cost of maintaining a cow. In fact, the value of any cow is dependent not only on her current and future remaining productivity, but also upon the market value of that production. Which brings us to the next part of the question of what a cow is worth: how much will her calves be worth?

EXPECTED FEEDER CALF PRICES AND COW VALUES

The preceding discussion indicates that age is the primary factor affecting a cows productivity and hence her value. But the price of the calves she produces will also greatly influence her profitability and hence her worth. More important the expected future pattern of feeder cattle prices will not only influence the net worth of any one cow, but the relative worth of cows of different ages. In the discussion that follows, a series of calculations will be made to show that in a period of generally rising prices 4- and 5-year-old cows are to be most highly valued, and in a period of falling prices, younger cows are likely to be worth more in the long run.

Calculating the expected value of a cow involves determining the expected revenue flow from her future calf sales and her expected value if sold. From these revenues the cost of maintaining the cow and calf must be subtracted to get a net value. A cow's calculated, expected net worth is dependent upon her current age and how many years one plans to keep her. For example, the exercise below calculates the expected net worth of a 5-year-old cow kept for 1 year and then sold along with her calf. It is assumed the calf will sell for $.70 lb and the cow for $.40 lb. Expected weaning weights, calving success, the cow's weight, and cow death probability are all taken from table 1.

Calf Revenue
(Weaning wt x calving rate) x price
(488 x .945) x $.70   = $341.60

Cow Sales Revenue
(Cow wt x probability of living) x price
(1041 x (1.0 - .0235) x $.40   = $406.61

Maintenance Cost
Maintenance cost x (1.0 - death probability x
.5 - aborts x .25)
$350 x (1.0 - .0235 x .5 - .055 x .25)   = $-341.08

Net Worth $407.13

According to the above calculations a 5-year-old cow kept for 1 year, which produced one calf and was sold, has an "expected" net present value or net worth of $407.13. Stated alternatively a producer could affort to pay up to $407.13 to purchase a 5-year-old cow, keep her for a year, and then sell her and any calf she produced. The probability that she would die or fail to produce a calf has been injected into the determination of her value. Hence the value of $407.13 is the average value of what might be expected for 100 cows, i.e., 94.5% of them would have calves and 2.35% of them would die, etc.

Next, consider the alternative of buying a 5-year-old cow and keeping her for 2 years. Assume during the second year any calves produced could be sold for $.75 lb and the cow could be sold for $.42 lb.

Value Of A 5-Year-Old Cow Kept For 2 Years

Year 1

Calf Revenue
(488 x .945) x $.70   = $341.60
Maintenance Cost
$350 x (1.0 - .0235 x .5 - .055 x .25)   = -$341.08
Cow Revenue - Early Culls
(Cow wt x price) x Calving failures - Death rate
(1041 x $.40) x .055 - .0235   = $13.12

Year 2

Calf Revenue
Weaning wt x calving wt) x price / discount rate
(488 x .943) x $.75 / 1.15
                 $345. / 1.15
                 $300.12 x % cows remaining
                 $300.12 x .9215   = $276.56
Cow Sales Revenue
(Cow wt x probability of living) x price / discount rate
(1100 x (1.0 - .0245) x $.42 / 1.15
                 $450.68 / 1.15
                 $391.90 x % cows remaining
                 $391.00 x .9215   = $361.13
Maintenance Cost
Cost x (1.0 - death rate x .5 - aborts x .25) /
     discount rate
$375 x (1.0 - (.0245 x .5) - (.057 x .25) / 1.15
                 $365.06 / 1.15
                 $317.45 x % cows remaining
                 $317.45 x .9215   = $-292.53

Net Worth $358.80

The calculations for the second year are similar to those for the first year but with a few added considerations. It is assumed that any cows that fail to calve will be culled. Hence considering deaths and culls only 92.15% of the original 4-year-old cows will typically remain in the herd at the beginning of the second year. Some revenue is realized at the end of the first year from the sale of the 5.5% of the cows that failed to calve and did not die. All calculations in the second year are then based upon 92.15% of the total original number of cows. The result of this calculation indicates that even though calf and slaughter cow prices were assumed to rise during the second year, it appears more profitable to hold the cow for only 1 year. The reason for this is revealed by examination of the revenue calculations for the second year. With regard to calf revenue, the calving rate changes little for a 6-year-old cow versus a 5-year-old. The major cause of reduction in calf revenue comes from the fact that revenues earned 2 years into the future are discounted by 15%. This is done to reflect the impact of inflation. With a 15% rate of inflation, the value of a dollar earned 2 years from today is only $1.00/1.15 or $.87. Thus to make the earnings in the second year comparable to those in the first, second year earnings and costs are discounted. The resulting net worth figure is then in terms of current dollars and is referred to as a "net present value."

Obviously these types of calculations can become very cumbersome if very many ages or years of keeping a cow are considered. The general formula for making such calculations is given below.

$$NPV_s = \sum_{t=0}^{c}(1+r)^{-t}R_{s+t}(t)+(1+r)^{-c}M_{s+c}(c)-\sum_{t=0}^{c}(1+r)^{-t}+C_{s+t}(t)$$

where

NPC_s — net present value (worth) of a cow $\underline{s}$ years of age kept for $\underline{c}$ additional years.

$R_{s+t}(t)$ Calf revenue flow from a cow $\underline{s}$ + $\underline{t}$ years of age in period $\underline{t}$.

M_{s+c} The slaughter market value of a cow $\underline{s}$ + $\underline{c}$ years of age in period $\underline{c}$.

$C_{s+t}(t)$ The cost of maintaining a cow-calf unit for a cow of age $\underline{s}$ + $\underline{t}$ in period $\underline{c}$.

$\underline{r}$ Discount or interest rate.

$\underline{c}$ Number of future years the cow will be kept.

$\underline{s}$ Current age of the cow.

$\underline{t}$ Number of years into the future being considered.

This formula can be entered into a computer and calculated readily provided future prices, production costs, and the productivity data given in table 1 are made available. The

748

key question really is what prices and costs should be entered as expected prices and costs. In the examples to be
presented here expected prices have not been used. Rather
the typical feeder cattle and slaughter price patterns observed over the past 30 years have been used. Also the
general upward sloping cost trend observed over the past 30
years has been used. To the extent one believes that past
price patterns and cost trends will continue, these
generalized historical trends and patterns of the past can
be taken as expectations of the future. The patterns found
over the past 30 years for feeder cattle prices were in
essence a 12 year cyclical pattern trending upward at
approximately the same rate as costs of production were
rising. Feeder cattle prices were observed to typically
vary from about 16% below the trended average to 16% above
the trended average. In general, six years of "falling"
prices were observed to be followed by six years of "rising"
prices. Falling and rising are used here (and throughout
the rest of the discussion) to mean relative to the general
trend as opposed to absolute changes. Figure 2 shows feeder
cattle prices for the past 30 years, and the general trend
and cycle pattern deduced.

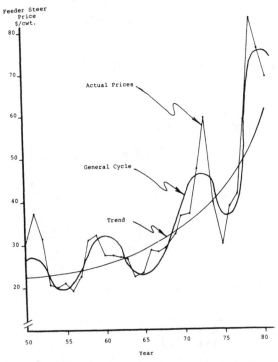

Figure 2. Actual and Generalized Trend and Cycle Patterns
for Feeder Cattle Prices, 1950-80

A computer has been used to calculate the expected value of cows of different ages kept for different lengths of time at four points in the feeder cattle price cycle. The points considered were the peak of the cycle, halfway through the down-phase of the cycle, at the bottom of the cycle and halfway through the up-phase of the cycle. The values found for these four cases are reported in tables 2 through 5.

The values reported in tables 2 through 5 are calculated as previously discussed. For example, in table 2 the value under the 11 in the top row represents the expected "net present value" of costs and revenue flows for a 2-year-old cow kept 11 years. All future net revenue flows have been discounted for inflation so that the values are in terms of 1982 dollars. Also calving rates, death rates, etc., as reported in table 1, have been incorporated. Table 2 indicates that at the peak of the cycle the most valuable cow is a 2-year-old planned to be kept for 11 or 12 years. If a slaughter cow price of $.50 lb is assumed for this point in the cycle, most cows in the herd will have an approximate slaughter value of $550. From the table it is noted that no cow over 6 years of age has the potential to be worth more than $550 as a breeding animal and hence should be culled. Younger cows are in general the most valuable at the peak of the price cycle. This is because they have the potential of still being productive when prices start to rise again in the next cycle up-phase—some 6 years into the future. Hence they not only can produce nearly half their calves in a rising market but also will be productive long enough to be culled and sold when market prices are relatively high. It should be noted, however, that if 2-year-old heifers are bringing $.70 on the market, their slaughter value is approximately $595, which is not far from their breeding herd value of $642.

Table 3 is similar to table 2, except it has been calculated for a situation midway through the down-phase of the cattle price cycle when deflated prices are expected to decline for 3 more years and then rise for about 6 years. The most valuable cow in the herd is now a 5-year-old cow planned to be retained for another 7 years. The culling age has now become 12 years of age. As will be shown later, once a cow has been placed into the herd and a declining market develops, the best strategy appears to be to keep the cow to a relatively old age.

Table 4 depicts cow values at the bottom of the price cycle. The most valuable cow is now a 4-year-old planned to be retained for 6 years. the culling age is 10. Note that in general the breeding herd value of a cow is at its peak when cattle prices are "bottoming." This is due to the fact that cows under the age of about 5 or 6 will be producing the majority of their calves around the peak of the cattle price cycle, which is expected to occur some 6 years in the future.

TABLE 2. ESTIMATED VALUE OF A BROOD COW – CYCLE PEAK

Current age	\<-- Years planned to be kept --\> 1	2	3	4	5	6	7	8	9	10	11	12	13	14	15
2	441	506	546	568	580	582	592	608	626	638	642	642	640	638	638
3	468	520	546	560	564	576	596	618	632	638	638	636	634	632	
4	474	504	522	526	542	566	594	610	616	616	614	612	610		
5	456	476	480	498	538	558	576	584	584	580	578	576			
6	428	432	452	484	520	542	550	550	546	542	540				
7	396	418	454	494	518	528	528	524	520	518					
8	386	428	474	504	514	514	510	504	502						
9	398	454	488	500	500	494	488	486							
10	414	466	480	482	474	468	464								
11	434	454	456	446	436	432									
12	438	440	426	412	406										
13	428	406	384	374											
14	396	354	336												
15	340	302													
16	304														

Market value
Heifer @ $.70/lb $595
Cow @ $.50/lb $550
Culling age   16

TABLE 3. ESTIMATED VALUE OF A BROOD COW MIDWAY THROUGH THE CYCLE DOWN PHASE

Current age	1	2	3	4	5	6	7	8	9	10	11	12	13	14	15
2	346	356	376	410	456	502	548	584	608	612	610	604	600	596	596
3	350	376	418	476	534	590	636	664	672	666	660	654	650	650	
4	370	418	488	558	626	680	714	722	716	708	702	698	696		
5	400	480	558	638	700	740	750	742	732	726	720	720			
6	440	528	618	688	732	744	736	726	716	712	710				
7	474	574	654	706	718	710	696	688	682	680					
8	508	600	658	672	662	648	638	630	628						
9	526	596	612	600	584	570	562	560							
10	524	546	532	510	494	484	480								
11	484	464	436	414	402	398									
12	432	390	358	340	332										
13	372	318	288	278											
14	310	254	234												
15	250	212													
16	220														

Market value
Heifer @ $.63/lb $517
Cow @ $.40/lb $440
Culling age   12

TABLE 4. ESTIMATED VALUE OF A BROOD COW - CYCLE BOTTOM

Current age	1	2	3	4	5	6	7	8	9	10	11	12	13	14	15
2	298	354	434	522	606	662	696	710	712	702	692	686	684	682	682
3	360	460	570	674	744	788	806	806	794	782	776	772	770	768	
4	442	576	700	784	836	858	858	844	830	820	816	814	814		
5	522	666	762	822	846	848	832	814	804	800	796	796			
6	576	686	754	780	782	764	766	734	728	724	724				
7	584	660	692	692	672	652	638	632	628	626					
8	562	598	600	576	552	536	528	524	524						
9	512	514	288	258	440	430	426	414							
10	454	420	382	360	348	342	340								
11	380	330	300	284	276	274									
12	321	276	254	244	240										
13	284	246	228	224											
14	264	232	222												
15	250	232													
16	270														

Market value

Heifer @ $.55/lb $468
Cows @ $.35/lb $385
Culling age   11

TABLE 5. ESTIMATED VALUE OF A BROOD COW MIDWAY THROUGH THE CYCLE UP PHASE

Current age	1	2	3	4	5	6	7	8	9	10	11	12	13	14	15
2	396	504	604	680	728	742	740	734	730	728	726	742	742	742	742
3	478	602	696	758	774	774	766	760	756	754	754	754	752	752	
4	548	660	734	754	752	744	738	732	730	728	728	726	626		
5	578	662	684	684	672	666	660	656	656	656	654	652			
6	564	590	588	576	568	562	558	558	556	556	554				
7	506	504	490	482	474	470	470	468	466	466					
8	442	426	416	406	402	402	400	398	398						
9	386	374	362	358	356	354	352	350							
10	352	338	332	330	328	326	342								
11	328	320	318	316	312	310									
12	328	326	322	316	312										
13	240	334	324	320											
14	350	332	322												
15	340	322													
16	346														

Market value
Heifer @ $.63/lb $517
Cow @ $.40/lb $440
Culling age   8

Table 5 depicts breeding cow values midway through the up-phase of the cycle. The most valuable cow is now a 3 year old planned to be retained 5 years. The culling age is 8. At this point breeding cow values are beginning to decline as the peak of the price cycle is approached. Younger cows are beginning to become relatively more valuable. They have the potential to remain productive through the down-phase and into the next up-phase.

PLANNING, CULLING, AND REPLACEMENT STRATEGIES

The preceeding section has indicated how cyclical and nonconstant prices in general will cause the value of a brood cow to vary and hence the economical culling age to vary also. Ideally a producer would like to have as many calves as possible to sell when prices are high and few or no calves to sell when prices are low and he is losing money. Given the preceding table, one might quickly conclude that the producer should buy all the 4-year-old cows he can possibly handle at the bottom of the price cycle (if and when the "bottom" can be ascertained), keep them 6 years, and then go out of business for 6 years. However, most producers plan to continually stay in business over the long run through at least several price cycles. In general, they also have a ranch with a significant capital investment in land and equipment that cannot afford to remain idle, i.e., they are equipped for a certain "normal" herd size. If they are to expand much beyond normal capacity, a certain amount of supplementary feed to augment stressed pastures will be needed. If the normal herd size is not achieved the pasture and equipment will be underutilized and cost per head will rise. A typical per head cost structure reflecting this situation has been developed for a firm with a normal cow herd capacity of 100 head and is reported in table 6. The total cost series is U-shaped and "bottoms" at 100 head. The typical practice is to produce your own replacement heifers. Thus a realistic question, if a rancher plans to stay in business over the long run through several price cycles and produce his own replacement stock, is what kind of a culling and replacement strategy should he use? Once again the computer is consulted for a generalized answer. The answer must be taken as general since a typical price cycle, cost curve, and cow productivity schedule are assumed. However, the strategy developed appears to be intuitive enough and stable enough to be used in a rather wide variety of cases.

Table 7 reports the general strategy developed with the computer. The values reported in table 7 are also displayed graphically in figure 3. Review of table 7 and figure 3 indicates that herd size should be varied over the price cycle. The average herd size found was 113.78 head. But

TABLE 6:   AVERAGE COST PER COW-CALF UNIT

Herd size	Variable cost	Fixed cost	Total cost
50	122.42	466.03	588.45
60	115.68	388.36	504.04
70	111.29	332.88	444.17
80	107.88	291.27	399.15
90	105.72	258.90	364.62
100	114.91	233.01	347.92
110	161.81	211.83	373.64
120	228.89	194.18	423.07
130	273.72	179.24	452.96
140	327.30	166.44	493.74
150	369.79	155.34	525.13

the strategy called for the herd size to vary from 104.45 head to 122.97 head, or by approximately 16%. The average herd size found in this general case is larger than the minimum cost herd size of 100 head due to the fact that average profit per head is positive at $29.60. If prices and costs had been entered so that an average profit of zero were obtained, the herd size would have varied in a similar pattern but averaged about 100 head over the 12 year cycle.

Interestingly the largest herd size is not called for when prices are peaking. Rather it occurs 3 years in advance of the price-cycle peak. This occurs to permit optimal timing of the age structure for the herd. To ensure that the herd will be mostly middle-aged and highly productive animals during the peak of the price cycle, 2-year-old heifers must be brought into the herd 4 to 6 years in advance of the price peak. Bringing a large number of replacement heifers into the herd during the up-phase of the price cycle ensures three desirable atributes: (1) many of the 2-year-old heifers will be entering the herd when their slaughter market values are low; (2) a large portion of the cows in the herd will be at or near their breeding productivity peak as prices peak; and (3) cows in the herd will be young enough when prices peak that they can be retained until the beginning of the next up-phase of the cycle. The latter point ensures that a large number of cows will be ready for culling when the next set of 2-year-old heifers should be brought in for the next cycle.

The preceding strategy leads to a relatively stable herd size pattern but a somewhat volatile culling and replacement pattern. This is the case for a good reason. Expanding the herd size is expensive in terms of rising costs per head (table 6). It makes sense to retain a large number of 2-year-old heifers in the herd when their slaughter value

TABLE 7. SIMULATED OPTIMAL HERD SIZE, REPLACEMENT AND CULLING PATTERNS

Period	Herd size	Profit per head	Replace-ments	Culls[a]	Optimal[b] culling age	Age of[b] oldest cow in the herd	Average age of cows in the herd	Feeder[c] steer price
(1)	(2)	(3)	(4)	(5)	(6)	(7)	(8)	(9)
1	114.84	-4.46	26.45	22.65	12	12	6.35	54.14
2	118.64	-1.42	42.38	39.11	10	12	5.94	56.39
3	121.91	-1.74	28.14	27.08	9	10	4.23	61.13
4	122.97	10.78	20.40	22.39	8	9	4.03	66.07
5	120.98	42.58	36.67	40.09	6	8	4.29	71.07
6	117.56	49.82	23.50	28.43	6	6	3.74	74.35
7	112.63	51.64	8.94	12.92d	9	6	3.92	75.00
8	108.65	51.32	7.96	11.35d	13	7	4.74	72.99
9	105.26	54.37	10.17	10.98d	13	8	5.48	68.49
10	104.45	40.30	14.39	12.14d	13	9	6.06	63.08
11	106.70	22.56	17.94	13.90d	13	10	6.28	59.03
12	110.74	6.49	20.62	16.52	12	11	6.39	54.76
Average	113.78	29.60	21.46	21.46	10.3	9	5.12	64.60

aCulls include cows culled due to enforcement of the culling rule and due to failure to produce
a calf. Based upon specified calving rates by age of cow, all cows who do not calve are
culled.
bAll cows are assumed to produce their first calf at age 2.
cFeeder steer prices are detrended and based at 1966 levels.
dCulling is due only to failure to produce a calf. The optimal culling age is not a constraint
since all cows in the herd are younger than the optimal culling age.

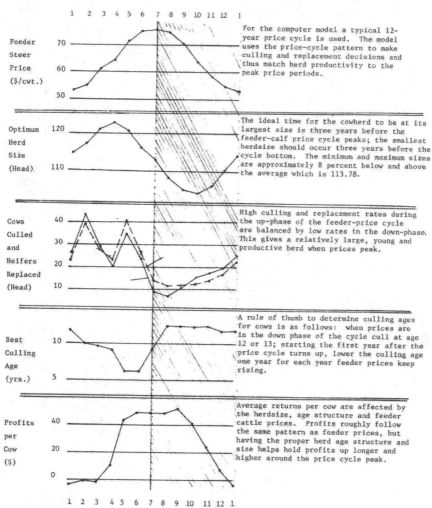

Figure 3. Simulated Optimal Herdsize, Replacement and Culling Patterns.

For the computer model a typical 12-year price cycle is used. The model uses the price-cycle pattern to make culling and replacement decisions and thus match herd productivity to the peak price periods.

The ideal time for the cowherd to be at its largest size is three years before the feeder-calf price cycle peaks; the smallest herdsize should occur three years before the cycle bottom. The minimum and maximum sizes are approximately 8 percent below and above the average which is 113.78.

High culling and replacement rates during the up-phase of the feeder-price cycle are balanced by low rates in the down-phase. This gives a relatively large, young and productive herd when prices peak.

A rule of thumb to determine culling ages for cows is as follows: when prices are in the down phase of the cycle cull at age 12 or 13; starting the first year after the price cycle turns up, lower the culling age one year for each year feeder prices keep rising.

Average returns per cow are affected by the herdsize, age structure and feeder cattle prices. Profits roughly follow the same pattern as feeder prices, but having the proper herd age structure and size helps hold profits up longer and higher around the price cycle peak.

is low and their potential "net present value" as a brood
cow is high; especially if a large number of old cows are
ready to cull. The strategy developed here recommends a
"flurry" of culling and replacement activities during the
up-phase of the price cycle followed by a relatively slow
period of culling and replacement in the down-phase of the
price cycle.

Several "rules of thumb" can be used to summarize the
strategy. With regard to replacement rates, the first year
a producer observes feeder cattle prices rising after a
series of down years (or whenever he believes the feeder
calf price cycle has "bottomed") he should increase the number of 2-year-old heifers added to the herd to about 25% of
the herd size. Such a replacement rate should be continued
as long as feeder calf prices continue to rise, but probably
for no more than 6 years. During the first year after a
producer has noted that feeder cattle prices have apparently
"bottomed," he should begin lowering the culling age for his
mature cows by approximately 1 year of age for each year
prices continue to decline, starting from an age of 12 and
ending at 6. After having added a large number of heifers
to the herd for 5 to 6 years, or upon seeing an apparent
peak in prices, a producer should reduce the number of
heifers added to the herd to about 12% of the herd size.
The culling age of cows during the down-phase of the cycle
should be relatively high, i.e., 12 to 13 years of age.
Most of the culling should be due to failure to produce a
calf; most replacements would be to replace cows culled for
poor performance.

LIMITATIONS AND CONDITIONS

It should be emphasized that this strategy is oriented
toward firms specializing in cow-calf production that produce their own replacement heifers. It is assumed that the
firms do not consider alternative uses of their land and
equipment, such as stocker grazing or forage-crop production. In addition, sporadic 2 to 4 year renting in or renting out of pasture land is not assumed to be possible.
Finally, use of the strategy assumes regularity of the feeder-cattle price cycle and an ability to determine one's
position in the cycle at any given time. This is a heroic
assumption and not always possible. Nevertheless the principles and reasoning presented are still believed to be useful and valid for a rather wide range of circumstances. The
same logic and principles can be used to rationally alter
the strategy given alternative conditions or expectations.
The computerized model can also be used to accomplish this.
However, the model is currently rather cumbersome to operate
and not generally appropriate for public use. Modification
of the model in the future for use by producers on personal
microcomputers is being considered.

REFERENCES

Bentley, J., R. Waters, and C.R. Shumway. 1977. Determining optimal replacement age of beef cows in the presence of stochastic elements. Southern Journal of Agricultural Economics. 8:13.

Kay, R. D., and E. Rister. 1977. Income tax effects on beef cow replacement strategy. Southern Journal of Agricultural Economics. 9:169.

King, C. S. 1979. A systems approach to the determination of optimal beef herd culling and replacement rate strategies. Masters Thesis. Oklahoma State University.

Rogers, L.,Jr. 1972. Economics of replacement rates in commercial beef herds. J. Anim. Sci. 34:921.

Rogers, L.,Jr. 1971. Replacement decisions for commercial beef herds. Washington State Agricultural Experiment Station Bulletin #736.

Trapp, J. N., and C. S. King. 1979. Cow culling and replacement strategies for cyclical price conditions. Oklahoma Current Farm Economics. 52:4.

82

PRODUCTION COST COMPARISONS
OF BEEF, PORK, AND CHICKEN

James N. Trapp

CHANGING PRODUCTION METHODS

Casual reflection upon the nature of the changes in the production processes for beef, pork, and chicken over the past 25 years brings to mind numerous changes and technological innovations. Perhaps the most notable changes have been in chicken production, followed by those in pork and beef. Chicken production has been transformed from a backyard, morning-and-evening, chore operation to a full scale, capital intensive enterprise. To a lesser degree a similar transformation has occurred in the pork production process. Capital intensive, totally confined to farrow-to-finish operations have become common. Beef production methods have also changed, but not as dramatically. Large scale commercial feedlots evolved during the 60s. Feed additives and improved breeding have increased feeding efficiency. These and other changes have not only affectd the manner in which beef, pork, and chicken are produced, but have also profoundly changed the cost of beef, pork, and chicken production. In fact, the conclusion drawn from the production costs presented in this paper is that changing relative costs of beef, pork, and chicken have been the dominate cause of dramatic increases in chicken production and consumption relative to beef and pork production and consumption. In the past 25 years, chicken consumption per capita has increased nearly twice as much as beef consumption and more than twice as much as pork. During the same time period beef production costs have tripled relative to chicken production costs, and pork production costs doubled relative to chicken production costs.

PRODUCTION COSTS: BEEF, PORK, AND CHICKEN

The cost impact of structural/technological change in the beef, pork, and chicken production process are reflected by the data in figures 1 and 2. Costs of production data, profit estimates, break-even price calculations, etc. such as those presented in figures 1 and 2 are always difficult

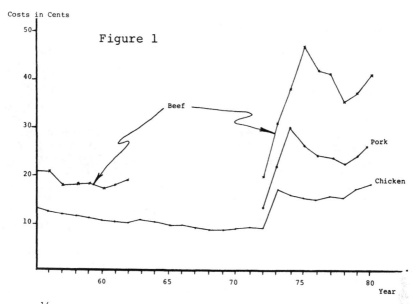

Costs in Cents

Figure 1

Beef

Pork

Chicken

Year

1/ Data collected from "Livestock and Meat Situation" and "Poultry Situation."

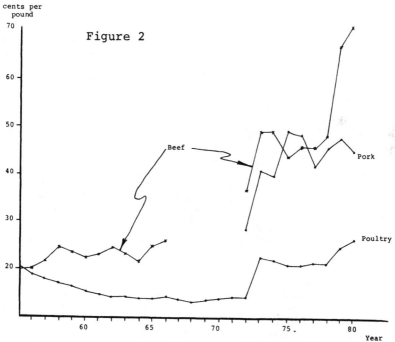

cents per
pound

Figure 2

Beef

Pork

Poultry

Year

1/ Data collected from "Livestock and Meat Situation" and "Poultry Situation."

Figure 1 (top) and Figure 2. Cost impact of
structural/technological change in the beef,
pork, and chicken production processes.

to obtain and define and thus should be viewed with caution. However, the cost data displayed here have consistent
definitions and collection procedures over time, which
allows valid comparisons of shifts in the relative costs of
production.

Figure 1 reports the feeding cost per pound of beef,
pork, and chicken. Only the feed used in the finishing
phase of production is considered--not the total feed for
breeding animals, backgrounding, etc. In the case of beef,
only grain-concentrate-type-ration feed costs are considered. The limited data available prior to 1972 depict a relatively stable feed cost situation. However, the beef/
chicken feed cost ratio did increase from an average of 1.6
over the 1955 to 60 period to 2.17 in 1972; i.e., a 36%
increase. In 1973, the cost of feed rose dramatically for
all three meat types as grain prices in general rose
sharply. However, beef ration costs continued to rise
longer and reached a new plateau relative to pork and
poultry.

TABLE 1. BEEF, PORK, AND CHICKEN FEED COST RATIOS PER POUND
OF GAIN

Feed cost ratio	Year			
	1972	Average 1975-80	1980	Percent change 1972 to 1980
Beef/chicken	2.17	2.53	2.30	+ 6.0
Pork/chicken	1.40	1.51	1.42	+ 1.4
Beef/pork	1.54	1.68	1.61	+ 4.5

Table 1 indicates that between 1972 and 1980 grain-fed
beef ration costs rose by some 6% relative to chicken feed
costs. These comparisons are spot comparisons between two
selected years. However, they appear to be representative
of the 70s. In summary, figure 1 provides casual evidence
that feeding efficiency in the broiler industry has tended
to improve more rapidly in the past than in the beef or pork
industry.

Figure 2 presents estimated historical break-even
prices (i.e., total cost of production estimates) for beef,
pork, and chicken. Problems also exist in calculating and
comparing the data series depicted in figure 2. The beef
and pork costs per pound include the market price of the
feeder animal as a cost. Hence any profits or losses in
beef and pork feeder-animal production are included in the
price of the feeder animal and thus also in the cost of producing a pound of beef or pork. Economic theory suggests
that if feeder-animal production activities are competitive,

profits/losses to producing feeder animals should be driven to zero in the long run. An additional comparison problem is that the beef budgets consider only the costs of producing grain-fed beef.

Figure 2 basically contains the same relationships that were found in figure 1. All break-even prices rose sharply in 1973. Unlike the feed cost case, the break-even prices for beef and pork stabilized immediately along with chicken break-even prices. This was largely due to feeder-animal prices dropping to off-set ration cost increases.

TABLE 2. BEEF, PORK, AND CHICKEN BREAK-EVEN PRICE RATIOS

Break-even price ratio	Year			
	1972	Average 1975-80	1980	Percent change 1972 to 1980
Beef/chicken	2.56	2.35	2.75	+ 7.4
Pork/chicken	1.96	2.02	1.70	- 13.3
Beef/pork	1.31	1.16	1.62	+ 23.7

The ratios of cost of production for beef, pork, and chicken present a mixed pattern when the 1980 and 1975 to 80 average ratios are compared to the 1972 ratio. One consistent relation does stand out. Pork production costs (break-even prices) have held the line against both beef and chicken. Figure 2 indicates that beef production costs remained in line with both pork and chicken production costs until 1979, and then rose sharply.

Despite the lack of comprehensive data over the period from 1955 to 1972, it is clear from figure 2 that beef production costs per pound have steadily increased relative to chicken. In 1955, beef is shown to have had a production cost below that of chicken. Furthermore, in contrasting the percentage changes in relative feed costs (table 1) and total costs (table 2), it is evident that increased cost efficiencies of chicken relative to beef have occurred in non-feed costs as well as feed costs.

COMPOSITION OF PRODUCTION COSTS: BEEF, PORK, AND CHICKEN

The composition of beef, pork, and chicken production costs in 1981 is considered next, with special emphasis given to analyzing the long-term investment cost of production versus short-run out-of-pocket expenses. The costs presented are not calculated to reflect national average costs; rather budgets for specific production systems have been used to develop the costs reported.

Beef production costs are based upon the combined costs of a 100-head cow-calf operation, a 100-head stocker backgrounding operation that carries calves from weaning to 600 pounds, and costs for feedlot finishing heifers to 950 pounds and steers to 1050 pounds. The pork production figures reflect the costs of a 90-sow, fully confined farrow-to-finish operation. The chicken costs are based upon a composite set of costs for a hatching-egg operation with a capacity for 8,000 layers and a broiler production system with four 15,000 bird houses. These budgets reflect typical competitive commercial production systems for beef, pork, and chicken. All cost summaries will be presented in cents per pound.

Table 3 describes capital, land, labor, and operating costs per pound for beef, pork, and chicken. Capital and land costs have been computed as either 5%, 10% or 15% of the total capital and land investment required.

TABLE 3. 1981 BEEF, PORK, AND CHICKEN CAPITAL, LAND, LABOR AND OPERATING COSTS PER POUND OF LIVE WEIGHT PRODUCED, ASSUMING THREE RATES OF RETURN TO CAPITAL AND LAND INVESTMENT

Meat type	Rate of return to capital and land		
	5%	10%	15%
	(cents/lb)		
Beef			
Capital	6.697	13.394	20.091
Land	16.284	32.568	48.852
Labor	8.779	8.779	8.779
Operating	46.131	46.131	46.131
Total	77.891	100.872	123.853
Pork			
Capital	1.750	3.500	5.250
Land	.282	.564	.846
Labor	2.471	2.471	2.471
Operating	46.334	46.334	46.334
Total	50.837	52.869	54.901
Chicken			
Capital	1.392	2.784	4.176
Land	.005	.010	.015
Labor	1.013	1.204	1.204
Operating	21.025	24.999	24.999
Total	23.435	28.997	30.394

Several insights into the nature of the comparative cost structures of beef, pork, and chicken are revealed in table 3. Relative to market prices existing in 1981, beef production costs are extremely high, i.e., market prices for beef, pork, and chicken averaged approximately $.64, $.45 and $.28/lb, respectively, during 1981.

Capital and land costs constitute a much larger portion of the cost of production for beef versus pork and chicken. Table 4 further illustrates this point.

TABLE 4. RATIOS OF CAPITAL AND LAND COSTS VERSUS OUT-OF-POCKET COSTS FOR LABOR AND OPERATION - ASSUMING A 10% INTEREST RATE

Meat type	Cost		Ratio
	Capital & land	Labor & operation	
Beef	45.962	54.910	.837
Pork	4.064	48.805	.083
Chicken	2.794	26.204	.107

Assuming a 10% charge for capital and land investment expense, the ratio of capital and land cost to labor and operating cost is .837 for beef, .083 for pork and .107 for chicken, i.e., the ratio of long-term investment costs to short-term out-of-pocket costs is nearly ten times greater for beef than for pork or poultry. This relationship causes, among other things, the cost of beef production to be much more sensitive to interest rates (i.e. the rate of payment to land and capital investment) than are pork and chicken costs of production.

Another observation concerns the cost of land used in beef production when a 10% interest rate is assumed. At this interest rate, land cost constitutes nearly one third of the cost of beef production. The cost of land in pork and chicken production by comparison is neglibible. It is perhaps useful when considering the impact of land prices upon beef prices to think of beef firms as producing two products--land and cattle. One has often heard of "land and cattle" companies but never of "land and hog" or "land and chicken" companies. Land, unlike other forms of operating and capital inputs, tends to appreciate in value over time. Hence part of, if not most of, the payment to land is expected in many cases to be covered by the land's appreciation in value. Therefore, when the land input for beef, pork, and chicken is "full costed" it is not surprising that the spread between beef production costs and its market price is the widest of the three cost/market price spreads. If land costs are removed from the cost of beef when a 10% interest rate is assumed, the cost of beef production per pound is reduced from slightly over $1.00/lb to $.68/lb. A cost level of $.68/lb is only a few cents above the market price received for beef in 1981.

A final observation regarding table 4 is that the costs reported in the table imply the capital and land investment required to generate a specified gross income in beef production is much greater than that required by pork and chicken. Table 5 has been developed from table 3 to illustrate this point. Table 5 indicates that the investment level required to generate a dollar of revenue from beef

TABLE 5. CAPITAL AND LAND INVESTMENT REQUIRED TO GENERATE ONE DOLLAR OF GROSS REVENUE FROM BEEF, PORK, AND CHICKEN PRODUCTION[a]

Meat type	Investment per dollar of gross revenue
Beef	71.82
Pork	9.03
Chicken	9.98

[a]In calculating this table a 10% interest rate and the following meat prices were assumed; beef $.64/lb; pork $.45/lb; and chicken $.28/lb.

production is some 7 to 8 times larger than that required for chicken or pork. While it is argued that not all land cost should be covered by the revenue generated from production, it must be recognized that the large investment for land and other capital required by the beef industry creates a distinct problem. The economic rule and observed practice that a firm should continue to operate in the short-run only as long as they can cover short-term out-of-pocket costs takes on new meaning in relation to tables 4 and 5. The tables indicate that total costs and out-of-pocket costs are not much different for pork and chicken, but are substantially different for beef. The result is that pork and chicken supplies tend to be cut back very quickly by this economic rule as prices fall. On the other hand, according to this rule, beef supplies do not respond until prices fall substantially. In the beef industry, as opposed to the pork and chicken industry, longer, more severe periods of losses are usually required to obtain production cut backs that lead to higher prices.

SUMMARY AND CONCLUSIONS

The price trend relations and cost of production data presented here provide significant evidence that over the past 25 years greater improvements in production efficiency have occurred in the chicken and pork industries than in the beef industry. This has caused beef production costs and market prices to rise relative to pork and chicken. It is argued that these relative changes in production costs have

been the primary cause of changing meat consumption patterns. Fundamental tastes and preferences for beef, pork,
and chicken appear to have changed very little. Consumers
are, however, dealing with a new and, probably permanently,
different set of relative meat prices. Their response to
this new set of meat prices is primarily responsible for the
changes in consumption patterns observed. Producers have
produced what is profitable and consumers have eaten it.

The problems created in the beef industry by its failure to improve its cost efficiency as rapidly as chicken and
pork are compounded by and partially due to the composition
of its input costs. The beef industry is a relatively extensive user of land and capital. The chicken and pork industries have transformed themselves into intensive, highly
capital-efficient industries. As a result their long-term
capital investment to short-term operating capital ratios
are much lower than beef's. This allows the pork and
chicken industries to shut down production operations quickly when market prices fall. In so doing only small losses
are encountered since fixed costs upon capital investments
are relatively small. In the beef industry, fixed costs
upon land and capital investment constitute nearly half of
all production costs, and short-run termination of production during periods of low prices is not an option for the
industry as a whole. This inability of the beef industry to
reduce production temporarily in the short-run is further
hindered by the longer biological production periods required by beef versus pork and poultry. The ability of the
chicken industry to rapidly curtail production, because of
its low, fixed-capital overhead cost, and then rapidly expand production again because of its short biological production period, places it in a commanding position in the
meat market. This command becomes stronger and stronger as
chicken captures a larger and larger share of the meat
market.

83

EMERGING TRENDS IN THE COMPETITIVE POSITION OF AUSTRALIAN BEEF IN THE U.S. MARKET

A. John De Boer

INTRODUCTION

Australian beef exports to the U.S. market have become a significant factor in the U.S. meat trade. Total U.S. domestic beef and veal production in both 1981 and 1982 is estimated at about 10.35 million t per annum. The 1981 beef and veal imports (preliminary estimates) were 600,000 t, or about 17% of U.S. production. The Australian proportion of these imports dropped to 43% in 1981 compared to 52% in 1980. Despite this drop in Australian exports to the U.S., there seems little doubt that Australian beef production and exports will continue to play a dominant role in the U.S. and world beef trade. Therefore, a close examination of the Australian industry seems in order.

First, the international trade in beef and veal is examined and the volatile nature of the market is highlighted. Second, Australian participation in this market is described. Next, the structure of Australian beef production, marketing, pricing, and exporting is set out. Long-term prospects for expansion of the industry and its international cost competitiveness conclude the paper.

INTERNATIONAL TRADE IN BEEF AND VEAL

International trade in beef and veal can be characterized by the following:
- A few dominant exporters (Australia, Argentina, New Zealand).
- A dominant importer (USA).
- Many minor exporters and importers whose total trade exceeds that of the major traders above.
- Total trade is a minor proportion of total production.
- The presence of cattle cycles in many of the producing countries.

These conditions tend to result in a market with considerable instability; a major price setting role by the large, high-income importing countries; and little possibility for buffer stock accumulation and/or speculation to even out demand-supply imbalances. The 1970s were a particularly unstable period for beef prices. Demand, particularly in the high-income importing countries, rose steadily throughout the 1960s and peaked during 1972/73. This latter period coincided with a simultaneous downswing in the cattle cycle in both the major producing and exporting countries and a worldwide shortage of feed that reduced supplies of competing livestock products. Beef prices reached record levels during 1972/73, herd buildup started, and investments in beef production expanded.

By the end of 1973, a combination of factors led to sharply reduced demand, a cyclical recovery of beef production, and import restrictions in several countries. Beef prices plunged while production costs, which were driven up by high levels of inflation, continued to increase. By 1979, cattle numbers had been reduced significantly in the U.S., Australia, New Zealand, Argentina, and several other major participants in the market. The combined beef output of Argentina, Oceania, and North America in 1979 was 15.4 million t, down 10% from 1978 and down 12% from 1977. Trade in beef and veal actually fell in 1979. Tables 1 and 2 summarize these changes. These tables show that only about 7% of production enters international trade and much of this is subject to quotas and other trade restrictions, particularly in Japan, the U.S., Canada, and the EC-9 countries (Belgium, France, West Germany, Italy, Luxembourg, Netherlands, Denmark, Ireland, and United Kingdom).

TABLE 1. WORLD BEEF PRODUCTION BY SOME MAJOR PRODUCERS AND CONSUMERS (1,000 t)

	1970	1975	1978	1979	1980
World	39,086	44,121	48,211	45,831	45,350
Argentina	2,624	2,439	3,193	3,092	2,923
Australia	1,010	1,534	2,134	2,018	1,557
Brazil	1,845	2,157	2,300	2,106	2,200
Canada	851	1,116	1,060	946	950
China	1,433	1,487	1,623	1,668	1,683
EEC	5,750	6,608	6,427	6,729	6,870
Japan	278	353	403	402	418
New Zealand	387	508	555	501	470
USA	10,103	11,271	11,283	9,925	10,002
Uruguay	379	350	330	312	337
USSR	5,393	6,473	7,086	7,029	6,700

Source: FAO Production Yearbooks.

Since 1979, modest herd rebuilding has started in Australia, New Zealand, Argentina, and the U.S. In the U.S., total cattle numbers on January 1, 1982, were estimated at 115.7 million head, up from 110.87 million head in

1979, but well under the record inventory of 132 million head in 1975 and below the 1972 inventory of 117.86 million head (USDA, 1982). However, profitability has been limited by declining consumer purchasing power, resistance to beef prices, and plentiful supplies of chicken, pork, and turkey. Although international beef prices have increased gradually over the 1979 to 1982 period, the real (inflation adjusted) prices are still well below the 1970/74 peak period (table 3). The World Bank (1980) projections department has forecast sluggish demand growth in the major importing group, little demand from the highly protected European community market, and ample supplies from the major exporters. Therefore, real prices received by exporters are forecast to drop back to 104 cents/kg (1977 prices) for 1985 and 1990 (top, table 3). While these prices are well above the averages for the depressed period of the mid-1970s, they are still less than those obtained over the 1960/64 period.

These estimates are consistent with those of the USDA, FAO, and the Australian Bureau of Agricultural Economics. Following the moderate recovery we are now experiencing, the mid-1980s will be characterized by stagnant demand by major importers, plentiful supplies from traditional and smaller exporters, and downward pressure on beef prices. The import requirements of several rapidly growing Asian countries will help pick up some of the slack, but these increments will be minor compared to total world trade (De Boer, 1982).

Australian meat exports to the U.S. are dominated by beef. Over the 1965/67 period, an average of 171,700 t of Australian beef was exported to the U.S. compared to total average Australian exports over this period of 287,300 t of beef and veal. By 1975/76, U.S. imports of Australian beef had increased to 307,300 t compared to total Australian beef and veal exports of 483,000 t (McCalla et al., 1979). Australian exports peaked at over 1 million t in 1978, or about 25% of total world beef and veal exports. A relatively new factor in Australia-U.S. beef trade is the U.S. Meat Import Act of 1979 (Simpson, 1982). This law was designed to help adjust imports relative to U.S. domestic supply situations by permitting greater meat imports when U.S. supplies were low and consumer prices for beef were high. The law also is supposed to reduce meat imports when domestic supplies are large and prices low. As Simpson (1982) illustrates, the law may not have the desired countercyclical effect under quite reasonable assumptions about changes in U.S. beef cattle inventories. So far, however, the law has not had a major impact on U.S. beef imports because depressed demand in the U.S. and reduced supplies in many major exporting countries have dampened U.S. beef and veal imports.

For 1982, U.S. imports of beef and certain other meats subject to import constraints under the 1979 Act will be well under the trigger point of 1.3 billion pounds (589,670 t). Current USDA estimates place 1982 imports at 1.23 billion pounds (557,919 t) which is only slightly above 1981

TABLE 2. WORLD BEEF TRADE BY SOME MAJOR PRODUCERS AND CONSUMERS (1,000 t)

	1970		1975		1978		1979		1980	
	Imports	Exports	Imports	Exports	Imports	Exports	Imports	Exports	Imports	Exports
World	2,094	2,096	2,477	2,368	3,033	3,193	3,249	3,444	3,235	3,378
Argentina	-	351	-	75	-	340	-	338	-	204
Australia	-	328	-	416	-	1,147	-	835	-	580
Brazil	1	98	24	5	113	755	110	2	64	46
Canada	61	47	58	14	66	10	56	38	53	46
Central America	25	111	32	92	33	31	30	138	33	103
EED	877	523	951	1,115	1,180	1,126	1,240	972	1,179	1,199
Iran	-	-	15	-	28	-	20	-	30	-
Japan	23	-	45	-	100	-	130	-	122	-
Korea	1	-	-	-	40	-	57	-	2	-
New Zealand	-	178	-	192	-	226	-	245	-	216
Uruguay	-	131	-	79	-	93	-	61	-	95
USA	527	9	557	21	673	52	715	54	642	64

TABLE 3. ACTUAL AND PROJECTED INTERNATIONAL TRADE PRICES FOR BEEF IN 1977. PRICES (U.S. CENTS/kg)

A. Average real prices using 1977 prices

			Adjusted					Projected	
1960/64	1965/69	1970/74	1975	1976	1977	1978	1979	1985	1990
110	148	168	67	78	76	75	129	104	104

B. Average of actual prices paid

			Actual					Projected	
1960/64	1965/69	1970/74	1975	1976	1977	1978	1979	1985	1990
43	60	96	61	72	76	87	169	201	216

Source: World Bank (1980).

imports subject to the law of 1.22 billion pounds (553,383 t) but about 75 million pounds (34,019 t) below the trigger level. A recent analysis of the effect of the U.S. import quota was conducted by Chambers et al. (1981) who found that removal of the quota would reduce total U.S. expenditures on beef imports, reduce beef prices by about 2% and increase the quantity of imported beef by 1.9%.

Much of the beef trade from Australia to the U.S. is composed of frozen, boneless cow meat used primarily in manufacturing by blending this lean meat with fat and trimmings from the U.S. slaughter industry. Much of the growth of Australian sales to the U.S. over the past 20 years has followed the spectacular growth of the fast-food/convenience-food sectors that are heavy users of hamburger. In addition, consumer resistance to high beef prices led to substitution for lower-priced cuts and hamburger.

If recent trends in the U.S. industry are an indication of the future, there will be far fewer heavy, overfed cattle going to market, and there will be a shortage of fat and fatty trimmings to blend with the lean meat that Australia traditionally exports to North America. Another factor that may serve to depress U.S. demand for hamburger and cheaper cuts is the depressed state of the fast-food industry as a result of the economic recession in the U.S. Currently, the U.S. consumes from 39% to 45% of its beef in the "ground" form (Conner and Rogers, 1979). This compares with 33% in 1972 and some industry estimates for 1985 that put the figure at between 50% and 65%. This ground beef comes from the following sources (Conner and Rogers, 1979): 13% from imported deboned beef, 35% from the block beef trade, 11% from nonfed steers and heifers, and 41% from U.S.-supplied boneless manufacturing beef. On balance, most estimates indicate that there is a continuing trend toward more ground beef consumption despite the current slowdown in the growth of imported boneless beef. Conner and Rogers (1979) also indicate that possibilities for increasing the U.S.-supplied lean meat by growing out more lean steers and heifers will not be economic, a conclusion also made by Standaert et al. (1980).

The impact of Australian beef imports on the U.S. beef sector has been examined in a number of studies (Freebairn and Rausser, 1975; Chambers et al., 1981; Ryan, 1980; Houck, 1974). The most recent estimates by Ryan (1980) examined the impact that a 200-million-pound increase in the U.S. meat import quota would have on U.S. prices. The estimated reduction in hamburger price was $1.47/lb, a $.74/cwt reduction in farm level prices for cows and a $.02/cwt reduction in farm prices received for choice steers.

AUSTRALIAN PARTICIPATION IN THE WORLD BEEF AND VEAL TRADE

The Australian beef cattle sector has, until recently, been of much less importance than are sheep, wheat, and sugar. However, the industry has expanded rapidly since WW II and by 1977/78 had the highest gross output of any agricultural industry. In the three years ending in 1977/78, the sheep industry provided 26% of total rural export income followed by wheat (20%), beef and veal (13%), and sugar (12%). About 50% of Australian beef and veal output is exported. However, the proportion exported varies considerably in response to economic factors.

The growth of the beef export trade has been most pronounced since the mid-1960s. Exports as a proportion of total production averaged only 14% over the 1951 to 1955 period and 21% over the 1956 to 1960 period. In 1961 to 1965, the proportion reached 27% and was 44% by 1968. Export percentage peaked in 1973 when 61.5% was exported in response to very high export prices. These increasing proportions were on top of increasing livestock numbers; thus, total exports grew even more.

Another important feature of the export industry is its concentration by production system and by state. Most beef production in the southern regions is oriented toward the domestic market. The south is where most of the population is concentrated and young, grass-fattened cattle are produced. In Queensland, the Northern Territory, and the northwest parts of western Australia, a high proportion of output is exported. Since beef prices in Australia are highly correlated with U.S. prices, producers in these areas suffer most when U.S. prices drop since there is no alternative enterprise. The breakdown of statewide export proportions is given in table 4. Exports by state are highly seasonal with peak exports from the southern states in the summer-autumn months of January–June, while exports from the northern areas peak in the winter months of May through September. Overall, slaughtering capacity of Australia's approximately 110 licensed export abattoirs ranges from a low of 28% to 30% in September-October to highs of 60% to 65% in May-June.

STRUCTURE OF THE AUSTRALIAN BEEF CATTLE INDUSTRY

This section draws heavily on Longworth (1979).

The outstanding feature of the Australian beef cattle industry is the dependence on grazing of pastures and crops. Almost no beef is lot fed, although a fledgling feedlot industry was started in the early 1970s to supply the Japanese market with high-quality beef. This collapsed following the 1974 Japanese beef embargo and has not revived. Another feature is that the decline of the dairy industry has resulted in only about 15% of the national output coming from dairy breeds. The Australian Bureau of Agricultural Economics has carried out surveys on beef producers and has defined four major production systems that encompass about 75% of the cattle population (BAE, 1976). These four systems are now briefly outlined.

Breeding and fattening vealers. Calves are produced for sale between 6 to 12 months of age with an average carcass weight of 160 kg. This is predominant in the south and produces most of Australian table beef.

Breeding and fattening older cattle. This system produces primarily steers from 1 to 3 years of age. These are termed yearlings, steers, and bullocks, depending upon age and weight. These are sold on the domestic market as well as for export, and better quality animals may be exported as specific cuts rather than as whole boneless carcasses.

Breeding and selling store (feeder) cattle. This is carried out where growing out calves to slaughter weights is not practical or takes too long. The store cattle are sold to higher potential grazing areas for growing out.

Fattening store cattle for sale. This system specializes in purchasing 1- to 2-year-old cattle (primarily steers) from the breeding areas and grass fattening them for a maximum period of one year.

Despite identification of the producers with the four systems, one should not get the impression that beef production in Australia can be treated as groupings of large, specialized cattle producers. Cattle are often raised on mixed enterprise farms and a high proportion are produced on farms with less than 50 head. In 1971, there were 61,500 beef producers who had 50 or more head while in 1978, a total of 120,073 farmers had at least one head of beef cattle. In the 1970/71 beef industry survey, only 20% of all beef producers (those having 50 or more beef cattle) were classified as "beef only" (where the producer had at least 85% of total gross returns from beef production). A further 26% were beef/sheep, 10% were beef/sheep/cereal, and 34% were beef/other.

Australian beef moves into the export market through a variety of marketing systems, the most common of which is open auction on either a "per head" or live weight basis. In the major export states of Queensland and the Northern Territory, sales by private treaty in the paddock are still common. About 95% of the beef is exported as cartoned meat, bone out. Most meat is frozen although certain markets take only chilled beef. Japan is virtually the only market for table-quality Australian beef although Korea and Taiwan are growing markets for this better-quality, grass-fattened beef.

Competitive Aspects

Because of the nature of Australia-U.S. beef trade, it is not correct to think in terms of direct Australian competition in the U.S. market in the same manner that a Toyota

automobile sale displaces a potential sale by Ford or GM.
The majority of U.S. beef consumed has passed through a
feedlot and is consumed as a retail cut. The vast majority
of Australian beef enters the manufacturing trade where the
domestic supply of commercial grade animals plus trimmings
is nowhere near sufficient to provide the hamburger and
processed meat requirements of the U.S. market. While there
is an effect, it is indirect and influences retail cut
prices through substitution effects. Also, it is difficult
to compare relative production costs, partly because
Australia produces and exports a somewhat different product
than that of the U.S. beef industry, but also because it is
conceptually very difficult to estimate beef production
costs in Australia. This is because so much of the output
is produced under extensive range conditions where there are
relatively few direct-cash costs and also because over 80%
of cattle are held under mixed farming situations so it is
very difficult to allocate many of the input costs specifically to the beef enterprise.

TABLE 4.   BEEF AND VEAL EXPORTS AS A PROPORTION OF STATEWIDE
           OUTPUT IN AUSTRALIA, 1967/68-1977/78 (PERCENTAGES)

Financial years	N.S.W.	Vic.	Qld.	S.A.	W.A.	N.T.	Tas.	Total
1967/68	23	41	66	20	42	50	30	44
1968/69	22	37	66	19	43	48	31	44
1969/70	28	48	77	26	52	52	35	50
1970/71	28	47	79	24	42	48	32	49
1971/72	41	51	76	34	49	55	44	55
1972/73	51	58	84	46	54	57	44	62
1973/74	48	54	74	37	55	52	47	56
1974/75	31	39	60	28	43	39	38	42
1975/76	41	40	64	26	43	32	43	46
1976/77	45	42	72	34	53	22	48	51
1977/78	50	48	71	48	50	50	50	55

Source: Production-Australian Bureau of Statistics
          Exports-Australian Meat and Livestock Commission.

Another way to assess the competitiveness is to compare
the Australian saleyard prices to the landed cost in the
U.S. This analysis was carried out by De Boer (1979) and
was based heavily on an earlier study by Forrest (1977).
The procedure takes the Australian Meat and Livestock Commission saleyard price quotations plus the processing costs
given below plus the charges for handling, freight, and
import duties in the U.S. Table 5 presents the result of an
August 1977 survey (Forrest, 1977) and covers costs up to
the point of delivery for domestic retail outlets and to the
point of loading aboard ships (FAS) for export beef. The
costs are given in cents per kg carcass weight and since
most beef is exported on a boned-out basis, the original

carcass weight must be converted to a boned-out basis. On average, the boned-out weight is two-thirds the weight of the original carcass. The original carcass represents about 55% of the original weight of the live animal so a live animal weighing 500 kgs would provide about 183 kgs of boned-out beef for export. The AMB converts the shipped weight of Australian beef exports into a carcass weight equivalent by multiplying by a factor of 1.5; for mutton or lamb the factor used is 2.0.

TABLE 5. PROCESSING COSTS FOR BEEF FOR LOCAL TRADE AND EXPORT TRADE

Local trade: Kill, chill, and deliver to retail butchers[2]

Livestock description	Carcass wt (kg)	¢/kg High	Cost carcass wt Low	wt Avg.	Avg. total cost/carcass ($)
Yearling	160	11.4	7.8	10.9	17.44
Trade steer	200	8.4	4.7	7.8	15.60
Medium ox	250	6.2	2.2	5.5	13.75
F.A.Q. cow[1]	215	7.3	3.5	6.7	14.40

Export trade: Kill, chill, bone, pack, freeze, deliver to ship for export[3]

Livestock description	Carcass wt (kg)	¢/kg High	Cost carcass wt Low	wt Avg.	Avg. total cost/carcass ($)
Japanese trade, Chilled ox	280 plus	21.7	11.7	17.4	48.72
Heavy ox	280 plus	18.2	10.2	14.7	41.16
Medium ox	250	19.4	11.2	15.9	39.75
F.A.Q. cow[1]	215	20.7	11.8	17.0	36.55
Third grade cow	180	21.9	14.0	18.9	34.02
Boner cow	150	23.3	16.6	20.5	30.75
Bull	250	24.6	13.3	19.9	49.75

Notes: [1] F.A.Q. = fair average quality
[2] local trade is in chilled carcasses
[3] Export trade is in chilled halves or quarters for the Japanese chilled beef market. All other costs are for frozen, boned-out beef.

Source: D. Forrest, (1977).

The figures in table 5 include Commonwealth levies and state inspection costs but do not include documentation charges, bank charges, or rejection insurance. By adding freight rates for containerized meat packed in cartons, it is possible to make a rough breakdown of landed meat in the U.S. Given 1979 saleyard prices, this exercise is carried out below for one full container of frozen, deboned beef packed in cartons of about 28 kg:

1. Cost of cattle for a container load equivalent of 16.33 metric tons of frozen boneless beef: 16.33 t x 1.5 = 24.5 t carcass beef. 24.5 t carcass beef based on 300 to 320 kg carcass weight export quality ox at $0.50 per kg carcass weight=$12,250.

2. Processing cost for heavy ox, based on average cost (kg from table 5) of $0.147/kg = 24,500 kg carcass weight x $0.147/kg = $3,600. This would require about 75 to 80 head for one container load.

3. Freight, Australia to United Kingdom, as of June 1976 based on full-container load packed by exporter and unpacked by importer, $30.50 per container.

4. Total cost per container $12,250 meat
                                         3,600 processing
                                         3,050 freight
                                        $18,900

5. Average cost per kg boned-out beef, frozen in cartons, delivered duty-free to importer: $\frac{\$18,900}{16,330 \text{ kg}}$ = $1.16 Australian per kg, or about $1.30 U.S. per kg.

On a per kg basis, the cost breakdown is as follows:

	$/kg
Cost of meat (boned-out equivalent)	0.75
Processing	0.22
Shipping	0.19
Total landed cost	$1.16

The figures in table 5 can be easily updated using 1982 or 1983 prices and costs. The analysis is perhaps even more relevant going from U.S. beef prices back to the price the Australian producer receives since, to a very large extent, Australia is a price taker in the international beef market. Several factors have worked against the Australian producer in this regard. First, costs of marketing, processing and handling beef in Australia have increased rapidly as inflation has continued, although at a lower rate than in the U.S. The boom in coal and mineral exports, although it has slowed since the mid-1970s, has kept the value of the Australian dollar high, and this reduces Australian producer receipts since most export sales are in U.S. dollars. High interest rates, as in the U.S., have had a deleterious effect on beef producers. However, freight rates have softened considerably.

SUMMARY AND CONCLUSIONS

The Australian beef industry experienced the same roller coaster effects as the U.S. industry during the turbulent 1970s. Those who survived are now slowly expanding production and exports in response to better short-run and long-run price expectations in the export market. Although Australian exports have recently shown greater diversification away from the U.S. market, we remain the bread and butter customer for the major item of the beef trade--boneless, frozen, lean meat.

The long-run prospects are closely tied to developments in the U.S. beef market. Although there are few prospects for overall U.S. beef consumption to surge back to levels of the early 1970s, a continued shift in total beef consumption from retail cuts to hamburger would continue to increase the demand for Australian imports. This would inevitably lead to imports hitting the import-quota trigger levels. If this happened during an inflationary period of high beef prices, pressure by importers and consumer groups could lead to legislative action. If, however, this occurred during a period of depressed beef prices, little pressure for reform would result.

There seems little doubt that, given the right combination of prices and export market demand, the Australian beef industry could undergo another major expansion phase. However, given the climatic instability inherent in many of the beef exporting areas, producers may not always be able to capitalize on these factors.

REFERENCES

Australian Bureau of Agricultural Economics. 1976. Production Systems in the Australian Beef Cattle Industry. Beef Research Report No. 18, Canberra.

Chambers, R. G., R. E. Just, L. J. Moffitt and A. Schmitz. 1981. Estimating the impact of beef import restrictions in the U.S. import market. Aust. J. Ag. Econ. 25:123.

Conner, J. R. and R. W. Rogers. Ground beef: implications for the southeastern U.S. beef industry. Southern J. Ag. Econ. 11:21.

De Boer, A. J. 1979. The short-run and long-run position of Australian beef supplies and the competitiveness of Australian beef in international trade. Working Paper No. 5, Center for Research on Economic Development, University of Michigan, Ann Arbor.

De Boer, A. J. 1982. Livestock development prospects in Asia: Implications for the U.S. livestock industries. Paper presented to International Stockmen's School, San Antonio, Texas. January 1983.

Freebairn, J. W. and G. C. Rausser. 1975. Effects of changes in the level of U.S. beef imports. Am. J. Ag. Econ. 57:676.

Forrest, D. 1977. Range Management Services/Graphic Marketing Services. The Australian Beef Situation. Brisbane, Australia.

Houck, J. P. 1974. The short-run impact of beef imports on U.S. meat prices. Aust. J. Ag. Econ. 18:60.

Longworth, J. W. 1979. The Australian beef industry. Food Pol. Stud. 19:26.

McCalla, A. F., A. Schmitz and G. G. Storey. 1979. Australia, Canada, and the United States: Trade partners or competitors. Am. J. Ag. Econ. 61:1,022.

Ryan, T. J. 1980. A note on bias in the estimated effect of beef imports on U.S. meat prices. Aust. J. Ag. Econ. 24:60.

Simpson, J. R. 1982. The countercyclical aspects of the U.S. Meat Import Act of 1979. Am. J. Ag. Econ. 64:243.

Standaert, J. E., L. L. Blakeslee and R. J. Folwell. 1980. Price and demand constraints on lean beef production in the U.S. College of Agriculture Research Center, Technical Bulletin 0095, Washington State University.

USDA. 1982. Livestock and meat situation and outlook. Economic Research Service, LMS-244, United States Department of Agriculture.

World Bank. 1980. Price Prospects for Major Primary Commodities. Commodity and Export Projections Division Report No. 814180. World Bank, Washington, D. C.

84

DIRECT DELIVERY OF MARKET INFORMATION THROUGH RANCHER-OWNED MICROCOMPUTERS: A RESEARCH REPORT

Harlan G. Hughes, Robert Price,
Doug Jose

Ranchers' needs for marketing information have changed dramatically since the early 1970s. Increasing price variability, rapid inflation, higher interest rates, and closer ties to world supply-and-demand conditions for agricultural commodities have resulted in increased needs for short, intermediate, and long-run marketing information. Also, ranchers continually have fewer market outlets available so that they must do a better job of marketing their product. The net result is that many ranchers are unable to adequately evaluate marketing alternatives and, thus, are often unable to make good marketing decisions.

In late June 1981, Cooperative Agreement Number 12-05-300-522 was signed between the USDA Extension Service and the Colorado State University Cooperative Extension Service on behalf of the Western Livestock Marketing Information Project to give ranchers decision assistance. The agreement was to conduct a pilot study concerning the feasibility of direct electronic delivery of marketing and management information to farm and ranch families.

Ranchers base marketing decisions on information from both internal and external sources. Accounting records, herd performance records, and budgets are examples of internal information used. Market news, outlook reports, price forecasts, weather forecasts, and research reports are examples of external information. Internal and external information are required for almost all short, intermediate, and long-run marketing decisions.

Needs for short-run market information commonly relate to selling decisions. There are sometimes substantial risks associated with selling agricultural commodities today rather than waiting a few days, or vice versa. Short-run decisions are relatively simple to analyze in a budgeting sense as the costs are readily predictable. The difficult element is the probability of price increases and decreases.

Typically, university and government outlook specialists have not provided short-run market information. It has generally been left to the commodity brokerage firms and other private organizations to provide short-run market information. These sources tend to discount the risk and uncertainty aspects.

Intermediate-run needs for market information relate to such decisions as purchasing of stocker and feeder cattle, crop selections, fertilizer application, feed choice, and other decisions that do not result in immediate revenue. These decisions are generally more complex as the information needed to evaluate possible outcomes is more complicated and has more chance of error. University and government outlook specialists generally have been most active in providing intermediate-run information.

Examples of long-run market-information needs include land purchases, irrigation development, machinery selection, cattle herd expansion, and the construction of livestock production units. These decisions, although not made as frequently as the previous types, require significant information to allow for success of a farm business. Although farm management economists have devoted much time and effort to investment analysis, outlook specialists in the university and government realm generally have concentrated very little on this long-run arena.

Because of variability in agricultural prices and production, as well as high financing requirements, producers may risk bankruptcy before profits from an investment can be realized. Long-run market information can also be useful in assessing the amount of risk that a specific producer can afford when making investment decisions.

A comprehensive marketing-information system, used properly, could play a major role in stabilizing or increasing net ranch income during the 1980s. In the coming years, ranchers are going to need more marketing information, delivered faster, and available in an easy-to-use form. Computers can and should play a major role in such a marketing -information system and the associated educational needs. The rapid development of electronic technology also presents an exceptional opportunity for the Cooperative Extension Service to assume an even greater role in the delivery of timely market information.

DELIVERY OF MARKET INFORMATION IN THE WEST

The problem of delivering timely market information in the western U.S. is compounded by the vast geographical dispersion of producers. The extension specialists and county agents must travel extensively to accommodate the needs of farmers and ranchers. Most newspapers carry very little, if any, current market data. Farm magazines are major sources of intermediate-run market information, but timeliness of that information does not meet the standards necessary for decision making in today's economic environment.

The Western Livestock Marketing Information Project (WLMIP) was created over 25 years ago in recognition of the void that existed in the delivery of timely market information to livestock producers in the West. The proven record of WLMIP as a major source of useful intermediate-run market information for the region has been well documented (WLMIP, 1977; Bolen, 1949). However, the changing complexities of the livestock market, combined with the rapid growth in computer technology, present the need and opportunity for WLMIP to expand services. These opportunities include direct delivery of market information at the producer level, as well as increased service to professional economists and others in the West. It also presents the opportunity for WLMIP to go from almost exclusive emphasis on intermediate-run market information to expanding short-run information.

AGNET--AGRICULTURAL COMPUTER SYSTEM

AGNET is a time-sharing computer network headquartered in Lincoln, Nebraska. There are over 2500 subscribers to the network with a total yearly connecttime of over 75,000 hours. This averages out to 8.5 users per hour concurrently on a 24-hour 7-day-a-week basis. AGNET is being utilized for problem-solving and information networking. The system is very "user-friendly" and is designed for use by people with no computer background. Ranchers are allowed to subscribe to the AGNET System by paying variable costs associated with operating the system.

PREVIOUS STUDIES

In 1979 a survey was sent to state extension service administrators inquiring about the priority of marketing extension programs. (Watkins and Hoobler, 1980). Thirty-seven of the 44 state administrators returning the survey placed extension marketing programs in the range of "important" to "of highest importance." The following summary statement was taken from the report:
- "It is recommended that each state Cooperative Extension Service administration, in cooperation with their marketing specialists and representative clientele groups, examine the results of this national study, analyze their state's specific needs, determine where a cooperative effort is needed with other states, and develop plans for renewing and/or initiating programs to effectively manage the problems."
Brown and Collins (1978), University of Missouri, conducted a national study in 1977 on the information needs of large commercial farms. Their study revealed that:
- Commercial family farmers and ranchers perceive marketing information as their number one need.
- Extension and universities were rated the most important source of production technology, but only of minor importance as a source of marketing information.

- Farmers, agribusiness, extension, and the agricultural media all expressed the belief that marketing information is critical now and will continue to be critical in the future. They also agreed that present sources of market information are inadequate.

A joint USDA/NASULGC study (1968) committee recommended in 1968 "that extension increase its emphasis on marketing and farm-business management while reducing the percentage of effort in husbandry and production." The study goes on to say, "Extension should gradually shift towards giving more in-depth training to producers and to wholesaling information through supply firms."

Most extension marketing-program-appraisal studies generally include recommendations for experimentation with the latest electronic and computer innovations. For example, New York dairymen in a 1977 telephone survey felt that extension could improve its effectiveness by placing more emphasis on the use of the computer as an educational tool. (Ainsle et al., 1977).

In spite of the emphasis placed more than 10 years ago on changing extension priorities, little progress has been made to implement these program shifts. This is mainly due to extension administrators' reluctance to changing priorities of their extension programs. We are hearing the same priority requests coming from producers today as we did a decade ago. This paper reports on one pilot project that attempted to respond to some of these priority requests.

PILOT PROJECT

The state of Wyoming piloted a basic electronic market information system on the AGNET computer network during 1978-79 (Skelton, 1980). Four objectives of the pilot system were:

1. To collect price information of interest to Wyoming producers.
2. To provide county extension offices with the ability to retrieve market information that allowed them to put together today's, yesterday's, last week's, last month's, or last year's markets of interest for use by their producers.
3. To provide simple, down-to-earth interpretations of what market prices and associated outlook mean to Wyoming producers.
4. To provide price forecasts for extension personnel to use with producers in planning.

The Western Livestock Marketing Information Project piloted some initial work in computerized market-information delivery in 1980. Major livestock reports (Cattle on Feed, Hogs and Pigs, etc.) were placed on AGNET, complete with analysis and interpretation. WLMIP was instrumental in sub-

stantially increasing listings of producers with hay for
sale and making these listings available to areas hardest
hit by the drought of 1980. In addition, WLMIP served as a
clearing house for drought conditions in many areas of the
western plains region. This information was collected and
transmitted throughout the region and forwarded to the office of the Secretary of Agriculture in Washington, D.C.

OBJECTIVES OF THIS STUDY

In an effort to provide an evaluation of electronic delivery of market information, a cooperative agreement was
signed between the Colorado State University Extension Service on behalf of WLMIP and the USDA Extension Service.
Subsequent cooperation was obtained from the University of
Wyoming and the University of Nebraska. Four of the six objectives of the pilot study reported in this report are:

1. Research and develop mechanisms for direct
   producer access to AGNET via farmer-owned
   microcomputers and computer terminals.
2. Add current livestock market news information
   on AGNET for retrieval by producers and
   others.
3. Improve the documentation of marketing information and other pertinent information on
   AGNET and make it available to producers.
4. Evaluate the effectiveness and efficiency of
   this new delivery system in providing useful
   information to farmers.

RESULTS

Objective 1: Research and develop mechanisms for direct producer access to AGNET by farmer-owned microcomputers
and computer terminals.

The technology of communication between computers of
different brands and types is an involved science of its
own. Different hardware requires different communication
protocols and procedures. Much additional work is needed in
this area that is receiving a lot of interest at the current
time.

The importance of networking between microcomputers and
mainframe computers housing networks such as AGNET is becoming increasingly obvious. As an example, of the 12 producers that participated in the pilot study, 9 accessed
AGNET through the use of microcomputers. The other 3 used
"dumb terminals" for communication. It is the authors'
opinion that microcomputers will become more the norm in
producer hardware than dumb terminals. The reason is that
the microcomputer can also be used to solve on-the-farm
types of production and marketing problems, handle production and accounting records, and handle other applications.

Current technology is readily available to enable a microcomputer to operate as a dumb terminal in communicating with AGNET. Generally, all that is required is a modem (telephone coupler) and software for the microcomputer, which is generally included with the hardware coupler. Such packages for microcomputers generally run in the price range of $500 or less. However, the technology involved in making a microcomputer into an "intelligent terminal" with AGNET is more complex. A high degree of interest in this type of software appears evident throughout the western region.

The important of operating a microcomputer in an intelligent mode with AGNET arises from the tremendous potential savings in telephone costs. A large part of the user's time during any terminal session is now spent typing in the needed information to respond to the AGNET questions. If such files could be developed on the microcomputer before the telephone call is actually made to AGNET, much of the telephone cost could be eliminated.

The authors have experienced substantial savings in telephone costs (50% or more) when using microcomputers as intelligent terminals. The capability to access AGNET as a central warehouse for information, download the information to the microcomputer, hang up the telephone, and work with the information that has been accessed results in even more cost savings.

In summary, it is a relatively easy procedure to turn a microcomputer into a dumb terminal for communicating with AGNET. It becomes a little more difficult to operate in the intelligent-terminal mode, but software has been developed in this study that makes this possible for most brands of microcomputers. The idea of interfacing farmer-owned microcomputers and a regional computer, such as AGNET, could be one of the most significant thrusts in extension service activities for computer applications to agriculture in the coming years.

Objective 2: Add current market-news information on AGNET for retrieval by producers and others.

This objective has been pursued heavily since the beginning of the pilot project. The system has been expanded so that 17 different market-price files are going onto AGNET daily. In addition, weekly and monthly analyses are going onto AGNET. During the six-month study period, 19,873 market-price files were retrieved by all AGNET users.

In addition to providing information for its current market value, most of the information included in the price files is captured by the computer and put into historical data files. By building such a data bank, files are in place for retrieval by the user in various programs for management and marketing decisions. Most of the captured information is already available for use in various retrieval and charting programs. However, retrieval programs for AGNET market information are still under development.

In an effort outside the scope of this study, the Foreign Agriculture Service (FAS) of USDA has begun a test using AGNET markets for distribution for much of their information. The response from users of the FAS information has been very enthusiastic, and several new users have subscribed to AGNET just to receive the FAS information.

The addition of market news and other information on AGNET will be a continuing process. Feedback from participating county agents and producers during this pilot study resulted in several files being added. Additional feedback on the final end-users' evaluations points to the need for even more types of files.

Market information that is currently being placed on AGNET is almost exclusively done by volunteer labor. Therefore, relatively little money is allocated to staffing explicitly for placing market information on AGNET. Several staff hours weekly are being devoted from numerous offices to place the information on AGNET.

By relying so heavily on manual labor to provide the information to the computer network, costs are magnified and the chances for errors arise. USDA Extension Service is working with Agricultural Marketing Service for direct electronic transfers of market information from AMS to the AGNET computer. If such a system could be put in place, cost savings for staff time would be tremendous and the timeliness of the availability of the information could be much improved.

Objective 3: Improve the documentation of marketing information and other pertinent information on AGNET and make it available to producers.

One of the developments coming out of this pilot study has been an AGNET Market Information Users Guide. The guide is intended to be just that, a guide to help the new user know what type of market information is available and how to access that information. In addition to documenting the market information, management decision tools are also referenced in the guide with a brief explanation of how to access and use those tools.

AGNET is a very "user-friendly" computer system. The major part of any documentation needed by the user is accessible directly from the computer with the use of "HELP" commands built into the system. Such HELP commands are unique to AGNET. Consequently, most of the needed documentation and aids are available at any time during a terminal session and preclude the necessity of having a manual available for reference while the user is online.

Objective 4: Evaluate the effectiveness and efficiency of this new delivery system in providing useful information to farmers.

The study tested two methods of market-information delivery. The first method was actual direct delivery to farmer-owned microcomputers or computer terminals located on the farm or ranch. The second method tested was "wholesaling" market information through county agents or trained agricultural professionals. These agricultural professionals used marketing bulletin boards in the county

agent's office or in the financial institution center and
then used frequent mailings of market information from these
offices to selected producers in their area. This course of
delivery was used mainly to acquaint producers with the type
of information that could be obtained from the computer network and to test their responses to see if the information
delivered was useful. The authors were also interested in
seeing whether, after receiving the information in this manner, producers would be more interested in obtaining their
own computer hardware for direct delivery of the information.

EVALUATION OF USERS DIRECTLY ACCESSING INFORMATION

An evaluation form was sent to the producers who were
directly accessing the information from their own hardware.
Evaluation forms were returned from 12 direct-access users.
It should be noted that this group is a representative sub-
set of farmers and not all AGNET users.

Users were asked to evaluate six general types of market information they could access from the computer. The
results of that evaluation are listed in the following
table:

Evaluation By Direct Users

Information	Very useful	Slight-ly useful	Not useful	No response	Total
Futures prices	3	4	1	4	12
Cash prices	5	3	1	3	12
Commentary & interpretation	6	3	0	3	12
News releases	2	5	1	4	12
Retrieval programs	3	2	1	6	12
Conferences	1	4	0	7	12

The files that contain commentary and interpretation of
factors influencing the market were very well received by
the users who were accessing them directly. Various comments received on this question included: "Good insights."
"Comments really helped to get a feel for the market."
"More of this type of information needed."

The files on various cash prices were also very well
received by the users who were directly accessing the information. AGNET is very unusual in that several files contain
localized information for various areas within a state that
is not available anywhere else in a condensed, summarized
form. Comments included such things as: "We need more
local prices on the system." "Used these the most." "Often
AGNET is the only source of this information." "Excellent."

The files on futures prices were not perceived by the end users to be as useful as the two previously discussed categories. One of the main reasons for this is that AGNET only offers each day's open and close of the futures. Producers who are active in the futures markets find that they need more current quotes, which they obtain from their farm radios or from their brokers. Also, not many producers use the futures market. Some comments on the futures included: "Would be better if we had a detailed report on weekly futures price movement." "Information didn't fit our area completely as there were no sugar futures." "Out of date by the time the producers really need this information." "Useful if picked off daily."

The retrieval programs were not used as heavily as the authors hoped they might be. One of the main reasons was that perhaps the users did not feel they were sufficiently versed in the correct technical aspects to use the program. Typical comments for this information included: "Did not use." "What are these?" "We need a lot more information on how to run these programs." "I liked these very much and accessed them regularly."

NEWSRELEASE items were also not rated very highly by the end users. This was not too surprising as the NEWSRELEASE program on AGNET is generally considered to be more consumer oriented, although there is much good useful information for livestock and grain producers. Typical comments included: "Very few used." "Checked only on occasional basis." "Some good, some bad." "Especially liked the ones on economic issues." "Some were excellent."

The lowest-rated information source by the end users was the electronic CONFERENCES. Again, this was not too surprising. CONFERENCES are of more use to people other than farmers. It is the responsibility of an individual AGNET user to link with the electronic CONFERENCES. Although the authors had sent out the procedure for doing this via U.S. mail, it is doubtful that many of the users took the time to go through the procedure to link up to the CONFERENCES. Typical responses included: "Did not use." "So what?" "Helped sometimes." "Need more information on how to use." "Not enough conferences sales or prices."

The users were asked to evaluate the timeliness of information delivered by AGNET. The response broke down as follows: very timely--5; average timeliness--5; too late to be useful--1; no response--1.

The users, who were all paying their own computer and telephone costs, had a very high expectation of when the information should be available on the computer. Many times the information for a given day would not be available until the following morning because of the manual transfer of the information onto the system. Once again, this points out the high desirability of automatic linkages with the AMS teletype system so that the information can be available much more quickly. Typical responses to the timeliness question included: "Most information was available from other sources at lower costs like newspapers and radio; however, this service shines in the fact that information is available on demand." "Many times it is hard to check information everyday. Why not put on a program that records daily futures-prices information and then on Friday evening we could pull them off for our records."

Users were asked to report costs. Very few had kept records of their costs, but those who did report indicated that $50 to $75 a month was a normal combined telephone and computer cost for accessing the AGNET information. A good share of the users responded that they did take advantage of nonprime-time telephone and computer costs by calling early in the morning or late in the evening.

Only three users indicated any problems from trying to access the information on AGNET. They also indicated that a workshop on operating technique would have been helpful. Most indicated that they felt it was quite easy to use the system. However, six respondents indicated that a workshop on how to apply the information being received from AGNET would be extremely useful.

Users were asked to give suggestions for improving AGNET delivery of market information and whether they felt it was worthwhile to continue providing information across the system. Most of the respondents who indicated that additional information would be desirable were looking for more localized cash prices and more commentary with specific projections for what the markets might do in the future. The overwhelming response was that the direct delivery of market information was extremely worthwhile and that the project should be continued. Only two users indicated that they did not intend to continue accessing AGNET information regularly.

Although very few respondents put a dollar value on the information received, the majority indicated that the cost-benefit ratios for accessing the information were highly favorable.

SUMMARY

The need for better information to be used by agricultural producers in making agricultural marketing decisions has been well documented. The thrust of this study has been to evaluate the feasibility of direct electronic delivery of this needed market information. The development of mechanisms for direct producer access to AGNET via farmer-owned microcomputers and computer terminals was one of the main objectives.

The best evaluation of this project lies in the large increase in retrievals of market information. During the time period of the study there was over a three-fold increase in the number of times that AGNET was accessed for market information.

Of the cooperators in this study, nearly three-fourths accessed AGNET through the use of microcomputers, while the remainder used dumb terminals for communication. It was found that operating a microcomputer in an intelligent mode with AGNET becomes increasingly desirable due to the tremendous potential in telephone savings. Savings of 50% or more resulted from using microcomputers in this manner.

It is relatively easy to turn a microcomputer into a dumb terminal for communicating with AGNET. It becomes much more difficult, however, to operate in the intelligent-terminal mode. This study uncovered software that makes this possible for most brands of microcomputers. The idea of interfacing farmer-owned microcomputers and a regional computer such as AGNET could be one of the most significant thrusts in extension service activities for agricultural computer applications in the future.

The pilot study was successful in providing current market news information to AGNET for retrieval by producers and others. A wide variety of new files has been made available in the MARKETS section on AGNET. Many of these files were the direct result of this pilot study. These files and others will continue to be available on AGNET.

Feedback from the final end user indicated the need for several more types of files, particularly of a regional type. The timeliness of the market information provided on AGNET was a concern to the producers involved in the study. Although the majority of the participants felt that the material was very helpful to them, they also expressed a desire for more timely information. This end-user evaluation points toward a critical need for direct electronic transfers of market information from AMS to the AGNET computer.

In summary, this pilot study provided much needed background information on the electronic delivery of market information. This study found that there is, indeed, a demand for the direct electronic delivery of marketing and management information to farm producers. There exits a distinct opportunity for the extension service to assume an even greater role in the delivery of this timely market information. In addition, the study provided the documentation of the need for increased development of computer applications to agriculture in information networking and evaluation of marketing alternatives. This information provides a base from which the extension service can evaluate and plan their activities in the computer arena for the future.

REFERENCES

Ainsle, et al. 1977. An evaluation of cooperative extension dairy programs. Specialist Report. Cornell University.

Bolen, Kenneth R. 1979. Economic information needs of farmers. Report of ESCS and SEA/Extension Study.

Brown, Thomas R. and Arthur Collins. 1978. Large commercial family farms information needs and sources. A Report of the National Extension Study Committee.

Skelton, Irvin. 1980. Wyoming agricultural extension service accomplishment report for FY-1980.

Watkins, Ed and Sharon Hoobler. 1980. Report of ECOP Subcommittee on agriculture forestry, and related industries extension marketing program and priorities survey. SEA/Extension.

USDA. 1968. A people and a spirit. Report of the joint USDA/NASULGC study committee on cooperative extension. Colorado State University.

WLMIP. 1977. Evaluation of the western livestock marketing information project. Report of WLMIP Technical Advisory Committee Survey of Users.

Part 13

COMPUTER TECHNOLOGY AND USE

85

A USER'S VIEW
OF IMPORTANT ATTRIBUTES
OF A COMPUTER SYSTEM

Jay O'Brien

Often when purchasing a new system or equipment, one should start small and work up. Due to the nature of computers and computer programs, it is best to go after a computer and programs that can offer the complete solution from the start. Little more work is required to establish a complete solution than to adapt to a system that will meet only part of one's needs. Due to the greatly reduced price of hardware, a first-class system is not much more expensive; however, operating a second-class system is definitely more expensive, as one will actually save little labor with an incomplete system.

You might compare the decision with that of picking a sound reliable horse that costs more money initially as compared to a horse that can do special jobs but is unable to do all of the jobs required of a good ranch horse. After you have supported the questionable horse for a while, you will wish you had chosen on the basis of dependability and quality.

WHAT IS A FIRST CLASS SYSTEM?

Hardware

The hardware should not only handle one's present needs; it should be capable of being expanded to handle future needs. All machines will break down and need service; therefore, dependability and repairability are extremely important. If one lives outside of a major metropolitan area, service should be the most important product. Go with the firm that can offer a service contract and has a history of support. Systems that cannot be repaired in one day by local representatives should be avoided. The good companies will even come to a ranch for the same monthly fee they charge me to come to my office. Check into the history of the manufacturer of the machine and be sure the company will be around when parts are required in three of four years. Also, check with someone who is running on the hardware and ask them about the service. I cannot stress dependability enough.

A machine capable of handling word processing should also be considered. This capability can allow someone with no secretarial skills to produce letters that look as if they have been written by the best secretary in the world. I also keep my common contracts on it and adjust them for new trades.

Software

Dependability is as important in software as hardware. Find a system that is presently operating. Unless you have a great deal of money and patience, do not try to design your own programs if good software is available. Also, inquire about support and see if the firm providing your software is capable of enhancing it to meet your needs. If at all possible, visit an installation of the system you are considering.

Quality software has the same concept whether it is for a feedlot, a rancher, a horse-farm operator, or a purebred herd. Do not settle for a glorified general ledger on an underpowered machine. These systems can take as much work to operate as good handbooks with little more product. With your general ledger, you should produce complete management reports by profit center (whether that be ranch, pasture, or pen in a feedlot). Profit centers do not start and end with tax years; therefore, profit-center books should be completely separate from tax books.

Be positive that a one-time entry will go to both profit-center and cash accounting. This one entry should probably write your check as well. A normal transaction in an office requires writing a check, addressing an envelope, two entries into the general ledger, and one entry into the profit-center books at a bare minimum. This means five chances for mistakes and numerous chances for further mistakes when these figures are totaled, multiplied, or divided. It also provides numerous opportunities to put the entries off, keeping you from having current books.

Be sure that your system will keep track of original costs of cattle, regardless of how many times the cattle are moved, and that it will make automatic general ledger entries when the cattle die or are sold. This appears to me to be one of the hardest areas of general ledger accounting for cattle.

If you have a one-time entry system, all of the books have to reconcile. You will review the check before it goes out, and if it is correct, all of the entries are correct. The second audit comes when you reconcile the bank statement. Once the bank account reconciles, you can rest assured that both the general ledger and profit-center books are current and correct.

A good system will allow evaluation of each individual operation separately. This means that it will not allow any expenses to get away unnoticed. Our system not only allocates our administrative expenses to all of the cattle we own, it also allocates the administrative expenses particular to one ranch to just the cattle on that ranch. That means that those poor cows have to be responsible for their part of my truck, gas, cowboy, cake, utilities, and windmilling expenses. A system such as this is not very forgiving to unprofitable enterprises.

How many of you know, or are, the type of rancher who takes the cost of all of your cattle each year and adds your expenses for those cattle and subtracts them from income to check your profits? Many ranchers do not even add in all of their overhead expenses. This type of system cannot analyze the profit or loss of any certain area. Worse than this, this type of system is usually done in coordination with tax returns and does not even show an actual profit or loss on any one group of cattle. A quality computer system can pay for itself in what it will save you on producing an accurate general ledger, but the profit will come from being able to analyze your profit/loss centers and discontinue or improve the loss areas.

A good system should allow you to keep track of interest at the rate you pay. Even if you are fortunate enough to operate out of your hip pocket, you have an opportunity cost in that money and the individual profit centers should have to be able to support that opportunity cost, or you will be better off lending the money to the bank and letting someone else hassle with the ranch.

A good system should keep track of hedges. Whether you use this tool or not, it is available and it would be short-sighted to buy a system that does not have this capability. The hedges should carry their profit or loss to the group of cattle they cover and the margin money should be charged interest just like other money invested in the cattle.

A good system should allow you to inventory your branded calves at branding time and assign all of their mothers' costs to them at weaning time.

A quality system should allow you to transfer cattle from one pasture or profit center to another and carry the cost and original investment with them.

I want a system that allows me to keep histories of the cattle I have owned. One of the most valuable attributes of my system is the ability to go back and look at my profit-center close-outs by a combination of twenty-two different variables. This means that I can analyze where the best and worst jobs are being done in different areas of cattle performance: the cattle buyers, the feedlots I feed in, caretakers or farmers on my wheat pasture, or cowboys at a ranch.

Computer systems can produce too much information as well as too little. Reports should be concise as possible. Often programmers ask if certain information is desired and the producer automatically answers yes. Be sure that any information you get on a report has a value. Computers can produce more data than can be assimilated and too much information often clouds or obscures the important information. When I get a long report in the mail, I often set it aside to examine it later. I might or might not ever get to it. If I get a concise report, I spend some time with it analyzing it. Figure 1 is an example of that type of financial report.

You should have the ability to call up lot sheets, close-outs, inventories, general ledger, and financial reports on demand and have as much or as little detail as needed. These reports should be user designed as much as possible. Detail that I feel is important on the lot sheets and close-outs (figure 2) include any transaction (other than allocated expense) that involves money or cattle movement. We use this detail to check back on financial or head-count discrepancy in the case of a mistake in inventory or a discrepancy with a caretaker or feedlot. The inventory (figure 3) should contain summary information that allows you to know about the groups of cattle without knowing the detail of each transaction. For each group, you would want to know the number of head on hand, death loss, number sold, cost and expenses, and futures contracts applicable to that lot.

Documentation should be complete and easy to understand. With a computer system, one standardizes bookkeeping practices so they are not subject to the styles of individual bookkeepers; but new bookkeepers should be able to learn the system easily.

So look for a system that covers your present needs, but look for a system that is adaptable enough to cover your future needs as well. It doesn't cost any more to maintain a good horse than a bad horse, the only difference is when you go to get the job done.

Figure 1

5/31/82   SHAWGO CATTLE CO. MONTHLY FINANCIAL REPORT

EXPENSES (RANCH)	MONTH	YR TO DATE
RANCH OVERHEAD	9661.22	48409.94
RANCH FEED	4937.04	56854.78
GENERAL OVERHEAD	2410.64	12543.42

CASH INCOME STATEMENT

INCOME	MONTH	YR TO DATE
CATTLE SALES	1008.56	445332.49
COMMODITY HEDGES	123.26	-8766.74

EXPENSES		
COST OF CATTLE SOLD/DEAD	-2931.69	-223827.43
FEED	-128380.18	-218677.38
PASTURE	.00	-133291.34
FREIGHT	.00	-14243.07
VET & MEDICINE	.00	-4023.25
INTEREST	.00	-28731.50
OTHER EXPENSES	-2410.64	-12543.42
RANCH FEED	-4937.04	-56854.78
RANCH VET & MEDICINE	-3218.51	-11432.07
RANCH LEASE	-4001.08	-78202.38
RANCH SUPPLIES & REPAIR	-529.49	-3808.98
RANCH TRUCK EXPENSE	-113.06	-2590.04
RANCH FUEL EXPENSE	-992.40	-6177.42
RANCH OTHER EXPENSES	-4807.76	-24601.43

INCOME TO DATE		-376061.32

BREEDING HERD SALES	12768.19	64201.75
COST OF BREEDING HERD SALES	.00	.00

CAPITAL INCOME TO DATE		64201.75

FINANCIAL FACTS

CASH IN BANK	1484.18
NOTE PAYABLE/PCA	-1841546.09
NOTE PAYABLE/SBA	-629933.30

Figure 2

SHAMGO CATTLE CO.   CLOSE OUT

```
AVG IN DATE  31882
CLOSE OUT DATE 82/07/08

Location: BBRNCH     Pen or PASTURE: R.HILL      Lot #: 39004     Sex: MALES      SOURCE: PNA TRN VR  RMG DAYS ON FEED  88
>>>>>>>>>>>>>>>>>>>>>>>>>>>>>>>>>>>>>>>>>>>>>>>>>>>>>>>>>>>>>>>>>>>>>>>>>>>>>>>>>>>>>>>>>>>>>>>>>>>>>>>>>>>>>>>>>>>>
Present Count                                  Total Cost      54204.23     Over Head                93.24
Total Hd In        160.00       Total Wt In    81927.00        Total Int              2235.94
Avg Wt             512.04       Avg Cost/Cwt      66.16        Tot Fd/Exp             7598.83
>>>>>>>>>>>>>>>>>>>>>>>>>>>>>>>>>>>>>>>>>>>>>>>>>>>>>>>>>>>>>>>>>>>>>>>>>>>>>>>>>>>>>>>>>>>>>>>>>>>>>>>>>>>>>>>>>>>>
Tot Hd Sold        160.00       Hd Transferred Out    .00      Tot Deads                .--        .0%
Tot Wt Out      101255.00       Avg Wt Out       632.84        Futures P/L           1808.00
Tot Recpt        67840.00       Avg Price         67.00        Shrink                    .00
>>>>>>>>>>>>>>>>>>>>>>>>>>>>>>>>>>>>>>>>>>>>>>>>>>>>>>>>>>>>>>>>>>>>>>>>>>>>>>>>>>>>>>>>>>>>>>>>>>>>>>>>>>>>>>>>>>>>
```

CATTLE PURCHASES (#=ACTUAL PAID)

Date	Source	No Hd	Tot Wt	Avg Pay Wt	Avg In Wt	Shrink	# Amount	#/Cwt	Ck No	Inv No
32082	FRAZIER PENA 81008	54	29160	540.00	540.00	.00	19099.80	65.50	31582	TRANSFER
32082	81001 PROTEIN PENA	37	20000	540.00	540.00	.00	13100.00	65.50	32082	TRANSFER
40882	HERB RAMAGE	17	8847	520.00	520.00	.00	5882.96	64.24	40882	3900N/17 STRS
50282	FUNNY FARM	52	23920	460.00	460.00	.00	18321.47	68.23	50282	TRANSFER

CATTLE SALES (#=ACTUAL RECEIVED)

Date	Buyer	No Hd	Tot Wt	Avg Pay Wt	# Amount	#/CWT
83082	FRIONA	160	101255	632.00	67840.00	67.00

EXPENSES (#=ACTUAL PAID)

Date	Desc	Ck No	Inv No	# AMOUNT	LBS FEED
33182	J.R. STEELE	1837	FREIGHT	341.69	0
40982	ACCO FEEDS	1843	615238	731.61	0
40982	EDDY COLLIE	1844	FENCE REPAIR	60.00	0
41282	GEBO'S	1866	193789	17.38	0
41282	ACCO FEEDS	1870	615514	1075.25	0
43082	JOHNNY JOHNSON	1900	CARE	104.45	0
61082	GLEN SPILLER	1900	CARE/MAY	104.45	0
61082	JOHNNY JOHNSON	1964	CARE/MAY	81.00	0
61082	GLEN SPILLER	1964	CARE/MAY	81.00	0
62582	JOHN BALDIKE	62582	RANCH LEASE	5000.00	0

FUTURES TRANSACTIONS

Date	Description	No. of Month	Fat-1. Fdr-2.	# Contracts	Trade Price
52682	HEDGE 2 AUGUST	08	2	-2	86.300
62982	AUGUST BUY BACK	08	2	2	64.200

```
                    Deads In     Deads Out     Overhead In
                   19328.00     19328.00
                     120.80       120.80          39.79
                      39.30        39.30          51.36
                      50.87        50.87
```

	Lot Totals	Per Head	Per CWT		
WEIGHT GAIN	67840.00	424.00	67.00	DAYS ON FEED	88
HD WT GAIN	64130.24	400.81	63.34	AV DAILY GAIN	1.37
GAIN COST W/O INT	3709.76	23.19		AVG FEED CONS	.00
FEED GAIN COST+INT	5517.76	34.49		FEED CONV. CST	.00
				TON FEED COST	.00

```
                              CATTLE COST    54204.23
                              EXPENSES        7690.07
                              INTEREST        2235.94
                              TOTAL COST     64130.24
```

				Lot Totals	Per Head
WEIGHT OUT				101255.00	632.84
WEIGHT IN				81927.00	512.04
HEAD IN				160.00	
HEAD OUT				160.00	
HEAD DEAD				.00	

	Lot Totals
SALES	67840.00
COST	64130.24
CATTLE PROFIT	3709.76
FUTURES PROFIT	1808.00
CAT+FUTS PROFIT	5517.76

Figure 3

L O T   I N V E N T O R Y

SHAWGO CATTLE CO.

DATE LAST UPDATE 06/29/82 DATE PRINTED 02/07/08

LOCATION YARD LOT PEN	SEX	HEAD CT	DAS ON FD	AVG PAY WT	PUR CST CWT	DAILY CONSUMPTION PER.	CONSUMPTION YTD	AVG DALY COST	CATTLE SOURCE	TOTAL CATTLE COSTS	EXPENSES TO DATE	NO. DED	NO. SOLD	TOTAL RECEIPTS	FUTURES P/L
FSW 6260 E6E5 STEERS		198	92	589	84.35	23.40	21.27	1.4	AZL	75765.95	21771.69	.5%	1	415.50	-405.00
FAW 6270 E13 STEERS FAT		336	92	652	64.35	20.80	20.86	1.4	AZL	141055.21	29699.86	.0%	0	.00	-587.00
FUT.MO NO C$ MARG FAT							-2 62.15 -2108.09								
FSW 6272 61 STEERS FAT		70	91	719	64.70	26.80	23.41	1.6	FSW	32582.92	6768.15	.0%	0	.00	-162.00
FUT.MO NO C$ MARG FAT		08			-2 67.00 -1083.95										
HERFRD 7252 H3 STEERS		366	118	586	52.75	25.20	31.68	1.9	LEW RNC OZK TRN	113451.12	58379.51	1 .3%	0	.00	.00
1/2 INTEREST HRFRD															
HERFRD 7278 XW0 HEIFER		57	100	564	57.55	22.10	29.59	1.9	AZL	18508.08	7054.33	0 .0%	0	.00	.00
HERFRD 7285 I8Q1S5 STEERS		585	108	605	65.22	25.10	32.51	2.0	PNA TRN OZK KM	232291.76	84832.58	3 .5%	1	188.21	-3386.00
CONT MBP 63.5 AUG															
HERFRD 7295 519T4 STEERS		332	106	613	65.50	26.30	35.12	2.1	PNA TRN OZK KM	133230.26	50038.09	0 .0%	0	.00	-2184.00
CONT MBP 63.5 AUG															
HERFRD 7399 W7 STEERS		207	47	697	54.12	29.90	24.96	2.0	OZK VR ALX TRN	78109.32	4962.33	0 .0%	0	.00	.00
HERFRD 7400 106 HEIFER		51	46	562	52.90	23.70	19.41	1.7	OZK VR ALX TRN	15172.04	962.97	0 .0%	0	.00	.00
ALEX 38003 WHEAT HEIFER		206	1	310	71.39	.00	.00	.3	OZK	57458.40	3376.55	3 1.2%	51	15172.04	.00
FNYFRM 39003 MALES		52	108	427	71.31	.00	.00	.1	OZK	31668.59	470.21	0 .0%	52	16321.47	.00
CONT FRIONA 65 OCT															
BBRNCH 39004 R.HILL MALES		160	87	512	86.16	.00	.00	.2	PNA TRN VR RMG	54204.23	2890.07	0 .0%	0	.00	.00
CONT FRIONA 67 JUL															
YRRNCH 50009 HR5PAS HORSES		9	369	989	69.60	.80	.10	.2	LIT	17900.00	2015.41	0 .0%	17	7898.78	.00
YRRNCH 50011 BULPAS BULLS		46	336	999	74.07	.00	.00	.0	TRN CHR PAN NH	74501.04	-863.13	3 4.2%	23	17152.20	.00
YRRNCH 70101 LOOPHL STEERS		382	106	530	86.35	.00	.00	.2	PNA TRN OZK KM	135836.40	6114.86	4 1.0%	0	.00	.00
YRRNCH 70104 1 SECT COWS		19	108	712	63.17	.00	.00	.1	PNA XL TRN CLV	8550.00	302.01	.0%	0	.00	.00
YRRNCH 70301 6 SECT STEERS		299	117	383	75.03	.00	.00	.4	V-R OZK TRN	83293.29	13796.31	7 2.3%	0	.00	.00
YRRNCH 70401 S CLAY HEIFER		417	108	361	61.26	.00	.00	.4	OZK MSB	93488.38	17835.03	4 .9%	1	166.07	.00

Figure 3a

LOCATION YARD LOT	PEN	SEX	HEAD CT	DAS ON FD	AVG PAY WT	PUR CST CWT	DAILY CONSUMPTION PER.	YTD	AVG. DALY COST	CATTLE SOURCE	TOTAL CATTLE COSTS	EXPENSES TO DATE	NO. DED	NO. SOLD	TOTAL RECEIPTS	FUTURES P/L
YRRNCH 70420 N GAUT	STEERS		179	87	520	72.00	.00	.00	.1	PHL	66991.68	2222.82	0 .0%	0	.00	.00
YRRNCH 70430 BAIRD	STEERS		233	87	441	72.00	.00	.00	.1	PHL	74231.28	2899.30	1 .4%	0	.00	.00
YRRNCH 70462 N CLAY MALES			550	104	375	71.84	.00	.00	.4	OZK V-R MSB RMG	193082.88	25766.97	10 1.4%	157	43871.15	.00
								CONT FRIONA 65 OCT								
YRRNCH 70492 CORSNO	COWS		152	248	714	66.28	.00	.00	.4	TRN	98501.67	12083.38	8 3.9%	44	21470.41	.00
YRRNCH 70498 CORSNO	STRCLV		65	150	0	.00	.00	.00	.0	BRN BB	.00	.00	0 .0%	0	.00	.00
YRRNCH 70499 CORSNO	HFRCLV		71	150	0	.00	.00	.00	.0	BRN BB	.00	.00	0 .0%	0	.00	.00
YRRNCH 70522 BUTTE	COWS		225	226	805	85.31	.00	.00	3.3	TRN	167547.32	19180.59	19 7.6%	0	.00	.00
YRRNCH 70523 BUTTE	STRCLV		92	273	0	.00	.00	.00	.0	CHR BB	.00	.00	0 .0%	0	.00	.00
YRRNCH 70524 BUTTE	HFRCLV		82	273	0	.00	.00	.00	.0	CHR BB	.00	.00	0 .0%	0	.00	.00
YRRNCH 70542 W RIVR	COWS		202	235	999	42.02	.00	.00	1.8	TRN	141265.20	13898.33	1 .4%	35	13927.26	.00
YRRNCH 70548 W RIVR	STRCLV		53	122	0	.00	.00	.00	.0	CHR BRN	.00	.00	0 .0%	0	.00	.00
								SOLD FI AT 65 OCT								
YRRNCH 70549 W RIVR	HFRCLV		33	122	0	.00	.00	.00	.0	CHR BRN	.00	.00	0 .0%	0	.00	.00

HEAD			CATTLE--$			EXPENSES			TOTAL	
FEEDYARD	2094		FEEDYARD	840166.68		FEEDYARD	264269.51		HEDGE INV	-10376.25
STEERS	2094									
HEIFERS	108									
PASTURE			PASTURE	1298522.16		PASTURE	121890.71		INTEREST	153471.93
STEERS	1093									
HEIFERS	623									
INTACT MALES	762									
COWS-BULLS	844									
STEER CLV	210									
HEIFER CLV	186		COST	2136688.84		EXPENSES	386160.22		INVESTMENT	2388989.97
	5720									

802

86

AGNET: A NATIONAL COMPUTER SYSTEM FOR CATTLEMEN

Harlan G. Hughes

In the early 1970s, two professors at the University of Nebraska conceived the idea of an agricultural computer system designed specifically for farmers and ranchers. They developed the computer system now known across the country as AGNET--The Agricultural Computer Network. In 1977, the governors of five states (Nebraska, South Dakota, North Dakota, Montana and Wyoming) jointly funded a pilot project to test if farmers and ranchers in their respective states would use a computer system to make better management decisions. AGNET has now developed so that over 400,000 calls a year are being made to the AGNET computer. AGNET is, indeed, a management tool for agriculture.

Wyoming now has computer terminals in all 23 county extension offices, and county extension agents are now receiving training on how to use these terminals with their farmer and rancher clientele.

AGNET is one of three computers in the Wyoming computer center. This operator controls the AGNET computer from the central station. If we have done the job right, the operator should not have to do much. Due to the speed of the computer, we prefer that the machine do as much of its own operation as possible. This operator, however, can and does take over control of the machine whenever necessary.

AGNET is a mass storage system. Behind the dark windows in AGNET storage units are stacks of phonograph-like records used for storage of data and programs. All of AGNET's programs are stored on disks so that when you type in the name of a program, the computer can immediately go to the appropriate disk and find the requested program. We do not have to wait for an operator to mount a tape or to do any manual intervention. AGNET is one of the largest mass storage systems in the world.

Farmer advisors on the AGNET payroll are very special persons to AGNET. George, one of the advisors, is a real character whom I wish everyone could met. George's role is to help make sure that what we have on AGNET will work for farmers and ranchers. I have heard George say, "Harlan, that is the dumbest #%&"* thing I have ever heard!" Or I have heard George say, "That may be well and fine in your ivory tower, but out on the farm we do not have that kind of data." We have two half-time farmers on the AGNET payroll and they play a very important and unique role in the design and operation of the total AGNET system.

AGNET is equipped so that we can have over 200 telephone calls coming in to the computer at one time. We are now averaging a phone call into AGNET every four minutes, seven days a week, 24 hours a day. That is over 400,000 phone calls a year.

The AGNET computer in Nebraska is located in the basement of the State Capital. By design AGNET is not on a university campus computer (and probably never will be on a campus computer) because of our computer needs and demands. A university computer is set up for research and administrative data processing. We need a service-oriented computer center that can consult us before the system is changed or shut down. Our users are not computer science PhD's and become frustrated with computer down time or off time. AGNET is often our user's first contact with a computer and since they are paying for the computer time, we place some stringent demands on the computer center.

THE FULL PARTNER STATES AND THE STAFF

In 1977, five states previously mentioned became full partners in the AGNET system. The best way to describe a full partner state is to say that each state has a member on the AGNET Board of Directors.

In July of 1980, the state of Washington joined as a full partner state and in October 1980 the state of Wisconsin joined AGNET. In July of 1981, Wisconsin withdrew, which leaves six full-partner states in the AGNET system, but other states currently are considering partner status.

The concept behind AGNET is to share the development and operating costs among the full-partner states. There are approximately 17 people on the AGNET payroll. Of the 17 people, Wyoming is paying for two. Each state pools its resources with the other states so that each state can take advantage of the total efforts of the total 17 people.

I have a goal in life and it is to dissolve state boundaries when it comes to information dissemination and use. We are proving that states have information and computer programs to share across state lines, and as we share our extension resources, the winners are our clientele.

AGNET PROGRAM LIBRARY

The six partner states in the AGNET system have developed the world's largest agricultural and home economics computer-program library in the world. Today there are over 200 programs available to AGNET users. With a library of this size, no one is expected to use or even know how to use all the programs.

Our goal is not to have users able to use all the programs in the library, but rather to have a large enough library so that every user can find at least one program of interest. This large smorgasbord of programs means that users should be able to find several programs of special interest. Appendix A provides a partial list of the programs available on AGNET. I have grouped the programs by subject matter to facilitate user interests.

The AGNET library has been put together with approximately 35 man-years of programming effort. In addition, each program development is supervised by a subject matter extension specialist who is responsible for the content of the program. Each subject-matter program is owned by the subject matter specialist and not by AGNET.

AGNET is exceptionally well equipped for the livestock producer. There are livestock ration-formulation programs available for range cattle, feedlot cattle, hogs, sheep, and poultry. There are programs available that will let you simulate on paper what your cattle will do in the feedlot given a description of your cattle and the ration that you are going to feed them. There are livestock budgets and planning prices stored in selected programs.

AGNET also has programs for the crop farmer. Machinery-cost calculators and crop budgets are available. In addition, there are whole farm or ranch budgeting programs designed to help you make long-run business investment decisions. There are many, many more programs designed to help you make better management decisions.

HARDWARE USED TO ACCESS THE AGNET LIBRARY

Touch-tone telephone. The first computer terminal that I installed in a county extension office in 1972 was a touch-tone telephone. It cost us $14 per month. We used the number pad to send information to the computer and the computer sent back the information over the special loud speaker attached to the phone. We would send in the input numbers by typing them into the telephone. The computer would talk back and say, "Answer number 1 is 420." We printed the answer onto a preprinted form that explained the interpretation of the number. This touch-tone terminal served us very well as a low-cost computer terminal. Industry still uses this type of small, low-cost terminal.

Execuport terminal. It soon became evident that we would like to have terminals in our county extension offices that would print out the computer information. We now have five of the Execuports in the Wyoming AGNET inventory. These cost $1,400 for reconditioned terminals.

Texas Instrument 745 Terminal. We have installed small portable TI-745s in most of our extension offices. The TI-745 weighs 13 pounds, has a clamp-on lid and a handle. It is the size of a small briefcase and weighs about half as much. Wyoming agents transport their terminals all over their counties. The TI-745 costs approximately $1,400 new.

North Dakota's CRT Terminal. Terminals come in all sizes and shapes. The Animal Science Department at North Dakota has a CRT terminal with a TV screen where one can read the data. It also has a printer that can be used to generate a printed copy of the output when desired. These dual-purpose units cost more money, but the flexibility is convenient and does reduce paper costs.

A Decwriter terminal. A Decwriter terminal is used by the Department of Agriculture in Alberta, Canada. I used their terminal to check my electronic mail. Alberta Agriculture subscribes to the AGNET system. This terminal costs around $2,000.

Teletype 43 terminal. My secretary and I use Teletype 43 terminals. Obviously this is the terminal that I like best. The TT-43 gets used more hours than any of our terminals and is virtually a maintenance-free terminal. The only problem is that it is not portable. It weighs 45 pounds and has the terminal plus the telephone coupler and the paper to move. This terminal also costs approximately $1,400.

Terminal with TV screens. We have one special terminal that drives 23-inch TV screens for demonstration and teaching purposes. These are the same TV screens that you see in airports with flight schedules. We use these screens so that clientele and students can see exactly what we type on the terminal and exactly what the computer sends back to the screens. These screens have helped to promote AGNET in Wyoming. The screens work so well that I will not give a demonstration without these screens. The special terminal and the two screens cost approximately $4,000; therefore, we have only one in Wyoming.

MICROCOMPUTERS FOR FARM AND RANCH

Let's now boil this all down. What does it mean for you on the farm or ranch?
Agriculture is going to have some serious challenges in the 80s. During the 60s and 70s your challenge was production, but the challenge in the 80s is going to be financial management.
Yes, the computer has the potential to improve your financial management. Let me make a prediction. Those of you that will be farming in 1990 will be using computers. Those of you that do not want to use a computer will not be farming in 1990. I often hear, "No damn computer is going to tell me how to run my ranch!" I predict that that person

won't be farming in 1990. Many will have retired and others will have gone out of business. Computers are going to become commonplace on U.S. farms and ranches during the 1980s. Producer owned microcomputers can be useful in relation to AGNET. As I travel around the country talking to farmers and ranchers, I hear them expressing interest in three applications of microcomputers. The three applications are:
- Business accounting.
- Herd performance reporting.
- Financial management.

In an accompanying paper, I have discussed microcomputers, their use, and purchase.

Information Networking on AGNET

If you have a telephone coupler for your micro, you can access the following from AGNET:
- Current commodity market prices.
- Current USDA, Foreign Ag and Wyoming news releases.
- Agricultural outlook and situation reports.
- Western Livestock Market Information Project livestock analyses.
- Hay for sale.
- Sheep for sale.
- Certified pesticide applicators in Wyoming.
- People interested in judging county and state fairs.
- Horticultural tips during the summer.
- Home-canning tips during the canning season.
- Emergency information such as drought tips, Mount St. Helen's emergencies, etc.

You can even use your micro to access the UPI and AP news services for news stories dealing with, for example, the Farm Bill or "beef." The AP and UPI news services are available from two commercial time-sharing companies. You can do all this today with your micro if it has a telephone coupler on it.

Marketing Information On AGNET

We are putting about 17 different daily market-price files on AGNET, including the futures opening and closing prices. We have Chicago, Kansas City, and Minneapolis futures going onto AGNET. In addition, we have both national and selected local cash markets. We are reporting local feeder-cattle sales in Wyoming, Northeastern Colorado, and Western Nebraska. Local grain and cattle markets are being put on weekly for Nebraska. Feedlot reports for the major cattle feeding states are going periodically. Export data is also going on weekly. AGNET is becoming a major source of market information for agricultural producers.

This appears to be the major reason for most of our producer subscriptions to AGNET. They want current market information.

SUMMARY

In Wyoming we are using the AGNET system to provide Wyoming farmers and ranchers with their first contact with computers for:
- Record keeping such as beef herd performance.
- Problem solving for computer-aided decision making.
- Information networking such as daily market information.
- Electronic mail to speed up the delivery of research and extension information to clientele.

Computerized Management Aids (CMAs) are not new to agriculture. They are just new to the west. Leading midwest farmers have been using CMAs for over 10 years.

APPENDIX A

PARTIAL LIST OF AGNET PROGRAMS AVAILABLE

Livestock Production Models on AGNET:

BEEF   Simulation and economic analysis of feeder's performance.

BHAP/BHPP   Beef herd performance program and beef herd analysis program.

COWCULL   Package to help determine which dairy cow to cull and when.

COWGAME   Beef genetic selection simulation game.

CROSSBREED
            Evaluates beef crossbreeding systems & breed combinations.

FEEDMIX   Least cost feed rations for beef, dairy, sheep, swine, & poultry.

FEEDSHEETS
            Prints batch weights of rations including scale readings.

RANGERATION
            Ration balancer for beef cows, wintering calves, horses & sheep.

SWINE   Simulation and economic analysis of feeder's performance.

TURKEY   Simulation and economic analysis of turkey's performance.

VITAMINCHECK
            Checks the level of vitamins & trace minerals in swine diet.

WEAN   Performance testing of weaning-weight calves.

YEARLING   Performance testing of yearling-weight calves.

AG Engineering Models on AGNET:

BINDRY   Predicts results of natural air & low temp. corn drying.

CONFINEMENT
            Ventilation requirements & heater size for swine confinement.

DRY   Simulation of grain drying systems.

DUCTLOCATION
            Determines ducts to aerate grain in flat storage bldg.

FAN   Determination of fan size and power needed for grain drying.

FUELALCOHOL
            Estimates production costs of ethanol in small-scale plants.

GRAINDRILL
            Calculates the lowest cost width for a grain drill for your farm.

PIPESIZE   Computes most cost-effective size irrigation pipe to install.
PUMP   Determination of irrigation costs.
SPRINKLER Examines feasibility of installing sprinkler irrigation.
STOREGRAIN
           Cost analysis of on-farm and commercial grain storage.
TRACTORSELECT
           Assists in determining suitability of tractors to enterprise.

Crop Production Models on AGNET:

BASIS   Develops "historical basis" patterns for certain crops.
BESTCROP   Provides equal return yield & price analysis between crops.
CROPINSURNACE
           Analyzes whether to participate in crop insurance program.
FLEXCROP   Forecasts yields based on amount of water available for crop growth.
IRRIGATE   Irrigation scheduling.
RANGECOND Calculates the range condition and carrying capacity.
SEEDLIST   Lists seed stocks for sale.
SOIL LOSS Estimates the computed soil-loss (tons/acre/year).
SOILSALT   Diagnoses salinity & sodicity hazard for crop production.
SOYBEANPROD
           Demonstration soybean production management model.

Home Economic Models on AGNET:

BEEFBUY   Comparison of alternative methods of purchasing beef.
BUSPAK   Package of financial analysis programs.
CARCOST   Calculates costs of owning & operating a car or light truck.
DIETCHECK Food intake analysis.
DIETSUMMARY
           Summary of analysis saved from DIETCHECK.
FIREWOOD   Economic analysis of alternatives available with wood heat.
FOODPRESERVE
           Calculates costs of preserving foods at home.
MONEYCHECK
           Financial budgeting comparison for families.
PATTERN   Helps select a commercial pattern size & type for figure.
STAINS   Tells how to remove certain stains from fabrics.

4-H and Youth Models on AGNET:

CARCASS   Scoring & tabulation of beef or lamb carcass judging contest.

FAIR   Scoring and tabulation of judging contests.

JUDGELIST   List of judges available for fairs and contests.

PREMIUM   Compiles and summarizes fair premiums.

Farm and Ranch Planning Models on AGNET:

BUSPAK   Package of financial analysis programs.

CALFWINTER
     Analyzes costs and returns associated with wintering calves.

COWCOST   Examines the costs and returns for beef cow-calf enterprise.

CROPBUD   Prints out select Wyoming crop budgets.

CROPBUDGET
     Analyzes the costs of producing a crop.

DAIRYCOST   Analyzes the monthly costs and returns with milk production.

EWECOST   Analyzes the costs & returns of sheep production enterprise.

FARMPROGRAM
     Analyzes USDA Acreage Reduction Program.

GRASSFAT   Analyze costs and returns associated with pasturing calves.

LANDPAK   Package of programs to assist in land management decisions.

LSBUDGETS   Designed to print out stored livestock budgets.

MACHINEPAK
     Machinery analysis package.

PLANPAK   Package of programs designed to help analyze and plan aspects of the business.

PLANTAX   Income tax planning/management program.

Information Networking on AGNET:

CONFERENCE
     A continuing dialogue among users on a specific topic.

EWESALE   Lists sheep for sale.

FAS   Prints trade leads & commodity reports provided by USDA-FAS.

GUIDES   Prints available reports of reference material information.

HAYLIST   Lists hay for sale.

MAILBOX   Used to send and receive mail.

NEWS   Latest notifications about programs and user-related information.

NEWSRELEASE
     A program for rapid dissemination of news stories.

WHO IS   Retrieves name and company affiliation of in-
             dividual users.
WYOPROGS   List of specialized Wyoming programs avail-
             able only to Wyoming users.

Market Price Retrievals, Plotting, and Forecasting Models on AGNET:

CASHPLOT   Prints a plot of selected cash prices.
CORNPROJECT
             Projects avg U.S. corn price for various
             marketing years.
MARKETCHART
             Prints various charts on selected future and
             cash prices.
MARKETS   Various market reports and specialists' com-
             ments.
PRICEDATA   Prints selected historic cash and/or futures
             prices.
PRICEPLOT   Designed to plot market prices in graphic
             form.

Miscellaneous Programs on AGNET:

EDPAK   Demo programs illustrating computer-assisted
             instruction.
FILLIN   A "fill in the blank" quiz routine.
GAMES   Package of game programs.
INPUTFORMS
             Prints available input forms.
JOBSEARCH   Matches abilities and interests to occupa-
             tions.
MC   A multiple choice quiz routine.
MICROPROGRAM
             Lists programs for microcomputers.
TESTPLOT   Standard analysis of variance.
TREE   Summarization of community forestry inven-
             tory.

87

RANCHER-OWNED MICROCOMPUTER SYSTEMS: WHAT'S AVAILABLE

Harlan G. Hughes

In the fall of 1977, Radio Shack started advertising the TSR-80 microcomputer for Christmas. This was the beginning of general-public awareness of the personal microcomputer. Another highly advertised microcomputer is the Atari which can be hooked up to a regular TV set, but the Atari is a game computer and, to my knowledge, has no agricultural programs available yet.

CURRENT MICROCOMPUTERS FOR RANCH AND FARM USE

There are two levels of microcomputers being considered by farmers. For the lack of any other terminology, I will use Level I and Level II as the classifications. Level I micros are the lowest cost and most popular systems. The three most common Level I micros in agriculture are the Radio Shack, Apple, and Pet Commodore.

Level I Hardware

Radio Shack Models I, II, & III. The Animal Science Division at the University of Wyoming has a Model I Radio Shack microcomputer. As is typical of most microcomputers, it has a keyboard, a TV screen, a disk unit, and a printer. Dr. Schoonover from Wyoming has developed a herd performance program for the Radio Shack Model I and III microcomputers. This program keeps track of the cow/calf information that ranches have been keeping on 3- x 5-inch cards. Once the data is inside the computer, management reports can be quickly printed out to help the rancher determine the cows to keep and the cows to cull. The same herd performance program that Dr. Schoonover has on the Radio Shack microcomputer is also on the AGNET system. We have several Wyoming ranchers currently using these herd performance programs. The Radio Shack Model II has the disk drive built into the unit. Radio Shack refers to this as their small business machine. The Radio Shack Model III has two disk drives built into the unit and presents pictures and graphs of your data.

The Model III can present a bar graph to show how ranch profits have changed the last 5 years. It has been suggested by some ranchers that a graph is purely academic if it represents ranch profits since profits have disappeared rather than changed.

Apple Computers.

AGNET has an Apple computer with which we have one of our Teletype 43 AGNET terminals as a slave terminal. With proper connections, you can use your existing terminal as a slave printer on your microcomputer. Also, if you have a black and white TV you can back it up as the CRT on the Apple (and other brands as well). The resolution is not quite as clear as a regular monitor, but it is a cheaper way to get set up with a microcomputer.

Dr. Menkhaus, at the University of Wyoming, uses his Apple microcomputer in his Price Analysis class to teach undergraduate students how to use microcomputers.

The newest Apple is the Apple III. It has been out for about a year, but has had some technical troubles that has set its acceptance back. The Animal Science Division at Wyoming cancelled its order for the Apple III and ordered the Apple II Plus. This fast-growing company moved into a new product and forgot something called "quality control."

Pet Commodore Microcomputer.

The Pet Commodore is being used by Alberta Agriculture in Canada and the Ag Economics Department at Wyoming. The Canadians have written a fair amount of agricultural software for the Pet and have been willing to share it with Wyoming so that we do have several decision aids for our Pet Commodore.

Word Processor on Screen.

Micros also can be used for word processing. You can buy word processing programs for almost all micros that will let you use your micro to generate printed materials like letters and reports.

Word processing allows you to electronically add words, delete words, add paragraphs, move paragraphs, etc. When you have your paper like you want it, you can print out the letter or paper on the computer's printer. I now write all my papers on the word processor.

While word processing will not be a big thing for many farmers or ranchers, it might be of value to those of you that are officers of farm organizations. Dave Flintner, President of Wyoming Farm Bureau, could surely use word processing in his Farm Bureau business.

Level II Computers

The more common level II microcomputers that farmers are considering are: Northstar, Vector Graphics, Superbrain, Hewlett Packard, and Cromenco. There are also other brands but they tend to be less popular.

Northstar Microcomputer. One Level II microcomputer that is fairly popular is the Northstar. Country Side Data out of Utah is selling agricultural software for the Northstar computer.

Vector Graphics Microcomputer. Another Level II microcomputer is called the Vector Graphics. Homestead Computers out of Canada has several software packages for the Vector. In addition, Loren Bennett in California has a dairy-ration package for the Vector.

This microcomputer and others can be equipped with a "professional" printer that is used for word processing. If we had a letter typed with this type of printer, I could convince you that the letter was typed by my secretary on her IBM electric typewriter. Professional printers sell for around $3,000; however, if you are going to do word processing, a professional printer is preferred.

One purebred cattleman has a professional printer on his micro. He uses the word processor to write individual letters to his purebred cattle customers. He keeps a list of potential customers inside his computer. When he has a bull for sale, he then uses the word processor to generate and address personal letters to each customer. Each customer thinks the cattleman personally typed the letter to them. In reality, his microcomputer merged the names into the standard letter stored in the micro. This cattleman argues that this is a very cost-effective way to advertise his purebred cattle. The key is the professional printer and the word processing software.

Superbrain microcomputer. A Superbrain is used by South Dakota AGNET with disk drives that are built into the cabinet. This is extremely nice when you move the microcomputer around.

Hewlett Packard. Hewlett Packard recently announced the HP-125 as their small business machine. HP long has a reputation of producing high quality products, and we believe this is also true for their microcomputers. To date, I am not aware of any agricultural software available for the HP machines.

Cromenco Computer. Cromenco microcomputer is configured to be a fairly powerful microcomputer, yet there are several empty slots for future additions to meet your expanding needs. The Level II machines are considerably more flexible than the level I machines.

Comparing Level I and II Microcomputers

There are several differences in the Level II micros as compared to the Level I micros. The key differences are: 1) basic language compilers that are faster than Level I interpretors, 2) 80 character screens that make VISICALC and communications easier to use, 3) more standard operating systems such as CP/M (this means it is easier to exchange programs from one machine to another), 4) more error diagnostics for software and hardware, 5) and the S-100 buss (for more hardware exchangeability).

Hardware Accessories

Data cassette. In the past, we used cassettes for data and program storage. In fact, you can use your kids' cassettes and their tape recorder on your micro to record data and programs. While this is a very cheap storage device, by today's standards it is too slow and inflexible.

Floppy disk. The technology that has made microcomputers of value to agriculture is the floppy disk--a phonograph record with a paper covering around it. Instead of recording music on the disk, the micro records data and computer instructions on the disk. The floppy disk now provides the microcomputer with mass storage capability. Dr. Schoonover can store data for 500 cows in the beef program on one of these disks. If you have 1,000 cows, you simply use two disks. In fact, you can have as many of these disks as you want on the shelf. You just pull off the shelf the disk that you want and put it into your microcomputer.

Hard disk. The newest storage technology is the hard disk. Inside this little box is the ability to store 5 million characters of data. You could store all the management information that you would ever need or generate on your farm or ranch on one hard disk. Most farmers or ranchers do not have this kind of data storage need. The purebred cattleman I know with a Vector Graphics machine keeps all his pedigree information for his cow herd on the hard disk. He can go back to 1932 with his pedigree searches. He feels that the microcomputer has helped his purebred business out considerably.

Instructional Aids

A Radio Shack Teaching Center on our campus has 15 microcomputers hooked up to a sixteenth computer. The sixteenth computer can monitor the other 15 computers. Wyoming's Agricultural Extension Service needs one of these to bring 15 ranchers or farmers in for computer training. You learn more about microcomputers by hands-on experience than from lecture or books. Many high schools and vocational technical schools have such instructional centers but the university extension services are behind.

MICROCOMPUTERS FOR FARM AND RANCH

Let's now boil all this down--what do microcomputers mean for you on the farm or ranch? Agriculture is going to have some serious challenges in the 80s. During the 60s and 70s your challenge was production, but the challenge in the 80s is going to be financial management. And the computer has the potential to improve your financial management. Let me make a prediction. Those of you that will be farming in 1990 will be using computers, and those of you that do not want to use a computer will not be farming in 1990. I often hear, "No damn computer is going to tell me how to run my farm!" I predict that that person won't be farming in 1990. Many will have retired and others will have gone out of business. Computers are going to become commonplace on U.S. farms and ranches during the 80s.

As I travel around the country talking to farmers and ranchers, I hear them expressing interest in three applications of the microcomputer. The three applications are:

- Business accounting.
- Herd performance reporting.
- Financial management.

Top producers are recognizing that they need to keep better books. They are looking to the microcomputer as a means to make bookkeeping easier and more flexible. They want current cashflow situations several times during the year. Today's profit margins do not allow the management errors that you could get by with in the 70s.

Top ranchers know the benefits of good cow-calf records and they have been keeping them on the 3- x 5-inch cards; however, it takes a lot of time to sort them into useful management reports. A herd performance system fits well onto a microcomputer and once the data is in the computer, management reports can easily be printed out. We even know of one rancher that takes his micro right out to the scales and enters the calf weights as they are weighed. When the last calf is weighed, he pushes the button and identifies the cows to be immediately culled. By not having to wait for culling data, this rancher argues that the dollar amount saved from not rounding up cattle the second time will pay for his microcomputer.

Bankers are requiring more and more financial information before they will make loans to producers. Top producers are starting to see the potential of being able to use the microcomputer to help generate these needed reports: financial statements, profit and loss statements, cash flow projections, five-year plans, etc.

VISICALC - a financial management tool. One of the most powerful financial management tools available is VISICALC. It is designed so that you don't have to be a programmer to program your own financial management programs. There is nothing equivalent on AGNET! Since I don't know how to describe in words what VISICALC can do, I suggest that you stop into a computer store and ask for a VISICALC demonstration.

Disk oriented system. In order to have sufficient capacity to handle your agricultural applications, producers should buy a disk-oriented system. It should contain:
- Dual-disk drives.
- A good 80-column printer.
- 32K to 48K memory (the horsepower of the computer).
- 80-column screen (preferred over a 40-column screen).
- Telephone coupler.

The system will cost between $4,000 to $5,000 for the hardware and about $2,000 to $3,000 for programs (software) for your farm or ranch.

Telephone coupler. One of the extremely useful attachments that you can purchase for your microcomputer is a telephone coupler. This will allow you to use your micro as a terminal to large mainframe computers such as AGNET, TELEPLAN, and CMN. You can call the mainframe on the telephone and type in your information on your micro's keyboard and have the output printed out on your micro's printer. The cost of a phone coupler is around $300 and you can access:
- Current commmodity market prices.
- Current USDA, Foreign Ag, and Wyoming news releases.
- Agricultural outlook and situation reports.
- Western Livestock Market Information Project livestock analyses.
- Hay for sale.
- Sheep for sale.
- Certified pesticide applicators in Wyoming.
- People interested in judging county and state fairs.
- Horticultural tips during the summer.
- Home-canning tips during the canning season.
- Emergency information such as drought tips, Mount St. Helen's emergencies, etc.

You can even use your micro to access the UPI and AP news services such as news dealing with the Farm Bill and "beef." The AP and UPI news services are available from two commercial time-share companies. You can do all this today with your micro if it has a telephone coupler on it.

HOW TO BUY A SMALL COMPUTER

What should a farmer and rancher do if he is thinking about buying a small computer?

There are two newsletters that I recommend that you subscribe to on computers in agriculture. Successful Farming publishes one newsletter for $40.00 per year. They make useful evaluations of hardware and agricultural software.

The second newsletter is published by Doane-Western Agricultural Service out of St. Louis, Missouri. Their subscription rate is $48.00 per year. If you are seriously considering a microcomputer, I strongly recommend that you subscribe to one or both of these newsletters.

The second thing I recommend that you do if you are considering purchasing a computer is attend one of the computer seminars that are being held around the country. Almost every state extension service is holding these seminars specifically for farmers and ranchers interested in learning more about microcomputers and the potential agricultural applications. Contact your local county extension agent or extension advisor for information on these seminars.

Books and magazines on microcomputers and how to use and program them are also helpful when selecting a microcomputer. I strongly encourage farmers and ranchers who are thinking seriously about purchasing a computer to get one or two magazines or books on microcomputers. Farmers and ranchers read several agricultural-related magazines, so why not read at least one computer-related magazine.

I personally subscribe to BYTE. It is a good magazine to read to find out what kind of hardware is available and to learn the jargon of computers.

I also subscribe to the Personal Computing magazine. It has stories written by people who are familiar with microcomputers for people like you and me who are not familiar with microcomputers.

SUMMARY

Microcomputers are the new farm- and ranch-management tools and innovative producers are buying them. More and more farmers and ranchers are going to own one or more microcomputers.

If you buy a microcomputer, be sure and buy the telephone coupler so that you can access the agricultural information networks being set up across the country. You will need to spend around $4,000 to $5,000 for a microcomputer with enough horsepower and flexibility to do your farm or ranch applications. I assure you that we are going to see considerably more farm and ranch purchases in the next five years.

88

SIX STEPS FOR A CATTLEMAN
TO TAKE IN BUYING A COMPUTER

Harlan G. Hughes

INTRODUCTION

Today's low profit margins and high interest rates place a premium on a cattleman's management-information system. Automation of that system lends itself to the microcomputer. Microcomputers represent a relatively new farm and ranch-management tool that farmers and ranchers are investigating. Purchasing one may prove to be one of the few profitable equipment purchases of the 1980s. One study indicates that as high as 64% of the producers interviewed were planning to buy a microcomputer as a management tool in the next five years. Twenty-seven percent indicated they would purchase a microcomputer in one to two years. These producers ranked business record keeping as the number one management function they wanted to perform on the microcomputer. The preparation of financial balance sheets and income and cash-flow statements ranked second. Breakeven analysis of individual enterprises and crop-production records ranked as the third and fourth management functions, respectively.

An Alberta, Canada, study of producers owning microcomputers indicated they were using the microcomputers for 1) farm planning, 2) financial record keeping, 3) physical record keeping, and 4) analysis of records (cash flow, breakeven analysis, and costs of production).

What kind of microcomputers do producers own? Sixty percent of the Canadian producers owned Radio Shack and the rest owned Apple, Pet Commodore, Vector Graphics, and others.

A recent Successful Farming magazine survey indicated that 46% of the respondents owned Apples, 34% owned Radio Shacks, 4% owned IBMs, 4% owned Commodore or Pet and the remaining percentage covered all other brands.

SIX STEPS FOR A COST-EFFECTIVE INVESTMENT

A producer-owned microcomputer should pass the same cost/benefit analysis as any other machinery investment.

Costs can be easily identified and documented; however, the benefit of improved management is considerably more difficult to document. What is clear, however, is that benefits received depend heavily on the preparation that the cattleman makes before purchasing the microcomputer.

Step 1

Before purchasing a microcomputer, study your management-information needs. Collection and analysis of management information requires time and money. You cannot afford to collect management information that you do not use or need. Some questions that you should ask are: What are the most important and significant decisions that I need to make? What information is needed to make these decisions? Can the generation of the needed information be scheduled? Can a microcomputer make this information collection easier? Studying your information requires some time and effort. It may well be worth your time to hire a consultant or visit with your university extension service and get a second opinion.

Step 2

Identify computer programs (software) that are available that might meet your management-information needs. As a cattleman you have four potential ways that you can obtain needed software. You can (1) buy it from a commercial vendor, (2) obtain it from the extension service, (3) hire it custom programmed, or (4) program it yourself.

If the software needed is available from a commercial vendor, this may well prove to be the most satisfactory method of acquiring software. Sometimes, however, you'll need software that is not available from a commercial vendor. The local extension service may have what is needed. Occasionally the only viable alternative is to hire a program custom-programmed or to program it yourself. Unless you have special training or a lot of spare time, I cannot recommend that you program the software on your own. Obtaining software tailored to your specific needs will be the most difficult and time-consuming task.

Step 3

Determine the hardware specifications required to execute the needed software. The size of the business affects the volume of management information needed and this, in turn, determines the size of hardware needed. Microcomputers come in different sizes (memory units), have different storage capabilities on the diskette (floppy disk), and have different add-on capabilities (80 column screens, upper and lower case characters, computer languages, CP/M operating systems, telephone modems, word processing software, etc.) Again, it is recommended that you contact a consultant or the extension service. Computer dealers are not necessarily the best information sources for determining specific hardware needs. Generally, they promote what they have to sell.

Step 4

Contact local hardware dealers and determine the viable hardware alternatives. Cattlemen should use the same criteria that they would use for any other equipment purchase: dealer knowledge of his own hardware, quality of the service department, apparent financial stability of the dealer's business and, in general, compatability with the dealer. Since cattlemen have purchased equipment before, they should feel reasonably comfortable with this step.

Step 5

Estimate the cost/benefit of the proposed computerized management-information system. A dealer can tell the purchaser exactly what the hardware will cost; and the cattleman already should have an estimate of what the software will cost. Remember that the cattleman-buyer can take investment credit and depreciation on computer hardware just like any other piece of machinery.

The clerical cost of collecting and processing the management information should also be included. This frequently is your time or that of your spouse. Collecting and typing data into the computer is time-consuming and boring. You might even consider hiring a person to be specifically responsible for the data processing of the management information.

While determining the cost/benefit, cost of the total management information system should be projected. A Michigan State University study indicates that it may cost $500 to $600 a year to process a producer's business records through his own microcomputer. Again, an outside consultant can be useful.

Estimating the dollar benefit of having a computerized management-information system is difficult for most cattlemen to do. Today's high costs of production and high interest rates do not leave much margin for management errors. Just preventing one management error a year may well pay for the microcomputer system. As could the ability to experiment with a decision on paper before implementation.

Step 6

The final step is to make the decision whether to set up a computerized management-information system. You should consider talking to other cattlemen that already own microcomputers. Many states are offering educational seminars for ranchers and farmers to learn more about how microcomputers can enhance a producer's decision-making process. The final decision rests with you the individual. There is no blank recommendation that will fit all situations. Microcomputers can, however, be an effective management tool.

Microcomputers are becoming a more common management tool for cattlemen. Innovative producers are purchasing microcomputers to enhance their personal management-information systems. This article summarizes six recommended steps that you should go through in making the decision to purchase a microcomputer. If these six steps are followed, you will have a higher probability for a successful experience with your first microcomputer.

REFERENCES

Engler, Verlyn, E. A. Unger and Bryan Schurle. 1981. The potential for microcomputer use in agriculture. Contribution 81-412-A. Department of Agricultural Economics, Kansas State Univ.

Nott, Sherrill. 1979. Feasibility of farm accounting on microcomputers. Agricultural Economics Report No. 336. Michigan State Univ.

Successful Farming. 1982. Successful Farming farm computer news. A special survey summary. Successful Farming.

1981. A survey of on-farm computer use in Alberta. Alberta Farm Management Branch, Olds, Alberta.

89

THE MICROCOMPUTER AND THE CATTLEMAN

Donald R. Gill

THE FUTURE

The use of the microcomputer on farms, ranches, and feedlots will probably have as large an impact on the cattle business over the next few years as did the introduction of hybrids on corn production in years past. The possible uses of computers by cattlemen are unlimited and are now available because of the rapid decline in the cost of processors and storage units. Today a unit that costs less than $400 can do more computing than a unit that cost over a million dollars in the early 1960s, and that's using four hundred 1983 dollars and a million 1960 dollars. In the 60s, computers took extensive training to use, but have become easier to use every year since their introduction. Today, while they scare hell out of a 50-year-old rancher, most eight-year-olds find running a computer and developing programs about as hard as learning to ride a bicycle.

WHAT CAN A COMPUTER OFFER A CATTLEMAN?

What can computers do for a cattleman? This list ranges from adjusting the fuel mixture on his pickup to keeping his records--the possibilities are endless. I believe that the greatest computer impact on the cattle will be the production decision-making programs. The reason for this lies in the normal variation in the many inputs that go into the beef business. No cattleman knows how much grass he will grow next year (unless it's none) or how much his steers will gain. For years, because of the wide range of variables that affect the business, cattlemen have used this variation as a good excuse not to budget and project the impact of different management procedures.

WHY THE CATTLE BUSINESS HAS SO MUCH TO GAIN FROM USING COMPUTERS

The real reason for the cattleman's reluctance to figure costs and projected returns is that there are at

TABLE 1
PASTURE GAIN ANALYSIS; NOTE: IF YOU HAVE BEEN USING IMPLANTS IN THE
PAST ENTER THE COST, BUT USE 0 RESPONSE.

	(INPUTS)	TOTAL COST	COST /DAY
CATTLE COST $ PER CWT.	69		
PURCHASE WEIGHT LBS.	450		
DAYS PASTURED	140		
	(INPUTS)	TOTAL COST	COST /DAY
CATTLE INTEREST (RATE) %	16.50	19.92	0.14
PASTURE COST $ / CWT / MO.	2.00	42.00	0.30
MEDICAL COST / HEAD ($)	5.00	5.00	0.04
DEATH LOSS (%)	1.00	3.11	0.02
LABOR COST ($) PER HEAD DAY	0.05	7.00	0.05
MARKETING COST PER HEAD ($)	0.00	0.00	0.00
FIXED FEED COST ($) HEAD	5.00	5.00	0.04
*RUMENSIN 0=NO, 1= RUMENSIN	0*		
*IMPLANTS 0=NONE, 1=IMPLANT	0*		
*IMPLANT (S), COST ($)	0.00*	0.00	0.00
*FEED 0=ENERGY, 1=PROTEIN	0*		
*POUNDS PER HEAD PER DAY	0.00*		
*FEED COST PER 100 LBS.	0.00*	0.00	0.00
OPERATING CAPITAL INTEREST	16.50	1.99	0.01
	TOTAL $	84.02	0.60

COST OF GAIN DEPENDING ON RATE OF GAIN

DAILY GAIN #		COST OF GAIN		SALE WEIGHT		BREAK EVEN $	
BASE	*EST.	BASE	*EST.	BASE	*EST.	BASE	*EST.
0.50	0.50	1.20	1.20	520.00	520.00	75.87	75.87
0.75	0.75	0.80	0.80	555.00	555.00	71.08	71.08
1.00	1.00	0.60	0.60	590.00	590.00	66.87	66.87
1.25	1.25	0.48	0.48	625.00	625.00	63.12	63.12
1.50	1.50	0.40	0.40	660.00	660.00	59.78	59.78
1.75	1.75	0.34	0.34	695.00	695.00	56.77	56.77
2.00	2.00	0.30	0.30	730.00	730.00	54.04	54.04
2.25	2.25	0.27	0.27	765.00	765.00	51.57	51.57

DEVELOPED BY DONALD GILL, OKLAHOMA STATE UNIVERSITY, 1982

NOTE; BASE GAIN ETC. DOES NOT INCLUDE THE AFFECTS OF DAILY FEED, IMPL-
ANTS, OR RUMENSIN. THE EST. GAIN ETC. IS EFFETED BY THE ITEMS MARKED BY
THE (*). WHEN IMPLEMENTING THESE OPTIONS KEEP IN MIND THAT; IMPLANTS IN-
CREASE GAIN 12%, FEED (.09# GAIN /LB FEED), PROTEIN (.15 LB GAIN), AND
RUMENSIN (.2 LB /DAY ADDITION). ***FEED MUST BE FED IF A 1 IS PRESENT
FOR EITHER RUMENSIN OR PROTEIN. SEE YOUR OSU LIVESTOCK ADVISOR.
BPASTURE/VC

826

least 10 to 15 variables needed to figure a reasonable production budget for stocker steers. It takes about 2 hours to figure one of these on a desk calculator and when completed the cattleman might wonder what would happen if he changed one of the input items a little bit. Because most of these inputs interact with one another, a change means about 2 more hours on the calculator. Table 1 is an example of a simple stocker budget. It was done on a TRS-80 Model II® microcomputer using a software package called Visacalc©.

THE CATTLEMAN DOES NOT NEED TO BE A PROGRAMMER

It was not necessary that I know how to program to develop this budget. The Visacalc© package takes about 20 minutes to learn to operate and do the projection in table 1. The screen of the computer is set up into rows and columns by the software package and labels (i.e., CATTLE COST $ PER CWT) can be added at any place. The calculations are made by using the proper mathematical function with the appropriate row or column. The computer screen can display all the information shown in table 1. The user may change any number in the column labels (INPUTS). Cattle costs change and affect costs and break-evens, as do most of the 17 input items.

HOW A SIMPLE PROGRAM CAN HELP

Table 2 is the same as table 1 except that the cattle were fed 4 lb/day of a grain mixture costing $7 cwt. For the winter-wheat-pasture cattle, that was an unprofitable condition. It takes about 2 seconds to change an input and get a new solution on the screen. Table 3 shows the results of cutting the grain to 1 1/2 lb/day and adding Rumensin® to the grain mixture. This condition shows a profit. Most experienced cattlemen recognize that feeding for lowest-cost break-even is a step in the right direction, but cattle weights and selling prices add another dimension. Table 4 illustrates even more forces acting on the cattleman's profit picture: stocker-cattle selling prices to account for the change in price (usually lower) as cattle become heavier; cattleman's equity per head for calculating return on his investment.

There are possibly a million or more combinations of inputs that might best describe a given cattle situation on a particular day. Through trial and error on the screen (instead of in the feedlot) you can control the numbers in your operation and in a very few minutes check out the key opportunities and avoid many of the pit-falls that often turn a profit into a loss.

TABLE 2
PASTURE GAIN ANALYSIS;   NOTE: IF YOU HAVE BEEN USING IMPLANTS IN THE
                           PAST ENTER THE COST, BUT USE 0 RESPONSE.

	(INPUTS)	TOTAL COST	COST /DAY
CATTLE COST $ PER CWT.	69		
PURCHASE WEIGHT LBS.	450		
DAYS PASTURED	140		
	(INPUTS)	TOTAL COST	COST /DAY
CATTLE INTEREST (RATE) %	16.50	19.92	0.14
PASTURE COST $ / CWT / MO.	2.00	42.00	0.30
MEDICAL COST / HEAD ($)	5.00	5.00	0.04
DEATH LOSS (%)	1.00	3.11	0.02
LABOR COST ($) PER HEAD DAY	0.05	7.00	0.05
MARKETING COST PER HEAD ($)	0.00	0.00	0.00
FIXED FEED COST ($) HEAD	5.00	5.00	0.04
*RUMENSIN 0=NO, 1= RUMENSIN	0*		
*IMPLANTS 0=NONE, 1=IMPLANT	0*		
*IMPLANT (S), COST ($)	0.00*	0.00	0.00
*FEED 0=ENERGY, 1=PROTEIN	0*		
*POUNDS PER HEAD PER DAY	4.00*		
*FEED COST PER 100 LBS.	7.00*	39.20	0.28
OPERATING CAPITAL INTEREST	16.50	3.25	0.02
	TOTAL $	124.48	0.89

COST OF GAIN DEPENDING ON RATE OF GAIN

DAILY GAIN #		COST OF GAIN		SALE WEIGHT		BREAK EVEN $	
BASE	*EST.	BASE	*EST.	BASE	*EST.	BASE	*EST.
0.50	0.86	1.20	1.03	520.00	570.40	75.87	76.26
0.75	1.11	0.80	0.80	555.00	605.40	71.08	71.85
1.00	1.36	0.60	0.65	590.00	640.40	66.87	67.92
1.25	1.61	0.48	0.55	625.00	675.40	63.12	64.40
1.50	1.86	0.40	0.48	660.00	710.40	59.78	61.23
1.75	2.11	0.34	0.42	695.00	745.40	56.77	58.36
2.00	2.36	0.30	0.38	730.00	780.40	54.04	55.74
2.25	2.61	0.27	0.34	765.00	815.40	51.57	53.35

DEVELOPED BY DONALD GILL, OKLAHOMA STATE UNIVERSITY, 1982

NOTE; BASE GAIN ETC. DOES NOT INCLUDE THE AFFECTS OF DAILY FEED, IMPL-
ANTS, OR RUMENSIN. THE EST. GAIN ETC. IS EFFETED BY THE ITEMS MARKED BY
THE (*). WHEN IMPLEMENTING THESE OPTIONS KEEP IN MIND THAT; IMPLANTS IN-
CREASE GAIN 12%, FEED (.09# GAIN /LB FEED), PROTEIN (.15 LB GAIN), AND
RUMENSIN (.2 LB /DAY ADDITION). ***FEED MUST BE FED IF A 1 IS PRESENT
FOR EITHER RUMENSIN OR PROTEIN. SEE YOUR OSU LIVESTOCK ADVISOR.
BPASTURE/VC

TABLE 3
PASTURE GAIN ANALYSIS;

NOTE: IF YOU HAVE BEEN USING IMPLANTS IN THE PAST ENTER THE COST, BUT USE 0 RESPONSE.

	(INPUTS)	TOTAL COST	COST /DAY
CATTLE COST $ PER CWT.	69		
PURCHASE WEIGHT LBS.	450		
DAYS PASTURED	140		
	(INPUTS)	TOTAL COST	COST /DAY
CATTLE INTEREST (RATE) %	16.50	19.92	0.14
PASTURE COST $ / CWT / MO.	2.00	42.00	0.30
MEDICAL COST / HEAD ($)	5.00	5.00	0.04
DEATH LOSS (%)	1.00	3.11	0.02
LABOR COST ($) PER HEAD DAY	0.05	7.00	0.05
MARKETING COST PER HEAD ($)	0.00	0.00	0.00
FIXED FEED COST ($) HEAD	5.00	5.00	0.04
*RUMENSIN 0=NO, 1= RUMENSIN	1*		
*IMPLANTS 0=NONE, 1=IMPLANT	0*		
*IMPLANT (S), COST ($)	0.00*	0.00	0.00
*FEED 0=ENERGY, 1=PROTEIN	0*		
*POUNDS PER HEAD PER DAY	1.50*		
*FEED COST PER 100 LBS.	7.15*	15.02	0.11
OPERATING CAPITAL INTEREST	16.50	2.47	0.02
	TOTAL $	99.52	0.71

COST OF GAIN DEPENDING ON RATE OF GAIN

DAILY GAIN #		COST OF GAIN		SALE WEIGHT		BREAK EVEN $	
BASE	*EST.	BASE	*EST.	BASE	*EST.	BASE	*EST.
0.50	0.84	1.20	0.85	520.00	566.90	75.87	72.33
0.75	1.09	0.80	0.66	555.00	601.90	71.08	68.12
1.00	1.34	0.60	0.53	590.00	636.90	66.87	64.38
1.25	1.59	0.48	0.45	625.00	671.90	63.12	61.02
1.50	1.84	0.40	0.39	660.00	706.90	59.78	58.00
1.75	2.09	0.34	0.34	695.00	741.90	56.77	55.27
2.00	2.34	0.30	0.30	730.00	776.90	54.04	52.78
2.25	2.59	0.27	0.27	765.00	811.90	51.57	50.50

DEVELOPED BY DONALD GILL, OKLAHOMA STATE UNIVERSITY, 1982

NOTE; BASE GAIN ETC. DOES NOT INCLUDE THE AFFECTS OF DAILY FEED, IMPL-ANTS, OR RUMENSIN. THE EST. GAIN ETC. IS EFFETED BY THE ITEMS MARKED BY THE (*). WHEN IMPLEMENTING THESE OPTIONS KEEP IN MIND THAT; IMPLANTS IN-CREASE GAIN 12%, FEED (.09# GAIN /LB FEED), PROTEIN (.15 LB GAIN), AND RUMENSIN (.2 LB /DAY ADDITION). ***FEED MUST BE FED IF A 1 IS PRESENT FOR EITHER RUMENSIN OR PROTEIN. SEE YOUR OSU LIVESTOCK ADVISOR.
BPASTURE/VC

TABLE 4
PASTURE GAIN ANALYSIS; WITH PROFIT OR LOSS

NOTE: IF YOU HAVE BEEN USING IMPLANTS IN THE PAST ENTER THE COST, BUT USE 0 RESPONSE.

	(INPUTS)
CATTLE COST $ PER CWT.	69
PURCHASE WEIGHT LBS.	450
DAYS PASTURED	140

CATTLE SELLING PRICE ---->>>>>

PRICE STRUCTURE AT SALE WTS

WEIGHT	$ PER CWT
450	69.00
500	68.50
550	68.00
600	67.00
650	66.00
700	65.00
750	64.00
800	63.00
850	62.00
900	61.00

	(INPUTS)	TOTAL COST	COST /DAY
EQUITY IN $ PER HEAD	75.00		
CATTLE INTEREST (RATE) %	16.50	15.11	0.11
PASTURE COST $ / CWT / MO.	2.00	42.00	0.30
MEDICAL COST / HEAD ($)	5.00	5.00	0.04
DEATH LOSS (%)	1.00	3.11	0.02
LABOR COST ($) PER HEAD DAY	0.05	7.00	0.05
MARKETING COST PER HEAD ($)	0.00	0.00	0.00
FIXED FEED COST ($) HEAD	5.00	5.00	0.04
*RUMENSIN 0=NO, 1= RUMENSIN	1*		
*IMPLANTS 0=NONE, 1=IMPLANT	1*		
*IMPLANT ($), COST ($)	2.65*	2.65	0.02
*FEED 0=ENERGY, 1=PROTEIN	1*		
*POUNDS PER HEAD PER DAY	1.50*		
*FEED COST PER 100 LBS.	12.00*	25.20	0.18
OPERATING CAPITAL INTEREST	16.50	2.89	0.02
TOTAL $		107.95	0.77

COST OF GAIN DEPENDING ON RATE OF GAIN

DAILY GAIN #		COST OF GAIN		SALE WEIGHT		BREAK EVEN $		PRICE OF CATTLE		PROFIT OR LOSS		RETURN ON EQUITY	
BASE	*EST.	BASE	*EST.	BASE	*EST.	BASE	*EST.	BASE	*EST.	BASE	*EST.	BASE	*EST.
0.50	1.05	1.17	0.74	520.00	596.30	75.47	70.17	68.50	68.00	-36.24	-12.97	-125.99	-45.08
0.75	1.33	0.78	0.58	555.00	635.50	70.71	65.85	68.00	67.00	-15.04	7.33	-52.30	25.49
1.00	1.61	0.59	0.48	590.00	674.70	66.52	62.02	68.00	66.00	8.76	26.85	30.44	93.33
1.25	1.89	0.47	0.41	625.00	713.90	62.79	58.61	67.00	65.00	26.31	45.58	91.45	158.45
1.50	2.17	0.39	0.36	660.00	753.10	59.46	55.56	66.00	64.00	43.16	63.53	150.02	220.85
1.75	2.45	0.33	0.32	695.00	792.30	56.47	52.81	66.00	64.00	66.26	88.62	230.32	308.06
2.00	2.73	0.29	0.28	730.00	831.50	53.76	50.32	65.00	63.00	82.06	105.39	285.24	366.37
2.25	3.01	0.26	0.26	765.00	870.70	51.30	48.06	64.00	62.00	97.16	121.38	337.73	421.95

DEVELOPED BY DONALD GILL, OKLAHOMA STATE UNIVERSITY, 1982

NOTE; BASE GAIN ETC. DOES NOT INCLUDE THE AFFECTS OF DAILY FEED, IMPL-ANTS, OR RUMENSIN. THE EST. GAIN ETC. IS EFFETED BY THE ITEMS MARKED BY THE (*). WHEN IMPLEMENTING THESE OPTIONS KEEP IN MIND THAT; IMPLANTS IN-CREASE GAIN 12%, FEED (.09% GAIN /LB FEED), PROTEIN (.15 LB GAIN), AND RUMENSIN (.2 LB /DAY ADDITION). ***FEED MUST BE FED IF A 1 IS PRESENT FOR EITHER RUMENSIN OR PROTEIN. SEE YOUR OSU LIVESTOCK ADVISOR.
EPASTURE/VC

HOW TO SOLVE THE PROBLEM

The mathematical formulas necessary for this calculation are generally available from the state universities. No program is better than the logic that goes into it and some are very poor in this respect. For this reason it is somewhat better to use software like Visacalc© and put your own method of solving the problem into play. A complex program is unnecessary; a single or series of simple programs give satisfactory solutions. The value of the computer is to solve large numbers of repetitive calculations quickly and its ability to save and repeat a problem solving method for future use or expansion when needed.

LIMITATIONS OF THE PROBLEM-SOLVING COMPUTER

But what is a computer unable to do? It is important to remember that you do need to know how to get the answer before asking the question. When we do not know how to get the answers, we rely on the qualified people who write the computed programs. This can be a trap! Although the computer rarely makes errors, the people who write programs have been known to multiply when they should have divided.

GENERAL RECOMMENDATIONS

None of us here today can afford to put off buying a personal microcomputer very long. The competitive advantage in the business world and in personal finances has always belonged to those who could figure all their options. Not many years ago many a sharp young cattleman worked out a good business deal on the back of a napkin while having coffee in the local cafe. Today the napkin and the ball-point pen just can't keep up with a microcomputer that can do thousands or even millions of calculations before the coffee is gone. Few can use the excuse that computers are too expensive. A $3000 unit can do all the problem solving, record keeping and tele-processing on a good-sized beef-cattle operation today. A $400 unit can keep track of all the household expenses, do a lot of problem solving, and after the work is done give you time to play the latest video games.

THE COMPUTER IS JUST A TOOL

When dealing with people, getting advice or buying hardware or software, some salespeople think the world revolves around a computer. Run from them!!!! Those who will be helpful to you will want to understand your business, and instead of trying to change your way of operating to fit their ideas, they will help you use a computer to do better those things you already do well. The computer, like the claw hammer, is a tool. With the right application it is amazing how the proper tool can make a difficult job easy and vice versa.

90

PROGRAMMABLE CALCULATORS FOR BEEF CATTLE MANAGEMENT

John W. McNeill

Electronic devices are a vital part of modern America. They send off an alarm to wake us up, cook our breakfast, open the garage door, and entertain us. Video games such as Pac-Man occupy a prominent location in practically all stores. Computers were used to make your flight reservations to to check you into the hotel.

Electronic devices are available that could be valuable tools if beef producers utilized them with the same zeal as that shown by the video games enthusiast. Electronic calculators made mechanical adding machines obsolete. Computers have made hand recording of information a rarity.

Critics of such devices as computers say they can do the same thing as a $10,000 computer with a 15¢ pencil and a 50¢ pad of paper. This is true since a computer is simply a machine that handles numerical data. A single-bottom, mule-drawn walking plow tilled the soil in the early part of this century. A forty-foot-wide, off-set disc drawn by a 500 horsepower, turbocharged, four-wheel-drive tractor tills the same soil today. Both the mule and the tractor performed the same task. The only differences are in their speed, convenience, timeliness, and efficiency. Not all farmers can justify the four-wheel-drive tractor, but must utilize equipment that meet their specific needs.

The same principle applies to the adaptation of computer technology to beef cattle production. Computers come in all sizes, shapes, costs, and colors. They are merely fast calculators with electronic filing cabinets. Programmable calculators are smaller versions with more limited filing resources that limit their use as devices to handle accounting programs and extensive record-keeping systems. However, they have proven to be very effective for routine calculation purposes such as budget projections. They are an intermediate step between the lead pencil and the microcomputer. The Texas Instrument Model T.I. 59 has received widespread use in agriculture. It is a small, handheld, programmable calculator that can be programmed to perform many routine mathematical programs to aid beef producers. The machine is even more useful when attached to a printer so that a copy of the data can be stored for reference.

Such a calculator with a printer can be purchased for under $400. Programs that tell the system what to do can be purchased from a variety of sources. Many Cooperative Extension Services provide them at a nominal cost. Programs are generally recorded on a small magnetic card. Some programs are grouped into a package, called a module, that is inserted into the machine. Many producers have learned very quickly how to write programs to meet their particular needs. You don't have to understand anything about computers to operate a programmable calculator. I like to use the analogy that you don't have to be a mechanic to drive a car or be an electrical engineer to turn on the lights. All you need to know is how to operate the device and rely on others to service your needs and make repairs. There are many good programs available with self-explanatory instruction manuals.

For years, beef production saw fairly low operation costs and a relatively stable market. In the last decade we have experienced fluctuating prices and escalating production costs. The economic well-being of beef producers will depend upon their ability to make sound management decisions. They need some method to examine all of the management alternatives available to them to develop a management program that will result in the best return on their investment.

Programmable calculators are extremely useful for budgeting. Every producer does some type of budget analysis. Some managers do this in their head and others on a napkin at the coffee shop. Once a budget has been developed on a programmable calculator, one can quickly change a single variable, or combination of variables, and complete another budget. This allows an examination of the financial consequences of many "what if" situations to hopefully develop the best combination of management variables to maximize profits before real dollars are committed.

I would like to demonstrate the use of the T.I. 59 Programmable Calculator with the Stocker Cattle Budget we developed. To enter the program, follow these steps in the instruction manual:

1. Turn on the calculator.
2. Enter 3, 2nd, 9, 1, 7 (719.29 should appear in the register).
3. Touch CLR.
4. Enter side 1 of magnetic card (1 should appear in the register).
5. Touch CLR.
6. Enter side 2 of magnetic card (2 should appear in the register).
7. Touch CLR.
8. Enter side 3 of magnetic card (3 should appear in the register.
9. Touch CLR.
10. Touch A.

The calculator is now ready to execute a stocker cattle budget. This is a self-prompting program that asks you for all of the economically important variables that should be considered in a stocker cattle operation. Simply enter the appropriate information and touch R/S after each entry. Shown below are the prompts, an explanation of the prompts, and a sample entry.

Prompt	Explanation	Sample Entry
PW	Payweight of animals, lb	400
C/C	Cost per cwt (include freight, commission, etc).	$70.50
PCT	Pasture cost per cwt per month	$2.00
MOP	Months on pasture	5
VET	Vet and medicine costs	$4.00
%DL	Death loss, %	2%
LAB	Labor cost per head per month	$2.00
SMN	Salt and mineral cost per head for the season	$3.00
HAY	Lb of hay fed per head for the season	25
C/T	Cost per ton of hay	$75.00
ADG	Expected average daily gain from purchase weight to gross sale weight, lb	1.5
SUP	Lb of supplement fed per head per day	0
DAY	Days supplement fed	0
F/G	Expected lb of feed per lb of additional gain	0
C/T	Cost per ton of supplement	0
INT	Interest rate, %	17.5%
SEL	Sell price per cwt	$67.50
%SH	Pencil shrink on sale weight, %	2%
MKT	Marketing and miscellaneous costs	$3.25

It takes approximately 30 seconds to enter all of the above data. In less than 20 seconds you get all of the information shown below.

612.5 (lb)	Average sale weight
$413.44	Income per head
$282.00	Cost per head
40.00	Pasture cost
10.00	Labor cost
4.00	Vet cost
5.64	Death loss cost
.94	Hay cost
3.00	Salt and mineral cost
0.00	Supplemental feed cost
3.25	Marketing & misc. cost
22.82	Interest cost
$371.65	Total investment
$41.79	Profit per head
212.5 (lb)	Net gain per head
$42.19	Cost per cwt of gain
$60.68	Break-even per cwt

This budget shows a potential profit of $41.79 per head. Alternative budgets could be run on other weights, grades, and classes of cattle or the same set of cattle subjected to a modified management program to see if they offer a greater net return.

Research shows that cattle implanted with a growth promotant will generally gain 12% more than nonimplanted cattle. This would result in an average daily gain of 1.68 lb per day instead of 1.5. The cost of the implant would be approximately $1.00. The cattle would be 27 lb heavier and would likely bring less per pound.

If we enter the same data as in the previous budget with the following changes, we get the budget shown below.

Variable		New Entry
VET	$4.00 + $1.00 for implant	$5.00
ADG	1.5 lb x 112%	1.68
SEL	$67.50 - 1.50	$66.00

639 (lb)	Average sale weight
$421.74	Income per head
$282.00	Cost per head
40.00	Pasture cost
10.00	Labor cost
5.00	Vet cost
5.64	Death loss cost
.94	Hay cost
3.00	Salt & mineral cost
0.00	Supplemental feed cost
3.25	Marketing & misc. cost
22.89	Interest cost
$372.72	Total investment
$49.02	Profit per head
239 (lb)	Net gain per head
$37.96	Cost per cwt of gain
$58.33	Break-even per cwt

In this comparison, the additional $1.00 for the implant returned an additional $7.23 of net profit. Other variables such as supplemental feeding, additional hay feeding, and an improved health program could result in more net dollars. Oftentimes management alternatives that offer the greatest return are not all that obvious on the surface. Too often we get into a routine of buying the same weight, grade, and class of cattle from the same order buyer each year and subject them to identical management regimes without exploring alternatives.

When it came time to sell the cattle projected in the last budget, the owner had the option of selling them as feeder cattle or putting them in a feedyard and retaining ownership. He could explore the economic impact of this alternative by simply entering the Feeder Cattle Budget into the calculator and entering the appropriate data.

Payweight, lb	639
Break-even per cwt	$58.33
Feed efficiency, lb	8:1
Average daily gain, lb	2.85
% death loss	0.5
Processing cost	$4.25
Ration cost per ton	$110.00
Sale price per cwt	$62.00
Interest rate, %	17.5
Final gross weight, lb	1,100

It takes less than 40 seconds to enter the data and get the budget shown below. Routine factors such as a 4% pencil shrink on the gross sale weight are entered into the program to determine sale weight.

1,056	Net sale weight, lb
417	Net gain, lb
162	Days on feed
3,688	Lb of ration
$654.72	Income per head
$372.72	Cost of animal
6.11	Processing, vet & death loss cost
202.84	Ration cost
37.76	Interest cost
$619.43	Total investment
$35.29	Net profit
$48.64	Feed cost per cwt of gain
$59.16	Total cost per cwt of gain
$58.66	Break-even per cwt

Examination of this budget shows a net profit per head of $35.29. At the conclusion of the stocker phase the projected profit was $49.02 per head, so the producer would reduce the overall net profit by $13.73 by retaining ownership through the feeding phase of production. The sound management decision would be to sell the cattle at the conclusion of the stocker phase.

There are many other programs that are available to beef producers. Shown below are available programs we have written.

Stocker Cattle Budget
Feeder Cattle Budget
Cow/Calf Budget
Beef Cow Supplement
Yearling Heifer Supplement

Projected Gains Based Upon Feed Intake
Cattle Diet Analyzer
Feedlot Break-evens
Feedlot Closeouts
Ingredient Inventory

There are many other programs available from other universities and commercial companies. There are people who will write programs to meet the specific needs of individual producers.

The programmable calculator is not as powerful a management tool as the microcomputer, but it is an intermediate management aid that can be of tremendous benefit in helping design the optimal production system to maximize returns from any type of beef cattle operation.

91

COMPUTERS FOR
STOCKER CATTLE MANAGEMENT

John W. McNeill

Each year, millions of calves are weaned off cows and placed in some type of stocker operation prior to entering feedlots for fattening. Most of these calves encounter a change of ownership at least once during this phase of beef production. When stocker cattle are purchased, the owner requires a management system that will provide the greatest return on the money invested in the operation.

The two basic resources any stocker operator has are forage and his labor. How efficiently he converts the forage into net profit determines the net return he receives for his managerial skills.

There are many production variables that must be considered when purchasing stocker cattle such as: weight, grade, class, and sex. Should they be purchased at an auction or through an order buyer? Other decisions: How should the cattle be processed, fed, and marketed? Trying to resolve these factors and numerous others into a viable management program becomes a challenge.

Microcomputer programs have been developed that can quickly, accurately, and easily help in deciding which combination of factors appear to be most economically rewarding.

Management is defined as "manipulation of production variables." The success we have in manipulating the economically important variables in a positive fashion will determine how financially rewarding the venture will turn out.

Shown below is a sample data input form with a hypothetical set of stocker cattle that could have been purchased last year to graze wheat pasture. The costs and input items are typical of those reported by producers.

1. Average payweight   400 lb
2. Cost per cwt $ 70.00
3. Freight cost/head $ 5.00
4. Head Purchased   200
5. Drylot phase
   a. Days held in drylot   14
   b. Pounds of ration/head/day   6
   c. Cost/ton or ration $125.00
   d. Pounds of hay/head/day   6

e.	Cost/ton of hay	$ 75.00
f.	Expected daily gain in drylot	.25
6.	Pasture cost/cwt per month	$ 2.00
7.	Months on pasture	5
8.	Processing cost	$ 5.00
9.	Expected death loss	2%
10.	Labor cost/head/month	$ 2.00
11.	Salt & mineral cost	$ 3.00
12.	Pounds of hay/head	50
13.	Cost/ton of hay	$ 75.00
14.	Average daily gain	1.5 lb
15.	Equity in cattle	30%
16.	Interest rate	17%
17.	Shrink on sale weight	2%
18.	Expected sale price/cwt	$ 64.00

Shown below is the type of economic and performance data that is generated by the computer program.

INCOME	PER HEAD	196 HEAD SOLD
Sale weight	616.	120,722.
Expected cash income	$394.20	$77,262.30
EXPENSE		
Stocker	$285.00	$57,000.00
Drylot	8.40	1,646.40
Pasture	40.35	7,908.60
Death loss	5.92	$0.00
Labor	10.00	1,960.00
Salt and minerals	3.00	588.00
Hay	1.88	367.50
Processing	5.00	1,000.00
Interest	18.62	3,650.36
TOTAL EXPENSES	$378.17	$74,120.90
Net income	$16.03	$3,141.40
Dry lot gain	4	
Pasture gain	225	
Total gain	229	
Days owned	164	
Net gain	216	40,722.00

- Gross av. daily gain (deads out; before shrink)	$1.39
Net av. daily gain (deads in; including shrink)	$1.27
Break-even net rate of gain	$1.11

- Feed cost per cwt gain	$25.81
Interest cost per cwt of gain	$8.96
Labor, processing, market & misc. cost per cwt gain	$7.27
- Total cost per cwt gain	$42.04
Annual return on investment of $17,100.00	40.3%
Income per dollar of expense	$1.04
- Buy price of $70 the break-even sell price is	$61.40
- Sell price of $64 the break-even buy price is	$73.62

The budget shows a projected net income of $16.03 per head on a total investment of $378.17 that amounts to a return of $.04 for each dollar invested. If a producer had purchased the cattle and subjected them to the management system described without prior examination of the potential return, he would likely be disappointed. However, if he examined the budget prior to buying the cattle, he might be able to manipulate some of the management variables to develop more net return. There are some variables over which producers have very little control and others that he can manipulate. Shown below are seven variables over which producers have some managerial control. If they vary either positively or negatively by the units shown in the middle column, they have the impact on the net income by the amount shown on the right.

Production Variable	Unit of change	Influence on Net Income
Purchase price/cwt	± $1.00	± $4.30
Pasture cost/cwt/month	± $ .25	± $5.23
Death loss	± .5%	± $1.61
Average daily gain	± .1 lb	± $9.41
Interest rate	± 1%	± $1.10
Sale shrink	± .5%	± $2.02
Sale price/cwt	± $1.00	± $6.16

The above chart assumes the production variables described change by the units shown and all other variables remain constant. In the previous budget, the projected net income was $16.03. If the stocker operator could shop around and buy the same cattle for $1.00/cwt cheaper from another source, he could expect to make a net profit of $20.33 instead of $16.03. Producers should evaluate this kind of data and determine which areas of management are subject to manipulation and which factors offer the greatest potential to improve the net return. For example, if the producer could 1) buy the same cattle $1.00/cwt cheaper, 2) make certain that they have plenty of pasture and be implanted so that they gain 1.6 lb/day (instead of 1.5), and

3) do a good job of marketing so as to receive $1.00/cwt more than was projected--he could anticipate a net income of $35.90 ($16.03 + 4.30 + 9.41 + 6.16) if all other factors remained constant.

If he were able to influence all of the factors 100% positively, the net income would be $45.99; whereas the net loss would be $-13.62, if the factors were influenced negatively by the same degree.

It is not likely that all of the factors described could be influenced 100% positively. However, if each could be influenced to some degree through proper managerial skills, the cumulative effect would be reflected in better profits. The key to doing this is to determine which areas of management are economically more important than others and to concentrate on those for which management has the expertise and resources.

This is where computers offer tremendous potential to stocker cattle operators. Initial planning prior to purchasing cattle and periodic updates to determine the current status are essential to a successful operation.

Computers come in all sizes, shapes, colors, and capabilities. If you are considering purchasing a computer, first determine what you want it to do for you and find programs (software) that will perform those functions. Then find a computer (hardware) that will run the programs. There are many good companies tht have these resources, and There are commercial firms that provide this service on a fee basis. In some states the Cooperative Extension Service provides this service to producers.

92

COMPUTERS FOR COW/CALF PRODUCERS

John W. McNeill

No segment of agriculture is more steeped in heritage than is the cow/calf industry. Much of the tradition developed due to practical adaptation to environmental production conditions. Early day ranching involved turning cows out on open rangeland and then gathering them in the spring to see how many calves made it through the winter. A mild winter and timely spring rains meant more pounds of beef to sell. Pioneer ranchers never heard of AI, estrus synchronization, the futures market, or investment credit.

Today's rancher is a businessman who uses cows to convert a limited amount of forage into a saleable product -- pounds of beef. Land costs, high interest rates, wages for labor, and an erratic market add up to a high economic-risk occupation.

For years, the beef industry has been promoting maximum production. For example, we have adopted new technology and ideas that maximized weaning weight, such as proper nutrition of the cow so that she milks well, and selecting replacements that have greater genetic ability to grow.

I don't know how often I have been asked "What's a performance tested bull worth?" "How much protein supplement does my cow need during the winter?" and "Does it pay to creep feed calves?"

The answers to these questions are: "It varies," "That depends" and "Maybe."

EVALUATING PRODUCTION ALTERNATIVES

Too often we have been imposing a management regime on our cowherd with little concern for evaluating the economic impact of production alternatives. Every rancher knows yu always need one more bull per 100 cows than you can afford, and that a cow needs cottonseed cake until you feed the last sack! We must ask ourselves if it would pay to buy more bulls if we could increase the conception rate and wean more calves. Would prolonged supplementation in the spring improve cow productivity?

With the current cost/price squeeze, we must closely scrutinize each cost input to determine the economic benefit of its implementation. There are certain cost items that can't be eliminated. For example, adequate bull power is necessary to get an adequate conception rate so that a high percentage calf crop can be weaned. But, if we think that this is a management problem, do we buy more bulls or confine the cows in a smaller area so that a bull can breed more cows? We must analyze the economics of all alternatives and determine which alternative to pursue. It may well be that the cost of doing either is greater than the financial return. If we decide that more bulls would be the answer to our problem, we might have to consider whether or not we need performance-tested bulls. If so, what traits should we stress, what level of performance must we demand, and what can we afford to pay for them?

Cow/calf producers face many questions such as these. Resolving the optimal management program is a mental encounter that most finally discard -- and then continue as before.

Computer programs have been developed to facilitate logical evaluation of production alternatives. One of the greatest benefits of using such an approach is that if forces us to list all of the parameters that influence the profitability of our operation. Oftentimes operations costs that we consider to be incidental are major items when we add them all together.

Let's look at how a computer program we developed can aid cow/calf producers in answering some of these questions. In this system, you simply turn the machine on, insert the program, and answer the following questions that appear on the monitor.

Question	Data Entry
1. NUMBER OF COWS IN HERD	350
2. COST TO PURCHASE A COW	$425.00
3. EXPECTED YEARS OF COW PRODUCTION	6
4. ESTIMATED COW SALVAGE VALUE	$350.00
5. ACRES OF PASTURE PER COW	20
6. PASTURE COST PER ACRE PER YEAR	$4.00
7. TEMPORARY PASTURE COST PER COW	$0.00
8. POUNDS OF HAY PER COW PER YEAR	200 LB
(A.) COST OF HAY PER TON	$75.00
9. POUNDS OF SUPPLEMENT PER COW PER YEAR	240 LB
(A.) SUPPLEMENT COST PER TON	$175.00
10. SALT & MINERAL COST PER COW PER YEAR	$4.50
11. VET & HEALTH COST PER COW PER YEAR	$5.00
12. NUMBER OF BULLS IN HERD	14
13. AVERAGE COST PER BULL	$1,000.00
14. EXPECTED YEARS OF USE	3
15. ESTIMATED SALVAGE VALUE	$750.00
16. PERCENT CALF CROP	90%
17. PERCENT OF HEIFERS KEPT FOR REPLACEMENT	20%
18. WEANING WEIGHT OF STEERS	425 LB

19. WEANING WEIGHT OF HEIFERS	400 LB
20. SALE PRICE OF STEERS - DOLLARS PER CWT	$72.00
21. SALE PRICE OF HEIFERS - DOLLARS PER CWT	$62.00
22. PERCENT SHRINK ON SALE WEIGHT	2%
23. POUNDS OF CREEP FEED PER CALF PER YEAR	0.0 LB
24. MARKETING COST PER CALF	$0.00
25. LABOR COST PER COW	$0.00
26. VEHICLE AND EQUIPMENT COST PER COW	$10.00
27. MISCELLANEOUS COST PER COW	$2.00
28. INTEREST RATE ON BORROWED MONEY - %	17.5%

This budget assumes that the rancher is managing the herd himself with no outside help so that the return is for his own labor. Shown below is the type of data generated.

PERFORMANCE SUMMARY

===

NUMBER OF COWS IN HERD	350.0
NUMBER OF STEERS SOLD	157.5
NUMBER OF HEIFERS SOLD	126.0
NUMBER OF HEIFERS KEPT FOR REPLACEMENT	32.0
POUNDS SOLD, STEERS	65,599.0
POUNDS SOLD, HEIFERS	49,392.0
POUNDS SOLD, TOTAL	114,991.0
POUNDS PRODUCED PER COW	371.3
POUNDS SOLD PER COW	328.5

COW/CALF BUDGET

INCOME	PER HEAD	TOTAL
CASH SALES	$222.44	$77,854.20
REPLACEMENT HEIFERS	$22.22	$7,777.28
TOTAL INCOME	$244.66	$85,631.40

EXPENSES		
COW COST PER YEAR	$12.50	$4,375.00
PASTURE COST	$80.00	$28,000.00
HAY COST	$7.50	$2,625.00
SUPPLEMENT COST	$21.00	$7,350.00
SALT & MINERALS	$4.50	$1,575.00
VET COSTS	$5.00	$1,750.00
BULL CHARGE	$3.33	$1,166.67
CREEP FEED COST	$0.00	$0.00
MARKETING COST	$0.00	$0.00
LABOR COST	$0.00	$0.00
EQUIPMENT COST	$10.00	$3,500.00
MISCELLANEOUS COST	$2.00	$700.00

INTEREST	$88.23	$30,880.20
	-------	-----------
TOTAL EXPENSES	$234.06	$81,921.90
NET EXPECTED	$10.60	$3,709.55

PERCENT OF HEIFERS KEPT FOR REPLACEMENT	20.0%
PRICE SPREAD EXPECTED BETWEEN HEIFERS & STEERS	$10.00
BREAK-EVEN PRICE FOR STEERS	$75.54
BREAK-EVEN PRICE FOR HEIFERS	$65.54
INCOME PER DOLLAR OF EXPENSE	$1.05

This budget shows a net of $10.60 per head. If the rancher wants to look at methods to potentially increase the net, he can simply change the appropriate variables and in a matter of a few seconds create another budget. For example, if he wondered if he could make more money if he were to creep feed the animals, the following items would be changed and all other variables could remain the same.

23. POUNDS OF CREEP FEED PER CALF PER YEAR	480.0 LB
A. COST PER TON	165.0
(ASSUMES 12:1 FEED EFFICIENCY IF FED 4 LB PER DAY FOR 4 MONTHS)	
18. WEANING WEIGHT OF STEERS	465.0 LB
19. WEANING WEIGHT OF HEIFERS	440.0 LB
20. SALE PRICE OF STEERS - DOLLARS PER CWT	$70.00
21. SALE PRICE OF HEIFERS - DOLLARS PER CWT	$60.00
(ASSUMES THE PRICE PER LB WOULD BE LESS BECAUSE HEAVIER)	

With these changes, the following budget is generated.

PERFORMANCE SUMMARY

==

NUMBER OF COWS IN HERD	350.0
NUMBER OF STEERS SOLD	157.5
NUMBER OF HEIFERS SOLD	126.0
NUMBER OF HEIFERS KEPT FOR REPLACEMENT	32.0
POUNDS SOLD, STEERS	71,772.8
POUNDS SOLD, HEIFERS	54,331.2
POUNDS SOLD, TOTAL	126,104.0
POUNDS PRODUCED PER COW	407.3
POUNDS SOLD PER COW	360.3

COW/CALF BUDGET

INCOME	PER HEAD	TOTAL
CASH SALES	$236.68	$82,839.70
REPLACEMENT HEIFERS	$23.65	$8,279.04
TOTAL INCOME	$260.34	$91,118.70
EXPENSES		
COW COST PER YEAR	$12.50	$4,375.00
PASTURE COST	$80.00	$28,000.00
HAY COST	$7.50	$2,625.00
SUPPLEMENT COST	$21.00	$7,350.00
SALT & MINERALS	$4.50	$1,575.00
VET COSTS	$5.00	$1,750.00
BULL CHARGE	$3.33	$1,166.67
CREEP FEED COST	$35.64	$12,474.00
MARKETING COST	$0.00	$0.00
LABOR COST	$0.00	$0.00
EQUIPMENT COST	$10.00	$3,500.00
MISCELLANEOUS COST	$2.00	$700.00
INTEREST	$91.35	$31,971.70
TOTAL EXPENSES	$272.82	$95,487.30
NET EXPECTED	-$12.48	-$4,368.66

PERCENT OF HEIFERS KEPT FOR REPLACEMENT	20.0%
PRICE SPREAD EXPECTED BETWEEN HEIFERS & STEERS	$10.00
BREAK-EVEN PRICE FOR STEERS	$80.03
BREAK-EVEN PRICE FOR HEIFERS	$70.03
INCOME PER DOLLAR OF EXPENSE	$0.95

This clearly shows that under the described conditions that it would not be a wise economic decision to creep feed the calves. On another set of cattle or in another year it might pay to do so, but not under these conditions.

Let's look at another alternative such as implanting the steer calves. Research has shown that implanted steers generally increase weaning weight by 15% over nonimplanted cattle. The implant will cost approximately $1.00, so we must increase the vet and health cost by this amount and make the other appropriate changes in the budget to reflect the weight and value of the steer calves.

11. VET & HEALTH COST PER COW PER YEAR	$6.00
20. WEANING WEIGHT OF STEERS	489.0 LB
21. WEANING WEIGHT OF HEIFERS	400.0 LB
22. SALE PRICE OF STEERS - DOLLARS PER CWT	$69.00
23. SALE PRICE OF HEIFERS - DOLLARS PER CWT	$62.00

PERFORMANCE SUMMARY

===

NUMBER OF COWS IN HERD	350.0
NUMBER OF STEERS SOLD	157.5
NUMBER OF HEIFERS SOLD	126.0
NUMBER OF HEIFERS KEPT FOR REPLACEMENT	32.0
POUNDS SOLD, STEERS	75,477.0
POUNDS SOLD, HEIFERS	49,392.0
POUNDS SOLD, TOTAL	124,869.0
POUNDS PRODUCED PER COW	400.1
POUNDS SOLD PER COW	356.8

COW/CALF BUDGET

INCOME	PER HEAD	TOTAL
CASH SALES	$236.29	$82,702.30
REPLACEMENT HEIFERS	$22.22	$7,777.28
	-------	---------
TOTAL INCOME	$258.51	$90,479.60

EXPENSES		
COW COST PER YEAR	$12.50	$4,375.00
PASTURE COST	$80.00	$28,000.00
HAY COST	$7.50	$2,625.00
SUPPLEMENT COST	$21.00	$7,350.00
SALT & MINERALS	$4.50	$1,575.00
VET COSTS	$6.00	$2,100.00
BULL CHARGE	$3.33	$1,166.67
CREEP FEED COST	$0.00	$0.00
MARKETING COST	$0.00	$0.00
LABOR COST	$0.00	$0.00
EQUIPMENT COST	$10.00	$3,500.00
MISCELLANEOUS COST	$2.00	$700.00
INTEREST	$88.32	$30,910.80
	------	----------
TOTAL EXPENSES	$235.15	$82,302.50
NET EXPECTED	$23.36	$8,177.06

PERCENT OF HEIFERS KEPT FOR REPLACEMENT	20.0%
PRICE SPREAD EXPECTED BETWEEN HEIFERS & STEERS	$7.00
BREAK-EVEN PRICE FOR STEERS	$68.68
BREAK-EVEN PRICE FOR HEIFERS	$61.68
INCOME PER DOLLAR OF EXPENSE	$1.10

This shows that implanting the steers would be an economical management factor to consider under the described circumstances.

Other variables could be considered such as the use of performance-tested bulls, pregnancy palpation, additional supplemental feeding, etc. Individual variables and all sorts of combinations of variables can be evaluated. The computer is an excellent tool to use to apply technology to practical production and determine the economic impact of implementing various factors into our current management system.

There are many other programs available that relate to cow/calf producers, such as ration analysis, calculating adjusted 205 day weaning weights, herd record systems, and determining supplement requirements.

Before buying a computer, decide what you want it to do for you. List functions that you need it to perform, shop for software that will do them, then purchase a computer that will provide computer access services on a fee basis.

Not all producers can justify the use of a computer, but for many it has become an essential piece of equipment in their cow/calf operation. It allows you to make mistakes on paper before you commit real dollars in a high risk enterprise.

93

COMPUTERS FOR FEEDYARD MANAGEMENT

John W. McNeill

Commercial feedyards probably have utilized computers to a greater degree than any has other area of beef cattle production. Feedyards were the only segment of beef production that had a high enough concentration of capital to be able to justify the use of the macrocomputers when they first became available. These machines were large, complex, and costly tools that were not as well received as are the microcomputers that are now available. Microcomputers have many potential uses in feedyards; these compact, relatively inexpensive machines have adequate storage capacity to handle the record keeping and inventory control duties of the large machines. Many of the early systems were on-line, time-sharing systems. Immediate access to the system and lack of personalization of the programs to meet unique needs of feedlots made them less than optimal.

Computers are just high speed calculators with the ability to file large volumes of information. They just add, subtract, multiply, and divide, and have the ability to store data. They don't do anything you couldn't do with a 15¢ pencil and a 50¢ pad of paper. They vary according to the speed with which they process data and the volume of information they can store for reference.

Computers will not solve the problem of poor organizational management. For a computer system to be an asset to management, office management practices must be efficient. The computer is able to process, store, and display only that data which it is given. Systems must be designed to meet the specific needs of a feedlot. Management needs to decide what types of information would aid them in doing a more effective job in feeding cattle. Some items are needed on a daily basis, while others might be adequate if summarized on a weekly or monthly basis. Once output needs are determined, an efficient data collection system must be developed to facilitate those needs. Then computer programs must be developed to be able to process the data into the form needed by management.

The benefits of such a system must be more than the costs associated with it or it is not a sound, economical decision. If some type of electronic device to calculate "break-evens" is all that is desired, you could do that with a programmable calculator at a much cheaper cost.

If you have adequate resources to justify a more detailed system, you should consider developing a total management program. This allows examination of each segment of the feedyard operation that contributes to the overall management efforts. Individual programs such as the following can be developed to meet the specific needs of the feedyard:

Break-even projections
Ration balancing
Inventory control
Customer billing
Employee payroll

Feed mill batch cards
Current pen performance
   projection
Closeouts
Machinery & equipment
   purchases

Computers can aid greatly in creating effective customer relations. Budget projections can be run on various groups of cattle to aid producers in determining what weight, grade, class, and sex of cattle to place on feed. Programs that allow periodic update of cattle performance to see if they are on target with original projections are useful in maintaining communications with customers. There are programs that project current live weight and anticipated marketing date based upon current ration consumption and the ration energy content. Word processors are extremely useful to personalize communication to individual customers. Periodic newsletters on cattle performance, market outlook, and other subjects help maintain good rapport with minimal effort.

Programs are available that can aid managers in making day-to-day and long-range management decisions. As new feed additives, implants, and alternative feedstuffs become available, decisions must be made as to whether or not to incorporate them into the feeding program. For example, the following information was entered into a program we have developed:

1.	Cattle description	Medium frame steers
2.	Origin of cattle	Amarillo auction
3.	Date purchased	7-10-82
4.	Number purchased	200
5.	Average Payweight	650 lb
6.	Cost per cwt.	$62.00
7.	Freight cost per cwt	$1.50
8.	Processing cost per head	$4.00
9.	Death loss	.5%
10.	Average daily gain	2.85 lb
11.	Lb of feed per lb of gain	8.00 lb
12.	Ration cost per ton	$110.00
13.	Misc cost per head	$1.00
14.	Interest rate	17.0%
15.	Average final sale weight	1,050 lb
16.	Sale price per cwt	$64.00
17.	Pencil shrink	4%
18.	Equity in cattle	30%

Shown below is the type of printout that is produced by the computer.

FEEDER CATTLE BUDGET

INCOME	PER HEAD	199 HEAD
SALE WEIGHT, LB	1,056	210,144
GROSS INCOME	$675.84	$134,492.00

EXPENSES		
PAY WEIGHT, LB	650	130,000
FEEDER CATTLE	$412.75	$82,550.00
RATION COST	$198.00	$39,402.00
MISC.	$1.00	$200.00
HEALTH AND PROCESSING	$4.00	$800.00
DEATH LOSS	$2.10	$0.00
INTEREST	$38.53	$7,667.47
TOTAL EXPENSES	$656.38	$130,619.00
NET EXPECTED	$19.46	$3,872.70

PERFORMANCE SUMMARY

ITEM	QUANTITY
HEAD PURCHASED	200.0
NUMBER OF DEADS	1.0
HEAD SOLD	199.0
DAYS ON FEED	157.9
READY FOR MARKET	12/14/82
NET GAIN, PER HEAD	406.0
NET GAIN, PEN TOTAL	80,144.0
FEED CONSUMED, PER HEAD	3,600.0
FEED CONSUMED, PEN TOTAL	716,400.0
FEED CONSUMED PER POUND OF NET GAIN	8.9
FEED COST PER CWT OF GAIN	$49.16
TOTAL COST PER CWT OF GAIN	$59.98
BREAK-EVEN PER CWT	$62.16
DOLLAR OF INCOME PER DOLLAR OF EXPENSE	$1.03
WITH $62 FEEDERS @ 30% EQUITY, ANNUAL RETURN IS	36%

This budget shows a projected net return per head of $19.46. Let's assume that a new feed additive became available that claimed to improve feed efficiency by 10% and average daily gain by 5%. The cost per ton of feed to supply the additive at the recommended level is $6. The following changes could be made in the budget in approximately 10 seconds and a new budget could be produced in less than 30 seconds to determine whether or not feeding this additive would be a sound business decision:

10. Average daily gain   3.0 lb
    (2.85 x 105% = 3.0)
11. Lb of feed per lb of gain   7.2 lb
    (8 x 90%) = 7.2
12. Ration cost per ton $116.00
    ($110 + 6) = $116.00

FEEDER CATTLE BUDGET

INCOME	PER HEAD	199 HEAD
SALE WEIGHT, LB	1,056	210,144
GROSS INCOME	$675.84	$134,492.00

EXPENSES		
PAY WEIGHT, LB	650	130,000
FEEDER CATTLE	$412.75	$82,550.00
RATION COST	$171.72	$34,172.30
MISC.	$1.00	$200.00
HEALTH AND PROCESSING	$4.00	$800.00
DEATH LOSS	$2.10	$0.00
INTEREST	$35.44	$7,052.56
TOTAL EXPENSES	$627.01	$124,775.00
NET EXPECTED	$48.83	$9,717.33

PERFORMANCE SUMMARY

ITEM	QUANTITY
HEAD PURCHASED	200.0
NUMBER OF DEADS	1.0
HEAD SOLD	199.0
DAYS ON FEED	149.0
READY FOR MARKET	12/06/82
NET GAIN, PER HEAD	406.0
NET GAIN, PEN TOTAL	80,144.0
FEED CONSUMED, PER HEAD	3,240.0
FEED CONSUMED, PEN TOTAL	644,760.0
FEED CONSUMED PER POUND OF NET GAIN	8.0
FEED COST PER CWT OF GAIN	$42.64
TOTAL COST PER CWT OF GAIN	$52.69
BREAK-EVEN PER CWT	$59.38
DOLLAR OF INCOME PER DOLLAR OF EXPENSE	$1.08
WITH $62 FEEDERS @ 30% EQUITY,	
ANNUAL RETURN IS	95%

These comparisons show that implementation of the new product into our existing feeding program would net $29.37 ($48.83 - $19.46) per head above the previous budget. We would need to determine the additional costs associated with handling the product such as storage requirement, minimal purchases required, and withdrawals to make the final determination.

There are many such programs that are very useful in determining profitable management alternatives such as whether or not to change grain processing techniques, and whether or not additional grain storage is feasible.

Collection and analysis of management information requires two valuable resources--time and money. Management information is of no value unless you use it to upgrade the effectiveness of your operation.

The following steps should be followed to determine whether or not a computer fits into your operation:

1. Determine your management information needs. Be sure to include the information that you are getting now by other methods and the data that you feel would help you be a better manager.

2. Determine what programs (software) are available to process the raw data into useful information.

3. Determine the cost/benefit ratio of implementing the program.

94

MICROCOMPUTERS: TOOLS FOR MANAGEMENT IN REGISTERED BEEF HERDS

Bill Borror

Management decisions in all segments of the business world require vast amounts of information. Now is an age of information: we are bombarded with it by mailmen, newscasts, extension service programs, and we attend the Stockmen's School for even more! What do we do with all this information? What types of data information are needed for managing a cowherd?

It is obvious to me that we cannot keep all the data we need for managment decisions neatly organized in our brain. There is just too much that we need to know. There are theories that we never really forget anything, that all of the experiences and knowledge acquired in a lifetime are permanently stored. We just can't find it when we want it. You know the feeling? Computers are no smarter than our cows--certainly no smarter than we. But they do have nearly unlimited retrievable memories. Furthermore, the computers memory is almost infallible; once it gets the message it will repeat it over and over again. But the computer is too dumb to attempt any further handling of those bits of information unless prompted by some outside means. Thus we can expect reliable information from a vast area of accessible, retrievable memory from our computer. With a good system this information will be organized in a manner so that we have access to it fast. Information accessibility--that's the name of the game in making management decisions. I'm convinced my microcomputer serves a useful purpose in the game plan for my ranch operation.

Getting into the specifics of computer systems, I'll explain what I want from my system.

Easy operation. I want to be able to operate it without a lot of technical computer experience. Most farmers and ranchers are not going to be hiring professional people to operate the equipment, therefore we need a system that is user friendly.

One-time data entry. The system should be designed so that all input data is entered only one time. For example, once a birthdate is entered on an individual animal, it should never have to be entered again. If we enter that a group of calves was weaned on October 1, 1982, the machine will store it in all the corresponding records.

Discretionary sorting and selecting. I want to be able to sort and select records at my discretion, and the computer is a speed demon at this task. How about a list of cows that has taken more than 2 breedings per conception? How about a list of cows with Maternal Breeding Value Ratios (MBVR) above 102 sorted in order of MBVR to help in making breeding decisions? With a good flexible system you can let your imagination run wild. So my system must be flexible, i.e., I want to be able to set up, select, and sort modes at my discretion--without the help of a professional programmer.

The reports and forms must be in a usable format. My reports are all run on 8 1/2" by 11", 18 lb paper using the page lengthwise so that we can get as much information as possible on one page. This size sheet fits in any standard binder which should be sturdy enough for field conditions.

Accessible information. One of the primary reasons for my buying a computer is that I want the information available when I need it. Most breed associations have a good track record on turn-around-time on processing records, but the mail service doesn't. With the system I have developed I have weight reports completed minutes after the last calf is weighed. Not hours or days--minutes.

I would like to describe each of the reports I am generating and relate how they are used in making management decisions. The listings titled "Form" are used for entering data; those titled "Report" and "Summary" are used for analyzing processed data.

BREEDING DATA FORM   FALL 1982

COW#	SIRE	DAM	WBVR	YBVR	MBVR	BULL#1	DATE 1	BULL#2	DATE 2
A128	B12	C847	101	102	100	------	------	------	------
A146	K40	D203	104	105	104	------	------	------	------

After each cow to be bred in a particular breeding season, I list the cow's ID, the sire and dam of each cow, and her corresponding breeding values. These values are used in making mating decisions. There is blank space for the bull that I will assign to each cow and a space for the date of my actual breeding. I breed all cows AI so that is the basis of my breeding records.

BREEDING DATA REPORT   FALL 1982

Cow#	Bull#1	Date 1	Bull#2	Date 2	Bull#3	Date 3
A128	174	11/20/82				
A146	207	11/22/82				
B456	174	11/22/82	174	12/11/82		

The above is a printout of the breeding data information that is stored in computer memory after the information from the Breeding Data Form is entered. I like to enter the breeding data in the machine at 21 day intervals during the breeding season and run this report after each entry session. One can tell at a glance whether the breeding operation is successful.

CALVING ORDER REPORT   FALL 1983

Cow #	SIRE	EXCLFDATE
A128	174	8/27/83
A146	207	8/29/83
B456	174	9/18/83

After the completion of the breeding season I process the breeding information to calculate an expected calving date and update each cow's individual record with the no. of breedings to reach conception. The Calving Order Report lists the cows in order of expected calving date. I put this with the next report, the Calving Form (example follows), so that at calving time I can merely cross off the the cows that have calved and then can tell at a glance any overdue cows that might require special attention. The computer is a management tool that saves calves.

CALVING FORM   FALL 1982

COW#	BULL	DATE	CALFID	BDATE	SEX	SIRE	BW	CE	BC	COMMENT
A197	005	9/04/81	--
B356	MOOS	9/06/82	--
B492	KING	9/06/81	--

The Calving Form (above) also lists all cows of the particular calving season by cow ID with the recorded expected calf's sire and expected calving date with blanks for the information collected at calving time. I have purposely asked for sire information again just to double check that our expected calving date is within proper parameters. If it isn't, we can check our original Breeding Data Form for possible error and take appropriate action. This form is particularly helpful in cutting down error in recording birth data.

DAM CALVING REPORT   FALL 1981

COW#	CALFID	SEX	SIRE	BDATE	BW	CE	GEST	BC	COMMENT
K485	M412	C	PP	8/20/81	66	1	275		
K488	M626	C	567	9/30/81	55	1	276		
K493	M459	B	419	8/29/81	90	2	277		

The Dam Calving Report (above) lists each cow ID giving all the calving information stored in memory supplied by the Calving Form with the addition of a calculated gestation length. This report is very useful when you want to know the data of the calf of any particular cow, without having to thumb through pages of calf data to find it. An example of readily accessible information!

CALF SIRE SUMMARY   FALL 1981

CLFID	SEX	DAMID	SIRE	BDATE	BW	CE	GEST	BC	COMMENT
M439	B	H520	MOOS	8/24/81	64	1	276		TWIN
M473	B	D163	MOOS	8/30/81	81	1	278		
M485	B	G239	MOOS	8/31/81	83	1	280		
3 B from MOOS ave.					72		278		
M406	C	C681	MOOS	8/23/81	78	1	277		
M421	C	H520	MOOS	8/24/81	66	1	276		TWIN
2 C from MOOS ave.					72		276		
5 calves from MOOS ave.					72		277		

The Calf Sire Summary gives the same information listed on the dam Report except this report is sorted by sex within sire groups, and averages are calculated for birth weight and gestation length. This report gives the number and sex of calves from each sire. It also gives more information on which to base management and breeding decisions!

WEAN DATA FORM   FALL CALVES 1981   WEIGHED

CALFID	S	DAMID	SIRE	BDATE	BW	WWT	WMGT	BC	COMMENT
M401	B	H345	419	8/20/81	77	---------------------			
M402	B	G239	567	8/20/81	80	---------------------			
M404	B	K456	PP	8/21/81	74	---------------------			

The Wean Data Form (above) lists all the calves for collection of weaning data. I list them in calf ID order and sorted by sex since I normally weigh the bull and heifer calves in different groups. Here again by using a pre-printed listing we can catch any entry errors that might have been made at calving time. Any tags that are misread while weighing are easily picked up before the calves are turned out!

WEAN SIRE SUMMARY   FALL 1980 CALVES   4/21/81

CALFID	S	BDATE	SIRE	DAMID	BW	GES	CE	WWT	205W	205R	MGT
L011	C	8/19/80	005	G263	70	274	1	456	403	97	1
L022	C	8/21/80	005	G164	65	271	1	540	476	114	1
L029	C	8/23/80	005	D151	70	276	5	495	433	100	1

The Wean Sire Summary (above) gives birth and weaning data sorted by sex within sire groups and with averages calculated. This is essentially the same information that we get from the breed association. I have not tackled the problem of calculating breeding values, so I am plugging them into the system from breed association reports. Here we can analyze what bull or bulls are superior for weaning weights at a glance by looking at the progeny averages.

WEAN SELECTION REPORT   FALL 1980 CALVES   4/21/81

CALFID	S	DAMID	SIRE	BDATE	BW	205W	205R	COMMENT
L005	B	D076	105	8/15/80	78	597	114	
L050	B	G141	KING	8/27/80	78	593	114	
L043	B	C686	005	8/26/80	86	580	111	
L071	B	E355	TIT	8/29/80	90	579	111	

The Wean Selection Report (above) lists all weaned calves in order of descending adjusted weights at 205 days, sorted by sex. I use this form for making culling selections at weaning time. My keep/cull decisions on the bull calves are largely made on adjusted weaning weights. This form puts that information right at my finger tips!

YEARLING DATA FORM   SPRING 1982 CALVES   WEIGHED   --/--/--

CALFID	S	DAMID	SIRE	BDATE	BW	205W	205R	WWT	YEARWT	COMMENT
N801	B	J881	REX	2/27/82	85	650	111		678	---------------
N804	B	F887	REX	2/28/72	89	643	110		654	---------------
N806	B	C683	REX	3/01/82	90	556	103		610	---------------

The Yearling Data Form is used for collecting yearling data--listed in calf ID order sorted by sex and showing the birth and weaning data already collected.

YEARLING SIRE SUMMARY   FALL CALVES 1980 9/25/81

CALFID	S	BDATE	SIRE	DAMID	BW	205W	205R	365W	365R	COMMENT
L011	C	8/19/80	005	G263	70	403	97	691	101	
L022	C	8/21/80	005	G164	65	476	114	695	102	
L028	C	8/23/80	005	D931	67	394	94	669	98	
L139	C	9/04/80	005	H735	74	488	117	714	105	
					--	---	---	---	---	
AVE. for 4 C from 005					69	440	105	692	102	

The yearling Sire Summary (above) is similar to breed association reports showing calves listed in calf ID order sorted by sex within sire groups. Again, we are showing all birth and weaning data in addition to the new yearling information.

YEARLING SELECTION REPORT   FALL CALVES 1980   9/25/81

CALFID	S	BDATE	SIRE	DAMID	BW	205W	205R	365W	365R	COMMENT
L207	B	9/16/80	174	H501	100	646	124	1272	117	
L155	B	9/07/80	105	G373	78	610	117	1271	117	
L209	B	9/16/80	94	E451	94	642	123	1260	116	
L256	B	10/13/80	TIT	F716	100	606	116	1214	111	
L174	B	9/09/80	174	H639	90	596	114	1213	111	

A listing of calves sorted by sex in descending order of 365 day adjusted weight is on the Yearling Section Report (above).

YEARLING DATA REPORT   FALL 1980 CALVES

CALFID	S	DAMID	SIRE	BDATE	BW	205W	205R	WBVR	365W	365R	YBVR	MBVR
L003	B	C681	005	8/13/80	63	522	100	97	1076	100	101	100
L004	C	B484	KING	8/13/80	55	442	91	92	640	94	9 7	98
L005	B	D076	105	8/14/80	78	597	114	105	1133	101	103	101

The Yearling Data Report is the final report of a particular calf crop. It is a listing of all calves in order of calf ID. This is a permanent office copy showing all information on each calf.

SIRE OF DAM REPORT   FALL CALVES 1980   11/20/81

CALFID	S	DAMID	SIRE	205R	365R	WBVR	YBVR	MBVR	DAM'S SIRE
L199	C	E397	KING	102	94	102	98	99	R72
L126	B	F733	SSN	88	101	100	101	101	R72
L117	C	E368	SSN	112	104	103	105	106	R72
L211	B	F730	TIT	117	112	109	107	100	R72
L234	B	F816	TIT	113	99	108	103	103	R72
L123	C	F825	TIT	100	99	100	101	99	R72
				---	---	---	---	---	
6 CALVES FOR R72 AVE				107	101	103	103	101	
L249	C	H540	005	103	103	98	101	99	SH52
L054	C	H711	105	104	98	99	101	99	SH52
L059	C	H631	174	116	120	105	111	100	SH52
L051	C	H673	KING	104	88	101	96	98	SH52
				---	---	---	---	---	
4 CALVES FOR SH52 AVE				104	102	101	102	99	

The information on a calf crop is sorted by sire of dam and averaged for each of the sires represented. This can be done for a single calf crop or for all progeny of the current cow herd. You can read through this report and get a feeling for which bull or bulls are putting the high-producing females in the herd. A future program can be developed to quantify the information in this report to aid in making proper breeding decisions.

DAM PROGENY REPORT   JULY 23, 1982

DAM ID	CALF ID	S	SIRE	BDATE	GES	BW	CE	205R	365R	WBVR	YBVR	MBVR
E511	H741	B	450	11/02/77	279	75	1	95	100	95	99	99
E511	J231	B	STAR	10/09/78	280	90	1	104	100	101	99	99
E511	K569	C	F31	9/12/79	278	89	1	107	110	105	107	99
E511	L166	B	SSN	9/08/80	281	70	1	102	103	101	104	104
E511	M652	B	BRAV	10/12/81	285	98	1	99				
					---	--		---	---	---	---	---
AVE OF 5 PROG OF E511					281	85		101	103	100	102	100
F718	H523	C	105	9/01/77	275	65	1	110	112	108	110	103
F718	J140	C	F31	9/12/78	280	75	1	128	118	109	112	104
F718	K552	B	F31	9/10/79	282	94	1	106	101	104	104	101
F718	L252	B	TIT	10/18/80	280	77	1	100	99	101	104	98
					---	--		---	---	---	---	---
AVE OF 4 PROGENY OF F718					279	78		110	107	105	107	101

The Dam Progeny Report (above) shows the records of all cows in the herd. I list birth weight, calving ease, birth code, 205 ratio, 365 ratio, all BVRs, and gestation length. Can run this report at any time after any new information is added to the calf file. Each time it is run the cow's own record is updated with the progeny averages and the no. of calves reported.

COW MAIN RECORD						
COW ID . . . D203 DAM ID . . .A128 SIRE ID . . .DYO						
REGISTRATION NO 76740001						
NAME . . . . .TEHAMA EISA ERICA D203						
AGE	DATE	WEIGHT	RATIO	BVR	PROG RATIO	AVE GESTATION
BIRTH	9/10/73	67				6-278
WEAN	4/20/74	450		104		
205 DAY		467	105		6-105	
YEAR	10/14/74	785		105		
365 DAY		789	107		5-108	
MATERNAL				106		
FIRST CALVING DATE. . . 9/12/75			COW GESTATION . . .278			
LAST CALVING DATE . . . 9/15/81			BREEDING COEF. . . .1.22			
EXPECTED CALVING DATE . 8/29/82			CALVING INTERVAL . .366			
COMMENT						

The Cow Main Record (above) lists the information being recorded on each individual cow. All of the identification data and her performance records are transferred electronically from the calf file at the time the decision is made to put the heifer into the cow herd. The progeny ratio information and calving and breeding information at the bottom of the record are updated electronically after each calving. These cow records can be called up on the computer screen one at a time, can be edited, or printed out on hard copy.

CALF MAIN RECORD					
CALF ID . . K456	DAM ID . .	D031	SIRE ID . .		105
SEX C	REG# . . .	9456322			
NAME . . . TEHAMA LOBELIA K456					

AGE	DATE	WEIGHT	CODE/MGT	RATIO	BVR
BIRTH	790918	77	1		
WEAN	800421	504			105
205 DAY		486		108	
YEAR	801014	803			104
365 DAY		789		109	
A.D.G.					
MATERNAL					106

CALVING EASE 1	SORT OPTION 2
COMMENT:	GESTATION . . . 281

The Calf Main Record (above) has the information on
each calf. This record can be called up individually for
observation, editing, or printing.

The various forms supplied by breed associations can be
filled out from computer memory. This is a real time saver
when it comes to registering calves or sending in weight
data. It is also cutting down on transcribing errors. Once
the information is in the machine correctly there is just
no way that our reports can have errors or omissions.

The 18 reports I have listed are the ones that have
proven useful to me so far--I would like to emphasize the
"so far." I doubt if anyone has even scratched the surface
as to what can really be done with these microcomputers for
cattle management, breeding decisions, or keeping and
analyzing our other farm records for maximum returns. The
mechanical technology is here for us to use. The production
records I have outlined here are a beginning. There is much
left to be done.

Part 14

FEEDLOT AND RANCH MANAGEMENT AND PRODUCTION SYSTEMS

95

USING TECHNOLOGY
IN BEEF REPRODUCTION

R. A. Bellows

PRESENT TECHNOLOGY

One of the most effective techniques for beef cattle improvement is use of genetically superior bulls through artificial insemination (AI). Only limited use was made of this technique in beef cattle until frozen semen became available, and it is estimated that still only about 6% of the beef cattle in the U.S. are bred artificially. This low percentage is due to many factors--farm or ranch location, size of land area, terrain, feed-forage base, personal preference of the owner--to name a few. Possibly the main deterrents to use of AI in beef cattle are the increased labor involved and a suspected lower conception rate from AI (The latter point may or may not be valid; some studies support this contention and others do not.)

One of the potential problems in using semen that has been frozen is the relatively low number of sperm that survive the freezing process--this is often only 20 to 40%. A new technique has recently been developed in which the sperm are processed before freezing, resulting in 40 to 60% live sperm following thawing. Preliminary studies suggest that use of the processed semen results in improved conception rates.

The increased labor requirements for AI are very real. In natural service breeding conditions, it is simply a matter of turning the bull into the pasture and letting him handle it from there. But in AI, the detection of heat and breeding depends on the operator, his management capabilities, and his cattle-handling facilities. I am convinced that the majority of beef cows will continue to be bred by natural service, but we will see an increase in AI in the future. If, for no other reason, prices for genetically superior sires are rapidly reaching levels that exceed what can be paid by many producers--whether seedstock or commercial.

Use of heat synchronization as a labor-saving tool will grow in the future. Several techniques are available for experimental use and prostaglandin $F_2\alpha$ is now on the market. The advantage of successful heat synchronization is that it reduces the labor requirement for detection of heat and allow concentration of labor involved for AI over a short (4- to 7-day) time period.

Much work is also underway to synchronize when the egg is shed from the ovary--synchronization of ovulation. If these efforts are successful, detection of heat will not be necessary and breeding can be done at a predetermined time following treatment with the synchronizing drugs.

REPLACEMENT HEIFERS

One of the critical phases of the reproductive cycle is development of the replacement heifer. This heifer must reach puberty, be potentially fertile, and conceive early in the breeding season. This will set the stage for her to produce more, earlier maturing, and heavier calves throughout her lifetime. This requirement means the heifer must be fed nutrients at a level to meet all of her body requirements for growth and development, but she must not be overfed. Overfeeding is costly and can cause the heifer to be a poor milk producer. Adequate growth is also important since the size of the pelvic opening or birth canal of the heifer can be reduced if nutrient intake is not adequate. This is important since it leads to more problems at calving, resulting in increased calf deaths and more labor requirements. Selection of replacement heifers that have conceived early in the breeding season--in addition to selection for growth and maternal traits--must be considered important technology in routine breeding and selection practices.

Induction of puberty in replacement beef heifers has been successful. Various hormone treatments and management schemes have been studied or developed to produce heat and ovulation in the prepuberal heifer. Other studies have shown that treating heifers with zeranol (Ralgro®) at various times following weaning will produce increased growth and increased pelvic size. But, a word of caution. These systems involve use of compounds and drugs. The use of these are controlled by the Food and Drug Administration, and these drugs have not been cleared for use in replacement heifers. Some studies also have shown a reduced pregnancy rate in the treated heifers; therefore much work is needed before we can recommend use of these compounds for these purposes in the industry.

CALVING

Calving time is a very critical period, and it is the time we "harvest the fruit" of our breeding and management decisions--since losses at calving result in a major reduction in the net calf crop. These decisions range from feeding for adequate growth rate in the replacement heifer, to wise selection of the bull to breed the heifer, through having adequate obstetrical facilities and equipment available for assisting and correcting problems that occur at calving. Studies on calf losses at calving have shown that calf birth weight is the most important factor, with pelvic area of the dam ranking second in importance. As mentioned in the discussion on puberty, adequate growth in the replacement heifer is important to assure that growth of the pelvic opening is not retarded.

Technological implications are apparent. Selection of bulls that sire calves with low birth weights and a low incidence of calving problems is mandatory. However, low birth weight and slower growth rate of the calf after birth appear to be highly associated. Various selection indexes are available to slow the increase in birth weight and calving difficulties while resulting in continued increase in growth as measured by higher weaning and yearling weights. Use of these indexes must increase. In addition, culling heifers with small pelvic openings must be considered also as available technology.

Methods for inducing calving are available. Research indicates that gestation can be shortened by up to 10 days without detrimental effects on the calf, although birth weights are reduced. To date, the shortened gestation length and lower birth weights have not resulted in a lower incidence of calving difficulty; but, in most instances, increased calving problems are encountered and the incidence of retained placenta increased. These factors are clearly the result of not exactly duplicating the hormonal changes that normally occur near the end of gestation. Drugs under study at the present time show promise of alleviating these problems.

Induced calving allows accurate prediction of when the calf will be born. This factor has major advantages for allowing planning of labor and facility availability for an intense calving period. Thus, scientists are developing techniques to synchronize breeding and calving.

Our data show that diseases such as scours and pneumonia are the greatest cause of calf deaths during the 6 weeks after birth. New vaccines and drugs have been and will be developed to prevent or treat these conditions. But, good management will still be a key technological factor in any successful disease control program.

REBREEDING

The interval from when the calf is born until the dam comes in heat is termed the postpartum period. The length of this interval is affected by breed, age of dam, and nutritional status of the dam before and after calving. This interval is important because, if it is prolonged, the female may exhibit heat very late in the breeding season or, in some instances, not until after the breeding season is over. In the first case, the female would conceive late in the breeding season, if at all, and in the later instance would not conceive. Nutrition management is the technology of key importance, but breeding programs involving selection of dams that conceive early in the breeding season would tend to eliminate dams that have long postpartum intervals.

Suckling has a marked effect on length of the postpartum interval. If the calf is weaned at birth, the dam will be back in heat within 3 to 4 weeks after calving. This compares to 40 to 90 days in dams that continue nursing their calves. Hormonal treatments have been developed to shorten the postpartum interval, but early weaning of calves to assure that the dam is potentially fertile early in the breeding season is technology that may become more commonplace in future years. Many research studies have involved treatment to synchronize heat and short-term (48 hour) weaning of the calf. This practice has the double benefit of concentrating labor for breeding and causing more cows to respond to the synchronizing treatment.

BULLS

Again, it is my opinion that the majority of beef cows will continue to be bred by natural service, and bull selection must consider reproduction. The bull must produce sperm and have the ability to breed. If either of these conditions is lacking or even partially lacking, conception rates in the herd will suffer. Recent studies have shown that scrotal circumference and testicular consistency in the bull are highly correlated with sperm production. Other studies involve actually determining the mating capacity or libido of the bull when he is exposed to a cow in heat. Bulls that are more active give higher pregnancy rates. One study found that bulls that had greater scrotal circumference had half-sisters that reached puberty at younger ages. These important findings pinpoint technology for selection of breeding bulls based on growth, scrotal circumference, and mating capacity.

"HORIZON" TECHNOLOGY

Semen sexing or sexing of embryos has been a goal of reproduction research for many years. At present, sexed semen has shown little change from the 50:50 sex ratio. Recently, an announcement was made indicating semen was available that would result in a slightly altered sex ratio--approximately 58% bull calves to 42% heifers. Tests of this claim are underway. Sexed semen will be available in the future, but "when" is still a question.

Sexing of embryos involved in embryo transfer is now being accomplished. The method requires removing some cells from the embryo and typing them to determine the structure of the chromosomes. This cell removal and embryo manipulation appears to damage the embryo since pregnancy rates following transfer of sexed embryos is reduced. The procedure is very laborious.

Embryo sexing by observing to determine the presence of male- or female-specific proteins is another promising research area. All male cells produce a protein called the H--Y antigen. This protein appears to be coded by DNA on the Y chromosome, and thousands of copies are inserted into each male cell membrane. The presence of the protein can be detected by antibodies. A method of determination of embryo sex with anti H-Y antibody and a flourescent technique is in preliminary stages of development. If this procedure is successful, it will enable rapid sex determination without killing or damaging the embryo.

EMBRYO TRANSFER

Embryo transfer has exciting possibilities--not only in terms of the growth of that industry noted in recent years, but also in terms of the potential use of this technique. Structures called oocytes are the predecessors of ova (eggs) in the cow. Each ovary contains many thousands of oocytes during fetal life, but production of new oocytes cease before the heifer is born. There is constant degeneration of oocytes after birth. In some species, about 90% of the oocytes degenerate and fewer than 1% are ovulated during the life of the cow. This wastage of ova could be decreased if satisfactory means for harvest could be developed.

The common scheme for increasing ova harvest is by superovulation through hormonal treatment with fertility drugs. These treatments can increase ova production and ovulation up to 50 or 100X, but the average increase ranges from 3 to 10X. We have used this to produce twins and triplets in beef cows. The major problem with hormonal superovulation is its unpredictable animal-to-animal response--with response to the same treatment ranging from 0 to 20 ova. Harvest of ova can be dramatically increased by mincing the ovary and adding enzymes. However, success would require in vitro (in the test tube) fertilization and storage techniques that have not been perfected.

Early embryo transfers involved extensive and expensive surgical procedures. Techniques for nonsurgical collection of fertilized ova from the donor and nonsurgical transfer to suitable recipients are rapidly becoming more successful. These techniques open the door for rapid, safe, economical, experimental, or on-the-farm transfer of ova.

Viable embryos have been obtained from prepuberal heifer calves. Studies to date have involved surgical collection of embryos from prepuberal donors; but as in vitro fertilization techniques advance, ova may be harvested from a prepuberal donor, fertilized in vitro, and then transferred to a suitable recipient. But, remember, selection of the donor would be on pedigree only--not performance.

NEW TECHNIQUES

Selection for high fertility in males is imperative, and techniques appear promising for screening semen in vitro to determine fertilizing capabilities. This procedure could be accomplished by collecting ova from ovaries obtained from slaughtered females and using them in an in vitro fertilization test. This technique, however, would do nothing to determine the libido and breeding ability/capacity of the male in question.

In vitro fertilization and fertilization outside the body are techniques that are definitely on the horizon. This procedure has been accomplished, and interesting "oviduct" factors have been identified as facilitating successful maturation and fertilization of the egg. This technology would provide the possibility of inserting sperm directly into an oocyte. To date, live offspring have been obtained through this technique in laboratory species, but not in cattle. It would also mean that ova could be harvested from ovaries and stored--either fertilized or unfertilized--for future use. This would mean that unfertilized ova from a particularly valuable cow could be kept frozen for various time periods and thawed and fertilized by exceptional bulls as they become available, thus adding greater flexibility to mating plans and breeding systems.

Other "futuristic" techniques that might become available are parthenogenesis and cloning. This might be accomplished by fertilizing one oocyte with another, as has already been done in the mouse. Although none of these embryos have gone to term, all individuals would be female. Another intriguing possibility would be to fertilize an ovum with two sperm and then remove the female pronucleus. Two Y-bearing sperm would result in death; one X- and one Y-bearing sperm would result in a male and two X-bearing sperm, a female. Thus, one could obtain females with DNA provided only by sperm. This would be the equivalent of fertilizing one sperm with another and would permit crossing of two bulls instead of a cow and a bull. This might result in greater genetic progress since the genetic merit of bulls is often known more precisely than that of cows. If one could extend this to use of two sperm from the same bull--in effect crossing a bull with himself--you would be approaching an identical copy of the individual.

Sexual reproduction is a game of chance resulting in a nearly infinite number of possible gene combinations. Breeders have succeeded in removing some of the change by progeny testing, planned matings, developing inbred lines, careful selection, etc. However, progeny of outstanding individuals are often less outstanding than their parents, although they are generally above average. Cloning, defined as asexual reproduction and splitting of an embryo to make two offspring from one, is a technology that could be used to produce genetically identical copies of outstanding animals. The ability to clone embryos through several generations of identical clones would provide a tool for accelerating genetic changes that is potentially more powerful and effective than artificial insemination.

Methods of early pregnancy diagnosis are on the horizon. Palpation of the embryo per rectum is of value as early as 40 days postbreeding, and determining progesterone levels in milk or blood appears to have merit by day 24. Detection of HCG--a pregnancy-specific protein in humans--is the basis for pregnancy detection in humans. It is quite likely that a pregnancy-specific protein will be found in blood, mucus, or milk of cattle that can be used as an early indicator of pregnancy. Early pregnancy diagnosis would be of value for both management and selection purposes. Pregnant animals could be placed on diets suitable for pregnant dams; open animals could be culled. It would be a breakthrough for selection for pregnancy occurring early in the breeding season. In heifers, it could be used to eliminate those animals with delayed puberty, and lactating dams with excessive postpartum intervals could be identified and culled.

In the discussion of superovulation, I mentioned we have produced twins and triplets in beef cows. These studies combined synchronization of heat and superovulation and resulted in calf crops of 119% in the treated dams, compared to 70% in the untreated. Other scientists are studying the possibility of developing lines of cattle that naturally produce a high percentage of twin births. Some researchers have successfully produced twins by embryo transfer techniques. These studies have major potential for beef producers of the future.

GENERATION INTERVAL

Another point we are often not fully aware of is the long generation interval in beef cattle. For example, under good conditions, a bull used for breeding as a yearling is about 480 days old at the end of the first breeding season. A replacement heifer will be about 760 days old when her first calf is born, and that calf will be almost 400 days old when it reaches the Choice slaughter grade. We cannot be satisfied with these long time periods--time is money-- and this factor shows little evidence of changing very quickly. We must develop replacement heifers more quickly and find methods to develop and measure genetic merit of bulls at younger ages. Gestation must be shortened, as must the postpartum interval. The feeding period can be reduced through more effective rations, feeding bulls and appropriate carcass grade changes. All these factors must be given careful consideration in beef breeding-production programs of the future. But--let there be no doubt--we are on the threshold of a technological revolution in the beef industry. Are you ready for it?

96

USING GROWTH SIMULATION MODELS
IN BACKGROUNDING AND FEEDLOT
MANAGEMENT DECISIONS

James N. Trapp

Making informed decisions about buying, selling, and feeding cattle are critical to success in today's cattle industry. Sooner or later, all cattlemen find themselves sitting down with paper and pencil, asking some hard questions, and taking a stab at budgeting a feeding or grazing activity to see if it will pay. This budgeting process is likely to be repeated several times during the life of the cattle in light of new information. For every decision made, a producer probably has (and should have) considered several alternative production activities. All this adds up to a lot of pencil pushing and guesswork about growth performance on different rations or pasture systems, time on feed etc.

Computerized beef-growth simulation models have been developed that can aid in the decision-making process for feedlot and backgrounding operations. The programs can assist the decision-making process in several ways. First, they take the drudgery out of pencil-pushing by computerizing the calculations required. This speeds up the process and allows one to easily look at many more alternatives than he may have previously taken the time and trouble to consider. Second, the growth simulation models provide a method for removing some of the guesswork about animal performance. The models have been designed and tested using data from many experimental and commercial feeding experiences. They are capable of accurately describing growth response over time for a fairly wide variety of feeds, pastures, cattle types, and seasons. An experienced producer familiar with his system of feeding and certain types of cattle can probably out-perform the model in projecting the results of certain systems. When it comes to considering new systems or situations where one is lacking in experience, the simulation models and a basic familiarity with cattle feeding and/or backgrounding can greatly reduce guesswork. Third, the simulation models provide an organized and consistent way to obtain information with which to make decisions. The models force their users to explicitly consider the same comprehensive set of information each time a production activity is considered. All calculations are made in the same manner each time and are free of math errors. Finally, the output or results are presented in neat and consistent forms for the user to keep a record of what he has considered and what was found.

Two beef-growth simulation models designed to aid in management decision-making are presented here. The first of these is a stocker-growth model designed to simulate and budget the performance of stockers on various types of pasture and forage systems. The second is a feedlot-growth simulation model designed to perform the same activities for cattle on feed. Emphasis is on outlining the information required for their operation and the information generated. A more detailed discussion can be obtained from publications cited in the reference list.

STOCKER CATTLE GRAZING MODEL

The stocker-cattle simulation model outlined here was developed to predict stocker-cattle performance and to compare the profitability of alternative stocker cattle production choices. (The model was developed by B. W. Borsen and O. L. Walker of Oklahoma State University. Copies of the model that will function on either a Tandy or Apple computer can be obtained from Dr. Ted Nelson, Oklahoma State University, Stillwater, OK 74078. A nominal charge is made for the disk and documentation provided.) The information required by the program is listed below:

1. Cattle data
   - Sex
   - Purchase weight (lbs)
   - Purchase price ($/cwt)
   - Planned selling weight (lbs)
   - Expected selling price ($/cwt)
   - Death loss (%)
   - Shrink from purchase weight to on-feed weight (%)
   - Shrink from production weight to sale weight (%)
2. Growth adjustment data
   - Is rumensin used? (Yes/No)
   - Is an implant used? (No, Synovex or Ralgro)
   - Expected weight to grade choice (lbs)
   - Previous average daily gain (lbs)
   - Percent reduction from normal feed intake for first day after purchase and shipment
3. Financial data
   - Commissions ($/head)
   - Trucking ($/cwt)
   - Veterinary/medicine ($/head)
   - Interest rate (%)
   - Equity ($/head)

- Labor ($/head/month)
- Equipment ($/head/month)
- Pickup ($/head/month)
- Minerals ($/head/month)
- Pesticides ($/head/month)
4. Pasture selection
- Pasture type (overseeded Bermuda, Bermuda, tall native grass, short native grass, Sudan Grass, Lovegrass, fescue, wheat)
- Beginning month (Jan 1...Dec 12)
- Days to graze
- Annual pasture cost per head
5. Supplementary feed data
- Pounds per head of corn, milo, soybean meal, cottonseed meal, wheat, alfalfa hay, prairie hay and wheat straw
- Nutrient content of each feed
- Price of each feed type ($/cwt)

The computer program prompts the user to enter each value required by printing a line of abbreviated headings under which the data are entered. If the user is not certain of what to enter for any data item, or is content to use a default value reflecting a standard condition, the data slots in question may be left blank. This will cause a default value to be automatically entered. Once all the desired data have been entered, a display of the entries is made that allows the data to be checked and changed if desired. The default values provided are also displayed. Figure 1 shows a typical display for a set of entered data. The display shows a summary of the input data as well as the output/results generated by the program.

The first section of the output lists the nutrients specified as available for the type of pasture selected. In this case overseeded Bermuda grass pasture was selected. The assumed total digestible nutrient (TDN) and crude protein (CP) content of the dry matter (DM) produced by the pasture are reported in percentage terms. The pounds of dry matter consumed per acre each month also are listed. Dry matter consumed per acre is a function of forage quality and does not necessarily equal dry matter available. In the case cited, no dry matter would be consumed by animals on overseeded Bermuda pasture during November through February, even though some might be available. Any of the listed default nutrient values may be altered by the user.

The next four lines of the output indicate the headings and data entered for various cattle data, growth data and financial data. Following this, a tables shows the projected growth pattern for 400 lb steer placed on overseeded Bermuda grass on March 1. The respective columns across the table report projected weight in 15 day increments, dry matter consumed per day, gain per day, grazing capacity per acre, pounds of supplementary feed fed per day, marginal

OVERSEEDED BERMUDA

	JAN	FEB	MAR	APR	MAY	JUN	JUL	AUG	SEP	OCT	NOV	DEC
TDN	35.6	37.6	68.0	66.7	63.9	56.9	55.2	52.1	54.9	50.1	42.8	41.9
CP	5.6	6.6	25.0	24.2	20.6	16.9	10.0	9.8	10.0	12.1	8.2	7.1
DM	0	0	265	1000	810	925	1030	970	950	220	0	0

```
SX   BUYWT  BUYPR   RUM   IMPLANT  PADG  CHWT
S 400.00  95.01   1.00   1.00    1.00  1050
COMM  TRKRT VETMED OTH/DY  INTRT $EQUITY
3.50   0.34   4.85   0.07   0.12   0.00
```

STOCKERS ON OVERSEEDED BURMUAGRASS

DATE	WEIGHT	FD/DY	GAIN/DY	HD/AC	LB.SUP.	MR-MC	PROF/DY
3 0	400.00						
3 15	435.03	12.21	2.34	0.72	0.00	0.30	0.30
3 30	471.18	13.28	2.41	0.67	0.00	1.11	0.70
4 15	507.67	14.66	2.43	2.27	0.00	1.00	0.80
4 30	545.17	15.80	2.50	2.11	0.00	0.90	0.83
5 15	577.10	16.14	2.13	1.67	0.00	0.61	0.78
5 30	607.92	16.81	2.06	1.61	0.00	0.48	0.73
6 15	622.10	14.61	0.94	2.11	0.00	-0.01	0.63
6 30	635.80	14.82	0.91	2.08	0.00	-0.04	0.54
7 5	639.31	14.48	0.70	2.37	0.00	-0.13	0.52
7 10	642.80	14.54	0.70	2.36	0.00	-0.14	0.49
7 15	646.29	14.61	0.70	2.35	0.00	-0.14	0.47
7 20	649.76	14.67	0.70	2.34	0.00	-0.15	0.45
7 25	653.24	14.74	0.69	2.33	0.00	-0.15	0.43
7 30	656.71	14.81	0.69	2.32	0.00	-0.15	0.41

```
STEER CLOSEOUT AFTER  150 DAYS.            POUNDS     $
ADG = 1.71 LB/DAY... INTAKE=14.76  LB/DAY
AVG HD/AC = 1.79   MIN HD/AC= 0.67
CATTLE AT $  95.01/CWT..........      400.00   380.04
MISC. COSTS AT $ 0.07/DAY...(LABOR= 0
EQU = 1.5 PICK-UP= 3 MIN= 4 PEST= 1.5 )        10.00
INTEREST @ 12 PERCENT............             20.91
COST OF SUPPLEMENT AT $12.10/CWT........  0.00   0.00
PASTURE COST AT $ 1.43/CWT D.M......   2,213.98  31.67
D.L = 7.10 +MED= 4.85 +COM= 3.50 +TRK= 1.36    16.81
TOTAL SPECIFIED COSTS                        459.43
SALE VALUE @ $  79.25/CWT.............  656.71  520.46
NET RETURNS TO $ 0 EQUITY,MGMT,RISK.
          & UNPAID LAND & LABOR                61.02
BREAKEVEN SALE PRICE..............             69.96
```

NUTRIENT REQUIREMENTS TDN=% DP=% DM=LB/ACRE

	JAN	FEB	MAR	APR	MAY	JUN	JUL	AUG	SEP	OCT	NOV	DEC
TDN	0.0	0.0	68.0	66.7	63.9	56.9	55.2	0.0	0.0	0.0	0.0	0.0
DP	0.0	0.0	7.4	6.6	5.9	5.0	4.8	0.0	0.0	0.0	0.0	0.0
DM	0	0	382	457	494	441	439	0	0	0	0	0

Figure 1. Computer output from simulation of stocker cattle performance on overseeded bermuda grass

revenue minus marginal cost or net profit per day, and average profit per day to date.

The growth projection table is followed by a projected "closeout statement." In this case the closeout is for July 30, after 150 days of grazing. The closeout data is relatively self-explanatory. Average daily gain, average dry matter intake per day, and average and minimum grazing capacity per acre are reported. The minimum grazing capacity per acre figure is of interest because it allows one to figure the acres of pasture required at the most stressed point in the grazing period. If this quantity of pasture is not available, then supplemental feed will be required to achieve the gains reported. If the use of supplemental feed is specified, then the minimum stocking rate will vary accordingly. In the case reported here, no supplemental feed was used. The stocking rates calculated and reported are those that will produce maximum growth, i.e., the animal has all the grass it will eat.

The closeout table also reports the purchase cost per head, miscellaneous cost including labor, equipment, pickup, mineral and pesticide costs. Interest and supplemental feed cost are calculated when appropriate. Pasture cost is calculated on a per head basis and per cwt of dry matter basis, as determined from the stocking rate, dry matter consumption rate, and average annual pasture charge per head. Finally, other miscellaneous costs are calculated, including a death loss allowance, medical costs, commissions, and trucking expenses. All costs are then totaled and compared to the sales values as determined by the selling price, sell-shrink factor, and simulated weight. A net return figure is calculated from this comparison. Finally a break-even sales price is calculated that would yield a zero net return.

The final output generated is a table indicating the actual nutrients used per head per month during the feeding period. In this case, only pasture nutrients were used and reported. If supplemental feeds had been used, their use would have also been reported in a separate table.

In summary, the stocker grazing model simulates stocker growth on alternative pastures and presents an estimated financial closeout statement for the simulated activity. Alternative pastures, supplemental feedings, in-weights, out-weights, prices, equity positions, etc., can be entered into the model and analyzed in a matter of minutes.

FEEDLOT-GROWTH SIMULATION MODEL

The feedlot-growth simulation model is similar in concept and operation to the stocker-cattle grazing model. The major difference is that the model is oriented toward predicting growth from concentrate rations rather than from roughage grazing. This requires different calculations within the model and a somewhat different set of information. The output generated is quite similar in that the data provided is summarized, a growth simulation table is reported, and an estimated closeout summary is given. The data required are listed below:

1. Cattle data
   - Sex
   - Purchase weight (lbs)
   - Purchase price ($/cwt)
   - Expected sales weight (lbs)
   - Expected sales price ($/cwt)
   - Expected death loss (%)
   - Shrink from purchase weight to on-feed weight (%)
   - Shrink from production weight to sales weight (%)
2. Growth adjustment data
   - Gainability code (1-6 with 4.5 as normal)
   - Percent reduction in normal feed intake for first day on feed
3. Financial data
   - Commissions ($/head)
   - Trucking ($/head)
   - Veterinary/medicine ($/head)
   - Yardage/labor ($/day/head)
   - Interest rate (%)
   - Equity ($/head)
4. Ration description
   - Feed types (corn, milo, soybean meal, alfalfa hay, silage, wheat, hulls, wheat straw)
   - Nutrient value of each feed type (energy for maintenance, energy for gain, protein, moisture)
   - Price of each feed type ($/cwt)
   - Ration composition and number of days a specified ration is to be fed

As with the stocker program, the computer prompts the user to enter each required piece of data. If the data are not entered a standardized default value is automatically entered and used. The data used in the simulation process are then summarized and printed along with the simulated growth pattern and an estimated closeout statement. Figure 2 depicts a typical set of program output/results.

The first line summarizes the input data beginning with the sex code and proceeds to list the buy weight, buy price, sell weight, sell price, gainability code, commissions, trucking fee, veterinary/medicine cost, yardage, interest rate and equity per head. The second line reports the specified rations. In this case, to simplify the illustration, only one ration is specified to be fed for 300 days (or to sell weight, which ever comes first). The ration contained 87 Mcal/cwt of ration of energy for maintenance,

SX	BUYWT	BUYPR	SELLWT	SELLPR	GNGRADE	COMM	TRKRT	VETMED	YDG/DY	INTRT	$EQUITY
S	575.00	72.50	1050.00	68.25	4.30	0.00	0.40	4.85	0.10	0.13	400.00

ENDINGDAY	EN MNT	EN GN	$/CWT	%MOIST	PRT INT
200.00	87.00	56.00	5.00	15.00	30.00

DAY	WEIGHT	FD/DY	GAIN/DY	FD$/DY	FD#/#GN	PRICE	PROF/DY
5	560.50	14.98	2.10	0.75	7.13	67.78	-11.13
9	569.39	15.88	2.25	0.70	7.07	68.16	-5.71
13	578.94	16.80	2.39	0.84	7.04	68.25	-3.72
17	589.04	17.75	2.53	0.89	7.03	68.25	-2.69
21	599.68	18.71	2.66	0.94	7.03	68.25	-2.04
51	682.38	19.50	2.76	0.97	7.07	68.25	-0.41
81	762.19	20.89	2.66	1.04	7.85	68.23	-0.06
111	838.06	21.83	2.53	1.09	8.63	66.64	-0.05
141	909.34	22.37	2.38	1.12	9.42	64.98	-0.11
171	975.61	22.57	2.21	1.13	10.22	68.25	0.10
189	1,012.26	22.48	2.04	1.12	11.04	68.25	0.08
199	1,031.59	22.32	1.93	1.12	11.54	68.25	0.07

```
STEER CLOSEOUT AFTER  199 DAYS                    POUNDS      $
CATTLE AT $ 72.5 /CWT.........                     575.00   416.88
LOT CHARGE @   1 PER DAY..............                       19.80
INTEREST @  12.5 PERCENT..............                       17.47
FEED COST @ $ 5 /CWT.................            4,179.21   200.96
D.L=13.11 +MED= 4.85 +CCM= 0.00 +TRK=2.30                    20.26
TUTAL COST OF SLAUGHTER ANIMAL                   1,031.59   683.36
TOTAL COST OF GAIN...(ADG=2.24#/DAY)               456.59    56.79
SALE VALUE @ $ 68.25 /CWT.............            1,021.28   697.02
NET RETURNS..........................                 1.34    13.66
BREAKEVEN SALE PRICE...................                      66.91
AVE. FEED/LB. GAIN @ 90%DRYMATTER =...                        7.78
PERCENT RETURN TO EQUITY OF $ 400 =...                        6.26%
```

FEED DATA

	EN.M /CWT.	EN.G /CWT.	PROT. /CWT	WEIGHT /PUR.UN.	MOIST %	ASIS-PRLBS. IN % IN /CWT.	RAIN	IN % IN RAIN
1 CORN	102.00	67.00	10.00	0.56	15.00	4.29	500.00	50.00
4 ALFY	57.00	27.00	20.60	20.00	17.00	3.50	100.00	10.00
5 SILGE	73.00	46.00	8.40	20.00	66.67	1.00	400.00	40.00
9 TOTALS	57.81	36.85	7.00	4.56	35.87	2.89	1000.00	100.00

Figure 2. Computer output from simulation of fed cattle performance

56 Mcal/cwt of energy for gain, 15% moisture, and cost $5.00/cwt. These ration data are a summary of the more specific ration specified at the bottom of the output where the ration is shown to consist of 50% corn, 10% alfalfa, and 40% silage.

Following the ration summary line, a projected growth table is presented. The table shows the weight of the animal after a specified number of days on feed. Subsequent columns in the table show the feed fed per day, gain per day, feed fed per pound of gain, estimated current price at the current weight (based on purchase price, sales price, and sales weight), and average profit per day to date.

After the growth projection table, an estimated closeout summary indicates the purchase cost of the animal and expenses in raising it. All costs are then totaled and compared to the sales value to obtain a net return figure. A break-even sales price, which would have yielded a net return of zero, also is calculated and reported. Finally, computations are made of average pounds of feed per pound of gain and the rate of return to equity.

In summary, the feedlot growth model simulates feedlot cattle growth from a variety of rations and reports the estimated financial closeout summary of the simulated activity. The program is capable of quickly reconsidering many alternatives, including different purchase weights and prices, different selling weights and prices, alternative rations, different quality of cattle as reflected by the gainability code and different interest rates and equity positions.

CONCLUSION

Growth simulation models provide a practical means for making informed decisions in feedlot and stocker grazing operations. The programs do not make decisions themselves, nor do they provide all the information needed. But they do provide a systematic, organized way of considering the available alternatives, and of summarizing the results of alternatives considered.

Use of the program is not difficult. The information required to run the programs is readily available from normal cattle feeding records and activities. Learning the mechanics of operating the computer takes only a few hours and will save days of pencil and paper calculating in the future. Microcomputers capable of executing the program cost approximately $1,000 to $4,000. These computers are also capable of performing many other tasks such as record keeping, billing, and payroll.

The decision-making process in many feedlot and back-grounding operations can be enhanced by the use of computer-ized growth-simulation models. The primary benefit of using such models would appear to be providing an organized, quick, and consistent way of evaluating alternative feeding and grazing operations. Use of the computer makes the traditional pencil and paper budgeting process easier--more options are available with less effort. But perhaps, most important, the computer consistently requires that all necessary information be considered in each analysis, thus methods of calculating cattle performance and the financial results are performed consistently and accurately. In addition, all of the information used and the results are neatly summarized in standard printed forms.

REFERENCES

Brorson, B. W. 1980. Economic analysis of stocker cattle production alternatives using a computer simulation model. Unpublished M.A. Thesis. Oklahoma State University.

Brorson, B. W., O. L. Walker, G. W. Horn, and T. R. Nelson. 1981. Economic analysis of alternative stocker cattle production systems using a minicomputer. Professional paper. Department of Agricultural Economics, Oklahoma State University.

Gill, D. R. 1979. Time-tuning management with computer assisted decisions. Proceeding, 15th Annual Cattle Feeders' Seminar. Oklahoma State University.

Gill, D. R. 1975. Beef gain simulator. Department of Animal Science, Oklahoma State University.

Nelson, T. R. 1980. Description and sample illustrations of OSU microcomputer programs. Department of Agricultural Economics, Oklahoma State University.

97

PRODUCING A POUND OF CARCASS BEEF PER POUND OF CONCENTRATE FED

Donald M. Kinsman

INTRODUCTION

The preamble to the Declaration of Independence of the United States begins with the stimulating verbiage: "When in the course of human events, it becomes necessary . . . it is the right of the people to alter or abolish . . . to prove this, let facts be submitted to a candid world." This famous passage could be paraphrased to read "When in the course of human events--such as when, for various reasons, we need to conserve energy, reduce the use of food-type grains for livestock use, and lower the fat content of our meat supply--it becomes necessary to feed our livestock and produce our meat more efficiently and economically, we must alter or abolish our less-efficient, less-economical, more wasteful ways. Let facts be submitted to a candid world." Our cattle are finished fatter, both externally and internally, than generally is required by our clientele. Witness the distribution of quality and yield grades and the proposed beef grade changes, although it must be recognized that beef grading, as a voluntary service, paints a biased picture for all beef produced in the U.S. (table 1). Only the beef that has the greatest demand is graded and then only in sufficient quantity to meet the requirements. In brief, we need to discuss how we can produce more Choice 1 and 2 beef and reduce the proportion in the 3, 4, and 5 yield grades.

CORN SILAGE OFFERS ECONOMY OF GAINS

To reduce fat content of our beef, yet maintain quality, it is necessary to select cattle with the frame size and growth potential to utilize forage and still finish at a young age to meet the USDA beef quality grading requirements for Choice and yield-grade specifications for 1 or 2. Then we must furnish them with sufficient high energy but economical feedstuff to make rapid and efficient gains at the proper stage of the growth curve to accomplish this goal. To do so, it will necessitate feeding optimum protein requirements plus appropriate mineral supplements to satisfy the maintenance and growth of a young, rapid-gaining animal.

TABLE 1. U.S. BEEF GRADED IN 1980 WITH COMPARISONS TO 1979

Quality grade	% of each yield grade					1980 Total %	% Change from 1979
	1	2	3	4	5		
Prime	0.5	18.1	57.9	17.3	6.2	6.00	−0.1
Choice	1.3	28.2	59.7	9.7	1.1	89.00	0.0
Good	7.0	53.4	38.0	1.4	0.1	4.30	+0.2
Standard	21.6	56.0	15.8	0.4	6.2	0.20	0.0
Commercial	6.4	39.6	46.5	5.9	1.0	0.10	−0.1
Utility	12.5	53.8	29.6	2.6	1.6	0.35	0.0
Cutter & Canner	41.5	41.5	14.6	1.4	0.0	0.05	0.0
1980 total	1.6	28.9	58.4	9.7	1.4	100.00	
% Change from 1979	−0.2	0.0	−0.4	+0.4	+0.2		

TABLE 2. POUNDS OF BEEF PER ACRE FROM FEEDER CALVES AT DIFFERING LEVELS OF GAIN AND YIELDS OF FEEDABLE SILAGE PER ACRE

Tons of feedable silage per acre	Average daily gain per head						
	2.75	2.50	2.25	2.00	1.75	1.50	1.25
25.0	3,053	2,775	2,498	2,220	1,943	1,665	1,388
22.5	2,750	2,500	2,250	2,000	1,750	1,500	1,250
20.0	2,444	2,220	1,998	1,776	1,554	1,332	1,110
17.5	2,140	1,945	1,750	1,556	1,362	1,167	973
15.0	1,832	1,665	1,499	1,332	1,166	1,000	833
12.5	1,529	1,390	1,251	1,112	973	834	695

Based on consumption of 45 lb of corn silage and 2.5 lb of soybean oil meal per head per day for 200 days.

By producing and properly harvesting and preserving high-yielding corn silage, it is possible to provide an excellent, low-cost ration for bovine of this description. Select the right kind and maturity of corn seed for your area, feed it well, harvest it at the proper stage (late dent), chop it fine, ensile it well (packed), and supplement it adequately, either into the silo or as fed, to assure a balanced ration for these feeders. Table 2 demonstrates the relationship between the average daily gain per head with the tons of feedable silage per acre to yield a given number of pounds of beef per acre from feeder calves. This is all predicated on top-quality corn silage fed free choice and topped with 2.5 lb of soybean oil meal per head daily for a 200-day period. These feeder calves starting at a 500 lb average and finishing at approximately 1000 lb liveweight would average 45 lb of corn silage consumed daily. This would amount to 18 lb of corn silage per pound of gain for cattle gaining 2.5 lb daily or about 6 lb of feed on a dry-matter basis. Table 3 indicates the feed costs per pound of gain determined by the gain ability of the cattle and the cost of corn silage and soybean oil meal (SOM). In general, it is feasible to produce at least a ton of live weight beef per acre on this program if the cattle gain at least 2 lb per head daily and the producer realizes at least 20 tons of corn silage per acre. The 500 lb of gain could realistically be accomplished in 200 days at 2.5 lb ADG (Average Daily Gain). With corn silage valued at $30 per ton and SOM at $225 per ton and the cattle gaining at 2.5 lb per head per day, the feed costs per pound of gain would be 38.4¢. These capabilities have all been demonstrated in our corn silage--SOM feeding trials with Angus, Hereford, Shorthorn, Charolais, and Holstein steers at varying times during the past 25 years. In all cases, the steers were fed under cover from June to November and in open sheds from November through May, which included at least three months of severe weather that reduced feed utilization and efficiency of gain.

TABLE 3. FEED COST (CENTS) PER POUND OF GAIN

	Average Daily Gain Per Head							
	3.00	2.75	2.50	2.25	2.00	1.75	1.50	1.25
Silage at $20/ton soybean oil meal at $150/ton	21.2	23.2	25.6	28.4	32.0	36.4	42.6	51.0
Silage at $25/ton soybean oil meal at $250/ton	29.3	31.8	35.0	39.0	43.8	50.0	58.3	70.0
Silage at $30/ton soybean oil meal at $300/ton	35.2	38.4	42.0	47.6	53.2	60.8	71.2	85.2

Based on consumption of 45 lb of corn silage and 2.5 lb of soybean oil meal per head per day for 200 days

This program has produced very acceptable beef of between 30¢ and 40¢ per pound of gain depending on costs assigned to the corn silage. It can be done for no more than the off-farm direct costs for protein supplement (chiefly SOM) at the rate of 2.5 lb per head daily for 200 days--or 500 lb of purchased grain. This 500 lb of SOM corresponds with the total weight gain of 500 lb or one pound of SOM per lb of live weight gained.

WHAT ABOUT THE CARCASS?

On this type of feeding regime, it would be desirable to produce at least a 500 lb dressed carcass. It should have attained 1.25 lb of carcass weight per day of age, which translates into a 400-day-old (13.2 months) steer yielding a 500 lb carcass. For example, a 15-month-old (450-day-old) steer would produce a 563 lb dressed carcass or at 18 months (550 days) a 685 lb carcass. Furthermore, it would be preferable that these carcasses grade Choice and fall within the yield grade 1 or 2 categories. It would also be preferable to have no more than 0.05 in. of fat cover per hundredweight of dressed carcass, which would total 0.25 in. on a 500 lb carcass or 0.30 in. for a 600 lb carcass. The ribeye area should make at least 2.0 sq. in. per hundredweight or produce 10.0 and 12.0 sq. in. for 500 and 600 lb carcasses, respectively.

It is possible to attain all of these goals on a corn silage-soybean oil meal regime fed to cattle of the right genetic growth potential, feed efficiency, and carcass merit to achieve this end. Where feed costs represent such a large proportion of production input, it is extremely important to maintain a feed cost as low as possible without restricting the growth potential. The accompanying pictures illustrate the carcass characteristics of 20 to 30 steers of various breeds used in feeding trials over a 20-year period. Thus, some 500 steers have proven the value of this system even under New England winter conditions. If we can do it, so can you. We recommend it most highly.

Photo 1. A typical steer (h-66) fed on corn silage-soybean oil meal ration for 200 days at University of Connecticut.

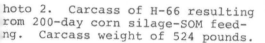

hoto 2. Carcass of H-66 resulting rom 200-day corn silage-SOM feed-ng. Carcass weight of 524 pounds.

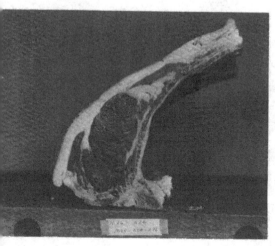

Photo 3. Rib cut from H-66. REA=10.25 square inches (1.96 sq. in./ cwt.). Fat cover=0.54 inch. Quality grade= good plus. Yield grade=3.2

Photo 4. Carcasses of five steers fed on corn silage-SOM
ration for 200 days at University of Connecticut.

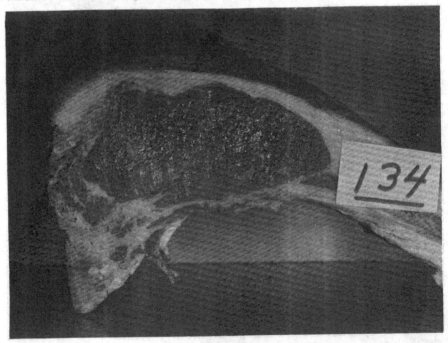

Photo 5. Rib cut from a typical steer fed on corn silage-
SOM ration for 200 days at University of Connecticut.

98

SPAYING HEIFERS: A LOOK AT THE DATA

R. A. Bellows

"Varying economic conditions and changes in the demands of the meat-consuming public have been responsible for the turns that have taken place in the beef industry during recent years. The world is experiencing a period of overproduction of beef cattle. Prices of feeder and fat cattle are low. Female stock of all kinds are being unloaded on the markets in excessive numbers, and prices of all animals of this sex are low. With each period of overproduction, there is a tendency on the part of the range man to return to the practice of spaying at least a portion of each year's heifer crop." These statements have a familiar ring don't they? They were copied from a Nebraska Experiment Station Bulletin published in 1930. The beef cycle and changes in demand for beef are going to be with us for a long time, so let's take a look at some of the data on spaying.

SPAYING

First of all, spaying is the surgical removal of the ovaries--or simply, female castration. This removes the primary source of estrogen, which is the hormone that produces estrus or heat. It also removes the source of ova or eggs which combine with sperm cells to initiate a pregnancy. This means that two things will be accomplished by properly spaying a heifer: 1) she will not come in heat and 2) she will not become pregnant.

There are numerous techniques for spaying and all involve opening and penetration of the abdominal cavity. Perhaps the most common procedure is penetration of the abdominal cavity through an incision made in the mid- to upper-left flank region. First, the hair is clipped in the area where the incision is to be made and surgical disinfectants are applied. Local anesthesia is given as desired; and an incision, 5 to 6 inches in length, is made through the hide. The exposed muscles are separated and the lining of the abdominal cavity (peritoneum) is opened by the surgeon's hand. The ovaries are then grasped, clamped to prevent hemorrhage, cut off with scissors, and removed from the body cavity. Total ovary removal is important since, if even a small piece of the ovary remains, the heifer may continue to come in heat. This is also true of dropping an ovary into the body cavity. That ovary will often simply transplant and grow at a new site in the body cavity, and the heifer continues to come in heat. The incision is dressed with suitable antibiotic dressings, and the incision is closed with metal clips or surgical sutures. An intramuscular injection of an antibiotic preparation can also be given as further precaution against infection. Heifers are back on feed and show little discomfort within 48 hours after a successful surgery.

Recently, Drs. Kimberling and Rupp at Colorado State University, College of Veterinary Medicine, designed an instrument for spaying heifers that does not require the external incision in the flank. This instrument is inserted into the vagina to a position forward and above the cervix. The vaginal wall and peritoneum are carefully perforated and, one at a time, the ovaries are manipulated into the cutting chamber of the instrument. The ovaries remain in the instrument; and upon completion of the surgery, the instrument is removed from the vagina. Routine antibiotic therapy is used as desired. This technique is useful but must be used by trained and qualified operators since the entire process is done internally by palpation manipulations.

These descriptions of the spaying process are presented to emphasize that spaying is not a simple procedure. Anyone attempting this procedure must be adequately trained. The opportunities for hemorrhage and infection are many; and if these occur, the heifer may die or be damaged in such a way that gains are not normal. These chances must be considered before you embark on a spaying program. Veterinarians charge from $3 to $12 per head for spaying, and one dead heifer will destroy any profit potential realized through spaying.

GROWTH AND GAINS

What about performance of the spayed animal? A number of studies have been made, and the results are summarized briefly in table. 1.

In general, steers outgained heifers in the feedlot regardless of whether the heifers were spayed or not. The rates of gain of the two heifer groups show spayed heifers gained more slowly in the feedlot than did the open, intact heifers in all 11 of the studies--regardless of age of the animal. The differences ranged from as little as .06 lb per day in the 1956 study at North Dakota to .31 lb per day in the 1960 Wyoming study. Some of these differences were not statistically significant; but if you calculate the overall average daily gain of intact and spayed heifers in the 11 studies, you obtain a value of 1.89 lb for average daily

TABLE 1. SEX DIFFERENCES AND EFFECTS OF SPAYING ON FEEDLOT PERFORMANCE

Scientists, state and year [a]	Animal class at start of study	Avg daily gain (lb) in feedlot			Total feed per 100 lb gain		
		Steers	Open, intact heifers	Spayed heifers	Steers	Open, intact heifers	Spayed heifers
Gramlich and Thalman, Nebraska; 1930	Two-yr-olds	2.12	—b	1.99	1110	—b	1154
	Yearlings	2.10	2.15	1.89	1042	1005	1103
	Calves	2.07	1.92	1.66	789	854	974
Hart, Guilbert and Cole, California; 1940	Yearlings	—b	1.89	1.82	—b	1047	1085
Dinusson, Andrews and Beeson Indiana; 1950	Yearlings	—b	1.84	1.67	—b	991	1082
Langford, Douglas and Buchanan North Dakota; 1955	Yearlings	1.78	1.66	1.46	2880	2866	3274
Clegg and Carroll, California; 1956	Yearlings	2.38	1.87	1.80	1137	906	857
Langford and Douglas North Dakota; 1956	Calves and yearlings	2.01	1.92	1.86	2196	2276	2457
Smith, Richardson, Koch, Cox and Stitt Kansas; 1956	Calves	—b	1.74	1.45	—b	1759	2154
Smith, Richardson, Koch, MacKintosh and Stitt Kansas; 1956	Calves	—b	1.74	1.53	—b	1290	1384
Kercher, Stratton, Schoonover, Gorman and Hilston Wyoming; 1958	Calves	—b	1.96	1.79	—b	837	853
Kercher, Thompson, Stratton, Schoonover, Gorman and Hilston, Wyoming; 1960	Yearlings	—b	1.93	1.62	—b	1029	1172
Ray, Hale and Marchello, Arizona; 1969	Yearlings	2.21	2.04	1.82	966	988	1030

[a] Some studies conducted at the same location have been combined to simplify the table
[b] These groups not studied in the experiment

gain of the intact, open heifers and 1.72 lb for the spayed heifers. Comparing these two values, we find that the average daily feedlot gains of the spayed heifers averaged 9.9% lower than for the open, intact heifers.

Table 1 also shows the total feed required per 100 lb gain. You can again see that steers were generally more efficient than heifers, though not in every case. However, in 10 of the 11 studies comparing open, intact heifers to spayed heifers, the intact heifers gained more efficiently in the feedlot than did the spayed heifers.

One cannot compare the total amount of feed consumed among studies, because the rations differed widely. Some animals were on a hay-grain ration, while others were on different types of silage, and the feed weights shown in the table include differences in moisture content of the rations. However, if we calculate the percentage differences within the various studies, we find that the open, intact heifers consumed 8.5% less feed per 100 lb gain in the feedlot than did the spayed heifers. This means the gains made by the spayed heifers were less efficient.

This all sounds very negative. Were there any advantages for spaying? There was a trend in some of the studies for the spayed heifers to have slightly higher carcass grades than the open, intact heifers and to have slightly higher dressing percentages. The spayed heifers were slightly fatter.

The studies summarized in table 1 were conducted on heifers in the feedlot. What about heifers on range forage? Table 2 shows figures from a 1960 Wyoming study and a 1977 Montana study that involved heifers on range. The results of the Wyoming study show an average daily gain of 1.47 lb for the open, intact heifers and 1.28 lb for the spayed - a decrease of 14.8%. The Montana results are similar. Average daily gains of 2.08 lb for the open, intact heifers and 1.95 lb for the spayed--a decrease of 6.6%.

The study at Montana included determining the effects of implanting the heifers with zeranol (Ralgro®) or Synovex-H®. The results are interesting. Implants improved the rate of gain in all heifers regardless of whether they were intact or spayed. Some of those differences were not statistically significant, but implanting the spayed heifers with either zeranol or Synovex-H resulted in higher gains than either open, intact, or spayed heifers.

Scientists at South Dakota State University have also found that implanting spayed heifers brought their gains back to a level comparable to those of the open, intact heifers.

These studies clearly indicate that any producer considering spaying heifers must also consider implanting growth stimulants in order to recover losses due to the gain depressions experienced from spaying. But, even with gain restored by implants, the producer should vigorously pursue a premium market for spayed heifers, and the premium must return the cost of spaying plus the labor and cost involved in implanting.

TABLE 2. EFFECTS OF SPAYING AND IMPLANTS ON GAINS OF HEIFERS ON RANGE

Scientists, state and year	Animal class at start of study	Grazing period (days)	Avg daily gain (lb) by group						
			Open intact heifers	Spayed heifers	Bred heifers	Intact + Ralgro	Intact + Synovex	Spayed + Ralgro	Spayed + Synovex
Kercher, Thompson, Stratton, Schoonover, Gorman and Hillston, Wyoming; 1958	Yearlings	120	1.47	1.28	1.36	----	----	----	----
Cameron, Thomas and Brownson, Montana; 1977	Yearlings	159	2.08	1.95	----	2.10	2.16	2.13	2.17

99

PROJECTING FEEDLOT GAINS

Don Williams

One of the basic decisions in projecting profits or losses in cattle feeding is the anticipated gain of your cattle on feed. The science of projecting cattle gains has made tremendous improvements over the last 25 years, but there are still many different methods to project gains.

A major improvement in gain projections has occurred with the union of the computer and the California Net Energy System. The California Net Energy System has calculated the calories for each feed ingredient available for maintenance and gain. It is thus possible to take the ingredients and respective percentages of any ration and calculate the calories for maintenance and gain for the ration as a whole. Once a person knows what his cattle weigh and how much they eat, it is possible to calculate the theoretical gain each day. By building the correct computer program, a feeder only has to enter the feed consumption each day and obtain the theoretical gain and new weight of each group of cattle on a daily basis.

The term theoretical gain is used because there are many variables other than caloric intake that can affect gain. Some of the variables are listed below.

Frame size. The reclassification of USDA Feeder Grades is an attempt to put some parameters on the variables. Of course, a steer that reaches its mature size at 1050 will deposit more fat between the weights of 900 lb and 1050 lb than will a steer that reaches its mature size at 1300 lb. Since fat has 2.25 times more calories than muscle, the gain on the small steer will not be as great, in this rate range, as the large steer if they are eating the same amount of feed.

Condition. An animal that is put on feed in thin flesh will out-gain an animal that is put on feed carrying moderate to heavy flesh. The industry refers to this as compensatory gain. Compensatory gain is a phenomena whereby an animal's body seems to be more efficient when it is playing catch up. It is probably closely related to the type of tissue that is deposited in the gain. The thin-fleshed animal is probably putting on more pounds of muscle and bone while the composition of the better-fleshed animal contains more fat.

Weather. Energy is expended by animals to either dissipate excess heat or to produce extra heat in cold weather. Neither of these functions is a direct function of ambient temperature since they are influenced by humidity and wind. When cattle have excessive heat, they cool themselves through sweating, evaporation, or loss of heat from the body surface. They must also depend on expelling body heat from the respiratory tract by panting--a process that requires energy. A breeze makes the process more efficient, while high humidity lowers the efficiency greatly. Likewise, cattle can stand low temperatures quite well with their winter coat if they are dry, but let their coat become wet and expose them to 20 to 30 MPH winds and their heat loss becomes significant. To overcome excess heat loss in these conditions, we see animals shiver, which is actually small contractions of muscle groups. These muscle contractions produce heat to keep animals warm, but they also require energy that could be used by the body to produce gain. Thus, in either hot or cold weather, the body may use excess energy to regulate body temperature. The use of this energy reduces gain.

Pen conditions. A fourth factor that alters gain is the amount of mud that is in a pen. Good research verifies that feedlot gains are reduced when an animal spends all day wading mud. The extra energy expended as animals pull themselves through a muddy lot decreases the amount available for gain. Unfortunately, muddy lots are more frequent in winter weather as the low ambient temperatures do not produce good weather for the mud to dry. Hence, muddy lots and chilling temperatures have a detrimental effect.

In an attempt to find some method that would give us a better guideline as to feedlot gains, Hitch Feedlot decided to take a small group of animals and weigh them at the same time every morning. We have taken approximately 50 animals that were mid-way through their feeding period and have put them in a small pen close to the scales. Every morning we weighed them between 6:30 and 7:00 a.m. We found that the cattle actually lost weight for a day or two as they were becoming accustomed to their new surroundings. By the end of the first week, the animals were in such a routine that the same animals came out of the gate first every morning and, as you could expect, the same animals each day were dragging along behind. Needless to say, we were concerned as to what effect these daily weighings were having on gain. We have run three groups through this procedure and have been gratified to find that the animals weighed every morning outgained their pen mates that were left in the original pen. We assume that the stimulation of the morning stroll increases their appetite slightly and results in a slight increase in feed consumption. Another factor is probably that there is less stress on the animals from the small pen on the day of the final weighing because they are trained to walk to the scales. The cattle being driven from the pen to the scale are not accustomed to being taken from the pen and this extra stress of excitement. So actually, we may be getting to the scale with more fill and pounds in the experimental cattle than the cattle that stayed in the original pen.

We feel that these daily weighings helped us to explain variations in the performance of cattle as the conditions varied on the morning of the final weighing. Subconsciously, we all think of the weight of cattle on feed as increasing in a straight line when, in reality, the increase in gain is steady by jerks as seen in Chart No. 1.

The change in weights is seen more graphically at the bottom of Chart No. 2; the weights increased as much as 21 lb/head one day, only to be followed by an 18 lb/head loss the following day. If these animals had been shipped at the end of a 100-day feeding period under these variable conditions, it would have influenced the average daily gain by .2 lb/day.

The average daily feed consumption was computed by taking the total lb fed that day and subtracting the estimated lb of feed left in the bunk the next morning by an experienced feed caller. The total feed consumed was then divided by the number of animals in the pen that day. A comparison of feed consumed per day against daily gain does not show the correlation that one would expect. Obviously, the only other factor that would effect gain was water consumption, and I feel that one must credit water consumption with the primary change in weights.

I would criticize our data in that the maximum temperatures and daily rainfall do not accurately reflect the weather conditions at the time of weighing. If the rainfall occurred during the early morning hours, it could have lowered water and feed consumption; conversely if it had been relatively hot all night and the day before and with rainfall at 4:00 a.m. or 5:00 a.m., the cattle could have eaten heavily the first part of the night and that would have caused them to drink more water and weigh heavier. If they are wet at the time of weighing, their coat could contain 7 to 10 lb of water.

Chart No. 3 is an attempt to summarize the performance of these steers over the period of the trial. Grain prices were quite high during this period as was interest. The upper graph of cost of gain (COG) was computed with all changes plus interest and demonstrates the effect of interest costs at the end of the feeding period when the accumulated costs must be financed at 21% interest. The lower graph labeled "COG Without Interest" demonstrates the added costs incurred at the end of the feeding period when the feed consumed must furnish the maintenance for more pounds of body weight. At the end, the animal is not gaining as fast because more of his gain is fat and this requires more energy and thus more feed. These same factors will, of course, serve to increase the dry matter conversion (the pounds of absolute dry pounds of feed required to produce a pound of gain) toward the end of the feeding period.

We have found several advantages to having a set of animals that are weighed each day. Some of these are:

- If fat animals to be sold are weighed up on a morning when the animals weigh extremely heavy, these data allow you to explain it to the buyer of the cattle. He can then alert the packing plant that the dressing percentage on any cattle that are weighed that morning will be lower than he had estimated.
- If cattle are weighed on a morning when animals are weighing light, the data explain to the owner of the cattle why his gain was not as good as he expected. It also offers an opportunity to point out to the fat-cattle buyer that weighing conditions don't always go against him and that he can inform the packing plant that the dressing percentage will be higher than he had estimated.
- In one instance, we shipped cattle to a packer the day after we had a slow, freezing rain. When the cattle were slaughtered, there was a high percentage of dark cutters. The packer was concerned because our cattle were the only ones that had the dark-cutter problem. Further investigation revealed that all of the other cattle slaughtered that day came from feedlots further south that had not experienced the freezing rain. We were able to present the data from our daily weighings to show a decrease in weight from the day of the rain and substantiate our claim that the stress which produced the dark cutters was due to the weather and beyond our control.
- Probably the biggest advantage that we derive from our daily weighings is that we are able to monitor daily gain under existing weather conditions and adjust our estimates accordingly. Not only does weighing allow us to alter our performance projections for the cattle that we are feeding, it also allows us to project the performance of all the cattle in the area since other feedlots will be experiencing similar effects from the weather. With this information, we can better anticipate the beef tonnage that will be marketed from our area in the next 60 to 90 days.

896

CHART 1   WEIGHT OF FED STEERS

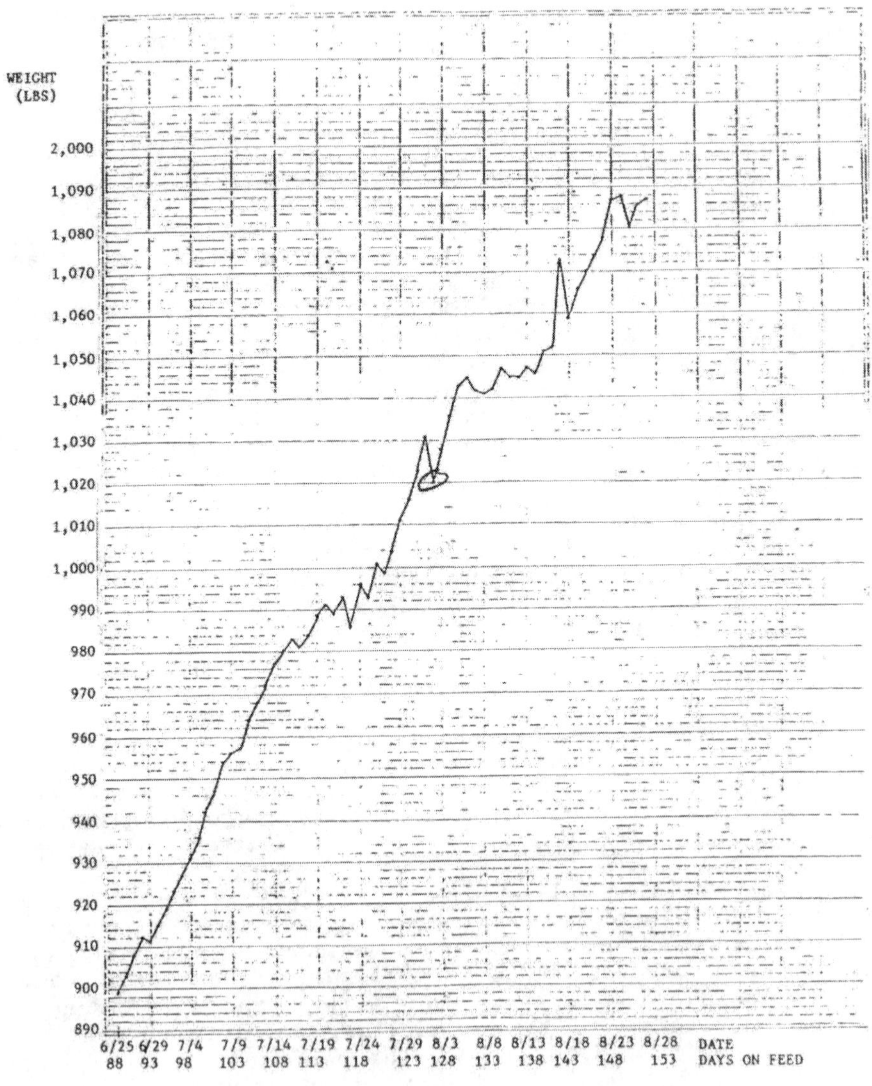

897

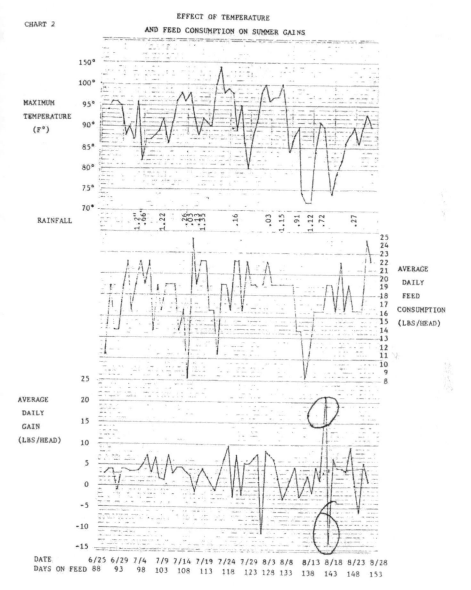

CHART 2

EFFECT OF TEMPERATURE
AND FEED CONSUMPTION ON SUMMER GAINS

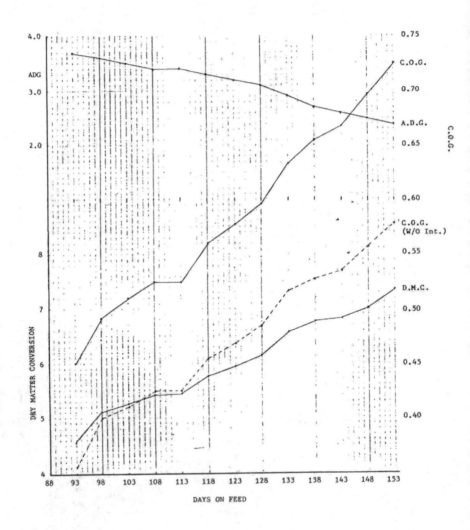

CHART 3   CATTLE PERFORMANCE

100

LET'S FEED BULLS FOR PROFIT

Don Williams

U.S. livestockmen have shown increasing interest in feeding young bulls in recent years. As with anything new, this concept has some opportunities under proper management practices, but it also has some pitfalls and misconceptions. This discussion will attempt to recognize the pros and cons of feeding young intact males.

The concept of feeding bulls has been an accepted practice in Europe for years. Of course, many of the cooking practices in France, Italy, etc., are designed to serve beef that has not been grain fed--thus it is unlike the beef commonly sold to the American consumer. A possible corollary is the fact that the per capita consumption of beef in Europe is much below that of the U.S. Hence, one of the prime considerations in a discussion of feeding young bulls in the U.S. is: Will it be accepted by the American consumer, and if so, under what conditions? In June 1982, Kansas State University held a National Symposium on the feeding of young bulls and the most significant statement made at the conference was to the effect that no one should start feeding bulls until they know where they will sell them.

Let us look at the question of palatability and consumer acceptance and the various research projects that have been completed in this area. As a result of health concern, recent surveys tell us that the American consumer is desirous of a more lean beef product. The beef from young bulls certainly fits these specifications. Studies indicate there is 35% less fat in carcasses from young bulls, but there are factors in consumer acceptance other than fat content. The two primary components of palatibility are flavor and tenderness. It has long been recognized that beef from grain fed beef has a preferred flavor. This is the basis for establishing quality grades in our meat grading system that is based on the premise that the more marbling a carcass has, (small specks of fat within the muscle) the better the flavor. Recent work tends to discount this to some extent. Research at Texas A & M has shown that an animal that has been on a high grain ration for 100 days will have acceptable flavor regardless of how much marbling the carcass may show. Some of these researchers have suggested that government graders might be better used traveling from feedlot to feedlot certifying length of time on feed rather than looking at carcasses.

Tenderness in beef is basically determined by the amount of cold shortening that occurs as the carcass is chilled and the amount and type of connective tissue in the carcass. Cold shortening is due to the contraction of muscle fibers when the fibers are chilled too quickly after slaughter, resulting in the muscle fiber being short and thick and the meat less tender. If the carcass has 0.3 in. of fat over the carcass, the fat acts as insulation and the muscles chill slowly enough to prevent cold shortening. On bull carcasses, beause they have less fat cover on the carcass, cold shortening is definitely a problem. Fortunately, new technology has been developed that overcomes this problem by the use of electrical stimulation of the carcass at the time of slaughter. The use of voltages from 250 to 3,600 volts has been shown to stimulate muscle contraction, resulting in faster postmortem degeneration of the muscle that prevents cold shortening. Therefore, cold shortening is not a factor in tenderness of young bull carcasses.

The other factor of tenderness, i.e., the connective tissue, is not thought to be improved by electrical stimulation. Connective tissue is the supportive tissue of the body that surrounds the muscles, attaches the muscles to the skeleton, and furnishes general support to the body. The connective tissue varies in thickness and density in different parts of the carcass; the heavier accumulation becomes gristle when meat is served. Light strands of connective tissue surround each muscle bundle and are an integral part of any cut of meat. As bulls reach puberty, all of the connective tissue becomes stronger. For this reason, comparison of steer carcasses to bull carcasses has shown that young bull carcasses are slightly less tender than steer carcasses.

The question thus becomes, is this decrease in tenderness great enough to effect consumer acceptance? One attempt to answer this question was conducted by an experiment supported by the International Charolais Association in which half of the male calves were castrated shortly after birth. Through supermarkets they distributed over 5,000 meat cuts equally divided between steer and bull carcasses and were fortunate to obtain over 3,000 returned questionaires from consumers. Though preference was shown for two cuts from the steer beef and for two cuts from the bull beef, the researchers concluded that there was no difference in consumer acceptance between steer and bull carcasses. This distribution was all in one city, and it would be presumptuous to say that the preferences of these shoppers represented the preferences of all consumers in the U.S. However, judging from current data, there does not seem to be a problem with consumer acceptance of beef from young bulls.

Perhaps we should interrupt our discussion here to discuss the term "young bulls" as young, intact males. If bull feeding becomes widely accepted, one of the more difficult problems for the industry will be to have an effective quality-control program to ensure that the carcasses merchandised are from young animals not more than 15 or 16 months of age. As noted in the discussion of connective tissue, shortly after puberty the increased production of testosterone causes a change in the chemical nature of the connective tissue. As the males mature, the connective tissue becomes thicker and stronger. Numerous trials indicate that the critical age for carcass tenderness is 15 to 16 months of age. There is probably some difference between breeds, and even within breeds, but all the remarks made in this discussion apply to animals less than 15 to 16 months of age. Another related factor is the tendency for the Brahma breed to have heavier and denser connective tissue. With this in mind, and also to avoid carcasses that might have heavier necks that could be a cause of discrimination, we try not to put any animals into our bull-feeding program that are over 1/4 Brahma (and prefer to keep them under 1/8 Brahma).

Now, if we are convinced that the consumer will accept beef from young intact males, the next question is: Will it be accepted by the retailer and the packer? Probably the biggest problem that the beef industry faces here is the strong prejudice against carcasses from mature bulls. A creative, innovative designation is needed for the carcasses from young intact males (YIM) that would be acceptable to the industry. "YIM carcasses" (from young intact males) would be a handy acronym, but how do you get it established? The problem is that the retailer is afraid to handle these carcasses for fear of bad publicity if the word should spread that he is pawning off bull meat on his customer. The packer has some of the same fear as he sells to the retailer. All of this pattern leads to quite a frustration as the advantages are proven for both the retailer and the packer.

For the retailer, the advantages include:
- Less fat to trim so that the retail cuts will be acceptable to the consumer.
- The reduction of "seam fat" in the chuck that usually makes these cuts so difficult to move.
- A leaner, more acceptable, hamburger product.
- More pounds of retail cuts on the meat counter from the carcass or box beef that he buys because of the greater muscle-to-bone ratio and larger rib eyes of young intact male carcasses. This is in the range of 6% from boxed beef and 13% for carcass beef.

For the packer, the advantages include:
- No yield grade 4 carcasses to try to sell from his coolers.

- A higher percentage of yield grade 1 & 2 carcasses even though there will not be as many choice carcasses.
- If he is selling his production as boxed beef, he will obtain: more pounds of meat in the box and less waste as trim, and less fat content in trimmings that come off his fabrication line, providing a higher value and more dollars from his trim.

Thus, in summary, the big problem in merchandising carcasses to the packer and retailer stems from the beef industry's strong prejudice against use of bull carcasses. New terminology could make such carcasses more desirable.

Previously in our discussion we have been concerned with merchandising these carcasses; now let us consider the advantages and disadvantages in producing the product. Various feedlot trials have shown that bulls will gain from 7% to 23% faster than steers and that they will make this gain from 5% to 16% more efficiently. They will also be heavier at weaning. The difference at weaning, of course, will vary depending on the breed and the nutritional regime used in production of the calves. However, it seems that this might be the ideal disposition for calves on excellent pasture to be weaned at 600 pounds under an intensive crossbreeding program.

As compared to steers, young bulls show different social behavior and require some changes in management for efficient handling. First, bulls are, of course, very aggressive in establishing their social order. When a group of bulls is put together, there will be a period of time when gains and feed intake are reduced due to their physical activity. For this reason, many people feeding bulls will purchase the bulls, assemble and sort them while on pasture, and take great care not to add additional animals to the pen after they are brought to the feedlot. Assembling young bulls is much easier before puberty and can be accomplished easily with weaning calves. Thus a good pasture program is a nice complement to a bull feeding program.

The second point related to the bulls social behavior is the high incidence of "bullers," both on pasture and in the feedlot. Since allowing the cow to mount him is part of the normal sexual behavior of the male, it seems that pens of bulls more frequently select one of their sex to ride. We have had as high as 12% to 15% bullers at times. After we remove those animals and put them in a buller pen for a few days, we have been fairly successful in placing the bullers back in their home pen and having no further trouble with that particular animal.

There is one controversy that is yet to be resolved in bull feeding. Many of the comparisons of bulls versus steers have been university trials using only a few animals per pen. While commercial feeders do experience increases in rate of gain and feed efficiency, they are ususally not as great as those reported in research trials. The U.S.

Meat Research Center at Clay Center, Nebraska, has experienced reduced gains when 60 young bulls were fed per pen as compared to 30 per pen, even though the 60-head pens were exactly twice as big as the 30 head pens. Each animal had the same bunk space and square footage in the pen in these trials. It is thought that if this is truly a factor in bull feeding, it probably reflects more physical activity in the pen as more animals are congregated in a group. In an attempt to evaluate this hypothesis, we made observations 5 to 7 times a day of pens of bulls that were paired with similar pens of steers. Using the criteria of how many animals were standing or lying down and how many animals were playing with another animal, our study did not detect any difference in physical activity of bulls versus steers.

Yet another management factor that must be considered in a bull feeding program is the effect of cold weather on bulls. Since bulls deposit less fat beneath the skin, they have less insulation and therefore less resistance to cold stress. Experience has shown that there is more sickness and more death loss in bulls than in steers during winter weather. Some commercial feeders have altered their program so that no bulls are placed on feed from mid-December until early spring. The disadvantage is that these young bulls are more easily available for feedlot purchase in the fall and their discount purchase price as compared to the price for steers is usually wider. Whether this discount will continue as bull feeding becomes more popular is a matter for speculation.

Other decisions required in establishing a bull feeding program are those regarding the type of implanting system to use and which breeds are most advantageous to feed. The data on the use of implants in young bulls is conflicting. Much of the older research does not show a significant advantage for implants; however, much of this work was done with D.E.S.--the other implants have been improved since this research was conducted. In the summer of 1982, we ran a trial on three breeds and four implant regimes. The bulls were fed for 112 days before slaughter; their average weight at the end of the trial was 1,024 lbs. Their average daily gain was:

	Hereford	Charolais X	Angus X	Average
No implant	3.83	3.57	3.47	3.62
Ralgro (reimplanted)	4.16	3.91	3.70	3.92
Synovex (reimplanted)	3.79	3.70	3.90	3.80
Compudose	4.04	4.15	3.86	4.02
Average	3.95	3.83	3.73	3.84
Average final weight	1043	1043	984	1024

We obtained a positive response to all three implants used in this trial. Research done in Kansas in recent years has given a positive response to Ralgro. I personally feel that I would use implants even if there was no growth response because it seems to prevent these young bulls from becoming as masculine as nonimplanted bulls. Present data indicates that implanted bulls develop smaller testicles and produce less testosterone. The carcasses seem to marble a little better and not show as much neck and muscle development as that characterizing bull carcasses.

My present preference for breeds is for an animal that is at least 5/8 British breed; not over 1/8 Brahma, with the balance representing one of the large-muscled Continental breeds. My prejudices are based on beliefs that 1) more than 1/8 Brahma may effect tenderness, 2) some of the larger-framed European breeds may be too slow in maturing and not develop the fat cover needed, and 3) the British breeds will marble a little better than do the other breeds, giving assurance that the carcasses grade at least U.S. Good.

In summary, young bulls should gain faster on less feed to produce a carcass that is leaner but will not grade much better than U.S. Good. Since young bulls can usually be purchased at a discount to steers, the total cost of the finished animal is less than that of a steer; however, the profitability of feeding young intact males is dependent on being able to market these animals for their value as retail cuts on the meat counter of the supermarket. At the present, this is very difficult.

101

CATTLE FEEDLOT MANAGEMENT

Pat Shepherd

FEEDLOT ACCOUNTING AND RECORD SYSTEM

The key to efficient cattle finishing in the feedlot is management--animal health care, proper nutrition, orderly marketing, and risk management are crucial to produce the most economical pound of beef and to making a profit in the cattle finishing operation.

The financing of a feedyard operation is even more important now that cattle, most feedstuffs, machinery, labor, and expenses are inflated greatly over similar items of 20 years ago, and the cost of money itself has now become a very important input in a breakeven analysis. Risk management through hedging, contracting, and projections is another key to profits, a perfect job of purchasing, feeding, and animal health can be wiped out with a bad or erratic market. Cattle finishing by nature has to be a continuous operation year-round, but markets are not always profitable nor predictable. Risk management achieves a reasonable profit by working around the market highs and lows.

A record system that can cover all the important aspects of cattle feeding and finishing is an important tool to the cattle owner, feedlot manager, financial institution, and investor. Records are only as accurate as the information recorded. They should not be so cumbersome that the feedyard becomes overburdened with paperwork and neglects the primary purpose--efficient and profitable beef production.

In a typical commercial cattle-finishing feedyard, the records needed by management are cattle performance, feedyard operations, risk management, financial statement, regulatory, and miscellaneous.

Cattle Performance

A pen of cattle is treated as an accounting entity from the time it arrives until all cattle and costs are accounted for through records. This is important whether the feedyard is large or small, or whether the feedyard owns all the cattle or not. All costs, performance, profit or loss are kept separate by the lot number. Records here are more meaningful if a group of cattle of similar weight, origin, and quality are started at the same time.

On each lot of cattle a file of records is kept intact even after the cattle are closed out for audit and future reference purposes. These individual lot records include the following:

- In sheet
- Processing record
- Medicine card
- Report of death loss
- Feed statement
- Performance record

All these records collectively account for the feeding costs of the lot of cattle until they are finished. The proper completion of these records while each activity is being accomplished by the feedlot personnel becomes one of the real challenges of management. A well-documented set of files/records will provide management information for future planning and explain to customers exactly what their cattle did or failed to do.

Feedyard Operations

A commercial custom-cattle feedyard is in the business of selling feed and performing services such as cattle purchasing, marketing, financing, and feed procurement for customers who are the cattle owners. The feedyard company or ownership may also own and feed cattle for their own account. In either case, the feedyard operation should stand on its own feet as a separate entity and make a profit.

Shipping and Receiving Recap and the Daily Yard Sheet are records used by management and personnel to keep track of the movement of the cattle. Both of these records keep up with total cattle inventory and the yard sheet shows average feed consumption, daily cost, death loss, days on feed, total pen inventory, and cattle in hospitals and in buller pens. For the milling operation, a Feed Inventory Record and Feed Production Sheet total the individual pen's feed cards and give a control on amount of feed being purchased in the mill compared to the amount being manufactured and fed to customer cattle. These records account for any losses or shrinkage due to handling and for gains made in mixing.

The most important record for management and owners is the financial statement which summarizes all the input and output data that determine a profit or loss. The balance sheet of the company gives the complete accounting for each lot and operation at the end of each accounting period. This document is extremely important for making long-range plans for debt service, depreciation, income taxes, and other important business decisions. On the income statement, costs are kept separate by department. In a feedyard, departments such as feed mill, feed yard, feed and rolling stock (trucks), office, and farm should keep expenses separate for proper analysis of each area. These breakdowns become important for budgeting costs for another year and comparing costs from previous time periods.

Risk Management

Risk management is the method used to handle or avoid the price fluctuations of commodity markets such as cattle or feed through the use of futures markets, contracting, or hedging. Risk management becomes extremely important in the cattle feeding business where the margins are close and price fluctuations have been more nearly normal since about 1973. Because of the high investment involved, a feedyard must keep feeding cattle year-round, purchasing feeder cattle and grains, and selling finished cattle. This, in turn, involves the risks of losing large amounts of money, unless risk management is used. Finished cattle are a perishable product in the sense that they are at their optimum for marketing for a 2 to 3 week period and then begin to get over-fat if fed any longer. A cattle feeder cannot simply wait a month or two until the market gets better.

Feedyards project a breakeven cost on each lot of cattle fed. This breakeven point will be used to hedge a profit on the futures market or in forward contracting with a packer. Projections also are made using the futures or current cash markets and working backwards to figure the price you can pay for feeder cattle to make a profit feeding. Projections are made with different weights of cattle, growing vs finishing programs, and pasture vs finishing on grain. Analyses are also made on feeder cattle from different parts of the country, different order buyers, times of the year, and weights of cattle. Many of these comparisons are put on the modern computer systems with great effectiveness. Good records are very useful in making these comparisons.

Another area of hedging and risk management is in the purchasing and protection of feedstuffs from erratic market price fluctuations. Feedyards use more grain than any other feed ingredient. Hedging is done by forward contracting with elevators or farmers; or it is done with long or short hedging on the futures market. Long hedging is buying a supply of grain on the futures such as corn, that cannot be bought now or is not available. This is always a consideration when prices are at extreme highs or lows.

Another form of risk management is for feedyards to be in the grain or cattle-grazing business so that they grow some of their own feed supplies. Vertical integration is a method used by large corporations as well as the midwestern farmer who grows his own feed supply, feeds cattle or hogs on the farm with family labor, and markets livestock when the feed supply is used up. Risk management has extended to the cow-and-calf producer who now is looking at owning his production all the way from start to finish. We aid a cow-calf producer or a grazer in hedging his cattle at the most profitable time by finishing the cattle in the feedlot after they come from grass.

Regulatory

Cattle feedyards are subject to several state and federal regulatory agencies, as are most businesses. Cattle feedyards are especially regulated because we are producing food for human consumption and we use chemicals and feed additives that could be harmful to humans if used improperly. A Food and Drug Administration (FDA) file is kept with clearances for all the feed additives used in our feeds. An annual registration, periodic assay reports from feed samples, a log of daily use of registered drugs or feed additives, and master formulas of all ration formulas used in our mill are kept in the FDA file.

Other regulatory files necessary to keep are Occupational Safety and Health Administration (OSHA), social security, unemployment compensation, workers compensation, and other required by local, state, or federal authorities or agencies.

Miscellaneous

Many miscellaneous records are needed for specific feedyard customers or ownership. Grain-credit records are kept for customers who purchase grain or other commodities at the end of a year to feed their cattle the next few months of the new year. This is done for tax planning as well as to take advantage of a harvesttime low price. Our South Plains Feed Yard, Inc. does a lot of work with investors who feed cattle to take advantage of the tax-shelter aspects of cattle feeding that are very good. Many planning, financing, hedging, and special records are necessary to obtain the desired tax shelter.

Summary

All records listed and discussed in this paper are used in the cattle management program at South Plains Feed Yard, Inc., Drawer C, Hale Center, Texas 79041.

Figure 1

SOUTH PLAINS FEED YARD, INC.

AC 806-879-2104
P.O. DRAWER C
HALE CENTER, TEXAS 79041
AC 806-293-4078

NAME	: RESALE	DATE	:	08/31/82
ADDRESS	:	OWNER	:	498
ADDRESS	:	DATE STARTED	:	06/23/82
CITY, STATE	:	LOT NO	:	297 PEN: 211
ZIP	: 00000	SEX	:	SI

INVOICE

DESCRIPTION	QTY	RATE	CHARGES	CREDITS	BALANCE
BALANCE FORWARD					16,359.98
RATION #4	63,500	129.000	4,095.77		
HOSPITAL FEED			43.50		
BULLER PEN FEED			784.40		
MEDICINE			17.44		
INSURANCE			6.72		

****

TOTAL CURRENT CHARGES		4,947.83

****

*PAY THIS AMOUNT   21,307.81

****

DAYS ON FEED	70	HEAD SHIPPED TO LOT	168
COST PER HEAD PER DAY	2.05	HEAD SOLD	
FEED PER HEAD PER DAY	26.8	HEAD CURRENTLY IN LOT	168
P-T-D HEAD DAYS	2,234	HEAD DEAD	
P-T-D BULLER PEN DAYS	424	% DEATH LOSS	.0
P-T-D SICK PEN DAYS	30		

All invoices are net payable in cash within 15 days from ending date of billing period at the office of South Plains Feed Yard, Inc., or Drawer C, Hale Center, Hale County, Texas 79041 unless provisions have been made for financing feeding costs. Interest charges for past due invoices will be 10% ANNUAL PERCENTAGE RATE starting on the 16th day after the end of the billing period. Any discounts offered for prompt payment will be show in the statement.

Figure 2

(806) 879-2104 (806) 293-4078

SOUTH PLAINS FEED YARD, INC.

DRAWER C

Hale Center, Texas 79041

PERFORMANCE RECORD

_____

Date

Owner: _____ Lot No. _____ Pen No. _____

Head In _____	Dates In _____
Deads _____	Off Truck Shrink _____
Head Sold _____	Dates Out _____
Total Out Weight _____	Aver. Out Weight _____
Total In Weight _____	Aver. In Weight _____
Total Gain _____	Aver. Gain/Head _____
Total Head Days _____	Aver. Days on Feed _____
Total Feed Cons. (lbs., as fed) _____	Aver. Daily Cost $_____
Total Feed Consumed $_____	Aver. Daily Gain _____
Processing & Medical $_____	Aver. Daily Feed Cons. _____
Insurance $_____	Cost/lb. Gain $_____
_____ $_____	Conversion (as fed) _____
Total Feeding Costs $_____	Conversion (dry matter) _____
	Aver. Selling Price/cwt. $_____
	Breakeven Price/cwt. $_____

SALES RECAP:

    Total Cattle Sales $_____

    Laid In Cost $_____

    Total Feeding Cost $_____

    Total Expenses $_____

        Total Income (Loss) For Lot $_____

        Aver. Income (Loss) Per Head $_____

Sex _____	Type _____	Background _____
1. Steers	1. Angus	1. Native Grass
2. Heifers	2. Hereford	2. Wheat Pasture
3. Bulls	3. English Cross	3. Silage Grow
4. Cows	4. Okies	4. Local Auction
	5. Brah. or Brah. X	5. Southeast Auction
Condition _____	6. Exotic Cross	6. Cent. & S. Tex. Auction
1. Thin	7. Holstein	7. E. & NE Tex. Auction
2. Medium	8. Santa Gert.	8. Preconditioned
3. Fleshy	9. _____	9. _____

Figure 3

(806) 879-2104 (806) 293-4078

SOUTH PLAINS FEED YARD, INC.

DRAWER C

Hale Center, Texas 79041

Date________________

Owner____ Pen____________________

Kind____ No. of Head____________

Breakeven Estimate:

1. Laid-in Weight & Cost ______________________________
2. Estimated Feeding Cost ______________________________
3. Total Feeding Cost ______________________________
4. Estimated Interest ______________________________
5. Other Costs ______________________________
6. Total Feeding Cost & Weight ______________________________
7. Estimated Break-even Price Per Cwt. ______________________________
8. Estimated Sale Date ______________________________

Hedging Potential:

1. Futures Month & Price ________________ AT ____________________
2. Less Price Differential ____________________
3. Other Differentials or Discounts ____________________
4. Expected Sale Price ____________________
5. Breakeven Price Per Cwt. ____________________
6. Approx. Hedged Profit (Loss) Per Cwt. ____________________
7. Approx. Finished Weight ____________________
8. Approx. Hedged Profit (Loss) Per Head ____________________

Comments:

____

____

____

Part 15

BEEF PRODUCTION
IN THE TROPICS

102

BEEF CATTLE PRODUCTION
ON PASTURES IN THE AMERICAN TROPICS

Ned S. Raun

Cattle play a prime role in the overall development of countries and regions in the Latin American tropics. Cattle provide the mechanism for the rational use of extensive pasture lands not suitable for crop production, as well as for the immediate use of potentially arable lands for grazing--until the necessary infrastructure for crop farming develops. Beef cattle transform these pastures and forages, as well as other cellulosic crop residues and by-products, into high-quality animal protein for human consumption. In many countries, expanded cattle production could also supply additional animal protein for human consumption and could provide beef for export above effective local demand.

LAND USE

Ecosystems vary greatly in the tropical lowlands: i.e., new fertile soils and old infertile ones; desert regions and areas with an annual rainfall as high as 10,000 mm; open savannahs and dense rain forests; and level as well as hilly topography.

In the tropical belt countries of Latin America, there are 150 million ha (10.2%) in crops, 370 million ha (25.3%) in permanent pasture, and 944 million ha (64.5%) in forest crops (FAO, 1980). Of the 370 million ha in pasture, approximately half are located in infertile soil areas and the other half in more fertile soil areas.

Red-yellow latosols, or oxisols and ultisols, predominate in the infertile savannah areas. They are highly weathered, quite acid (pH 4.5 to 5.0), highly aluminum saturated, and very low in bases (Ca, Mg, K) and phosphorus. Since most of these lands cannot support sustained crop production without sizable fertilizer inputs, grazing by ruminants appears to be the only feasible method of use in the future.

In contrast, alluvium soils predominate in the fertile savannah areas, and support the growth of more productive and nutritious species than those in the infertile soil areas.

LIVESTOCK NUMBERS AND PRODUCTION OF ANIMAL PRODUCTS

There are large numbers of livestock in the tropical belt countries of Latin America/Caribbean. Estimated numbers (in thousands) in 1980 were: cattle, 196,739; sheep, 56,803; goats, 24,291; and swine, 70,662.

Production of animal products (in thousands) was: beef, 4,740; milk, 25,819; sheep meat, 116; goat meat, 82; pork, 2,212; poultry meat, 2,733; and eggs, 2,087 (FAO, 1980).

CATTLE PRODUCTIVITY

Animal productivity is low in the lowland tropics of the Americas. Most cows calve every 2 years, giving a calving rate of approximately 50%. Slaughter animals are marketed at 3 to 5 years of age when they weigh 350 to 450 kg. Current extraction rate is 13%, with an annual production of carcass beef from the total beef herd of only 24 kg per head (FAO, 1980). This compares with 52 kg in Argentina and 59 kg in Australia where predominantly pasture-based production systems are used. In the U.S., however, where forages provide approximately 73% of the total feed units and 23% are derived from grain, the annual production of carcass beef from the total national herd is 90 kg per head.

Productivity per hectare is low. Combining all types of land used for cattle grazing (approximately 370 million ha), the stocking rate averages 1.9 ha per animal, and carcass beef production is only 12.8 kg/ha/year (FAO, 1980).

The Centro Internacional de Agricultura Tropical (CIAT) conducted surveys in the Colombian llanos, a region comprising open savannas with infertile soils, and on the north coast of Colombia, a generally fertile soil area. In the llanos, calving rates ranged from 15% to 81%, calf mortality from 0% to 60%, and adult mortality from 0% to 6% (Stonaker et al., 1975b; CIAT, 1975). In the more fertile north coast area, mean production levels were higher than in the Colombian llanos. Calving rates ranged from 57% to 74% and mortality from 2.4% to 7.4%, and animals were marketed at 400 kg at 3 to 4 years of age--at least 1 year less than those marketed from the Colombian llanos (Rivas, 1974). This reflects a higher plane of nutrition attained as a consequence of higher soil fertility supporting growth of more nutritious pasture forage.

PRODUCTION CONSTRAINTS

There are nutritional, disease, and germ-plasm constraints to achieving the production potential of the feed-animal continuum. Of these, inadequate nutrient intake from grazed forages, particularly during the dry season, is the principal constraint to the attainment of high levels of productivity. The severity of this constraint varies with the distribution and amount of rainfall and is accentuated in the low-fertility soil areas.

In infertile soil areas, the unique problems of beef cattle production arise from the low nutritive value of grazed forages, which is principally a consequence of low soil fertility. The most serious consequences are calving percentages of 40% to 50% and slow growth rate resulting in the marketing of animals at 4 to 5 years of age. In the short-term, the remedy lies in improved herd and pasture management practices combined with mineral supplementation, and in the establishment of improved pastures needed to offset inadequacies in nutrient intake during critical phases of the production cycle (Stonaker, 1982).

Mineral supplementation is especially critical in the acid infertile soil regions. The phosphorus content of grass species in the llanos, for instance, ranges from .09% to .14% (Lebdosoekojo et al., 1980). Even the tropical legumes that are adapted to these allic soils are often deficient in phosphorus; the phosphorus content of Stylosanthes guyanensis is approximately 0.15%, which is marginal or below that required for satisfactory growth and reproduction (CIAT, 1973).

In fertile soil areas, the increased competition for land use, expanded market possibilities, and greater production potential call for more intensive production systems and practices. Particular attention should be given to the selection of pasture species that are not only nutritious and high-yielding, but also can withstand grazing on a continuing basis and be competitive with weeds. The greater frequency of milking cows in beef cattle herds in the fertile-soil areas and the higher incidence of disease and parasitism associated with increased concentrations of cattle also require improved managerial and preventive medicine programs.

Disease and parasitism constraints are related principally to failure to apply available technology, not to the unavailability of prevention and control methods and agents. However, there are two major technological gaps. Currently used foot-and-mouth vaccines confer immunity for only 4 to 5 months (new gene-spliced vaccines produced in Echerichia coli are purported to provide immunity for up to 1 year). Dipping or spraying with chemicals for the control of ticks is costly, laborious, and inefficient, but is necessary because biological control methods have not yet been developed.

Genetic potential places an ultimate limit on productivity. In many instances animals are not producing up to the potential of the current feed-management continuum. Genetic potential will become increasingly limiting as other constraints are removed and as higher production levels are sought.

POTENTIAL TO INCREASE PRODUCTIVITY

Productivity can be markedly increased through the application and adaptation of technology, both in the acid-infertile and fertile savannah regions. The collaborative Instituto Colombiano Agropecuario (ICA) and Centro Internacional de Agricultura Tropical (CIAT) beef production systems experiment tested several improved practices in the acid infertile Colombian llanos region. Using grade Zebu cattle grazing native pasture, mineral supplementation increased calving rate from 52% to 78%, weaning weights at 9 months of age from 159 kg to 163 kg, 18 month weights from 160 kg to 199 kg, and markedly reduced calf and cow mortality (see 9-month weaning figures in table 1). This

TABLE 1. EFFECTS OF MINERALS AND EARLY WEANING ON CALVING RATE, GROWTH RATE, AND MORTALITY IN THE ICA/CIAT BEEF PRODUCTION SYSTEM EXPERIMENT IN THE COLOMBIAN LLANOS, 1972-77

Weaning	Herds 2,3 non minerals		Herds 4,5,6,7,8,9 minerals	
	3 month*	9 month	3 month*	9 month
% Born	92	52	107	78
% Mortality				
3 months	11	22	3	8
9 months	13	25	10	10
18 months	23	27	15	12
% Raised**				
3 months	82	41	104	72
9 months	80	39	96	70
18 months	71	38	91	69
Cow mortality (%)	0	16	0	12.5
Calf weights (kg)				
3 months	66	68	78	81
9 months	112	129	112	163
18 months	151	160	151	199

Source: Stonaker (1982).
* All early weaned calves received minerals following 3 months of age.
** Due to termination of experiment, death loss is underestimated.

resulted in an overall doubling (7,720 kg increase) of beef production of cow herds receiving mineral supplementation during the course of the 5-year experiment (table 2). On a market basis, cost:benefit ratio could easily be as much as 1:20. Another innovation, early weaning, increased calving rate to as high as 107% (table 1). Early weaning removes the nutrient drain for lactation and enables the cow on a marginal plane of nutrition to replenish weight losses and

TABLE 2. CATTLE LIVE WEIGHT PRODUCED PER HERD 1972-77 IN THE ICA/CIAT BEEF PRODUCTION SYSTEMS EXPERIMENT. INCLUDES LIVE WEIGHT OF ALL COWS AND CALVES IN HERDS MAY 1977 AND 18 MONTH WEIGHTS OF STOCKERS LEAVING THE EXPERIMENT, LESS INITIAL 1972 WEIGHTS OF BREEDING HEIFERS ENTERING THE EXPERIMENT

Herds	M[1]	U[2]	Calves no.	Calves kg	Stockers no.	Stockers kg	End cows-1977 no.	End cows-1977 kg	Beginning heifers-1972 no.	Beginning heifers-1972 kg	Net weight produced per herd (kg)	Wt ratio relative to herd 3
1	0	0	10	1185	21	4066	24	7363	34	(6468)	6146	.75
2	0	+	27	2883	20	3197	31	9900	34	(6606)	9374	1.14
3	0	0	21	2124	27	4280	27	8378	35	(6574)	8208	1.00
4	+	+	32	4235	43	7852	32	11402	34	(6467)	17022	2.07
5	+	0	38	4317	38	7228	33	10862	33	(6408)	15999	1.95
6	+	+	30	3456	51	10248	33	10389	35	(6709)	17384	2.12
7	+	0	36	4929	45	9044	35	11858	34	(6486)	19345	2.36
8[3]	+	0	38	3064	49	10032	25	7458	36	(7013)	13541	1.65
9	+	+	28	3699	48	10651	32	10277	37	(7093)	17534	2.14

Source: H. Stonaker (1982).

1 Mineral effect (M) = +7720 kg/herd.
2 Urea effect (U) = 2036 kg/herd.
3 Urea effect on molasses grass only = 3993 kg/herd.

get into body condition to rebreed. These are highly acceptable production levels that can be achieved quickly and economically, with minimal cash outlays and using existing farm management skills and resources.

Similarly, in fertile soil regions, major increases have been attained. At the ICA Turipana station in the Sinu River valley in the north coast of Colombia, calving rates of 86% have been achieved using simple improved production systems (ICA, 1972). Even higher production levels (with calving rates as high as 91%) have been obtained in Ecuador at the Pichilingue station of the Instituto Nacional de Investigaciones Agropecuarios (INIAP, 1974).

RECOMMENDED PRACTICES TO INCREASE PRODUCTIVITY

In the lowland tropics of Latin America, there are two predominant cattle production systems based on pasture, i.e., beef cattle and dual-purpose (beef/milk) production systems. In crop-based farms having livestock (crop/livestock systems), dual-purpose-cattle production systems predominate. This paper discusses only recommended practices for beef cattle and dual-purpose-cattle production systems on pasture.

Herd-Breeding Practices

Only bulls that have shown good growth and fertility should be used for breeding. One bull will be required for every 20 to 30 cows. To maintain the bulls in good breeding condition, a high plane of nutrition is essential and it is often desirable to alternately use and rest bulls where seasonal breeding is used.

Breeding herds should not exceed 100 cows. Nonbreeders and old cows should be rigorously culled and not kept around for "another chance" to get bred.

Seasonal breeding for periods of 3 to 4 months facilitates the programming of breeding and calving at the most propitious times of the year, simplifies management, and reduces labor. Breeding at the beginning of the rainy season (which lasts 6 to 8 months) is the most advantageous because the high nutritive value of new pasture growth puts cows in breeding condition. This results in the majority of the calves being dropped in the dry season (lasting 4 to 6 months), which maximizes calf weaning weight and minimizes mortality. Even without a controlled breeding season, most cows breed in the early part of the rainy season because of improved body condition/ weight. As a general rule of thumb, Zebu cows will not breed unless they weigh at least 300 to 320 kg.

Despite the advantages of controlled breeding, continuous breeding should be used where management skills are deficient. Also, continuous breeding is advantageous in dual-purpose herds where cows with nursing calves ("vaca con ternero") are milked throughout the year.

Under typical circumstances, only Bos indicus and Bos taurus breeds that are adapted to the lowland tropics should be used. These include all of the Zebu (Bos indicus) breeds and the locally adapted Criollo (Bos taurus) breeds that were introduced into Latin America by the Spanish and Portuguese colonists. Unadapted European Bos taurus breeds are generally not suited for the tropics under pasture-management conditions. In some instances, it may be advantageous to infuse some exotic Bos taurus breeding into local Zebu stocks; but this practice should be approached with great caution, and no more than one-fourth to three-eighths exotic bloodlines should be introduced.

There are, however, major advantages in crossing locally adapted Criollo breeds with Zebus. Heterosis levels (more than 10%) are generally higher than those obtained in crossing Bos taurus breeds in temperate climates, and off-spring have even greater livability than either of the parent breeds (Plasse et al., 1974). In the ICA/CIAT beef-production-systems experiment, crossbred calves were 10% heavier at 18 months of age (Stonaker, 1982).

Early Weaning

When cows are on high planes of nutrition, nutrient intake is generally adequate to support lactation and to maintain adequate body weight/condition to enable rebreeding. However, nutrient intake for cows on many tropical savannahs, particularly in the acid infertile soil areas, is barely adequate for reproduction and necessary weight gain without producing milk. Under these conditions, early weaning will remove the nutrient drain for lactation, enabling the cow to get up to the threshold breeding weight of 300 kg or so and to rebreed.

In the ICA/CIAT beef production systems experiment, weaning at 3 months of age increased calving rate 29% to 40% over the 5-year experimental period as compared to traditional weaning at 9 months of age (table 1). Also cow mortality was reduced from 12% to 16% to zero. And available information indicates that growth rate of early-weaned calves will not be seriously reduced if high-quality pastures and concentrate are provided. In this experiment, calves weaned at 3 months of age were given limited concentrate (less than 1 kg/head/day). One group was placed on high-quality pasture and the other group on low-quality pasture. Early-weaned calves on high-quality pasture weighed as much at 18 months of age as calves weaned at 9 months, while those on low-quality pastures weighed somewhat less (table 1). These results emphasize 1) the feasibility and desirability of the early weaning of calves to enable the rebreeding of cows on marginally deficient pastures and 2) the prime need for high-quality pastures/forage and some concentrate for the weaned calves to maintain satisfactory growth rates.

Pasture Management

Pastures should be seasonally grazed to take advantage of their maximum production potential. Higher, well-drained areas should be preferentially grazed in the rainy season when soil moisture is adequate to support plant growth, and the lower, high-water-table areas should be grazed in the dry season.

Invasion of weeds is generally a consequence of incorrect range management or unsuitable pasture species, and should be treated accordingly. Routine use of chemicals should generally be limited to troublesome weeds that are difficult to control and(or) eradicate through management and cultural practices (Doll, 1975).

Setting a stocking rate for the efficient utilization of forage that does not result in deterioration of the pasture is often delicate and highly location-specific. While overgrazing is generally the primary cause of pasture degradation, undergrazing also can be disadvantageous. An example is the replacement of a less desirable species (Trachypogon vestitus) by a more desirable species (Axonopus compressus) as stocking rate increases (Paladines, 1974).

Under extensive management conditions where pasture forage production exceeds consumption, periodic burning can be an important management tool (Paladines, 1974). Burning near the end of the rainy season removes low-quality, mature forage and permits the regrowth of new, higher quality pasture going into the dry season. However, burning should not be used under conditions where pasture forage is limiting, nor should it be used in improved pastures--particularly legume-grass associations--because burning usually kills the legumes.

Improved Pastures

Practically every beef cattle enterprise--whether small or large, intensive or extensive, in fertile or infertile soil areas--should have a program for the establishment of improved pastures. The extent of improved pasture will generally be determined by the feed nutrient requirements for the critical phases of the production cycle, i.e., breeding, weaning, and fattening. Improved pastures will assume greater importance as grazing land becomes limiting, as higher production levels are sought, and as land use becomes intensive.

Many adapted grass species that will withstand grazing can be used over a wide range of situations for the establishment of improved pastures. Some, like pangola (Digitaria decumbens), para (Brachiaria mutica), star grass (Cynodon nlemfuensis), and guinea grass (Panicum maximum), are suitable for the more fertile areas. Others, like Brachiaria humidicola, Brachiaria decumbens, Andropogon gayanus and Hyparrhenia rufa, and Melinus minutiflora (in some instances), are suited to infertile soil areas where there is little or no application of fertilizer.

Forage legumes included in grass swards offer the possibility of increasing productivity, particularly in the dry season, because of their high protein content, drought resistance, and capacity to fix nitrogen in the soil in association with soil rhizobia. In some areas, native legumes abound in association with grasses. But very few tropical forage legume varieties are available for commercial use in Latin America. However, experimental evidence from the ICA/CIAT experiments indicates that live weight gains of growing/fattening cattle on Andropogon gayanus in association with the legumes Stylosanthes capitata, Zornia latifolia, or Pueraria phaseoloides can be as high as 201 to 214 kg live weight gain/animal/year as compared to 128 kg on Andropogon gayanus alone and 90 to 118 kg on other grasses. This relationship also generally exists on live weight gain/ha/year (CIAT, 1981). More important still, cattle grazing the grass-legume associations registered weight gains as high as 303 g/animal/day during the dry season, while cattle on grass pastures suffered weight losses of 50 to 167 g/animal/day.

Another approach is the use of standing legume "protein banks" to provide a protein supplement for low-protein, native grasses during the dry season. Native savannah plus 10% Kudzu (Pueraria phaseoloides) tripled weight gains/ha/ year (22 kg vs 74 kg) and gave weight gains in the dry season (126 g/animal/day) vs weight losses (167/g/animal/ day) on native savannah alone (CIAT, 1981).

However, it should be emphasized that these legume species are yet to be tested under commercial management conditions. The feasibility of these tropical forage legumes will depend upon their persistence under grazing and costs of establishment and maintenance relative to benefits derived (Raun, 1982).

Mineral Supplementation

While the formulation of optimized mineral supplements will be area-specific, it is possible to make general recommendations that can be adapted to local conditions. A mineral supplement composed of one part salt and one part of dicalcium phosphate or other good phosphorus source provided ad libitum is recommended in all tropical regions (Raun, 1976). Fortunately, ruminants consume minerals commensurate with need, and excessive consumption is uncommon. To protect against imbalances in consumption, the mineral mix should not be diluted with salt because this prevents animals from consuming adequate phosphorus and other minerals-- and they may consume excessive salt in obtaining needed minerals. Conversely, salt alone should be provided apart from the mineral mix to enable animals to consume only salt when desired. The mineral mix should be fortified (ppm in the total diet) with copper (5 to 10), cobalt (0.5), iodine

(0.5), and zinc (10). It generally will not.be necessary to add manganese and iron, because levels are high in most tropical soils. Need for the addition of other minerals, e.g., selenium, is unclear and should be approached with caution.

Potential benefits from mineral supplementation are very high, particularly in the acid, infertile soil regions, and probably provide the greatest overall benefit of any intervention that can be applied in the short-term. Results of the 5-year ICA/CIAT beef-production-systems experiment clearly indicate that total productivity per animal unit is nearly doubled (59 vs 95 kg/year) through mineral supplementation (Stonaker, 1982). Mineral supplementation increased calving rate from 52% to 78%, cow weights from 304 kg to 331 kg, calf weights at 9 months from 131 kg to 169 kg, and lowered cow mortality from 16% to 1% and calf mortality from 27% to 12%. Although benefits from mineral supplementation would be markedly less in more fertile, alluvial soil regions, a free choice mineral mix should be provided because animals generally consume minerals only in amounts needed. Consumption, and cost of supplement, will be less than in infertile soil regions.

Protein Supplementation

In situations where the dry season is severe and long, it may be advisable to provide a protein supplement to prevent undue weight losses, long calving intervals, and delayed marketing of feeder cattle. However, compensatory gain in the rainy season must be considered in evaluating the economics of supplementation.

But where the dry season is less severe and of shorter duration, protein supplementation will often not be economical. In the ICA/CIAT beef-production-systems experiment in the Colombian llanos, supplementation of cows with 80 g urea, 500 g molasses, and 3 g sulphur per head per day during the last 3 months of a 4-month dry season reduced weight losses but did not improve calving rate of cows on native pasture (Stonaker, 1982). Similarly, dry season supplementation of Zebu steers had a limited effect on annual weight gains--although supplemented animals weighed more than nonsupplemented animals at the end of the dry season, the latter recovered 50% to 71% of this weight difference with higher weight gains during the subsequent rainy season (Centro Internacional de Agricultura Tropical, 1974, 1975).

Herd Health Programs

Only simple, standardized health programs are required. Foot-and-mouth disease is endemic in South America but not in Central America and the Caribbean. In South America, cattle should normally be vaccinated against foot-and-mouth disease (aftosa) every 5 to 6 months if outbreaks are frequent in the area. If there have been no outbreaks, vaccination can be less frequent. It may be necessary to vaccinate calves against blackleg and anthrax. If brucellosis is a problem in the area, heifer calves (not males) should be vaccinated with Strain 19 brucellosis vaccine. Cattle should be treated for gastrointestinal parasites only when fecal egg counts indicate parasitism. Normally, problems with gastrointestinal parasites are limited mostly to calves, with a few problems in adult animals. Routine scheduled use of antihelminthic agents for gastrointestinal parasites is uneconomical and unnecessary. Babesiosis and anaplasmosis should not be problems for animals born and raised in endemic areas. Dangers arise only when animals from noninfected zones are moved into infected zones, e.g., from high elevation zones (above 2,000 m or so) to lower zones (below 2,000 m). Cattle in tick infested zones (below 2,000 m) should be dipped/sprayed with locally available acaricides every 3 weeks, and if infestation is low, treated again when tick populations build up. Excessive spraying/dipping to keep animals absolutely free of ticks is inadvisable as animals will lose the antibodies protecting them against babesiosis. Some ticks are necessary to periodically reinfect animals and thus maintain protective antibody levels. In the intermediate elevation areas, it may be necessary to dip animals against nuche (Dermatobia hominis). Lungworms (Dictyocaulus viviparus) in calves may be a problem.

DUAL-PURPOSE-CATTLE PRODUCTION SYSTEM

Many so-called beef cattle herds are used for both beef and milk production. A recent evaluation of the Colombian livestock industry indicates that 46% of the milk that enters commercial channels comes from beef cattle herds, most of which are located in tropical climates (Ministerio de Agricultura, 1974). The milk produced represents a principal source of income, particularly where the livestock industry is more developed and where there is a demand for milk in the urban centers. An example is the north coast of Colombia where Rivas (1974) found that on farms of up to 200 ha, sales of milk accounted for 33% of total income. The corresponding figures for farms of 200 ha to 500 ha and over 500 ha were 30% and 13%, respectively. This is especially significant when one considers that milk production averaged only 2.5 liters per cow per day in the dry season and 3.1 liters in the rainy season.

The usual practice in beef/milk herds is to separate the calves from the cows in the evening, milk the cows in the morning, leave some milk for the calf, and allow the calf to be with the cow during the day. Continuous rather than seasonal breeding generally is used in these beef/milk systems.

The increased demand for fluid milk in urban centers has resulted in the establishment of dairy-oriented enterprises in some lowland tropical areas. These systems are strictly pasture-based, like the traditional systems. Cattle breeding schemes have been largely based on the infusion of some exotic Bos taurus dairy breeds into existing Bos indicus (Brahman) and Criollo Bos taurus (Costeno con Cuernos) breeds. These crossbreds typically have one-fourth to three-eighths of exotic dairy breeding (e.g., Holstein), possess sufficient adaptability and resistance to the tropical environment, and produce 50% to 100% more milk than do either Brahman or Criollo breeds.

Particular attention must be given to the dry season feeding of the cow and to the nutrition of the calf. Provision for nutritious dry-season pasture is crucial, e.g., grass-legume associations. Browse legumes such as Leucaena leucocephala might be considered (Hutton, 1974). Fresh chopped sugarcane might serve as an effective energy supplement (Preston, 1974). Conservation of forage produced during the rainy season as hay or silage and other methods of supplementation might also be possible approaches to resolving dry-season nutrient deficiencies.

REFERENCES

Centro Internacional de Agricultura Tropical. 1973. Annual Report 1973. Cali, Colombia. p 254.

Centro Internacional de Agricultura Tropical. 1974. Annual Report 1974. Cali, Colombia. p 260.

Centro Internacional de Agricultura Tropical. 1975. Annual Report 1975. Cali, Colombia.

Centro Internacional de Agricultura Tropical. 1981. Annual Report 1981. Cali, Colombia. p 112.

Colombia. Ministerio de Agricultura. 1974. Programas ganaderos 1974-75. Bogota. p 245.

Doll, J. 1975. Control de malezas en cultivos de clima calido. Centro Internacional de Agricultura Tropical. Boletin serie ES-16. p 10.

FAO. Production yearbook, 1980. Rome.

Hutton, E. M. 1974. Tropical pastures and beef production. World Animal Review 12:1-7.

Instituto Colombiano Agropecuario. 1972. Programa nacional de ganado de carne. Informe anual 1972. Bogota. p 33.

Instituto Nacional de Investigaciones Agropecuarias. Estacion Experimental Tropical Pichilingue. 1974. Programa de pastos y ganaderia bovina. Informe tecnico anual 1974. Quito. p 24.

Lebdosoekojo, S., C. B. Ammerman, N. S. Raun, J. Gomez and R. C. Littel. 1980. Mineral nutrition of beef cattle grazing native pastures on the eastern plains of Colombia. J. Animal Sci. 51:249.

Paladines, O. 1974. Management and utilization of native tropical pastures in America. Seminario. Potencial para incrementar la produccion de carne en America tropical. Cali, Colombia.

Plasse, D., S. O. Verde, B. Muller-Haye, A. M. Burguera and J. Rios. 1974. Comportamiento productivo de Bos taurus y Bos indicus y sus cruces. VII. Estimacion de heterosis en crecimiento. Memoria: Asociacion Latinoamericana de Produccion Animal. 9:61.

Preston, T. R. 1974. Intensive beef fattening systems for the tropics. Potential to increase beef production. Seminario. Potencial para incrementar la produccion de carne en America tropical. Cali, Colombia.

Raun, N. S. 1976. Beef production practices in the lowland American tropics. World Animal Review. 19:18.

Raun, N. S. 1982. Tropical forage legumes in the American tropics. International Stockmen's School. Winrock International, Morrilton, Arkansas.

Rivas, L. 1974. Some aspects of the cattle industry on the north coast plains of Colombia. Tech. Bull. No. 3. Centro Internacional de Agricultura Tropical. Cali, Colombia. Technical Bulletin No. 3.

Stonaker, H. H., J. Villar, G. Osorio and J. Salazar. 1975b. Reproduccion en vacas en los llanos orientales de Colombia. Quinta Reunion Latinoamericana de Produccion Animal. Maracay, Venezuela. (Abstr.).

Stonaker, H. H. 1982. Beef cow-calf production in the Colombian llanos as affected by early weaning, minerals, crossbreeding, urea supplementation, and improved pasture. ICA, CIAT, Winrock International (in press).

103

TROPICAL FORAGE LEGUMES
IN THE AMERICAN TROPICS

Ned S. Raun

There are 12,000 species of Leguminosae in the world; of these, 7,000 are tropical and 4,000 are native to the Americas. Stylosanthes, Desmodium, Centrosema, Macroptilium are among the principal genera in the Americas. Although these legumes are widely distributed throughout tropical America, only limited exploitation has been made of this germ plasm base in the establishment of legume-based pastures.

This paper first discusses productivity of legumes and legume-grass associations, followed by commentary on considerations in developing improved pastures based on tropical forage legumes. Principal sources of information are from the Centro Internacional de Agricultura Tropical (CIAT), the Instituto Colombiano Agropecuario (ICA) in Colombia, and the Instituto Veterinario de Investigaciones Tropicales y de Altura (IVITA) in Peru.

PRODUCTION OF LEGUME AND LEGUME-GRASS BASED PASTURES

Legumes

Legumes alone would generally be used under two situations, i.e., as fresh-cut forage and as a standing forage for deferred grazing.

An example of a fresh-cut forage is Desmodium distortum. This type of forage is generally used as a supplement to low-protein diets during the dry season and for dairy cattle. These legumes are not suitable components of perennial pasture mixtures, being biennial and annual plants, respectively.

Standing-legume forage is sometimes used during the dry season as a feed reserve for deferred grazing to provide a high-protein forage to supplement low-quality, low-protein grass pastures. Leucaena leucocephala is one example. It is a browse plant, deeply rooted, withstands drought, and is suitable for medium- to higher-fertility soils. It is generally not used under continuous grazing. Kudzu (Pueraria phaseoloides) is widespread throughout the wet tropics, has a wide range in soil adaptability, and is excellent for standover feed for the dry season.

Legume-grass Pastures

Although optimal ratios of grasses and legumes will vary widely depending upon species and environment, available information would indicate that approximately 50% of legume grass is desirable in the tropics and subtropics. On the other hand, it is generally felt that, if pastures are used all the year round, some grass mixed with the legume would be preferable to a pure legume pasture. This is in consideration of the generally higher dry matter production of the grasses during the rainy season, their faster growth following early rains, and their tendency to provide more ground cover that controls weeds and minimizes erosion and water runoff. An exception would be standing "protein banks" of pure legumes to be used as protein supplements in the dry season, e.g., Kudzu.

Legumes contribute in two principal ways to increasing productivity when incorporated in a grass pasture. First, the legume forage component, being higher in protein than the grasses, acts as a protein supplement particularly in the dry season when grasses are often very deficient in protein, and also for grasses that are marginally deficient in protein in the rainy season. In this instance, the supplemental legume increases digestibility and intake of grass forage. In a series of CIAT investigations, addition of Stylosanthes guyanensis to dry season Melinis minutiflora (3% crude protein) increased total intake and digestibility of consumed dry matter, and increased intake of digestible dry matter from 60% to 100% of maintenance requirements (CIAT, 1973). Secondly, legumes and soil Rhizobia symbiotically fix nitrogen in the soil, improving nitrogen status and providing nitrogen for enhanced growth of both grass and legume.

In CIAT's tropical pastures program, principal emphasis is placed on developing improved legume-grass pastures. Much of the field work in this program is carried out in colaboration with the Instituto Colombiano Agropecuario (ICA) at their Carimagua Station in the heart of the Colombian Llanos. Climates and soil type are similar to those found in other savanna grassland areas, i.e., Venezuelan Llanos and Campo Cerrado of Brazil. The average rainfall is 2,000 mm, the dry season lasts 4 to 5 months, the mean temperature is 27° C, soils are acid (pH 4.5), phosphorus is low (2 ppm), other bases are low (Ca, Mg, K), and exchangeable aluminum is high (3.5 m.e/100 g) (Spain, 1974).

Predominate native grass genera are Trachypogon, Andropogon, Paspalum, Axonopus, and Leptocoryphium. All of these species are regarded to be of low nutritive value and slow growth. Principal naturalized grass species include Melinis minutiflora, Hyparrhenia rufa, Brachiaria decumbens, and more recently Andropogon gayanus. During the rainy season, total weight gains/ha are considerably higher on the naturalized grass pastures as compared to the native grass pastures, but weight gains/animal are increased only slightly. During the dry season, weight losses are encountered on all grasses. The only exception would be in the bajos, low areas that remain moist during the dry season, and where growth of pasture plants continues.

Adapted improved "naturalized" grasses are available, and establishment methods and management systems are reasonably well-defined. These include Andropogon gayanus, Brachiaria humidicola, and Brachiaria decumbens. However, much less technology is available for legumes. In many instances, adapted varieties that will persist and are insect and disease resistant are either not available or are just being identified. Further definition is needed on fertilizer requirements, establishment methods, required pasture management, and seed production systems. Serious errors have been made in assuming that "overseas" selected varieties and technology would be directly usable in the Latin American tropics.

Table 1 summarizes productivity (expressed as weight gains) of different types of pastures in Carimagua, an agricultural research center in the Colombian eastern llanos that is operated by ICA and CIAT (CIAT, 1981). In native savanna, animal productivity per year increases only slightly when improved grasses are introduced alone. However, when improved grasses are accompanied by legumes, the yearly per-animal productivity is approximately doubled. In the same way, by using protein banks in improved pastures, per-animal productivity is twice that on native savanna and, with only 1/20 of the area in a small Kudzu bank, it improves as much as 50%. With respect to yearly weight gain per hectare, Brachiaria decumbens alone as compared to native savanna increased yearly weight gain/ha by seven times and Andropogon gayanus 20 times. When Andropogon gayanus is associated with a legume, the area productivity is augmented approximately 15 times, somewhat less than with Andropogon gayanus alone. When protein banks were used in native savannas, weight losses did not occur during the dry season, animal productivity increased by 60%, and per-hectare productivity tripled.

Similar results have been obtained by researchers in the Instituto Veterinario de Investigaciones Tropicales y de Altura (IVITA) at their tropical station in Pucallpa, Peru. This station is located in the Amazon basin. Terrain is undulating, vegetation is of the high rain forest type. Annual rainfall is 1,600 mm and is fairly well distributed, with June, July, and August being relatively dry. The mean temperature is 30.8°C. Soils are lateritic clay loams, acidic (pH 3.8 to 5.3), low in phosphorus (1.5 to 4.5 ppm), low in Ca, Mg, and K, and high in exchangeable Al (0.20 to 7.60 m e/100 g). Several legume species (Macroptilium atropurpeum, Glycine wightii, Lotononis bainsii, Desmodium

TABLE 1. PASTURE PRODUCTIVITY EXPRESSED IN WEIGHT GAINS.
SUMMARY OF EVALUATIONS, CARIMAGUA, COLOMBIA

Type of pasture	Average pasture productivity (weight gains)		
	Drought	Rains	Total / year
	g/animal/day		kg/animal kg/ha
Grasses only*			
Savanna with improved herd			
& pasture management	-167	449	90	22
Melinis minutiflora	-445	508	97	43
Brachiaria decumbens	- 50	506	118	147
Andropogon gayanus	- 97	567	128	457
Associations**			
A. gayanus +			
S. capitata	303	656	201	330
A. gayanus +			
Z. latifolia	163	765	214	357
A. gayanus +			
P. phaseoloides	290	696	210	380
Protein banks***			
B. decumbens + P.			
phaseoloides in blocks	317	625	191	303
B. decumbens + P.			
phaseoloides in rows	540	606	213	341
Savanna + 1/20 of			
P. phaseoloides	52	468	121	30
Savanna + 1/10 of			
P. phaseoloides	126	537	147	74

Source: Centro Internacional de Agricultura Tropical
(CIAT) Report 1981.
* More than three years of observation.
** Two years of observation.
*** One year of observation.

intortum, Stylosanthes humilis, Stylosanthes guyanensis, Pueraria phaseoloides, Centrosema pubescens) were observed with associated grass species at Pucallpa. Of these, Stylosanthes guyanensis, Pueraria phaseoloides and Centrosema pubescens appeared to be suitable legumes to develop grass-legume mixtures in that region (Santhirasegaram et al., 1972). Productivity of the legume-grass association was markedly higher than that of the pure grass pastures. Santhirasegaram (1974) obtained live weight gains of 352 kg/ha/yr with two yearling Nellore heifers/ha on Pueraria phaseoloides-grass pastures.

Santhirasegaram (1974) also observed that direct phosphorus supplementation is necessary for animals grazing legume-grass associations. Although the legumes have a higher phosphorus content than the grasses, they are still deficient in phosphorus. Spain (1974) observes that, although the phosphorus fertilization level will influence the phosphorus content of tropical forage legumes, it generally will not be economical to fertilize to the level necessary to make the legume forage nutritionally adequate in phosphorus. The fertilization level should be adjusted to achieve optimum dry matter productivity, which in Carimagua, Colombia, would generally be no more than 20 to 25 kg P/ha at time of establishment with maintenance levels of no more than 5 to 10 kg P/ha/yr thereafter. Deficiency of phosphorus in the forage can be corrected by ad libitum feeding of a mineral supplement (Lebdosoekojo, 1980).

Calving percentages are low in most lateritic soil areas, averaging probably no more than 45% to 50% with cows calving every two years (Raun, 1976). Inadequate nutrient intake is the principal cause of low reproductive performance and often is accentuated during the dry season when forage protein content declines sharply, often to as low as 2% to 3%. In these situations, the legume could be invaluable in correcting the marked protein deficiency of dry-season pasture grasses, in acting as a protein supplement in meeting metabolic needs for protein, and in increasing forage digestibility and intake, thereby raising the nutritional plane sufficiently to enable the lactating cow to rebreed.

Although no reproduction data are yet available in CIAT, growth data indicate that legume-grass pastures would provide a plane of nutrition adequate for lactating cows to rebreed. Nutrient requirements established by the National Research Council indicate that a ration that would support live weight gains of 500 g/day in yearling steers would also provide nutrient intake adequate to enable the lactating cow to rebreed. Current live weight gains in Carimagua on legume-grass pastures are well above this amount on a year-round basis. However, gains on the native savanna-Kudzu-protein-bank pastures are marginal.

Developing Pasture Improvement Programs Based on Tropical Legumes

These results on tropical forage legume pastures and other results obtained in the Latin American tropics, when reinforced by those of the Australian experience, are highly encouraging and illustrate their tremendous potential to increase beef cattle production. But there are still some gaps in workable, legume-based pasture systems at the field level. Some of the gaps are technological, some are socio-economic, some relate to availability of inputs. These problems are critical; they must be resolved before the legume package will be salable and workable. Principal constraints are 1) adapted varieties that will persist for at least 3 years to break even on costs and 6 years for satisfactory returns on investment, 2) availability of phosphorus fertilizers, 3) availability of seed, and 4) economical and practical methods of establishment.

Adapted Varieties

Native tropical forage legumes abound in the Americas. Of these, the Australians have collected ecotypes and developed varieties that will perform under their conditions, which are generally less tropical than in much of the Latin American tropics. With the success obtained in Australia with these varieties, many organizations were prompted to directly use these varieties in the Americas. Results have been disappointing. The Australian commercial varieties have generally not been adapted to the tropical conditions in the Americas.

Susceptibility to disease, insect attack, and inability to compete with other species have been common problems. This perhaps seems strange since these varieties were developed from ecotypes collected in the Americas. However, the American ecotypes were selected for Australia and not Latin America.

But little has been done in the Americas to understand and exploit this germ plasm. There is a job to be done in adequately sampling and evaluating native species. Natural variability within legume species appears to be quite high. Consequently, the probability of identification of adapted-- and superior--ecotypes from the native germ plasm pool must be fairly good.

The identification of varieties that are adapted to the major tropical ecosystems in the Americas--and which will persist in association with grasses using simple management practices--is an absolute requirement for the success of legume-grass associations.

Phosphorus Fertilization

Establishment of legume-based pastures in most humid tropical areas will require the application of some phosphorus fertilizer. Although amounts required for establishment and maintenance are moderate, sizable quantities of phosphorus fertilizer will nonetheless be required in any pasture improvement program of significant dimension. High cost of phosphorus fertilizers is a serious constraint and emphasizes the need to seek varieties/establishment procedures that minimize requirements for phosphorus fertilizers.

It is encouraging that for many legume species, phosphorus fertilization required is fairly low (no more than 20 to 25 kg/P/ha for establishment, and maintenance requirements of 5 to 10 kg P/ha/yr).

TABLE 2. PRINCIPAL ADVANTAGES AND DISADVANTAGES OF PLANTING METHODS AND SPATIAL DISTRIBUTION IN PASTURE DEVELOPMENT, CARIMAGUA, COLOMBIA

Systems	Advantages	Disadvantages
Conventional seeding (broadcast)	-Can be done manually or with scattering devices -A traditional method	-Greater seed requirement -More weed problems -Low fertilizer efficiency
Row seeding	-Requires less seed -Greater fertilizer efficiency -Better initial establishment for each component -Reduces initial competition -Reduces shade	-Requires more complicated machinery -Slower than broadcast
Spatial distribution (species planted in separate bands)	-Results in more stable and persistent associations of some species than in intimate mixtures -Permits association between otherwise incompatible species -Keeps the advantage of association; avoids some of the problems of protein banks	-More complicated than traditional planting -Wide bands do not favor efficient use of nitrogen by associated grasses
Low density methods	-Initially low labor, seed, and fertilizer are required (as little as 5% of that received for conventional seeding) -Well accepted on small farms -Results in very strong and persistent mother plants -Reduces risk of inherent failure in establishing pastures	-May need more time to become established -Not suitable for all species -May not work where weed potential is high

Source: Centro Internacional de Agricultura Tropical (CIAT) Report 1981.

Seed Production

After adapted varieties are identified, seed production will be limited first by the initial quantity of seed; second, by the time needed to produce a seed crop that often is as much as a year; third, by yield, which is about 50 to 70 kg harvested for every kilo of seed planted in the case of S. capitata; and, fourth, by economic feasibility of commercial seed production. Although recommended seeding rates are only 1 to 5 kg/ha, total area required annually in seed production would average 2% of the area to be seeded. These combined considerations all emphasize the importance of developing comprehensive seed production packages that are efficient, economically feasible, and attractive to commercial seed producers. Any large-scale pasture-improvement program based on tropical legumes will require sizable quantities of commercially produced seed.

Apart from commercial seed production, sizable quantities can and should be produced in "on farm" seed production plots. For instance, three kilograms of Stylosanthes capitata will seed one hectare, which in 12 months will produce 180 kg of seed or more that could seed 60 ha. If every livestock farm could seed such an area annually with an adapted variety that would persist, the beef cattle industry would be revolutionized and beef production increases would be phenomenal.

Establishment

To be successful and accepted by producers, pasture-establishment methods and maintenance practices must be practical at the farm/ranch level; reasonable in cost, i.e., cost attractive; and must provide persistent, productive, and stable legume-grass associations. Principal methods with advantages and disadvantages are cited in table 2.

CONCLUSION

Tropical forage legumes are not new. But, even so, we have not yet exploited this resource. A concentrated and systematic effort is needed to develop comprehensive tropical forage-legume packages, i.e., the varieties, fertilizer, and the seed production systems to enable us to put tropical forage legumes to work at the field level.

936

REFERENCES

Centro Internacional de Agricultura Tropical. 1973. Annual Report, CIAT, Cali, Colombia.

Centro Internacional de Agricultura Tropical. 1981. Annual Report, CIAT, Cali, Colombia.

Lebdosoekojo, S., C. B. Ammerman, N. S. Raun, J. Gomez and R. C. Litell. 1980. Mineral nutrition of beef cattle grazing native pastures on the eastern plains of Colombia. J. Anim. Sci. 51:1,249.

Paladines, O., E. Alarcon, J. Hilton, J. M. Spain, B. Grof and R. Perez. 1974. Development of a pasture program in the tropical savanna of Colombia. Proceedings - XII International Grassland Congress.

Raun, N. S. 1976. Beef production practices in the lowland American tropics. World Animal Review No. 19:18.

Santhirasegaram, K., V. Morales, L. Pinedo and J. Diaz. 1972. Interim report on pasture development in the Pucallpa region. FAO-IVITA. Pucallpa, Peru.

Santhirasegaram, K. 1974. Legume-based improved tropical pastures. Proceedings of seminar on tropical America: Potential to increase beef production. Centro Internacional de Agricultura Tropical, Cali, Colombia.

Spain, J. M. 1974. La fertilizacion fosforica de praderas en suelos alicos. Suelos Ecuatoriales. Proc. Tercer Coloquio Sobre Suelos. Sociedad Colombiana de la Ciencia del suelo. Volumen VI, 1:234.

Stonaker, H. H. 1982. Beef cow-calf production in the Colombian Llanos as affected by early weaning, minerals, crossbreeding, urea supplementation, and improved pasture. CIAT, ICA, Winrock International (in publication).

104

SUGAR CANE MOLASSES
FOR FATTENING STEERS

Manuel E. Ruiz

Ruminants by virtue of their rumen microorganisms, can break down fibrous feeds and survive on poor-grade forages. These animals have the capacity to synthesize good-quality microbial protein from simple nitrogenous compounds such as urea. However, this requires the availability of rapidly fermentable carbohydrates such as sugar or starch. In the U.S. and Europe, intensive fattening systems have been developed based on the use of high levels of cereals. These systems, in turn, are based on the fact that the most effi- cient feed conversion for meat is obtained with the highest possible proportion of cereals in the ration. However, the application of such systems to the southeastern region of the U.S. and in developing tropical countries is constrained by low cereal-crop yields, or high import cost of grains. An exception is rice, which is grown for human nutrition and not for animal feeding. Thus, given the inadaptability of grain-based feeding systems for the subtropical and tropical regions, alternatives must be found.

The tropical and subtropical areas of the world abound in energy-feed resources: for example, sugarcane, molasses, culled commercial bananas, cassava roots, taro, sweet pota- toes, and other lesser-known roots and by-products. Sugar- cane molasses is one of the most easily available feeds derived from a highly efficient plant, as indicated in a comparison with other energy-rich crops (table 1).

THE 10%-MOLASSES TABOO

The use of molasses as feed for cattle has been known for many years. However, the proportion in which it has been used has rarely exceeded 5% to 10% of the total ra- tion. The reasons for this limitation have included:
- Higher levels cause diarrhea due to its laxative effects (Morrison, 1956).
- Higher levels cause difficulties in the preparation of rations due to its viscous nature.

- Storage and transport difficulties arise because of the increased humidity that higher levels of molasses impact to rations.
- The net energy value of molasses decreases when it is used in levels above 10%, as shown in table 2.
- High intake of molasses may cause toxicity resulting in polyencephalomalacia (Verdura and Zamora, 1970).
- Molasses is very poor in protein value.

TABLE 1. TDN YIELDS OF ENERGY-RICH CROPS IN SELECTED COUNTRIES, METRIC TONS/HA

Country	Grain	Sorghum grain	Cassava roots	High-Test molasses[a]
Mexico	0.96	2.00	-	7.48
Jamaica	0.96	-	0.40	8.28
Peru	1.28	1.36	2.07	17.40
Taiwan	1.92	1.28	2.80	8.80
India	0.80	0.40	2.35	5.72

Source: B. Gohl (1981).
[a] Assumes that 100 tons of sugarcane produces 18.4 tons of high-test molasses

TABLE 2. NET ENERGY OF MOLASSES AS A FUNCTION OF ITS PROPORTION IN COMPLETE RATIONS

Level of molasses %	Net energy kcal/g
10	1.518
25	0.833
40	0.773

Source: G. D. Lofgreen and k. K. Otagaki (1960).

Against this set of constraints, research in Cuba and Costa Rica has shown that molasses can be consumed at very high levels without causing toxicity problems. This permits the preparation of liquid feeds and provides a solution to the problem of handling and mixing complete diets. Initially, however, the outlook for such a solution was not encouraging as indicated in the following discussion of molasses-feeding-systems research.

THE DEVELOPMENT OF HIGH-MOLASSES-FEEDING SYSTEM

To avoid feed-mixing problems, molasses must be provided free-choice in open containers, thus ensuring minimum trough space requirements (Preston, 1972). An initial experiment consisted of allowing the animals free consumption of molasses under two sets of conditions: free-choice feeding of grains or forage (table 3).

TABLE 3. EFFECT OF GIVING MOLASSES/UREA FREE CHOICE TOGETHER WITH GRAIN OR FORAGE TO BRAHMAN BULLS IN DRYLOT

	Molasses/urea	plus
Item	Ad libitum forage	Ad libitum grain
No. of bulls	246	80
Initial weight, kg	218	194
Final weight, kg	398	368
Daily gain, kg	0.59	0.97
Daily DM intake, kg		
Molasses	4.0	0.75
Forage	6.6	-
Grain	-	5.50
Daily intake of ME, Mcal	25.3	18.2
Conversion per kg gain		
Molasses, kg DM	6.8	0.75
Forage, kg DM	11.2	-
Concentrates, kg DM	-	5.50
ME, Mcal	43.0	18.8

Source: T. R. Preston and M. B. Willis (1970).

While performance was best when molasses was used together with sorghum grain, the intake of molasses was very low, contributing no more than 30% of the total metabolizable energy (ME). When forage was the basic ingredient, molasses intake was much higher and contributed up to 42% of the total ME (Preston and Willis, 1970). However, this resulted in a poor rate of gain and inefficient rate of feed conversion. It was observed that when the animal was given a choice between molasses and forage or grain, it favored either forage or grain. This finding led to the proposition of restricting these feeds and allowing ad libitum consumption of molasses as a means of increasing molasses intake and, hopefully, improving performance. This hypothesis proved to be correct as shown in table 4.

When the forage was restricted to 1.5% body weight (as-fed basis), the molasses intake increased, and animal performance also improved both in weight gain and feed conversion. With the restricted forage intake, molasses contributed some 75% of the total ME.

Experiments in Costa Rica have shown that the source of fiber does not affect animal weight gain as long as the intake of the fibrous ingredient is greater than 230 g/DM/100

TABLE 4. PERFORMANCE OF BRAHMAN BULLS GIVEN AD LIBITUM MOLASSES WITH 3% UREA AND EITHER AD LIBITUM FORAGE (MAIZE) OR RESTRICTED FORAGE (1.5% OF LW) AND CONCENTRATE SUPPLEMENT (0.4% OF LW)

	Forage	
Item	Ad libitum	Restricted
No. of bulls	24	23
Initial weight, kg	220	210
Final weight, kg	385	390
Daily gain, kg	0.63	0.76
Feed intake, kg (as-feed basis)		
Molasses	3.36	5.6
Forage	30.6	4.3
Concentrate	-	1.12
Feed conversion, Mcal		
ME/kg gain	32.8	20.6
Killing out, %	51.7	52.3

Source: J. L. Martin et al. (1968).

kg LW and less than 800 g/DM/100 kg LW. At rates between these levels, molasses intake decreases if green forage is used in increasing quantities; but if the fiber source is bagasse, the molasses intake increases as the level of fiber increases (figure 1).

The negative relationship between green-forage intake and molasses intake is presumably the result of a substituting effect, both feeds being energy sources. In the case of bagasse, its effect may be the result of a stimulation of rumen motility and more rapid rate of passage of rumen contents.

Figure 1 shows that levels of fiber less than 230 g DM/kg LW are associated with bloat and symptoms of molasses toxicity (drunkenness). Figure 1 also shows that very high levels of molasses intake can be achieved at either the minimum level of green forage or at the high level of bagasse. For example, the minimum use of green forage, a 300 kg steer may consume up to 10 kg of molasses daily.

If the source of fiber is bagasse, the logical decision is to use the lowest safe level so as to minimize transportation and storage expenses and to keep to a minimum the ad libitum consumption of molasses without affecting animal performance. If the fiber source is green forage, its availability and relative cost will determine the most appropriate level of use.

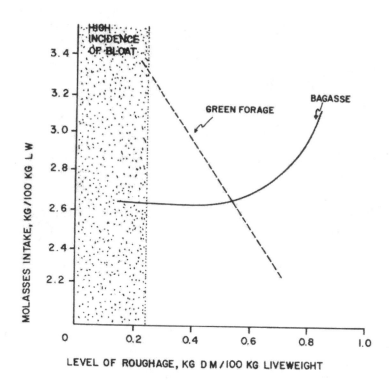

Figure 1. Effect of daily roughage consumption on daily intake of molasses (80° Brix)(M. E. Ruiz, 1973)

THE 1%-UREA TABOO

The most critical element in a feeding system based on molasses is the protein, since molasses contains only 3% CP on a dry-matter basis. Moreover, because molasses is a liquid, rate of passage and rate of microbial protein synthesis may be significantly different from that found with dry, solid rations. Thus, a first step in the practical definition of the amount and composition of the CP to be given was to establish the relationship between the level of a high-quality protein supplement and animal response. Figure 2 shows the results of a study with Peruvian fish meal as the supplementary protein.

Biologically, the response of the animal to protein is very evident up to a certain point after which there is no appreciable response to additional amounts of protein. The point at which the near maximum response is obtained is around 700 g CP/100 kg LW, however, the protein level necessary to achieve 1 kg of daily weight gain is around 400 g/100 kg LW. It is important to note that in this experiment a high-quality protein source was employed. It is likely that more protein will be necessary as the quality of the protein decreases, as would be the case with cottonseed meal, or more obviously, if urea were used.

Table 5 provides a comparison between fishmeal and chicken litter as supplements providing 30% of the total N consumed. As a carrier for urea, molasses has the advantage of its liquid nature, which makes it possible to obtain an even distribution of this source of NPN. This, in turn, reduces the risk of urea toxicity due to inefficient mixing. Another advantage of molasses is its masking effect on the unpalatable urea, which allows for the incorporation of urea levels higher than the traditional ceiling of 1%. Finally, because of its high concentration of soluble sugars, molasses is a convenient energy substrate for microbial activity, especially when a highly hydrolizable N-source, such as urea, is used.

TABLE 5. EFFECT OF PROTEIN SOURCE ON PERFORMANCE OF BULLS GIVEN HIGH LEVELS OF MOLASSES/UREA

| | Protein source | |
Item	Fish meal	Poultry waste
No. of bulls	36	33
Initial weight, kg	281	277
Final weight, kg	429	280
Daily gain[a], kg	0.98	0.68
Feed intake		
DM, kg/day	7.47	8.21
ME as molasses, %	71.0	70.9
Conversion, Mcal ME/kg		
daily gain	20.8	34.3

Source: T. R. Preston et al. (1970).
[a] Adjusted for differences in final weight

As expected, the higher the proportion of the total N as urea-N, the slower the weight gain rate will be. This is clearly shown in figure 3, the data having been obtained from an experiment involving young bullocks weighing initially 300 kg LW.

Even though there will be a decrease in animal performance as the level of urea is raised, the ultimate criterium for deciding what level of NPN should be used, will be the profitability. This, obviously, has to be worked out for every specific location and situation. Application of molasses/urea-based feeding systems at the commercial farm level in Costa Rica (Clavo, 1974) has proven that rates of gain of Brahman crossbred steers can be as high as 841 g/day, with urea providing 72% of the total CP (total CP:350 g/100 kg LW/day).

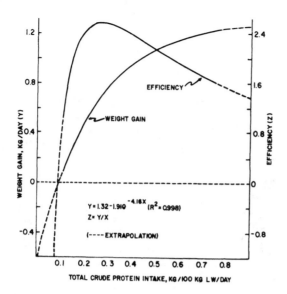

Figure 2. Weight gain and protein efficiency as a function of protein intake in young bulls (300-400 kg LW) fed high levels of molasses (M. E. Ruiz, 1976)

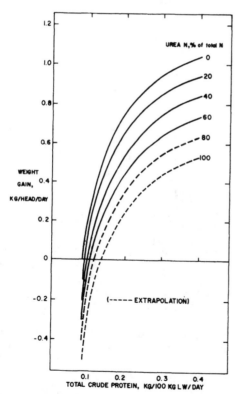

Figure 3. Weight gain of steers fed variable levels of total crude protein and urea (M. H. Ruiloba et al., 1978)

TWO SYSTEMS

Examples of intensive feeding systems based on molasses will be given. Both stress the need for a gradual build up of molasses and urea intakes and a gradual reduction, or change, of the roughage ingredient. The first example concerns a procedure developed in Cuba (Preston and Willis, 1970), as shown in table 6.

TABLE 6. RECOMMENDED PROCEDURE FOR FEEDLOT FATTENING OF CATTLE ON A MOLASSES-BASED DIET

Day	Fresh forage kg/head	Molasses/urea[a,b] kg/head	Protein supplement[c] g/100 kg LW	Minerals[d]
1	ad libitum	1.0	120	ad libitum
2	"	1.5	"	"
3	"	Gradual increase by	"	"
⸱	"	0.5 kg/day until	"	"
⸱	"	ad lib levels are	"	"
⸱	"	reached	"	"
⸱	"		"	"
14	"	ad libitum	"	"
15	30	"	"	"
16	25	"	"	"
17	20	"	"	"
18	15	"	"	"
19	10	"	"	"
20	1.5 kg/100 kg LW	"	"	"
⸱	"	"	"	"
⸱	"	"	"	"
⸱	"	"	"	"
⸱	"	"	"	"
to slaughter				

Source: T. R. Preston and M. B. Willis (1970).
a 2% urea, 4% water, 0.5% salt, and 93.5% molasses.
b Equal parts of water added during the first 4 weeks.
c Fish meal (65% CP) of any other equivalent source of insoluble protein. It contains, additionally, 80,000 IU of vitamin A and 20,000 IU of vitamin D per kg.
d Contains, in g/kg, 500 $CaHPO_4$, 400 NaCl, 20 $ZnCO_3$, 0.1KI, 19.8 ground corn, 27 $FeSO_4$; $5H_2O$, 23 $MnSO_4$; H_2O, 10 $CuSO_4$; $5H_2O$, 0.1 $CoSO_4$; $7H_2O$.

The other example comes from Costa Rica and is designed for fattening operations near sugar mills, as both bagasse and molasses are used. Notice the gradual replacement of green grass by bagasse (table 7).

TABLE 7. A SYSTEM OF INTENSIVE FEEDING BASED ON MOLASSES, BAGASSE, AND UREA. ALL FIGURES ARE EXPRESSED IN INDICATED UNITS/100 KG LW/DAY

Day	Green forage[a]	Bagasse DM	Molasses[b]	Urea	Meat and bone meal
	kg	g	kg	g	g
1-3	5	0	0.3	15	50
4-6	2.5	100	0.6	30	100
7-9	1.0	200	1.2	45	200
10-12	0.5	300	2.4	60	300
13	0	400		75[c]	341
'	"	"	"	"	"
'	"	"	"	"	"
'	"	"	"	"	"
'	"	"	"	"	"
To slaughter					

Source: M. E. Ruiz (1974).
[a] Usually containing 26% DM.
[b] Contains 75% DM.
[c] Rejection at this level may ocur; if so, the increase may be modified to 60 -> 65 -> 70 -> 75 every 3 days.

CARCASS CHARACTERISTICS

Animals fattened with molasses-based rations tend to produce carcasses with less fat than is the case with corn-fed animals. Table 8 shows data from both Cuba and Costa Rica. The differences observed in carcass yield and fat content may be explained by the fact that the animals in the Costa Rican experiment had been fed the molasses-rich diet from the time of weaning.

IRONING OUT THE SYSTEM WITH STARCH

More recently, various experiments at CATIE (Herrera, 1974; Ruiz and Ruiz, 1978) have shown that molasses/urea diets can be greatly improved by substitutions of starch for molasses. Figure 4 shows that improvements of up to 30% in rate of gain and feed conversion can be obtained by iso-caloric substitutions of molasses by starch. The most like-ly reason for such beneficial effects of the starch is a more efficient utilization of urea for microbial protein synthesis, as demonstrated by p32 - uptake experiments (Olivo, 1978). This finding opens a new array of possibil-ities, especially for tropical and subtropical countries where there is an abundance of starch-rich agricultural pro-ducts and by-products.

946

TABLE 8. CARCASS CHARACTERISTICS OF BULLOCKS FATTENED WITH DIETS BASED ON MOLASSES, UREA, AND PROTEIN SUPPLEMENT WITH RESTRICTED FORAGE

	Cuba[a]	Costa Rica[b]
Number of bulls	47	58
Final LW, kg	387	401
Carcass weight, kg	203	226
Yield, %	52.4	56.4
Total meat, %	73.0	77.0
Excess fat, %	8.6	3.0
Bone, %	18.4	18.6
Trimmings, %	−	1.4

Source (Cuba): A. Elias et al. (1969).
Source (Costa Rica): M. E. Ruiz (1974).
[a] DM intakes: molasses 5.1 kg; green forage 1.5 kg; concentrate 0.4 kg; of the total N, 59.7% was NPN.
[b] DM intakes: molasses 4.9 kg; bagasse 1.1 kg; of the total N, 30% was NPN.

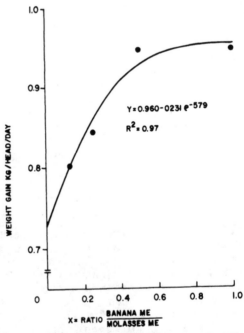

Figure 4. Effect of the substitution of ME derived from bananas for ME derived from molasses on weight gain in cattle fed high levels of urea (E. Herrera, 1974)

The findings discussed here indicate that molasses-based systems rival those based on corn and soybean meal in terms of live weight gain. Although feed efficiency of molasses-based diets requires improvement, this can be achieved by appropriate use of a good-quality, insoluble source of supplementary protein and a relatively small proportion of starch.

REFERENCES

Clavo, N. 1974. Respuesta a diferentes niveles de urea por novillos alimentados con melaza y bagazo de cana de azucar. M.S. Thesis. Turrialba, Costa Rica, IICA-CTEI.

Elias, A., T. R. Preston, and M. B. Willis. 1969. Subproductos de la cana y produccion intensiva de carne. 8. El efecto de la inoculacion ruminal y de distintas cantidades de forraje sobre el comportamiento de toros Cebu cebados con altos niveles de miel/urea. Revista Cubana de Ciencia Agricola 3:19.

Gohl, B. 1981. Tropical feeds. FAO Annual Production and Health Series No. 12. FAO, Rome.

Herrera, E. 1974. Engorda de vacas de desecho con subproductos de la cana y diversos niveles de almidon de banano. M.S. Thesis. Turrialba, Costa Rica, IICA-CTEI.

Lofgreen, G. D. and K. K. Otagaki. 1960. The net energy of blackstrap molasses for fattening steers as determined by comparative slaughter technique. J. Anim. Sci. 19:392.

Martin, J. L., T. R. Preston, and M. B. Willis. Intensive beef production from sugarcane. 6. Napier or maize as forage sources at two levels in diets based on molasses/urea. Revista Cubana de Ciencia Agricla (English ed.) 2:175.

Morrison, F. B. 1956. Feeds and Feeding, 22nd. ed. Cornell Univ., Ithaca, New York.

Olivo, R. 1978. Evaluacion del crecimiento microbial in vitro con diferentes relaciones amilosa/amilopectina y almidon/sacarosa en el sustrato energetico. M.S. Thesis. Turrialba, Costa Rica, UCR-CATIE.

Preston, T. R. 1972. Fattening beef cattle on molasses in the tropics. World Animal Review 1:24.

Preston, T. R. and M. B. Willis. 1970. Intensive beef production. Pergamon Press, New York.

Preston, T. R., M. B. Willis, A. Elias, and J. Garcia. 1970. Intensive beef from sugarcane. II. Effect of vitamin E and selenium, type of molasses and method of administration of the protein supplement. Revista Cubana de Ciencia Agricola (English ed.) 4:175.

Ruiloba, M. H., M. E. Ruiz, and C. Pitty. 1978. Produccion de carne durante la epoca seca a base de subproductos. II. Niveles de proteina y substitucion de proteina verdadera por urea. Ciencia Agropecuaria (Panama) 1:77.

Ruiz, A. and M. E. Ruiz. 1978. Utilizacion de la gallinaza en la alimentacion de bovinos. III. Produccion de carne en funcion de diversos niveles de gallinaza y almidon. Turrialba 28:215.

Ruiz, M. E. 1974. Desarrollo de sistemas intensivos de produccion de carne en confinamiento para el tropico. CATIE, Turrialba, Costa Rica.

Ruiz, M. E. 1973. Intensive beef production with sugarcane molasses and bagasse. In: Activities at Turrialba. CATIE, Costa Rica 1:3.

Ruiz, M. E. 1976. New animal feeding systems based on the intensive use of tropical by-products. In: P. V. Fonnesbeck, L. F. Harris, and L. C. Kearl (Eds.) First International Symposium, Feed Composition, Animal Nutrient Requirements, and Computerization of Diets. pp 660-666, Utah State University.

Verdura, T. and I. Zamora. 1970. Cerebrocortical necrosis in Cuba in beef cattle fed high levels of molasses. Revista Cubana de Ciencia Agricola (English ed.) 4:209.

105
CALF REARING
IN TROPICAL ENVIRONMENTS

Manuel E. Ruiz,
Arnoldo Ruiz

In dairy-oriented herds of tropical America, two milk production systems are used: 1) the specialized dairy and 2) the dual-purpose cattle production system in which both milk and beef are important outputs. Each system is associated with a corresponding calf-rearing procedure.

<u>Artificial calf rearing</u> describes the method used when man, and not the cow, is directly and intensely involved in feeding and caring for the calf. This is almost always associated with specialized dairy operations where the calf is separated from the dam at birth, or not later than five days when colostrum production comes to an end.

<u>Natural calf rearing</u> is the typical method whereby the cow raises her own calf until weaning with minimum participation by man. This method shows variations, but all are typical of the dual-purpose-cattle production system. In tropical Latin America, this term implies a management system by which both milk and beef are produced; the milk is obtained through once-a-day milking and the beef from the sale of weaners, yearlings, and culled cows.

The specialized form of milk production (dairy breeds, two milkings per day, artificial rearing of calf) can be found near the large cities of tropical Latin America (for example, Lima, Peru) or in high-altitude areas such as the central plateau of Costa Rica and the highlands of Guatemala. Such farmers include small, intermediate, and large landholders.

In contrast, the dual-purpose enterprise is found mostly in outlying, less-developed areas, with hot and humid climate; usually these are operated by small farmers.

Thus, both milk-producing systems are found within the group of small farmers, which makes up about three-fourths of the total population of farmers. A diagnosis of farms having less than 25 ha of land and/or less than 50 head of cattle revealed a number of management characteristics that serve to differentiate between the two systems. Some of these factors are shown in table. 1.

ARTIFICIAL REARING

In most artificial-rearing operations, the calf is kept indoors for up to 7 mo and milk is sometimes fed for the total period. The trend, however, is to reduce the total amount of milk and to utilize the main feed resource--the forages. Obviously, in this respect the research organizations are in the forefront, although important differences may be noted. CATIE, for example, has pushed for minimization of milk use and maximization of forage and agricultural by-product utilization. Experimentally, the total amount of milk fed has been reduced 120 kg/calf (about 5 weeks of milk feeding at a rate of 3.5 kg/day for small and medium-sized breeds); however, to retain a safety-margin, the recommendation is 180 kg of milk (Ruiz et al., 1980).

TABLE 1. FREQUENCY AND CHARACTERISTICS OF THE ARTIFICIAL AND NATURAL METHODS FOR REARING CALVES ACCORDING TO THE DAIRY SYSTEMS PRACTICED BY SMALL FARMERS IN COSTA RICA[a]

| | System | |
	Dual purpose	Specialized dairying
No. and frequency of small farms	182(83.5%)	35(15.2%)
Milking, % of farms		
- Stimulation by calf	Yes	No
- Once a day	89	3
- Twice a day	11	97
Calf feeding, % of farms		
- One quarter left	59	NA[b]
- Residual milk	33	NA
- Other related practices	8	NA
- Milk or whey in bucket	NA	88
- Milk substitutes	NA	12

[a]Unpublished results, IDRC-CATIE Project, Turrialba, Costa Rica.
[b]Not applicable.

Other features of CATIE's calf-feeding program include:
- Colostrum feeding by the dam for the full 4 to 5 days during which it is produced. If milk-letdown problems are likely to occur because of the separation from the calf, the colostrum may be removed from the mother at birth and hand-fed. It has been shown that only one day of colostrum feeding (10% LW) is sufficient to impart protection to the calf (Ruiz et al., 1981).

- Initiation of limited grazing at the first or second week of age. This practice has been approved cautiously because tropical pastures, especially in the humid tropics, are often not a safe environment for calves due to parasite infestation and diseases. Although this risk must be taken, the younger the animal when first put out to pasture, the more resistance it seems to develop to infectious diseases (Ruiz et al., 1980). However, a judicious parasite-control program is necessary. For example, some degree of cattle-tick infestation is beneficial (to develop resistance against anaplasmosis and piroplasmosis) but no infestation by lungworm can be tolerated. Weight gains of calves initiated to grazing as early as two weeks of age do not differ from those of calves initiated at a later stage (table 2).

TABLE 2. THE EFFECTS OF AGE OF CALF AT FIRST GRAZING AND OF THE RESTING PERIOD OF PANGOLA GRASS PADDOCKS ON WEIGHT GAIN, G/HEAD/DAY[a]

Resting period days	Age when grazing starts, months				
	2	6	10	10	Avg
21	314	296	262	399	318
42	401	282	248	296	307
Average	358	289	255	348	312

Source: M. E. Ruiz et al. (1980).
[a]Milk feeding was terminated at 10 weeks of age.

- Use of rations based on local feeds, mainly byproducts. To reduce costs and import dependency, dry rations are offered ad libitum from the fifth day of age. Intake is noticeable when the calf is 2 weeks old. At 3 weeks, the intake of the ration will be around 0.5 kg/day. The starter rations should have a source of high-quality protein, such as fishmeal or soybean meal, and a high-energy value. An example of a ration used at this stage would include corn (30%), fishmeal (20%), meat and bone meal (25%), cane molasses (12%), tallow (10%), salt (1%) and a vitamin/mineral premix (2%). Rations like this are used until the calf is 4 months old. Later, simpler supplements are used that contain a high proportion of molasses, small amounts of a protein source, a starch source, and urea.

- Animal performance can be very satisfactory if sanitary conditions are maintained. During the first 2 mo, gains of 350 to 400 g/day can be expected for calves of Jersey, Criollo, Ayrshire, and crossbred extractions; from 3 to 6 mo of age, gains of 500 g would be normal results; from 6 to 12 mo, weight gains would vary between 550 and 700 g/calf/day. This implies that, for small breeds, the heifer may receive her first service at 12 to 14 mo of age.

The total cost of a program with the above characteristics will be $150 to $200 per animal raised to breeding weight. (Estimated for Costa Rican conditions.)

Figure 1 provides a summary of such a calf-breeding program.

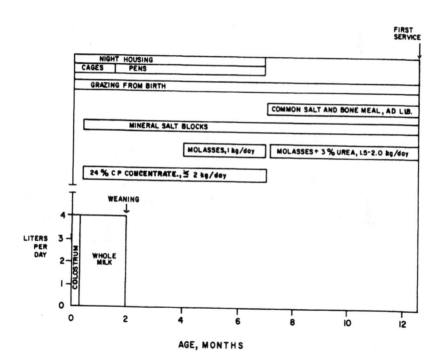

Figure 1. Diagram of calf feeding and management program at CATIE

Another calf-rearing method involves complete confinement but with minimum use of milk. Both the INCAP Nutrition Institute for Central America and Panama in Guatemala (Cahezas and Sahli, 1976) and CEDA Agricultural Development Center in El Salvador (Cahezas and Sahli, 1976) have developed similar programs. CEDA's program is described below.

- The calf is separated from the dam 5 days after birth and housed in covered pens.
- Weaning (withdrawal of milk) occurs when the calf becomes 2 mo old.
- Whole milk is used during the first 4 weeks and then replaced by skimmed milk for the following 4 weeks.
- An 18% crude protein starter and hay are offered ad libitum from the first week of age.
- The schematic representation of the feeding program up to 17 weeks of age is shown in table 3.

TABLE 3. FEEDING PROGRAM FOR DAIRY CALVES AS DEVELOPED IN EL SALVADOR

Age weeks	Whole milk (2 portions)	Skimmed milk (2 servings)	Commercial starter (18% CP)	Hay
1	10% LW		Ad libitum	Ad lib
2				
3				
4				
5		10% LW		
6				
7				
8 (weaning)			Intake limited to 2.3 kg	
17				
Total, kg	136	159	191	93

Source: M. T. Cabezas and E. Sahli (1976).

- Animal performance has been satisfactory and comparable to NRC (1971) recommendations, as evidenced in figure 2.
- According to the authors (Cabezas and Sahli, 1976) of the described program, the feeding costs during the first 4 weeks is $1.60 (US), per kilogram of weight gained. This figure is reduced to $0.90 (US) after milk is withdrawn.

NATURAL REARING

In the tropics, the use of European dairy breeds has been seriously limited, not only by a lack of adaptation of

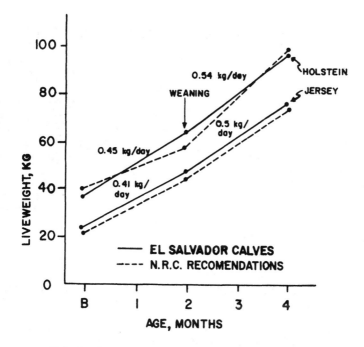

FIG. 2 GROWTH OF HOLSTEIN AND JERSEY CALVES
IN EL SALVADOR.

**Figure 2.   Growth of Holstein and Jersey Calves in El
Salvador**

these animals to tropical conditions but also as a result of
the high plane of nutrition required by these breeds--a diet
which cannot be easily obtained with tropical feedstuffs.
This situation, and the lack of a specialized tropical dairy
breed, is forcing the farmer to use Zebu x European breed
animals, which may not have the genetic potential of European breeds, but are more adapted to tropical conditions.
    The milking of Zebu crossbred cows in the absence of
their calves results in milk let-down problems, short lacta-
tions, and low levels of production. Table 4 shows data on
the productive behavior of crossbred cows that were milked
without their calves to stimulate milk let-down (Alvarez et
al., 1980).
    The above data indicate that 40% of the cows had termi-
nated their lactation before 150 days, with an average lactation length of 34 days. Consequently, these cows did not
produce sufficient milk to raise their calves, and thus
showed a deficit of 257 kg.
    Selection for milk let-down in the absence of the calf
is of no practical value. The low level of repeatability of
this behavior was demonstrated in the same study (Alvarez et

956

TABLE 4. PERFORMANCE OF ZEBU x EUROPEAN COWS MILKED
WITHOUT THEIR CALVES (3 YEAR DATA ON 309 COWS)

	Lactation length, days	
	>150	<150
Number of cows	185	124
Average number of days milked	305	34
Total milk production, kg	1571	103
Milk fed to calves, kg[a]	360	360
Salable milk, kg	1211	-257

Source: F. J. Alvarez et al. (1980).
[a]Calves artificially raised with 360 kg of milk.

al., 1980) when cows that had not shown any problems when
milked in absence of the calf during a previous lactation
were subjected to the same management in the following lac-
tation--approximately 50% of these cows had milk let-down
problems.

The need for calves to stimulate milk let-down in Zebu
crossbred animals became evident when a group of cows that
had produced hardly any milk in the absence of their calves
noticeably improved their performance during the following
lactation, as a result of having calf stimulation. (Alvarez
et al., 1980). If, in addition to the previous considerations, account is taken of the high technology and investment required for a successful artificial raising of calves,
it becomes clear that there is a need to find new calf-rearing methods for the tropics with the restricted suckling
being the most promising alternative.

The use of restricted suckling is not unknown to the
Latin American small farmer. He practices it in his dual-
purpose system by allowing the calf to suckle its dam for 6
to 8 hours a day. However, the system, as applied today,
implies the use of large quantities of milk in calf raising. A possibility for overcoming this situation would be
to further restrict the time the calf has access to its
dam. Several studies have been carried out in relation to
this possibility, and the data presented in table 5 are representative of the results obtained (Gaya et al., 1977).

As expected, cows that were suckled produced less milk
at the parlor than did those whose calves were raised artificially. However, the situation is reversed, when total
milk production is considered, indicating that the restricted suckling had a stimulating effect on milk production.
Several explanations have been given to this phenomenon.
Some have related it to a more efficient udder evacuation
which increases available space for milk synthesis and re-

TABLE 5. MILK PRODUCTION AND CALF PERFORMANCE AS A RESULT
OF DIFFERENT CALF RAISING METHODS

	Criollo		Holstein x Hereford	
	RS[a]	AR[b]	RS	AR
Milk production, kg	(n = 20)		(n = 16)	
at milking	7.9	8.8	4.5	4.9
obtained by calf	2.7	4.0	3.9	3.9
total produced	10.6	8.8	8.4	4.9
saleable	7.9	4.8	4.5	1.0
Calf weight gain, g/day	317	433	497	353
Milk conversion	8.4	9.3	7.8	11.4

Source: H. Gaya et al. (1977).
[a]Restricted suckling during 30 minutes after milking.
[b]Artificial raising.

duces the incidence of mastitis (Ugarte and Preston, 1972).
Others claim that it is the result of maintaining more appropriate hormone levels, an explanation that is reinforced
by the fact that the tendency towards higher total milk production is maintained even after the calf is weaned (Ugarte
and Preston, 1973). Table 5 also shows the beneficial effect of restricted suckling on the amount of milk available
for sale.

In relation to calf response, data in table 5 indicate
that weight gain is partially a function of the amount of
milk the calf receives (compare milk intake and weight gains
in the Criollo cattle). In addition, when equal amounts of
milk are consumed, those calves under restricted suckling
tend to gain more weight and are more efficient than those
artificially reared (compare data on Holstein x Hereford
cattle). This latter finding is the result of a higher nutrient content of the milk received by the calves that
suckle, because residual milk has a higher fat content (more
energy), and this milk constitutes a large proportion of the
milk obtained by the suckling calves.

Some of the barriers found when trying to implement restricted suckling at the farm level are related to 1) the
possible negative effects of this practice on reproduction
and 2) the procedures to be followed. In relation to possible negative effects, it has been demonstrated that restricted suckling has no detrimental effects--on either heat
presentation or fertility. Breed effects seem to be more
important (Ugarte and Preston, 1972).

In relation to management, it is not possible to blind-
ly recommend either once- or twice-a-day suckling since its
application would depend on the amount of milk present in
the udder after milking. If the cow maintains a large quan-

tity of milk in its udder, once-a-day suckling would be enough to obtain adequate weight gains and higher milk production. On the other hand, if the calves are not growing at the desired rate, this could be the result of insufficient milk intake and, consequently, twice-a-day suckling would be more appropriate. All this means that when deciding whether to use once- or twice-a-day suckling, the final decision will be based more on the farmer's judgment than upon theoretical grounds.

Another question concerns the interval between final milking and the initiation of the daily suckling. This interval should be no longer than half an hour, and the calf should have access to its dam for half an hour. This reduces to a minimum the amount of newly synthesized milk that the calf will receive and ensures that most of the milk consumed will be residual milk that cannot be extracted even when careful hand-milking techniques are applied (Ugarte and Preston, 1972).

TOWARDS A MORE APPROPRIATE CALF-REARING METHOD FOR THE TROPICS

The procedures discussed so far in this paper apply to both general systems of calf rearing mentioned earlier. The advantages and disadvantages of using each method have been discussed also. A final question deals with how to combine the benefits into a recommendation that could offer the safety of natural rearing and the efficiency of artificial rearing.

With this purpose in mind, a study (Vargas, 1980) was designed to combine various levels and characteristics of natural and artificial rearing to determine their effect on calf and cow performance. Table 6 describes the different combinations applied. In the cows to be suckled, one of the quarters was not milked in the morning, and the calf had access to its dam only in the morning for one hour. In the afternoon, all quarters were completely milked. The milk intake of the suckling calf was determined by weighing the calf before and after the suckling period.

Table 7 shows the data for a total of 25 cows (5 cows/treatment) studied during 145 days.

According to these results (Vargas, 1980) restricted suckling had no negative effects on milk production as long as the total suckling period was less than 70 days. However, in view of the incidence of mastitis, suckling should not exceed 21 days (plus the 5-day colostrum feeding). Note that the incidence of mastitis was low during the period the cow was suckled. Extraordinarily high weight gains were obtained with all methods, which reflects not only excellent nutrition but could perhaps be the result of using new housing facilities and the strict application of a good sanitary program.

TABLE 6. MILK FEEDING OF CALVES UNDER VARIOUS REARING
METHODS

Method	Milk intake
1. Specialized milk	No suckling. Milk given twice a day at a rate of 4 kg/day, completing 180 kg/calf (straight artificial rearing)
2.	7-day suckling. Afterwards milk given twice a day at a rate of 4 kg/day, completing 180 kg/calf
Progressive change	
3.	21-day suckling. Afterwards milk given twice a day at a rate of 4 kg/day, completing 180 kg/calf
4.	70-day suckling. No additional milk
5. Dual purpose	140-day suckling. No additional milk

Source: H. E. Vargas (1980).

TABLE 7. COW AND CALF RESPONSE AS A FUNCTION OF VARIOUS
CALF-REARING METHODS

	Mastitis incidence, %		Weight change		
Method[a]	During suckling	After suckling	Salable milk kg/145 days	Cow kg/145 days	Calf g/day
1	–	6.3	1068	25.4	737
2	5.0	20.0	1048	8.0	677
3	0.0	8.2	1030	-0.8	715
4	5.0	23.5	1096	-9.9	744
5	3.3	–	815	-15.6	788

Source: H. E. Vargas (1980).
[a]The methods are described in table 6.

It is evident from these results that it is possible to design different calf-rearing methods that combine most of the advantages of both artificial and natural rearing. This opens a totally new area of research, and the final recommendations should be based on economic considerations and social acceptability of the new procedure.

960

REFERENCES

Alvarez, F. J., G. Saucedo, A. Arriaga and T. R. Preston. 1980. Efecto sobre la produccion de leche y el comportamiento de los becerros al ordenar las vacas cebu/europeo con o sin apoyo del becerro y amamantamiento restringido. Produccion Animal Tropical 5:27.

Cabezas, M. T. and E. Sahli, 1976. Crianza y alimentacion de la hembra de reemplazo en los hatos lecheros. Revista Pecuaria de Centroamerica 62:42.

Gaya, H., J. G. Delaitre and T. R. Preston. 1977. Effecto del amamantamiento restringido y la alimentacion en cubo sobre la tasa de crecimiento de becerros y la produccion lechera. Produccion Animal Tropical 2:293.

National Research Council. 1971. Nutrient requirements of domestic animals. 3. Nutrient requirements of dairy cattle National Academy of Sciences, National Research Council, Washington, D. C.

Ruiz, M. E., G. Cubillos, O. Deaton and H. Munoz. 1980. A system of milk production for small farmers. In: Animal Production Systems for the Tropics, Proceedings. pp 246-264. International Foundation of Science, Provisional Report No. 8, IFS, Stockholm, Sweden.

Ruiz, M. E., E. Perez and R. Medina. 1981. Efecto del periodo de amamantamiento con calostro sobre el comportamiento de terneros de lecheria. Turrialba 31:21.

Ugarte, J. and T. R. Preston. 1972. Amamantamiento restringido. 1. Efectos del amamantamiento una o dos veces al dia sobre la produccion de leche y el desarrollo de los terneros. Revista Cubana de Ciencia Agricola 6:185.

Ugarte, J. and T. R. Preston. 1972. Amamantamiento restringido. 2. Efecto del intervalo de tiempo entre el ordeno y el amamantamiento sobre la produccion de leche y el comportamiento del ternero. Revista Cubana de Ciencia Agricola 6:351.

Ugarte, J. and T. R. Preston. 1973. Amamantamiento restringido. 3. Efecto de disminuir a una vez diaria el amamantamiento, despues de la cuarta semana, sobre la produccion de leche y el desarrollo del ternero. Revista Cubana de Ciencia Agricola 7:151.

Vargas, H. E. 1980. Influencia del amamantamiento post-ordeno sobre el crecimiento de terneros y la produccion de leche. M.S. Thesis. Turrialba, Costa Rica, URC-CATIE.

106

FATTENING STEERS UNDER GRAZING CONDITIONS

Manuel E. Ruiz

Although beef production systems in the developed countries usually include the feeding of high-energy rations to young animals, most of the beef produced in the developing countries still comes from extensive systems of production. These systems take two general forms: 1) small farmers, most of whom have so-called dual-purpose cattle enterprises that include the once-a-day milking of Zebu-type cows, where beef production consists of the sale of weaned calves and culled cows, and 2) ranchers, whose operations emphasize extensive grazing of beef-bred herds whether the operation is cow-calf or growing-finishing of purchased steers. In any of these three situations, beef output may be as low as 7 kg/ha/year on a carcass basis, and acceptable carcass weights (equal to or above 150 kg) can be achieved only when steers are 4 to 5 years old or older (Auriol, 1974; Pearson, 1970).

THE SEASONALITY OF PASTURE AND ANIMAL PRODUCTION

The lengthy period required to produce marketable steers can be attributed to the climate-dependent curves of pasture and animal growth, which consist of periods of good animal weight gain followed by periods of weight loss because of feed scarcity. This situation also has been described for the Australian tropical region (Evans, 1976) and represented in figure 1.

Variations in the availability of pastures occur in both the dry/humid tropics and the humid areas, the two principal ecosystems in Latin America, as illustrated in figure 2.

Thus beef production in the dry/humid tropics is confronted with a serious seasonal lack of pastures that causes substantial losses in animal-product output, including death losses. During this time, farmers sell cattle in an attempt to reduce stocking rates, but suffer loss in profits due to market flooding. Obviously, any research and development program designed to improve the nutritional well-being of

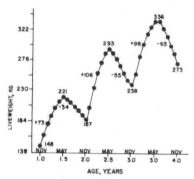

Figure 1. Seasonal Weight Changes of Cattle from 1 to 4 Years of Age Grazing Native Grasses in the Australian Tropics (Source: T. R. Evans (1976).

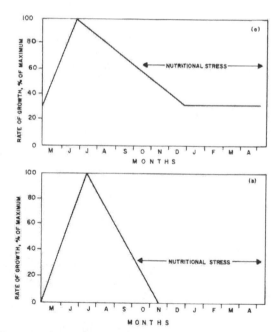

Figure 2. Seasonal Rate of Growth of Grasses in (a) The Humid Tropics and (b) The Dry/Humid Tropics of Central America (Source: G. Cubillos et al. (1975).

the herds during the critical dry period is of utmost importance--although nutrition is just one component of a complex production system that includes biological factors, as well as physical, social, and economic factors.

In the humid tropics, in spite of the almost constant rainfall, grass fluctuates in both quantity and quality. For example, on the Atlantic Slope of Costa Rica, total pasture production may range from 100% (maximum relative biomass during the most productive months) to only 20%, with the lower percentage caused by diminished rainfall, lower temperatures, and reduced daylight (Cubillos et al., 1975). These variations in forage availability cause fluctuations in animal productivity that are reflected in lower efficiency levels in the tropics as compared to those in countries with more sophisticated technology.

PROTEIN AND ENERGY LIMITATIONS OF TROPICAL PASTURES

In addition to fluctuations in pasture productivity, tropical pastures have two other constraints to grazing systems. One constraint is the seasonal reduction of the protein content in grasses, such as Jaragua grass (Hyparrhenia rufa) as the dry season settles in. Protein levels reach lows of 2% CP (Tergas et al., 1971) and such levels will obviously restrict feed intake and animal productivity. The other constraint is the generally low or modest level of digestibility that partially explains the lower weight gains found in cattle-grazing tropical grasses (table 1).

TABLE 1.   COMPARATIVE WEIGHT GAIN OF CATTLE GRAZING
TROPICAL OR TEMPERATE GRASSES OR CONCENTRATES

Diet	DM Digestibility %	Weight Gain Kg/Head/Day
Tropical pasture		
1. Immature	60 - 65	0.7 - 0.9
2. Semimature	50 - 55	0.4 - 0.5
3. Mature (dry season)	33.5	-0.225
Temperate pasture	70 - 80	0.9 - 1.2
Concentrate rations	80 - 85	1.2 - 1.4

Sources: R. Cerdas (1977), H. Roux (1966), Stobbs and
Thompson (1975).

BASIC STRATEGY FOR INCREASING BEEF PRODUCTION UNDER GRAZING CONDITIONS

On the other hand, tropical grasses grow vigorously and far exceed temperate grasses in total biomass production. Under high humidity and fertilization conditions, tropical grasses produce six times as much material as temperate grasses. As water availability decreases, the differences become less evident (table 2).

TABLE 2. ESTIMATES OF THE TOTAL ANNUAL DRY MATTER PRODUCTION (TONS /ha) OF GRASSLANDS IN THE MAIN CLIMATIC ZONES OF THE WORLD

Temperature	Water Supply			
	Wet	Humid	Semiarid	Arid
Subartic	4	8	9	–
Temperate	25	15	10	4
Subtropical	120	40	10	4
Tropical	150	70	12	4

Source: R. W. Snaydon (1981).

Therefore, the primary strategy to increase beef production in tropical areas is to saturate the capacity of the grassland to supply feed for the herd. This capacity depends on the nutritive quality of the forage and the total biomass produced. The nutritive quality will be reflected in the animal's rate of weight gain, while the forage yield will determine the stocking rate--this in terms of animal units/ha or total animal weight/ha. The product of these two factors will be the amount of beef that is produced per hectare. These relationships have been theoretically considered in a well-known model (figure 3).

In general, figure 3 indicates that at low stocking rates the maximum individual performance is obtained (the extent of which is dependent upon the forage quality and the animal's genetic ability). As stocking rate increases (i.e., forage availability per animal diminishes) a point will be reached after which the individual performance will decrease. Nevertheless, beef output per hectare will continually increase until a maximum is reached, then at this extreme stocking rate animal production per hectare will fall rapidly. Table 3 provides an example of how the model operates.

REFINEMENTS OF THE BASIC STRATEGY

Now, given a suitable pasture species and knowing the optimum carrying capacity, there are three ways by which beef production can be further increased. These are:
- The use of fertilizer
- The use of associated legumes
- The use of supplements

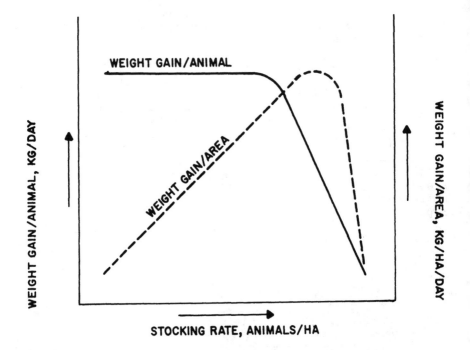

Figure 3. The Effect of Increasing Stocking Rate Upon Productivity Per Animal and Per Unit Area
Source: G. O. Mott (1960).

TABLE 3. PRODUCTIVITY OF SAVANNAH GRASSES IN TERMS OF CATTLE WEIGHT GAINS PER HEAD AND PER HECTARE

Country	Stocking rates animal/HA	Weight Gain in 203 days	
		KG/Head	KG/HA
Nigeria and savannahs	0.42	21	9
of Northern Guinea	0.50	26	12
	0.62	24	16
	0.83	19	15
	1.25	1	3

Source: P. N. DeLeeuw (1971).

Fertilization

Normally, N is the limiting growth factor of tropical grasses even when the humidity and temperature are adequate. The extent of the response to N applications will depend chiefly on the specific pasture. Thus, Elephant grass (Pennisetum purpureum) will show extraordinary responses to N up to 800 kg N/ha/year, while grasses like molasses grass (Melinis minutiflora) will hardly respond to fertilization.

Fertilization with N will modify, in most species, the growth pattern so that the growing period is extended well into the dry season when N is applied towards the end of the rainy season. However, this will not prevent the normal loss of the nutritive value due to the dry season (Tergas et al., 1971).

Moreover, contrary to common beliefs, fertilization with N will not cause obvious increases in the crude protein values. Rather, the main effect is a significant. increase in total biomass production and, therefore, an increase in the pasture's carrying capacity. Table 4 shows a sample of results directly related to the above concepts.

TABLE 4. BEEF PRODUCTION ON FERTILIZED PANGOLA GRASS (DIGITARIA DECUMBENS) IN PUERTO RICO

		Weight gain	
Fertilizer level KG N/HA/year	Stocking rate animals[a]/HA	KG/head/day	KG/HA/year
64	2.0	0.56	404
168	3.2	0.55	642
382	4.9	0.55	990
535	5.9	0.49	1070

Source: J. Vicente-Chandler et al. (1974).
[a]300-kg steers

Grass/Legume Associations

The rising cost of fertilizers has forced researchers to readdress the search for suitable, highly productive, and persistent legumes in association with grasses. Recently, this search has included the evaluation of legume trees and shrubs. Where legumes have been successfully introduced, dramatic increases in beef production have been obtained. One important feature of legumes is their significant contribution to the alleviation of the nutritional stress during the dry seasons; in fact, they can support weight gains throughout this critical period as may be seen in table 5.

Table 6 compares the production potential of various types of pastures both in temperate and tropical climates. Native tropical grasses have serious limitations for production, although this situation can be vastly improved through the introduction of legumes and the application of phosphorus. Temperate grass/legume associations are more productive than tropical associations and N-fertilized tropical grasses are superior to N-fertilized grasses in temperate climates.

Supplementation

Supplementary feeding is justified by the seasonal variations in both pasture and animal production (figures 1 and 2). Also, at times, a farmer may find himself with too many animals and a rapidly decreasing productivity per animal and per hectare (figure 3), and he may be forced to look for additional feed resources.

Supplementation is undoubtedly beneficial if a premium price is paid for high quality beef that can be obtained only through sustained weight gains and proper nutrition. Finally, the practice of supplementation allows the farmer to program the sale of fattened steers at times when prices are most favorable.

TABLE 5. BEEF PRODUCTION BASED ON GRASSES OR GRASS/LEGUME ASSOCIATIONS IN AREAS WITH MORE THAN 1500 mm RAINFALL

Country	Pasture	Stocking rate animal/ ha	Weight gain	
			Kg/head/ day	Kg/ha/year
Australia	Panicum maximum	4.2	0.22	378
	P. maximum + C. pubescens	4.2	0.30	460
Fiji Islands	Discanthium caricosum	1.5	0.22	110
	D. caricosum + 10% area with Leucaena leucocephala	1.5	0.30	170
	D. caricosum + 20% area with Leucaena leucocephala	1.5	0.50	270
Peru	Hyparrhenia rufa	1.2	0.16	70
		1.8	0.23	149
		2.1	0.17	130
	Hyparrhenia rufa + Stylosanthes guyanensis	2.1	0.40	309
		2.4	0.40	351
		2.7	0.34	335
		3.0	0.34	378

Source: P. C. Whiteman (1976).

TABLE 6. BEEF PRODUCTION POTENTIAL OF NATIVE AND IMPROVED
GRASSES WITH OR WITHOUT LEGUMES OR FERTILIZATION

| | | Tropical climate | |
Pasture	Temperate climate	Monsoon (5-6 dry months)	Humid
Native grass			
- Unimproved	100-400[a]	10-80	60-100
- Associated with legumes plus P. fert.	200-500	120-170	250-450
Improved grass			
- Associated with legumes plus P. fert.	400-1200	200-300	300-800
- Fertilized with N	700-1400	300-500	400-1800

Source: J. R. Simpson and T. H. Stobbs (1981).
[a]Weight gain in kg/ha/year

The success of supplementation, aside from economic considerations, depends on the following factors (Bailey et al., 1972):
- If pasture availability is not limiting, animal performance will be determined by the voluntary intake of energy and protein that, in turn, depends on the digestibility and protein content of the pasture.
- If pasture quality is low, energy and/or protein supplements will increase total intake and may even increase the consumption of pasture.
- When pasture quality is high, grazing animals will respond to supplements only if pasture availability is low, thus limiting the amount of pasture that the animal can harvest. Little response will be obtained at low stocking rates.
- At high stocking rates, not only is the available pasture reduced but the opportunity for selection by the animal is also reduced. In this situation, supplementation allows for a more efficient use of the available pasture.

Work at CATIE has led to a model illustrating expected behavior of young bulls grazing Guinea grass (Panicum maximum) and supplemented with molasses/cottonseed meal (figure 4). Figure 4 illustrates some of the principles stated above and, in addition, presents two other features of supplementation effects. One is what may be called an "additive effect," "which simply means that the consumption of small amounts of a supplement does not interfere with the

970

normal consumption of the forage. This, of course, would translate into higher animal productivity. In figure 4, this would be represented by a 33% increase in rate of gain.

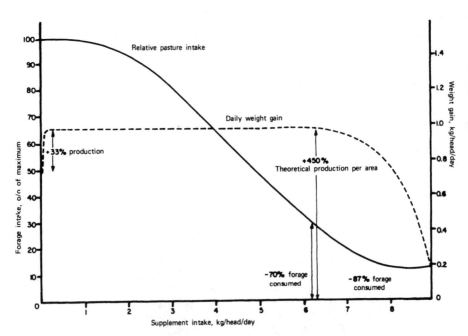

Figure 4. General Relations Between Supplement Intake, Forage Intake, and Weight Gain of Grazing Young Bulls

A "substitution effect" of the supplement indicates a competition between the intake of the supplement and the intake of the forage. When this occurs, the individual performance may not be affected. However, if the substitution is too large, specific nutrient deficiencies may appear, thus reducing animal performance. In general, substitution effects should be avoided since the cost of the supplement is usually higher than the cost of the forage; if this effect occurs, it can be offset by increasing the stocking rate, thus improving productivity per hectare. An example of such manipulation is offered in table 7.

Clearly, there is a tremendous potential for improving beef production in tropical environments. What is needed is the building of road networks, marketing facilities, carcass-grading systems, price incentives, and a well-run technical assistance and credit program. No economic analyses were given here because they are difficult to relate to data from other countries. However, such data are definitely the ultimate criterium for discerning which, if any, alternative is most appropriate to technify beef cattle production systems in tropical areas.

TABLE 7. MAXIMUM BEEF PRODUCTION PER HECTARE IN HUMID TROPICAL PASTURES VARYING THE STOCKING RATE AND LEVEL OF SUPPLEMENTATION WITH GREEN BANANAS

Green bananas[a] Kg/head/day	Stocking rate Kg/Ha/day	Weight gain Kg/Ha/year
0	1000	694
2.5	1100	748
5.0	1500	821
7.5	1850	949
10.0	2500	1168

Source: G. Cubillos et al. (1975).
[a]As-fed basis, 20% DM

REFERENCES

Auriol, P. 1974. Intensive feeding systems for beef production in developing countries. World Animal Review 9:18.

Bailey, P. J., J. H. L. Morgan and A. H. Bishop. 1972. Supplementary feeding of beef cattle in the high rainfall zone of southern Australia. Australian Vet. J. 48:304.

Cerdas, R. 1977. Cambios en el valor nutritivo de los pastos Jaragua (Hyparrhenia rufa (Nees) Stapf) y Estrella Africana (Cynodon nlemfuensis) durante la epoca seca del tropico. Ing. Agr. Thesis, San Jose, Costa Rica, University of Costa Rica.

Cubillos, G., K. Vohnout, and C. Jimenez. 1975. Sistemas intensivos de alimentacion del ganado en pastoreo. In: El potencial para la produccion de ganado de carne en America Tropical. CIAT, Cali, Colombia, Serie CS-10, pp 125-142.

De Leeuw, P. N. 1971. The prospects of livestock production in the northern Guinea zone savannas. Samaru Agricultural Newsletter 13:124.

Evans, T. R. 1976. The establishment and management of tropical pastures for beef production. In: Seminario Internacional de Ganaderia Tropical, Acapulco, Mexico. Memoria. Secretaria de Agricultura y Ganaderia and Banco de Mexico, S. A. Vol. 4. Produccion de Forrajes. p 51-86.

Mott, G. O. 1960. Grazing pressure and the measurement of pasture production. Proceedings, VIII International Grassland Congress, pp 606-612.

Pearson, L. S. 1970. The role of livestock in developing economies. Advancement of Science 26:289.

Roux, H. 1966. Estudio preliminar sobre el uso de la urea en la alimentacion del ganado bovino en Panama. University of Panama, Technical Publication No. 3.

Simpson, J. R. and T. H. Stobbs. 1981. Nitrogen supply and animal production from pastures. In: F.H.W. Morley (Ed.) Grazing Animals. pp 261-268.

Snaydon, R. W. 1981. The ecology of grazed pastures. In: F.H.W. Morley, (Ed.) Grazing Animals. pp 261-268.

Stobbs, T. H. and P. A. C. Thompson. 1975. Milk production from tropical pastures. World Animal Review 13:27.

Tergas, L. E., W. G. Blue and J. E. Moore. 1971. Nutritive value of fertilized Jaragua grass (<u>Hyparrhenia rufa</u> (Nees) Stapf.) in the wet-dry Pacific region of Costa Rica. Tropical Agriculture (Trinidad) 48:1.

Vicente-Chandler, J., F. Abruna, R. Caro-Costas, J. Figarella, S. Silva and R. W. Pearson. 1974. Intensive grassland management in the humid tropics of Puerto Rico. University of Puerto Rico, Agricultural Experiment Station, Bulletin No. 233.

Whiteman, P. C. 1976. Beef and milk production from legume-based tropical pastures. In: Seminario Internacional de Ganaderia Tropical, Acapulco, Mexico, 1976. Memoria. Secretaria de Agricultura y Ganaderia and Banco de Mexico, S. A. Vol. 4. Produccion de Forrajes. pp. 190-122.

Part 16

PASTURE, FORAGE, RANGE, AND AGROFORESTRY

107

UNDERSTANDING RANGE CONDITION
FOR PROFITABLE RANCHING

Martin H. Gonzalez

INTRODUCTION

In many parts of the world, improvement of the livestock industry is believed to be dependent on two primary factors related to animals: genetics and disease control. Another nonanimal consideration seems to be rain! It is common to listen to ranchers talking about a new bull they have or plan to import, about the last rain--or complaining about the lack of it even when the normal rainy season is still months ahead--or about prices of livestock on the market and the latest government regulations. However, less often we find a group of ranchers here or in Mexico, discussing range conditions, the forage production on the ranch, or range improvements.

Perhaps because we are overgrazing and because our pasture and forage production is poor, we manage to put the blame on the rain, the market, or the government, we do not like to recognize that a great part of our attention should be devoted to the basic things on the ranch: soil, grass, and water.

In the U.S. and Canada, many ranchers might be familiar with the range condition concept and its implications, but in other countries the situation is different. Even when the common sense, the personal ability, and the experience of a producer is outstanding, he will require a good and basic knowledge of the condition of the range because it is the key to profitable ranching. And this is true, particularly, today, when the livestock industry needs to reduce the cost of production--and can by using desirable range plants--still one of the cheapest sources of forage.

SOIL DEVELOPMENT, PLANT SUCCESSION, AND RANGE MANAGEMENT

Range management as an art and as a science has its roots in ecological principles. A range user has to deal with the basic components of the ecosystem around him and his operation, and should understand how they function and how they interact. Profitable ranching requires an understanding of basic ecology.

978

Figure 1 will help us to understand how the basic ecological processes relate to range management, exemplified by
a rangelands ecosystem.

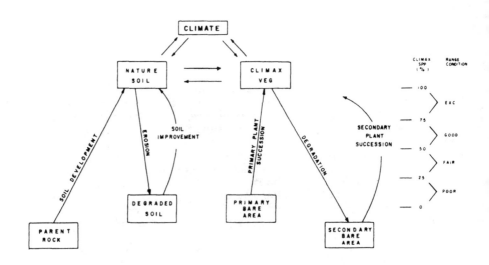

Figure 1. Relationship between soil, vegetation, climate,
and range condition

Soil has developed for thousands of years from parent
material. After a long process under various climatic
conditions, it developed freely and reached maturity in its
most productive phase. During that process, plant life had
a parallel development, beginning from a primary bare area
until reaching its climax stage (or maximum degree of
development). This series of natural vegetative changes
toward the climax stage are known as the primary plant succession. At this point, a fully developed soil and climax
vegetation were in equilibrium with the climate for that
particular area.
However, mainly due to man's actions, the soil has
often been eroded, caused by degradation of the vegetative
cover, and the pristine or climax condition has been
destroyed. That which took centuries to build was destroyed
by man in a much shorter period: by plowing of rangelands,
by excessive fire, by timber exploitation, by over-grazing,
by erosion, etc. It was then that the mature soil was converted into a degraded soil, and the climax vegetation was

converted into a secondary bare area very far from its productive potential.

But man had not always been abusive of his resources. After he realized the serious damage caused to soil and vegetation and, of course, to the total productivity of an area, he started a series of soil improvement practices, together with some actions to accelerate the rehabilitation of the vegetative cover (secondary plant succession). These man-activated or "artificial" improvement practices (usually very expensive) were particularly needed in those areas where disturbances had been so severe that a spontaneous recovery was impossible.

And it is here, during different stages of this secondary plant succession, that most of our ranches are found today. The degree of disturbance they once experienced (or still are) and the degree of advancement within this succession, relate directly to range condition, according to the percentage of climax (desirable) forage species in the botanical composition found in the different pastures. It is impossible, under practical grazing conditions, to expect a range to remain in its primitive, climax state, and very few pastures can be found in excellent condition. There are many well-managed ranches in "good" condition, but most of the land, particularly in Latinamerican native ranges, would be characterized as in "fair" and "poor" condition. Many of these lands have the potential--under some grazing restrictions and good management--to improve and to recover productivity. However, on the lands, some agronomic improvement practices (soil conservation, brush control, range reseeding, etc.) must be used to reach their potential. We must pay this high price for the misunderstanding and mismanagement of our range resources.

QUANTITATIVE ECOLOGY AND RANGE MANAGEMENT

The Range Condition Concept

Range condition reflects the health of a range or pasture. It is based on the relation between the present vegetation and the vegetation that, potentially, a given site should have.

Dyksterhuis (1948) classified range plants in three categories according to their response to use by animals, and determined range condition based on the quantity of each of these species when sampling the range. These species are decreasers, increasers, and invaders. Figure 2 shows these relationships that every rancher should keep permanently in his mind.

Extensive field work has been done to classify plant species as decreasers, increasers, or invaders. Based on this work, range condition guides have been developed for every different site within any given vegetative type.

These include most of the plants that fall within each category and the percentage of each that is permitted in the botanical composition--based on what the climax vegetation should be. This permitted percentage, particularly for the increaser species, has been determined by experience and comparisons of the same species in different sites and under different grazing pressures.

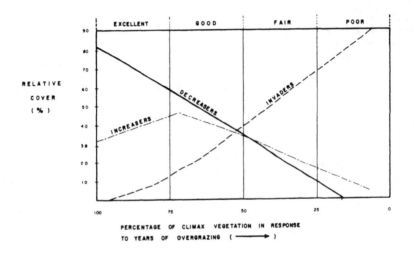

Figure 2. Quantitative basis for determining range condition (Dyksterhuis, 1948)

The decreasers are usually components of the climax community and are perennial, highly productive forage plants that are palatable and desirable. They are permitted in the percentages found in the botanical composition. The invaders are not permitted at all in the composition, so the percentage they accomplish for is substracted from the total. The increasers, most of which belong to the climax community, are permitted in varying percentages according to the range site and the associated species. The permitted percentage of each increaser species for the different sites has been calculated according to their response to grazing. This percentage has a maximum that will not interfere with the most desirable species--and a maximum that will not benefit the expansion of invaders. These changes in composition take place only when overgrazing occur over long periods, thus causing the deterioration of the range.

Dyksterhuis established a quantitative basis to determine range condition as follows:

Excellent Condition describes the botanical composition including more than 76% of climax or desirable forage plants. In other words, the vegetation is a mixture of decreasers and allowable increasers. There is almost no erosion and plant density is high. Almost no invaders are present. Litter is abundant.

Good Condition describes the biological composition that contains desirable species. The proportion of decreasers is less than that found in excellent-condition composition and few invaders may be evident. Litter is not as abundant but does occur.

Fair Condition describes the composition that has only 26% to 50% of the plants of climax or desirable species. In this condition, the signs of deterioration are more noticeable, particularly in marginal areas. Invaders appear in the same proportion as do decreases and increases; erosion has begun to be evident and density is generally low.

Poor Condition describes a badly degraded range with less than 25% of the plants in the composition of desirable or climax species. Forage production is very low. Erosion is a problem and invaders constitute a majority in the botanical composition. Almost no litter is evident. Spontaneous revegetation may not be economical under this condition and rehabilitation can be achieved only through range improvement practices.

We must realize that these conditions cannot be compared among different sites. Range site separation must be done before determining condition since their characteristics and potential are different.

Forage Production and Range Condition

The condition of the range and the amount of desirable forage produced are directly related for all types of range vegetation. However, from a production standpoint, a disclimax condition (just below the climax) may present a better diet for the grazing animals since some of the "increasers" in a given site could be more palatable or have a higher nutritive value than do some "decreasers" or climax species.

Figure 3 shows comparisons in the productivity of five selected rangeland types in northern Mexico, each for different condition: excellent, good, fair, and poor (Gonzalez, 1969).

It is observed that, even when there are substantial differences in kg/ha of usable forages between good an excellent conditions, there are greater differences between fair and good conditions and still greater differences between the poor and the fair conditions. This is true for all the vegetative types represented. Such large differences can serve as an index in estimating the effort, the

982

expense and the time that may be needed to bring back the
poorer conditions to a more productive stage.

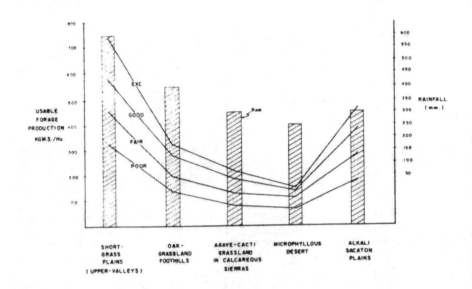

Figure 3. Forage production in five vegetative types under
different range conditions in northern Mexico
(Gonzalez, 1969)

RECOGNIZING RANGE CONDITIONS ON A RANCH

There are now sufficient range-condition guides for
most of the rangeland areas of North America. For example,
the Soil Conservation Service in the U.S. and COTECOCA-SARH
in Mexico represent the agencies that developed such guides
and that are using them extensively, along with many other
government and educational institutions. It is not difficult to learn how to use these guides, but a basic knowledge
of the terrain and the vegetation is needed.

Hoffman and Ragsdale (1974) from Texas A&M recommend
for following steps in using the range-condition guides:
    - Determine the range site.
    - List the plants found on the site.
    - Estimate the percentage of each plant that could
      be in the composition.

- Refer to the guide to determine the percentage of each plant that could be in the composition in the vegetational areas and site for which you are judging.

Carrying Capacity

Each guide includes the carrying capacity that is estimated for the different conditions in every site because range condition directly reflects the forage production in that particular site. A good condition means a stable, well-managed range; fair and poor conditions are indicators of heavy or severe use. On the other hand, if a site is in excellent condition, perhaps it is under-utilized. These guides provide an efficient, easy way to determine the carrying capacity (AU/ha) that the ranch should have--as compared with that under current management.

After the condition of the range has been determined, some adjustments may be necessary to keep the ranching operation flexible and to give the range the opportunity to recover. Adjustments may be necessary on:
- Livestock numbers (usually reducing them if fair and poor conditions are dominant).
- The present grazing system.
- The most economically convenient improvement practices to accelerate rehabilitation (brush control, range reseeding, etc.)--if condition is poor and hope for natural recovery is low.

REFERENCES

Dyksterhuis, E. J. 1948. Guide to condition and management of ranges based on quantitative ecology. Amer. Soc. Agron. App. Sec. Mimeo. p 25 (Abstr).

Gonzalez, M. H. 1969. Coeficientes de agostadero para el estado de Chihuahua. Memoria COTECOCA-SARH. Mexico.

Hoffman, G. O. and B. J. Ragsdale. 1974. How good is your range? Tex. Agr. Ext. Service Bull. Texas A&M Univ., College Station, Texas.

108

RANGE IMPROVEMENT PRACTICES
AND COMPARATIVE ECONOMICS

Martin H. Gonzalez

Grazing lands of the world's rangelands have suffered abusive management that has, in turn, caused reduced productivity. This low productivity often is attributed to critical drought periods; however, the direct effect of a drought is to aggravate the poor management of the land.

The world demand for livestock products is increasing but the production from rangelands is decreasing because: faulty management has reduced the condition of the land and consequently the productivity; other demands for the land (agriculture, industrial, urban, highways) are deminishing the number of hectares available.

World evidence confirms the desertification of our ranges and the destruction of millions of highly productive hectares that have become unproductive and denuded.

A study conducted in Mexico (CFAN-CID, 1969), which included a survey in nine central and northern states and covered around 100 million hectares, indicated that overgrazing was evident in 85% of the land; light or advanced erosion was a problem in 87.5% of the overgrazed area, and 49.7% of the land was infested by undesirable plants, mostly brush. Overgrazing is a problem in almost all countries where degradation of rangelands has been so severe that economically it is impossible to expect a natural recover. Before the rangeland reaches this point, various range improvement practices, basically agronomic, are needed to facilitate and accelerate the secondary plant succession for returning those lands to a productive stage.

In this paper, emphasis will be given to the principles governing the most common improvement practices. More specific information for a particular problem may be obtained from neighboring experimental stations, the extension and university people, or the county agent.

MAIN RANGE IMPROVEMENT PRACTICES

Of the four elements applied in the rehibilitation of grazing lands (climate [rainfall], soil, vegetation, and grazing management), climate is the only one that man <u>cannot</u>

manipulate. However, manipulation of water from rainfall once it hits the ground is one of the most important aspects of range improvement. On the other hand, with correct planning and management of the other three elements, man may rehabilitate deteriorating lands. Some of the most common (and needed) improvement practices on rangeland are the following:

Water Conservation

Efficient use of rainfall enhances all the other improvement practices. When properly managed, even excess water can be beneficial. The principle in water conservation is to supply the forage species with moisture in such a way and at the right time so that the effects last the longest.

Water conservation can be managed in two ways: (1) by retaining each drop of rain where it falls and (2) by diverting excessive surface runoff to the highest-producing sites on the ranch. These two steps can be accomplished by either simple or complicated structures. However, the best way to conserve water for plant use is to have a good vegetative (grass) cover so that rain infiltrates into the soil and runoff and erosion are avoided.

Figure 1 shows how different grass covers on the rangelands of Chihuahua affect infiltration of water (Martinez, 1959). This figure averages data for tests on different short-grass ranges where bluegrama (Boutelouagracilis) was dominant. It was found that soils in an exclosure where they had been protected from grazing for 7 years had an infiltration rate 118% higher than bare ground, 63% higher than an overgrazed area, and 25% higher than a moderately grazed area. In turn, the soil that was moderately grazed absorbed 30% more water than the overgrazed site and 74% more than the bare area.

Table 1 shows the results of similar tests comparing different types and density of vegetative cover to conserve rainfall on different soils. On short-grass plains, sandy loam soil under excellent condition absorbed 450 mm more rain in a period of 105 minutes than a bare area; 270 mm more than range in poor condition, and 170 mm more than the soil with a cover in good to fair condition. The results were the same for the other two vegetative types--the oak-bunchgrass in the stony foothills and on the alkali flats with their heavy clay and deep soils (Sanchez, 1972).

Among the commonly used water conservation practices on ranches are contour furrows, range pitting (small, medium, or large pits), subsoiling, and low-retention fences. The conservation practice(s) to use depends on soil type, topography, slope, plant cover, costs, and storm intensity and frequency.

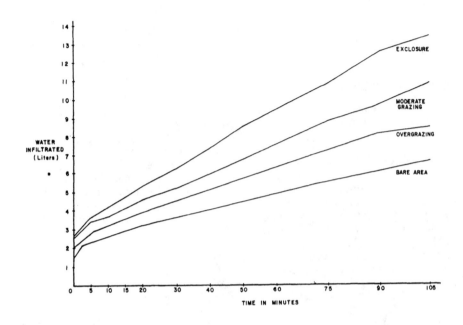

Figure 1. Water infiltration in short grass range under different covers. Chihuahua, 1964.

TABLE 1. INFLUENCE OF RANGE CONDITION ON WATER INFILTRA-TION[1] IN THREE VEGETATIVE TYPES AT LA CAMPANA, CHIHUAHUA

Range condition	Shortgrass plains	Oak bunchgrass foothills	Alkali flats bottom lands	$\overline{X}$
	mm	mm	mm	mm
Excellent	650	450	150	417
Good-fair	480	375	120	325
Poor	380	230	81	230
Bare area	200	190	73	154

[1] Mm of rain or equivalent absorbed in 105 minutes.

Soil Conservation

The most effective practice for soil conservation on rangeland is the same as for water conservation--a permanent cover of desirable plants. All soil conservation practices have a double purpose: to conserve water and to avoid erosion. Contour furrows, gully control by using different structures, terracing, and wind breakers are all effective soil conservation practices. As for water control, the practice to be selected will depend on the terrain. Heavy, sophisticated equipment for soil conservation is easily replaced by using some of the natural materials found on the ranch or in the pastures. For example, a gully can be controlled by filling an area with rocks, yucca plants, palmilla (sacahuiste) plants, old trees, or the undesirable vegetation nearby. Gullies can serve as dump sites for those materials. In a very few years the plant material, rocks, and soil eroding from upstream will compact, cause a dam to form, and erosion downstream will be reduced.

Control of Brush and Other Undesirable Plants

Unfortunately, most of the world's rangelands in use are infested, by varying degrees, with undesirable vegetation of which the woody species are the most common. This invasion is the direct result of the degradation caused by the different activities of man while using his grazing resources. Control of undesirable brush can be done by different methods: mechanical, chemical, and biological.

Mechanical control. Mechanical control is usually done with heavy or with light equipment. Heavy equipment includes dozing, chaining, disking, and heavy shredders or crushers. Light equipment includes the use of smaller tractors and some manual tools for shredding and disking. the equipment to use depends on the type of vegetation and its density, conditions of the terrain, availability of equipment, and costs. When using mechanical brush control, try to move or remove as little top soil as possible.
Research and rangeland experience all over the world demonstrate the competition between undesirable brush and grasses. The woody species of brush requires nearly four times more water to produce one kg of aerial growth and provides much lower forage quality than some of the best perennial grasses.
On Rancho Experimental La Campana, Chihuahua, Mexico, mechanical control of shrubs in a short-grass range invaded by Acacia, Mimosa, Eysenhardtia, and Brickellia resulted in a 102% range increase in forage production (Gomez and Gonzalez, 1978). Chemical control of "chaparrilo" (Eysenhardtia spinosa) produced 428 kg DM/ha more than the untreated areas, which had only 171.1 kg DM/ha. This represents 250% increment in forage production (Gomez and Gonzalez, 1976).

Chemical control. The chemical control of undesirable plants has developed intensively during the last two decades. The chemical products (herbicides) on the market for almost any brush-control program can be applied either by aerial spraying or ground spraying. Ground applications can be foliar, on the trunk or stumps, or directly on the ground in the form of pellets.

Herbicides work in different ways:

- contact herbicides are those that kill the plant or parts of the plant directly exposed to the chemical. This type is used on annual weeds. Examples: diesel oil, diquat, and paraquat.
- Translocated herbicides (also called hormonal or systemic)--are used in low concentrations. The toxic substance is carried through different parts of the plant by the plant's own liquids. Examples: 2,4-D, 2,4,5-T, silvex, dicamba, and picloram.
- Selective herbicides are the ones that kill a particular species or group of species without damaging others. If heavy dosages are used these may act as nonselective. This type of herbicide is the most widely used for range improvement because it does not affect grass or grass-like plants (monocots). Examples: 2,4,5-T picloram, TCA, and MCPA.
- Nonselective herbicides kill or damage all plants where applied. Example: AMS, amitrol, PCP, diesel oil, kerosene.
- Soil sterilants are generally in the form of pellets and are applied directly to the soil. When the pellets are diluted by rain, they are absorbed by the roots. Examples: Atrazine, bromocil, dicamba, diuron, and fenuron. The effect may be permanent or temporary, selective or nonselective. A detailed treatise on herbicidal plant control including products, methods, advantages, problems, etc., is found Range Development and Improvements (Vallentine, 1971).

Phenological stage of the plants, time of year, and meteorological conditions are three of the important factors to consider in brush-control programs.

Biological control. Another, but complicated, way to control undesirable plants and shrubs is through biological control. This can be achieved by insects or domestic or wild animals. There are insects that specifically live on certain plants, feeding from them, and eventually damaging them. Examples are the mesquite twig girdler (Oncideres rhodosticta) and two species of leaf-feeding beetles (Chrysolina gemellata and C. hypericy) that attack the klmath weed (Hypericum perforatum).

Although biological control of shrubs and other undesirable plants by insects is complicated, the use of domestic animals is not. Studies in Central Chihuahua (Fierro, et al., 1979) used goats over a period of 3 years to defoliate five different woody species. Results indicated the goats' preferred shrubs to grasses. In moderate vs intensive browsing and in shredded and not shredded areas, intensive defoliation by goats caused a 36% mortality of "Chaparrilo" plants (Eysenhardtia spinosa); moderate defoliation killed 14% of the plants. A similar trend was observed for Acacia and Mimosa.

An additional benefit of using goats for biological control of noxious shrubs was their milk production--an average 30 ml per goat per day for 4 mo a year. The production of perennial grasses increased from 94 kg/ha to 360 kg/ha in the first year (1977) and to 950 kg/ha 3 years later with moderate browsing.

The main factor to consider if planning to control shrubs with sheep or goats is the type of vegetation. On many ranges invaded by shrubs and other plants that cattle will not eat, grazing systems combining small and large herbivores have been established. Even in some areas where toxic plants for cattle existed, changing to goats (or perhaps sheep) allowed those areas to be utilized.

Fire. The cheapest and most effective brush control tool is fire. However, there are certain rules that have to be followed: (1) the amount of fuel (grasses and associated vegetation) must be high enough to carry the fire at the desired intensity; (2) wind direction, velocity, and soil moisture content determine the season and time of day to use fire. Extreme care must be taken to avoid letting the fire run wild and damage adjacent areas. Vallentine (1971) in his book, Range Development and Improvements, describes in detail the procedures of burning for brush control and its consequences.

Numerous controlled burning experiments have been done to control undesirable vegetation. Glendening and Paulsen (1955) obtained 52% kill of young mesquites having basal diameters of 0.5" or less; however, only 8% to 18% of the taller trees were destroyed by fire. In other areas, only 9% were destroyed by burning (Reynolds and Bohning, 1956), but White (1969) reported a wild fire that killed 20% of the trees with moderate and severe burns. Cable (1965) reported that variable results were obtained from different fuel covers: 4500 kg/ha of fuel (ground cover) killed 25% of the trees when burned; areas with 2200 kg/ha of fuel killed only 8% of the trees. Other species damaged by fire were Ocotillo (Fouqeria splendens) with 40% to 67% killed; Sotol (Dasylirion texanum) had 97% dead plants, and creosote bush (Larrea tridentata) was reported with a "very high" percentage killed.

June fires killed 44% of the choya cactus (Opuntia fulgida) and 28% of the pricklypear (O. engelmanii) (Reynolds and Bohning, 1956).

Range Reseeding

Reseeding of denuded areas is an improvement practice that obtains fast results. However, many risks are involved in this type of operation, and extreme care must be exercised to establish and produce good forage.

Range reseeding has proven successful in many areas in the U.S., Canada, Mexico, and some Latin American countries. In arid and semiarid lands it is combined usually with some water catchment devices or structures in order to assist in the establishment of the new plants. Commonly associated with land clearing and brush control, reseeding is a must in many parts of the world.

The following are some points to help in understanding, planning, and managing a range reseeding operation.

What is reseeding? Reseeding refers to the artificial, not natural, revegetation of the range. It includes planting native or introduced species with one or several of the following objectives: 1) to improve plant cover/density, 2) to improve forage production, 3) to improve forage quality, 4) to improve efficiency of water use, and 5) to improve the diet of grazing animals.

Why reseeding? Reseeding is initiated for two reasons:
- To rehabilitate unproductive areas and put them into production again. Nongrazed areas, because of lack of forage, are a waste of money.
- To accelerate secondary plant succession in areas where natural revegetation is too slow or almost impossible.

When is reseeding necessary and justifiable? Reseeding is needed and justified:
- when percentage of desirable vegetation in the botanical composition is very low (below 18-20%) and plant density is rare.
- when the range site to reseed has the agronomic potential to respond to the treatment (topography, soil quality, etc.).
- when climatic and meteorological conditions are favorable and meet the minimum safe requirements: amount and distribution of rainfall, and frost-free period.
- when the proper seed is available.

How to reseed. There are some basic rules that one has to follow to minimize risks when trying to improve a range by revegetation:
- Choose the best adapted species, native or introduced, based on experimental evidence or on regional experience.
- Select a high quality seed with a high PLS (pure live seed) percentage; do not sacrifice quality for a low price.

- Seed at the beginning of the <u>formal</u> rainy season--when you are sure the rains will continue with regular frequency. Do not plant late in the summer because the plants will not have a chance to get established before the first frosts.
- Use the recommended seeding rates on a PLS basis.
- Try to use a grass seed driller that will assure the best distribution of seed on the ground. If no machinery is available and you have to broadcast, cover the seed lightly with branches from shrubs.
- Broadcast a mixture of seeds separately according to size, i.e., the small seeds (weeping live, clover, bluepanic) should be mixed with fine sand for a better distribution and planted separately from larger or fluffy seeds (side oats, crested wheat, buffel). Drillers usually have different types of boxes for different sizes of seeds and can be calibrated separately for the desired rate.
- Plant grass seeds no more than 1 to 1.5 cm deep in a firm seed bed. If possible, press the soil down lightly on the seeds to make contact between seed and soil. If this is not possible, and broadcast is used, cover lightly with a rake made of shrub branches.
- Do not fertilize at the time of planting since many weeds will take advantage of the fertilizer and will compete strongly with the planted species. It is better to wait until there is an established stand of the seeded species.
- Protect the seeded area from grazing at least during the first growing season. Depending on the density of the stand, the area could be grazed lightly in the second year. Remember: reseeding is expensive and risky, and proper management is necessary for a long-lasting stand.

Range fertilization. Recent increases in prices of fertilizers have limited their use on rangelands. However, areas receiving annual rainfall above 350 mm have responded significantly in terms of forage production, forage nutritive quality, and carrying capacity so that it might still pay to fertilize.

Nitrogen is the most commonly used fertilizer for semi-arid and temperate rangelands; however, nitrogen plus phosphorous have demonstrated to be better in some rangelands. For instance, experiments in Central Chihuahua, Mexico, using different levels of N and P and their combinations, indicated that either N or P alone produces more forage than the nontreated areas, but the application of N and P together was much better. By applying 80 kg of N and 50 kg

P_2O_5/ha, forage production increased 132% the first year; crude protein per hectare changed from 31 to 107 kg and phosphorus per hectare increased from 0.42 to 1.46 kg. In this same experiment, 73% of the cost of fertilizers was recovered during the first year (Gonzalez, 1972).

In spite of the "energy crisis" and the high price for fertilizer, it is important to fertilize rangeland because of the growing demand for land and because of the need to intensify as much as possible by getting the maximum out of every hectare of range. That is why fertilizers have an important role in areas where ecological conditions are favorable for their use.

Type of fertilizer, apportionment/ha, time of application, and method of application are some of the aspects to consider once you have had a soil analysis and determined the need to fertilize.

PRIORITIES IN SELECTING IMPROVEMENT PRACTICES

Every ranch has its own problems and own needs for range improvement. All of the range improvement practices discussed are expensive, but each must be considered carefully when bringing rangeland back into production.

The following are some guides to help in the selection of which improvement practice(s) to use.
- First, evaluate the present range condition to determine if an improvement is necessary. Usually ranges in poor or fair condition need some help, but ranches in better condition may need only a change in management.
- Second, define the problem. Brush infestation? Low plant density? Poor botanical composition? Erosion? . . . Establish your priorities. Treat the best sites first because they respond better and faster to treatment. Define which improvement practice(s) to use and, if several are needed, establish your priorities.
- Third, in selecting the practice to improve your ranch, costs are basic. There must be an evaluation of costs in terms of the potential to improve forage production and pay back the expenses. One must be alert to prices and cost fluctuations.
- Finally, once the practice has been selected, plan how to do it: 1) what to use, 2) when to do it, and 3) who is going to do it. Before starting the program define the type of equipment and materials, the time of the year, the number of hectares to treat, and the people responsible for doing the job.

In analyzing all the parameters, a combination of two or more practices (if needed!) may be the most economically productive.

For example, figure 2 is a classic for understanding the priorities for improvement in northern Mexico in relation to the interaction of rain and vegetation. Of the 340 mm average annual rainfall, 40 mm, almost 12%, fall in the winter time when it is not used readily by plants. Of the remaining 300 mm about 100 mm are lost through evapotranspiration and surface runoff; only 200 mm are available for the vegetation. According to studies on over 100 million hectares of rangelands in Central and Northern Mexico (CFAN-CID, 1968), botanical composition in range pastures includes more than 50% of undesirable plants; therefore, at least 125 mm of rain are used by shrubs, forbs, and weeds, and the final 75 mm for desirable forage species--the equivalent of a mere 22% of the total rain recorded.

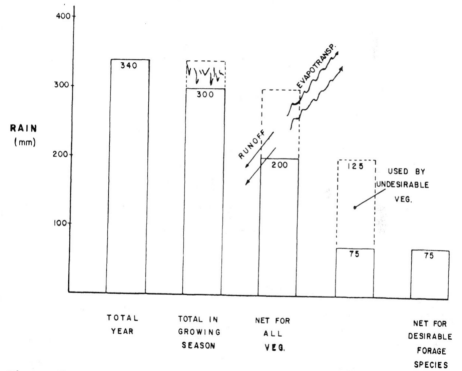

Figure 2. Used and wasted rainfall. Northern Mexico.

This simple rain-vegetation relation shows that range improvement practices should be directed towards: (1) obtaining a better conservation of water and (2) eliminating undesirable vegetation so the limited available rain is used by desirable plants.

ESTIMATED PRODUCTION INCREASE DUE TO IMPROVEMENT PRACTICES

Based on evidence accumulated by research at Rancho Experimental La Campana, in Chihuahua; CIPES, in Carbo, Sonora; and Vaquerias Experimental ranch, in Jalisco, as well as field data in different ranches in northern and central Mexico, the increase in range production due to several improvement practices is illustrated in table 2 (Gonzalez, 1977). This table includes averages of results of different vegetative types where improvement practices have been conducted. These indicate the benefits in a 4- to 6-year period (short-term) and the projected increases that can be obtained in 8 to 12 years (long-term).

TABLE 2. ACTUAL AND POTENTIAL INCREASE IN FORAGE PRODUCTION IN RANGELANDS OF CENTRAL AND NORTHERN MEXICO

| | | Potential | |
	Present	Short-term (actual)	Long-term (projected)
Forage Prod. kg DM/ha	210	420	580
% increase due		100	176
Grazing systems and intensities		40	46
Soil and water conservation		15	40
Brush control		30	60
Range reseeding		15	30
Carrying capacity Ha/AU	20.6	10.3	7.5

Source: M. Gonzalez (1977).

In the short-term period, the forage production of the range increased 100% from an average of 210 kg DM/ha to 420 kg DM/ha. This increment was due to the adjustment of the carrying capacity of the ranches and the modification of their grazing systems (40%); to soil and water conservation (15%); to brush control (30%); and range reseeding (15%). These improvements increased the yearly carrying capacity of the range from 20.6 ha/AU to 10.3 ha/AU.

The projected increase in production for the long-term period was based on the continuation of adequate management and improved range conditions. For an 8- to 12-year period, it was estimated that forage production would reach 580 kg DM/ha or 176% above the present. Carrying capacity would be 7.5 ha--almost triple of the present one.

The range improvement will be reflected by increased livestock production, both in quantity and quality, because of better forage in the pastures.

REFERENCES

Cable, D. R. 1965. Damage to mesquite, lehmann lovegrass and blackgrama by a hot June fire. J. Range Mgt. 18:326.

Fierro, L. C., F. Gomez and M. H. Gonzalez. 1979. Control biologico de arbustivas indeseables utilizando ganado caprino. Bol. Pastizales, Vol. X No. 3 Rancho Exp. La Campana INIP-SARH, Mexico.

Glendening G. and H. A. Paulsen. 1955. Reproduction and establishment of velvet mesquite and relation to invasion of semidesert grasslands. USDA Tech. Bull. 1127.

Gomez, F. and M. H. Gonzalez. 1976. Evaluacion de cinco mezclas de herbicidas en elcombate de chaparrillo, Eycenhardtia spinosa. Bol. Pastizales Vol III, No. 2 Rancho Exp. La Campana INIP-SARH. Mexico.

Gomez, F. and M. H. Gonzalez. 1978. Efecto del chapeo mecanico en el incremento de un pastizal invadido por arbustivas Bol. Pastizales, Vol. IX, No. 5 Rancho Exp. La Campana INIP-SARH, Mexico.

Gonzalez, M. H. 1972. Aumentos en la produccion de carne con la fertilizacion de un pastizal. Bol. Pastizales Vol. III, No. 1 Rancho Exp. La Campana, INIP-SARH, Mexico.

Gonzalez, M. H. 1976. El potencial de las tierras no cultivables en la produccion de alimentos. Ingenieria Agronomica, Vol. II, No. 2, Col. Ing. Agr. de Mex. AC. Mex.

Martinez, J. 1959. Pruebas de infiltracion en un pastizal mediano-abierto. Tesis. Esc. de Ganaderia, Univ. de Chihuahua, Mexico.

Reynolds, H. G. and J. Bohning. 1956. Effects of burning on a desert grass shrub range in southern Arizona. Ecology 37.

Sanchez, A. 1972. Efecto de la condicion de pastizal en la infiltracion del agua de lluvia Bol. Pastizales, Vol. III, No. 3. Rancho Exp. La Campana INIP-SARH, Mexico.

Vallentine, J. F. 1977. Range Development and Improvements. Brigham Young Univ. Press. Provo, Utah.

White, L. D. 1969. Effects of a wild fire on several desert grass land shrub species. J. Range Mgt. 22.

109

WHAT TYPE OF GRAZING SYSTEM FOR MY RANCH . . . ?

Martin H. Gonzalez

INTRODUCTION

Ranchers have developed a growing interest in grazing systems during the last decade. The introduction of the new, "intensive" systems has received more attention from producers and researchers than the traditional, continuous, and extensive rotation systems we knew.

A never-ending controversy could evolve from a group of ranchers discussing the pros and cons of each one's favorite system. All of the systems have been used under different conditions and have their particular advocates but, at the same time, everybody knows that a specific system cannot be universally implemented. Each ranch has its peculiar objectives, ecological features, market situations, financial capabilities, and the personal preference of its owner. As Penfield (1982) wrote, "The most important element in a grazing system involves a commitment by you to make it work." And at the same time the rancher must be committed to working with nature not against it--a principle to be kept in mind when selecting a grazing system.

Both native rangelands and cultivated forage communities and their biological and economic ecosystems lead ultimately to the production of livestock products used by man. Most lands under livestock or wildlife production have unbalanced ecosystems in which one or more of the major constituents (plant, animal, or man) has a chronic or seasonal lack of nutrients, water, or other environmental requirement (Whythe, 1978). It is the responsibility of the soil and crop scientists to help overcome these deficiencies. But it is the rancher, the manager of the range, who has to decide whether the proposals of the scientists and the new "discoveries" are economically acceptable; the rancher, after all, is part of the ecosystem in which he lives with his domestic livestock.

TYPES OF GRAZING SYSTEMS

The number and variation of grazing systems is almost infinite. Every ranch has its peculiar characteristics--

even adjoining ranches with similar ecological conditions differ in objectives, management, and financial capabilities. For this presentation we will consider some of the most common grazing systems, knowing that within each of them the variations are numerous. Any rancher can identify with one of these models and design his own system.

Continuous Grazing

Continuous grazing is the constant use of forage on a given area, either throughout the year or during most of the growing period. This type of grazing does not always result in range decline (overgrazing). Light, continuous use results in range improvement, but returns per hectare are lower than those obtained from a four-pasture, deferred-rotation grazing (Merrill, 1980).

Under continuous grazing, stocking rates are set with only minor adjustments through the year. The number of animals does not vary greatly except during long drought periods (Kothmann, 1980). Under continuous grazing, stocking rate is the only variable the manager can adjust; this allows little flexibility in responding to drought seasons (Vaughan-Evans, 1978).

Rotation System

A rotation system moves animals from one pasture to another according to a fixed schedule. Within this system are the extensive (or non-intensive) and the intensive systems. In the extensive system, several pastures are grazed and one is resting for a certain number of months. The rest period is rotated so that the same pasture has the rest period during different months each year. In the intensive system, all the animals graze one pasture (for a short period) while all the other pastures rest.

Deferred rotation. Under deferred rotation, each pasture is deferred during each season (Ambolt, 1973). The design includes a fixed number of pastures for each herd of livestock. The goal is to improve the vegetation condition of the resting unit, but the length and intensity of grazing on the remaining units should not be so long as to cause range deterioration (Merrill, 1980). Carrying capacity is determined by the total land area in the system and should be set conservatively. Deferment periods usually vary from three to six months, but may be longer (Kothmann, 1980).

The purpose of such a system is two-fold: to improve the vigor and production of forage, which in turn increases the desirable forage species, the carrying capacity, and the animal response to the improved forage. The result is higher returns per unit area. Figure 1 shows the Merrill 4-pasture deferment-rotation grazing as a good example of this system.

998

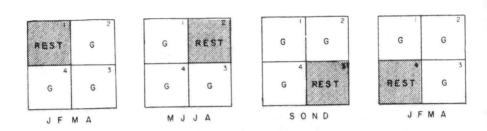

Figure 1. Merrill's four-pasture, deferred-rotation grazing system

Rest rotation. In the rest-rotation system, a pasture is not grazed for a full year and the rest period rotates among pastures. Deferments are provided for seed production and for seedling establishment. The number of pastures in the system may be from two to five (Hormay, 1970). The carrying capacity is based on that portion of the range that is grazed each year. Generally a much smaller percentage of the total area is available to grazing during the growing season (Kothmann, 1980).

These two basic nonintensive rotation systems sometimes are combined to take advantage of both the rest and the deferment on a varying number of pastures.

A disadvantage for some ranches is that pastures should be reasonably uniform in both size and carrying capacity. This often requires additional fencing and watering (figure 2).

Intensive systems. The two most common types of intensive grazing systems are the high intensity-low frequency, and the short-duration grazing. Both of these systems are based on grazing all the animals in one pasture while the other pastures rest. The difference between the two systems is the length of time given to the grazing and rest periods. The advantages of the long rest period are: it provides maximum vegetation improvement, eliminates seasonal rotations, and concentrates the livestock in one area which makes handling and working easier. The disadvantages of the intensive systems are: an increase in the need for water storage and daily output on individual pastures, generally lower livestock gains per head, and a possible parasite problem if sheep and goats are present (Merrill, 1980).

High intensity-low frequency (HILF)--When designing an HILF system, the forage species present are evaluated and a

system that favors the plants we want to produce is designed. The time plants need for recovery from grazing determines the rest periods needed in the system (figure 3).

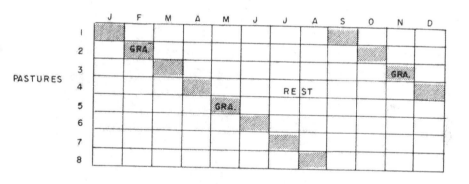

Figure 2. Rest rotation system with five pastures

Figure 3. High-intensity--low-frequency grazing system

The HILF systems are based on the use of intensive
grazing periods with relatively long rests. At least three
pastures are required per herd. The grazing periods are
generally more than two weeks, rest periods longer than 60
days, and grazing cycles over 90 days. Carrying capacity is
based on the total land area in the system and moderation in
length of grazing cycle to prevent an excessive decline in
animal production.

Short duration--Like the HILF, this system also must
have three or more pastures per herd, but the grazing and
rest periods are shorter. Grazing periods should be less
than 14 days (preferably 7 days or fewer and rest periods
should vary from 30 to 60 days, but never exceeding 60 days
(figure 4).

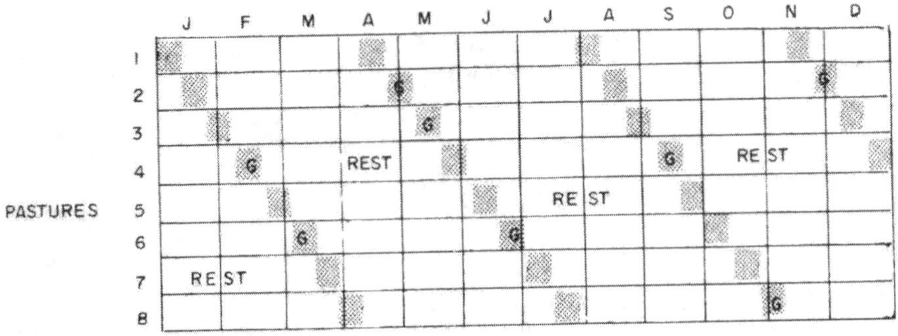

	J	F	M	A	M	J	J	A	S	O	N	D
1	G							G			G	
2	G			G				G				G
3		G			G			G				G
4		G	REST		G				G	REST		G
5			G			G	REST		G			
6			G			G						
7	REST		G					G			G	
8				G				G			G	

PASTURES

Figure 4. Short-duration grazing system

Grazing cycles are generally short enough to allow six or more full rotations per year (Savory, 1979). Carrying capacity is based on the total land area in the system. Because of the high degree of control over both frequency and intensity of defoliation (greater than under other systems), a higher degree of use in this system is possible without detriment to plant or animal production (Kothmann, 1980).

PARAMETERS FOR THE EVALUATION OF GRAZING SYSTEMS

Objectives

Apart from the specific objectives that a rancher might have, there are a number of general objectives for selecting his grazing system. Some of the most important are the following:
- To obtain a better livestock grazing distribution and, consequently, a more uniform utilization of the range or pasture.
- To maintain or improve forage density.
- To maintain or improve the botanical composition.
- To meet the nutritional needs of the grazing animals.

All of the objectives are affected by the major objective--to increase the overall efficiency of the operation.

Considerations

Thee are some basic premises that the rancher must consider when choosing his grazing system. The following deserve his attention:
- Viability of the system. Will all factors combined in its planning make it work?
- Cost of additional infrastructure (fencing, watering, pens, etc.) and labor involved.
- Minimizing the effect of adverse weather.
- Stocking rate.
- Pasture size.
- Location of water, salt, and supplements.
- The number of pastures in the system.
- The number of herds per system, if more than one system is to be used.
- The length of the grazing cycle and the rest period--the heart of the system.

Consequences

A well-planned and implemented grazing system will have some beneficial consequences. On the contrary, if the development of the system is not in agreement with the basic objectives and considerations, the results are likely to be poor. Some of the positive effects of a well-designed and managed system are:

Effects on the vegetation.
- The system will produce more forage.
- Seed production is increased.
- Botanical composition of pastures is improved.
- Chemical composition of forage plants (animal's diet) is improved.
- Over-all range resources are improved by the rapid increase in range condition.
- The amount of forage consumed by livestock due to nonselective grazing is increased.

Effects on the animal.
- Minimal or no stress on animals.
- Improved animal performance.
- Less handling of animals in general.
- Reduced supplemental feeding.
- Minimized labor costs.
- Higher individual weight gain.
- Better calving intervals.
- Increased stocking rate.
- Interrupted disease and parasite cycle.

A possible added benefit would be the increase in wildlife population and an improvement in their condition.

PLANNING THE SYSTEM FOR YOUR RANCH

Assuming that you have decided to develop your own grazing system or to adopt one of many models available, a logical sequence of action is recommended.

Inventory and study carefully your range resource.
- Vegetative types
- Range sites within vegetative types
- Your pastures:
  a Number
  b Size
  c Forage production
  d Season of use
  e Present range condition

Reaffirm your objectives for:
- Type of livestock operation
  a Cow-calf
  b Feeder cattle
  c Combination
- Type of grazing system
  a Extensive rotation
    1. Deferred rotation
    2. Rest rotation
  b Intensive rotation
    1. High intensity-low frequency
    2. Short-duration grazing

Design your grazing system according to:
- Your objectives
- Present number and size of pastures
- Projected number and size of new pastures
- Expenses involved in additional proposed infra-structure
  a Fencing
  b Water development
  c Labor

Expand the management plan.
- For the range-pastures
  a Grazing cycle
  b Range improvement practices needed
- For the livestock
  a Breeding and calving
  b Nutrition program
    1 Range supplements for cows
    2 Possibilities of creep feeding
    3 For replacement heifers
  c Sanitation program
  d Marketing
- For monitoring
  a Forage utilization
  b Range condition and trend
  c Animal production

1 Percentage calf crop
2 Weaning weights
d Market situations
We must emphasize that the best grazing system is the one best adapted to your particular interest and the conditions on your ranch. Even if economic resources and infrastructure are limited, a grazing system can be improved through better management. Work with what you have and follow the basics for the grazing system that best suits your ranching business.

REFERENCES

Hormay, A. 1970. Principles of rest-rotation grazing and multiple use land management. USDA, F. S. Training Test 4(2200).

Merrill, L. B. 1980. Considerations necessary in selecting and developing a grazing system. What are the alternatives? Proc. Grazing Mgt. Systems for S. W. Rangelands Symposium. The Range Impr. Task Force. New Mexico State Univ., Las Cruces, N. M.

Kothmann, M. M. 1980. Integrating livestock needs to the grazing system. Proc. Grazing Mgt. Systems for S. W. Rangelands Symposium. The Range Impr. Task Force. New Mexico State Univ., Las Cruces, N. M.

Penfield, S. 1982. Are you ready for a grazing system? Rangelands 4(1). Soc. Range Mgt.

Savory, A. 1979. Range management principles underlying short-duration grazing. Beef Cattle Sci. Handbook. Agr. Services Found., Clovis, Ca. 16:375.

Vaughan-Evans, R. H. 1978. Short-duration grazing improves veld conservation and farm income in the Oue group of I.C.A.'s Cenex reports. Mimeo. Rhodesia.

Wambolt, Carl L. 1973. Range grazing systems. Coop. Ext. Serv. Bull. 340. Montana State Univ.

110

DETERMINING CARRYING CAPACITY
ON RANGELAND TO SET STOCKING RATES
THAT WILL BE MOST PRODUCTIVE

John L. Merrill

Every rancher has to decide the stocking rate for his pasture--how many animals for how long. If he makes the wrong decision, the problem will be compounded by continuing indecision or unsound decisions.

Successful range management depends upon proper use which is defined as "stocking according to forage available for use--always leaving enough for production, reproduction, and soil protection." This is easier said than done. Other important proper-use factors are the species of livestock and wildlife to be grazed, the proper distribution of grazing, and the grazing method employed, but a key concern is setting and adjusting the stocking rate as determined by carrying capacity.

Experience and research have helped the grazing-land manager estimate and/or calculate the carrying capacity of either range or tame pasture for making better stocking-rate decisions and adjustments. By using several different methods simultaneously, the grazing-land manager's judgment and accuracy are improved.

Over a period of many years the USDA Soil Conservation Service (SCS) has developed range site descriptions and range condition guides for determining carrying capacity for almost every area of privately owned rangeland in the U.S. The safe-starting stocking rates derived by the SCS are fairly conservative so as to avoid recommendations that might cause trouble.

A range site is a distinctive kind of rangeland of relatively homogeneous climate, soils, and topography that will support a typical group of plants and level of production. Range condition refers to its current productivity relative to its natural capability to produce in climax condition. Range condition is classified by the percentage of climax plants present and usually is expressed as excellent, good, fair, and poor by 25% increments. For each classified condition there is a corresponding carrying capacity or safe-starting stocking rate range expressed in acres per Animal Unit Year Long (AUYL) or Animal Unit Month (AUM). The range allows for varying forage production from year to year within a given range condition.

Climax plants include decreasers, which are those plant species dominant in the original-climax vegetation that decrease with continued overuse. Increasers are secondary-climax plants that increase for a period of time as decreasers decline from overuse, but they in turn will also decrease from continued overuse. Invaders are those plants that are not found (or found in small numbers) in the climax situation that begin to cover bare ground as the climax plants decline and, therefore, are not counted in range-condition classification.

For many years SCS range-condition estimates were based on a visual estimate of percentage by ground cover (basal density and/or canopy cover) of climax plants with percentages of decreasers counted, but with no greater percentage of increasers counted than were estimated to be present in climax condition. This system was easy for both range technicians and ranchers to learn and use and provided a reasonably accurate track of changes in plant composition if recorded year after year (appendix A).

In an attempt to refine that system, SCS range-condition guides now assign a percentage of each plant or group of plants similar in production and reaction to grazing pressure. The percentage present is determined by estimating the pounds of air-dry forage per acre of these plants or groups divided by the total pounds per acre (appendix B).

This system is much more difficult and complicated to learn and use and confuses annual forage production (which fluctuates widely from year to year) with actual changes in plant composition (which are much slower to change) so that neither is tracked accurately and separately. For these reasons, the present system seems to be more complicated and time-consuming, less useful, no more accurate than the former--and the estimate of total forage production per acre is not used per se in estimating the carrying capacity.

For the past twenty years, the author has developed and used a third method, proven accurate in practice, that is fast and easy to learn and use in a year-long grazing area. It is based on estimating the total forage production per acre in the fall at or near the end of the growing season to determine the carrying capacity through the dormant season. The pounds of forage per acre may be estimated as an average per acre for the whole pasture. In either case the pounds per acre are multiplied by the number of pasture acres to determine the total forage available in the pasture.

To determine the amount of forage available for animal intake, the total forage present should be divided by four to allow for the amount already grazed during the growing season, the amount lost to weathering and trampling, and the amount that should be left for soil protection at the end of the grazing season before the following year's growth is initiated. To determine the amount of usable forage required per animal unit, multiply 30 lb/day (3% of body weight) by the number of days to be grazed (dormant season plus 30 to

45 days to allow for a late spring or a shorter period if livestock are to be rotated). If the pasture is being grazed yearlong, the winter stocking rate will approximately equal the yearlong stocking rate.

The eye of the technician or rancher can be "set" for visual estimates by clipping all the forage from a 21 in. radius circle. Weigh in grams, multiply by 10 to convert to pounds per acre; adjust for moisture content to determine total pounds of air-dry forage per acre. The moisture adjustment can be made by using SCS tables on forage type and stage of growth or by carefully heating the clippings (Avoid fire!) in the kitchen oven and adding 10% to get air-dry forage amount. Usually, only two or three clippings are necessary to condition the eye to recognize typical areas before proceeding with visual estimates. This procedure comes very naturally to those accustomed to estimating livestock weights (appendix C).

A fourth method for estimating carrying capacity involves using the total forage producer per year required to support an animal unit (but provide for forage loss and soil protection in areas of differing annual rainfall) and dividing by the actual total annual forage production per acre as determined by clipping or ocular estimate to determine acres per animal unit on a yearlong basis. This method was developed independently by Hershel Bell, H. L. Leithead, and a number of other workers based on their field observations and experience. The following table, published by the Texas Agricultural Extension Service, uses this method.

Forage requirement per animal unit of domestic livestock

Average annual rainfall	Annual forage requirement
More than 30 inches	30,000 pounds
20 to 30 inches	24,000 pounds
15 to 20 inches	20,000 pounds
Less than 15 inches	15,000 pounds

A fifth method of annually setting or adjusting stocking rate is used by Dick Whetsell on the Oklahoma Land and Cattle Company in the Osage area of northeastern Oklahoma—an area of 32 in. to 34 in. average annual rainfall. Since no one can predict future forage production accurately, he bases his adjustments on the previous year's forage production. In an average area where 8 acres are required per animal unit throughout the year, if the previous year has been above average in forage production, the allowance would be reduced by 1A to 7A. If another good year followed in succession, the allowance would be reduced to 6A but never below 6A. If that succeeding year had been average, the stocking rate would remain at 7A; if below, it would have returned to 8A. In years of reduced forage production, reductions would be made in 1A increments. To adjust this method for areas of less rainfall larger increments of adjustment would be needed. An area requiring 16 A/AU would require adjustment in 2A increments.

A sixth method uses the net-energy method to estimate carrying capacity by balancing animal energy needs with energy available in the amount of forage per acre with energy provided by supplements deducted to arrive at the number of acres required per animal (appendix D).

A seventh method, simple and widely used, is to make adjustments of stocking rate based on the degree of use and overall range trend. The present degree of use is the indicator of range trend, followed by plant vigor, production and reproduction, with litter, organic matter, ground cover, and livestock production following. A stable trend would indicate leaving livestock numbers as they are with concomitant additions of livestock if the trend is upward and decreased numbers if the trend is down.

Data for all of these methods can be obtained rather rapidly for the purpose of original inventory or subsequent monitoring by moving through each pasture and recording data in abbreviated form site by site and/or pasture by pasture on aerial photographs or other maps. Site can be noted by initials, as can classification of condition (E, G, F, or P), the amount of forage present in pounds per acre, current degree of use (L--light, M--medium, or H--heavy)and trend (+, 0, or - to represent upward, stable, or downward). An example map notation follows:

Fair condition   Moderate use

   Deep upland site   DU/F/2500/M/+

   2500 pounds of forage   Upward trend

Calculations of an example by methods one and three might be done as follows:

North pasture - 328 acres   November 1 Observations

        AUYL by Forage   AUYL by Condition

28 ac. S/F/1500/H/- = 42,000   @ 18 AU = 1.56
200 ac. CL/G/2500/M/0 = 500,000   @ 10 AU = 20.0
100 ac. BL/G/3000/M/+ = 300,000   @ 12 AU = 8.33
                        _____   ____

                        842,000   29.89 = 30 AU
                        ÷ 4 = 210,500
                        ÷ 5400 = 38.98 = 39 AU
(30#/AU/DA x 180 DAS = 5400#/AU)

If your stocking rate the previous year had been 35 AU and this had been an average forage producing year with a total forage production of 3500 lb/A, a crosscheck by method four would indicate 38 AU and by methods five and seven would be

35 AU. With these calculations as background, you might confidently decide to stock on the basis of 37 AU in that pasture.

Before determining the number of domestic livestock to be grazed, be sure to deduct the animal unit equivalent of grazing wildlife, if numbers are significant. Species and classes of livestock should be chosen according to forage and water resources, topography, markets, facilities, and knowledge available. Stocking rate usually can be increased if rotation grazing and/or multiple species are used to increase efficiency.

Animal Unit Equivalents Guide

Kind and classes of animals	Animal-unit equivalent
Cow, dry	1.00
Cow, with calf	1.00
Bull, mature	1.25
Cattle, 1 year of age	.60
Cattle, 2 years of age	.80
Horse, mature	1.25
Sheep, mature	.20
Lamb, 1 year of age	.15
Goat, mature	.15
Kid, 1 year of age	.10
Deer, white-tailed, mature	.15
Deer, mule, mature	.20
Antelope, mature	.20
Bison, mature	1.00
Sheep, bighorn, mature	.20
Exotic species	(to be determined locally)

Stocking-rate calculations are exactly the same situation as calculating a feed ration. They are used to decide on a reasonable course of action to be followed by continuous close observation of results; in this case, the response of both forage and livestock are observed and the necessary adjustment made to achieve the desired results. No amount of calculation can take the place of good judgment, keen observation, and timely action, but used properly, they certainly can contribute to wiser decisions and less costly errors.

Since years are seldom alike in forage production, good ranchmen plan for adjustments in stocking rate to match forage available without wrecking breeding management, cash flow, and tax management. Understocking, overstocking, and even constant stocking at the same rate increases costs and reduces profitability. A constant, knowledgeable effort to stock properly will pay good dividends.

APPENDIX A

TECHNICAL GUIDE TO RANGE SITES AND CONDITION CLASSES (WORK UNIT--FORT WORTH)

KEY CLIMAX PLANTS	Plant Percent Allowable by Sites			INVADING PLANTS
	DU	RP	VS	
Little bluestem	-	-	-	Annuals
Big bluestem	-	-	-	Texas grama
Indiangrass	-	-		Tumblegrass
Switchgrass	-	-		Hairy tridens
Perennial wildrye	-			Scribner panicum
Sideoats grama	10	15	0	Fall witchgrass
Hairy dropseed		5	-	Halls panicum
Vine-mesquite	5	5		Sand dropseed
Meadow dropseed	5	15		Roemer senna
Silver bluestem	5	10	15	Gray goldaster
Texas wintergrass	10	5	5	Dyschoriste
Texas cupgrass	5	5		Buckwheat
Buffalograss	Inv.	5	10	Curly gumweed
White tridens	5	5		Mealycup sage
Deep muhly	Inv.	5		Nightshades
Tall hairy grama	Inv.	5	10	Milkweeds
Rough tridens		Inv.	15	Western ragweed
Perennial threeawn	Inv.	Inv.	10	Baldwin ironweed
Climax forbs	5	5	10	Texas stillingia
Woody canopy *	5	5	5	Prickly pear
				Mesquite
				Yucca
				Sumac

Maximum Total Allowable Increasers	20	30	35

Range Condition	Safe Starting Stocking Rates - Ac./AUYL			% Climax Vegetation
Excellent	7	10	16	76-100
Good	9	12	20	51-75
Fair	14	16	28	26-50
Poor	20	23	35	0-25

LEGEND

Blank	=	Not significant
-	=	Decreaser: all allowed
5, etc.	=	Increaser: allowable percentage
Inv.	=	Invader

Range Sites and Major Soils:

DU= Deep Upland (GP) 2-San Saba clay; 2X-Denton, Krum clay; Lewisville clay loam.

RP= Rolling Prairie (GP) 18c-Denton-Tarrant complex; 18-Denton clay, shallow phase.

VS= Very Shallow (GP) 24c-Tarrant stony clay, unfractured substrata.

* Use open canopy method to determine percentage of all brush species. On savannah sites, determine range condition by the percentage composition of understory vegetation. To estimate available grazing, reduce the acreage in the site by the percent that climax brush exceeds the indicated allowable. On prairie sites, estimate percentage of invading brush species by loosely compacting the canopy to simulate total shading. The percentage figure thus obtained is "counted" against range condition.

APPENDIX B

RELATIVE PERCENTAGE

Grasses	90%	Woody	5%	Forbs	5%
Little bluestem	35	Elm		Engelmanndaisy	
Big bluestem	20	Hackberry	5	Maxmilian sunflower	
		Pecan		Yellow neptunia	
Indiangrass	15	Plum		Catclaw sensitive-	
Switchgrass		Liveoak	T	briar	
Virginia and	5			Prairie clovers	
Canada wildrye				Scurfpeas	
				Gaura	
Sideoats grama	10			Heath aster	5
Texas wintergrass				Trailing ratany	
				Blacksamson	
Tall dropseed				Golden dalea	
Vine mesquite				Wildbeans	
Texas cupgrass	5			Tickclovers	
White tridens				Gayfeather	
Silver bluestem				Prairie bluets	
Hairy grama				Bundleflower	

b. As retrogression occurs, big bluestem decreases rapidly followed
by Indiangrass, little bluestem, and switchgrass. Sideoats
grama, tall dropseed, and Texas wintergrass increase initially
and then decrease as retrogression continues. Buffalograss,
Texas grama, tumblegrass, red threeawn, western ragweed, Baldwin
ironweed, queensdelight, mesquite, sumac, lotebush and common
honeylocust invade the site.

c. Approximate total annual yield of this site in excellent
condition ranges from 3000 pounds per acre in poor years to
6500 pounds per acre of air-dry vegetation in good years.

4. WILDLIFE NATIVE TO THE SITE: Dove and quail inhabit this site.

5. GUIDE TO INITIAL STOCKING RATE:

a.
Condition Class	Climax Vegetation	Ac/AU/YL
Excellent	76-100	6-9
Good	51-75	8-12
Fair	26-50	10-16
Poor	0-25	14-20

b. Introduced Species

Species	Percent of the Area Established			
	100-76	75-51	50-26	25-0
King Range bluestem	8-10	10-14	14-18	18+
Common bermudagrass	7-10	10-13	13-18	18+
Kleingrass	7-10	10-13	13-18	18+

APPENDIX C

YIELD DETERMINATIONS

CIRCLES

RADIUS	UNIT OF WEIGHT	CALCULATION
3.725 ft. (3 ft. 8 3/4")	pounds	x 1000=lbs./ac.
2.945 ft. (2 ft.11 3/8")	ounces	x 100=lbs./ac.
1.75 ft. (1 ft. 9 in.)	grams	x 10=lbs./ac.

SQUARE

LENGTH OF SIDE	UNIT OF WEIGHT	CALCULATION
6.6 ft. (6 ft. 7 1/4")	pounds	x 1000=lbs./ac.
5.22 ft. (5 ft. 2 5/8")	ounces	x 100=lbs./ac.
3.1 ft. (3 ft. 1 1/4")	grams	x 10=lbs./ac.

Percentage of Air-Dry Matter in Harvested Plant
Material at Various Stages of Growth

Grasses	Before heading; initial growth to boot stage	Headed out; boot stage to flowering	Seed ripe; leaf tips drying	Leaves dry; stems partly dry	Apparent dormancy
	Percent	Percent	Percent	Percent	Percent
Cool season............	35	45	60	85	95
wheatgrasses					
perennial bromes					
bluegrasses					
prairie junegrass					
Warm-season					
Tall grasses..........	30	45	60	85	95
bluestems					
indiangrass					
switchgrass					
Mid grasses..........	40	55	65	90	95
side-oats grama					
tobosa					
galleta					
Short grasses.........	45	60	80	90	95
blue grama					
buffalograss					
short three-awns					

Trees	New leaf and twig growth until leaves are full size	Older and full-size green leaves	Green fruit	Dry fruit
	Percent	Percent	Percent	Percent
Evergreen coniferous.......	45	55	35	85
ponderosa pine, slash				
pine-longleaf pine				
Utah juniper				
rocky mountain juniper				
spruce				
Live oak.................	40	55	40	80
Deciduous................	40	50	35	85
blackjack oak				
post oak				
hickory				

APPENDIX C (con't)

Percentage of Air-Dry Matter in Harvested Plant
Material at Various Stages of Growth

Shrubs	New leaf and twig growth until leaves are full size	Older and full-size green leaves	Green fruit	Dry fruit
	Percent	Percent	Percent	Percent
Evergreen................ big sagebrush bitterbrush ephedra algerita gallberry	55	65	35	85
Deciduous................ snowberry rabbitbrush snakeweed Gambel oak mesquite	35	50	30	85
Yucca and yucca-like plants.................... yucca sotol saw-palmetto	55	65	35	85

Forbs	Initial growth to flowering	Flowering to seed maturity	Seed ripe; leaf tips dry	Leaves dry; stems drying	Dry
	Percent	Percent	Percent	Percent	Percent
Succulent................. violet waterleaf buttercup bluebells onion, lilies	15	35	60	90	100
Leafy..................... lupine lespedeza compassplant balsamroot tickclover	20	40	60	90·	100
Fibrous leaves or mat...................... phlox mat eriogonum pussytoes	30	50	75	90	100

Succulents	New growth pads and fruits	Older pads	Old growth in dry years
	Percent	Percent	Percent
pricklypear and barrel cactus....................	10	10	15+
cholla cactus...............	20	25	30+

APPENDIX D

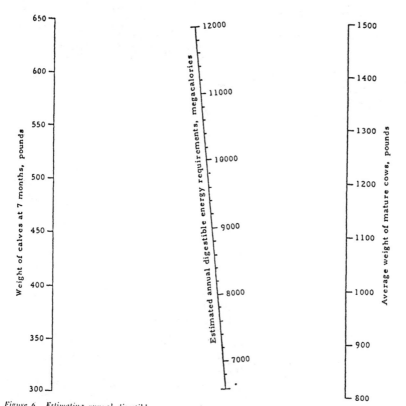

Figure 6. Estimating annual digestible energy requirements for a cow and her calf (2 miles travel).

			Example		Your herd	
1*	Energy requirement for cow and calf (1,000 lb. cow—500 lb. calf)		8,817		1000# 8800	1300# 11300
2**	Minus energy in supplemental feed (300 lb. CSC 1.3 megacalorie) (150 lb. hay 0.9 megacalorie)	390 135 525	−525	−390	8410	10910
3	Energy needed from pasture forage		8,292			
4	Megacalorie per pound of forage		+1.0			
5	Total pounds of air-dry forage required for each cow and calf		8,292		8410	10910
6	40 to 65% of the forage used by cow and calf (40 to 50% for western native pastures) (55 to 65% for eastern improved pastures)		+.40			
7	Total pounds of air-dry forage required per cow		20,730		21025	27275
8	Pounds of air-dry forage produced per acre		1,300		2000	2000
9	Number of acres per cow and calf		.16		10.5	13.6 ac

*Use your average cow weight and 7 month calf weight and determine the digestible energy requirements from figure 6.
**Use your planned supplemental feeding program to calculate this figure. If grain is fed, use 1.60 megacalories for this fraction.

111

OPPORTUNITIES AND CONSTRAINTS IN INCREASING FARM AND RANCH INCOME THROUGH FOREST MANAGEMENT FOR TIMBER PRODUCTS

Evert K. Byington,
J. Rick Abruzzese

Farmers and ranchers own a substantial portion of the forested land in the U.S.; however, most of the forested land on farms and ranches is not being adequately managed to develop the forest-product potential. Lack of forest management is robbing landowners of needed income and can deprive the nation of vital economic growth. This paper presents an overview of the basic management and marketing information needed by private landowners to start developing the forest product potential on their farms and ranches. Five aspects of forest management and marketing within the farm and ranch context will be discussed:
- Characteristics of forest resources on farms and ranches.
- Factors affecting the value of timber products.
- Considerations in marketing timber products.
- Requirements for timber management.
- Opportunities to integrate timber-product management and marketing into overall farm and ranch management.

FARM AND RANCH FOREST RESOURCES

In 1977, farmers and ranchers owned 117 million acres of U.S. commercial timberland, i.e., forested land capable of producing at least 20 cu. ft. of wood biomass per acre per year. This acreage represents about 25% of total U.S. forest lands (USFS, 1977). Two out of every five farms and ranches have at least some commercial timberland. On the average 9% of all U.S. farm and ranch land is forested but regional variation is considerable. For example, over 30% of the farm and ranch land in the southeastern states is forested (USDC, 1978).

In general, farmers and ranchers are not capturing the potential economic returns from their forested lands. Less than one farmer in ten who owned forested land marketed forest products in 1974. Total gross earnings were only $232 million or about $2.50 per acre of farm and ranch forested land (USDC, 1974). Such low rates of return help explain why the area of forested land owned by farmers and ranchers dropped by nearly 60 million acres over the last 30 years even though the total forested area in the U.S. remained nearly constant (USFS, 1977). However, the $2.50 rate of earnings is misleading since most farms and ranches earned nothing from the sale of forest products, while the 72,000 farmers and ranchers who did sell forest products had gross earnings per farm of $3,200 on the average (USDC, 1974).

Forested land can be a source of additional income particularly when forest management can be integrated into the overall operation of the farm and ranch. Although the value of many forest products is currently down due to the national economic slowdown, the long-term view is favorable. Even now earning can be made from forested land through good management and marketing practices.

FACTORS AFFECTING THE VALUE OF TIMBER PRODUCTS

Forest products continue to be an important part of our economy despite the tremendous increase in the use of synthetic materials. About 13 billion cu. ft of wood was used by the forest industry in 1977 to make a broad range of products including lumber, plywood, particle board, hardboard, and insulation board for use in the construction of homes and apartments; commercial, industrial, and public buildings; utilities and highways; and other construction projects. The rest of the wood was used to produce paper and pulp, shipping materials (pallets, containers, dunnage, etc.), cooperage, pilings, poles, posts, mine timbers, furniture, etc. An additional 1.4 billion cu. ft of wood was used as fuel wood. The nation's forests also supplied Christmas trees, naval stores, pine needles for mulch, landscaping materials, and other forest products (USFS, 1980).

The Forest Service expects the demand for timber from domestic forests to rise from 12 billion cu. ft in 1976 to 25 billion cu. ft by 2030--an increase of 107%. The greatest growth in demand will be for hardwood roundwood, which will increase some 224% between 1976 and 2030. This compares to a 71% rise for softwoods. Much of the increased production to meet this expected demand for forest products must come from privately owned forests (USFS, 1980).

Seventy-two percent of all commercial timberland was privately owned in 1977. More specifically, farmers, ranchers, and other nonindustrial owners control about 70% of the nation's hardwood inventory and 27% of all softwood inventories. Thus farm and ranch forested land is going to become an increasingly valuable resource but even now there are opportunities to earn additional income through the management and marketing of forest products (USFS, 1980).

ECONOMIC VALUE OF VARIOUS FOREST PRODUCTS

Few private landowners realize the value of their forested land, and even fewer are able to fully capture that value through the management and marketing of forest products. The complexity of marketing forest products is largely inherent in the industry and is difficult to change. However, the factors controlling the prices paid for forest products can be understood by landowners and used to help ensure that a fair price is received.

On a specific timber stand, the two factors controlling the value of forest products are 1) the structure of the forest-processing industry that buys the forest products from the landowner and 2) the characteristics of the forest stand including how it was managed.

The Structure of the Forest Industry and its Impact on Timber Prices

The forest processing industry is machine-intensive and, in general, the larger the processing plant, the more economical it is to process a given volume of raw materials. However, large processing plants require major capital investments that only a few companies can make. Also, the larger the plant, the greater the daily flow of raw materials that must be processed. Because of the high cost of transportation of timber, processing plants must acquire most of their raw materials from the surrounding area. The result is frequently a very limited number of potential buyers for the forest manager's output. In fact, there may be no local buyers for some speciality products because local landowners are not producing a sufficient volume to support a processing plant.

The presence of near-monopolies in the timber-processing industry creates problems and opportunities for buyers and sellers of timber products. One problem is the lack of information on current prices being paid for various timber products delivered at the plant gate or on the stump. Acquiring such information from published sources or even by word of mouth is nearly impossible. The lack of public price information helps the forest processing industry avoid violations of antitrust laws and charges of price-fixing. Prices are arrived at through direct negotiations between a buyer and a seller. As long as prices are never made public and competing buyers do not discuss prices, price-fixing charges can be avoided. And as long as sellers can negotiate with more than one buyer, competition is maintained.

Negotiating the price of timber products on a case-by-case basis can cause problems for the landowner with limited experience in forest management and marketing. First of all, the landowner may not even realize that he has a valuable product to sell and never attempt to manage or market his forest resources. In this case, everyone loses. Or the landowner may sell his timber products but for much less than they are worth because he lacks the knowledge to negotiate effectively. In this case, the landowner is the loser.

Farmers and ranchers often have substantial timber holdings, but forestry is not their primary occupation. They can seldom hope to master all of the factors that control prices for timber products delivered at the factory gate, much less those additional factors affecting the value of the unharvested timber stand. Fortunately, the landowner can get professional guidance to help perform these marketing functions along with help in managing the timber stand to get it ready to market. Nevertheless, some general understanding of management principles and the marketing process can aid the landowner in more effectively using the professional resources available.

Site Characteristics Affecting Value of Forest Products

Although negotiations determine the price finally paid for timber products, biological potential and stand management determine the intrinsic worth of the forest resource. Negotiating skills can never compensate for poor management or a poor site.

Biological and management factors influencing the value of the timber stand include:

- Tree species composition and quality. Stands can be mixtures of hardwoods and softwoods, or stands of either hardwoods or softwoods, or a monoculture of just one species. Pines and other softwoods tend to be more marketable than the hardwoods because of their variety of potential uses. However, high quality hardwoods can be much more valuable than softwoods if there is a local furniture or veneer industry to create a demand. In the past, there has not been a market for many hardwood species but with the rapid growth in demand for firewood, these species will increase in value.
- Size class composition. In general, large, quality logs are of greater value than are logs suitable only for pulpwood. A stand with many large pulpwood stems should be studied closely to evaluate rate of growth and local wood prices to determine if greater earnings can be made by allowing the trees to grow to sawlog size. Similarly, sawlog-size trees could be allowed to grow into more valuable specialty products, such as veener peeler logs or poles and pilings. Postponing a sale to allow trees to grow into the next product class does involve risks, particularly if there are very few local markets for the larger logs.

Stands that are homogeneous in size and species are easier to harvest and will bring a higher price in most cases. However, if many larger logs are mixed with smaller ones, or if species are mixed, the landowner may earn more by marketing timber products rather than by marketing the stand as a whole. This may require selling to several buyers who must selectively harvest the various tree classes; or the stand can be valued for each product, with buyer allowed to separate the products.

- Stand accessibility and size. A stand's value increases with ease of access. Rough or wet terrain and heavy underbrush increase the time required for harvesting and may require specialized equipment: such factors will be reflected in the offering of a lower price for the stand. Stands that are identical except for tract size and location can bring significantly different prices. Large, contiguous tracts are more valuable because the logger need not relocate crews and equipment as often as on many small, scattered tracts. Also, the farther the timber stand from the nearest mill, the less valuable the stand.

CONSIDERATIONS IN MARKETING TIMBER PRODUCTS

The marketing of timber products, particularly those still on the stump, is more complex than most other agricultural marketing because the product for sale is not as readily apparent. Forest landowners must first determine exactly what they have to sell before they find potential buyers, negotiate the sale, and finally have the product harvested.

Estimating the Timber Products Available in a Stand

The first step in marketing a timber stand is to determine the quantity and quality of timber products found in the stand. As indicated in previous paragraphs, a number of factors influence the value of a stand and evaluating these factors requires considerable training and experience.

A forester "cruises" a timber stand to indentify the types of timber products available and the "volume," i.e., quantity of each that can be harvested. A description of the stand in terms of accessibility, location, and size is also prepared.

"Cruise volumes" are reported in various units including:

- Thousand board feet (MBF) "log scale" for saw-timber, veneer peeler logs, poles, and piling.
- Cords for chip-n-saw logs and pulpwood.

- Tons for pulpwood and fuelwood.
- By piece or linear foot.

Unfortunately, there are three commonly used "log scales" for estimating volume and they do not give the same results. Doyle log scale will calculate fewer boardfeet in a given tree than Scribner Decimal C, which will calculate fewer than International 1/4-inch. All three scales improve in accuracy as logs increase in diameter at breast height (DBH), which is 4 1/2 ft above the ground. The International scale is the newest of the three scales and it better reflects the advanced sawmill technology that has reduced slab and sawdust waste. Information in table 1 can be used to compare or convert volumes in the three scales if the board foot volume and stand mean DBH are known. Due to the close relationship between weight and volume and the ease of measuring wood weight on the truck, some prices are now being calculated in dollars per ton so that scale conversion will be less of a problem.

TABLE 1. A TABLE OF CONVERSION RATIOS BETWEEN THREE MAJOR LOG SCALES STATED IN TWO-INCH-DIAMETER CLASSES

DBH inches	Doyle to Inter	Inter to Doyle	Scribner to Inter	Inter to Scribner	Doyle to Scribner	Scribner to Doyle
10	2.56	.39	1.325	.755	1.93	.52
12	2.07	.48	1.210	.825	1.70	.59
14	1.76	.57	1.170	.855	1.50	.67
16	1.60	.63	1.130	.885	1.36	.75
18	1.46	.69	1.115	.895	1.31	.76
20	1.36	.73	1.095	.915	1.25	.80
22	1.27	.79	1.080	.925	1.18	.85
24	1.20	.83	1.070	.940	1.13	.89

Source: USFS. Service Foresters' Handbook (1975).

Many timber stands are sold at a fraction of their value because landowners did not obtain an objective appraisal from a qualified source. Since most farmers and ranchers lack the skills needed to cruise their timber they need to seek help from government or private consultants. Many states have forestry agencies that will cruise a stand and develop a timber management plan for the private land-owner. The cruise will produce an estimate of volume and quality for the timber products and usually a description of accessibility, topography, and other related factors. However, state agencies normally do not estimate an economic value for the stand or implement the management plan. Private forestry consultants provide a full range of services from developing and administering management plans to cruising the stand and administering the sale. Fees are usually a fixed annual amount for managing a stand and a percentage of the harvest value for administering a sale.

The consultant's integrity is extremely important and it is frequently advisable to use a local person who has ties to the community.

CONSIDERATIONS IN MARKETING TIMBER PRODUCTS

After farmers and ranchers understand the quantity and quality of timber products and have some idea of the local value of these products, they are prepared to market them to earn the greatest return. Seeking bids from various potential buyers is the key step in marketing.

The Marketing Process

The first marketing step is to advertise that a stand is available and allow buyers to cruise the stand and make bids. The buyers may offer a total price for the standing timber in the stand or offer a set price per unit for each product in the stand. The buyers' cruise volumes and bids must be compared carefully with those volumes and estimated values obtained from a consultant. One landowner in Alabama received several bids from local buyers for a stand and the highest bid was $40,000. But the landowner's consultant had estimated the value to be $90,000. The landowner advertised the stand in some other counties and was ultimately able to get $80,000--longer hauling distances reduced the value from the original estimate.

Care should be taken to ensure that the cruise volume logscale and the payment logscale are the same. Money will be lost if one sells Doyle scale volumes for Scribner scale prices.

Once the highest suitable bid is received, the landowner should investigate the purchaser. Not all buyers are operators of forest-processing plants. Many buyers are middlemen who assemble the timber from many small stands at a woodlot for sale to a large processor. Reputable loggers will protect land or trees that are not to be harvested. Local state forestry representatives and neighbors can be helpful in selecting a suitable buyer.

The preparation of a written sales contract is the next step after a buyer is selected. A written contract is required no matter how small the sale (Stuart and Walbridge, 1978). Points to be addressed in the contract include:
- Definition of the harvest area indicating logging roads, log loading areas, boundary lines, buffer strips along streams and ponds, service roads, internal fences, and restricted areas, if any. Set responsibility for trees that are accidentally removed from adjacent lands; replacement of damaged fence; penalties for removing buffer trees or other trees marked to remain; and damage to stream and pond banks.

- Specify merchantable limits on trees that are to be cut and penalties for merchantable wood that is left on-site. State the requirements that will make a tree a cull. These points are particularly important when a sale is based on scale and the owner is not paid for unharvested wood. Also, specify who is responsible for residue wood since a cost is frequently incurred to clear such wood prior to reforestation. The contract may require that tops, limbs, and cull trees be felled, lopped down, or chipped depending on future management plans.
- Specify weather-related restrictions and conditions in the stand. Wet or dry conditions may expose the site to unreasonable soil compaction, erosion, fire hazards, etc.
- Clearly define the timber's ownership during all phases of the sale--severance, natural resource, and other taxes are accessed against the timber owner at specific times. A knowledge of tax laws is critical. Ownership also influences liability for accidents and losses due to fire, theft, or storm.
- Specify equipment restrictions to control damage, particularly to the soil. There are two important equipment-related factors: type of equipment used and the skill of the operator in using the equipment. Operator damage can be controlled through stiff penalties for careless damage. Equipment can usually be controlled by limiting weight and(or) horsepower--but, in some instances, specialized equipment also should be specified.

Although forest product marketing can be complex and require time and effort, the potential rewards are great. Also, not all forest products require marketing through the forest-processing industry. Firewood, fence posts, and Christmas trees are examples of forest products that the landowner can sell directly.

Examples of Timber Product Prices for Various Locations in the U.S.

As noted, the landowner can have considerable difficulty in estimating the current market value of his timber products. Tables 2, 3, 4, and 5 are examples of prices that were being paid in various parts of the country for sawtimber, pulpwood, firewood, and Christmas trees in early 1982. A review of these tables indicates the tremendous variations in price--not only between locations and products, but also for the same product in the same state. Also note the values of firewood. Firewood may be the best way to enter the timber-products business since marketing and management are less complex and there is a rapidly growing market across the country.

TABLE 2. EXAMPLES OF STUMPAGE PRICES FOR HARDWOOD AND SOFTWOOD
SAWTIMBER IN SELECTED STATES IN 1982.*

State	Species	Scale	Dollars/1,000 board ft		
			High	Low	Average
Connecticut	Mixed hardwoods	International ¼"	150	80	120
	White pine	International ¼"	55	40	48
Ohio	Pine	Doyle	80	40	54
	Poplar	Doyle	150	40	84
	Walnut	Doyle	540	90	311
Arkansas	Yellow pine	Scribner	160	129	153
	Oak	Doyle	75	54	65
Florida	Yellow pine	Scribner	148	70	109
	Poplar	Doyle	126	50	88
	Hardwood	Doyle	68	27	48
Virginia	Yellow pine	Scribner	90	62	76
	Oak	Doyle	79	43	61
Arizona	Douglas fir	Scribner	175	150	170
	Ponderosa pine	Scribner	200	125	170
Oklahoma	Yellow pine	Scribner	101	58	79
	Oak	Doyle	64	45	53
Texas	Yellow pine	Scribner	170	161	166
	Oak	Doyle	58	46	51
	Mixed hardwood	Doyle	53	48	50
Colorado	Ponderosa pine	Scribner	185	120	160
	Lodgepole pine	Scribner	180	100	120
Montana	Ponderosa pine	Scribner	131	32	--
	Spruce	Scribner	84	3	--
	Red cedar	Scribner	65	3	--
Utah	Ponderosa pine	Scribner	140	120	140
California	Ponderosa pine	Scribner	350	120	165
	Douglas fir	Scribner	380	165	170
Washington	Ponderosa pine	Scribner	250	120	170
	Douglas fir	Scribner	375	150	180

* Data obtained from state forestry agencies or printed, public
sources.

TABLE 3. EXAMPLES OF STUMPAGE PRICES FOR HARDWOOD AND SOFT-
WOOD PULPWOOD IN SELECTED STATES IN 1982.*

State	Species	Price per cord		
		High	Low	Average
Maine	White pine	11.00	2.50	5.50
	Tamarack	10.00	3.00	6.50
Minnesota	Pine	-	-	6.80
	Mixed hardwoods	-	-	4.50
Ohio	Mixed hardwoods	8.00	4.00	6.00
Arkansas	Yellow pine	15.00	11.75	14.00
	Mixed hardwoods	4.75	3.25	4.00
Florida**	Yellow pine	24.93	15.71	20.32
	Mixed hardwoods	6.22	3.08	4.68
Virginia	Yellow pine	12.25	8.00	10.00
	Mixed hardwoods	4.00	2.50	3.25
Oklahoma	Yellow pine	14.00	9.00	12.00
	Mixed hardwoods	3.50	2.00	2.75
Texas	Yellow pine	17.50	9.75	14.50
	Mixed hardwoods	4.00	2.00	3.00

* Data obtained from state forestry agencies or printed,
  public sources.
** 1980 prices.

TABLE 4. EXAMPLES OF PRICES FOR DELIVERED FIREWOOD IN
SELECTED CITIES IN 1982. MOST DATA OBTAINED
FROM NEWSPAPER ADS.

City	Species	Unit	Price
Pittsburgh, PA	Hardwood*	Cord**	95-105
Burlington, VT	Hardwood	Cord	90
Milwaukee, WI	Hardwood	Face cord	50
Montgomery, AL	All	Cord	80-100
Little Rock, AR	All	Pickup load	40-65
Orlando, FL	All	Face cord	40
Raleigh, N.C.	All*	Cord	85
Dallas, TX	Mesquite	Cord	125
Houston, Tx	Oak	Cord	150
Kansas City, KS	All	Cord	80
Omaha, NE	Hardwood	Cord	115
Grand Forks, N.D.	Hardwood	Cord	110
Urban areas in AZ	Ponderosa pine	Cord	80
	Mesquite*	Cord	200
Oakland, CA	Hardwood	Cord	120
Riverside, CA	Orange wood	Face cord	80-100
Boise, ID	Mixed	Cord	65-75
Salt Lake City, UT	Pinon-juniper	Cord	79
Spokane, WA	Fir	Cord	70

* Wood has been seasoned.
** Two ricks or face cords in a cord.

TABLE 5. EXAMPLES OF PRICES FOR CHRISTMAS TREES IN SELECTED STATES IN 1982.*

State	Species	Unit	Price			Remarks
			High	Low	Average	
Connecticut	Scotch Pine	Tree	20.00	12.00	16.00	6'-7' & "you cut" retail
Indiana	White Pine	Tree	--	--	8.25	6'-7' & wholesale
New York	Douglas Fir	Tree	10.00	6.00	8.00	6'-9' & on stump
Alabama	Virginia Pine	Tree	--	--	8.00	Wholesale
Mississippi	Scotch Pine	Tree	22.00	18.00	20.00	8' & "you cut" retail
Missouri	Scotch Pine	Foot	1.50	1.00	1.25	Wholesale
California	Scotch Pine	Foot	3.50	1.50	2.50	Cut & choose
California	Douglas Fir	Foot	1.20	0.70	1.00	Cut & choose
New Mexico	Douglas Fir	Foot	--	--	1.50	Wholesale
Texas	Virginia Pine	Tree	--	--	8.00	Wholesale
Texas	Virginia Pine	Foot	3.00	2.50	2.75	"You cut" retail

* Data from state forestry agencies and local producer associations.

REQUIREMENTS FOR TIMBER MANAGEMENT

Good marketing practices help ensure that landowners receive full value for their timber products, but one can only sell what one has, and product availability is controlled by production management. Unfortunately, many people think that high-quality timber stands just "happen." Early settlers did find superb timber stands of timber all across this country but these stands evolved over many centuries and most have been cut over. Modern society cannot wait for hundreds of years for natural processes to regenerate the forests. Timberlands must be managed to increase productivity while shortening the time for forest regeneration.

Forest management is similar to other agricultural activities that grow plants for food and forages. Forest productivity per given area and period of time can be increased through insect and weed control, use of genetically improved species, soil and water conservation, fertilization, and well-thought-out management strategies. The main difference is that the time between planting and harvesting is decades instead of months.

Forest management is a highly developed profession based on knowledge gained from years of experience and research. Farmers and ranchers seeking to improve the productivity of their forested lands should seek professional help in developing and implementing management plans. The forestry professional can develop many alternative timber management plans for a site, so landowners must have an idea of their objectives in timber management and of how forestry will integrate with other farming and ranching activities. There are two broad strategies to forest management: even-age stand management and uneven-age stand management. Each strategy has its advantages and disadvantages depending on the broader land-management objectives of the farm or ranch.

Even-Age Timber Management

The objective of even-age management is to produce a forest dominated by trees in one age class. Even-age stands are usually created by man when reforesting an abandoned field or a forest that has been clearcut. Such stands also can occur naturally following a catastrophe such as fire or high winds. There are four basic phases of even-age management:

1. Site preparation is necessary prior to establishment of most tree species. The land can be prepared through mechanical, chemical, fire, and(or) biological methods. Each method has its own advantages and disadvantages, so landowners must consider which method or combination of methods will best meet local conditions. In general, mechanical methods are

giving way to the use of chemicals and fire because of expense and damage to the soil. Use of livestock, particularly goats, has potential as a method of site preparation but because goats have not been extensively used, knowledge is limited about this method. How intensive site preparation must be depends in part on the method of tree regeneration that will be used.

2. The importance of <u>regeneration</u> cannot be stressed too strongly. The seeds or seedlings planted today will largely determine the timber products that can be marketed in the future and how long the growing cycle will take. Regeneration can be done naturally or artificially.

Natural regeneration uses a seed source on the site or from adjacent land. Seed tree and shelterwood seed sources require planning before a harvest operation and care during harvest. Genetically superior trees should be left as seed-source trees. These trees should be adequately marked so that they are not accidentally harvested, and care should be taken not to damage them during harvest and site preparation.

Natural regeneration is less expensive than artificial regeneration and requires less site preparation (Williston and Balmer, 1974). However, genetically improved varieties or new species cannot be introduced, and frequently the seed trees cannot be harvested following regeneration; therefore some of the most valuable timber is lost.

Artificial regeneration is more expensive than natural regeneration but offers much more control over stand development. Direct seeding is done from the air or the ground. Seeds are treated with fungicides and animal repellents and spread over a prepared site. This method is relatively inexpensive but can have mixed results; for example, when rain washes seeds into dense clumps that later require expensive precommercial thinning (Williston and Balmer, 1977).

Seedlings can be planted either mechanically or by hand to ensure proper spacing and to improve the certainty of regeneration. Mechanical planting uses a tractor-pulled machine to plant bare-root seedlings or cuttings. Mechanical planting is faster and less expensive than hand planting, but it can only be used on intensively prepared sites or old fields on mild terrain.

Hand planting is the most commonly used procedure because it is so adaptable to variations in site conditions. Bare-root seedlings, containerized seedlings, and cuttings can be hand planted using very simple tools; but this method is more expensive than machine planting and spacing is less uniform.

The tree spacing selected at the time of planting has important implications for later cultivation and harvesting operations. The distance selected between rows and trees in the rows should be based on desired product, growing time, and labor input. Widely-spaced trees will develop full crowns, retain lower limbs, have a large diameter but be short in height, and will be largely unmarketable unless artificially pruned during growth.

By contrast, closely spaced trees elongate faster to compete for light and have few lower limbs, smaller crowns, and less diameter. Closely spaced trees also compete below the ground for water and nutrients. Such stands must be thinned periodically to ensure that trees continue to grow (Smith, 1962). With proper planning, the tree biomass removed during thinning can be sold for pulpwood or as fence posts and rails. However, if a stand's initial spacing is too close, the first thinning may not produce products of commercial value (Balmer and Williston, 1973).

Sullivan and Matney (1980) state that wider spacings are now being recommended in pine plantation establishment. In the South, for example, 6 x 8-ft or 7 x 8-ft spacings are used for integrated management to produce pulpwood, poles, and sawtimber over the life of the plantation. If sawtimber is the primary goal, 8 x 8-ft or more often 10 x 10-ft plantings are used. The number of seedlings required per acre for various spacing patterns is indicated in table 6.

A management plan should be based on a spacing that will produce desired products with a minimum input of material and labor costs in a reasonable time period (Feduccia and Mann, 1976). Usually this calls for a spacing that will negate the need for a precommercial thinning, but with the trees sufficiently close to eliminate the need for pruning. However, such a stand will require periodic thinning to prevent stand stagnation.

TABLE 6.   THE NUMBER OF TREES PER ACRE FOR GIVEN
DISTANCES BETWEEN ROWS AND TREES

No. of ft between trees	No. of ft between trees	Number of trees per acre
6	6	1,210
6	8	907
6	9	807
6	10	726
8	8	681
8	9	605
8	10	545
8	12	454
9	9	538
9	10	484
9	12	403
10	10	436
10	12	363
10	15	290
20	20	109

3. A timber stand must be _cultivated_ over its
   growth cycle just like any other crop. Thinning and pruning are only two of the important
   activities in forest cultivation. Cultivation
   to ensure stand protection, "weed" control, and
   nutrient availability also are important.
   The greatest threats to a timber stand are
   fire, disease and insects, and wind. The
   landowner can do little to control wind damage
   beyond using proper thinning practices. Thinning also helps control forest pests by maintaining tree vigor. However, if pests do
   become established in a stand, quick action
   must be taken to control the spread of the
   insects or disease. Therefore the stand must
   be regularly monitored to ensure rapid and
   accurate identification of the initial outbreak. The most frequent treatment is to
   harvest or destroy infected trees. Chemical
   control can be used but is expensive and often
   gives mixed results.
   Fire prevention, control of unwanted vegetation (weeds), and nutrient availability are
   all closely related. Controlling unwanted
   vegetation, particularly shrubs, noncommercial
   tree species, and herbage, reduces 1) feed
   build-up that can carry fires and 2) competition for water and nutrients between the

desired trees and the other vegetation. Interestingly, fire is one of the best tools for controlling this unwanted vegetation. Controlled burns are frequently used, particularly in even-age pine plantations to reduce the build-up of unwanted plant material. The burning not only kills unwanted plants but also releases nutrients back into the soil (Hough, 1981). However, fire is a tool that requires considerable skill to use safely. Chemicals and mechanical tools can also be used to control vegetation but are more expensive. Additionally, livestock is a useful vegetation-control tool if herbage and low shrubs dominate the understory vegetation.

4. Harvesting is usually thought of as the final step in even-age management, although the various thinnings performed in cultivating the stand often produce a commercial product that can be marketed. The final harvest or "clearcut" marks the end of the cycle and the land can be reforested or shifted to another use. The landowner should decide how the land is to be managed prior to the final harvest so that trees can be left to support natural regeneration or other land-use values such as wildlife or recreation.

Uneven-Age Timber Management

The objective of uneven-age management is to manage a forest with many age classes of trees. The forest can be managed as a mix of all-aged trees or as a mix of even-aged stands that represent all ages. The main advantages of uneven-aged forests are steadier income flow after the forest is established; greater ability to support other activities such as wildlife and recreation; and better protection of fragile sites and watersheds; and makes it possible to maintain some hardwood forest types. The main disadvantage is that uneven-age stands require greater management and are, therefore, less profitable.

The basic management practices that apply to even-age stands also apply to uneven-age stands, but all may be required at the same time in the uneven-age stand.

INTEGRATING TIMBER PRODUCT MANAGEMENT WITH OTHER FARM AND RANCH ACTIVITIES

Farming and ranching and forestry need not be an either/or situation. A well-managed forest can increase farm and ranch income through the marketing of timber products while providing improved opportunities for recreation and wildlife (Duerr et al., 1975). Recreation and wildlife resources can improve the quality of life of the rancher and farmer and also serve as a source of additional income (particularly near large urban areas). Often, another important benefit of good forest management is an increase in forage production--forage that can be used by livestock and wildlife. Forest forages are one of this nation's largest, untapped resources. Developing and utilizing this forage potential demands good management of both the livestock and the forests. An increased forage supply may more than justify improved management. (The principles of forest-grazing management are discussed in another presentation in this book.)

Farm and ranch forests are a valuable asset to the landowner as well as to the community and the nation. Taking full advantage of this asset requires additional effort, but the rewards can be great.

REFERENCES

Balmer, W. E. and H. L. Williston. 1973. The Need For Precommercial Thinning. Southeastern Area State and Private Forestry, Atlanta, Georgia.

Duerr, W. A., D. E. Teeguarden, S. Guttenberg and N. B. Christiansen. 1975. Forest Resource Management. Oregon State Univ. Book Stores, Inc.

Feduccia, D. P. and W. F. Mann, Jr. 1976. Growth following initial thinning of loblolly pine planted on a cutover site at five spacings. USDA Forestry Service Research Paper SO-120. Southern Forestry Experiment Station, Pineville, Louisiana.

Hough, Walter A. 1981. Impact of prescribed fire on understory and forest floor nutrients. USDA Forestry Service Research Note SE-303. Southeastern Forestry Experiment Station, Asheville, North Carolina.

Smith, David M. 1962. The Practice of Silviculture. John Wiley & Sons, Inc., New York.

Stuart, W. B. and T. A. Walbridge, Jr. 1978. Contract Terminology in the Pulpwood Industry. Industrial Forestry Operations Program, School of Forestry and Wildlife Resources, VPI & SU, Blacksburg, Virginia.

U.S.D.C. (United States Department of Commerce). 1978. 1978 Census of Agriculture. Bureau of the Census, Washington, D.C.

U.S.D.C. (United States Department of Commerce). 1974. 1974 Census of Agriculture. Bureau of the Census, Washington, D.C.

U.S.F.S. (United States Forest Service). 1980. An Assessment of the Forest and Rangeland Situation in the United States. Dept. of Agriculture, Washington, D.C.

U.S.F.S. (United States Forest Service). 1977. Forest Statistics of the U.S., 1977. Dept. of Agriculture, Washington, D.C.

Williston, H. L. and W. E. Balmer. 1977. Direct Seeding of Southern Pines: a Regeneration Alternative. Southeastern State and Private Forestry, Atlanta, Georgia.

Williston, H. L. and W. E. Balmer. 1974. Managing for Natural Regeneration. Southeastern State and Private Forestry, Atlanta, Georgia.

112

PRINCIPLE OF LIVESTOCK GRAZING ON FORESTED LAND: WHAT WE KNOW AND DON'T KNOW

Evert K. Byington

This presentation has three objectives. First, to increase landowners' awareness of the large, underutilized livestock-forage resource that exists on the nation's forested lands. Second, to briefly examine obstacles that must be overcome before livestock producers can efficiently utilize forest forages. Third, to review general principles of management that can aid in the development and use of forest forages, particularly on farms and ranches.

FOREST FORAGES AS A NATIONAL RESOURCE

Over 475 million acres, or about 25% of the continental U.S., is commercial forest land, which is defined as forested land (or land that will be reforested) with the potential to produce at least 20 cu ft of wood per acre per year. This land is distributed across the U.S. with the largest concentration in the southern states.

Region	Commercial forest land (millions of acres)
Northern states	170.8
Southern states	188.4
Rocky Mountain states	57.8
Pacific Coast states	59.6

About 72% of the U.S. commercial forest land is privately owned with farmers and ranchers owning 116 million acres compared to 68 million acres owned by the forest industry (USFS, 1977).

America's forested land provides many goods and services essential to our national well-being. The most obvious products coming from forested lands are timber products that include lumber; plywood and other veneer products; pulp and paper; pilings, poles, and posts; and shipping materials such as pallets and dunnage. But the forests also provide other equally important but less obvious products: much of the nation's clean air and water, recreation for millions of people, and hunting and fishing.

A major challenge facing the world is how to maintain the forested land while increasing food production. One possibility for resolving this dilemma is to manage the forest understory as a source of forage for livestock. Such an approach would result in greater food supplies while maintaining other forest values.

U.S. forested land has the potential to produce substantial amounts of range forages, i.e., native grasses, forbs, and browse. Forest land has its highest forage-producing potential when there is little or no true cover. Such a condition occurs following a total timber harvest or a natural disaster such as a fire. Such forest land temporarily cleared of trees or with trees in the early stages of growth is called "transitory range." Average potential production for most transitory range is between 1,000 and 2,000 lb of native forage per acre (USFS, 1980). As the trees grow and the forest canopy forms, forage production can drop to less than a tenth of this value. Since only a portion of the forest land is transitory range at any given time, and much of the forest land is not available for grazing because of ownership patterns, it is difficult to estimate how much forage could actually come from the forests. One study estimated that the privately owned forests of the eastern U.S. could supply between 35 and 90 million AUMs (Animal Unit Months) depending on management intensity; at the same time the National Forest System could supply 16 million AUMs of forage through intensive management of three of its larger and more productive forest types (Byington and Child, 1979).

Forested lands are the greatest untapped range-forage resource in the U.S. Forested lands supplied only 22 million AUMs of forage in 1976 even though a fourth of the country is forested. Twenty-two million AUMs were only 10% of the range forage used that year and less than 3% of the 976 million AUMs of grazed forages provided from all sources including pasture and crop lands (USFS, 1980).

If the nation's forests can supply so much forage why isn't this resource being utilized? Unfortunately there is no simple answer because several factors are involved. Among the more important of these are ownership, lack of management, and availability of alternative livestock forage and feed sources.

As previously stated, most of our forested land is privately owned. Thirty years ago farmers and ranchers owned half of the nation's privately owned forests, but today they own 56 million fewer acres of forest. Only a small portion of the forest acres given up by farmers have gone to the government or timber companies. Some 44 million acres have gone to rural home owners, nonforest industry corporations, real estate interests, and people who moved to the city and are absentee landowners. Thus, the people most likely to practice forest livestock grazing (farmers and ranchers) no longer have ownership of millions of acres of forests, and the new owners frequently are opposed to forest grazing.

However, millions of forest acres are still owned by farmers and ranchers who graze livestock on them so that low forage off-take from forested lands is only due in part to ownership patterns. In 1978 farmers and ranchers were running livestock on 48 million acres of their forested land (Census of Agriculture, 1978) and millions of additional acres leased from the Forest Service and the forest industry. The problem is that much of this grazing received little or no management input and in some cases resulted in forest destruction.

HISTORY OF FOREST FORAGE USE

Forest livestock grazing is not a new idea. Ponce de Leon brought cattle into the southern states in 1521 and by 1650 Spanish "ranchos" were established in northern Florida with cows running in the pine forests. Early settlers in the Ozarks found large, fairly open woodlands and prairies with abundant bluestem grasses and found livestock grazing to be the most profitable occupation (Byington, 1980). Much of the western sheep and cattle industry evolved around summer grazing in the mountain forests. Thus, forest grazing in the U.S. originated during the era of open range and cut-and-run forestry. There was no concern for integrated management that would ensure sustained livestock grazing and timber production on the same land.

But by the 1930s, most of the nation's original timber stands were gone and reforestation programs began, particularly in the South. The forest plantation system and open-range livestock grazing quickly came into conflict. Further north in the hardwood forests of the Midwest, foresters and livestock producers were also coming into conflict. As more and more of the prairie went under the plow, farmers moved their livestock into the trees and stocking rates became so high that the land was often stripped of vegetation and erosion began (Byington, 1980). Unfortunately these conflicts continue in both the South and Midwest today but at a greatly reduced level (Croft and Cutshall, 1980; Shiflet, 1980).

The forest-livestock conflict was not resolved through better management to integrate the two land uses. Instead the two land-use activities separated and went their own way, particularly in eastern U.S. The abandonment of millions of acres of farmland before and during World War II, combined with the introduction of exotic grasses and cheap fertilizers, made it possible to establish and maintain the pasture resource that now supports the majority of the nation's cattle. The demand for range forages was further lessened by the huge grain surplus of the postwar years that provided most of the nation with an abundance of low-cost livestock feeds.

The lack of a need for more forages may be the main reason there has been little effort to fully develop and manage the forest's forage potential in harmony with timber production. But what about the future? Will we need the forest forages to support tomorrow's livestock population?

FUTURE FORAGE DEMAND AND SUPPLY

Long-term population growth, higher costs for feed grains, and consumer demand for leaner beef are expected to increase U.S. demand for grazed livestock forages. USDA has projected that the demand for beef, veal, mutton, and lamb will increase from 27 million lb carcass weight in 1976 to between 35 and 49 million lb in the year 2030. This would require that livestock forage grazing increase from 976 million AUMs to between 1,400 and 1,700 million. The increased demand for grazed forage would be greatest in the South where there is the largest undeveloped forage-resource base (USFS, 1980).

Increasing forage production will require that all forage resources from range, forest, pasture, and crop lands be more fully utilized. The most straightforward way to increase forage production is through more intensive management of the nation's 134 million A of pastureland and 414 million A of range (Soil Conservation Service, 1981). Less than one in five of the pasturelands was being fertilized in 1978, so that forage production was dramatically increased on existing acreage by increasing management inputs. There are two potential problems with this approach. First, 76 million A of the pastureland can be used for crop production with minimum effort; if demand for crops continues to grow, many acres of pastureland will be lost. Second, most of the management inputs for increased pasture production are energy-intensive and use fertilization and irrigation. Thus, our pasture resource could be caught between higher energy-related input costs and decreasing area.

By contrast, U.S. rangelands can continue to supply forages at current levels or increase production with more intensive management. The Forest Service (USFS, 1980) estimates that range forage production on the nation's 414 million A of rangeland could be increased by 150 to 200 million AUMs. The relatively small increase for such a large area is due to the arid nature of most rangeland that limits potential productivity.

If conditions further reduce pasture acreage and it is not economically feasible to use high-energy inputs to increase production on the remaining pasture, then the untapped forage potential of the forested land will be needed. The prospect of increasing the use of forested land for food production is an important issue to the nation, to the individual landowners who own the majority of the forest land, and especially to the farmers and ranchers who must manage the livestock and forage resource to increase production. But, do we know enough about managing livestock production on forest forages to develop this largely untapped resource in an economically efficient and environmentally sound manner?

ASSESSING THE TECHNICAL KNOWLEDGE AVAILABLE FOR MANAGING FOREST FORAGE-PRODUCTION SYSTEMS

During the 1970s, the Department of Agriculture conducted several assessments of the nation's forage resources produced on range, forest, crop and pasture lands. One of those assessments was a cooperative research program between Winrock International, U.S. Forest Service, Agricultural Research Service, and the Soil Conservation Service. The objective of this research program was to assemble the information available on forest forage utilization to determine if it was adequate to aid landowners in developing this resource. Emphasis was put on the southern states because they have the largest forest-forage potential.

The findings of this research program were mixed. Although most of the nation's forests have been grazed by livestock at one time or another, much of the forest grazing was opportunistic with little regard for other forest uses. Paradoxically, a sizable amount of forestry and wildlife research has been directed at proving forest grazing unworkable.

There have been two notable exceptions to this general pattern. The Forest Service has maintained small but active research programs in forest grazing on the pine forests of the intermountain West and the southern Coastal Plain. Also both regions have documented examples of forest forage management that illustrate workable production systems that have integrated forestry, livestock, wildlife, and other forest uses. Unfortunately, these research findings and management experiences reveal that forest forage management is relatively complex. There are many options for the production of multiple products and numerous approaches to production management that have evolved out of the need to adapt to local social, economic, and biological conditions.

Several factors make it difficult to generalize about forest forage-production systems based on the work in these two regions. Three of the most important are:

1. There are two philosophical approaches to land management and both find application in forest forage management: the agronomic approach and the natural resource approach.

   The agronomic approach is based on the use of many mangement inputs into a unit of land to maximize the production of the desired product. This approach uses fertilizers, irrigation, pesticides, improved genetic material, and machinery to elicit the full potential of the land to produce the desired product. The agronomic approach is normally used on the more productive soils found in favorable climates so that per acre yields cover the cost of the inputs. A well-managed, irrigated, and fertilized stand of Bermuda grass producing 4 or 5 tons of hay per acre is an example of the agronomic approach.

The natural resource approach focuses on the management of natural plant communities to produce a mix of desired products. Physical inputs, such as fertilizer or improved genetic material, are limited by lack of profitability. However, management can be very intensive by using natural forces to regulate production. An example of the natural-resource approach is using grazing-management systems to produce livestock and wildlife on a sustained basis.

Traditionally, forest management has followed the natural-resource approach, but in recent years more and more land, particularly privately owned land on the more productive sites, is being managed using the agronomic approach. This shift is forcing a reevaluation in the approach to forest forage utilization. Forest forages have traditionally been viewed as a "range" resource to be managed using the principles of range management. This approach is still the most suitable, particularly in the more arid and mountainous locations in the West. But on more productive sites, such as those found in the South, more intensive management is possible. For example, even-aged pine plantations consisting of slash pine growing in nearly pure stands of Bahia grass (that is fertilized yearly) are highly productive and managed more like an Iowa cornfield than rangeland.

In many locations and for many landowners, forest forage management should be based on a mixture of these two approaches to land management. Unfortunately, our research, education, and extension activities tend to be organized along these two separate approaches. The first step many landowners must take is to break down the compartmental barriers in their own minds and start thinking as land managers and not just as foresters or livestock producers. The next step is to contact the various specialists to get as much information as possible on both approaches and then combine the best of both to meet your own needs.

2. Land ownership patterns and objectives make it difficult to formulate a general management strategy for an area. For example, a large timber company with a million acres of forested land will require a very different strategy from the farmer with 100 acres of woods on his 260 acre farm. Even two farmers with 100 acres of forest land will need different strategies if one farm has its land in one block while the other farmer's forest land was in small, scattered stands. A farmer's age will also affect the strategy pursued. For example, a farmer in his 60s would likely be less interested in planting trees that are going to take 30 years to grow than would a man in his 20s.

    One of the key factors influencing many landowners in their management of forest forages is their objective for wildlife. The more important wildlife values are, the more important the natural-resource approach becomes, and in most cases the lower the carrying capacity for livestock.

3. Variations in such ecological factors as soils, terrain, climate, tree and forage species, and animal species make it even more difficult to formulate standard management strategies for forest forage use. Even in the same county, one landowner can have a very difficult resource base to work with. This variation is particularly important when considering forest forage management in hardwood stands as contrasted to pine stands. Most research and intensive management experience has been in pine forests. Much of this work is not applicable to hardwood forests.

Despite the limited amount of information and the limitations in using it to solve specific management problems, there is some good news. The work performed in the South and intermountain West has demonstrated that it is both technically and economically feasible to combine forestry and livestock production, particularly in pine forests. This work has also amassed a considerable amount of information on how to manage such integrated production systems. While the specific management details are mainly useful only to the landowners where the research and management were performed, they are useful to aid landowners to develop a strategy for using their forest forage resources.

GENERAL PRINCIPLES IN DEVELOPING AND MANAGING FOREST FORAGE POTENTIAL

Development of the Forest's Forage-Producing Potential Can Enhance Other Land Uses and Create New Production Opportunities on the Farm and Ranch

There is evidence that many farmers and ranchers do not actively manage their forested lands; thus the forest is untapped land that could be used to increase farm income. Forest management, particularly for livestock forages, can be an integral part of the overall farm and ranch operation by providing forages to increase livestock numbers, freeing land for more valuable uses, utilizing surplus labor and equipment, and enhancing other land uses.

Examples are many. Livestock grazing in forests with heavy undergrowth opens up the understory to improve hunting and reduce fire hazards. By shifting summer grazing to forested areas, additional livestock numbers can be put on the freed pasture, or pastureland can be used for hay production or even converted to cropland. Many forest-management practices can be performed in the winter in the more southerly states so that labor and equipment used in the summer for crop production can be shifted to the forest.

Unfortunately, some farmers and ranchers have used the hardwood forests of the Midwest to an excess. The grain farmers who have large quantities of crop residues available for winter grazing use their small woodlots to hold their cattle herds during the summer months and overstocking is common. This results in serious damage to trees, forages, and the soil.

Forest Forage Resources Can Seldom Be Fully Used in Isolation From Other Forage Sources

This principle mainly applies to eastern farms and ranches and forested land in the western states. In these cases forest forages are usually not adequate to support the livestock enterprise on a year-round basis. Generally, forest grazing depends on native forages and has many of the characteristics of rangeland grazing. Production per unit area is low, and forage quality and quantity decline during the winter months.

Forage quality and quantity problems in winter are solved similarly on both range and forested lands. Supplemental feed and (or) grazing are provided. This is done by moving the livestock to other grazing locations such as annual-grass pastures in the South or lower elevation ranges in the West, or by bringing supplemental hay or feed concentrates to the animals.

The low forage production per acre is solved in the West by having access to vast areas of rangeland. Ranches with thousands of acres of grazing land are the norm. However, the situation is different on most farms in the eastern half of the nation. The typical farm in the South has 100 A to 200 A of forested land on which it can only support 5 to 20 head because of its low carrying capacity. For example, 10 A to 20 A of longleaf forest on the Gulf Coast is required for each cow-calf unit/year. Therefore pasture or crop grazing land must be used to increase livestock numbers to an economic unit. As a result forest grazing is usually just one component in the livestock enterprise's forage picture.

Utilization of Forest Forages Should Be Managed in Conjunction With the Development of Other Forest Products and Services

The philosophy of land management that stresses maximizing the production of a single product on a piece of land has blinded many landowners to the advantages of managing land for multiple products. The logic behind product maximization per unit area is that this approach will maximize income. The fact is that profit maximization is seldom the goal of farmers and ranchers in forest management.

Timber products are generally the most valuable products that can be grown on forested land, yet many farmers and ranchers do not manage to take advantage of them. There are several explanations for this seemingly irrational behavior. The main problem is cash flow. Forestry can result in sizable earnings, but many farmers and ranchers would rather eat a little everyday than wait 30 years for a tree crop to mature so that they can get one big paycheck. Another problem is that forestry does not appear related to other farm activities. Why learn a new set of management practices just to manage a few acres that produce one crop of trees in a life time?

The solution according to the product maximization approach would be to clear the forest and establish pasturefor possible income. But this ignores the noncommercial values the landowner may be getting from the forest if he hunts or enjoys a quiet walk in the woods.

Fortunately, landowners need not be trapped by the dilemma of single-product maximization from forested land. Forested land can be managed to produce several good services simultaneously. Trade-offs are involved. Each of the goods and services cannot be maximized. Some timber production must be sacrificed to ensure a steady supply of forage for livestock. And some livestock forage and timber must be forfeited to maintain wildlife habitat, but all three can coexist together. However, all of this comes at a cost. A high level of management is continuously required.

Multiple Product Management on Forested Land Should Focus on Interactions Between the Major Plant, Animal, and Abiotic Components

A frequent criticism of the livestock grazing industry is that many farmers and ranchers have concentrated their efforts on the animal's well-being and have not given adequate attention to the forage resource and management of the interactions between animal and plant. Past abuse of western rangelands is the most striking example of failure to adequately balance the grazing animal and the grazing resource. Modern rangeland management illustrates the benefits of understanding how animal, plant, and the physical environment of soil, terrain, and climate interact to determine the long-term productivity of the land. Managing such complex interactions does take more knowledge and greater effort but our better ranchers realize there is no other way on our rangelands.

The interactions that must be managed in forest grazing include many of those in range management plus the tree component. If the forest manager is to fully develop and protect the land, it is necessary for him to manage the following.

Forages and trees. Trees and forages compete for the essential factors for plant growth--sunlight, water, nutrients, and space. During the early stages of forest regeneration, the forage plants, particularly perennial grasses, have the advantage over the young tree seedlings. However, once the trees are fully established and the forest canopy forms, the forages are at such a disadvantage that production can drop to such low levels that grazing is not economical. Maintaining equilibrium between the overstory and understory requires careful management of forage-tree interactions--interactions that change over the life cycle of the forest.

Grazing animals and trees. The interaction between trees and grazing animals can be very positive or a total disaster, depending on management. The tree canopy can provide shade for animals in the summer and protection during the winter. Also trees can extend the grazing season for a few weeks in the fall by protecting forage from early frost. In turn, the grazing animal helps control the build-up of forage and other plants on the forest floor. This reduces the competition between plants and trees for water, nutrients, and sunlight during tree establishment and reduces the danger of fire, because less fuel is available.

On the negative side, grazing animals can seriously damage and kill trees, particularly during establishment. Most of the conflict between the forester and livestock producer stems from the destruction of tree seedlings by livestock. Control of stocking rates, season of grazing and use of livestock feed supplements can substantially reduce livestock damage to young trees to acceptable levels, particularly in pine regeneration.

Livestock-wildlife. The forest is the natural habitat for a variety of wildlife and many of these animals depend on the grasses, forbs and browse growing on the forest floor as their food source. Managing the forest for livestock forages can either have a positive or negative impact on wildlife. Year-round forest grazing at high stocking rates with livestock that utilize the same forages as local wildlife can deplete the wildlife's food supply. For example, heavy cattle grazing on mountain meadows can reduce grass availability for elk; heavy goat grazing in forested land can reduce browse availability for deer in the winter. Grazing systems should be selected that balance livestock and wildlife objectives.

On a more positive note, livestock management can improve habitat and feed sources for many wildlife species. Maintaining a more open forest favors the increased production of forages for both livestock and wildlife so a well-managed forest grazing program may permit substantial increases in both. Also, certain game birds such as quail and turkey prefer more open forests and can move into previously unoccupied areas following forest modification for grazing. Grazing improvements such as water developments, and salt and supplement feeders benefit wildlife as well as livestock.

A Broad Range of Tools are Available to the Landowner to Manage Forest Grazing

Forest grazing involves a combination of forestry and forage production and use. Forest grazing managers have all of the tools of forestry, range, pasture and animal science and management at their disposal. Although an individual would have to spend a lifetime mastering the use of these tools for a particular location, the greater the manager's knowledge of these tools, the greater the chances for efficiently solving problems and capturing opportunities.

Many of the more important tools available are for vegetation control, and they can be placed into one of four categories:

Fire. Controlled burning is the oldest management tool for forest forage management and was used by the Indians to keep the forest open to improve forage production for wildlife and to facilitiate movement for hunting and travel. Fire continues to be an important management tool particularly in the fire-adapted, yellow pine forests of the southern U.S. Fire removes brush, pine needles, and unused forages to improve access to new growth and facilitiate animal handling. Burning warms the soil in the early spring and releases nutrients to help promote early spring grazing. Controlled burning also reduces wildfires by reducing fuel buildup, and aids in controlling insect pests.

Fire is a relatively inexpensive tool, but does have several disadvantages. Many tree species, particularly hardwoods, are killed easily by fire, so fire cannot be used if these species are desired. Smoke from fires can be a problem and limits the use of controlled burning near urban areas. Finally, the use of controlled burning requires considerable skill--without adequate preparation, a controlled burn can turn into a wildfire that can damage surrounding ownerships.

Mechanical. There is an almost endless variety of mechanical tools and vehicles, ranging from the ax to the bulldozer, used in forest and livestock management. Many of the more important tools are small and hand operated. These tools have the advantages of being easy to use and maintain, inexpensive to purchase, and usable on steep or wet terrain. However, most of these tools require high labor inputs and can be overly expensive on large land areas.

Machine-powered tools and vehicles have become the backbone of the forest industry for planting, harvesting, and transporting forest products. Machine-powered vehicles are also important in transporting equipment, feed, and animals in the livestock industry. These tools and vehicles save time and labor, and as long as energy was inexpensive, they found broad application. However, with higher energy prices, many operations can no longer be economically performed with machine-powered tools and vehicles. Brush control is seldom affordable any longer by mechanical means over large land areas.

Chemical. Chemical herbicides for controlling unwanted vegetation to increase tree and forage production became the main replacement for mechanical methods during the 1970s. However, environmental concerns and higher energy costs have made chemicals a less useful tool in many conditions. Considerable skill is required to use chemicals both safely and effectively.

Biological. For a time it appeared that cheap energy in the form of mechanical and chemical tools would eliminate the need for biological tools for controlling vegetation and transporting products. Now one of the "new" developments is the promotion of cattle and goats as brush-control tools in forestry. Perhaps, in a few years, mules will be rediscovered as a vehicle for moving logs in rough terrain just as cattle drives are being reintroduced in some locations to move cattle at lower costs.

Factors that should be considered in selecting vegetation control tools include cost, user experience, terrain and other local features, environmental constraints, and availability. But perhaps the key factor is the management strategy in which the tool is to be applied. In turn, the management strategy should reflect the landowners overall goals and objectives.

Landowners Require a Long-Term, Comprehensive Management Strategy for Forest Grazing and Other Forest Products to Ensure all Objectives are Adequately Met.

Simultaneous production of timber products, livestock forage, wildlife, and other goods is possible--with management. Multiple-use management is complex and involves many trade-offs among alternative uses. The management strategy, with its associated tools, formulated to manage the forest and livestock will determine the nature of these trade-offs. Thus, landowners should select strategies that most nearly achieve their overall objectives for their land.

The management strategies must be long-term and sufficiently flexible to adapt to changing conditions. In nearly all cases, timber-product cycles run at least 20 years, so the overall management plan must be for at least this length. Once the timber cycle begins, the landowner has lost a number of options for shifting the strategy to managing for an alternative set of trade-offs. However, there are still some opportunities to build in flexibility to shift the product mix and these should be identified in the plan for the management strategy. For example, planting space between trees can be selected to allow the landowners several opportunities to control the mix between pulp wood and sawlogs, and the amount of forage available. The more the stand is thinned to promote sawlog growth, the more forage there will be under the trees.

The management strategy must contain at least two components, one for forest products and one for livestock. The forestry component must indicate species of trees to be managed, timber products to be produced, management approach (even-age vs uneven-age stand), etc. (An overview of forestry management is presented in this volume in another article by the author, so forest management is not discussed in detail here.) The livestock component must contain the same information required in any livestock operation plus information to manage interaction with other forest products, particularly timber products. Attention must be given to adjusting stocking rates of livestock per acre as the forest matures and forage production drops. Plans must also be made for handling livestock in wooded areas.

The management strategy should also contain plans for other components indicated by the landowner's objectives. Wildlife and outdoor recreation are commonly included.

Finally, forest management strategy should reflect the management of the rest of the farm or ranch. For example, row-crop production can provide a source of winter feed for livestock, but on the other hand the landowner may have little time during spring planting or fall harvest to manage the livestock grazing in the forest.

SUMMARY

Forages produced on forested lands are one of this nation's largest untapped resources for food production. Farmers and ranchers own much of our forested land and can develop this forage resource while they manage the forests for other products including timber.

Forest grazing has been practiced for generations but generally with little regard for other forest values. Research and management experience have demonstrated that forest grazing need not be detrimental and can even benefit other uses. However, a very high level of management is required.

The basic principles for efficiently managing forest grazing in a multiple-use framework are available, particularly for pine forests. However, many specific management details are not available and landowners may have to work out day-to-day management practices on their own.

Multiple-use forest management is complex and will require considerable effort and imagination on the part of the landowner. But the results can be greater earnings, economic diversity, and a sense of accomplishment.

REFERENCES

Byington, E. K.   1978.   Livestock grazing on the forested lands of the eastern United States.   Final report to Office of Technology Assessment for contract 033-4570. Winrock International, Morrilton, Arkansas.

Byington, E. K. and R. D. Child. 1979. Forage for domestic livestock on the national forest system. Study of RPA Issue Area #13 for U.S. Forest Service. Winrock International, Morrilton, Arkansas.

Census of Agriculture. 1978. 1978 Census of Agriculture for the United States. Bureau of the Census, Washington, D.C.

Craft, M. D. and J. R. Cutshall. 1980. The Le Blanc pasture, pp 215-221.   In: E. K. Byington and R. D. Child. Proceedings of a Symposium: Southern Forest Range and Pasture Resources. Winrock International, Morrilton, Arkansas.

Shiflet, T. N. 1980. What is the resource. In: E. K. Byington and R. D. Child. 1980. Proceedings of a Symposium: Southern Forest Range and Pasture Resources. Winrock International, Morrilton, Arkansas.

Soil Conservation Service. 1981. Soil and water resources conservation act, 1980 appraisal. Soil Conservation Service, Washington, D.C.

U.S. Forest Service. 1980. An Assessment of the Forest and Range Land Situation in the United States. U.S. Forest Service, Washington, D.C.

U.S. Forest Service. 1977. Forest Statistics of the U.S., 1977. U.S. Forest Service, Washington, D.C.

113

PRACTICAL INTEGRATION OF LIVESTOCK, RANGE, AND WILDLIFE RESOURCES ON CHAPARROSA RANCH

Patrick O. Reardon

The 70,000 acre Chaparrosa Ranch located west of La Pryor, Texas, in Zavala County, is owned by Belton Kleberg Johnson, great-grandson of Captain King, who started the famous King Ranch. Additional properties that enhance this ranching enterprise include: the 6,000 acre Mangum Division south of La Pryor, Texas; the 10,000 acre La Puerta Division near Agua Dulce, Texas; and the 4,200 acre Carmel Ranch Division in the Carmel Valley, Carmel, California.

Over 20 years ago Mr. Johnson and his family purchased and moved to the Chaparrosa Ranch and began accumulating what is today one of the best purebred Santa Gertrudis herds in the world. This herd, which features the Masterpiece and Pico bloodlines, includes about 1,400 mother cows on four ranch divisions. The unmatched quality of this herd is documented by the long list of National and Regional Grand Champion awards. Also, evidence of their quality is the growing reputation of the "Cowman's Choice" production sale that annually offers over 300 head of outstanding purebred Santa Gertrudis bulls and females, crossbred heifers, and registered quarter horses. Both purebred and commercial cowmen from all across North America gather each year at this "one of a kind" production sale.

Although Chaparrosa Ranches are best known for their Santa Gertrudis cattle, the whole operation is a diversi- fied, integrated agricultural entity that includes a 7,500 head capacity feedyard, over 9,000 acres of irrigated and dryland farmland, 3,000 head of commercial crossbred cows, a progressive quarter horse operation, and an intensive range and wildlife improvement program. Each division of the overall operation has a manager who is responsible for making his unit as efficient and profitable as possible. It is also each manager's responsibility to coordinate his pro- gram and activities with the other division managers.

It would not be realistic to say that an operation such as this is typical in South Texas. There are many ranches much larger, but few large ranches with such a uniquely integrated program. The basic principles that make this operation successful will be discussed and should be helpful to any size ranching enterprise.

As previously stated, the overall goal of this ranch is to be as efficient and profitable as possible. Simple, proven, and practical principles that are utilized to integrate our livestock, range, and wildlife resources will be discussed. Some may seem elementary, but even today many proven principles of agricultural production are still not utilized. It has been said that agricultural research is still 5-10 years ahead of what most ranchers will accept and try.

LIVESTOCK

Maximum beef production from minimum number of cows. Our management works toward maximum beef production from a minimum number of cows. Why is it that today many of our farm products such as milk, milo, and cotton sell for less than it costs to produce? Basically, it is because farmers have over-produced. Even our ranch-raised beef does not bring what it should because of periodic oversupply. At Chaparrosa we try to keep range stocking rates low enough to allow cows to perform according to their genetic potential. Because South Texas has more dry weather than wet weather, this stocking rate has to be conservative.

Cost of dry cows and sterile bulls. A dry cow or sterile bull costs us just as much to run as a productive one. We feel that maximum individual animal performance means more beef per cow and per acre. Annual culling based on age, condition, and palpation records for cows, and age, condition, and fertility tests for bulls are utilized to keep individual animal production at its best. It is better to be able to brag about a "true" large calf crop than about the size of the calves. Experience has shown that up to 10% of a bull herd can be sterile at any given time. A fertile bull is especially critical in small herds of cows.

Supplemental feed. Except in extreme conditions, if we run out of forage, we have too many cows. Supplemental feed is just that, a supplement to what the rangeland provides. South Texas is blessed with good quality brush, as well as grass and forb, and all are necessary to a balanced diet. Here at Chaparrosa, supplemental protein and energy are provided in a high-quality liquid feed when forage is plentiful and by grain cubes when forage is limited.

Practically all Texas soils and forages are deficient in phosphorus. We find it absolutely necessary to feed a high-phosphorus supplement as well as salt and other trace elements. Sufficient consumption and price are two major factors in selecting a product.

The feedyard cover up. Having a feedyard on the ranch sometimes makes it too easy to cover up mistakes. Our feedyard operation is used to condition purebred animals for sale or breeding, to fatten ranch-raised calves, or to fatten customer cattle. Hay, grain, and silage are all raised by the farm division and utilized by the feedyard to feed the ranch or customer cattle. Under certain conditions it has been used to feed purebred or commercial cattle when range forages were depleted from drought or overstocking. In other words, it becomes too easy to keep too many cows, knowing we can always feed them at the feedyard.

A health program. A good health program is a continuous effort. A health program that includes vaccination for such things as blackleg and lepto-vibro, control for fly and grub, and supplement of vitamin A is standard procedure. Also, importance is placed on recommended procedures for vaccination and proper care of the vaccine itself.

Budgets and records. Budgets and records are essential. Annual budgets are prepared by each division manager. Through the use of computers, monthly statements are prepared comparing actual vs budgeted income and expenses. With this arrangement, monthly and yearly comparisons can be made to assist in decision making. Purebred cattle records also are computerized for registration, production records, and selection of outstanding individuals.

RANGE

Best rangeland feed. Our cheapest and best feed is produced by grasses, forb, and brush from our rangelands. Even though supplemental feed and mineral are necessary, we could not operate long without our native range forages.

Grazing management. Grazing management is essential in maintaining sufficient nutritious forage for both livestock and wildlife. Long-term Soil Conservation Service records, current conditions, and experience are all used in determining a stocking rate for the rangelands. Getting caught with too many cows in a dry period is costly to the livestock operation and extremely detrimental to the white-tailed deer, quail, and turkey.

Pastures are given a periodic deferment from grazing whenever possible and some are deferred systematically through 4-pasture deferred rotation (Merrill) systems. The Merrill system is utilized primarily because movement of the cattle can be done during spring, summer, and fall workings.

Brush control. Brush control is designed to benefit both livestock and wildlife. Any type of brush control is expensive, so it has to economically benefit both cattle and wildlife. When mechanical brush work is done, the cleared areas are seeded to introduced grass species. As a rule of thumb, clay and clay loam soils are seeded to kleingrass;

sandy loan, and sandy soils are seeded to buffel grass; and eroded, shallow, or steep soils are seeded to Kleberg bluestem. In most cases only one grass is seeded in an area although sorghum-alum is sometimes seeded on the edge of cleared strips for game birds. Different grasses are seeded in the same pasture, but not on the same cleared strip.

Brush clearing. Brush is always cleared in long narrow strips. After many trials, we feel the 300 to 400 foot wide cleared strip is best. Normally we clear 300 to 400 feet, then leave 300 to 400 feet of brush. With strips this narrow, they can be as long as necessary. We prefer to clear the strips in a checkerboard fashion. A 600 to 900 foot wide strip is always left along creeks or major drainages. Normally, with this design no more than 40 to 50% of a total pasture is cleared. No doubt deer utilize cleared areas no matter how big they are. However, if the strips are too wide, they will utilize only the edges until after dark. By having many narrow strips, deer are much more accessible to the hunter, and the hunter is the one who makes it profitable.

Improved hunting. It would be hard to justify mechanical brush work if it did not improve hunting. Brush is extremely thick in most parts of South Texas; this is one reason we have maintained the excellent quality in the deer herd. Hunters pay a lot for the privilege to hunt these trophies and their chances of getting one are greatly enhanced with a well-designed brush-clearing pattern.

Mechanical brush work. With mechanical brush work, it is necessary to treat whole pastures, and never part of a pasture. Deferment is necessary the year of treatment and the following spring. If the grass is well established by frost the first year, it is grazed for a short period and then deferred again the following spring.

Chemical brush work. Chemical brush work is used whenever possible. To date there still is not a good economical brush chemical on the market for the mixed-brush complex of South Texas. We use chemical sprays on areas with heavy mesquite infestation or pellets on heavy white brush. We do use strip treatment but treat 900 to 1200 feet and leave 300 foot strips. This treats about 75 to 80% of the pasture and does not have a detrimental effect on wildlife.

Improved grass maintenance. Maintenance of improved grass stands is required. Resprouting brush is periodically controlled with a small root grubber or brush wiper. If the regrowth brush has a 30% canopy cover or greater, the area is rootplowed and disked. Roamdisking, fire, and deep chiseling are used to maintain a vigorous improved grass stand.

Grass seed production. Grass seed production is often overlooked. Most ranches need some areas where introduced grass seed can be harvested. It does, however, cost about 75% more to clear rangeland good enough to harvest seed on. This means if it cost $60 per acre to rootplow, roller chop, and seed, it will cost over $100 to get it smooth enough to harvest seed. From experience we have learned that if the pasture to be treated is small enough to allow annual deferment and has good soil, it is worthwhile to spend the extra money.

WILDLIFE

Commercial hunting. On the same land it takes to run one cow (24 acres) we can take in $130 from hunting. No commercial buck hunting is done on Chaparrosa Ranches, but the value is still there. Ranches of this size with equal quality and quantity game species typically lease for $5 to $6 per acre per year. Some may wonder why we don't just manage the wildlife and not even mess with the cattle. But cattle grazed conservatively actually improve wildlife habitats, so why not get something from both?

Wildlife control. Everyone wants to shoot a trophy buck but not a spike or doe. Our most difficult task is getting someone to harvest does and spikes or inferior bucks. We conduct annual deer counts and try to harvest enough to maintain a herd density of about 25 acres per deer. In good years, such as in 1981, we try to kill every buck that is a spike (2 points). Most does are killed by paid hunters.

Quail hunting. The quail hunter normally has more expendable income than the deer hunter. Quail, turkey, dove, javelina and feral hogs can bring in bonus revenues because sometimes these animals are just thrown in as a bonus to deer hunting. Good quail hunting should and does bring in as much revenue as deer hunting. Split seasons and separate charges for each game animal will add greatly to gross income.

Main farming wildlife populations. You can round up cattle, move them to a certain pasture, and keep them there, but you cannot make wildlife stay there until they have sufficient food to sustain life and reproduction. Areas on this ranch that were overstocked for several years practically lost their wildlife populations and are extremely slow in recovering. All our wildlife species benefit from the periodic deferments and moderate stocking rates. Generally, good grazing management for cattle is also good for wildlife. Good grazing management should maintain a sufficient quantity and balance of desirable grass, forb, and brush species.

114

MORE GRASS = MORE BEEF = MORE PROFIT

Dick Whetsell

Grazing livestock on natural or native vegetation is one of our oldest and most accepted meat-producing practices. A few thousand years ago, Esau looked the range over and told his brother Jacob that there wasn't enough grass in the Haran area for all their livestock, so he would move his operation farther north. Since that time, this plan of moving to better range has been practiced more than any other grazing system. The most recent example is the exodus of Texas ranchers to Oklahoma, Arkansas, South Dakota, and other lush areas during the severe drought of the 1950s. The moves were dictated by the premise that it is more economical to buy or lease other grazing land and move than to stay and wait for recovery practices to produce more vegetation on the home place.

Hendrick Von Loon once said, "The history of man is the story of a hungry animal in search of food." That being true, we might say history is the story of man's constant search for grass. We have searched the known areas of our world well, and I hope we now have decided that it is time to start a sound, scientific, and economical grazing program that will again cover the land with a sea of grass. Yes, it can be done. We have the knowledge about plants and animals and the needed operator skills to produce more grass--more beef--and more money. Originally, I said "more profit" instead of "more money," however, after the experience accumulated during the past 8 to 10 years, I have changed the wording to say more money since it is not necessarily more profit. In fact, some years a good program just means less money lost.

Oklahoma Land and Cattle Company has been in the ranching business since 1954 and is a family-owned company that now owns and operates ranches in Oklahoma, Kansas, and New Mexico. All of our 175,000 acres are native vegetation, mostly grass, and I am sure our operation presents a fair sample of common problems.

After spending my life in this busines, I am convinced that many of our failures, reverses, and disappointments have been the product of one-practice programs. This may be the result of scientific training in only one field, of continuing an out-dated operation, or of being led by a strong but limited technician; and sometimes we even get misled at special-information field days, seminars, and other technical meetings.

Combining grass-cattle-people into a profitable operation requires skills that many of us do not possess. Several times over the past few years I have been driven by fear of loss rather than attracted by hope of profit. So, we are confronted with the problem of fitting all parts of our total-operation machinery into a smooth-running, efficient, productive unit.

Range management that will produce more grass and grazing management that will harvest more high-quality forage most efficiently no doubt offer the best ways to increase production and profit from most ranches.

Range plants are designed to withstand grazing--but not continuously. You must select a system that will keep the grass healthy and vigorous while providing lush nutritious forage for livestock. Some type of graze and rest system is needed. In most cases, a short grazing period of 5 to 8 days followed by 25 to 30 days rest in the 20 in. to 40 in. rainfall belt is very productive.

Relatively long grazing and rest periods require less moving, but usually result in slow movement. Most graze-rest systems carry more cattle than does the practice of continuous grazing on the same area. However, a majority of reliable tests show a bit lower gain per animal with the greater numbers. Stocking according to forage available is essential for success with all systems.

When the range is covered with productive palatable native plants and grazed uniformly when the grass is the most nutritious and free of undesirable brush and weeds, your chances for a profit are improved. Let's assume you have accomplished the above goals for forage production and you are now ready to stock with cattle. The class and quality of cattle used will have profound effect on your bottom-line figure. Twenty-five years ago, we had purebred Hereford and Angus, and in 1955 we weaned our steer calves at 460 lb. In July 1982, from the same range, we weaned 711 lb steers from crossbred cows and crossbred bulls. For the past 5 years, we have summered from 15,000 to 20,000 head of yearlings. Our records show as much as 100 lb per head difference in gain while on the ranch. This past summer-grazing season, we had strings of cattle purchased in March and April that varied as much as 75 lb in gain when weighed in August.

Yes, we like good grass grazed with genetically sound crossbred cattle or top-line purebreds.

The other essential, especially in larger operations, is good people. Not at the top or the bottom, but all the way through. Many good programs have been short-circuited by disinterested, uninvolved, uninformed people. Working people are especially important, because most programs call for more activity, not less.

So stock the range with productive, palatable plants, graze it with good-gaining cattle, and place it in the hands of informed, interested, intelligent working people, and you will have a more grass-more cattle-more money operation.

115

RANCHING INTERNSHIP: A GOOD INVESTMENT

Dick Whetsell

After investing 28 years of my life in our present working-ranch operation, I am firmly convinced of the value of an internship program. Over the years, we have hired a rather large number of working hands, several of whom have been college trained. For example, I recall one eager young man who really wanted to be a ranch hand and came to our ranch his first job out of school. On his third day of work, we were holding 600 steers in the corner of a pasture and were sorting on horseback. After noting some confusion on his part, I said, "Bill, I would appreciate some help." Bill said, "If I had any idea what you are trying to do, I would sure help." Another new hand couldn't understand why I insisted that he get on a certain side of the gate when sorting cattle in the alley. These are just two simple examples of things learned only by doing the work.

I could make a list of at least 50 separate job skills that a person should possess to do an effective job working with ranchers. These skills are invaluable to a person as he goes to work as a federal or state employee, ranch manager, private-industry salesperson or energy-related field representative.

This internship program is becoming more important each year as fewer students and fewer professors come to college from a working ranch. If all the students are exposed only to textbooks, what kind of a hand will they make in the industry?

Many ranch skills are known only to the older, long-time ranch hands, and I am really concerned that we do the necessary planning to see that these skills are passed on to our younger hands in a sound manner. There is an old saying, "We never miss the skills we don't have." It is also true we can't pass on or consider very valuable those skills that we don't have.

The importance of intern-type training has been recognized for many years by other industries and professions. The best example is in the field of human medicine, but veterinary medicine, architecture, construction engineering, forestry, and many others have intern requirements as an integral part of their basic training.

Several of our leading universities have excellent intern programs. Three of the better programs are at Texas Tech, Texas A & M, and Oklahoma State University. I am sure there are others and the number is growing each year.

We have been in the Animal Science Intern Program at Oklahoma State University since it was initiated several years ago and are happy with the results. We have had several students that were so lost that they could not stay out of the way as we worked cattle but who, 3 months later, looked and worked like one of the regular hands.

There are dividends to be gained for the rancher, student, school, and certainly industry in these ranching-internship programs.

We work directly with the university and student to set up the summer work schedule, pay scale, living quarters, transportation, and other details related to the assignment. If your school doesn't have a program, then help them set one up and be the first intern's employer. It is not complicated; all you need is an industry-connected progressive school; an intelligent, working student, and a cooperative, willing rancher.

Part 17

ANATOMICAL STRUCTURE
AND FUNCTION

116

SIMILAR ANATOMICAL AND DEVELOPMENTAL CHARACTERISTICS OF SHEEP, CATTLE, AND HORSES

Robert A. Long

Each species of animal with which we concern ourselves in animal science is constructed according to a fixed plan. Therefore, we consult a textbook on the anatomy of the bovine and have confidence that it applies to all cattle. There are variations, of course, due to differences in genetics, nutrition, age, sex, and even disease. However, the two basic tissues of bone and muscle are always present in the same general pattern. This constancy of structure or organization of tissues is of great value to us in the evaluation of cattle both alive and dead.

We should also realize that there is great similarity in this overall plan of structure among different species. All mammals have essentially the same skeleton and the same muscles attached to the skeleton. The similarity between cattle, sheep, and horses is very pronounced and, even though small differences exist, the factors that affect the growth and development of the three species are identical. Let us examine these common traits and discuss how they might help us in evaluating livestock.

THE SKELETON

All cattle, sheep, and horses are made according to the same general plan or design. Their skeletons are composed of the same number of bones, and the general shape of each bone is the same in all three species. Also, the percentage of total weight or linear size that each bone represents of the whole skeleton is constant. Butterfield (1964), Kauffman (1973), Ramsey (1976), Heird (1971), and Mukhoty (1982) are in agreement with this statement. This simply means that, for all practical purposes, if we can measure one bone in the skeleton, we can determine the dimensions of the whole skeleton. In fact, in the case of the long bones of the limbs, the measurement of one bone is much more accurate than an overall measurement of height. Most people currently measure frame size (or think they do) by measuring height at the withers and/or hips in cattle and sheep or hands at the withers in horses. Figure 1 illustrates how

Figure 1. Identical skeletons showing the effect on height
at the withers and hips of changing the angle of
movable joints of the long bones of the legs.

misleading this can be. Note that these are identical
measurements of these two skeletons, which is exactly the
case. If one bone is longer, every other bone in the skeleton will be longer and proportionately so. Remember, this
is only valid if the cattle are of the same age and sex.
You must compare bulls with bulls, steers with steers (castrated at the same age), and heifers with heifers, because
at puberty the level of sex hormone production changes
greatly; this results in closure or calcification of the
epiphyseal groove, and the length of the leg bones stops increasing. Thus, steers and wethers generally are taller and
shorter bodied than their identical twins that have been
left sexually intact. We do not notice this as often in
horses since they are usually castrated after sexual
maturity. Of course, nutrition has an effect on the age at
which animals reach puberty; thus, for comparisons to be
legitimate, they must be made under uniform conditions of
nutrition (or for that matter under uniform total
environment).

This is of less concern in castrates such as steers and
wethers. Skeletal growth or bone formation in growing animals takes priority for nutrients over that of muscle growth
and fat deposition. Therefore, regardless of plane of nutrition, if we compare animals of the same age, their frame
size has probably increased according to genetic potential
and is a good measure of what their mature size will be.
When compared at the same age, the larger the frame, the
larger it will be at maturity and the longer it will take to
reach that point. Also, we know that as an animal approaches maturity, he begins to deposit fat in the muscle, which
is the marbling that gives his carcass a quality factor.
This is the very basis of the USDA feeder cattle grades that
are discussed by the author in another presentation at this
school.

About ten years ago, a national field day was jointly
sponsored by a different breed association and the University of Wisconsin in each of three consecutive years. Each
breed selected different "types" of steers which were placed

on feed and were slaughtered when ready for market. A field day was built around the data collected. Some good things came out of these sessions but, unfortunately, the most attention was received by the profile drawings in figure 2, which is entitled "Body Types." Note that "body type #1" is shortbodied and lowest and shows heavy development in the dewlap, brisket, and belly, and great proportional depth of body. Also, observe that "body type #5" is tall and long and is trim-fronted and tight-middled. The implication here is that all small-framed cattle are wasty and fat and that all large-framed cattle are trim and desirable. Nothing could be farther from the truth.

I do not believe that such a thing as a body type exists. I believe, and will offer evidence to prove, that every frame size of beef animal can and does occur with every possible combination of fat and muscling. Some small-framed cattle are highly desirable in composition--some are not. Some large-framed cattle are desirable in composition---some are not. The same can be said for any frame size (Long, 1982).

I want you to look at the data from three steers in table 1. Their weight is very different but their skeletons are practically identical in size, which is, of course, their frame size. Now examine the dissection data in table 2. Not only were their skeletons identical in linear measurements, but their skeletons weighed the same. However, here the similarity stops. Note the tremendous difference in muscle both in total weight and as a percentage of the carcass of the #1 steer. This gives a muscle:bone ratio of just twice as much for the heavily muscled steer as for the thinly muscled one. Fat varies only a little in this case, but keep in mind that it would be easy to put together a large group of steers with identical skeletons that vary widely in fat and muscle composition. Table 3 lists the conventional carcass measurements. This table makes two major points.

1. The Yield Grade formula ranked these three steers essentially the same, which is obviously in error. This is because the formula was constructed with conventional British breeds that did not offer the range in muscling that we have here. It underevaluates the heavily muscled #1 steer, overevaluates the thinly muscled #3 steer, and does a good job on #2.

2. The frame size or skeletal size of these steers had nothing to do with the desirability of their carcasses.

I would hope that your conclusion would be something like mine that simply stated is: Why anyone would use frame size in the evaluation of cattle for slaughter is beyond me. Yet, that is exactly what takes place in the majority of steer shows in this country--they put the tall ones up. Think what this means. The cattle are shown by weight and

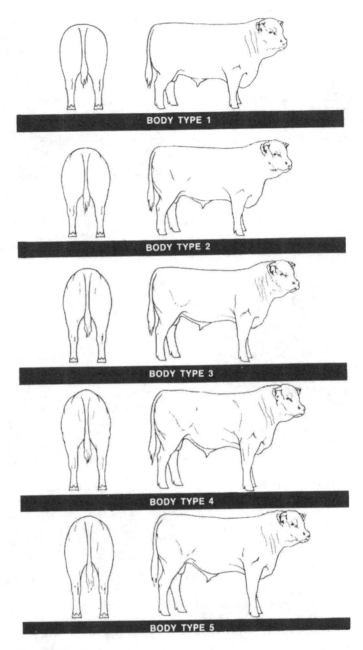

Figure 2. Body types.

TABLE 1. MUSCLE:BONE RELATIONSHIPS AMONG SLAUGHTER STEERS
LIVE MEASUREMENTS

Steer #	1	2	3
Live wt (lb)	1450	1300	1005
Length of body (in.)	60.23	60.23	59.84
Rump length (in.)	20.07	20.07	20.47
Ht. withers (in.)	51.96	51.57	52.36
Ht. hips (in.)	53.54	53.14	53.93

TABLE 2. MUSCLE:BONE RELATIONSHIPS AMONG SLAUGHTER STEERS
DISSECTION DATA

Steer #	1	2	3
Lb of bone	64	68	67
% bone	13.1%	16%	23%
Lb of muscle	320	262	168
% muscle	66%	63%	59%
Lb of fat	104	81	53
% fat	21%	19%	18%
Muscle:Bone	5.01	3.88	2.52
Mucle:Bone IM fat included	5.16	3.94	2.61

TABLE 3. MUSCLE:BONE REALTIONSHIPS AMONG SLAUGHTER STEERS
CARCASS MEASUREMENTS

Steer #	1	2	3
Carcass wt	976	820	570
Dress %	67%	64%	57%
Maturity	A75	A50	A75
Marbling	Small30	Slight80	Slight60
Quality grade	Ch$^-$	Gd$^+$	Gd$^\circ$
Fat thickness (in.)	.3	.3	.12
Rib eye area (sq in.)	18.1	14.3	9.9
% KHP	3.0%	2.5%	2.5%
Yield grade	1.8	2.3	2.3

most of them have been fed and managed in such a way that
they are not excessively fat. Therefore, placing the tall,
big-framed steers up in class and the small-framed ones down
means that selection is against muscle or meat, which makes
no sense at all in the beef-production business. The placing of the tall ones of the same weight on top of the class

further complicates the situation. Large-framed cattle mature later, which fact decreases the chances of the large-framed steer making the choice grade.

Unfortunately, it is commonly believed that size of the skeleton is of great value in evaluating other classes of livestock as well. As already stated, it tells us something about expected time of maturity but nothing about composition. Sheepmen are particularly prone to make the statement, "He has a longer hindsaddle with more weight in the high-priced cuts." Sheep are like cattle. They are all in the same proportion. Therefore, if one has a longer hindsaddle the rest of his skeleton is also longer. However, this does not mean that there is more meat or muscle on it (or more or less fat, for that matter.)

THE DETERMINATION OF MUSCLING

Now let us devote our attention to differences in muscling. First, dispose of the old, often-used phrase, "More weight (or more meat) in the high-priced cuts." This statement originated years ago when some cattlemen decided that more muscle in the rib, loin, and round--and less in the rest of the carcass--would be a great thing. It would be a good thing but, unfortunately, it isn't possible. The research data of Butterfield of Australia, Berg of Canada, and several people in this country show that different breeds of cattle (British, European, Brahman, dairy breeds, wild cattle, and water buffalo) have essentially the same relationship between the various muscles. This does not mean that we cannot increase their muscle, but it does mean that we cannot increase one muscle or group of muscles in a steer without increasing all of them. This should not be discouraging; indeed, it is most fortunate. If we can measure the amount of muscle in one part of the animal, we can depend on proportional development in all other parts. Therefore, we can appraise the muscling of a steer by looking at the forearm, at the muscle working in his top as it moves, and at the thickness through the lower quarter or stifle in proportion to that through the rump. Wide stance both in front and from the rear tells us the steer has a lot of muscle.

We know that the approximate shape of the forearm bone is that of a cylinder or piece of pipe which is approximately uniform in circumference and width. Since this is true, any change in shape or increase in width of the forearm region comes from muscle, because observation and experience have shown us that little fat is deposited here. Likewise, practically no fat is deposited over the outside, lower round; so, as we stand behind a steer, a horizontal plane that would pass through the stifle joint should be the thickest place in the steer. If it is not, he lacks muscling, or has heavy deposits of fat over the rump and on top of the round, or a combination of both.

We also know by looking at the bovine skeleton that the foreleg is attached to the rest of the skeleton only by muscle. The amount of muscular development determines the space between the leg bones and the rib cage and, therefore, how far the front legs are held apart. Likewise, the muscular development between the hind legs determines how far apart the steer stands as viewed from the rear. Therefore, the heavily muscled animal will stand wide when viewed both from the front and from the rear.

We hear a great deal about the "kind" of muscle on cattle, sheep, and horses, and the favorite terms are "the right kind of muscle" or "that good, long, smooth muscle." Fortunately, there is only one "kind" of muscle. It is composed of muscle fibers bundled together by connective tissue and attached by connective tissue and tendons to other muscles and to the skeleton. The "length" of the muscles is determined by the size of skeleton since each muscle is attached to the skeleton at the identical spot in each specie. Therefore, animals of equal frame size have the same length of muscle. "Smooth Muscle" is a term used to describe animals that have a layer of subcutaneous fat, or are thinly muscled, or both.

Just as the skeleton is in the same proportion, each muscle in its anatomic entirety represents a constant percentage of the total muscle mass. This is well-established by both Butterfield (1964) and Kaufman (1976) and is the basis for estimating total muscle by examining an animal for degree of muscling over the forearm or through the stifle. Let's face it, a steer or wether cannot produce an excellent carcass without being well muscled. This, of course, adds to his weight and when finish is constant the heavily muscled beast far outweighs the smooth-muscled one of the same frame size. Therefore, a large-framed animal will be considerably heavier than the packer wants if his composition is correct.

You will recall that we have pointed out that both the skeleton and musculature occur in essentially the same proportion in each specie. This results in a near constant percentage of carcass weight in each of the wholesale cuts. For example, a heavily muscled Limousin steer has the same percentage of hindquarter as the thinnest-muscled Jersey steer. Cattle just don't possess "more weight in the high priced cuts." The difference is in the percentage of meat, fat, and bone in each cut. The data that illustrate the constant proportionality of skeleton and muscle have often been misinterpreted to mean that all cattle are the same-- and, if you measure them, the longest or largest are the best. This is in complete error. You must know muscle:bone ratio and degree of fatness to know composition. This is also true in sheep and horses.

REFERENCES

Butterfield, R. M. 1964. Relative growth of the musculature of the ox. In: D. E. Tribe (Ed.) Carcass Composition and Appraisal of Meat Animals. The Commonwealth Scientific and Industrial Research Organization, Melbourne, Australia.

Heird, J. C. 1971. Growth parameters in the quarter horse. M.S. Thesis. Univ. of Tennessee.

Kaufman, R. G., R. H. Grummer, R. E. Smith, R. A. Long and G. Shook. 1973. Does live animal and carcass shape influence gross composition? J. Anim. Sci. 37:1142.

Kaufman, R. G., M. D. Van Ess and R. A. Long. 1976. Bovine compositional interelationships. J. Anim. Sci. 43:102.

Long, R. A., and C. B. Ramsey. 1982. Visual scores and linear measurements of slaughter cattle as predictors of carcass characteristics. Proceedings, Western Section, American Society of Animal Science. Vol. 33.

Mukhoty, H. and H. F. Peters. 1982. Influence of breed and sex on muscle weight distribution of sheep. Proceedings, Western Section, American Society of Animal Science. Vol. 33.

Ramsey, C. B., R. C. Albin, R. A. Long and M. L. Stabel. 1976. Linear relationships of the bovine skeletons. J. Anim. Sci. 42:221 (Abstr.).

117

FRAME SIZE AND ITS RELATIONSHIP
TO FEEDLOT AND CARCASS PERFORMANCE

Craig Ludwig

TRAITS TO USE IN DECISION MAKING

Cattle producers should carefully choose traits to be used in selection decisions. In most nations where the beef cattle industry is a major industry, production of certain types and sizes is dictated by domestic demand and requirements. Since demand for the finished product and the method of production vary considerably from one country to another, it is virtually impossible to design an animal that fits all needs. Consequently, each geographic area must determine its own inherent requirements and structure its seedstock segment to produce cattle that meet those specific requirements. The following remarks regarding optimum frame size are made to reflect the current industry needs in the U.S. They do not necessarily indicate that breed goals established by the American Hereford Association (AHA) will apply to those requirements noted in other countries. However, we do feel that it is our responsibility to influence the structure of the beef business in our country and to guide our breeders towards the development of Hereford cattle that best meet these domestic demands. In determining such goals, frame size influences the performance of Herefords in meeting these U.S. requirements as well as other countries' demands, although the final goals may vary from one country to another.

Within the U.S. cattle industry cattlemen generally place emphaiss on:
- Traits of economic importance.
- Traits of medium to high heritability.
- Traits that are easy to measure.
- Traits that are correlated.
- Herd strengths and/or weaknesses.

In the early, mid, and late 1960s the consumer's purchases told the beef producers that the consumer wanted less outside fat and more lean, red, edible meat.

The cattle industry in the U.S. has:
- 1.8 million cattle farms and ranches. (The seedstock portion of the industry is represented in this segment of the industry and is represented by such organizations as the AHA.)

- 135,000 feedlots.
- 6,100 meat processors and meat packers.
- 265,000 meat retailers.
- 500,000 food service operations.

The majority of the U.S. cattle are finished in feedlots that handle over 10,000 cattle at a time. Some lots handle from 40 to 80,000 cattle at a time. Such feedlots feed a cross section of every breed and breed cross in the U.S. Feedlots of this magnitude do not concentrate on specific breeds. They work on averages, but knowledge about the strengths and weaknesses of specific breeds is most important.

Because over one-half of the cattle in the U.S. are Hereford or Hereford-cross cattle, our studies concerning the breed are significant to the U.S. beef industry in general. The AHA is primarily concerned with how straightbred Herefords perform from conception to slaughter. Some special studies of the AHA involve the permanent identification of individual calves from birth through weaning. These calves are followed into the feedlot and fed in progeny groups until slaughter. Their individual identification is carried into the packing house where they are slaughtered and cooled for 24 hours. The AHA personnel then collects four different and separate pieces of carcass information.

After 12 years of collecting such feedlot and carcass data, the AHA has become more aware that the real problem facing all commercial and registered breeders in the U.S. is that of finding a new bull each year that will produce a genetically superior product in the new calf crop even when used on the average cow. The AHA and most of the Hereford breeders in the U.S. approach this problem through the use of the association's Total Performance Records (TPR) program. This TPR program is broken down into two phases: 1) the cow-calf phase and 2) the feedlot and carcass phase.

Within the feedlot and carcass phase of TPR, AHA personnel have tested the progeny from nearly 2,100 herd sires and collected individual carcass data on nearly 20,000 steers and heifers.

With the American Hereford Association's own high speed computer we are able to summarize and analyze a sire's progeny for average:
- weaning weight
- yearling weight
- weight-per-day-of-age
- percentage retail yield (proportion of edible red meat in a carcass)
- efficiency or lb of feed/lb of gain

At an American Hereford Association-sponsored University of Wisconsin conference in 1969 (13 years ago) we made available the results of a research project dealing with framesize and its relationship to beef cattle profitability. In this study the shortest, most compact steers were designated as frame size 1. The longest and tallest steers were given a frame score of 5. Cattle varying in length and height between these two extremes were designated as 2, 3, and 4 frame sizes. Since this conference in 1969, the industry has added frame sizes up to 7.

The AHA developed measures and frame scores as follows (in mm and inches).

Age	3 MM	in.	4 MM	in.
7	101.6	= 40"	106.7	= 42"
12	114.3	= 45"	119.7	= 47"
18	121.9	= 48"	127.0	= 50"
24	125.7	= 49.5"	130.8	= 51.5"

Age	5 MM	in.	6 MM	in.
7	111.8	= 44"	116.8	= 46"
12	124.5	= 49"	129.5	= 51+
18	132.1	= 52"	137.2	= 54"
24	135.9	= 53.5"	141.0	= 55.5"

A frame size 1 steer at 13 months old weighs 840 to 880 pounds, has .25 in. fat over the rib eye between the 12th and 13th rib, and is approximately 42 in. in height over the hips.

A frame size 5 steer's statistics are: at 15 months of age it weighs 1,150 to 1,175 lb, has .25 in. of fat over the rib eye between the 12th and 13th ribs, and is approximately 50 1/2 in. in height over the hips.

GROWTH CHART

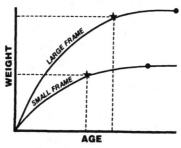

★ PSYCHOLOGICAL MATURITY
● MATURE SIZE

On the horizontal axes of the bovine growth chart (figure 1) we have the age of the animal. On the vertical axes we have the weight of the animal. The growth of the frame size 1 animal is plotted on line and the growth of the frame size 5 animal on the other line. Cattle with frame sizes between the 1 and 5 would naturally fall between these two lines. This graph visually shows the difference that frame size makes in the weight of the frame size 1 at 13 mo vs the frame size 5 steer at 15 mo of age.

Following the conference in 1969 the Association and its breeder members went about the task of finding herd sires that were both taller and longer than the ones used in the past. These larger herd sires matured a little later and had a higher proportion of lean meat to fat meat than those used in the past. It was not easy to find bulls that would produce steers from average cows that weighed 1,150 to 1,175 lb at 15 months, measured 50 1/2 in. tall, and had only .25 in of fat.

In the mid-1960s the kind of steer that was bringing the top dollar with most U.S. feeders weighed 440 lb at 7 months of age and was about 35 1/2" tall.

Results from our 1969 conference proved that this kind of a calf was too short and matured at too light a weight to be the most economically efficient for the various segments of our industry.

Why were we producing that kind of a steer? Because we had been selecting short, small, compact bulls as grand champions at our nation's largest Hereford shows. Back on our farms and ranches we were selecting for the same traits and were producing slaughter steers that by today's standards were too short, too compact, and too fat.

A grand champion steer of a large show in 1969 was termed an "industry ideal" if at 17 months of age he weighed 1,056 lb carried 0.5 in. of fat, and was 43.3 in. tall. By today's standards in the U.S. he is too short and compact but much better than the champions 5 years earlier. From the 1969 Conference an artist painted the ideal Hereford bull. At 30 to 35 mo of age he was 5 ft. 9 in. tall and weighed 2,020 to 2,255 lb. Following the 1969 conference, a majority of our breeders began:

- progeny testing their herd sires.
- keeping performance records that included weaning weight, yearling weight, and height measurements.

At the same time within our nation's show rings, our judges were subjectively evaluating our cattle for conformation and frame size. What we were doing was selecting the taller, longer breeding animals that were structurally correct.

By using objective and subjective measures in an 8-year period (two generations away from the 1969 Conference), our breeders were able to produce show winners like the 1977 Denver Grand Champion that at 27 to 30 months of age weighed 2,288 to 2,310 lb and was 59 in. tall. We found that sires like this one could be used on the nation's average cows and produce steers that at 15 months of age weighed 1,100 to 1,200 lb, had .25 in. of fat, and were 49 in. tall.

Our data further proved the bovine growth curve concept that differences in large vs small frame and physiological maturity are primarily all a function of frame size.

Further evidence as to the validity of the frame size concept and its affect on cattle slaughtered under U.S. conditions came from a recent study done by the AHA in cooperation with Colorado State University. This study included calves from the AHA National Reference Sire Evaluation Program (NRSEP). Among those calves we had 16 sires represented and 172 steers with the following body type

Steers
Body Type

	1	2	3	4	5
No.	3	40	96	31	2

A breakdown of the number of cattle for frame sizes 2, 3, and 4 as they went on test at 250 days of age was 40, 96, and 31, respectively. All the cattle in this study were slaughtered at a weight constant of 1,150 lb.

Age at slaughter by frame size:

2	3	4
508 days	483 days	449 days
17 mo	16 mo	15 mo

Days on feed by frame size:

2	3	4
201 days	177 days	142 days
7 mo	6 mo	5 mo

Average daily gain by frame size:

2	3	4
2.99 lb	3.10 lb	3.45 lb

Weight-per-day-of-age by frame size:

2	3	4
2.13 lb	2.24 lb	2.45 lb

Fat Thickness by frame size:

2	3	4
.65 in.	.57 in.	.45 in.

Cutability - percentage lean meat in a carcass by frame size:

2	3	4
48.8%	49.3%	51.0%

Marbling by frame size:

2	3	4
11.7	10.5	10.0

Correlation in beef traits shows that when we select for one trait, we see a related change in another trait. for example, as we select for yearling weight, we see weaning weights go up also.

Rank of correlation means: High, above .50; Medium, .30 to .50; Low, below .30. Correlations found in this study with on test hip height taken when the calves werre 250 days of age.

AND

Days on feed ....-.52
Age of slaughter.......................................-.52
ADG on test .... .32
WDA slaughter .... .43

Fat thickness..............................-.28
Retail yield............................... .33
Marbling..................................-.16

The CSU Growth Curve shows that cattle were slaughtered by weight at 1,150 lb. Small steers were carried too far and became too fat.

Yield Grade (1974-1978)

Weight	Gain	WDA	REA	BF	VG	MB
1,051	3.03	2.36	13.3	.39	2.3	5.25
1,069	2.70	2.20	12.7	.48	2.9	4.50

After the AHA and Colorado State University completed this study, the AHA decided to look at a cross section of our own feedlot and carcass data to see if the rank and file Hereford breeder had made any changes in a 4-year period or 1 cattle generation in the cattle they were producing. In comparing 1974 with 1978, there was 12% increase in gain, 7% increase in weight per day of age, 2% increase in red meat and 10% increase in quality grade!

These results proved to us that our breeders were now selecting for taller, longer cattle and were, in fact, changing that nation's commercial cattle to the benefit of the whole U.S. beef industry.

In a further study in cooperation with Oklahoma State University, the AHA further investigated frame size and its effect on slaughter cattle. The first steer observed in frozen cross sections was 16 months, 1,120 lb, 46.5 in. tall. His carcass data-fat = .65 in. REA = 11.2 sq in., % lean = 48%, and frame size = 3. This carcass is too wasty for normal U.S. beef trade.

The second steer of the study in frozen cross section was 16 months old, 1,150 lb, 50 in. tall. His carcass data: fat = .30 in, REA = 14.5 sq. in., and cutability = 53% red meat. This carcass is worth about $100.00 more than the one noted above.

The AHA and breeders have not placed all our selection emphasis on frame size and carcass traits. They have continued their selection pressures on other traits as well. 1) low heritability, 2) reproductive traits, 2) breed asset or liability (Hereford-is an asset), 3) high economic value.

REPRODUCTIVE TRAITS

Characteristics of these traits are 1) low heritability, 2) high economic value, and 3) an asset or liability for a breed (for Herefords an asset).

Breeders keep reproductive records for:
- Calving interval
- Most probable producing ability (MPPA)
- Maternal breeding value
- Birth weights

Bull and heifer replacements are selected from dams that calve regularly, have good weaning weights and yearling weights with moderate birth weights.

The AHA performance program measures reproductive efficiency for 1) calving, 2) maternal breeding value, and 3) MPPA.

Our breeders use such data along with measures of frame when selecting herd replacements for both bulls and heifers to maintain a biological equilibrium among all traits.

GROWTH TRAITS

Another group of traits that our cattlemen are concerned with are the growth traits: 1) heritability (moderate) 2) weaning weights, yearling weights, and gain (affect profits and efficiency of production).

Cattlemen look for sires that produce calves with the most potential for the growth traits but have moderate birth weights.

We use every available record on each animal to produce a large number of bulls superior to those we produced last year or several years ago so that their traits can filter down to the commercial cattlemen in large numbers and allow him to make the same improvement in production efficiency as have our seedstock breeders.

CARCASS TRAITS

Carcass traits are highly heritable and can change very rapidly through selection. Economic value is less important to the commercial cattleman than are the reproductive traits. But, if the commercial cattlemen in the U.S. cannot sell their calves because of buyer resistance to carcass traits, then frame sizes become economically important. Research and continued acceptance by America's housewife of the leaner beef we are producing has been very gratifying to America's cattleman. Fifteen years ago, the housewife in the U.S. told us what kind of product she wanted--and we used every available tool possible, including frame size, to genetically engineer an animal that would prove profitable for our producers and satisfy her needs and desires in our product. Beef remains the staple of the American family. We must produce it cheaply and keep it a good buy.

YES--we can model or mold our beef cattle to some extent to fit our needs as long as biological equilibriums are maintained between fertility, growth, and carcass traits. First, you need to know what you want your cattle to do. This is what prompted U.S. cattlemen to look at frame size.

Bulls of large frame size are now common in America-- rather than the exception as they were some 10 years ago. We see taller, longer bulls throughout America.

These bulls are being used on the nation's average cow herd and are improving the genetic merit of the nation's whole beef industry. By using frame size scores (in conjunction with other objective measures) we know the bull that fits our needs at 7 to 8 months of age will be 47 in. tall, weigh 770 lb and fit a U.S. frame size 5 to 6. We have found that they will mature into bulls that weigh 2,300 to 2,420 lb and stand 60 in. tall over the hips. Herd sires like this will produce calves at 7 to 8 months of age that weigh 550 lb and stand 43 in. to 45 in. over the hips. And when finished these steers will weigh 1,150 to 1,175 lb at 15 months of age, have .25 in. of fat, a 12.5 to 14.0 sq in. of REA and be 48 in. to 51 in. tall.

Why did AHA and its breeders use frame size and other records to increase the size of their cattle? Because for the past 15 years, there has been a struggle for survival for all breeds in the U.S.

Scientific, practical knowledge, and the success of our breed with the U.S. commercial cattleman has proven that a biological equilibrium between fertility, growth, and carcass traits have meant more profit for the commercial and registered breeders across America. We believe frame size has helped and is another objective measure of our cattle-- as are weight, most probable producing ability, yearling breeding value ratio, and maternal breeding value ratio.

118

SKELETAL SIZE AND DEGREE OF MUSCULARITY IN THE EVALUATION OF SLAUGHTER CATTLE

Robert A. Long

INTRODUCTION

The accurate evaluation of slaughter cattle is important in several phases of the beef cattle industry. Both the feeder and packer buyer can make more intelligent decisions if they are able to accurately predict the quality and cutability of the carcasses resulting from the slaughter of specific individuals or groups of cattle. Likewise, purebred breeders should evaluate seedstock in this manner because of the high heritability of carcass traits. Even in the showring, both breeding and slaughter classes should be evaluated largely for potential carcass characteristics. However, some breeders, feeders, packers, and judges of live-animals use evaluation criteria of doubtful accuracy. Examples are the various estimates of skeletal size--such as height and length of body often called "elevation," "stretch," and "scale,"--implying that a greater skeletal size is more desirable. Further measures of bone are such terms as "ruggedness" and "heavy bone," as determined by visual estimation of the circumference of the cannon bones and their overlying tissues. Here again the suggestion is that larger is more desirable. Muscling is described by such terms as "length," "smoothness," and "pattern,"--all terms that imply desirability but that have not demonstrated contribution to superior composition of bovine meat animals. Bush (1969) and Cundiff et al. (1967) reported that slaughter weight was a better predictor of carcass cutability than body measurements or subjective conformation scores. However, these researchers each worked with only one breed, in which case a heavier weight probably reflected later maturity. Good et al. (1961) found that circumference of round and cannon bone were positively correlated with rib eye area, while Orme et al. (1959) reported a negative association for cannon bone size and rib eye area.

Therefore, the study reported here was designed to determine the usefulness of various scores and linear measurements of slaughter steers as predictors of carcass characteristics.

EXPERIMENTAL PROCEDURE

In the experiment at Texas Tech, immediately before slaughter, 88 steers were grouped visually for trimness, muscling, and frame size. In evaluating for trimness, the steers with trimmer briskets and tighter dewlaps, flanks, twists, and underlines, along with concave foreflanks, rearflanks, and twists, were classed as "trim." Steers with opposite characteristics were classed as "wasty," and the intermediates were classed as "medium." For muscling evaluations, steers with bulging forearms, thicker stifles in relation to width at the hooks, and more muscle expression during locomotion were termed "heavy," the opposite—"light," and the intermediates—"medium." To evaluate from size, those steers with greater length of cannon bones and body, with some attention to height, were classed as "large," the opposite "small," and the intermediates as "medium."

Eight linear measurements were made on each steer before slaughter. A rigid metal caliper was used to determine to the nearest millimeter the height at the withers, height at the hips, length of rump (from hooks to pins), and length of body (from the anterior point of the shoulder to the posterior edge of the pins). A flexible metal tape was used to measure the length of the right metacarpus and metatarsus from the lateral condyle of the proximal end to the distal synarthrosis of each bone. The same tape was used to determine the circumference of these bones and overlying tissues at the smallest point. Each of the eight linear measurements was reported on two consecutive days and averaged.

Live weights were the average of three consecutive daily weights taken between 1400 and 1500 hours.

After collection of all measurements and scores, the steers were slaughtered at the Texas Tech Meats Laboratory. All carcasses were chilled for 48 hours before the determination of USDA yield grade (YG) and quality grades (QG). The right side of each carcass was weighed in air and water to determine carcass specific gravity (SG) as a measure of composition. The carcasses later were fabricated into boneless, closely trimmed (<1.0 cm) retail cuts plus ground beef (about 20% fat). The total amount of fat was recorded.

Simple correlation coefficients were computed between measurements and scores for carcass traits. One-way analysis of variance and Duncan's New Multiple Range Test (Steel and Torrie, 1960) were used to determine the significance of differences among trimness, muscle, and frame size groups for certain traits. The predetermined acceptable level of probability in analyses of variance was $<.05$, and the level used throughout this discussion. Stepwise multiple regression by backward elimination at a .10 entry level was used to apportion the effects of scores, live measurements, and certain carcass bone measurements on selected carcass traits.

RESULTS AND DISCUSSION

Visual Scores

Least-squares means and standard errors for trimness groups are presented in table 1. Carcasses from the trim steers were heavier than those from the wasty steers. The trim steers had the highest dressing percentage (DP, 63.8%). This pattern among groups is supported by the work of Kauffman et al. (1973), who reported greater carcass weight and DP for muscular cattle than for light-muscled cattle.

The trim steers produced carcasses with lower QG than that of the wasty and medium groups, but this difference was less than one-third grade. Such a difference may be of little practical importance because of reduced emphasis on marbling and the uniform maturity scores (Ao) among groups.

Measures of composition showed a definite pattern among trimness groups. The visual score for trimness accurately ranked the cattle for desirability in every measure of cutability. However, the means for fat thickness (FT) and kidney, pelvic, and heart fat percentage (KPH) were not statistically different among groups. All other measures of composition--rib eye area (REA), SG, fat trim percentage (FT%), bone percentage, and retail cuts percentage (RC%)--ranked the groups appropriately, and the trim and wasty groups were statistically different. The best measure of real value was RC%. A difference of 3.5% between the trim and wasty cattle represented a difference in retail value of $56 for a 320 kg carcass at an average retail price of $5.00 per kilogram.

Means for cutability measures showed a pattern among muscle groups (table 2) similar to that among trimness groups. The heavy-muscled group tended to be more desirable. Differences between the light- and heavy-muscled group means were significant for YG, REA, SG, and RC%. FT, KPH means and FT% did not differ among muscling groups.

Least-squares means and standard errors for traits of the steers and their carcasses by frame size groups (visually determined) are shown in table 3. Each measure of height and length revealed that the small-framed steers were significantly different from the medium and large framed ones. However, the groups of medium and large framed steers were not significantly different from each other except for right at the hips. This failure to separate these two groups, except for height at the hips, demonstrates the difficulty in visually evaluating size of frame on individual cattle without others for comparison. However, in all measurements, the large-framed group were tallest or longest of the three frame-size groups.

TABLE 1. LEAST SQUARE MEANS AND STANDARD ERRORS FOR CARCASS TRAITS BY TRIMNESS GROUP

Trait	Trimness group[a] Wasty	Medium	Trim
Carcass weight, kg	310 ± 6[g]	314 ± 8[gh]	335 ± 6[h]
Dressing percentage	62.4 ± .34[g]	62.6 ± .41[g]	63.8 ± .33[h]
Quality grade[b]	11.9 ± .28[g]	12.0 ± .34[g]	11.1 ± .27[h]
Maturity score[c]	14.2 ± .08	14.2 ± .10	14.2 ± .08
Marbling score[d]	4.9 ± .21[gh]	5.2 ± .25[h]	4.4 ± .20[g]
Yield grade	3.3 ± .14	3.0 ± .17	2.9 ± .14
Fat thickness, cm	1.3 ± .09	1.1 ± .10	1.1 ± .08
Kidney, pelvic and heart fat, %	2.9 ± .12	2.8 ± .14	2.7 ± .11
Rib eye area, cm2	75.94 ± 1.42[g]	78.52 ± 1.68[gh]	82.26 ± 1.36[h]
Conformation score	12.4 ± .25[g]	12.9 ± .30[g]	13.9 ± .24[h]
Specific gravity	1.0447 ± .0015[g]	1.0479 ± .0018[g]	1.0525 ± .0015[h]
Fat trim, %	21.0 ± .76[g]	19.1 ± .92[gh]	17.7 ± .74[h]
Bone, %	15.5 ± .27[g]	16.3 ± .32[gh]	16.6 ± .26[h]
Retail cuts[f], %	61.5 ± .65[g]	63.4 ± .78[gh]	64.9 ± .62[h]

[a] N = 32 for wasty, 22 for medium, and 34 for trim.

[b] Ch- = 12, Gd+ = 11.

[c] A- = 15, Ao = 14.

[d] Small = 5, slight = 4.

[e] Ch+ = 14, Cho = 13, Ch- = 12.

[f] Boneless, closely trimmed retail cuts plus ground beef.

[g,h] Means on the same line with different superscripts are different (P<.05).

TABLE 2. LEAST SQUARE MEANS AND STANDARD ERRORS FOR CARCASS TRAITS BY MUSCLE GROUP

Trait	Muscle group[a] Light		Medium		Heavy	
	mean	± SE	mean	± SE	mean	± SE
Carcass weight, kg	303	± 7	328	± 7^h	332	± 6^h
Dressing percentage	62.2	± .35^g	63.0	± .39^{gh}	63.6	± .34^h
Quality grade[b]	12.0	± .29	11.6	± .32	11.2	± .28
Maturity score[c]	14.3	± .08^g	14.0	± .09^h	14.2	± .08^{gh}
Marbling score[d]	5.1	± .22	4.8	± .24	4.5	± .21
Yield grade	3.3	± .15^g	3.1	± .16^{gh}	2.8	± .14^h
Fat thickness, cm	1.3	± .09	1.1	± .10	1.1	± .08
Kidney, pelvic and heart fat, %	2.9	± .19	2.8	± .13	2.7	± .11
Rib eye area, cm^2	73.55	± 1.29^g	79.36	± 1.42^h	83.87	± 1.23^i
Conformation score	12.1	± .22^g	12.9	± .24^h	14.3	± .21^i
Specific gravity	1.0446	± .0016^g	1.0505	± .0018^h	1.0505	± .0015^h
Fat trim, %	18.3	± .81	19.0	± .89	20.5	± .77
Bone, %	15.8	± .29	16.2	± .31	16.3	± .27
Retail cuts[f], %	61.9	± .69^g	63.5	± .76^{gh}	64.5	± .66^h

[a] N = 30 for light, 25 for medium, and 33 for heavy.

[b] Ch- = 12, Gd+ = 11.

[c] CA- = 15, Ao = 14.

[d] Small = 5, slight = 4.

[e] Ch+ = 14, Cho = 13, Ch- = 12.

[f] Boneless, closely trimmed retail cuts plus ground beef.

[g,h,i] Means on the same line with different superscripts are different (P<.05).

TABLE 3. LEAST SQUARE MEANS AND STANDARD ERRORS FOR LIVE MEASUREMENTS AND CARCASS TRAITS BY FRAME SIZE

Trait	Small	Medium	Large
		Frame size group[a]	
Live measurements			
Height, hips, cm	120.9 ± 1.79	125.6 ± .95h	128.6 ± .54i
Height, withers, cm	118.6 ± 1.79	121.8 ± .96h	123.8 ± .54h
Rump length, cm	44.1 ± .919	48.9 ± .49h	49.3 ± .28h
Body length, cm	136.8 ± 2.189	145.0 ± 1.19h	146.2 ± .68h
Forecannon length, cm	14.4 ± .539	16.3 ± .29h	16.4 ± .16h
Rearcannon length, cm	17.1 ± .679	19.4 ± .37h	19.5 ± .21h
Carcass traits			
Carcass weight, kg	254 ± 149	327 ± 8h	325 ± 5h
Dressing percentage	60.0 ± .759	63.5 ± .41h	63.1 ± .23h
Quality grade[b]	11.5 ± .7	11.6 ± .4	11.6 ± .2
Maturity score[c]	15.0 ± .29	14.1 ± .1h	14.1 ± .1h
Marbling score[d]	4.8 ± .5	5.1 ± .3	4.8 ± .2
Yield grade	2.8 ± .34	3.3 ± .19	3.0 ± .11
Fat thickness, cm	1.0 ± .20	1.4 ± .11	1.1 ± .06
Kidney, pelvic and heart fat, %	3.3 ± .26	2.7 ± .15	2.8 ± .08
Rib eye area, cm²	71.16 ± 3.299	79.48 ± 1.81h	79.61 ± 1.03h
Conformation score[e]	11.67 ± .609	13.55 ± .33h	13.13 ± .19h
Specific gravity	1.0510 ± .0037	1.0471 ± .0020	1.0487 ± .0012
Fat trim, %	19.0 ± 1.83	20.6 ± 1.0	18.8 ± .57
Bone, %	16.4 ± .64	16.0 ± .35	16.1 ± .20
Retail cuts[f], %	64.5 ± 1.58	61.9 ± .86	63.7 ± .49

[a] N = 6 for small, 20 for medium, and 62 for large.

[b] Ch- = 12, Gd+ = 11.

[c] A- = 15, Ao = 14.

[d] Small = 5, slight = 4.

[e] Ch+ = 14, Cho = 13, Ch- = 12.

[f] Boneless, closely trimmed retail cuts plus ground beef.

[g,h,i] Means on the same line with different superscripts are different (P<.05).

The small-framed steers produced significantly lighter carcasses, dressed over 3% lower, were more mature as judged by muscle and bone characteristics, had over 8 cm^2 less REA, and received lower conformation scores than the medium- and large-framed steers. The medium and large groups did not differ significantly in any carcass trait measured.

Simple correlation coefficients for both live and post-mortem linear measurements of frame size with carcass traits may be examined in table 4. All linear measurements of frame size or skeletal length were positively and significantly correlated with chilled carcass weight and DP--except for DP with height at the withers. Rump length was more highly associated with YG ($r = .43$), FT ($r = .31$) and FT% ($r = .35$) than with any other measurement. These positive associations no doubt were due to the fact that considerble fat is deposited on the anterior edge of the hooks and on the posterior surface of the pins (Stabel, 1974), thereby increasing the rump length as fattening progresses. When fat trim percentage was held constant, the association between rump length and yield grade was reduced from 18% to 7%. These results call attention to the widely held belief among cattle breeders and judges that long-rumped cattle are desirable. However, the study data makes this belief suspect, as does the finding that wither and hip and length metacarpel are significantly associated with percentage bone ($r = .40$, $.40$ and $.30$, respectively), whereas rump length and body length (both of which measure bone plus fat) are not significantly associated ($r = .08$ and $.19$, respectively).

Circumference of the metacarpal, taken on both the live animal and on the cleaned and dried bone, was positively associated with REA and conformation score.

Multiple Regression

The effects of visual scores and measurements of the live steers on selected carcass characteristics were apportioned by multiple regression analyses (table 5). Only rump length, forecannon length, and rearcannon circumference entered the model; QG was the dependent variable. These measurements accounted for only 18% of the variance in QG. Longer-rumped steers with shorter forecannon bones and smaller circumference of the rearcannon bones and overlying soft tissues tended to grade higher. As previously mentioned, fatter cattle tend to receive longer-rump measurements because of the inclusion of subcutaneous fat in the measurements. This probably explains the positive association between length of rump and QG in this model.

The model-predicting YG of the carcasses was more efficient than the one for QG, accounting for 36% of the variance in YG. Steers that were visually scored as thicker in muscling and larger in frame size tended to produce more desirable YG when rump, forecannon, and rearcannon lengths were constant. Longer-rumped steers tended to have less

1082

TABLE 4. SIMPLE CORRELATION COEFFICIENTS OF LIVE AND POSTMORTEM MEASUREMENTS WITH CARCASS TRAITS

	Live measurements[a]						Postmortem measurements[b]			
	Height at withers	Height at hips	Rump length	Body length	Fore-cannon length	Fore-cannon circumference	Meta-carpal length	Meta-carpal circumference	Meta-carpal area	Meta-carpal thickness
Carcass weight	.59***	.67***	.66***	.71***	.54***	.71***	.52***	.60***	.37***	.64***
Dressing percentage	.25**	.25**	.51***	.41***	.27**	.23*	.29*	.35**	.27*	.39**
Maturity score			-.22*				-.32*			
Marbling score									-.30*	
Yield grade	.31**	.22*	.43***	.27**	.22*					
Fat thickness			.31**							
Kidney, pelvic, and heart fat, %	.26**				.27**					
Rib eye area	.25*	.32***		.28**	.20*	.52***	.32**	.39**	.27*	.39**
Conformation score						.45***		.40*	.32**	.32**
Specific gravity	.27**								-.39**	.38**
Fat trim, %	-.40***		.35***	.21*			.30*			
Bone, %	-.40***	.40***			.43***					
Retail cuts[c], %	-.22*								.39**	.41**

aN = 88.

bN = 58.

cBoneless, closely trimmed retail cuts plus ground beef.

*P<.05.
**P<.01.
***P<.001.

TABLE 5. MULTIPLE CORRELATION AND REGRESSION COEFFICIENTS FOR INDEPENDENT VARIABLES AS PREDICTORS OF CARCASS CHARACTERISTICS[a]

Independent variable or R^2	Carcass characteristics[b]					
	Quality grade	Yield grade	Rib eye area	Specific gravity	Fat trim, %	Retail cuts,[c] %
R^2	.18	.36	.45	.25	.36	.31
				b_0		
Trimness group	[d]	...	...	.003**	-2.27***	1.60***
Muscle group	...	-.27**	.58***	...	...	...
Frame size group	...	-.29*	...	...	-1.68*	...
Height, hips	...	...	.09*	...	...	.52**
Height, withers	.31***	.17***	...	...	...	-.78***
Rump length	...	...	...	-.002***	.62**	...
Body length	-.52**	...	...	...	...	...
Forecannon length	...	-.31**	.29**	.003**	...	...
Rearcannon length	...	.24**	-.45**	...	.66*	...
Forecannon circumference	...	...	.46***	...	...	...
Rearcannon circumference	-.13*	...	...	...	...	...

[a]Stepwise regression by backward elimination.

[b]N = 88.

[c]Boneless, closely trimmed retail cuts plus ground beef.

[d]Did not meet .10 significance level for entry into model.

*P<.05.
**P<.01.
***P<.001.

desirable YG.   The fore and rearcannon lengths had opposite effects.   Longer rearcannon and shorter forecannon bones were associated with more desirable YG when the other three independent variables in the equation were held constant. The reason for these results is not known.   Note that trimness score did not enter the equation at the .10 level of probability.   Muscle score, wither height, and three cannon bone measurements were useful in predicting rib eye area (R^2 = .45).   When the other variables were held constant, steers that were thicker in muscling, taller at the withers, longer in the forecannon, shorter in the rearcannon and larger in forecannon circumference tended to have larger rib eyes. The positive effect of forecannon circumference on rib eye area may be due, at least partially, to the size of the tendons included in the circumference measurement.   Perhaps we should study the relation between a tendon and the muscle it serves.

Carcasses with a lower SG tend to be fatter.   Trimness score, rump length, and forecannon length were significantly associated with SG (R^2 = .25).   Steers that scored as less trim were longer rumped and shorter in the forecannon and were fatter.   However, of these three independent variables only trimness score remained in the equation predicting actual FT% resulting from the retail cut fabrication operations.   This model accounted for 36% of the variance in FT%.   Steers that were scored as less trim and smaller framed and that were longer rumped and longer in the rearcannon tended to be fatter.

Only three variables entered the equation predicting RC% (R^2 = .31).   Steers that scored as trimmer--were taller at the hips but shorter at the withers--tended to produce a higher percentage of boneless, closely trimmed retail cuts. This means that the most desirable cattle from the standpoint of cutability would be those that are low in front and high behind.   This finding supports the commonly held belief among cattlemen that cattle that are lower in front than behind are not mature and will continue to grow.   Therefore, such cattle are not mature, carry less fat than those of equal wither and hip height.   The lower fat content contributes to greater RC%.

CONCLUSIONS

No single visual grouping (trimness, muscling, or frame size) nor any single linear measure of frame size accounted for a sufficient amount of the variation in these cattle to be of value in predicting carcass characteristics.   However, a combination of traits appears to be of some value in predicting cutability.   Trim, heavily muscled, large-framed cattle that are low in front and high behind appear to be most desirable.

REFERENCES

Busch, D. A., C. A. Dinkel and J. A. Minyard. 1969. Body measurements, subjective scores, and estimates of certain carcass traits a prediction of edible portion in beef cattle. J. Anim. Sci. 29:557.

Cundiff, L. V., N. G. Woody, J. E. Little, Jr., B. M. Jones, Jr. and N. W. Bradley. 1967. Predicting beef carcass cutability with live animal measurements. J. Anim. Sci. 26:210 (Abstr.).

Good, D. L., G. M. Dohl, S. Wearden and D. J. Weseli. 1961. Relationship among live and carcass characteristics of selected slaughter steers. J. Anim. Sci. 20:698.

Orme, L. E., A. M. Pearson, W. T. Magee and L. J. Bratzler. 1959. Relationship of live animal measurements to various carcass measurements in beef. J. Anim. Sci. 18:991.

119

VISUAL SCORES AND SKELETAL MEASUREMENTS OF FEEDER CATTLE AS PREDICTORS OF FEEDLOT PERFORMANCE AND CARCASS CHARACTERISTICS

Robert A. Long

A trip to almost any feedlot in this country will establish that we are not now accurately sorting cattle as to yield outcome. Inspection of a pen of cattle being offered to packer buyers usually shows 3, 4, or even 5 yield grades represented. One can only conclude that the cattle are either not sorted or are sorted improperly.

Cattle feeders are not ordinary people. They are intelligent, experienced persons charged with both management decisions and financial responsibility. They are aware of the great variation in performance, carcass characteristics, and optimum slaughter weights among feeder cattle. These feeders are also aware of the fact that when widely divergent kinds of feeder cattle are fed in the same pen for the same length of time that part of the cattle may be fed too long, resulting in overfinished carcasses and poor feed conversion. They also know that others may not be fed long enough to meet the standards for the choice grade.

However, when you discuss these problems with feedlot managers they say "Yes, I know, but..."

Let us look at some of the reasons offered for continuing to operate as in the past.

REASONS GIVEN FOR THE STATUS QUO

- Uniform groups of feeders are not available.
  SHOULDN'T THEY BE?

- It is not practical to sort cattle after arrival at the feedyard because of too much shrink.
  COULDN'T THEY BE SORTED AT THE SAME TIME THEY ARE WORKED? I SERIOUSLY DOUBT THAT A 2- OR 3-WAY SORT AS CATTLE LEAVE THE WORKING CHUTE WOULD RESULT IN A SHRINK THAT WOULD BE PICKED UP 140 DAYS LATER.

- It is better to sort them off at the end of the feeding period as they reach market finish.

HOW CAN YOU OPERATE A FEEDLOT AT CAPACITY BY
SELLING HALF A PEN OF CATTLE AFTER 120 DAYS ON
FEED AND THE REMAINDER 30, 40, or 50 DAYS LATER?

I believe that everyone concerned with this problem
would agree that an accurate method of predicting outcome on
cattle would be a good thing. Such a system would not only
permit us to sort cattle into uniform groups according to
optimum slaughter weights but also result in more uniformity
as to cutability as well. We then could make intelligent
decisions as to the value of various groups and do a better
job of protecting the details of economic outlays and planning in general.

Feeders have not used the USDA feeder grades because
they simply do not work. As a result many local systems
have been developed, but when you observe the lack of uniformity within a pen of "Number 1 Okies" you can only conclude that these systems do not work either.

The big question is simply: Can a uniform sorting job
be done? I believe that we can do much better than we have
been doing. In particular, I believe that we should make
the effort to sort better by developing a system that will
get the job done. The increased efficiency resulting from
use of an improved sorting system is obvious.

No one buys feed without consideration of its moisture,
protein, fiber, or caloric density. Why then should we run
it through a group of cattle without regard for efficiency
of utilization?

Let us look at what we know about feeder cattle.

We know that cattle of similar age and condition vary
widely in size of frame. Conner (1974) and Nelson (1976)
have shown that height at the withers or hips at weaning age
is a good predictor of height at 12 months. Conner further
reports that in the population with which he worked, frame
size adjustments could be made by the following formula:

No. of days of age under 365 x .033 + actual height = adjusted height
No. of days over 365 x .025 + actual height = adjusted height

Tyler (1977) and Dikeman (1978) report that frame size at a
given age is highly correlated with mature size and, consequently, highly correlated with the weight at which, under
normal feeding practices, an animal will produce a carcass
of a given grade. The conclusion, of course, is that
small-framed cattle reach choice at lighter weights than do
larger-framed cattle. This, of course, all of us have observed.

The next question is how best to measure the size of a
skeleton. Ramsey (1976), Butterfield (1964), and Kauffman
(1973) have all observed that the bovine skeleton is in constant proportion. If you accept this constant proportion--
and I do--it is then more accurate to observe the length of
a single long bone, such as the cannon, and use it rather
than an inclusive measurement which crosses movable joints.
For example, a 10 degree change in the angle of the scapula

and the humerus will change height at the withers by 2 or 3 in. This is the reason that the casual observer reports seeing cattle of the same height that have shorter or longer bodies; thus, they fail to accept the constant proportion data. They should because the evidence is extensive and the conclusion logical.

In view of the observation that frame size predicts mature frame size and maturity end point, it follows that if cattle are sorted as to frame size and are in similar condition as they go on feed, they reach the choice grade at approximately the same time. Therefore, a sort based on frame size has helped us market pens of cattle of uniform quality grade. However, within each of these frame sizes of the same quality grade, there can be found wide ranges in cutability and here we must consider degree of muscling. Heavily muscled cattle not only produce carcasses with larger loin eyes but also tend to be leaner, and as a result we can observe wide differences in cutability within the same frame size. This suggests an additional criterium for sorting--that of degree of muscling, which, in turn, raises the question of how to identify degree of muscling. Here again we are fortunate since Butterfield (1964) and Kauffman (1973) have both reported a constant proportion between muscles. This fact allows us to measure degree of muscling in an exposed area of the animal's body and use it as a measure of total musculature because a single muscle represents a constant percentage of total muscle mass in all cattle.

These observations suggest a simple sorting criteria of frame size and degree of muscling that can be stated as follows:

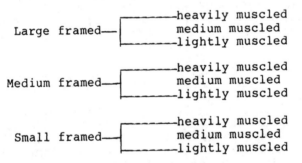

```
                       ┌───────heavily muscled
    Large framed──────┤        medium muscled
                       └───────lightly muscled

                       ┌───────heavily muscled
    Medium framed─────┤        medium muscled
                       └───────lightly muscled

                       ┌───────heavily muscled
    Small framed──────┤        medium muscled
                       └───────lightly muscled
```

This is exactly the system used in the USDA's Grades of Feeder Cattle. This method may not be perfect, but at least it is a step in the right direction--in my opinion; it is far better than what we have tried up to this point and certainly better than no sort at all.

REFERENCES

Butterfield, R. M. 1964. Relative Growth of the Musculature of the Ox. In: D. E. Tribe (Ed.) Carcass Composition and Appraisal of Meat Animals. The Commonwealth Scientific and Industrial Research Organization, Melbourne, Australia.

Conner, Fred. 1974. Beef cattle evaluation. National Livestock Testing Assn. Report.

Dikeman, M. E., M. D. Albrecht, J. D. Crouse and A. D. Dayton. 1976. Visual appraisal of bovine cannon size related to performance, carcass traits and actual metacarpus measurements. J. Anim. Sci. 42:1077.

Kauffman, R. G., R. H. Grummer, R. E. Smith, R. A. Long and G. Skook. 1973. Does live-animal and carcass shape influence gross composition? J. Anim. Sci. 37:1112.

Nelson, L. A. and R. e. Humsley. 1976. Evaluation of body type and muscling of cattle. The Simmental Journal. 3:18.

Ramsey, C. B., R. C. Albin, R. A. Long and M. L. Stabel. 1976. Linear relationships of the bovine skeleton. J. Anim. Sci. 43:221.

Part 18

ENVIRONMENT, FACILITIES, AND ANIMAL WELFARE

120

MEASURING AN ANIMAL'S ENVIRONMENT

Stanley E. Curtis

ASSESSING ANIMAL ENVIRONMENTS

Animal environments are characterized according to problems suspected of being associated with environmental stress on the animals and the effectiveness of control measures. Most of the elements of animal environments can be measured but, interpreting the results in terms of animal well-being, facility operation, and production economics often remains a dilemma because of the interaction of environmental factors. The effect of one stressful factor on an animal quite often depends on the nature of the rest of the environment.

ANIMAL ENVIRONMENT PROBLEMS

Troubleshooters often must engage in trial-and-error to identify animal-environment problems, but there are several points to be kept in mind:
- The environment results from all external conditions that the animal experiences, so all elements must be considered. Those which cannot be measured or controlled readily might influence animal health and performance nonetheless, so even they must be considered so far as is possible.
- The environmental complex acts as a whole on the animal, so interactions must be kept in mind. The combined effects of two or more environmental components may be difficult to evaluate, but they must be considered.
- Time and space affect the environmental factors. Environmental variables should be measured where the animal experiences them--in its micro--environment--taking into account the lack of spatial uniformity that occurs in all facets of the surroundings. Most important are vertical stratification of various parameters of the thermal environment and horizontal and vertical

variation in airflow due to design or mode of
operation of the ventilation system.
- Environmental elements change with time at a
  given place. Because weather and facility occu-
  pancy vary with time, control requirements do
  also. Thus, environmental assessments and control schemes must take daily and seasonal environmental cycles into account. Most animals adjust to environmental cycles readily as long as
  extremes of the excursions are not unduly
  stressful.
- The rate of environmental change is critical.
  Abrupt environmental changes tend to be more
  stressful than those occurring over a longer
  period. For example, preconditioning young animals to a cool environment before moving them
  from a warm, closed house to a cool, open one
  during cold weather reduces the stress. It is
  sometimes difficult to identify a single index
  of environmental stress or even an adequate multiple index or combination of indices. For example, daily temperature range per se may be of
  little consequence as long as the day's maximum
  and minimum values do not exceed or fall below
  respective trigger levels. Likewise, a certain
  rate of temperature change might be stressful if
  it occurs at extremes of temperature, but not
  within more moderate ranges. With modern statistical techniques, it is possible to develop
  many environmental indices, with relative ease
  but it is a very difficult job to determine
  their respective significances in terms of the
  environment's impingements on the animal.
- Animals modify their own environments by giving
  off heat, water vapor, urine and feces, disease
  causing microbes, and others. The animals' own
  processes help determine the nature of their
  microenvironment. Changes in age or number of
  animals in a facility alter these impacts and,
  therefore, the control measures required.
- Anthropomorphism is a common pitfall in the
  assessment of animal environments. A comfortable environment for a human is not necessary
  for an animal. Animals send signals of discomfort or uneasiness to alert caretakers. These
  behavioral indications are always useful signs
  that the environment could be improved. In some
  cases, the way animals behave is the only clue
  that stress is present.

MEASURING ENVIRONMENTAL FACTORS

A well-planned and organized approach to environmental-measurement programs is important. The means employed will depend on the amount of accuracy, the extent of detail required and the effort devoted to the actual work of measurement. Insights into sampling theory, as well as descriptions of some of the instruments and techniques that have proved especially applicable to measuring outdoor and indoor animal environments, are provided in the following sections.

Sampling Theory

Environmental assessments are based on interpretations of measurements of pertinent variables. Hence, the observations must represent the situation faithfully. To ensure this, environmental sampling programs must be planned carefully. The main reasons for this have been alluded to already: most environmental elements vary over time and space. Some of these variations are regular and predictable, others are not. The times and the places the environment is measured determine whether the resultant information reflects the character of the environment well enough to be of use. Of course, it is possible to gather more data than necessary, too.

Time considerations. To estimate the average impingement of an environmental factor over a period of hours or days, or to learn about excursions of these values over time, a continuous sample is needed. In short, continuous sampling from more or less permanent instrument stations, usually coupled with recording equipment, provides the data needed for detailed analyses of animal environments.

The most important consideration in regard to time is the length of the observation period. It should be a multiple of a well-established environmental cycle. In studying an animal facility in a temperate climate, for instance, seasonal periodicities must be accounted for to appreciate the facility's nature all year long. On the other hand, a particular problem may be limited to one season, in which case the observation period would be a multiple of the day, to account for diurnal cycles in meteorologic phenomena.

Of course, there is considerable variation among years and even among days within seasons. For example the number of cycles--the number of years or days--that should be observed to give meaningful results depends partly on the nature of this variation already known to occur in the facility's locale. Furthermore, interpretation of any results should include an historical perspective. For example, if observed values for air temperature are at the lower end of the acceptable range during a winter known to be relatively mild by local standards, the problem of a too-cold animal micro-environment might well be encountered during a more nearly normal winter.

The frequency of observations needed within a sampling period is another important decision. In general, measurements should be made as frequently as feasible, especially if automatic recording equipment is being used. Then, after observations have been completed, key periods of environmental extremes or change can be evaluated in detail, while less interesting periods can be ignored. Also, when observations are made too infrequently, errors can be made in characterizing both the ranges and the average values of the variables.

Time-averaging can obscure extremes of environmental factors that may trigger animal responses, hence it must be done only when warranted. For example, effective environmental temperature in an outdoor environment might range from -10° to 20°C one day, from 0° to 10°C another. Average temperature might be around 5°C on both days but the nature of the animals' thermoregulatory reactions would be different on these days, and thus averaging the environmental data over a day could lead to misleading impressions.

Nonlinearity in interactions among environmental factors causes additional difficulties so far as the time-averaging of environmental measurements is concerned. In other words, effects of combined factors must be calculated carefully, even when their relations for steady-state conditions are well-known. Take as an example the case of the wind-chill index (table 1). The average of the wind-chill indices for the three sets of conditions in the table is 1.6 x 10^3 Kcal m^{-2} hr^{-1}, while the wind-chill index for the average of the three conditions (namely, temperature -34°C and wind speed 9 m sec^{-1}) is 2.1 x 10^3 kcal m^{-2} hr^{-1}.

TABLE 1. AN EXAMPLE OF NONLINEAR INTERACTION AMONG ENVIRONMENTAL VARIABLES: WIND-CHILL INDEX

	Condition		
	A	B	C
Temperature (°C)	-18	-34	-51
Wind speed (m sec^{-1})	18	9	0
Wind-chill index (kcal m^{-2} hr^{-1})	1.8 x 10	2.1 x 10	.8 x 10

Time lags between environmental occurrences and animal responses also must be recognized--taking them into account could improve the probability of defining a connection between animal and environment.

Another approach is discontinuous sampling, sometimes called spot- or grab-sampling. This usually involves more portable equipment and is aimed at gaining information over short periods, such as hours. Discontinuous sampling is best suited to determining extremes in an environmental element when the basic nature of the variation of the factor is known. For example, air temperature might be measured only in the early afternoon on relatively hot days to gain information on upper values of air temperature, or concentrations of air pollutants might be measured in closed animal houses on relatively cold days when ventilation rate is relatively low.

Discontinuous sampling has several advantages. It requires less time and often less equipment. Further, fewer data are generated, so data-processing equipment requirements are less than when continuous sampling is practiced. Finally, equipment portability often facilitates economical use of a single instrument at many locations in a facility to learn more about environmental variation over space than might be feasible otherwise.

Space considerations. Two prime considerations should determine where environmental measurements are to be taken. In the first place, environmental factors generally should be measured in the immediate surroundings of the animals. For example, most elements of the environment vary consistently with distance above the floor, but most animals reside at discrete heights within a facility. Thus, the height at which the environment is measured is crucial if the measured values are to reflect the conditions to which the animals are being exposed.

In animal facilities it is sometimes tempting to sample the environment in an alleyway or some other place where the instruments will be relatively safe from animal damage. For the most part, these temptations should be resisted. The animals affect their own surroundings so greatly that most variables differ even from animal microenvironments to nearby areas where the animals are not permitted.

Second, the environmental and animal features known to affect the variable under study should be clearly in mind when the measurement sites are chosen. Major items include heat sources: air inlets and outlets; orientation of the facility to winds, the sun, and other structures; location of mechanical services, such as feed-delivery systems; and animal size and population density.

There also is the substantial problem of the effect of the measurement instrument itself, and its protective hardware, on the environment. Some equipment stations obstruct airflow, for example, and the air samplers may be drawing air from different heights and extracting components from the air around the sampler, thereby modifying it.

Instrument Choice

Dozens of instruments for measuring environmental factors are available on the commercial market today. The choice of instrument or set of instruments to be used in a given environmental-measurement program is based on several considerations: (1) the kind of information needed, (2) the relative efficiencies of the various instruments and their reliabilities under field conditions, (3) ease of use, cost, and availability, and (4) personal choice--often based on past experiences--is an important point.

These instruments have accompanying instructions. If the manufacturer's written advice proves inadequate, get in touch with a technical representative of the manufacturing firm directly; most will assist customers by mail or telephone with problems in specific applications of their product.

In the sections that follow, some instruments commonly used in measuring environmental factors in animal facilities are described in brief detail.

MEASURING AIR TEMPERATURE

Liquid-In-Glass Thermometers

A variety of commercially available mercury- or alcohol-in-glass thermometers are used widely to measure air temperature. As temperature rises, the liquid expands to occupy more of the capillary tube in which it is held. Of course, as in all thermometry, the temperature registered is actually that of the thermometer, not necessarily that of the environment, so factors apart from the temperature of the surrounding air that affect thermometer temperature lead to errors. Chief among these are solar and thermal radiations and air movement. Precautions must be taken to shield the thermometer against them. Furthermore, the thermometer may be placed in a spot where the air is stagnant or otherwise unrepresentative of the general area.

Lag time. When environmental temperature changes, the value required for a thermometer to reduce the difference between registered temperature and actual air temperature to 36.8% of the original difference is called the lag time of the thermometer. Lag time for most mercury-in-glass thermometers is around 1 min, while that for alcohol-in-glass is roughly 1.5 min. Some electronic thermometers have much shorter lag times. Especially in discontinuous sampling, lag time is an important consideration because a measurement can be made more quickly with the shorter lag time.

Maximum-minimum thermometer. A thermometer designed to register the maximum and minimum temperatures experienced during the measurement period often has been used in animal facilities. It consists of a U-tube with bulbs at both ends. One side of the U serves as the scale for maximum temperature, the other for minimum. The bulb on the maximum side serves as a safety reservoir and is partially filled with a liquid such as creosote solution. That on the other side is completely filled with the liquid. Between these two portions of liquid, in the bottom of the U-tube, is mercury. As the thermometer becomes warmer, the liquid in the filled bulb expands, pushing the mercury up on the maximum-scale side (the opposite side). Atop the mercury on both sides is an iron index (a sliver of iron), and as the mercury moves up the maximum side it pushes the index ahead of it. As the thermometer cools, the mercury retracts, but the index remains at its highest point due to friction with the inside of the U-tube that can be overcome by the mercury, but not by the other liquid. Of course, as the mercury column retracts upon cooling, it pushes the minimum-temperature side's index ahead of it, so this one registers the lowest temperature of the measurement period on the max-imum-temperature scale, which is upside-down. When maximum and minimum temperature have been observed, the thermometer is reset by replacing the indices atop respective mercury meniscuses by means of a magnet.

Although maximum and minimum temperature may be all the information needed in some situations, and despite the fact it is relatively inexpensive and straightforward in design, this instrument has serious drawbacks. Chief among them are the tendencies for the mercury column to become separated or broken and for the indices to become permanently lodged in the U-tube.

Bimetallic-strip thermometer. The sensor of a bimetal-lic-trip thermometer comprises a sheet of each of two metals having dissimilar coefficients of linear thermal expansion, which are joined along their faces. When such a strip's temperature changes, it bends. Lag time is relatively short--around 10 seconds.

Thermograph. The bending movement noted above for the bimetallic strip is usually magnified by shaping the strip appropriately--and in a thermograph the movement is recorded by affixing a pen to its free end and applying this pen to a piece of graph paper on a drum rotated by a clockworks. By this relatively inexpensive means, a permanent record of temperature variation is made. Further, it requires no electrical supply. Of course, a thermograph is so large that it affects the microenvironment it is used in and, be-cause of its construction, is prone to error due to radia-tion and stagnant air pockets.

Calibration of a thermograph is a critical matter. The recording element is very sensitive to physical shock, which often occurs when the instrument is moved from place to place, thus, after the instrument is moved, it is absolutely necessary to calibrate a thermograph by adjusting the re-cording pen several days in a row, preferably near the times of the daily high and low temperatures. An artificially ventilated psychrometer is commonly used as the standard in-strument for calibration of thermographs.

Electric thermometers

There are two general kinds of electric thermometers. One kind depends on the principle that as the temperature of a substance changes, so does its electrical resistance. The other kind--thermocouple thermometry--depends on the principle that when wires of two specific metals are joined at both ends, and when these two junctions are kept at two temperatures, an electromotive force is generated in that circuit. Voltage in such a circuit, when measured with a potentiometer, is directly proportional to the temperature difference between two junctions.

Thermistor. Certain semiconducting materials have negative temperature coefficients of electrical resistance: as temperature rises, resistance decreases. Resistance in the circuit to which a current has been applied is measured by an indicating unit. These sensing elements are called thermistors--parts of the electric thermometers most applicable to air thermometry in animal environments.

Thermistor probes, indicating units, and recording units are commercially available in a wide range of models. At one time, thermistor systems were notoriously unstable, but nowadays stable, calibrated probes are on the market and in recent years they increasingly have become the sensors of choice for routine assessment of animal environments.

Thermocouples. A variety of combinations of dissimilar wires are used for the two sides of the circuit in thermocouple thermometry. Copper and constantan are frequently chosen. One of the thermocouple junctions (the reference junction) is held at a constant or known temperature so changes in electromotive force measured reflect changes in the temperature of the measurement junction, which is placed in the environment to be monitored. The voltage generated can be used to drive a millivolt recorder or registered on a millivoltmeter.

In general, use of thermocouples for air thermometry in animal facilities has some disadvantages. The needed equipment, especially the constant-temperature bath for the reference junction, is relatively cumbersome, the physical integrity of the measurement and reference junctions critical and sometimes difficult to maintain, and careful calibration of the thermocouples very important.

Metal-resistance thermometer. In metals, as temperature rises, so does electrical resistance. Small-diameter platinum wire wound on a support having a small thermal expansion coefficient is a frequent choice. A small current is introduced into the wire, and resistance of the whole circuit is measured using a Wheatstone bridge circuit. The most sensitive resistance thermometers tend to be fragile, and for this reason alone they are of limited use in animal environments. In addition, the measuring equipment is relatively expensive.

MEASURING AIR MOISTURE

The measurement of water vapor in air is called psy-chrometry or hygrometry. Several principles have been used to measure air moisture, and three have been applied widely in quantifying animal environments.

Hair Hygrometer

Hair is hygroscopic, and its length is related directly with the amount of water it contains. Further, there is a nonlinear, direct correspondence between length of hairs and the relative humidity of the air surrounding them. Hair hygrometry is most accurate when relative humidity ranges between 20% and 80%. This principle has been employed extensively in hygrometry in animal facilities, partly because hair hygrometers require no electrical supply and they are affected little by other factors.

Hygrograph. Elongation and shortening of hair bundles can be magnified by an appropriate level system and trans-formed into movement of an arm holding a pen that is applied to graph paper on a rotating drum, giving a record of changes in relative humidity. Just as for the thermograph, hygrographs must be calibrated carefully over a period of several days after they have been moved to a new location before reliable measurements can be made.

Psychrometers

Psychrometers are a class of instrument by means of which the air's moisture content or relative humidity can be estimated indirectly. They use both a wet-bulb thermometer and a dry-bulb, and their measurement principle is based on the thermodynamic relation between the air's moisture content and wet-bulb temperature. (Wet-bulb temperature is affected by dry-bulb temperature and air pressure.) Once dry-bulb and wet-bulb temperatures of the air are known, the air's moisture content and relative humidity can be estimated from a psychrometric chart.

The dry-bulb thermometer can be of any type, while the wet-bulb thermometer is a similar instrument having a water-saturated wick closely surrounding the bulb or sensing the element. As the psychrometer is operated, evaporation from the wick occurs, and the temperature of the wet bulb is depressed. Of course, the drier the air, the greater the evaporation, and the greater the wet-bulb depression.

To give an accurate estimate of wet-bulb temperature, the thermometers must be ventilated adequately; maximum cooling of the wet bulb does not occur in still air. Also, temperature readings must not be taken until equilibrium has occurred. Other sources of error when using any psychro-meter are heat conduction down the wet-bulb thermometer (the wick is ordinarily extended up the stem), receipt of solar and thermal radiation (the latter even from the operator of the instrument), and wicks that are too thick or dirty (one way to minimize mineral crust is to wet the wick with distilled water only).

Wet-bulb temperature should always be read before that of the dry-bulb, as it will begin to rise as soon as ventilation ceases. It is also good practice to repeat the measurement several times to make sure the lowest wet-bulb temperature has been attained.

Sling psychrometer. The sling psychrometer, once the standard instrument for spot-sampling psychrometry, consists of dry- and wet-bulb liquid-in-glass thermometers in a frame that can be revolved around a handle. The thermometers are whirled--usually for a minute or more--to permit wet-bulb depression to occur. Larger sling psychrometers must be revolved at least two times per second to provide sufficient ventilation, and smaller ones five times. The movement of the operator mixes the air in the region.

Artificially ventilated psychrometer. Various models of psychrometer are now available in which air is drawn artificially past the temperature sensors. One popular model employs dry-cell batteries that supply a small fan, which pulls air at speeds up to 5 m sec^{-1} past the bulbs. Of course, this kind of instrument may draw air from as far away as 1 foot, hence it is not applicable to some microenvironmental measurements. Still, it is reliable, rugged, and portable.

Electrical Conductivity of Hygroscopic Materials

Salts such as lithium chloride are hygroscopic, and their electrical conductivity increases--thus, their resistance decreases--with increasing water content.

Electric hygrometer. Commercially available sensing units usually involve a film of lithium chloride on a nonconducting frame through which an electrical current is passed for the purpose of measuring electrical resistance changes. The logarithms of the resistance and the atmospheric humidity parameters are inversely related.

Dew cell. When a film of lithium chloride is applied to a heating-element frame, and the temperature is so high the salt is dry, the salt is also highly resistant to conducting electricity. The electrical circuitry of a dew-cell apparatus is designed so that when the salt is a conductor, the element is being heated, but the heating stops when the salt becomes warm enough to become dry. Thus, the lithium-chloride film is kept more or less at the same temperature and dry at all times. This equilibrium salt temperature is related directly with the air's dew-point temperature.

Electric hygrometers and dew cells make it possible to monitor air humidity continuously, but in animal environments they often become dirty, and this can lead to errors. Further, they remain operational for periods of only a few months under the best of conditions.

MEASURING AIR MOVEMENT

Drafts, stagnant spaces, and inadequate removal of moisture or noxious gases in animal houses are among the common symptoms of improper design or operation of a ventilation system. Air speed and distribution throughout an animal facility must be known if ventilation problems are to be remedied.

Anemometers

Several kinds of instruments to measure air speed are available. Each is best suited to a particular application in animal-environment measurement.

Pitot-tube measurement. When air moves into or across the mouth of an open tube, the air pressure in the tube changes; it increases in the former case, decreases in the latter. Such pressure changes are proportional to air speed and serve as the basis of pitot-tube anemometry.

The most common pitot-tube anemometer used today is the Velometer, a rugged and portable instrument that gives a direct reading of air speed and comes supplied with a variety of probes for different velocity ranges. This instrument is most adaptable to measuring air velocity at inlets, in areas of strong drafts, and in air ducts.

Because of its very nature, the pitot-tube is extremely directional; it is sensitive to air movement in one direction. Large errors can result when a probe is not properly oriented.

Hot-wire anemometer. Several modes of hot-wire anemometer are on the market. Some are directional and therefore applicable to measuring air speed in ducts and at inlets; others are more nearly omnidirectional. They operate on the principle that as air passes across a fine platinum or nickel wire that has been heated electrically, the wire tends to cool by an amount proportional to air speed. This instrument is designed so current flowing through the wire automatically changes so as to keep wire temperature nearly constant. The amount of current required to achieve this is thus related directly with air velocity.

Hot-wire anemometers are relatively sensitive and thus especially applicable to situations--such as animal microenvironments--where velocity can be as low as .5 cm sec-1. Another advantage is that some designs are very portable and fairy rugged, and so small they do not interfere much with the environment. They also respond very quickly to changes in air speed. However, the sensing wire can be affected over time by atmospheric pollutants, and for this and other reasons frequent calibration is necessary. Further, the sensing wire itself is exceedingly fragile and subject to damage. The hot-wire anemometer cannot be used outdoors during rainy periods or in any environments where water can reach the sensing wire.

Rotating-vane anemometer. There are various kinds of anemometers employing a lightweight propeller with eight or more blades. All are highly directional, and for applications where air direction changes greatly they are commonly attached to a vane device that keeps them aimed into the main flow of the air.
A variety of physical and electronic metering devices are used in conjunction with these anemometers. They are most useful at high air speeds, and thus are most applicable in outdoor settings. Indoor models are relatively fragile and subject to damage by corrosive gasses.

Cup anemometer. The device used to measure air velocity outdoors consists of several cups, each connected to a rod radiating from a rotor. A revolution of the assembly is directly related to air velocity. Cup anemometers are not suitable for measuring low air speeds. They are most applicable to estimating the average speed of the wind during periods of at least several hours.

Kata thermometer. A simple instrument was developed over a century ago to estimate the cooling power of the air. The kata thermometer is capable of measuring very low air velocities. It is simply an alcohol-in-glass thermometer with the stem marked at the 37.5° and 35°C levels.
The thermometer is warmed in a water bath to around 40°C, removed from the water and wiped dry, and placed in the environment to be measured. The time required for temperature to fall from 37.5° to 35°C is determined using a stopwatch, and this value substituted into a formula (which also includes air temperature and an individual instrument-calibration factor supplied by the manufacturer) for calculating air velocity.

Airflow-pattern measurement

Certain visible particles suspended in the air can be used as an aid to tracing how the air is distributed in an animal house. The source of the particulate matter is simply placed at the air inlet or in the microenvironment to be studied, and the course of the pollutant followed visually through the space of interest. Much cigar and pipe smoke has been used quite effectively for this purpose in the past. More reliable sources of larger amounts of visible tracers are now on the market.

In the absence of an anemometer, low and moderate air speeds can be estimated to a first approximation by briefly interrupting the tracer's flow and monitoring the movement of the turbulence so induced with the aid of a measuring tape and a stopwatch.

"Smoke"vials. When a solution of titanium tetrachloride is exposed to air, it gives off copious whitish fumes. Small glass vials of this solution areideal for use in studying airflow patterns in microenvironments or parts of a room. A vial is broken open and simply can be held in the location to be observed. An open vial is commonly placed in a cup, which may be attached to a pole in order to reach certain parts of the room of interest. Several devices are also available to increase the evolution of fumes by passing air over the solution. Titanium tetrachloride fumes in the concentrations used are not toxic to animals.

Talcum aerosolizer. Fine talcum powder is also used as a tracer in small-scale air-distribution studies in animal houses. Special talcum-powder aerosolizers are available commercially. Talcum tends to precipitate faster than does titanium tetrachloride, and it is generally more difficult to generate an adequate aerosol of the powder. For these reasons, it is inferior to titanium tetrachloride for this purpose.

"Smoke" pellets and candles. When long-term or large-scale observations are desired, more tracer might be needed than can be supplied by titanium tetrachloride or talcum aerosols. The commercial market has a wide range of pellets and candles that, when lighted, produce very large amounts of fumes. These are generally set in the way of the incoming air. The fumes of some of these devices are toxic and therefore must not be used in occupied houses.

Static-Pressure Measurement

In negative-pressure or exhaust-ventilation systems, adequate negative-static pressure must be maintained by the fans. Static pressure inside an animal house is usually measured by means of a manometer sloped 1:10 to increase accuracy. One end of the manometer is open to the outside atmosphere, the other to the inside, and the pressure difference is registered by the manometer.

Every animal house employing mechanical ventilation should be equipped with its own static-pressure manometer.

MEASURING SURFACE TEMPERATURE

The temperature of an animal's surface is an important determinant of convective and radiant exchanges of heat.

The temperatures of environmental surfaces likewise play a central role in determining the magnitude and direction of thermal-radiant flux. Accurate measurement of surface temperature is difficult, but there are two ways this can be approached.

Portable radiation thermometer. A variety of battery-powered instruments now available provide rapid measurement of surface temperature by a technique that doesnot involve contact with the surface. The portable radiation thermometer is simply aimed at the surface to be measured, and the surface's temperature is registered on the meter. Models providing different fields of view are available.

This kind of thermometer must be calibrated before every observation by means of a Leslie cube or some other temperature-calibration device. Solar radiation reflected by a surface leads to overestimation of that surface's temperature when measured with this instrument. Also, surfaces in the surroundings substantially cooler than that being measured tend to bias these estimates in the opposite direction. The instrument is reliable, reasonably rugged, and is an excellent means for spot-sampling animal and environmental surface temperatures.

Contact thermometers. Thermometers in a variety of styles involving contact with animal and environmental surfaces have been used sometimes to measure surface temperature. These include thermistor probes in hypodermic needles, contact discs (banjo probes), as well as thermocouples taped or glued to surfaces.

There are serious drawbacks to contact thermometers. First, the contact must be flawless--otherwise insulative air pockets, even very small ones, between thermometer and surface will introduce measurement errors--and this is rarely achieved on either animal or environmental surfaces.

There are other problems of introducing artifacts with animal surfaces in particular. It is a practical impossibility to achieve the necessary contact with covered areas of an animal's surface without disrupting that cover and altering surface temperature. Even on nude areas, affixing a thermometer to the skin in such a way as to ensure adequate contact alters cutaneous blood flow and surface temperature.

MEASURING SOLAR AND THERMAL RADIATION

Solar Radiation

The total direct and diffuse (sky) solar radiation received by a horizontal surface is measured by an instrument called a pyranometer. Several designs are available on the market. The standard instrument in the United States is the Eppley pyranometer.

Eppley 180° pyranometer. This instrument consists of horizontal concentric silver rings, one black and one white. The glass hemisphere that covers the rings transmits only radiation with wavelengths less than 3.5 e_m (solar radiation), for which the black and white rings have different absorptivities. The temperature difference between the two rings generates an electromotive force in a thermopile (a series of thermocouple junctions), alternate junctions of which are in thermal contact with respective rings.

The Eppley 180° pyranometer measures direct and diffuse solar radiation received by a horizontal flat surface from the upper hemisphere. When direct sunshine is blocked by a shade, only the sky radiation is measured.

This instrument is fragile and must be sited carefully. Its glass bulb must be cleaned daily.

Thermal Radiation

The rate of incoming thermal radiation is usually estimated as the difference between the rate of total incoming radiation having wavelengths between .1 and 100 e_m, and the rate of total incoming solar radiation as estimated by an unshaded pyranometer. The rate of total incoming radiation can be measured by any of several kinds of total radiometer.

Beckman and Whitley total radiometer. This instrument is ventilated to minimize errors due to variable convective heat loss and measures the rate of incoming radiation at wavelengths between .1 and 100 e_m. It essentially consists of three layers of plastic: the top one is painted black and exposed to radiation from the upper hemisphere; the middle one contains a thermopile; and the lower one is shielded to minimize its receipt of radiation. The temperature difference between top and bottom of the middle plate is related directly to the upper plate's rate of radiation absorption.

MEASURING LIGHT INTENSITY

Photovoltaic Meters

Light meter. A selenium photovoltaic cell comprises the basis of most light-intensity meters used by photographers. Illumination of the sensitive layer of selenium by visible radiation (wavelengths from .3 to 7 e_m set up a flow of current in an appropriate circuit. Photographic light meters are portable and very useful in measuring light intensity in animal environments.

In most commercially available light meters, sensitivity over the visible spectrum is trimmed to resemble that of the human eye; that is, it peaks at a wavelength of about .55 e_m instead of the .63 e_m wavelength radiation that drives photoperiodisms in animals.

Illuminometer. A recording hemispherical photometer called the Illuminometer is also on the market. It is particularly suited to stationary installations where light intensity is known to vary over time, such as it does outdoors.

MEASURING SOUND LEVEL

Sound-level meter. Several battery-powered models of sound-level meter are available commercially. All provide a simple, portable, and reliable means of measuring sound level in decibels. The microphones on such instruments are relatively nondirectional, and the operation of sound-level meters is straightforward. Necessary precautions include recognizing the possible presence of obstacles to sound waves; locating the microphone at the observer's side, not between observer and sound source; shielding the microphone from any moving air; and making sure interfering electromagnetic fields from other electrical equipment are accounted for.

MEASURING AIR PRESSURE

Aneroid barometer. Measurement of air pressure in conjunction with animal production is ordinarily accomplished by means of an aneroid barometer--an instrument in which the walls of an evacuated cell move as air pressure changes. The movements are transmitted to a pointer, which indicates air pressure. While not as accurate as a mercury barometer, the aneroid version is nonetheless quite useful in animal work. If moved, it must be recalibrated against a mercury barometer.

Barograph. Movements of an aneroid barometer cell's wall can be recorded on graph paper affixed to a rotating drum when a pen is linked to that wall.

MEASURING AIR POLLUTANTS

Measuring Aerial Gases and Vapors

Colorimetric indicator tubes. Several systems of the same general type are available commercially for the convenient and, when properly used, reasonable accurate measurement of aerial gases and vapors. These consist of an indicator tube and a precision piston or bellows pump operated manually to draw air. The detector tube contains a specific chemical that reacts with the gas or vapor being measured. These small detector tubes are available for all the major gases and some of the vaporous compounds commonly present in animal-house air.

When air is pulled through an indicator tube, the pollutant for which the tube's indicating gel is specific reacts with the chemical, resulting in discoloration. The extent of this change is related to the concentration of the pollutant in the air. One problem with such a measurement system is that gases associated with dust particles--for example, some of the ammonia in dusty air--are filtered out of the air before it reaches the colorimetric indicator. This tend to cause underestimation of the pollutant's concentration in the air.

The pump for this kind of system must be kept leakproof and must be calibrated. Also, the detector tubes must be handled carefully and their predicted shelf lives observed.

Measuring Aerial Dust

High-volume sampler. Several models of dust samplers in wide use draw through a filter made of cellulose paper, glass or plastic fibers, or organic membrane. Dust particles too large to pass the filter are collected on it.

In practice, a filter is dried and tared before sample collection, and the particle-laden filter is dried and weighed again at the end of a sampling period. The difference between the two is an estimate of the mass of the particles collected during the sampling period. This usually is divided by the product of sampling period and average airflow rate to give the concentration of the pollutant. Filters are commonly handled with tweezers and transported outside the laboratory in large, covered petri dishes.

Another critical factor is calibration of airflow rate. most high-volume air samplers are equipped with some sort of airflow meter, but these instruments should be calibrated frequently because errors in this estimate are perpetuated as errors in all concentration estimates.

Particle counting and sizing. Several dozen models of instruments to count airborne particles are on the market. Some are primarily collectors--based on the principle of impaction on a solid surface, impingement in a liquid medium, centrifugation, or settling--and used in conjunction with subsequent visual observation. Others are direct-reading instruments employing optics and electronics.

Impactors are most commonly used for discontinuous sampling in animal environments. Two popular instruments are the Anderson six-stage, stacked-sieve, nonviable sampler, and the four-stage cascade impactor. Both feature impaction of dust particles on pieces of glass, which are then inspected microscopically for counting. As for sizing of the particles, both instruments have several stages designed so that the polluted air is drawn through a series of jets with progressively smaller cross-sections. The result is that relatively large particles are impacted in early stages, smaller ones at later stages. The size ranges monitored by these instruments are pertinent to the site of deposition of the particles within an animal's respiratory tract.

MEASURING AERIAL MICROBES

Qualitative studies of airborne microbes in animal environments have long involved opening a petri dish of culture medium, permitting viable particles to settle out of the air onto the medium's surface. Of course, special media can be used when there is interest in particular kinds of microbes. When the aerial concentration of microbes or microbe-carrying particles must be determined, another method must be used.

All-glass impinger. One method of quantifying airborne microbes is to impinge them in an isotonic solution that can then be diluted appropriately, combined with nutrients, and cultured in preparation for counting. This method is well-adapted to situations where aerial microbic level is high, but has the disadvantage of disintegrating airborne particles containing more than one microbe so that, for instance, an airborne particle that would give rise to one colony in an animal's respiratory tract might give rise to hundreds to be counted in the culture dish.

Andersen six-stage viable sampler. The viable version of the Andersen stacked-sieve sampler holds special culture-medium plates instead of flat pieces of glass as in the nonviable model. Particles are impacted onto the solid medium's surface where colonies can grow and be counted.

The Andersen viable sampler is a very useful instrument for both counting and sizing airborne microbic particles in animal environments. Special media can be used when desirable, and both the size and the number results can be interpreted in terms of the challenge the aerial microbic particles present to the animals' respiratory tracts. On the other hand, because of its relatively high air-sampling rate, the Andersen sampler is less well-adapted to air environments in which microbic populations are very high, as in some closed animal houses during cold weather. In such case, the sampling period may have to be short as 15 seconds, and thus special care must be exercised to ensure accuracy in estimating the volume of air drawn through the instrument during the sampling period.

Andersen disposable two-stage viable sampler. A less expensive device is a disposable-plastic, two-stage sampler fashioned after the original Andersen six-stage model. Commercial petri dishes available in hospital microbiology laboratories and a variety of vacuum sources can be used with this system.

This instrument has a critical orifice providing an air-sampling rate of 1 ft^3 per min when a vacuum of at least 10 in. of mercury is maintained. When operated in this way, the colonies that grow on the upper stage have arisen from particles having an aerodynamic diameter greater than 7 em, and hence they would not have deposited in the lungs of an

animal. The particles that are impacted on the lower stage
are between 1 and 7 e_m in diameter, and many of these could
have reached the animal's lungs.

Like the Andersen viable sampler, the disposable sampler sometimes must be operated for a short sampling period
in commercial animal houses.

Also, the collection efficiency of the Andersen disposable sampler seems to be less than that of the standard version. Despite these drawbacks, the disposable model is an
inexpensive, relatively accurate means of estimating the
concentration of microbe-bearing particles in the air, and
whether they are of such a size as to directly threaten pulmonary health.

1112

REFERENCES

Anonymous. 1972. Air Sampling Instruments for Evaluation of Atmospheric Contaminants. Fourth Ed. Am. Conf. Gov. Indust. Hygienists. Cincinnati.

Curtis, S.E. 1981. Environmental Management in Animal Agriculture. Animal Environment Services, Mahomet, Illinois.

Gates, D.M. 1968. Sensing biological environments with a portable radiation thermometer. Appl. Optics 7:1803.

Hosey, A.D. and C.H. Powell (Eds.). 1967. Industrial noise--a guide to its evaluation and control. Pub. Health Serv. pub. 1572. U.S. Gov. Printing Off., Washington.

Johnstone, M.W. and P.F. Scholes. 1976. Measuring the environment. In: Control of the Animal House Environment. Vol. 7, Laboratory Animal Handbooks. Laboratory Animals, Ltd., London.

Kelly, C.F. and T.E. Bond. 1971. Bioclimatic factors and their measurement. In: A Guide to Environmental Research on Animals. Nat. Acad. Sci., Washington.

Munn, R.E. 1970. Biometerological Methods. Academic Press, New York.

Platt, R.B. and J.F. Griffiths. 1972. Environmental Measurement and Interpretation. Krieger, Huntington, NY.

Powell, C.H. and A.D. Hosey (Eds.). 1965. The industrial environment--its evaluation and control. Pub. Health Serv. Pub. 614, U.S. Gov. Printing Off., Washington.

Schuman, M.M., et al. 1970. Industrial Ventilation. Eleventh Ed. Am. Conf. Gov. Indust. Hygienists, Cincinnati.

Spencer-Gregory, H., and E. Rourke. 1957. Hygrometry. Crosby Lockwood, London.

Stern, A.C. (Ed.). 1976. Measuring, Monitoring, and Surveillance of Air Pollution. Vol. III, Air Pollution. Third Ed. Academic Press, New York.

Tanner, C.B. 1963. Basic instrumentation and measurements for plant environment and micrometerology. Soils Bull. 6, Univ. of Wisconsin, Madison.

Wolfe, H.W., et al.   1959.   Sampling Microbiological Aerosols.   Pub. Health Serv. Pub. 686. U.S. Gov. Printing Off., Washington.

121
MEASURING ENVIRONMENTAL STRESS
IN FARM ANIMALS

Stanley E. Curtis

Relations between agricultural animals and their surroundings always have been important. Those species recruited for domestication generally differ from their wild cousins in that the domesticated animals are adaptable to a wider range of environments than are their cousins (Hale, 1969). Hence, these animals we keep are more amenable to being confined and managed by the humans they serve (Bowman, 1977).

Ecology always has been at the heart of animal production. The shelter aspect of environmental management has been applied for a long time. shepherds kept their flocks in folds at night thousands of years ago. Only with the advent of widespread spacewise and time wise intensiveness in animal agriculture have animal-environmental relations become so important relative to other factors of production. And only with this intensiveness has major environmental modification been possible not to mention economically feasible. Now in addition to increasing the fit of the animals to the environment, we are coming closer to meeting the animals' needs by modifying their environments.

A DIGRESSION

"Measuring Environmental Stress in Farm Animals," calls to mind that which we should keep in mind. Let us examine the last five words of the title first and the first word last.

Environmental Stress in Farm Animals

Stress is of the environment, not of the animals (Fraser, et al., 1975; Curtis, 1981). Nevertheless, we measure stress in the animals, not in the environment. An animal is under stress when it is required to make extreme functional, structural, or behavioral adjustments in order to cope with adverse aspects of its environment. Thus, an environmental complex is stressful only if it makes extreme demands on the animal.

In other words, an environment is not stressful in and of itself; it is stressful only if it puts an animal under stress. And because animals differ in the ways they perceive and respond to the environmental impingements, the very same environment can be stressful to one animal and not to another.

An environmental factor that contributes to the stressful nature of an environment is called a stressor. When we "measure stress in an animal" we really measure the effects of the stress: the changes the stressor causes in the animal (such as the rise in body temperature when the animal is experiencing a net gain of heat from the environment) or the responses the animal invokes in an effort to establish a normal internal state in the face of a stressor (such as the rise in breathing rate when the animal needs to increase its heat-loss rate to bring body temperature back down to the desired point).

Measuring Environmental Stress

Scientists in a wide range of disciplines have been "measuring stress in animals" with increasing frequency over the past century and a half. Almost twenty years ago, The American Physiological Society published an epic tome of some 1056 pages called Adaptation to the Environment (Dill, 1964). The means of measurement have continued to develop as the scientific inquiry in animal ecology has blossomed profusely in the intervening two decades.

Yet the measurement of stress in animals is but the first step in applying ecological knowledge to animal production. The second step is the interpretation of the values. And of the two, the second step is by far the more difficult. In particular, it is necessary to determine where stress leaves off and distress (excessive or unpleasnt stress) begins.

Interpretation of stress parameters and indices is thus the real challenge as we continue to generate more knowledge and endeavor to use more completely what is already known for the purpose of increasing the fit between agricultural animals and their environments. And so it is this interpretation step on which we shall dwell.

STRESS RESPONSES: TRADITIONAL CONCEPTS

It is the unusual moment when an animal--in the wild or on a farm--is not responding to several stressors at once. Stress is the rule, not the exception. And nature has endowed the animals with a marvelous array of reactions to these impingements.

External environment comprises all of the thousands of physical, chemical, and biological factors that surround an animal's body. Each environmental factor varies over space and time. The animal's environment is therefore exceedingly complex.

The animal must maintain a steady state in its internal environment despite fluctuating external conditions. Claude Bernard (1957) said: "All vital mechanisms, however varied they may be, have only one object, that of preserving constant the conditions of life in the internal environment." This is the concept based on negative-feedback control loops that Walter Cannon later called homeostasis. More recently it has been called homeokinesis to emphasize its dynamic, yet consistent, nature.

All sorts of external environmental elements tend to modify corresponding internal environmental elements in an animal. Ultimately, if no homeokinetic mechanism acted, the internal environment would resemble the external, and life would cease.

Homeokinetic Control Loops

The homeokinetic animal attempts to control all aspects of its internal environment via adaptive responses similar in principle to a house's temperature-control system. Neural mechanisms participate in input reception and analysis, decision-making, and effector activation. Neuroendocrine mechanisms link neural and endocrine elements and activate effectors. Endocrine mechanisms take part in neural-endocrine and endocrine-endocrine linkages, as well as effector activation, and in some cases even effector action. These processes occur in specific configurations in the animal's many specific control loops. Muscles and glands are the body's chief effectors. Effector action is usually specific for the particular remedial reaction required.

Nonspecific stress response. In addition to specific stimulus/effector activation loops, Hans Selye (1952) has developed the concept of a nonspecific initial reaction to diverse stimuli. According to this facet of Selye's general adaptation syndrome, the rate of adrenal glucocorticoid secretion increases abruptly following any insult to the body. The teleological reason for this is that glucocorticoids promote mobilization of proteins from tissues. The amino acids liberated in this way can be used either as fuel or for synthesis of other proteins, such as immunoglobulins or scar tissue, that might be crucial at the moment.

This nonspecific reaction no doubt occurs, but specific impingements sooner or later require specific counterreactions. Further, for domestic animals, the nonspecific alarm reaction seems to be superfluous, if not counterproductive, whereas insult or injury might interfere with a wild animal's getting food, and thus crucial amino acids, food-getting is ordinarily not a problem for domestic animals. Finally, all productive processes involve protein synthesis, so a high glucocorticoid secretion rate can be detrimental to food-animal performance at least in the short term.

Adaptation: Stress and Strain

An environmental adaptation refers to any functional, structural, or behavioral trait that favors an animal's survival or reproduction in a given environment, especially an extreme or adverse surrounding. Rates of life processes are the criteria used most often to assess adaptation. Adaptation can involve either an increase or a decrease in the rate of a given process.

A strain is any functional, structural, or behavioral reaction to an environmental stimulus. Strains can be adaptive or nonadaptive. Many enhance the chances of survival, but others are seemingly of little consequence.

A stress is any environmental situation--and a stressor any environmental factor--that provokes an adaptive response. A stress might be chronic (gradual and sustained) or acute (abrupt and often profound). Thus, by definition, environmental stress provokes animal strain, or in other words environmental stress provokes a stress response.

Environmental stress occurs when a given animal's environment changes so as to stimulate strain (as when environmental temperature falls below the crucial level) or when the animal itself changes in relation to a given environment (as when shearing reduces a sheep's cold tolerance).

Kinds of adaptation. There are several categories of environmental adaptation. A given animal represents one stage in a continuum of evolutionary development. An animal's heredity determines the limits of its environmental adaptability. Hence, there are genetic adaptations to environment. Genotypic changes occur naturally due to genetic mutations. Environmental stress theoretically permits mutations having adaptive utility to be realized and ultimately to become fixed in animal populations. Artificial selection pressures for productive traits reduce such natural selection pressures. But individuals selected on the basis of productive performance are at least adequately adapted to the production environment; otherwise, they would neither perform at relatively high levels nor reproduce.

There are also induced adaptations. A given stressful environmental complex provokes various responses depending on the individual animal's current adaptation status, which is determined by heredity and by its life history, as well.

Acclimation is one kind of induced adaptation. It refers to an animal's compensatory alterations due to a single stressor acting alone, usually in an experimental or artificial situation, over days or weeks. A hen in a layer house might acclimate to altered day length, for example.

Acclimatization, on the other hand, refers to reactions over days or weeks to environments where many environmental factors vary at the same time. A ewe at pasture acclimatizes to seasonal variations in day-length in conjunction with variations in other environmental factors.

Finally, an animal may become habituated to certain stimuli when they occur again and again. Sensations and effector responses associated with particular environmental stimuli tend to diminish when these stimuli occur repeatedly. A pig raised near an airport becomes habituated to the roars of jet airplanes, for example.

Level of Adaptation. An animal's environmental adaptation can be analyzed at several levels of organization. At one end of the spectrum, adaptations can take the form of enzyme inductions or of changes in other modifiers of catlyzed biochemical reactions. In the middle are changes associated with adaptive responses to environmental stress in sensory, integrative, and coordinative neural functions, in neuroendocrine and endocrine functions, and in effectors' outputs. At the other end of the range, the animal's behavior often changes in response to environmental stimuli. Malcolm Gordon (1972) said: "There is certainly no logical basis for any claim that understanding the nature of life at one level of organization is more fundamental to overall understanding than comprehension at any other level."

STRESS RESPONSES: PSYCOLOGICAL COMPONENTS

It is now generally recognized that the amount of stress an animal is under depends not only on the intensity and duration of the noxious agent (the traditional concept), but on the animal's ability to modify the effects of the stressor as well (Mason, 1975; Archer, 1979).

Lack of Control

A recent study of stress effects on tumor rejection demonstrated psycological components of stress responses (Visintainer, et al., 1982). Stressors such as mild electrical shock depress an animal's ability to reject certain tumors in experimental settings. In this particular experiment, individually held rats were inoculated with a standard dose of tumor-causing cells and assigned to three treatments: control (no shock), mild shock that could be stopped by pressing a switch (escapable shock), and mild shock that stopped anytime the escapable-shock rate in the trial pressed its switch but over which this rat itself had no control (inescapable shock).

In other words, the amount of physical impingement received by animals in the two shock treatments was the same, but those in one group (escapable shock) could control the duration, while those in the other group (inescapable shock) could not.

Fifty-four percent of the control rates rejected their tumors. Inescapable shock caused so much stress that tumor rejection occurred in only 27% of the rats, while 63% of those subjected to escapable shock rejected their tumors. The conclusion: the low rate of tumor rejection was due not to the shock itself, but to the animal's inability to control this stressor.

Alliesthesia Modification

Central perception ("alliesthesia") of stress intensity depends on the context within which it occurs. Alliesthesia in the form of comfort rating or pleasure rating is affected by the animal's internal state and, hence, by its external surroundings as well.

For example, a thermal stimulus can feel pleasant or unpleasant depending on the body's thermal status. Hypothermic humans find cold stimuli very unpleasant and hot very pleasant, while hyperthermic humans have the opposite perceptions (Cabanca, 1971). Similarly, gastric loading with glucose decreased the human subjects' pleasure rating of the sweet taste of sucrose in a thermoneutral environment (26°C), but this negative alliesthesia due to glucose loading was eliminated when ambient temperature was reduced to 4°C (Russek, et al., 1979).

These findings remind us that "variety is the spice of life" and suggest that "taking the bitter with the sweet" is pleasurable in the long run. Extrapolating the concept of alliesthesia modification to agricultural animals' lives, it would appear that stress of one sort often primes the animal to receive pleasure from some other aspects of its environment.

In any case, the fact that an animal's psychological state can modify its perception of stress makes it all the more difficult to interpret how a specific stressor is affecting a specific animal.

MEASURING AND INTERPRETING STRESS RESPONSES

The scientific literature stores report of hundreds of experiments purported to measure stress in food animals (Hafez, 1968; Hafez, 1975; Johnson, 1976a,b; Stephens, 1980; Craig, 1981; Curtis, 1981). It is a relatively simple task to subject experimental animals to a controlled stressor and measure a resultant change in some physiological, anatomical, or behavioral parameter. Hormonal, cardio-respiratory, and heat-production parameters have been studied most in the past. Behavioral and anatomical changes are being characterized more lately. But an objective index of stress in terms of animal health, performance, and well-being has been elusive.

As Graham Perry (1973) said: "Even marked physiological changes may indicate only that an animal is successfully adapting to its environment--not necessarily that it is succumbing to adversity." And, similarly, Ian Duncan (1981) said: ". . . it should be of no surprise that chickens behave differently in different environments. This may simply demonstrate how adaptable they are." Again: at what point does stress become dis-stressful.

As for methodology, it is very difficult to study the effects of specific supposed stressors on an animal without introducing artifacts due to the stressfulness of the investigative techniques themselves (Adler, 1976). This is especially so in real or simulated production situations. What is the baseline adrenal-glucocorticoid secretion rate of an animal? Will it ever be certain beyond a reasonable doubt that the experimental manipulation necessary to obtain the needed samples or observations is not itself so stressful as to compromise the results?

Also, interpretation of the results of this kind of research is hampered by the fact that, by and large, there is not yet consensus as to the meaning of data on specific behavioral and hormonal changes in responses to stressors. What does it mean when an animal increases breathing rate by 250% in one environment compared with another? Does a 65% increase in plasma glucocorticoid concentration indicate the animal is under stress? If so, is the stress mild or severe? Scientists still do not understand how findings such as these relate to an animal's well-being, its health, and its productivity; consequently, we cannot rely on physiological or behavioral traits as valid indicators of the amount of stress an animal actually perceives, let alone how these might be related to the animal's health and productivity.

STRESS AND PRODUCTIVITY

Environmental stress generally alters animal performance (Curtis, 1981). The stress provokes the animal to react, and this reaction can influence the partition of resources among maintenance, reproductive, and productive functions in one or more of five ways:

1. The reaction may alter internal functions. Many bodily functions participate in productive processes as well as in reactions to stress. Survival responses may thus unintentionally affect productive preformance. For example, increased adrenal glucocorticoid secretion in response to stress can impair growth.

2. The reaction may divert nutrients. When an animal resonds to stress, it in effect diverts nutrients to use in higher-priority maintenance processes. Adaptive reactions are implemented even at the expense of productivity.

3. The reaction may reduce productivity directly. The animal's response sometimes partly comprises intentional reductions in productive processes. This generally frees some nutrients for maintenance uses. For example, an animal might reduce its productive rate in a hot environment in an attempt to re-establish heat balance with its surroundings.

4. The reaction may increase variability. Individual animals within a species differ from each other in functions, behaviors, and structures by what have been called "individuality differentials." Individual animals therefore differ in their responses to the same environmental stressor. In other words, two animals in the same group might successfully cope with the same stressful situation by calling different mechanisms into play. Then, if the complements of mechanisms used by the two individuals differ in the energy expenditure required to achieve them, the amount of energy diverted from productive processes will be different for the two animals. The result of this is that the amount of variation in individual performance in a group of animals tends to be related directly with the environmental adversities to which the animals are subjected.

5. The reaction may impair disease resistance. Because the animal's reaction to stress can impair disease resistance, that reaction influences the frequency and severity of disease. Of course, infection itself is a stress, so once established it in turn can influence the animal's productive performance. The mechanisms involved in the relations between environmental stress and resistance against infectious disease are just now being elucidated.

Kelley (1980) identified eight stressors: heat, cold, crowding, regrouping, weaning, limit-feeding, noise, and movement restraint. He documented the fact that all of these have been accorded a central role in stress-induced alterations of resistance against infection.

Having developed a framework for analyzing relations between adverse environments and animal productivity, it would be unrealistic to leave the impression that the link between stresses and productive processes are clear and simple. Consider two examples.

Lactating dairy cows held in a natural subtropical summer environment and provided no shelter are obviously under severe stress at mid-afternoon. They have markedly higher body temperatures and respiratory rates than their herdmates under the shade. Yet there might be no significant difference in fat-corrected milk yield between the two groups of cows (Johnson et al., 1966).

Socially and physically deprived animals often grow faster than do their counterparts in more enriched environments (Fiala et al.) So there is a risk in assuming that an animal stressed by a specific environmental complex is necessarily unfit for productive use in that environment. While one often might be justified in presuming that strain against stress reduces animal productivity, the animal can still be putting out an acceptable amount of product per unit of resource input.

Robert McDowell (1972) refers to "physiological adaptability" to environment (measured by physiological traits such as breathing rate) that is associated with survival responses, and to "performance adaptability" to environment (measured by productive-performance traits such as growth rate). These two often bear little positive relation to each other.

Thus, it is not sufficient for an animal producer to be concerned only with physiological and behavioral indices of environmental adaptability. Producers are more interested in the size of decrement, if any, in production associated with an animal's living in a particular environment. And to learn the quantitative effects of a given environment on animal performance, the productive traits themselves must be measured. After all, knowing a hen's breathing rate tells one little or nothing about her rate of lay.

There has been unfortunate ambiguity on this point among researchers and producers alike. An animal exhibiting marked strain has generally been assumed to be having markedly depressed performance. This is not necessarily so. Indeed, visible strain signifies that the animal is attempting to compensate for an environmental impingement. These attempts might succeed, and they might interfere with production only slightly or not at all.

The marvelous homeokinetic phenomena they possess make for resilient beasts and birds on our farms and permit profitable performance in a wide range of circumstances. The response flexibility that animals demontrate in the face of myriad stressors seem more remarkable than those instances when defensive reactions are inadequate and the environmental complex drastically reduces health or performance.

REFERENCES

Adler, H. C. 1976. Ethology in animal production. Livestock Prod. Sci. 3:303.

Archer, J. 1979. Animals Under Stress. Edward Arnold, London.

Bernard, C. 1957. An Introduction to the Study of Experimental Medicine. Dover, New York.

Bowman, J. C. 1977. Animals for Man. Edward Arnold, London.

Cabanac, M. 1971. Physiological role of pleasure. Science 173:1103.

Cannon, W. B. 1932. The Wisdom of the Body. Norton, New York.

Craig, J. V. 1981. Domestic Animal Behavior. Prentice-Hall, Englewood cliffs.

Curtis, S. E. 1981. Environmental Management in Animal Agriculture. Animal Environment Services, Mahomet, Illinois.

Dill, D. B. (Ed.) 1964. Handbook of Physiology. Section 4: Adaptation to the Environment. American Physiological Society, Washington.

Duncan, I. J. H. 1981. Animal rights-animal welfare: a scientist's assessment. Poul. Sci. 60:489.

Fiala, B., F. M. Snow, and W. T. Greenough. 1977. "Impoverished" rates weigh more than "enriched" rats because they eat more. Devel. Phychobiol. 10:537.

Fraser, D., J. S. D. Ritchie, and A. F. Fraser. 1975. The term "stress" in a veterinary context. Brit. Vet. J. 131:653.

Gordon, M. S. 1972. Animal Physiology: Principles and Adaptations. (Second ed.) Macmillan, New York.

Hafez, E. S. E. (Ed.) 1968. Adaptation of Domestic Animals. Lea and Febiger, Philadelphia.

Hafez, E. S. E. (Ed.) 1975. The Behavior of Domestic Animals (Third ed.) Williams and Wilkins, Baltimore.

Hale, E. B. 1969. Domestication and the evolution of behavior. In: E. S. E. Hafez (Ed.). The Behavior of Domestic Animals. William and Wilkins, Baltimore.

Johnson, H. D. (Ed.) 1976a. Progress in Animal Biometerology, Volume 1, Part I. Swets and Zeitlinger, Amsterdam.

Johnson, H. D. (Ed.) 1976b. Progress in Animal Biometerology, Volume 1, Part II. Swets and Zeitlinger, Amsterdam.

Johnston, J. E., J. Rainey, C. Breidenstein, and A. J. Gidry. 1966. Effects of ration fiber level on feed intake and milk production of dairy cattle under hot conditions. Proc. Fourth Int. Biometerological Cong., New Brunswick.

Kelley, K. W. 1980. Stress and immune function: A bibliographic review. Ann. Vet. Res. 11:445.

Mason, J. W. 1975. Emotion as reflected in patterns of endocrine integration. In: L. Levi (Ed.) Emotions-- Their Parameters and Measurement. Raven, new York.

McDowell, R. E. 1972. Improvement of Livestock Production in Warm Climates. Freeman, San Francisco.

Perry, G. 1973. Can the physiologist measure stress? New Scientist 60 (18 October):175.

Russek, M. M. Fantino, and M. Cabanac. 1979. Effect of environmental temperature on pleasure ratings of odor and testes. Physiol. Behav. 22:251.

Selye, H. 1952. The Story of the Adaptation Syndrome. Acata, Montreal.

Stephens, D. B. 1980. Stress and its measurement in domestic animals: a review of behavioral and physiological studies under field and laboratory situations. Adv. Vet. Sci. Comp. Med. 24:179.

Visintainer, M. A., J. R. Volpicelli, and M. E. P. Seligman. 1982. Tumor rejection in rats after inescapable or escapable shock. Science 216:437.

122

HOW ENVIRONMENT AFFECTS REPRODUCTION IN CATTLE

Richard O. Parker

INTRODUCTION

These days much is written and spoken about the environment for humans and animals. Recently, some groups have concerned themselves with the quality of animal environment provided by producers. It may surprise these groups, but the producer's concern for an animal's environment precedes theirs by many years. Cattlemen are no exception. A favorable environment combined with heredity offers the surest way of improving performance. Improved performance means more dollars.

All too often environment is thought of only as that which can be measured, seen, or felt. This includes such things as temperature, humidity, light-dark, and nutrition. Sometimes extremes in these environmental conditions produce traumatic outward manifestations. Other times the effects of environment occur more subtly. To fully appreciate "environment" it must be understood from a different level. This understanding will aid in an awareness that allows cattlemen to make choices and adjustments in their programs to maximally "control"--or at least avoid--environmental conditions that would limit production, primarily reproduction.

REPRODUCTION

Successful reproduction is the process by which beef animals produce other animals like themselves that go on to market or to reproduce. Success of the reproductive process is the foundation of a successful operation.

Reproduction processes include: mating, conception, pregnancy, birth and lactation. Failure, or reduced performance, at any point reduces efficiency--and any time efficiency is down, so are profits.

When viewing the complexity of the reproductive processes, it is a miracle that reproduction succeeds as often as it does. As with anything complex, numerous factors can influence the outcome. In a broad sense, these factors can be divided into those attributed to genetics (inheritance) and those attributed to the environment.

To measure the outcome of reproduction, a variety of "yardsticks" are available; usually the following are considered: (1) calving interval, (2) birth weight, (3) weaning weight, and (4) cow maternal ability. Indirectly these traits measure fertility, embryo mortality, prenatal maternal influences, lactation and postnatal survival (table 1).

HEREDITY

Within every cell of each beef animal, there are 60 chromosomes. They are composed of the substance deoxyribonucleic acid (DNA) that, due to its composition, contains the coded master plans for producing a beef animal from the embryo to the adult. Each animal receives one-half of its DNA from each parent at the time of conception. In turn, breeding animals pass on one-half of their DNA again when they mature and breed.

The master plan contained within the DNA, in part, directs to formation and the function of beef cattle, including those traits relating to reproductive efficiency. Depending on the trait, heredity varies in its influence, leaving the action of the environment to "fill the gap" in influencing the formation and the function of beef cattle. For example, heredity largely determines the color of the coat, while it has only a small effect on the calving interval.

TABLE 1. REPRODUCTIVE TRAITS IN BEEF CATTLE AFFECTED BY HEREDITY AND ENVIRONMENT

Trait	Source of variation	
	Inheritance	Environment
Calving interval	10%	90%
Birth weight	35%	65%
Weaning weight	30%	70%
Cow maternal ability	30%	70%

ENVIRONMENT

Environment on a broad basis is all of the surrounding things, conditions, and influences affecting the growth, development, and production of beef cattle. For a better understanding, environment can be considered at three levels: external, internal, and uterine.

External Environment

To the producer, the external environment is the obvious and measurable conditions and influences. This environment includes heat, cold, wet, dry, light, dark, wind, and feed supply and quality. Producers relate to these and are aware of their changes, but the external environment is capable of altering an environment at a more subtle level--the level of the cells of the body.

Internal Environment

The internal environment is the environment to which the cells are exposed--the level of all life processes. Despite the insults of the external environment, beef cattle strive to maintain a constant internal environment. This is accomplished through two cooperating systems: the nervous system and the endocrine system. Unfortunately, producers can only measure and relate to the external environment. There is no way to measure the internal environment--before problems develop.

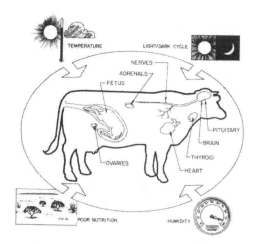

Figure 1. Two cooperating systems control the internal environment in response to the external environment: the nervous system and the endocrine system. The brain and nerves represent the nervous system, while the pituitary, thyroid, adrenals, and ovaries are some important components of the endocrine system. The heart and blood vessels distribute the hormones from the endocrines.

Uterine Environment

Uterine environment refers to that experienced by the embryo-fetus in the uterus. The conditions and influences affecting the developing calf can have profound and possibly long-lasting effects, though this environment is temporary. During residency within the uterus, the embryo-fetus is totally dependent upon the cow for an environment that is optimal for growth and development.

Heavier birth weights are associated with increased survival, and birth weight is largely determined by the uterine environment. Thus, through this environment, the maternal contribution to fetal size (birth weight) is greater than the paternal contribution.

One of the classical experiments demonstrating this effect of the uterine environment was reported by A. Walton and Sir John Hammond in 1938. They performed a reciprocal cross between the large Shire horse and the small Shetland pony. At birth, the crossbred foal from the Shire mare was three times larger than the crossbred foal from the Shetland mare. The environment provided by the Shetland mare limited the size of the foal. Even at four years of age, the cross-bred from the Shire mare was one and one-half times larger--demonstrating the possible long-lasting effects of uterine environment. Experiments in cattle have demonstrated similar effects on embryo-fetal growth. Since the uterine environment is nested within the internal environment it follows that changes in the internal environment affect the uterine environment.

GENETICS VERSUS ENVIRONMENT

Are genetics more important than environment in determining performance? This is an age-old question that will likely never be completely resolved or understood. When determining the effects of one or the other, the importance of individual variation becomes apparent. Often in individuals of similar breeding and environment there is a wide variation in their reproductive efficiency. Hence, stockmen and scientists usually talk in terms of averages.

Table 1 shows to what extent some reproductive traits in beef cattle are influenced by heredity (genetics) and environment. Regardless of the trait, environment (whatever it may be) exerts a greater influence on reproductive traits than does heredity. Hence, manipulation of the environment offers a tremendous potential for improving reproductive efficiency. Conversely, it provides a major avenue for inefficiency. No one should forget, however, that genetics and environment enjoy a complementary relationship.

INTERPRETATION OF THE ENVIRONMENT

Interpretation of the environment, or how the body of the beef animal reacts to the environment, is determined by the brain--which continually receives and interprets information relative to the internal and external environment. Then, through nervous signals, the brain alters the internal environment. While the whole brain is important, the hypothalamus area is most important to understanding how the external environment is capable of altering reproductive traits.

Environment-Brain-Hormone Connection

For years, stockmen and scientists were aware that the environment altered reproductive traits. For example, they found that hot weather lowered fertility. They also were aware of the nervous system, and of hormones controlling reproduction, but the connection between the two eluded scientists for years.

The brain is like a computer with many sensors, and it constantly receives information about temperature, humidity, amount of light, sounds, smells, pain, surroundings, etc. After interpreting this information, the brain decides what adjustments are necessary to ensure the well-being (survival) of the beef animal. In some cases, appropriate signals for adjustments are sent via the nerves to the various tissues, organs, or limbs. In other cases the signals must go via the endocrine system as a hormone. It is the hypothalamus of the brain that coordinates this activity. Small, but powerful, it responds to the various signals received by the brain and then acts like a "switchboard" by switching on or switching off the hormones needed to alter the internal environment. The hypothalamus is the bridge between the nervous and endocrine systems.

For the sake of perception, the nervous system may be likened to a "wired" system, while the endocrine system may be likened to a "wireless" system.

Power of Hormones

Endocrine glands secrete hormones. The word hormone comes from the Greek word hormon, which means "to spur on," "to set in motion," or "to excite to action." These phases are all very descriptive of a hormone. Hormones are "chemical messengers" that travel via the blood to specific organs or tissues and direct such processes as growth, reproduction, metabolism, and behavior. In the blood, they exist in extremely small quantities--millionths, billionths, and trillionths of a gram. Yet, their effects upon the body are profound. (Just as an example, to provide an appreciation of these tiny amounts of hormones, the fraction of one billionth equals one second out of the life of a 32-year-old individual, and one trillionth equals one second out of 320

centuries.) Extraction of the blood from about one million
animals may yield about a pound of many of the hormones.

Figure 2 illustrates the endocrine glands, their hormones, target organs, and the relationship of the brain and
hypothalamus.

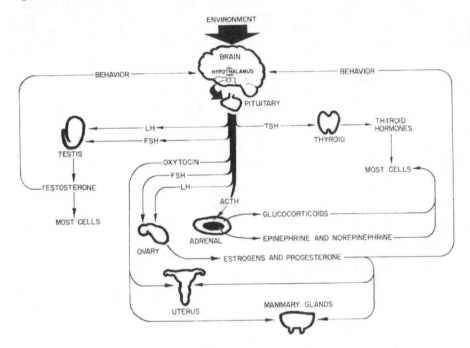

**Figure 2. Interaction of the environment, brain, endocrine
glands, hormones, and their target organs.**

Environments and Hormone Release

There are numerous hormones and no attempt will be made
to discuss each one—only those involved more directly in
the reproductive processes. Table 2 names the hormones,
lists environmental conditions affecting their release, and
briefly describes their actions and effects of a deficiency.

Secretion of the hormones should never be viewed as a
single event, but as a concert. As one hormone comes into
play, another may fade out; or one hormone may cause the
secretion of another; or the action of one may complement
the action of another. The brain or nervous system acts as
the conductor by signaling the proper time for increased or
decreased secretion of a hormone.

ENVIRONMENTAL PROBLEMS FOR STOCKMEN

It is apparent from table 2 that hormones have a wide-spread effect on reproduction, whether directly or indi-rectly. Also, many environmental situations are capable of altering hormone levels in the body of the beef animal. Some of the common reproductive problems include: silent heats, repeat breeders, anestrus, fertilization failure, embryonic mortality, newborn mortality, lactation failure, lack of libido, dystocia, and testicular degeneration. On the basis of table 2 many of these reproductive problems can be explained, or at least compounded by environmental fac-tors altering endocrine function. More and more, research suggests that environmentally induced alterations in the hormone levels may influence the physiological phenomena related to lowered reproductive performance in cattle.

PUTTING IT TOGETHER IN A SYSTEM

Genetics and environment complement each other. But genetics require considerable time to change due to the generation interval of cattle and the identification of superior animals. An awareness of the profound effects of environment on reproductive traits and the potential for improvements, offers stockmen an immediate method for improving reproductive efficiency. With a knowledge of the way cattle perceive the environment and changes it causes, stockmen can always be evaluating their system for potential problems or improvements. After all, stockmen are the best welfarists.

TABLE 2

HORMONES AND THE ENVIRONMENT

Hormone	Environments Affecting Release	Hormone Action	Effects of a Deficiency[1]
ACTH (Adrenocortico-tropic hormone)	Anxiety Injury Light-dark Temperature Toxins	Release of glucocorticoids. Liberates free fatty acids from fat tissue into the blood. Necessary for normal growth and lactation.	Inability to cope with stressful situations. Depressed growth or lactation.
Epinephrine and norepinephrine	Fright Pain Restraint Shock Surprise Unfamiliar surroundings	Alters heart function. Raises blood pressure. Dilates or constricts blood vessels. Increases metabolic rate. Increases blood glucose. Increases alertness.	Unknown
Estrogens	Humidity Light-dark Nutrition Temperature	Female secondary sexual characteristics. Estrous cycles. Female behavior. Development of uterus and mammary glands (udder).	Failure to exhibit estrus. Failure of function of accessory sex organs.
FSH (Follicle stimulating hormone)	Humidity Light-dark Nutrition Temperature	Secretion of estrogens. Follicle growth in females. Sperm formation in the male.	Failure to show heat in female. Failure to develop eggs. Lack of secondary sex characteristics. Depressed sperm formation in males.
Glucocorticoids	Anxiety Injury Light-dark Nutrition Pain Restraint Temperature	Balanced metabolism of proteins, fats, and carbohydrates. Anti-inflammatory.	Inability to cope with stressful situations. Abnormal protein and carbohydrate metabolism.
Growth hormone	Anxiety Restraint Unfamiliar surroundings	Growth of all tissues. Mobilization of fat for energy. Role in lactation.	Improper growth. General depression of metabolism.
LH (Luteinizing hormone)	Light-dark Nutrition Temperature	Testosterone production in the male. Progesterone production in the female. Release of egg from follicle.	Lack of libido in the male. Depressed sperm production. Pregnancy failure. Ovulation failure.
Oxytocin	Auditory or visual cues Mating Stressful situations Suckling	Milk letdown. Uterine contractions aiding birth process or sperm transport.	Possible disruption in sperm transport. Lack of milk ejection.
Progesterone	Light-dark Nutrition Temperature	Implantation and pregnancy maintenance. Mammary and uterine gland development.	Pregnancy failure. Lactation failure.
Prolactin	Variety of nonspecific environmental changes.	Promotes lactation, though high levels may inhibit. May prevent estrous cycles.	Cessation of lactation, but other hormones are involved.
Testosterone	Light-dark Nutrition Temperature	Necessary for sperm production. Develops and maintains male sex organs and masculine characteristics.	Depressed sperm production. Lack of libido.
Thyroid hormones	High temperatures Low temperatures Poor nutrition Unfamiliar surroundings	Increases metabolic rate. Increases nervous system activity. Stimulates protein synthesis. Increases motility and secretions of digestive tract.	Limited growth. Depressed metabolism. Depression or cessation of lactation. Improper development.

[1] Excesses may also be detrimental by carrying normal hormone action to an extreme or by preventing the action of another hormone.

123

LIVESTOCK PSYCHOLOGY
AND HANDLING-FACILITY DESIGN

Temple Grandin

Handling your cattle and sheep will be much easier if you learn a little livestock psychology. Many people do not realize that cattle and sheep have panoramic vision and they can see all around themselves without turning their heads (Prince, 1977; McFarlane, 1976). Sheep with heavy fleeces would have a more restricted visual field depending on the amount of wool on their head and neck. Both cattle and sheep depend heavily on their vision and are easily motivated by fear (Kilgour, 1971). Livestock are sensitive to harsh contrasts of light and dark around loading chutes, scales, and work areas. "Illumination should be even and there should be no sudden discontinuity in the floor level or texture" (Lynch and Alexander, 1973).

Solid shades should be used over the working, loading, and scale areas (Grandin, 1981). Slatted shades are fine for areas where the animals live and feel familiar. However, when the animals come into the handling areas they are often nervous. The zebra stripe pattern cast by the slatted shades constructed from snow fence or corrugated sheets suspended on cables will cause balking. The pattern of alternating light and dark has the same effect as building a cattle guard in the middle of the facility. Contrasts of light and dark have such a deterrent effect on cattle that in Oregon lines are painted across the highway to take the place of expensive steel cattle guards.

Shadowy stripes will cause balking problems with sheep. A single-file chute for sorting sheep should be oriented so that the sun does not form a shadow down the middle of the chute. The worst possible situation for sheep is to have half the floor of the chute in the shade and the other half in the sunlight. In shearing sheds and sheep holding areas, the wooden slats on the floor should face so that the sheep walk across the slats instead of in the same direction as the slats (Hutson, 1981). If you get down on your hands and knees and look at the floor, the floor appears more solid if you move across the slats. The floor should also be constructed to prevent sunlight from shining up through the slats.

A single shadow that falls across a scale or loading chute can disrupt handling. The lead animal will often balk and refuse to cross the shadow. If you are having problems with animals balking at one place, a shadow is a likely cause. Balking can also be caused by a small bright spot formed by the sun's rays coming through a hole in a roof. Patching the hole will often solve the problem. Handlers themselves should be cautious about causing shadows. Figure 1 illustrates a shadow that was formed when the handler waved at the cattle. The animals refused to approach the shadow of the waving handler cast at the entrance to the single-file chute.

Figure 1.   The handler's shadow cast on the entrance to the single-file chute caused the cattle to balk. This is just one of the many kinds of shadows which can cause balking problems in your cattle handling facility.

APPROACH LIGHT

Both cattle and sheep have a tendency to move towards the light. If you ever have to load livestock at night, it is strongly recommended that frosted lamps that do not glare in the animals face be positioned inside of the truck (Grandin, 1979). However, loading chutes and squeeze chutes should face either north or south; livestock will balk if they have to look directly into the sun.

Sometimes it is difficult to persuade cattle or sheep to enter a roofed working area. Persuading the animals to enter a dark, single-file chute from an outdoor crowding pen in bright sunlight is often difficult. Cattle are more easily driven into a shaded area from an outdoor pen if they are first lined up in single file.

Many people make the mistake of placing the single-file chute and squeeze chute entirely inside a building and the crowding pen outside. Balking will be reduced if the single file chute is extended 10 to 15 feet outside the building. The animals will enter more easily if they are lined up single file before they enter the dark building. The wall of the building should NEVER be placed at the junction between the single file chute and the crowding pen. Either cover up the entire squeeze chute and crowding pen area or extend the single file chute beyond the building. If you have just a shade over your working area, make sure that the shadow of the shade does not fall on the junction between the single file chute and the crowding pen.

PREVENT BALKING

Drain grates in the middle of the floor will make both sheep and cattle balk because the animals will often refuse to walk over them. A good drainage design is to slope the concrete floor in the squeeze chute area toward an open drainage ditch located outside the fences. The open drainage ditch outside the fences needs no cover and so it is easier to clean.

Animals will also balk if they see a moving or flapping object. A coat flung over a chute fence or the shiny reflection off a car bumper will cause balking. You should walk through your chutes and view them from a cow's eye level before moving or loading animals. You will be surprised at the things you may see. When cattle and sheep are being worked, the handlers should stand back away from the headgate so that approaching animals cannot see them with their wide angle vision. The installation of shields for people to hide behind can facilitate the movement of livestock (Kilgour, 1971; Freeman, 1975).

Problems with balking tend to come in bunches; when one animal balks, the tendency to balk seems to spread to the next animals in line (Grandin, 1980). When an animal is being moved through a single-file chute, the animal must never be prodded until it has a place to go. Once it has balked, it will continue balking. The handler should wait until the tailgate on the squeeze chute is open before prodding the next animal (Grandin, 1976). A plastic garbage bag attached to a broom handle is a good tool for moving cattle in pens. The cattle move away from the rustling plastic. When livestock are being moved, well-trained dogs are recommended for open areas and large pens. Once the animals are confined in the crowding pen and single-file chute, dogs should not be allowed near the fences where they still can bite at the cattle or sheep.

SOLID CHUTE SIDES

For both cattle and sheep the sides of the single-file chute, loading chute, and crowding pen should be solid. Solid sides prevent the animals from seeing people, cars, and other distractions outside the chute. A study with sheep showed that they moved more rapidly through a single-file chute that had solid sides (Hutson and Hitchcock, 1978). The principle of using solid sides is like putting blinkers on the harness horse. The blinkers prevent the horse from seeing distractions with his wide-angle vision. Cattle and sheep in a handling facility should be able to see only one pathway of escape--this is extremely important. They should be able to see other animals moving in front of them down the chute, when sheep are being sorted, the approaching animals should be able to see the previously sorted sheep through the end of the sorting chute.

Livestock will balk if a chute appears to be a dead end (Brockway, 1975; Hutson, 1980). Sliding and one-way gates in the single-file chute must be constructed so that your animals can see through them, otherwise the animals will balk (figure 2). The sides of the single-file chute and the crowding pen should be solid. The crowding-pen gate also should be solid so that animals cannot see through and will head for the entrance to the single-file chute (Rider, 1974). Mirrors could be used to attract sheep into pens and other areas that appear to be a dead end. The sheep are attracted to the image of sheep in the mirror (Franklin & Hutson, 1982).

HERD BEHAVIOR

All species of livestock will follow the leader and this instinct is strong in both cattle and sheep (Ewbank, 1961). Many people make the mistake of building the single-file chute to the squeeze too short. The chute should be long enough to take advantage of the animal's tendency to follow the leader. The minimum length for the single-file chute is 20 ft. In larger facilities 30 to 50 lineal ft is recommended.

Cattle and sheep are herd animals and, if isolated, can become agitated and stressed. This is especially a problem with Brahman-type cattle. An animal left alone in the crowding pen after the other animals have entered the single file chute, may attempt to jump the fence to rejoin its herdmates. A lone steer or cow may become agitated and charge the handler. A large portion of the serious handler injuries occur when a steer or cow, separated from its herd-mates, refuses to walk up the single file chute. When a lone animal refuses to move, the handler should release it from the crowding pen and bring it back with another group of cattle.

Figure 2. **The single-file chute to the squeeze should have solid sides to prevent the cattle from seeing distractions outside the fence. Sliding gates in the single-file chute must be constructed from bars so that the cattle can see through them. Solid sliding or one-way gates will cause balking.**

EFFECTS OF SLOPE AND WIND

To prevent livestock from piling up against the back gate in the crowding pen, the floor of the pen must be level. A 10° slope in the crowding pen will cause the animals to pile and fall down against the crowding gate. A small 1/4 in. to 1/8 in. slope per foot for drainage will not cause a handling problem. Livestock move more easily uphill than down, but they move most easily on a flat surface (Hitchcock and Hutson, 1979).

Research by Hutson and Mourik (1982) indicates that sheep will move more easily when they are heading into the wind. Heading into the wind can stimulate sheep to start moving along a chute.

WHY A CURVED CHUTE WORKS

A curved chute works better than a straight chute for two reasons. First it prevents the animal from seeing the truck, the squeeze chute, or people until it is almost in the truck or squeeze chute. A curved chute also takes advantage of the animal's natural tendency to circle around the handler (Grandin, 1979). When you enter a pen of cattle or sheep you have probably noticed that the animals will turn and face you, but maintain a safe distance (figure 3). As you move through the pen, the animals will keep looking at you and circle around you as you move. A curved chute takes advantage of this natural circling behavior.

Figure 3.   When you walk through a pasture the cows will turn and look at you. They will circle around you as you move about the pasture. Curved chutes take advantage of the cow's circling behavior.

Cattle can be driven most efficiently if the handler is situated at a 45° to 60° angle perpendicular to the animal's shoulder (Williams, 1978) (figure 4). A well-designed, curved single-file chute has a catwalk for the handler to use along the inner radius. The handler should always work along the inner radius. The curved chute forces the handler to stand at the best angle and lets the animals circle around him. The solid sides block out visual distractions except for the handler on the catwalk.

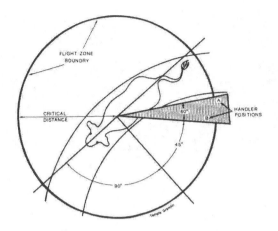

Figure 4. The shaded area shows the best position for moving an animal. To make the cow move forward the handler moves into Position B which is just inside the boundary of the flight zone. The handler should retreat to Position A if he wants the animal to stop. The solid curved lines indicate the location of the curved single-file chute.

The catwalk should run alongside of the chute and NEVER be placed overhead. The distance from the catwalk platform to the top of the chute fence should be 42 inches. This brings the top of the fence to belt-buckle height on the average person.

Figures 5 and 6 illustrate curved facilities for handling cattle and sheep. Curved designs are recommended by both Grandin (1980) and Barber (1977).

FLIGHT DISTANCE

When a person penetrates an animal's flight zone, the animal will move away. If the handler penetrates the flight zone too deeply, the animal will either turn back and run past him or break and run away. Kilgour (1971) found that when the flight zone of bulls was invaded by a mechanical trolley, the bulls would move away and keep a constant distance between themselves and the trolley. When the trolley got too close the bulls bolted past it. The best place for the handler to work is on the edge of the flight zone. This will cause the animals to move away in an orderly manner. The animals will stop moving when the handler retreats from the flight zone.

The size of the flight zone varies depending on the tameness or wildness of the animal. The flight zone of range cows may be as much as 300 ft whereas the flight zone

Figure 5. Cattle handling facility utilizing a curved single-file chute, round crowding pen, and wide curved lane. Up to 600 cattle per hour can be moved through the dip vat with only three people. The handlers work along the inner radius of the single-file chute and the wide curved lane (designed by Temple Grandin).

Figure 6. Sheep handling and sorting facility with a curved bugle crowding pen. The inner radius is solid to prevent the sheep from seeing the handler standing at the sorting gates (designed by Adrian Barber, Australia).

of feedlot cattle may be only 5 to 25 ft (Grandin, 1978). Extremely tame cattle or sheep are often difficult to drive because they no longer have a flight zone.

Many people make the mistake of getting too close to the cattle when they are driving them down an alley or putting them in a crowding pen. Getting too close makes cattle feel cornered. If the cattle attempt to turn back, the handler should back up and retreat to remove himself from the animal's flight zone instead, of moving in closer.

Cattle will often rear up and get excited while waiting in the single-file chute. The most common cause of this problem is the handler leaning over the single file chute and deeply penetrating the animal's flight zone. The cattle will usually settle down if the handler backs up.

When sheep are being handled in a confined area, pile-ups can occur if their flight zone is deeply penetrated. This is why dogs should not be used in the crowding pen or the single-file chute, because a dog, in a confined area, deeply penetrates the flight zone and the sheep have no place for escape. Dogs are recommended only for open areas and larger pens where there is room for the sheep to move away. During handling, minimize yelling and screaming so as to avoid enlarging the size of the animal's flight zone.

BREED DIFFERENCES

The breed of the cattle or sheep can affect the way it reacts to handling. Cattle with Brahman blood are more excitable and may be harder to handle than the English breeds. When Brahman or Brahman-cross cattle are being handled, it is important to keep them as calm as possible and to limit use of electric prods. Brahman and Brahman-cross cattle can become excited; they are difficult to block at gates (Tulloh, 1961) and prone to ram into fences. With this type of cattle it is especially important to use substantial fencing. If thin rods are used for fencing, a wide belly rail should be installed to present a visual barrier. Angus cattle tend to be more nervous than Herefords (Tulloh, 1961). Holstein cattle tend to move slowly (Grandin, 1980). Brahman cattle tend to stay together in a more cohesive mob than English cattle.

Brahman and Brahman-cross cattle can become so disturbed that they will lie down and become immobile, especially if they have been prodded repeatedly with an electric prod (Fraser, 1960). When a Brahman or Brahman-cross animal lies down, it must be left alone for about five minutes or it may go into shock and die. This problem rarely occurs in English cattle or European cattle such as Charolais.

There are distinct differences in the way various breeds of sheep react during handling (Shupe, 1978; Whately et al., 1974). Rambouillet sheep tend to bunch tightly together and remain in a group; crossbred Finn sheep tend to turn, face the handler, and maintain visual contact. If the handler penetrates the collective flight zone of a group of Finn sheep, they will turn and run past the handler.

Cheviots and Perendales are the easiest to drive into a crowding pen; the Romney, Merino-Romney cross, and the Dorset-Romney are the most difficult. The Romney tends to follow the leader but it is easily led into blind corners. Cheviots have a strong instinct to maintain visual contact with the handler and to display more independent movements than other breeds.

DARK BOX AI CHUTE

For improved conception rates, cows should be handled gently for AI and not allowed to become agitated or overheated. The chute used for AI should not be the same chute used for branding, dehorning, or injections. The cow should not associate the AI chute with pain. Cows can be easily restrained for AI or pregnancy testing in a dark box chute that has no headgate or squeeze (Parsons & Helphinstine, 1969; Swan, 1975). Even the wildest cow can be restrained with a minimum of excitement. The dark box chute can be easily constructed from plywood or steel. It has solid sides, top, and front. When the cow is inside the box, she is inside a quiet, snug, dark enclosure. A chain is latched behind her rump to keep her in. After insemination the cow is released through a gate in either the front or the side of the dark box. If wild cows are being handled, an extra long dark box can be constructed. A tame cow that is not in

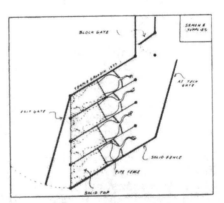

Figure 7. Chutes for A.I. can be laid out in a herringbone design. The two outer fences and the grates should be solid. The inner partition in between the cows should be constructed from bars. Cows will stay calmer if they know they have company.

heat is used as a pacifier and is placed in the chute in front of the cow to be bred. Even a wild cow will stand quietly and place her head on the pacifier cow's rump. After breeding, the cow is allowed to exit through a side gate, while the pacifier cow remains in the chute.

If a large number of cows have to be pregnancy checked or inseminated, two to six AI chutes can be laid out in a herringbone pattern (figure 7). This design is recommended by McFarlane (1976) from South Africa. The chutes are set on a 60° angle. They are built like regular dark box AI chutes except that the partitions inbetween the cows are constructed from open bars so the cows can see each other. The cows will stand more quietly if they have company. The two outer fences should be solid. If the cows are reluctant to enter the dark box, a small 6 in. by 12 in. window can be cut in the solid front gate in front of each cow.

LOADING CHUTE DESIGN

Loading chutes should be equipped with telescoping side panels and a self-aligning dock bumper. These devices will help prevent foot and leg injuries caused by an animal stepping down between the truck and the chute. The side panels will prevent animals from jumping out the gap between the chute and the truck.

A well-designed loading ramp has a level landing at the top. This provides the animals with a level surface to walk on when they first get off the truck. The landing should be at least 5 feet wide for cattle. Many animals are injured on ramps that are too steep. The slope of a permanently installed cattle ramp should not exceed 20°. The slope of a portable or adjustable chute should not exceed 25° (Grandin, 1979). Steeper ramps may be used for loading sheep but they are NOT recommended for unloading. Sheep will move up a steep ramp readily.

If you build your ramp out of concrete, stairsteps are strongly recommended. For cattle the steps should have a 3.5 to 4 in. rise and a 12 in. tread width. The surface of the steps should be rough to provide good footing. For sheep the steps should have a 2 in. rise and a 10 in. tread width.

On adjustable or wooden ramps, the cleats should be spaced 8 in. apart from the edge of one cleat to the edge of the next cleat (Mayes, 1978). The cleats should be 1 1/2 to 2 in. high for cattle and 1 in. by 1 in. for sheep.

Chutes for both loading and unloading cattle should have solid sides and a gradual curve (figure 8). If the curve is too sharp, the chute will look like a dead end when the animals are being unloaded. A curved single-file chute is most efficient for forcing cattle to enter a truck or a squeeze chute. A chute used for loading and unloading cattle should have an inside radius of 12 ft to 17 ft, the bigger radius is the best. A loading chute for cattle should be 30 in. wide and no wider. The largest bulls will fit through a 30 in. wide chute. If the chute is going to be used exclusively for calves, it should be 20 to 24 in. wide.

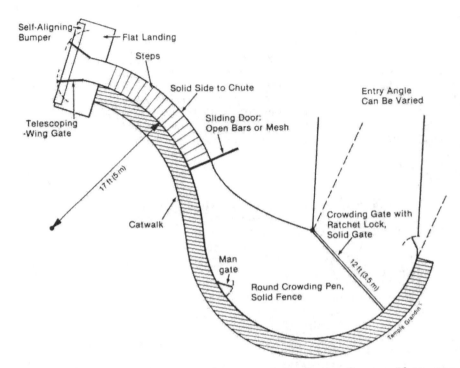

Figure 8. Curved loading chute with a round crowding pen. The sides of the chute are solid.

In auctions and meat packing plants where a chute is used <u>to unload only</u>, a wide straight chute should be used. This provides the animals with a clear path to freedom. These chutes can be 6 to 10 ft wide. A wide, straight chute should not be used for loading cattle.

SHEEP LOADING

Since most trucks have a 30 in. wide door, a good chute design for sheep is a 30 in. wide ramp that enables two sheep to walk up side by side. If a single-file chute is used, it should be 17 in. to 18 in. wide. The chute should be designed so that the animals walk up either in a single file or two abreast. Don't build a chute that is one and one-half animals wide--this creates jamming problems.

A wide ramp is recommended for loading sheep into shearing sheds or onto trucks that can be opened up the full width of the vehicle. In Australia sheep moved easily up ramps 8 to 10 ft wide when loading sheep onto ships for shipment to the Middle East (figure 9). The entrance ramp into a raised shearing shed should be 8 ft to 10 ft wide (Simpson, 1979).

Figure 9.   A wide ramp is used to load sheep onto a ship in Australia.   Once the flow of sheep was started the animals moved easily up the ramp.

1146

REFERENCES

Barber, A. 1977. Bugle sheep yards. Fact Sheet, Dept. of Agriculture and Fisheries South Australia, Adelaide, Australia.

Brockway, B. 1975. Planning a sheep handling unit. Farm Buildings Center, Nat. Agr. Center, Kenilworth, Warickshire, England.

Ewbank, R. 1961. The behavior of cattle in crises. Vet. Rec. 73:853.

Franklin, J. R., G. D. Hutson. 1982. Experiments on attracting sheep to move along a laneway. III Visual Stimuli, Appl. Animal Ethology 8:457.

Fraser, A. F. 1960. Spontaneously occurring forms of "tonic immobility" in farm animals. Canad. J. Comp. Med 24:330.

Freeman, R. B. 1975. Functional planning of a shearing shed. Pastoral Review 85:9.

Grandin, T. 1981. Innovative cattle handling facilities. In: M.E. Ensminger (Ed.). Beef Cattle Science Handbook, 18:117. Agriservices Foundation, Clovis, Calif.

Grandin, T. 1980. Livestock behavior as related to handling facilities design. Int. J. Stud. Animal Problems 1:33 etc.

Grandin, T. 1979. Understanding animal psychology facilitates handling livestock. Vet. Med. and Small Animal Clinician 74:697.

Grandin, T. 1978. Observations of the spatial relationships between people and cattle during handling. Proc. Western Sec. Amer. Soc. of Animal Sci. 29:76.

Grandin, T. 1976. Practical pointers on handling cattle in squeeze chutes, alleys, and crowding pens. In: M.E. Ensminger (Ed.). Beef Cattle Science Handbook 13:228.

Hitchcock, D. K., G. D. Hutson. 1979. The movement of sheep on inclines. Australian J. Exp. Agr. and Animal Husbandry 19:176.

Hutson, G. D., S. C. van Mourik. 1982. Effect of artificial wind on sheep movement along indoor races. Australian J. Exp. Agr. and Animal Husbandry 22:163.

Hutson, G. D. 1981. Sheep movement on slatted floors. Australian J. Exp. Agr. and Animal Husbandry 21:474.

Hutson, G. D. 1980. The effect of previous experience on sheep movement through yards. Appl. Animal Ethology 6:233.

Hutson, G. D. and D. K Hitchock. 1978. The movement of sheep around corners. Appl. Animal Ethology 4:349.

Kilgour, R. 1971. Animal handling in works, pertinent behavior studies. 13th Meat Industry, Res. Conf. Hamilton, New Zealand. pp 9-12.

Lynch, J. J. and G. Alexander. 1973. The Pastoral Industries of Australia. pp 371. Sydney University Press, Sydney, Australia.

Mayes, H. F. 1978. Design criteria for livestock loading chutes. Technical Paper No. 78-6014, Amer. Soc. Agr. Eng. St. Joseph, Michigan.

McFarlane, I. 1976. Rationale in the design of housing and handling facilities. In: M. E. Ensminger (Ed.). Beef Cattle Science Handbook 13:223.

Parsons, R. A. and W. N. Helphinstine. 1969. Rambo AI breeding chute for beef cattle. One-Sheet-Answers, University of California Agricultural Extension Service, Davis, California.

Rider, A., A. F. Butchbaker and S. Harp. 1974. Beef working, sorting, and loading facilities. Technical Paper No. 74-4523, Amer. Soc. Agr. Eng. St. Joseph, Michigan.

Shupe, W. L. 1978. Transporting sheep to pastures and markets. Technical Paper No. 78-6008, Amer. Soc. Agr. Eng. St. Joseph, Michigan.

Simpson, I. 1979. Building a modern shearing shed. Division of Animal Industry Bulletin A3.7.1. New South Wales Dept. of Agr., Australia.

Swan, R. 1975. About AI facilities. New Mexico Stockman. Feb., pp 24-25.

Tulloh, N. M. 1961. Behavior of cattle in yards: II. A study of temperament. Animal Behavior 9:25.

Whately, J., R. Kilgour and D. C. Dalton. 1974. Behavior of hill country sheep breeds during farming routines. New Zealand Soc. Animal Production 34:28.

Williams, C. 1978. Livestock consultant, personal communication.

124

DESIGN OF CORRALS,
SQUEEZE CHUTES, AND DIP VATS

Temple Grandin

CORRALS

A corral constructed with round holding pens, diagonal sorting pens, and curved drive lanes will enable you to handle cattle more efficiently because there is a minimum of square corners for the cattle to bunch up in. The principle of the corral layout in figure 1 is that the animals are

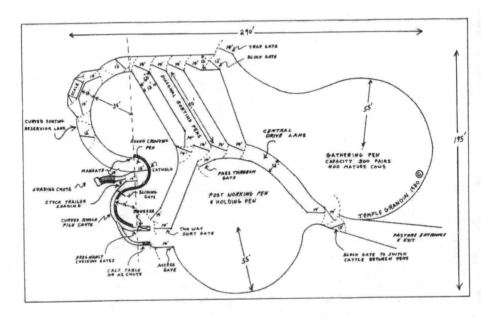

Figure 1. General purpose corral system for shipping, branding, sorting, and AI. It can handle 300 cow and calf pairs or 400 mature cows. Capacity can be increased by adding more diagonal pens and holding pen space (Grandin, 1981).

gathered into the big round pen and then directed to the curved sorting reservoir lane for sorting and handling. The curved sorting reservoir lane serves two functions: It holds cattle back into the diagonal pens, that are being sorted. It also holds cattle waiting to go to the squeeze chute, AI chute, or calf table.

Large Corral

The corral shown in figure 1 is a general-purpose system for shipping calves, working calves, sorting, pregnancy checking, and AI. It can handle 300 cow-calf pairs or 400 mature cows. It is equipped with a two-way sorting gate in front of the squeeze chute for separating the cows that are pregnant from cows that are open. Depending upon your needs, you can position either the squeeze chute, AI chute, or calf table at the sorting gate. If the cattle are watered in the large gathering pen, they will become accustomed to coming in and out of the trap gate. When you need to catch an animal, you merely shut the trap gate and direct her up the curved reservoir lane to the chutes. This is an especially handy feature for AI.

The curved sorting reservoir terminates in a round crowding pen and curved single-file chute. The crowding gate has a ratchet latch that locks automatically as the gate is advanced behind the cattle. To load low stock trailers, open a 8 ft gate that is alongside the regular loading chute. This provides you with the advantage of the round crowding pen for stock trailers. All fences in the curved single-file chute and the round crowding pen are solid. The ratchet crowd gate also should be covered with sheet metal or plywood.

Figure 1 can also be adapted for use with a prefabricated steel circle crowding pen and curved single-file chute. Since the prefabricated units have a 12 ft radius instead of the 16 ft radius shown in the drawing, you will have to move the sorting gate. If you plan to build the entire setup yourself, out of either wood or steel, keep the 16 ft radius, especially if you have large cows.

Diagonal Sorting Pens

When cows and calves are being separated, the calves are held in the diagonal pens and the central drive lane, and the cows are allowed to pass through one of the diagonal pens into the large post working pen. The diagonal pens and the central drive lane in figure 1 can hold 300 weaned calves overnight or 500 weaned calves crowded together. Each 70 by 12 ft diagonal pen holds 60 weaned calves overnight or 85 weaned calves crowded together. If the mother cows are put in the diagonal pens, each pen holds 40 cows overnight or 50 cows crowded. These capacities may vary depending on the size of your cattle.

To expand the corral system to handle more cattle, you can add more diagonal pens. Do NOT increase the length of the diagonal pens! If they are too long, the cattle will bunch up. You can increase the diagonal pen capacity to 1000 calves. It is NOT recommended to increase the size of the round gathering pen beyond the 55 ft radius shown. If the round gathering pen is too large, you may have difficulty getting the cattle into the curved reservoir lane. (Grandin, 1980a).

In order to increase the gathering area, you can build an additional round gathering pen at the pasture entrance. After the first 300 pairs are worked or sorted you can bring in 300 more pairs. The post working pen can be enlarged to hold cows after sorting or handling in the squeeze or AI chute.

Small Corral

The layout in figure 2 is designed for smaller ranches as a main working corral or a pasture corral on larger

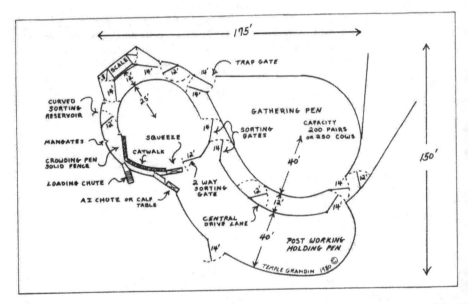

Figure 2. This is an economical corral system for a smaller operation or a pasture corral on a large ranch. It can handle 200 cow and calf pairs or 250 mature cows. It can be expanded to handle 300 cow and calf pairs (Grandin, 1980a).

operations. It is economical to build but still retains many of the features of the larger corral. It can handle

200 cow-calf pairs or 250 mature cows. By increasing the radius of the gathering pen to 55 ft and lengthening the central drive lane, it can be expanded to 300 pairs or 400 cows.

In figure 2 you can sort two ways out of the squeeze chute and three ways from the curved reservoir lane. Groups of cattle held in the curved reservoir lane can be sorted back into the post working pen, the central drive lane, or the round pen that is formed by the inner radius of the curved reservoir lane. When calves are being separated from the cows, the cows can be sorted into the post working pen and the calves into the central drive lane. For additional photos of corrals and a diagram of a minicorral, see Grandin, 1981, Beef Cattle Science Handbook 18:117.

Corral Construction Tips

Five foot high fences are usually sufficient for cattle such as Hereford and Angus. For Brahman cross and exotics a 5 1/2 ft to 6 ft fence is recommended. Solid fencing should be used in the crowding pen, single-file chute, and loading chute. If your budget permits, solid fencing should be used in the curved reservoir lane. If solid fencing is too expensive, then a wide belly rail should be installed. This is especially important if the corral is constructed from sucker rod (figure 3).

A V shaped chute can be built that will accommodate both cows and calves. It should be 16 to 18 in. wide at the bottom and 32 in. wide at the top. The 32 in. measurement is taken at the 5 ft level. If the single-file chute has straight sides it should be 26 in. wide for the cows and 18 to 20 in. wide for calves.

When a funnel-type crowding pen is built, make one side straight and the other side on a 30° angle. This design will prevent bunching and jamming. The crowding pen should be 10 to 12 ft wide (figure 4).

To prevent animals from slipping in areas paved with concrete, the concrete should be scored with deep grooves. The grooves should be 1 in. to 1 1/2 in. deep in an 8 in. diamond pattern. A diamond pattern should be used because it is easier to wash. If cattle are falling down when they exit from the squeeze chute in an existing facility, a grid constructed from bars will prevent falls. Construct the grid from 1 in. steel rods in 12 in. squares. Each intersection must be welded and the grid securely fastened to the concrete floor.

In areas with solid fence, small man-gates must be installed so that people can get away from charging cattle. The best type of man-gate is an 18 in. wide, spring-loaded steel flap. The gate opens inward towards the cattle held shut by a spring. A person can quickly escape because there is no latch to fool with. The man-gates can be constructed from 10 gauge steel with a rim of 1/2 in. rod.

Figure 3. Curved corral system shown in figure 1 constructed from steel with sucker rod fences. A
wide 24 in. belly rail has been installed in the
curved reservoir lane to provide a visually
substantial fence. The round crowding pen and
loading chute are covered with 10 ga. steel
sheets.

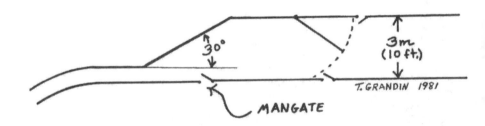

Figure 4. A funnel crowd pen should have one straight side,
and the other side on a 30 degree angle. Jamming
will occur if both sides are angled.

Many people have asked questions about how the corrals
should be laid out. It is really very simple. In figure 1
the curved single-file chute, round crowding pen, and curved

reservoir lane are laid out along the dotted line. The first step is to place a string on the site in the position of the dotted line. The radius points of the curved single-file chute, round crowding pen, and curved reservoir lane are all located along the string.

Layout steps for figure 1 (To be done in order.)
- Make a 16 ft 180° half circle for the single-file chute.
- Make a 12 ft 180° half circle for the round crowding pen.
- Make a 35 ft 180° half circle for the curved reservoir lane.
- Layout the diagonal pens on a 60° angle by placing a transit on the string.
- Layout gathering pen with a 55 ft radius. The radius point for the gathering pen is located 95 ft from the strike post of the last gate in the row of diagonal pens. The row of diagonal pen gates and the gathering pen radius point should be at a 90° angle relative to your string (dotted line).
- Layout the post working pen. The 55 ft radius point is found by measuring 55 ft from the hinge of the 14 ft sorting gate in front of the squeeze chute. The exact location of the hinge may vary depending on the length of the squeeze chute. Leave a 3 ft to 4 ft space between the end of the sorting gate and the headgate on the squeeze. This provides enough room so you can swing the sorting gate in front of the headgate without hitting the cow's head.
- After laying out the basics, finish laying out both sides of the lanes. The pasture entrance gates and the central drive lane will have to be laid out by eye.

Lay out everything in lime before building anything. This will prevent mistakes. Walk through the layout. If it looks like the drawing, then you have got it right. If there is a hill next to the site, look at the lime layout from there. If an aircraft is available, use it to check your layout.

SQUEEZE CHUTES

Herd health care is virtually impossible without a headgate or a squeeze chute for restraining animals. There are many headgates on the market and each type is especially suited for certain handling procedures. There are four basic types of headgates: scissors stanchion, full-opening stanchion, positive control, and self-catcher (Grandin, 1980b).

Scissors-stanchion headgates consist of two biparting halves that pivot at the bottom (figure 5). The full-opening stanchion consists of two biparting halves that work

Figure 5.   Scissors stanchion headgate with curved neck bars. The gate opens like a pair of scissors and has pivots at the bottom. The curved neck bars provide a good combination of head control and protection against choking.

like a pair of sliding doors. A positive-control headgate locks firmly around the animal's neck. This type of gate completely restricts up and down movement. The self-catcher headgate can be set like a trap. When the animal enters, its forward movement will close the gate automatically around its neck. The advantages and disadvantages of the four types of headgates are summarized in table 1.

Self catching, scissors stanchion, and full-opening stanchion are available with either straight or curved stanchion bars. A straight-bar stanchion headgate is extremely safe and will rarely choke an animal. The disadvantage of a straight bar stanchion is that an animal can slide its head up and down unless a nose bar or other restraint is used. The straight bar stanchion is recommended if the headgate is going to be used primarily for AI or pregnancy testing.

TABLE 1—Types of Manually-Operated Headgates Compared

	Self-Catcher	Scissors Stanchion	Positive	Full-Opening Stanchion
Recommended for	Hornless cattle, gentle cattle, one-man AI	General purpose, big feedlots, wild cattle, minimum maintenance, cattle of mixed sizes adjustment)	Dehorning, wild cattle, horned cattle, good head control, big feedlots. Requires less strength to operate than stanchion gates.	General purpose, vet clinics, mixed cattle sizes (because the gate seldom needs adjustment). Large bulls can exit easily.
Not recommended for	Wild cattle, big feedlots, horned cattle, groups of mixed-size cattle (because the gate has to be readjusted to catch animals of different sizes)	Very large bulls (because they may have trouble exiting due to the narrow space between the two bottom pivots)	Vet clinics where the animal is held in the headgate for a prolonged time. When AI and pregnancy testing are the primary uses of the headgate.	Big wild cattle, big feedlots (because many full-opening stanchion headgates are not sturdy enough to withstand constant heavy usage)
Warnings	Mechanism requires careful maintenance. Head and shoulder injuries may result if the animals are allowed to slam into the gate.	Be careful not to catch the animal's legs or knees between the two halves of the gate or the animal may be injured.	More likely to choke than a self-catcher, scissors, or full-opening stanchion.	Mechanism requires careful maintenance to prevent jamming. Animal may trip over the lower gate track if it becomes excited.

Self-catcher, scissors-stanchion, and full-opening stanchion headgates are available in models with either a straight or curved stanchion. Refer to the text for discussion on choking hazard *versus* head control.

Source: T. Grandin (1980).

A curved-bar stanchion is a good compromise between control of the animal's head and protection from choking. It is more likely to choke than a straight-bar stanchion, but it is safer than a positive type gate. A nose bar is not needed for ear implanting or tagging, if the animal is backed up in a stanchion headgate.

The problem of choking in a curved-bar stanchion or positive type headgate can be reduced by adjusting the squeeze sides of the chute. The V shape of the chute should support the animal. The proper spacing at the bottom of the squeeze sides is 6 in. for 250 to 400 lb calves, 8 in. for 600 to 800 lb animals, and 12 in. for cows and most fed steers. The space should be 14 to 16 in. for large bulls. The measurements are taken on the inside of the chute at the floor level. The best type of chutes have two squeeze sides that fold in evenly when the squeeze is applied.

Operator Skill Important

Results of a survey conducted by the author indicated that the main cause of handling accidents in hydraulic squeeze chutes was a careless operator handling cattle too fast. The skill of the operator affected the incidence of choking and escape from the squeeze chute (tables 2 and 3). Problems such as balking and falling while exiting from the squeeze chute were largely determined by conditions such as slick floors or shadows in the handling facility. You should be able to do a better job of operating your squeeze chute than that indicated by the percentages on tables 2 and 3.

TABLE 2—Effect of Cattle Breed and Operator's Skill on the Frequency of Handling Accidents in Hydraulic Squeeze Chutes

	Average for All 22 Groups (2150 Head)	Average for 12 Brahman Groups (1210 Head)	Average for 10 Non-Brahman Groups (940 Head)	Number of Groups with a Perfect Score	Single Worst Group Score	Observed Cause of the Worst Score
Mild choke	0.40%	0.17%	0.77%	17	3.00%	Rushing and carelessness*
Severe choke	0.30%	0.17%	0.40%	18	2.00%	Inexperienced operator
Entry balk	11.25%	10.30%	12.40%	0**	30.00%	Electric pole in front of the chute
Exit balk	15.20%	17.90%	12.00%	0	25.00%	Brahman cattle backed up after release
Partial escape	2.30%	3.90%	0.55%	10	11.00%	Long horns
Total escape	0.74%	1.00%	0.20%	14	5.00%	Carelessness
Falling	4.90%	7.30%	2.00%	5	15.00%	Slick, smooth concrete floor in front of chute
Headgate leg	2.50%	2.20%	3.00%	7	12.00%	Rushing

*All occurred in the same feedlot.
**Best balking score: entry balk 1.00%, exit balk 3.00%.

TABLE 3—Frequency of Handling Accidents in Hydraulic-Stanchion Headgate Squeeze Chutes*

	Ear Implant Only** (1230 Head Weighing More Than 600 lb)***	Full Processing[†] (920 Head Weighing 250 to 800 lb)*** 45% Castrated
Mild Choke	0.00%	0.88%
Severe Choke	0.11%	0.30%
Entry Balk	10.95%	12.70%
Exit Balk	13.90%	15.65%
Partial Escape	2.20%	1.39%
Total Escape	1.18%	0.07%
Falling	4.68%	3.15%
Headgate Leg	0.88%	3.77%

*Bowman and Trojan hydraulic squeeze chutes.
**Cattle received an implant of growth promotant in the ear. No other treatment was given.
***The sample consisted of both Brahman-cross and English-type cattle.
[†]In places where no Brahman or Brahman-cross cattle are handled the percentages should be lower. Full processing consisted of a minimum of two brands, two injections, an ear implant, and at least one other treatment such as deworming, or pour-on insecticide.

Source: T. Grandin (1980).

Different breeds of cattle react differently to handling (Ewbank, 1968; Tulloh, 1961). More handling accidents occurred when Brahman-cross cattle were being handled. The Brahman-cross cattle had more total escapes and partial escapes. A partial ecape was recorded when the animal was caught around the middle by the headgate.

Choking in the headgate occurs when the headgate applies excessive pressure to the carotid arteries or the wind pipe (White, 1961; Fowler, 1978). Excessive squeeze pressure in a hydraulic squeeze chute can also cause choking. A hydraulic chute is safe if the pressure relief valve is set correctly. In fact it may be safer for both man and animal because the dangerous levers are eliminated. At several feedlots, some cattle died several days after going through the hydraulic squeeze chute. The animals weighed over 600 lbs and appeared to have pneumonia. An autopsy revealed that the cattle had been ruptured internally due to excessive squeeze pressure.

Cattle can also be injured if a fast-moving animal is stopped suddenly by clamping the headgate around its neck. Examination of beef carcasses revealed old healed injuries to the back and neck. Even if the animal appears normal it can sustain a spinal injury if it slams into the headgate. A skillful squeeze chute operator can slow the animal down in the squeeze before it reaches the headgate. Rubber strips can be placed on the headgate to absorb shock and help reduce injury. Old, split motorcycle tires will work well.

Too many people try to set the world speed record for working cattle and they end up injuring a lot of animals. The survey indicated that a skilled crew can actually handle more cattle per hour by handling them gently and skillfully. A four-person crew using a hydraulic stanchion-type chute in a well-designed circular cattle working facility could catch an animal and place an ear implant every 15 seconds. The crew could also brand, vaccinate (up to four injections), and ear implant an animal every 45 seconds. In 60 seconds the crew could castrate, brand, vaccinate, ear implant, and give at least one other treatment such as pour on insecticide or clip a tail (Grandin, 1980b). When the crowding pen was filled with cattle, these timed procedures could be achieved without rushing. If a crew goes faster than these times, they will be doing a sloppy job and may be injuring the cattle.

DIPPING VAT DESIGN

Building a slide to make the cattle slide into the dip vat on their rear ends is wrong. The animal should be provided with good footing as it enters the water (Grandin, 1980c).

A 9 ft downward sloping, hold-down rack (figure 6) prevents the animals from leaping to the center of the vat.

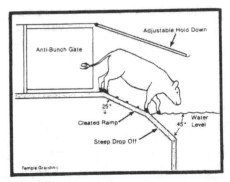

Figure 6. A grooved or cleated ramp with a nonslip surface enables the animal to enter the vat without hesitation. The steep drop off is hidden under the water. A 9 ft. adjustable hold-down rack makes the animal dive in and immerse its head. The distance between the hold-down rack pivot and the floor is 5 ft. The cleated portion of the ramp is 6 ft. long and on a 20 to 25 degree angle. The 45 degree angle portion is 4 ft. long (Grandin, 1980c).

The hold-down rack forces the animals to immerse their heads instead of being pushed under with a forked stick. It also helps prevent the chemicals from splashing out. Splashing can be further reduced by installing a 3 in. pipe along the inside and top edges of the dip vat. The pipe should be 3 to 4 ft above the surface of the water.

Each animal enters the vat by walking down a 6 ft gradual declining ramp that is on a 20° to 25° angle. This ramp has deep grooves in the concrete to provide the animal with good footing. The grooves should be 2 in. deep and 8 in. apart. The purpose of the ramp is to orient the animal's center of gravity towards the water. The steep drop-off is hidden under the water, but the ramp appears to continue on into the water (figures 6 and 7). When the animal steps out over the water it falls in. The ramp must have a **nonskid** surface or the animal may become scared and attempt to back out.

If both large and small cattle are going to be dipped, the entrance should be equipped with antibunch gates. These 7 ft gates allow only one animal to enter the vat at a time. The antibunch gates work on the same principle as the trigger trap one-way gates that are used to trap wild cows at the water hole. The opening between the ends of the two gates is adjusted to equal the width of one animal. The gates act as a valve to slow down incoming cattle. One of the antibunch gates can be spring loaded. The entrance should also be equippped with a semicircular block gate or a sliding gate to shut off the flow of cattle. A stanchion headgate powered by hydraulics also works well for a shut-off gate.

Figure 7. Cattle enter the vat and immerse their heads. They do not need to be pushed under with a forked stick. The hold down rack is easily constructed from 10 ga. steel and pipe.

Stairsteps are the best type of ramp for exiting from a dip vat. A steeper ramp can be used in a dip vat than for loading because the water supports the animal. Dip vat exit steps can have a 6 to 7 in. rise and a 12 in. tread width. The steps must be grooved to prevent slipping.

Drip Pen Design

A divided drip pen is recommended for dripping. When the cattle are drying on one side, the other side can be filled (figure 8). Figure 9 illustrates a dipping system layout with curved lanes and angled drip pens. Each side of the drip pen should be 30 to 40 ft long and 16 ft wide. The drip pens should be sloped 1/4 of an inch every foot towards the vat. The drip pen should be curbed with an 8 in. high

Figure 8. Divided drip pens with remote controlled exit
gates. The pipes go to cylinders to operate the
gates. The advantage of remote controlled gates
is they are labor saving and they allow the
handler to open the gates without entering the
flight zone of the cattle.

curb. To prevent the cattle from slipping, the drip pen
floor should be scored in an 8 in. diamond pattern. When
the wet concrete is scored, the last pass should be made
towards the vat so that the water will drain more easily.

Since often either hydraulic or air pressure is available at the site, placing cylinders on the exit gates is
recommended. This will let the cattle out by remote control. The exit gates can be opened without the handler entering the flight zone of the cattle. In conventional systems, the cattle often become agitated when the handler
walks up to open the gates manually. In some instances, the
cattle will ram the fences and attempt to jump back into the
vat when the handler approaches.

The drip pen exit gates and the dividing fence in between the two drip pens should be solid. This prevents the
cattle that are confined on one side from pushing on the
gate and attempting to follow the cattle that have just been
released. If Brahman-cross cattle are being dipped, a one-
way gate should be installed to prevent them from reentering
the vat.

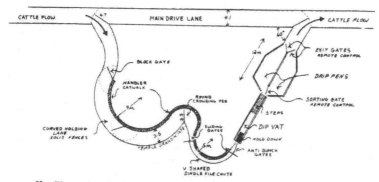

Handling system for dipping cattle with curved races.

Figure 9. Dipping vat system with curved lanes. A squeeze chute can easily be incorporated into this layout. A 14 ft gate positioned in front of the squeeze chute can be used to divert animals you do not want to dip. English measurements for metric measurements are: inner radius of curved holding lane 9m (30 ft); width of holding lane and length of crowd gate 3.5m (12 ft); and inner radius of the single file chute 5m (16 ft) (Grandin, 1980a).

To help keep the vat clean, the single-file chute and the crowding pen should be installed to prevent hair and manure from reentering the vat from the dip pens. The sump should have a valve to divert rain water that falls on the drip pen away from the vat. It is also wise to install a 2 ft wide curbed concrete apron alongside the vat.

To help keep the vat clean, the single-file chute and the crowding pen should have a concrete floor to prevent the animals from tracking dirt into the vat. This floor should be washed down after each dipping session.

Agitation and Aeration

Regular agitation and skimming off of floating debris will keep a dip vat cleaner. A vat that is left standing will quickly become stagnant and foul smelling. An easy way to agitate a vat is to install an airline in the bottom. The line should be a 1 in. pipe with 1/16 in. holes drilled at 5 in. intervals. The pipe is mounted 1 to 2 in. off the vat bottom. Two rows of holes are drilled on a 45° angle facing downward on both sides of the pipe (Saulmon, 1972). Connect the pipe to an air compressor capable of delivering 0.5 cu ft per minute at 40 psi per foot of pipe.

Frequent agitation of the water with the air compressor wil aerate the water and prevent it from becoming septic and foul. Good results have been obtained by connecting a timer and a solenoid to control the release of air. A 30 sec blast of air every 30 min almost completely eliminates bad odors.

Filtering and Cleaning

The installation of a filtering system can double the life of the chemicals in the vat. Use of a Hydrasieve sloping screen to remove solids from the vat has reduced chemical disposal requirements by 50% and pesticide usage by 30% (Sweeten, 1976; Miller, 1975). Almost twice as many cattle can be dipped before the vat requires recharging.

A cleaning system with a Hydrasieve is very simple. It consists of a 28 in. wide Bauer Hydrasieve screen with a .2 in. screen spacing. A 3 in. centrifugal trash pump brings the water to the top of the screen. Don't use a smaller diameter pump, it will clog up. The water runs back through the screen and the solids fall off the screen into a container. The water returns to the vat by gravity. When this system is built, the suction line should be exposed. Don't bury the suction line under the slab. If you get an air leak, you will not be able to fix it. The water-return lines and other pipes may be buried.

Another type of cleaning system is a sluice box or settling basin. This system can also double the life of the vat. The advantage of the sluice box or settling basin is that it is inexpensive. The use of this system and a Hydrasieve together may enable you to quadruple the life of the chemicals in the vat (Sweeten, 1982).

The Hydrasieve removes the bigger solids, and the sluice box or settling basin removes the fine dirt. Settling systems can vary in capacity from 300 to 100 ga. The bigger systems will remove more solids but they are harder to clean. An easy sluice box system to construct consists of a metal or concrete box 10 to 18 ft long and 2 ft deep. The box contains a number of removable baffles. The highest baffles should be located at the end where the dirty water enters. The box should be narrower where the dirty water enters and becomes wider as the discharge point is reached. Two ft wide at the entrance and 4 ft wide at the discharge point will work well. The idea is to make the water move very slowly at the discharge point so the solids will be left in the bottom of the box. The water should flow through a 300 gal box at a rate of 25 to 66 gal per min (Sweeten, 1982). The sluice box or sedimentation tank must be run at least 4 hours for every 1000 cattle dipped. It is of the utmost importance to keep the sluice box clean. If it is not cleaned out regularly, it will put dirt back into the vat. After each use, the baffles should be pulled out and the box cleaned out with a shovel.

A good way to design a cleaning system is to connect the Hydrasieve and the sluice box in a tee circuit. By using two valves, the larger portion of the pump output is directed to the Hydrasieve and the smaller portion of the output is directed to the sluice box. If the flow rate is too great the sluice box will not work. Sluice boxes or settling basins must be used with care when using wettable powders because they will remove the chemical. Check with the pesticide manufacturer.

Chemical Disposal

Haphazard dumping of used chemicals around a feedlot or ranch is not recommended. Used dip can be evaporated in a shallow concrete basin (Fairbank et al, 1980). For large feedlots in the Southwest the evaporation basin should hold three times the vat volume and the water level should be less than 2 ft deep. In high rainfall areas, an evaporation basin should have cover. Check with a local engineer to determine evaporation rates for your area. Each state has its own regulations on pesticide disposal. Check with your own state.

REFERENCES

Ewbank, R. 1968. The behavior of animals in restraint, In: M. W. Fox (Ed.) Abnormal Behavior in Animals. W. B. Saunders, Philadelphia, PA.

Fairbank, W. C., T. Grandin, D. Addis, and E. Loomis. 1980. Dip vat design and management leaflet. 21190 Division of Agricultural Sciences, University of California, Davis, Calif.

Fowler, M. E. 1978. Restraint and handling of wild and domestic animals. Iowa State University Press, Ames, Iowa.

Grandin, T. 1980a. Efficient curved corrals. Angus Journal, October 1980, pp 95-97.

Grandin, T. 1980b. Good cattle restraining equipment is essential. Vet. Med. and Small Animal Clinician. 75:1291.

Grandin, T. 1980c. Safe design and management of cattle dipping vats. Technical Paper No. 80-5518. Amer. Soc. Agr. Eng. St. Joseph, Michigan.

Grandin, T. 1981. Innovative cattle handling facilities. In: M. E. Ensminger (Ed.) Beef Cattle Science Handbook.18:117. Agriservices Foundation, Clovis, Calf.

Miller, D. 1975. Evaluation of Bauer separation equipment as a dip vat filtration system. (Unpublished). USDA, Beltsville, Maryland.

Saulmon, E. E. 1972. Ticks and scabies mites...dipping vat management and treatment procedures. Veterinary Services Memorandum 556.1, USDA/APHIS, Washington, D.C.

Sweeten, J. M. 1976. Results of Hydrasieve cattle dip recycling study. Field Day on Dip Vat Management Systems for Cattle Feedyards. Texas A & M University.

Sweeten, J. M. 1982. Dipping vat sedimentation tanks compared. Beef. March, 1982, p. 104.

Tulloh, N. M. 1961. Behavior of cattle in yards: II A study of temperament. Animal Behavior 9:25.

White J. B. 1961. Letter to the editor. Vet. Record 73:935.

125

HANDLING FEEDLOT CATTLE

Temple Grandin

Herrick (1979) reports that on a national basis 1 to 2% of all feeder cattle die within the first six weeks after they enter the feedlot. Losses in light (250 lb to 350 lb) calves arriving in southwestern feedlots can rise to 2 to 5% (Addis, 1980).

These calves are shipped in by truck from Florida, Tennessee, Kentucky, and other southeastern areas, to Arizona, Texas, and southern California feedlots. During the long trip, the calves become dehydrated, stressed, and shrunk. In many instances, they are just pulled off the mother cows and taken to the sale on the same day. About 30% to 45% of all southeastern calves are castrated on arrival at Arizona feedlots (Grandin, 1980). Calves that are castrated after arriving at the feedlot will have 18 lbs less gain (Addis et al., 1973). Many of the calves arrive at the feedlots with no vaccinations. If the producers in the Southeast would just vaccinate the calves, this would prevent a lot of health problems. In the Midwest, many producers fully precondition calves before selling them. Phillips (1982) suggests that handling at the farm and getting the calves used to mixing with strange animals would help to reduce stresses. He reported that calves that are accustomed to being handled have a lower corticoid level.

CARE FOR FEEDLOT CALVES

Why is there such a problem with stressed calves? The cattle industry is fragmented. Each person along the marketing chain tends to pass the losses on to the next person in the chain. The small rancher in the Southeast is not going to precondition his calves unless he gets a premium price. Full preconditioning consists of weaning, vaccinations, castration, dehorning, and bunk breaking.

Many preconditioning studies show conflicting results. Full preconditioning (consisting of 50% concentrate feeding and preweaning for 30 days) did not improve subsequent feedlot performance of calves in Texas, (Cole et al., 1982). Other studies conducted in the midwest indicated that full preconditioning was definitely recommended (Herrick personal communication). Most researchers agree that all calves should be vaccinated 2 to 3 weeks before they leave the ranch of origin. This will provide time for the buildup of protective immunity before they are shipped. The problem of shipping fever and stress is compounded when animals pass through many sale barns and markets on the way to the feed-lot. Sheldon (1981) found that if a load of calves is obtained from four or more sale barns, 80% or more of the calves will have to be treated for sickness after arrival at the feedlot. "Comingled calves not protected by any immunization, are prime candidates for Bovine Respiratory Disease Complex," (Herrick 1982).

Most of the sale barns in the Southeast do not provide water or feed for the calves (Phillips and Cole 1982). This is a situation which definitely needs to be corrected. A lack of feed and water may increase stress. Mills (1962) found that fasted sheep reacted with a greater output of corticoids in response to transit stress. After a calf has been starved, it takes two weeks for the rumen to recover (Cole 1982).

Fortunately, there is less cattle trading now, and most calves go through only one or two auctions before they reach the order buyer. When you buy calves, make sure that your order buyer has obtained them from less than three different auctions. Avoid buying cattle that have been traded by speculators. This is most likely to occur when the market is rising. Speculators will buy a load of calves and then sell them at another auction down the road in hopes of getting a better price. This is a practice that should be stopped. Hopefully, computer sales will reduce the need to physically move calves through auctions. Then the speculators and the trader can trade on the computer and live cattle would not have to be moved every time a transaction is made.

Feeder calves should be kept at the order buyer's barn for as short a time as possible (Cole 1982). Research has shown that feeding concentrates at the order buyer barn is helpful in super-stressed calves (Horn 1980). Feeding a high energy diet at the order buyer's barn improved the animal's ability to resist stress 3 weeks after arrival at the feedlot (Phillips 1982). The results of concentrate feeding research have been variable. The variability in the research results is probably due to some calves refusing to eat the feed. This could be due to inadequate trough space and competition between animals.

Receiving Procedures

Long-haul feeder calves become tired and stressed. Their rumen function will be impaired because transportation imposes an additional stress on rumen function over and above the stress of fasting (Galyean et al., 1980). New feeder calves should be processed as soon as possible after arrival at the feedlot (Addis et al., 1980). Processing consists of vaccinations, growth-promoter implant, horn tipping, dipping, and castration if needed. Prompt processing is essential to ensure that the unvaccinated calves develop protective antibodies before viruses incubating in their bodies attack them. The calves should be rested overnight, but processing should be conducted within 72 hours (Sheldon, 1981). Delayed vaccination for 2 to 3 weeks is not recommended. To reduce stress, many feedlots in Colorado and Texas are ear tagging their cattle instead of branding them. Some feedlots are still branding pen numbers in addition to the owner brand. Branding pen numbers is a practice that should be discontinued because branding on the rib causes a setback in gains (Addis et al., 1973). A hide with multiple brands will be discounted about $5 (Kilik, 1976).

Receiving-Pen Design

Each truck load of new feeder calves should be placed in a separate pen. This will help prevent the spread of disease. Each 450 lb calf needs 125 sq ft of dry space. These recommendations are for dry southwestern feedlots (Grandin, 1981).

If very small 200 lb to 350 lb calves are being received, it is advisable to split each truck load into groups of 15 to 20 head. This enables feedlot employees to observe them for sickness more carefully. Joe Clark, feedlot veterinarian for Miola Bros. Feedlot in the Imperial Valley is using this approach. Each small calf should have at least 25 sq ft for every 100 lb of body weight. Careful observation for sickness is essential. If a calf has lost 20% of its lung capacity to sickness, it still looks normal. Each day that it misses being treated, it will lose another 10%. Loss of lung function will ocur even faster in calves weighing less than 250 lb (Sheldon, 1981).

When calves arrive at the feedlot some may refuse to drink from an automatic water trough. When the float turns the water on, they get scared and back away. Calves have actually died of thirst because they have been afraid of the water trough. A good system for watering new calves is a long narrow water trough about 20 ft long. The water in the trough flows continuously to entice the calves. The flowing water will sound like the stream they drank from at the ranch of origin.

Large round stock tanks are another good water source. Many calves are accustomed to this system. It is advisable to feed new arrivals before they drink. This will keep thirsty calves from getting overloaded with water.

New calves need plenty of bunk space because they have a tendency to eat all at once. More mature cattle learn to take turns and can get by with very little bunk space. Ralph Durham, Texas Tech University, feeds his new arrivals from a series of large tubs on the ground. This helps to encourage eating and reduces competition over food. Feeding hay on the ground will also encourage eating.

New calves should get some milled ration into them promptly after arrival. Research conducted by Lofgreen et al., (1980) indicated that feeder calves stressed from transit regained purchase weight faster and had more efficient gains on a diet consisting of 50 to 75% concentrates compared to a diet consisting of 25% concentrates. Cole (1982) reported that providing 1.5% potassium and living lactobacillus in the receiving diet was helpful. You should consult a nutritionist for specific recommendations on diet.

Processing New Calves

Stressed calves need to be handled gently. Limit the use of electric prods. Electric prods should be used only when moving calves into the squeeze chute and not in the holding pens. A good tool for moving calves is a whip with a white flag on the end or a plastic garbage bag on the end of a stick. The animals will move away from the sound of the rustling plastic.

Cattle should not be left crowded in the crowding pen or alley more than 2 hours or the stress of waiting in line will elevate body temperatures and mask fevers (Gill, 1982). Ideally the calves should not have to wait in line for more than 30 minutes (Gill, 1982). This is especially important with light animals . Temperature taking and tail clipping should be done by the same person.

The various processing tasks should be evenly distributed among the crew members. Calves that are receiving vaccinations, ear implants, and ear tags can be processed in 20 to 30 seconds. A four man crew consisting of three processors and one drover can castrate a calf in 45 to 60 seconds.

Vaccinating the Calves

Another person should handle the syringes or needles to avoid contamination. Dirty needles can cause abscesss that remain in the muscle (figure 1). Abscesses have to be cut out at the meat packing plant. The crew must also be careful to give vaccines as specified by the manufacturer. Some vaccines are designed to be given only under the skin. If this type of vaccine is injected into the muscle, the vaccine may irritate the muscle and cause an abscess. Another bad practice that can cause abscesses is pouring vaccines into a bag or other container. If the bag becomes contaminated, abscesses may occur.

The processing area should be equipped with a clean table to hold all the vaccines and needles. A good system is to cut holes in a table top to hold stainless steel pans. The pans hold sponges that are soaked in disinfectant. The needle is rubbed on the sponge after each vaccination or ear implant (figure 2). This method is good for keeping ear-implant needles clean. A dirty needle can cause an infection that will wall off the pellet and the animal may not get the full-growth promoting effect.

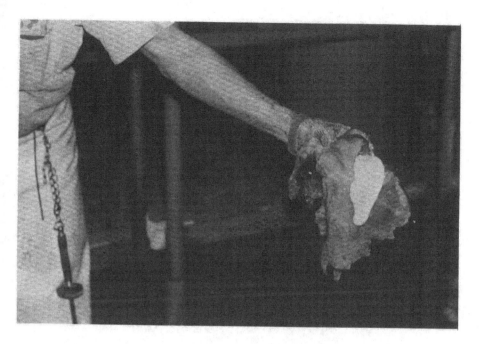

Figure 1.   Dirty needles and sloppy vaccinating practices
caused this abscess.   Cleanliness during processing is essential to prevent this.

Some precautions on vaccinating cattle need to be mentioned.   Chemical disinfectants should never be used to
clean syringes that have been used with live-virus vaccines.   The disinfectant will destroy the vaccine.   Live-
virus vaccines must be kept in the shade.   Sun and heat will
destroy the vaccine quickly.   Reconstitute one bottle at a
time.

Processing Facility Design

Well-designed handling facilities will help reduce
stress and injuries.   If your feedlot specializes in small
250 to 400 lb calves, a separate processing facility should
be built for small animals.   This will avoid the problem of
calves turning around and being difficult to handle in
chutes that are too large.   A single-file V chute for 200 to
400 lb calves should be 14 in. wide at the bottom and 28
in. wide at the top.   The height of the V chute is 54 in.
If larger (400 to 800 lb) animals are handled, the V chute
should be 16 in. wide at the bottom, 32 in. wide at the top
and 60 in. high.   Small hydraulic chutes are available for
handling calves and are recommended.   There are some really
nice hydraulic tilting chutes on the market.

Figure 2.   A table with pans for disinfectant-soaked sponges
will keep your implant needles and vaccination
needles clean.

Figure 3 (Grandin, 1981) illustrates a curved processing facility for calves and feedlot cattle. The single-file
chute should have a minimum inside radius of 12 ft. This
layout will accommodate a standard prefabricated round
crowding pen and a curved single-file chute. If you plan to
use this layout with large animals, it is advisable to increase the inside radius to 16 ft. Figure 1 is equipped
with a sorting gate in front of the squeeze chute headgate.
This is useful for diverting animals that have partially
escaped from the headgate back through the squeeze. Attempting to handle an animal that is caught in the headgate
by its hips can cause injuries to both the animal and a
person. (More specifications on facility design are found
in the author's other papers at this school.)

Make Your Work Area Washable

In many older feedlots, the processing and hospital
areas are almost impossible to wash because they do not have
proper drainage. When a new facility is being planned,
there must be a provision for draining the wash water away

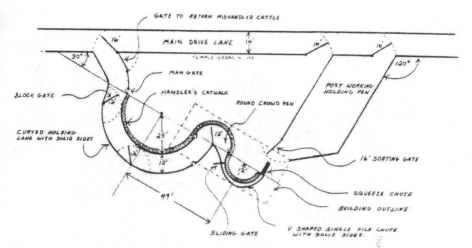

Figure 3. Curved layout for processing feedlot cattle.

from the facility. The squeeze chute, single-file chute, and the crowding pen should be washed down after each working day (figure 4). Concrete floors should be sloped 1/4 to 1/8 in. per foot towards the drain. The concrete slabs need to have curbs to contain the wash water and direct it to a drain. The curb prevents the water from running off the slab and making a mud hole.

The best type of drains are open ditch drains, which are constructed outside the areas where the cattle walk. For easy cleaning, the ditch drain should be slightly wider than the width of a shovel. Figure 5 illustrates a drain with a sloped sump, which is easy to clean out with a shovel. Another good drain design is to locate the drain directly under the squeeze chute. The squeeze has a grating floor. The single-file chute and area around the chute are sloped towards the drain. The work area surrounded by the curb must be small enough to wash easily.

All central processing areas and hospitals need a sink and running water for washing syringes and equipment. A refrigerator and a weather-proof storage space for medicines are also needed. Hoses for wash down must be large diameter fire hoses and plubmed into a 3 in. pipe. Small garden hoses are too small to be effective.

ANIMAL WELFARE

I was asked to discuss animal welfare at this conference. While working on building and designing cattle-handling facilities, I have witnessed many incidents of gross abuse of cattle. From my travels I estimate that 15% of

Figure 4. Processing facilities should be washed down after every working day. Well designed floors and drains are essential for easy wash down.

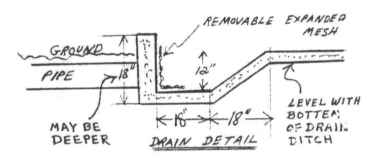

Figure 5. This sump can be installed in drain ditches. The sloped section makes it easy to clean out with a shovel.

U.S. feedlots and ranches are allowing gross abuse of animals to occur on a regular basis. I have seen the headgate of a chute slammed on a calf's head repeatedly and a cowboy try to poke an animal's eye out with his finger. During building projects, the construction crews have been shocked at the abuses that occurred almost every day on some feedlots or ranches. Abusive livestock handling is a problem that the beef industry must correct so that feedlots and ranches that are doing a good job are not penalized for the abuses of others.

Dirty processing and hospital areas are other feedlot problems. Approximately 50% of the feedlots have filthy cattle-handling facilities. Rough handling and dirty facilities are often located at custom feedlots. These feedlots make their money from selling feed for their customer's cattle. Feedlots that own a high percentage of their cattle often have better processing and doctoring practices. People take better care of things that belong to them.

I have visited many feedlots where everything in the yard is well managed and clean except the processing area. The processing areas might be an island of filth and roughness in the middle of a beautiful feedlot. Most of these operations would never allow their mill to be managed this badly. Part of the problem is the use of contract crews who are paid on a piece-work basis. One of the biggest animal welfare problems is rough handling and physical abuse of cattle.

1174

REFERENCES

Addis, D.G. 1980. Preventive medication for feedlot replacement calves, Study II. In: Stress Calf Studies. Riverside and Imperial Counties, Cooperative Extension, University of California.

Addis, D.G., J. Dunbar, J. Clark, G.P. Logfreen, and G. Crenshaw. 1980. Effect of time and location of processing on feeder calves. In: Stress Calf Studies. Riverside and Imperial Counties, Cooperative Extension, University of California.

Addis, D.G. et al. 1974. Stree study-castration branding. 13th Annual California Feeders Day. El Centro, California.

Cole, A. 1982. Nutrition-management interaction of newly arrived feeder cattle. Symposium on Management of Food Producing Animals. Purdue University, May 5-7.

Cole, N.A., J.B. McLaren, and D.P. Hutcheson. 1982. Influence of preweaning and B-vitamin supplementation of the feedlot-receiving diet on calves subjected to marketing and transit stress. J.Anim. Sci. 54:911.

Galyean, M.L., R.W. Lee, M.E. Hubbert. 1980. Influence of fasting and transit stress on rumen fermentation of beef steers. Agricultural Experiment Station, Research Report No. 426 New Mexico State University.

Gill, D. 1982. Effect of management practices upon performance. Symposium on the Management of Food Producing Animals. Purdue University.

Grandin, T. 1981. Feeder calf handling in southwestern feedlots. Amer. Soc. Agr. Eng. Technical Paper No. 81-6002.

Grandin, T. 1980. Good cattle restraining equipment isessential. Vet. Med. & Small Animal Clinician 75:1291.

Herrick, J.B. 1982. Bovine respiratory disease complex: Comingled calves, unprotected by immunization, are prime candidates. Beef 18:47.

Kilik, E.L. 1976. Current report on branding. Proceeding Livestock Conservation Institute, South St. Paul, MN.

Lofgreen G.P., L.H. Stinocher, and Kiesling. 1980. Effects of dietary energy, free choice alfalfa hay, and mass medication on calves subjected to marketing stresses. J. Anim. Sci. 50:590.

Phillips, W.A.   1982.   Factors associated with stress in cattle. Symposium on Management of Food Producing Animals. Purdue University.

Phillips, W.A., R.P. Wettemann, and F.P. Horn. 1982. Influence of preshipment management on the adrenal response of beef calves to ACTH before and after transit. J. Anim. Sci. 54:697.

Sheldon J. 1981. Cargill Feedlot Seminar, Kearney, Neb. In: Paul Andre (Ed.) Why Medicine Bills Skyrocket. Beef 17:12 No. 9 p 12.

126

REDUCING TRANSPORTATION STRESSES

Temple Grandin

Improved transportation and handling methods can save money. Seven to ten percent of all feedlot steers are bruised during handling, loading, transporting, and weighing. The discounting on the sale price for bruised (damaged) meat costs the producer an average of $57 per 100 head of fat cattle marketed (Rosse, 1974). Thirty percent of all the bruises occur in the valuable loin area. Livestock Conservation Institute estimates that the beef industry is losing $22 million annually from bruises that cause damage to the marketed beef. A feedlot with rough handling has twice as many bruises as a feedlot with gentle handling. A survey I conducted indicated that cattle sold live weight had 14% discountable bruises and the carcass cattle had only 8%. The producers were more motivated to handle cattle carefully when they were sold on a carcass basis because bruises showed as the carcass hung in the cooler and the producer received a penalty price. A study by Marshall (1977) indicated that people take better care of livestock that belong to them. Cattle hauled by contract truckers had more bruises than cattle hauled by a packer's own truckers. A few bad truckers often account for many of the bruises (Rickenbacker, 1958).

Horns will greatly increase the amount of bruising. Loads of horned cattle had twice as much bruise trim compared to loads of polled cattle (Meischke, 1974). Ramsey (1976) found that tipping the horns will NOT reduce bruising. The answer to the bruise problem is to dehorn your baby calves or to breed polled cattle. Thin cows have to be handled gently because they bruise more easily than steers (Wythes et al., 1979). Interior surfaces in livestock handling facilities must be smooth. Structural members such as posts should be on the outside of the fences to prevent bruises.

SHRINK

Range cattle that have been placed in unfamiliar pens will shrink more than cattle held in familiar pens

(Brownson, 1979). You can reduce shrink losses by handling livestock quietly and with little excitement. During hot weather it is advisable to ship during the night or in the early morning. Psychological stresses can contribute to shrink losses. Calves that have been preconditioned prior to shipping will shrink less than calves that have been weaned at shipping time (Woods et al., 1973). Calves that have become accustomed to transport will shrink less and animals that have become accustomed to handling procedures will be less stressed. Calves lost less weight the second time they were transported (Ried and Mills, 1962; Hails, 1978).

Transporting your livestock directly to the packer will help reduce shrink losses. Mayes, et al. (1980) found that feedlot cattle hauled for 52 mi had higher carcass yields than cattle hauled for 373 mi. Shrink can be reduced by providing water up until the time of transport. It is important to have water continuously available so that the animals do not engorge themselves shortly before loading. Tissue shrink can start soon after loading, especially if the animals get excited. Asplund (1982), found that in many instances loss of gut fill was a minor portion of the overall shrink. A portion of the shrink occurs early in the journey when excited animals urinate, sweat, pant, and defecate.

A continuous water supply prevents the animal's tissues from losing water. Tissue shrink starts quickly when the animal is being transported, but it takes time to regain the tissue weight loss after the animal is able to drink on arrival. After unloading, long-haul cattle should be fed and allowed to eat before they are allowed to drink. This will help prevent engorgement.

The type of feed the animals have been on also will affect the amount of shrink. Cattle that have been fed concentrates will shrink less than cattle that have been on green feed. Typical shrink figures for feeder calves shipped during the fall for 600 miles is 7.2% to 9.1% (Self and Gay, 1972).

WIND CHILL

Many people do not realize that the wind whistling through a truck can chill the animals. If a truck is traveling at 40 miles per hour on a 32° F. day the wind chill factor will be a chilly -20°F. (figure 1) for cattle with summer coats. During cold weather the nose vents in trucks should be closed. If the truck is being hit by a stong cross wind it may be advisable to cover one side. The important thing is to keep the animals DRY. If the hair becomes wet it loses its ability to insulate the animal from the cold. Wetting a calf has the same effect as lowering the outside temperature 40° to 50°F. Dry cold weather is usually less dangerous than wet weather around 32°F.

Even during very cold dry weather the coat retains its ability to insulate. Freezing rain is very hazardous and many animals have been lost due to wind chill under these conditions.

Ames Wind Chill Indexes

For Cattle with Summer Coats and Shorn Sheep (Dry Animals)

Wind speed mph	Actual Temperature (Fahrenheit)						
	−10	0	10	20	30	40	50
10	−20	−10	0	9	19	29	39
20	−37	−27	−17	−7	2	12	22
30	−53	−43	−33	−23	−13	−3	6
40	−60	−50	−40	−30	−20	−10	0

For Sheep with Full Fleece (Dry Animals)

Wind Speed mph	Actual Temperature (Fahrenheit)						
	−10	0	10	20	30	40	50
10	−10	0	9	19	29	39	49
20	−13	−3	6	16	26	36	46
30	−20	−10	0	9	19	29	39
40	−31	−21	−11	−1	8	18	28

Figure 1

There is no such thing as a single ideal temperature for an animal. The ideal temperature or thermal neutral zone, in which the animal feels neither hot nor cold is based on many factors, including wind speed, hair coat length, wetness, condition, and the level of nutrition. The amount of wool or hair will also affect the ability of the animal to withstand cold (figure 1) (Ames, 1974).

TRUCK TRAILER DESIGN

The design of the trailer can have an effect on the amount of bruising. In the West, side-loading trailers are used that unload through the side instead of through the rear. More bruises occur in this type of trailer because the cattle have to turn a 90° corner when they leave the trailer. If the animal becomes excited or rushed by the handler, it can easily bruise its loin on the door frame. The width of the door can have a significant effect on the amount of bruising. Replacing the standard 30 in. door with a door which was 42 in. wide at the top and tapered at the bottom reduced bruises. Tapering the door forced the animal to walk through the center and kept its hips from catching on the door frame (Grandin, 1980a).

There are still many unanswered questions about ventilation and the effects of exhaust fumes on livestock. Coleman and Sheldon (1974) reported that when the top of the exhaust stack was even with the trailer roof line, the calves on the top deck subsequently gained weight faster than the calves on the bottom deck. Calves loaded into trucks with low stacks 12 to 24 in. below the roof line gained faster if they rode on the bottom deck. Bottom-deck calves on trucks with high stacks may have been subjected to fumes sucked up in the vacuum that forms behind the trailer. Fortunately most trucks have higher stacks. Very few vehicles now have stacks 24 in. below the roof line. A recent study by Camp et al. (1981) indicated that there were no consistent differences in the average daily gains of calves transported on the bottom or top decks of a trailer. All the trailers surveyed were pulled by tractors with stack heights of 1 to 11.8 in. below the roof line. The elimination of the very low stacks could account for this result. Air movement through trailers should be researched. Muirhead (1982) is conducting studies on trailer streamlining to improve fuel efficiency and wind tunnel tests to determine air flow patterns inside the trailer.

Excessive vibration in a livestock trailer can be stressful to animals. Inflating the tires to 90 to 120 psi (to above mfg. specs.) prolongs the life of the tires but it greatly increases the vibration levels imposed on the livestock. Over-inflated tires also reduce the life of the trailer. In an aluminum trailer, the greatest amount of vibration occurs over the rear axles (Stevens and Camp, 1979).

SPACE REQUIREMENTS

Loading a truck correctly is essential for safe transport. Loading too few or too many animals can result in injuries. Livestock Conservation Institute has published recommendations for space requirements for different types of livestock (figures 2, 3, 4). Avoid overloading a truck with horned cattle. Too many animals can double the amount of bruising. If tiny baby ("bob") calves are being hauled, allow 2.9 sq ft per 100 lb calf and 3.5 sq ft for a 150 lb calf (Seubert, 1982).

TRUCKING TIPS

1. Use partitions to separate species transported on the same truck.
2. Clean trucks after each load to prevent the spread of disease.
3. Check truck load regularly enroute.
4. Load and unload animals quietly; limit use of electric prods. However, it is better to give a cow a small jolt than to break her

tail by twisting it. Abusive handling during loading and unloading is still a major problem.

Truck Space Requirements for Cattle

(Cows, range animals or feedlot animals with horns or tipped horns; for feedlot steers and heifers without horns, increase by 5 percent)

Av. Weight	Number Cattle per running foot of truck floor (92-in. truck width)
600 lbs.	.9
800	.7
1,000	.6
1,200	.5
1,400	.4

Examples (1,000 lb. cattle):

44 ft. single deck trailer—44 x 0.6 = 26 head horned, 27 head polled.

44 ft. possum belly (four compartments, 10 ft. front compartment; two middle double decks, 25 ft. each; 9 ft. rear compartment, total of 69 ft. of floor space)—69 x 0.6 = 41 head of horned cattle and 43 head of polled cattle.

Measure the total lineal footage of floor space in YOUR truck.

Figure 2

Truck Space Requirements for Calves

(Applies to all animals in 200 to 450 lb. weight range)

Av. Weight	Number Calves per running foot of truck floor (92-in. truck width)
200 lbs.	2.2
250	1.8
300	1.6
350	1.4
400	1.2
450	1.1

Examples (450 lb. calves):

44 ft. single deck trailer—44 x 1.1 = 48 head.
44 ft. double deck trailer—88 x 1.1 = 97 head.

Figure 3

Truck Space Requirements for Sheep

(Use for slaughter sheep, load 5 percent fewer if sheep have heavy or wet fleeces.)

Av. Weight	Number Sheep per running foot of truck floor (92-in. truck width)
60 lbs.	3.6
80	3.0
100	2.7
120	2.4

Example (120 lb. sheep):

44 ft. triple deck trailer—44 x 3 x 2.4 = 317 shorn sheep, 302 wooly sheep.

Figure 4

Source: Livestock Conservation Institute (1981).

5. Accelerate vehicle smoothly and avoid sudden stops.

6. Don't ship livestock that are full of green feed. Let them stand and empty out.

STRESS AND MEAT QUALITY

Stress can have a negative effect on meat quality in both cattle and sheep. Stress from fatigue, fighting, cold weather, or lack of food can use up the animal's store of glycogen (muscle energy). When the glycogen store is used up, the meat will be darker and drier than normal. This "dark cutting" meat is less palatable and has a shorter shelf life. Cattle that are "dark cutters" will receive USDA downgrading and discounting for the carcass.

The basic principle is that a long-term stress for 12 to 48 hours will use up the glycogen and make the meat darker than normal, but a short-term stress, such as excitement, immediately prior to slaughter will tend to make the meat tough. Stress-related meat quality problems are more likely to occur in the fall and spring when the days are hot and the nights are cool. When the weather turns more uniformly cold or warm the animal's body adapts and they become less prone to dark cutting.

To reduce stress, avoid mixing strange animals prior to loading or after they reach the meat packing plant. Everytime strange cattle are mixed they will fight to establish a new social order. Fighting is very stressful because large amounts of adrenalin are secreted. The rapid release of adrenalin that occurs during fighting uses up the animal's store of glycogen in the muscles. When this occurs, the pH of the meat rises and it becomes darker and drier.

When steers fight to establish a new social order during the 24 to 48 hours prior to slaughter, they increase the incidence of dark cutting meat (Grandin, 1980[b]) (figure 5). Fighting among bulls when unfamiliar animals are mixed will cause dark cutters within a few hours. When unfamiliar fed bulls were mixed overnight, prior to slaughter the resultant fighting caused 73% of the animals to become dark cutters (Tennessen and Price, 1980). Bulls fed for slaughter should travel to the packing plant with their penmates. Young bulls kept in penmate groups have a very low incidence of dark-cutters.

The shape of the pen that animals are held in can influence their behavior. Kilgour (1978) found that bulls tend to use the space more efficiently in a rectangular pen than in a square pen. A long narrow pen maximizes the length of fenceline space in relation to the floor space. Studies by Stricklin et al. (1979) indicated that cattle prefer to lie along the fenceline. A long narrow pen provides the animals with more fenceline space than does a square pen with the same area. The long narrow pen works on the same principle as that used by people when finding seats

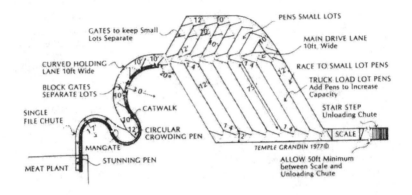

Figure 5. Stockyard for a slaughter plant with diagonal pens and curved chutes. One-way traffic flow and the elimination of sharp corners helps to keep animals calm (designed by Temple Grandin).

in a restaurant. People tend to prefer booths along the wall instead of center tables. The use of long narrow pens to hold livestock at the packing plant may help reduce stress. Figure 6 illustrates a layout for a packing plant with long narrow pens on a 60° angle. Each pen holds one truck load of cattle, and the livestock traffic flow is one-way. Pens on a 60° angle have been used for years in feedlots for shipping cattle. McFarlane (1976) was one of the first people to recognize the benefits of pens in a diagonal layout.

ROUGH HANDLING

Rough handling and the excessive use of electric prods is very stressful to livestock. An animal that becomes excited and agitated immediately prior to slaughter is more likely to produce tough meat. Poorly designed handling facilities and rough abusive handling can greatly increase the amount of stress. Efficient, quiet sorting of feeder calves resulted in an increase of heart rate of only 7 beats per minute. Rough handling in poor facilities resulted in an increase of 48 beats per minute (Stermer et al., 1981). A noisy environment, yelling, and dogs biting at livestock will increase stress (Kilgour and deLangden 1970; Pearson et al., 1977). Being bitten by a dog was more stressful to sheep than being chased by a dog. Tiny baby calves should not be shipped until they are 5 to 7 days old to avoid stress and death loss problems (Seubert, 1982).

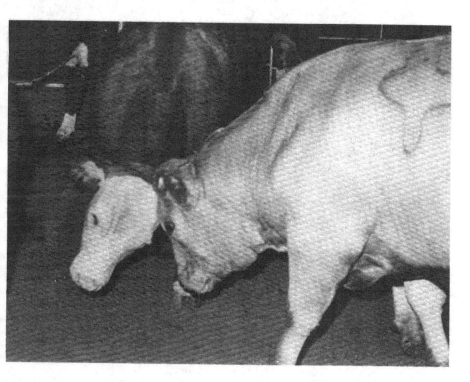

Figure 6. Fighting prior to slaughter is a major cause of dark cutting beef. When animals fight epine phrine (adrenalin) is secreted, which depletes the store of glycogen in the muscles. The big Charolais on the right, disrupted the entire pen. Extremely aggressive cattle observed fighting in the packing plant holding pens should be removed to reduce stress.

1184

REFERENCES

Addis, D.G. 1980. Preventive medication for feedlot re-
placement calves, Study II. In: Stress Calf Studies.
Riverside and Imperial Counties, Cooperative Extension,
University of California.

Addis, D.G., J. Dunbar, J. Clark, G.P. Lofgreen and G. Crenshaw. 1980. Effect of time and location of processing
on feeder calves. In: Stress Calf Studies. Riverside
and Imperial Counties, Cooperative Extension, University of California.

Ames, D.R. 1974. Wind chill factors in cattle and sheep.
Special Publications SP-174, International Livestock
Environment Symposium, Amer. Soc. Agr. Eng. St.
Joseph, MI. pp 68-74.

Asplund, J.M. 1982. Shrink: The silent rustler. Beef.
February, p 46.

Brownson, R. 1979. How to prevent cattle shrink. Montana
Stockgrower. Reprinted in Beef Digest. February 1979,
p 40.

Camp, T.H., D.G. Stevens, R.A. Stermer, and J.P. Anthony.
1981. Transit factors affecting shrink, shipping
fever, and subsequent performance of feeder calves.
J. Anim. Sci. 52:1219.

Grandin, T. 1982. Transportation of domestic animals, Symposium on Management of Feed Producing Animals. Purdue
University.

Grandin, T. 1981a. Bruises on southwestern feedlot cattle. Paper presented at 73d Annual Meeting, Amer.
Soc. Anim. Sci., July 26-29, 1981. (Abstr.).

Grandin, T. 1980. Bruises and carcass damage, Int. J.
Stud. Anim. Prob. 1:121.

Grandin, T. 1980b. The effect of stress on livestock and
meat quality prior to and during slaughter. Int. J.
Study of Anim. Prob. 1:313.

Grandin, T. 1978. Transportation from the animal's point
of view. Amer. Soc. Agr. Eng. Technical Paper No.
78-6013.

Hails, M.R. 1978. Transport stress in animals: A review.
Animal Reg. Stud. 1:289.

Kilgour, R. 1978. The application of animal behavior and
the humane care of farm animals, J. Anim. Sci.
46:1478.

Kilgour, R. and H. de Langden. 1970. Stress in sheep resulting from management practices. New Zealand Society of Animal Production proceedings 30:64.

Livestock Conservation Institute. 1981. Livestock Trucking Guide. By Temple Granding, South St. Paul, MN.

Marshall, B.L. 1977. Bruising in cattle presented for slaughter. New Zealand Vet. J. 25:83.

Mayes, H.F., M.E. Anderson, H.E. Huff, J.M. Asplund and H. B. Hedrick. 1980. Transport effects on carcass yield of slaughter cattle. Amer. Soc. Agr. Eng. Technical Paper No. 80-6509.

McFarlane, I. 1976. Rationale in the design of housing and handling facilities. In: M.E. Ensminger (Ed.), Beef Cattle Science Handbook. Agriservices Foundation, Clovis, CA 13:223.

Meischke, H.R.C. et al. 1974. The effect of horns on bruising cattle. Australian Vet. J. 50:432.

Muirhead, V.U. 1982. Interim report, an investigation of the internal and external aerodynamics of cattle trucks. Prepared for the Dryden Flight Research Center under Grant NAG4-8. University of Kansas Center for Research Inc., Lawrence, KS.

Pearson, A.M., R. Kilgour, H. deLangen and E. Payne. 1977. Hormonal responses of lambs to trucking, handling and electric stunning. Proc. New Zealand Soc. Anim. Prod. 37:243.

Ramsey, W.R. et al. 1976. The effect of tipping horns and the interruption of journey on bruising cattle. Aust. Vet. J. 52:285.

Reid, R.L. and S.C. Mills. 1962. Studies of the carbohydrate metabolism of sheep. Aust. J. Agr. Res. 13:282.

Rickenbacker, J.E. 1958. Causes of losses in trucking livestock. Marketing Research Report 261, Farmers Cooperative Service, USDA.

Rickenbacker, J.E. 1961. Loss and damage in handling and transporting hogs. Marketing Research Report 447, Farmer Cooperative Service, USDA.

Roberts, D.W. 1982. Yield in sheep meat processing. CSIRO Advances in Meat Science Conference, Brisbane, Australia.

Roose, J.C. 1974. Your stake in the $184,000,000 tangible farm to cooler loss. Proc. Livestock Conservation Institute, St. Paul, MN.

Self, H.L. and N. Gay. 1972. Shrink during shipment of feeder cattle. J. Anim. Sci. 35:489.

Seubert, T.J. 1982. Handling and transporting of bob calves and special fed veal calves. Official Proceedings, Livestock Conservation Institute, South St. Paul, MN.

Stermer, R.A., T.H. Camp, D.G. Stevens. 1981. Feeder cattle stress during handling and transportation. Amer. Soc. Agr. Eng. Technical Paper No. 81-6001.

Stevens, D.G. and T.H. Camp. 1979. Vibration in a livestock vehicle. Amer. Soc. Agr. Eng. Technical Paper No. 79-6511.

Stricklin, W.R. H.B. Graves and L.L. Wilson. 1979. Some theoretical and observed relationships of fixed and portable spacing behavior in animals. Appl. Anim. Ethol. 5:201.

Tennessen, T. and M.A. Price. 1980. Mixing unacquainted bulls: a primary cause of dark cutting beef. The 59th Annual Feeder's Day Report, Agriculture and Forestry Bulletin, University of Alberta, p 34.

Wythes, J.R., R.H. Gannon and J.C. Horder. 1979. Bruising and muscle pH with mixing groups of cattle pretransport. Vet. Rec. 194:71.

127

ANIMAL WELFARE:
AN INTERNATIONAL PERSPECTIVE

Stanley E. Curtis,
Harold D. Guither

Are today's intensive animal-production systems basically inhumane? This question is central to the issue of farm-animal welfare that has been developing in the U.S. for several years. Opinions vary across a broad spectrum. According to one extreme view, animals have the same mental experiences as humans ("anthropomorphism"), and thus they ought to be treated as humans. Ethical vegetarians believe we have no right to slaughter livestock or poultry for human consumption. A more widely held, more moderate position--and one where many animal welfarists and agriculturists find common ground--is that we should respect the animals we use and should not subject them to distress as a consequence of normal production pratices.

Most states have laws protecting domestic animals from neglect and abuse. But there is still debate about a third form of alleged cruelty to animals: depriving animals of opportunities to express certain supposedly necessary behaviors. Especially controversial are the practices of keeping laying hens in cages, gestating sows in stalls, and veal calves in crates. There are those who point out that some behaviors that animals express frequently in natural surroundings are not observed often in certain artificial environments (and vice versa), and they claim that this reflects undue stress on the animals in these unnatural surroundings. However, others say such differences in behavioral responses to the environments are to be expected because the array of behavioral triggers varies from one environment to another.

The general public's attention was first drawn to the welfare of food animals by the 1964 publication in England of the book Animal Machines by Ruth Harrison. As a result of the public interest this book generated, the British government appointed a committee that prepared a report on intensive animal-production systems. The report questioned the humaneness of several common husbandry practices. Since then, the debate over farm-animal welfare has spread all over northern Western Eruope, the U.S. and Canada, as well as Australia and New Zealand. Conflict has arisen because livestock and poultry producers--and indeed producers of feed grains too--have perceived their economic interests as being threatened by some of the policies and regulations proposed by animal-welfare groups, whose views are based on ethical judgments rather than scientific evidence or economic feasibilities.

The extent to which, and the ways in which, a society uses animals for companionship, recreation, power, or food are ethical decisions, and they are heavily influenced by social and religious traditions. Like other public policy matters, these decisions should be made by our political system and not by any one sector with strong opinions or interests. But such public decisions can be made wisely only after all of the facts and consequences of proposed policies and regulations have been fully explored and analyzed.

Scientific research can contribute to discussions of animal welfare by producing scientific evidence on animals' relationships with their environments. Current gaps in knowledge lie largely in the areas of perception and stress, and these are where many investigations are now being focused.

Perception is the immediate discriminatory response of an animal to energy-activating sense organs. What do we know quantitatively, and especially relative to the human experience, about conscious perceptions of comfort and discomfort, pleasure and displeasure, pain or the absence thereof, by farm animals? Little or nothing, if purely anthropomorphic musings are ignored.

We should also recognize that the design of accommodations for humans, about whom much more is known in this respect, is still hampered by the paucity of quantitative data available and by the practical impossibility of meeting an organism's needs so precisely over time that it will never experience discomfort. Added to this is the complicating fact that individuals' perceptions of comfort differ so greatly. One architect has suggested that, as a practical matter, even facilities for humans cannot be designed to achieve some "comfort" zone--rather a "lack of discomfort" zone is the best that can be hoped for.

We know more about stress and its consequences in farm animals than about perception. An animal is under stress when it must make extreme functional, structural, or behavioral adjustments to cope with adverse aspects of its environment. Thus, an environmental complex is stressful only if it makes extreme demands on an animal.

Interpretation of stress parameters and indices is the real challenge as we continue to generate more knowledge and endeavor to use more completely what is already known for the purpose of increasing the fit between agricultural animals and their environments.

It is the unusual moment when any animal--in the wild or on a farm--is not responding to several stressors at once. Stress is the rule, not the exception, and nature has endowed animals with marvelous arrays of reactions to these impingements.

It is now generally recognized that the amount of stress an animal is under depends not only on the intensity and duration of a noxious agent, but on the animal's ability to modify its perceptions and the effects of the stressor, as well. There is increasing evidence that an animal's emotional feelings depend to some extent on the predictability and controllability of its environment. When an animal's expectations are being fulfilled, or it is able to control its surroundings, it feels more comfortable, even if it is responding to stressors. But, according to this theory, the animal feels uncomfortable or even distressed when its environment is unpredictable or uncontrollable.

Furthermore, the concept of alliesthesia holds that central perception of stress intensity depends on the context within which it occurs. For example, a human whose body temperature is below normal usually finds cold stimuli unpleasant and hot pleasant. Extrapolating to agricultural animals, perhaps stress of one sort actually primes an animal to receive pleasure from a stressful aspect of its environment; the cool of the night might prepare the animal to take comfort from the heat of a summer afternoon.

Scientists still do not fully understand how findings such as these relate to an animal's welfare, health, and productivity. But it is accepted that we cannot rely on physiological or behavioral traits alone as indicators of the amount of stress an animal actually perceives, let alone how these might be related to the animal's welfare, health, or productivity.

Another set of indicators must be taken into account, too. Environmental stress generally alters animal performance and also elicits physiological and behavioral responses. The stress provokes the animal to react in some way, and this reaction can influence the partition of resources among maintenance, reproductive, and productive functions in one or more of five ways: the reaction might 1) alter internal functions involved in economically important processes as well as reactions to stressors, 2) divert nutrients from productive or reproductive processes to maintenance processes, 3) reduce productivity directly, 4) increase individual variability in performance, or 5) impair disease resistance.

Still, it would be unrealistic to leave the impression that the links between stresses and productive and reproductive processes are clear and simple. Consider three examples: 1) lactating cows can be under severe heat stress each afternoon, but so long as adequate feed is available they might not suffer any reduction in milk yield; 2) animals kept in relatively barren environments where specific social or physical stimuli are absent sometimes grow faster than do their counterparts in more enriched surroundings; and 3) there is not always a clear correlation between signs of physical and social trauma in individual hens and their respective individual egg yields. The situation is a complicated one which needs further scientific investigation before the fit between food animals and their environments can be optimized in terms of welfare, health, and performance of the animals.

As a result of public pressure, funds have been allocated in many countries for more research on intensive production systems for laying hens, veal calves, and swine and on the responses of these animals to various stressors. But sometimes the public pressure for governmental action has been so strong that political decisions have been made before scientific evidence was available to support such decisions. This is despite the fact that, in most countries, governmental officials have been sympathetic to the economic realities farmers face in terms of animal-welfare regulations' increasing production costs. Legislators have been especially reluctant to enact regulations that would put their nation's farmers at a competitive disadvantage with producers in other countries.

Animal-welfare laws and regulatios already have been established as part of public policy in several European nations. Through government-appointed committees or commissions, views of animal welfarists, scientists, and producers have been brought together. This is the first phase of bringing the public's awareness of the animal-welfare issue to the point that they can participate intelligently in the policymaking process. When such a group issues its report, the views and proposals it contains provide the bases for more media attention and broader public awareness. For example, the Report of the Technical Committee to Enquire into the Welfare of Animals Kept under Intensive Livestock Husbandry Systems in the U.K (1965), the National Council for Agricultural Research Committee of Experts' Report on Animal Husbandry and Welfare in The Netherlands (1975), the Council of Europe's European Convention for the Protection of Animals Kept for Farming Purposes (1976), and the House of Commons Agriculture Committee's report, Animal Welfare in Poultry, Pig, and Veal Calf Production in the U.K. (1981) have provided reference bases for public discussion of animal-welfare issues. What the public might not realize is that these reports may or may not represent a majority of public opinion. They are simply the reference base upon which further discussion and debate will be carried out before final policy decisions are made.

Once these decisions have been made, agencies of government are then established to oversee the new laws and regulations. In some European countries, the regulations have dealt with practices progressive producers would have applied in their livestock and poultry operations anyway. But, of course, with a rgulation in place they no longer have a legal choice in the matter.

In the U.S., a joint resolution introduced into the Congress by Congressman Ronald Mottl of Ohio in July 1981 called for the creation of a 16-member committee to study animal-production practices in this country. Those concerned with current animal-production practices viewed this legislative approach as a means to further promote discussion of animal welfare and broaden public awareness of the issue.

Producer groups generally felt that no such discussion or study was needed. They believed they were using the most advanced production methods that scientific research and practical experience offered for the most efficient and profitable production of food. Although no hearings on this resolution were scheduled in 1982, the idea of a commission to study animal-welfare issues in the U.S. probably will continue as a goal of some animal-welfare organizations for a long time.

The experiences with animal-welfare policies and practices in Europe suggest that policymakers in the U.S. weigh scientific evidence carefully and have the benefits of broad public discussion on the issues--including ramifications impinging on the food distribution and pricing system--before attempting to set up legislation that would regulate the ways livestock and poultry are produced.

128

ANIMAL WELFARE AND MEAT PRODUCTION IN GREAT BRITAIN

Harold Kenneth Baker

There is a growing awareness and interest in both environmental and welfare aspects of farming within Great Britain. Much of this concern by the consumer population is based on emotion rather than fact. However, it has to be recognized that some of the counter arguments from the agricultural community are also emotional and may be short on facts--particularly those facts relating directly to animal welfare. At the moment, only 2% of the U.K. population are directly involved in farming and just over half of these, or one per cent of the total population, are directly engaged in livestock production. Therefore, 98% of the British people have little, or no, detailed knowledge of current livestock practices. Their views are built up largely from what they read or see on television. Increasingly, welfare groups are gaining greater access to the media, and a higher degree of professionalism is being used in their campaigns. It must also be recognized that, while many of the activists in these movements are primarily concerned with welfare, there is also a hard core prepared, if necessary, to advocate vegetarianism as the ultimate weapon against those producers who appear immune to their arguments.

Whatever views the farming community have of the welfarists, they will ignore them at their peril. It is becoming increasingly clear that the population as a whole will not tolerate much longer what they believe to be cruelty being imposed upon farm animals. The increasing professionalism of the welfare lobby must, therefore, be worthy of a professional reply. Thus the population, as a whole, is becoming increasingly skeptical of the farmers' argument that when animals have a high level of performance (in terms of farming targets) they must, therefore, be thriving and contented. To the average consumer the terms intensive, or industrial, production of farm livestock must imply some degree of discomfort if not downright cruelty. Similarly to the welfarists, who within Britain will not be short of food, the economic arguments about producing more food cut little ice. They will often argue that if more humane methods of livestock production increase the cost of meat (for example) this is an acceptable price to pay.

In fact, within Britain, producers have a good track record in animal welfare and much of the emotional argument is caused by a genuine lack of communication between producers and the public. Currently, farmers and their representatives and legislators in Britain are devoting more time and thought to the welfare issues. In addition, the attitudes and legislation within the EEC also will have an increasing impact in the whole of this area.

LEGISLATION

There is already in Great Britain a solid base of legislation concerning animal welfare. The most important of these for meat producing animals are:
- The Protection of Animals Act 1911--designed specifically to prevent direct cruelty to domestic and captive animals. Acts amounting to cruelty include: "Cruelty to beat, kick, ill-treat, torture, infuriate, or terrify any animal" and "to cause unnecessary suffering by doing, or omitting to do, any act."
- The Slaughter of Animals Act 1958--controls the place and method of slaughter of farm animals.
- The Markets (Protection of Animals) Order 1964--provides for the prevention of unnecessary suffering of cattle, sheep, and pigs when exposed for sale live in markets and while awaiting removal after the sale.
- The Transit of Animals Order 1973 and 1975--are very detailed regulations concerning the transport of livestock. They include a number of general measures that are intended to safeguard a wide variety of animals during their carriage by sea, air, road, and rail.
- Codes of Recommendations for the Welfare of Livestock--are approved by Parliament and indicate the basic requirements necessary for the welfare of stock. These include water, feed, environment, housing requirements, etc. It is important to note the status of these codes in relation to other legislation.

Thus, to cause unnecessary pain or distress to a farm animal is an offense. However, failure to observe provision of a Code of practice, (i.e. size of pen) is not a legal offense. Such failure, however, can be used to support a charge of causing unnecessary suffering. Basic legislation is now available to prevent cruelty to animals in Britain. In general, however, this does not relate to systems of production as such; thus the 'intensive' form of livestock production, particularly with pigs and poultry that arouses the opposition of the Welfare Lobby, is completely legal.

Those areas that incite the greatest interest in the welfare area are described next.

VEAL CALVES

Veal production in Britain is very limited and is less than 1% of the total beef production. Nevertheless, the impact of the concept of veal production on the Welfare Lobby has been great and completely out of context with the small quantity of veal produced in Britain. Any talk of modern intensive beef production tends to be associated with veal production from calves housed in wooden crates and kept in darkened houses until they are slaughtered. Not only is very little veal produced, but the use of darkened houses has not been generally practiced in the U.K., although at one time it was common in Europe. The small number of veal producers in Britain have, in fact, already responded to the Welfare Lobby abandoning the individual crate system and moving to group housing in straw yards. This move to group housing should appeal to the Welfarists. However, it is interesting to reflect that when calves are acquired from a number of different sources and put into groups in a yard there can be higher levels of bowel and respiratory infection among the groups of young calves. The situation could in turn result in the regular and indiscriminate use of antibiotics with an associated build-up of residues in the meat of such calves. The yarded system could, therefore, require a higher-than-average level of stockmanship to safeguard the health of the young calves brought in from a varying number of sources. Thus it is concluded that 1) the yarded system must still be a specialist operation, not to be undertaken by an inexperienced farmer, and 2) calves will move more freely and establish social groups in a loose-house system with access to straw. The system would seem to be more humane than conventional pen-rearing methods, provided it is properly operated.

BEEF PRODUCTION

There are few welfare problems rising from the systems of beef production practiced in Britain. However, the frequent use of the term "intensive beef production," whether applied to the completely housed 12-month barley-beef system or intensive grazing of cattle on grass through heavy fertilizer rates, tends to raise the hackles of the Welfare Lobby. They automatically tend to think of industrial or factory farming without realizing the actual systems that are involved. Barley-beef production certainly involves the housing of calves from birth to slaughter. Only about 5% of Britain's beef is produced by this method and the housing conditions are normally excellent with plenty of pen space, usually on straw litter. "Intensive" grass-beef production really refers to high stocking levels, and there can be no question of welfare problems in this type of production.

A relatively new development in Britain is production using intact males, or "bull beef." Although this technique is widely practiced in some European countries, it is only slowly gaining ground in Britain. The bull-beef producer is restricted only by the need to observe a code of safety in the housing, fencing, and handling of bulls. The fencing requirements have tended to restrict this form of production to indoor systems based on intensive cereal and maize-silage feeding units. The housing together of beef bulls does require a high level of management, and failure to do so could lead to welfare problems. Thus, the careless mixing of bulls from different groups could lead to serious fighting and damage, or even death, to the stock.

One area that has not yet caused welfare problems, but could if not properly controlled by farmers, is that associated with difficult calvings. The introduction of Continental breeds to Britain has increased the incidence of calf mortality and dystocia, but so far the level of these is acceptable. However, it might be dangerous to move towards the situation in Belgium where the calvings associated the Belgium-Blue and White (a heavily double-muscled breed) invariably are associated with cesarean operations. Not only is this expensive, but if this were identified as a common practice in Britain it would almost certainly create problems with the welfarists.

In those areas where straw is difficult and expensive to obtain there has been a move to totally slatted floor areas. The overall cost is higher than purpose-built straw yards or cubicles, but it can be stocked more densely--thus competes with them in terms of unit cost. Provided the slats are properly designed and spaced, and the management of a high order, there should be no welfare problems. However, a low standard of management can produce miserable-looking animals in poor health under any condition, and if this is done on slats it would inevitably bring them into disrepute with the Welfarists.

SHEEP PRODUCTION

Within Britain, sheep have traditionally been kept completely on grass. Inevitably this presents a natural image to the public at large and so the Welfare Lobby and sheep producers have seldom clashed. It is interesting, however, to note that the new development of in-wintering ewes (either all the winter or simply for lambing) is generally referred to as intensification and hence leads the Welfarists to view this development with suspicion. In fact, of course, bringing the sheep indoors creates a much better environment, provided the correct tenants of ventilation, etc., are followed for both sheep and shepherds. Individual attention to individual animals is much better and lambing losses are generally considerably reduced.

PIG PRODUCTION

In the last 20 years there has been a move towards the keeping of large numbers of pigs in buildings--both feeding pigs and subsequently adult stock. Generally sows are housed individually or confined in farrowing crates in lactation, although there are some systems where sows and litters are mixed in small groups during lactation. After weaning, sows may be mixed or penned individually until service, when again they may be penned individually or run in small groups in yards. The two important trends have been the increasing use of partially or totally slatted flooring to facilitate the removal of dung and the close confinement of individual pigs in stalls or tethers, which do not allow the animal to turn around to assist the management and feeding of such pigs. After weaning, the growing pigs from litters are mixed into groups of 15 to 20 pigs, though the range can be much greater. Usually the pigs are moved at least once into another pen, at which time the number of pigs in the pens is reduced to 8 to 12, but again with a great range. The important trend in growing pigs has been to wean pigs at an earlier and earlier age so that the traditional weaning at 8 weeks has been reduced to 5 weeks--a great many herds wean at 3 weeks and a few even earlier.

There have been considerable developments in pig housing to assist in their indoor management. Particular attention has been paid to temperature and ventilation control and to the removal of dung. Good husbandry, especially in the housing and feeding of pigs, is necessary to control disease and to ensure that the pigs are kept clean. These requirements have led to the development of fully or partly slatted flat deck pens, both of which may be housed in a controlled environment building.

Welfare Considerations

The intensification of pig production has produced complaints that the welfare of pigs is not given enough attention. In particular the Welfarists have turned their attention to the totally confined sows (either tethered or in sow stalls) during gestation, which does not allow them to turn around. The large number of pigs kept in small areas may lead to the use of drugs to help make less successful management systems work profitably. (These aspects are considered further below.)

Effect of Close Confinement on Sows

The physical performance of sows totally confined has been compared with that of sows kept in yards during gestation. In general there is no difference in performance except for a trend for the weaning-to-service interval to be longer in sows totally confined. Comparisons of the general health of sows also have been made and here the reports are mixed. In some cases, there is no apparent difference, while in others it has been observed that the number of cases of farrowing fever, rectal prolapse, and lameness is increased in totally confined sows. There have, however, been few surveys in Britain. It should also be pointed out that conditions on some farms have produced serious problems because of poor housing design or a fault in adaptation of existing buildings, but such problems can be overcome by proper advice on construction. Because of the large variation in the general health of livestock on individual farms, it is not satisfactory to draw conclusions without surveying a large number of units.

In view of the importance of this matter, a study was made at the MLC station from September 1979 to December 1979 of the effects of using a neck tether system on dry sows. A total of 37 sows were examined on three occasions during this period, and the changes produced by the tether were classified as follows:

Grade 0 - no effect
1 - reddening of neck, wetness
2 - crust formation, serum exudation, redness
3 - severe inflammation with cuts in skin
4 - severe lesions on neck

Only one sow was found to show a Grade 3 change--and only on one occasion. No sow was found to have a Grade 4. About 50% of the sows showed no change, 32% improved, and in 16%, the change varied from time to time, while in only one sow did the condition become worse. Since December 1979, a further 132 sows have been examined (on one occasion only) for the effect of the tether with the following results:

	No.	%
Grade 0	99	75
1	22	17
2	11	8
3	0	-
4	0	-

The information indicates that the use of tethers does not produce changes that give cause for concern, but regular checking of the tether with adjustment or removal of the pig from the tether for a time is essential if a severe lesion is found.

However, lesions produced on the bony prominences of sows (possibly due to behavioral activity) in relation to the floor and equipment that confines the sow are more serious than those produced by tethers and occur, in particular, on the tail, head, and in the shoulder region. Their origin appears to be related to the contact between the sow and the concrete floor while suckling the litter. The lack of body fat in sows may be involved. Additionally, a lesion on the tail head may be produced by the sows actively rubbing against the metal crate for no apparent reason.

It is considered therefore that, although the main argument against close confinement of sows is that the animal cannot turn around and so may be the subject of mental and physical deprivation, the required physical conditions of confined sows can generally be met by good management, though it must be recognized that such a system is open to abuse. Ill-fitting tethers may cause pressure sores to develop, but by using the correct design these can be largely overcome, and any sows showing a continuing reaction against the tethers should be removed. These systems allow for the individual feeding of sows, which aids in keeping the sow in the best condition for each stage of her reproduction cycle and avoids the aggression and bullying that occur when sows are mixed together and run in groups. There is no satisfactory method of determining the mental welfare of confined animals, and until this is possible--which is not likely in the short-term--it is difficult to form objective opinions on this matter.

Although extensive systems tend to be more acceptable from the welfare viewpoint, serious problems can arise. Some involve competition for space, shelter, food, and water. The environment is less well controlled and there is considerable difficulty in supervision for a wide range of husbandry purposes; veterinary care is more difficult.

Early Weaning

Early weaning of piglets allows the production of more pigs per sow a year. However, because the piglets have special nutritional and housing requirements, good management is essential to allow these piglets to continue satisfactory growth without the losses caused by disease and high mortality. The flat-deck system allows this to be done quite satisfactorily. The levels of mortality, a good overall indication of the level of disease in young pigs, in such systems is low. Mortality of early weaned pigs from birth to slaughter is similar to those weaned later. The main cause for concern with early weaning is that many piglets are reared in a small space that appears to be overcrowded and this can lead to bad habits such as ear and tail biting. It is well recognized, however, that overcrowding leads to reduced performance but suitable management will correct such problems. Pigs are often reared on totally slatted floors. Originally some of these floors did cause leg injuries but avoidance of the use of unsuitable materials will prevent such a problem. With flat-deck or cage housing for young pigs, there may therefore be evidence of restlessness, or flank and tail biting, and variable growth, while in other units pigs appear highly content, healthy, and thriving, as in a well-operated deep straw system. Again, climatic environment, group size, stocking density, feeding space, and nutrition are critical.

Drugs

To keep pigs healthy, especially under intensive systems, good management is required, and where this is not optimal, there are fears that poor management may be masked by the widespread use of drugs. In addition, such widespread use of drugs may create a potential human health hazard by creating drug resistant strains of bacteria that are pathogenic to man. In pigs, the line dividing those diseases caused by an infection (which can only be treated by drugs) and those diseases produced directly by poor management can be very fine--in these cases, management changes often will not adequately control the disease. Responsible pig keepers have regular consultations with veterinary surgeons for guidance on this point. There are a number of drugs that are routinely fed to pigs as growth promoters. Such a usage as feed additive is restricted to drugs that are used at a low level. They are not used for therapy in either animals or humans in accordance with the policy decided after the Swann Committee's recommendations. So far there is no evidence of any development of bacterial resistance or risk to human health from such use of drugs in pigs.

STOCKMANSHIP AND WELFARE

Stockmanship has a more important influence on welfare than the actual systems of production, as good stockmanship can lead to a high degree of comfort and contentment in virtually all common systems of production. Conversely, average or below-average stockmanship imposed on a system that may appear acceptable from the welfare viewpoint can, nevertheless, give rise to welfare problems. The general trend in livestock production where one stock person tends more and more animals may have to be discouraged unless an improvement in labor-saving aspects of such systems allows ample attention to be given per animal. Labor-saving involves the elimination of chores but should not reduce the time spent caring for and observing stock. Very strong emphasis should be given to adequate training in the science and art of good stockmanship--along with experience. Eventually some form of licenses and qualifications may be involved, particularly in the pig and poultry sectors.

Poor housing coupled with bad management, can produce serious injury and disease problems. Adequate supervision to detect early signs of ill health and to correct such problems is essential at all times.

Finally, any new system of management is likely to have some "teething" problems, but this is not a reason for not continuing to develop systems to make them more profitable. Maybe it is unfair to compare such systems with those that have been in use for many years until such time as the snags have been ironed out.

NAMES AND ADDRESSES
OF THE LECTURERS AND STAFF

GEORGE AHLSCHWEDE
Sheep and Goat Specialist
Texas Agricultural Extension Service
Route 2, Box 950
San Angelo, TX 76901
--Sheep Specialist-Tour Coordinator

JOE B. ARMSTRONG
Associate Professor and
Extension Horse Specialist
Animal Science Department
New Mexico State University
Las Cruces, NM 88003
--Horse Geneticist

ROY. L. AX
Assistant Professor
Department of Dairy Science
University of Wisconsin
Madison, WI 53706
--Dairy Physiologist

HAROLD KENNETH BAKER
Meat and Livestock Commission
Box 44, Bletchley
Milton Keynes
MK2-2EF ENGLAND
--Beef Cattle Specialist & Geneticist

R. L. BAKER
Visiting Professor of Animal
Breeding & Genetics
Animal Science Department
University of Illinois
Urbana, IL 61801
--New Zealand Geneticist

R. A. BELLOWS
Location Research Leader
USDA-ARS
Route 1, Box 2021
Miles City, MT 59301
--Beef Cattle Physiologist

W. T. BERRY, JR.
Executive Vice-President
National Cattlemen's Association
P. O. Box 3469
Englewood, CO 80155
--Beef Cattle Organization Leader

HENRY C. BESUDEN
Vinewood Farm
Route 2
Winchester, KY 40391
--All-Time Great (Sheepman)

RONALD BLACKWELL
Executive Secretary-General Manager
American Quarter Horse Association
Amarillo, TX 79168
--Horse Organization Leader

BILL BORROR
Tehama Angus Ranch
Route 1, Box 359
Gerber, CA 96035
--Cattle Breeder

MELVIN BRADLEY
Professor of Animal Science and
State Extension Specialist
University of Missouri
Columbia, MO 65211
--Horse Specialist

B. C. BREIDENSTEIN
Director
Research and Nutrition Information
National Livestock and Meatboard
444 North Michigan Avenue
Chicago, IL 60611
--Meat Scientist

JENKS SWANN BRITT, D.V.M.
Veterinarian/Dairyman
Logan County Animal Clinic/J&W Dairy
Route 1
Russellville, KY 42276
--Veterinarian

HERB BROWN
Research Associate
Lilly Research Laboratories
P. O. Box 708
Greenfield, IN 46140
--Animal Scientist

O. D. BUTLER
Associate Deputy Chancellor
for Agriculture
Texas A&M University
College Station, TX 77843
--Animal Scientist

EVERT K. BYINGTON
Range Scientist
Winrock International
Route 3
Morrilton, AR 72110
--Range Scientist

B. P. CARDON
Dean
College of Agriculture
University of Arizona
Tucson, AZ 85721
--Animal Scientist

ARTHUR CHRISTENSEN
Manager
Christensen Ranch
P. O. Box 186
Dillon, MT 59725
--Sheep Producer

ROBERT L. COOK
Wildlife Biologist
Shelton Land and Ranch Company
P. O. Box 1107
Kerrville, TX 78028
--Wildlife Management Specialist

CARL E. COPPOCK
Professor
Animal Science Department
Texas A&M University
College Station, TX 77843
--Dairy Nutritionist

DICK CROW
Publisher
Western Livestock Journal
Crow Publications Inc.
P. O. Drawer 17F
Denver, CO 80217
--Journalist

STANLEY E. CURTIS
Professor of Animal Science
University of Illinois
Urbana, IL 61801
--Animal Scientist

A. JOHN DE BOER
Agricultural Economist
Winrock International
Route 3
Morrilton, AR 72110
--Agricultural Economist

WAYNE L. DECKER
Professor
Department of Atmospheric Science
University of Missouri
Columbia, MO 65211
--Meteorologist and Agri Weather Specialist

R. O. DRUMMOND
Laboratory Director
U.S. Livestock Insects Laboratory
P. O. Box 232
Kerrville, TX 78028
--Entomologist

ED DUREN
Extension Livestock Specialist
P. O. Box 29
Soda Springs, ID 83276
--Animal Scientist

WILLIAM EATON
Clear Dawn Angus Farm
R. R. 1
Huntsville, IL 62344
--Registered Cattle Breeder

WILLIAM D. FARR
Farr Farms Company
Box 878
Greeley, CO 80632
--All-Time Great (Cattleman)

H. A. FITZHUGH
Animal Scientist
Winrock International
Route 3
Morrilton, AR 72110
--Animal Scientist

MIGUEL A. GALINA
Professor
Universidad Nacional Autonoma de Mexico
A.P. 25, Cuautitlan Izcalli
Edo de Mexico, MEXICO
--Animal Scientist

HENRY GARDINER
Route
Ashland, KS 67831
--Cattleman

DONALD R. GILL
Extension Animal Nutritionist
Oklahoma State University
005 Animal Science
Stillwater, OK 74078
--Nutritionist

HUDSON A. GLIMP
Blue Meadows Farm, Inc.
Route 2, Box 407
Danville, KY 40422
--Sheep Producer

MARTIN H. GONZALEZ
President
ECO TERRA SA de CV
Fernando de Borja 208
Chihuahua, Chihuahua
MEXICO
--Range Scientist

TEMPLE GRANDIN
Livestock Handling Consultant
Department of Animal Science
University of Illinois
Urbana, IL 61801
--Livestock Facilities Specialist

SAMUEL B. GUSS, D.V.M.
Professor Emeritus
Veterinary Science Extension
Pennsylvania State University
2410 Shingletown Road
State College, PA 16801
--Veterinarian

JAMES C. HEIRD
Assistant Professor
Department of Animal Science
Texas Tech University
P. O. Box 4169
Lubbock, TX 79409
--Horse Specialist

A. L. HOERMAN
Extension Livestock Specialist
Texas A&M University Extension Center
P. O. Drawer 1849
Uvalde, TX 78801
--Beef Tour Coordinator

DOUGLAS HOUSEHOLDER
Extension Horse Specialist
Texas A&M University
College Station, TX 77843
--Tour and Clinic Moderator

HARLAN G. HUGHES
Agricultural Economist
University of Wyoming
Laramie, WY 82071
--Agricultural Economist

CLARENCE V. HULET
Research Leader
U.S. Sheep Experiment Station
USDA, ARS
Dubois, ID 83423
--Sheep Physiologist

HENRYK A. JASIOROWSKI
Director, Cattle Breeding Research
Warsaw Agricultural University
AGGW-AR 02-528 Warszawa
Rakowiecka Str. 26/30
POLAND
--Animal Scientist

DONALD M. KINSMAN
Professor, Animal Industries Department
University of Connecticut
Storrs, CT 06268
--Meat Scientist

JACK L. KREIDER
Associate Professor
Horse Program
Texas A&M University
College Station, TX 77843
--Physiologist

JAMES W. LAUDERDALE
Senior Scientist
The Upjohn Company
Performance Enhancement Research
Kalamazoo, MI 49001
--Physiologist

ROBERT A. LONG
Professor
Department of Animal Science
Texas Tech University
P. O. Box 4169
Lubbock, TX 79409
--Animal Scientist

CRAIG LUDWIG
Director of TPR
American Hereford Association
715 Hereford Drive
P. O. Box 4059
Kansas City, MO 64101
--Cattle Organization Leader

JAMES P. MC CALL
Director of Horse Program
Stallion Station
Louisiana Tech University
P. O. Box 10198
Ruston, LA 71272
--Horse Specialist

WILLIAM C. MC MULLEN, D.V.M.
Large Animal Medicine & Surgery
Texas A&M University
College Station, TX 77843
--Veterinarian

JOHN W. MC NEILL
Beef Cattle Specialist
Texas Agricultural Extension Service
6500 Amarillo Blvd., West
Amarillo, TX 79106
--Beef Cattle Specialist

DOYLE G. MEADOWS
Manager
Robinwood Farm
2822 East 2nd Street
Edmond, OK 73034
--Horse Specialist

JOHN L. MERRILL
XXX Ranch
Route 1, Box 54
Crowley, TX 76036
--Range Scientist and Rancher

BRET K. MIDDLETON
Animal Science Department
Iowa State University
Ames, Iowa 50011
--Computer Cow Game Coordinator

J. D. MORROW
Executive Vice-President
International Brangus Breeders Association
9500 Tioga Drive
San Antonio, TX 78230
--Cattle Organization Leader

HARRY C. MUSSMAN, D.V.M.
Administrator
APHIS/USDA
Washington, D.C. 20250
--Veterinarian

CHARLES W. NICHOLS
Manager
Davidson & Sons Cattle Company
Route 2, Box 15-0
Arnett, OK 73832
--Cattleman

J. DAVID NICHOLS
Anita, IA 50020
--Registered Cattle Breeder

MICHAEL J. NOLAN
Executive Secretary
Health and Regulatory Committee
American Horse Council, Inc.
1700 K Street, N.W., Suite 300
Washington, D.C. 20006
--Horse Specialist

JAY O'BRIEN
P. O. Box 9598
Amarillo, TX 79105.
--Cattleman

JERRY O'SHEA
Principal Research Officer
The Agricultural Institute
Moorepark, Fermoy Company
Cork, IRELAND
--Dairy Equipment and Mastitis Expert

RICHARD O. PARKER
Assistant to the President
Agriservices Foundation
648 West Sierra Avenue
P. O. Box 429
Clovis, CA 93612
--Physiologist

BRENT PERRY, D.V.M.
President
Rio Vista International, Inc.
Route 9, Box 242
San Antonio, TX 78227
--Veterinarian

GUSTAV PERSON, JR.
Extension Agent
Guadalupe County
Ag Building
Seguin, TX 78155
--Horse Tour Coordinator

L. S. POPE
Dean
College of Agriculture
New Mexico State University
Las Cruces, NM 88003
--Animal Scientist

DOUGLAS PRESLEY
Extension Agent
Bexar County
Room 310
203 W. Nueva
San Antonio, TX 78207
--Horse Clinic Coordinator

NED S. RAUN
Vice-President, Programs
Winrock International
Route 3
Morrilton, AR 72110
--Animal Scientist

PATRICK O. REARDON
Assistant General Manager
Chaparrosa Ranch
P. O. Box 489
La Pryor, TX 78872
--Range Scientist

RUBY RINGSDORF
President
American Agri Women
28781 Bodenhamer Road
Eugene, OR 97402
--Farm Organization Leader

DON G. ROLLINS, D.V.M.
Technical Veterinary Advisor
Mid-America Dairymen, Inc.
800 West Tampa
Springfield, MO 65805
--Veterinarian

BUDDY ROULSTON
Professional Horse Trainer
Brenham, TX 77833
--Horse Trainer

MANUEL E. RUIZ
Head
Programa de Produccion Animal
CATIE
Turrialba, COSTA RICA
--Nutritionist

CHARLES G. SCRUGGS
Vice-President and Editor
Progressive Farmer
P. O. Box 2581
Birmingham, AL 35202
--Journalist

RICHARD S. SECHRIST
Executive Secretary
National Dairy Herd Improvement Association
3021 East Dublin-Granville Road
Columbus, OH 43229
--Dairy Organization Leader

MAURICE SHELTON
Professor
Texas Agricultural Experiment Station
Route 2, Box 950
San Angelo, TX 76901
--Sheep Specialist

PAT SHEPHERD
Manager
South Plains Feed Yard
Drawer C
Hale Center, TX 79041
--Feedlot Manager

JOHN STEWART-SMITH
President
Beefbooster Cattle Ltd.
P. O. Box 396
Cochrane, Alberta
CANADA TOL OWO
--Cattle Breeder

GEORGE STONE
President
National Farmers Union
12025 East 45th Avenue
Denver, CO 80251
--Farm Organization Leader

JACK D. STOUT
Associate Professor
Extension Dairy Specialist
Oklahoma State University
Stillwater, OK 74074
--Dairy Specialist

MRS. BAZY TANKERSLEY
Al-Marah Arabians
4101 North Bear Canyon Road
Tucson, AZ 85715
--All-Time Great (Horsewoman)

MAURICE TELLEEN
Editor and Publisher
The Draft Horse Journal
Box 670
Waverly, IA 50677
--Horseman and Journalist

THOMAS R. THEDFORD, D.V.M.
Extension Veterinarian
Oklahoma State University
Stillwater, OK 74078
--Veterinarian

TOPPER THORPE
General Manager
Cattle-Fax
5420 South Quebec Street
Englewood, CO 80155
--Agricultural Economist

ALLEN D. TILLMAN
Rockefeller Foundation, Emeritus
523 West Harned Place
Stillwater, OK 74074
--Animal Scientist

JAMES N. TRAPP
Associate Professor
Agricultural Economics Department
Oklahoma State University
Stillwater, OK 74078
--Agricultural Economist

ROBERT WALTON
President
American Breeders Service
P. O. Box 459
DeForest, WI 53532
--All-Time Great (Dairyman)

RODGER L. WASSON
Executive Director
American Sheep Producers Council, Inc.
200 Clayton Street
Denver, CO 80206
--Sheep Association Leader

DOYLE WAYBRIGHT
Mason Dixon Farms
RD 2
Gettsburg, PA 17325
--Dairyman

GARY W. WEBB
Stallion Manager
Winmunn Quarter Horses, Inc.
Route 1, Box 460
Brenham, TX 77833
--Horse Specialist

RICHARD O. WHEELER
President
Winrock International
Route 3
Morrilton, AR 72110
--Agricultural Economist

DICK WHETSELL
Vice-Chairman
Board of Directors
Oklahoma Land & Cattle Company
P. O. Box 1389
Pawhuska, OK 74056
--Range Scientist

R. GENE WHITE, D.V.M.
Coordinator of the Regional College
of Veterinary Medicine
University of Nebraska
Lincoln, NE 68583
--Veterinarian

RICHARD L. WILLHAM
Professor of Animal Science
Iowa State University
Kildee Hall
Ames, IA 50011
--Animal Breeding Specialist

DON WILLIAMS, D.V.M.
Henry C. Hitch Feedlot, Inc.
Box 1442
Guymon, OK 73942
--Feedlot Manager and Veterinarian

JAMES N. WILTBANK
Professor
Department of Animal Husbandry
Brigham Young University
Provo, UT 84602
--Beef Cattle Physiologist

CHRIS G. WOELFEL
Dairy Specialist
Texas Agricultural Extension Service
218 Kleberg Center
College Station, TX 77843
--Dairy Tour Coordinator

JAMES A. YAZMAN
Animal Scientist
Winrock International
Route 3
Morrilton, AR 72110
--Dairy Goat Specialist and Nutritionist

Other Winrock International Studies
Published by Westview Press

Hair Sheep of West Africa and the Americas, edited by H. A. Fitzhugh and G. Eric Bradford

Future Dimensions of World Food and Population, edited by Richard G. Woods

Other Books of Interest from Westview Press

Carcase Evaluation in Livestock Breeding, Production and Marketing, A. J. Kempster, A. Cuthbertson, and G. Harrington

Energy Impacts Upon Future Livestock Production, Gerald M. Ward

Science, Agriculture, and the Politics of Research, Lawrence Busch and William B. Lacy

Developing Strategies for Rangeland Management, National Research Council

Proceedings of the XIV International Grassland Conference, edited by J. Allan Smith and Virgil W. Hays

Animal Agriculture: Research to Meet Human Needs in the 21st Century, edited by Wilson G. Pond, Robert A. Merkel, Lon D. McGilliard, and V. James Rhodes

Other Books of Interest
from Winrock International[*]

Ruminant Products: More Than Meat and Milk, R. E. McDowell

The Role of Ruminants in Support of Man, H. A. Fitzhugh, H. J. Hodgson, O. J. Scoville, Thanh D. Nguyen, and T. C. Byerly

Potential of the World's Forages for Ruminant Animal Production, Second Edition, edited by R. Dennis Child and Evert K. Byington

Research on Crop-Animal Systems, edited by H. A. Fitzhugh, R. D. Hart, R. A. Moreno, P. O. Osuji, M. E. Ruiz, and L. Singh

Bibliography on Crop-Animal Systems, H. A. Fitzhugh and R. Hart

[*]Available directly from Winrock International, Petit Jean Mountain, Morrilton, Arkansas 72110